Wolfgang Grünert, Wolfgang Kleist, Martin Muhler
Catalysis at Surfaces

Also of interest

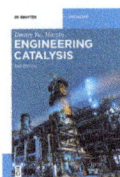

Engineering Catalysis
Murzin, 2020
ISBN 978-3-11-061442-8, e-ISBN 978-3-11-061443-5

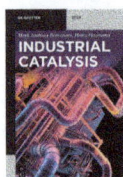

Industrial Catalysis
Benvenuto, 2021
ISBN 978-3-11-054284-4, e-ISBN 978-3-11-054286-8

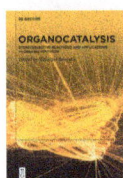

Organocatalysis.
Stereoselective Reactions and Applications in Organic Synthesis
Benaglia (Ed.), 2021
ISBN 978-3-11-058803-3, e-ISBN 978-3-11-059005-0

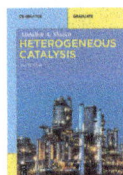

Heterogeneous Catalysis
Shaikh, 2023
ISBN 978-3-11-103248-1, e-ISBN 978-3-11-103251-1

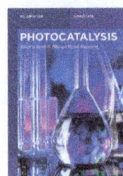

Photocatalysis
Pillai, Kumaravel (Eds.), 2021
ISBN 978-3-11-066845-2, e-ISBN 978-3-11-066848-3

Wolfgang Grünert, Wolfgang Kleist,
Martin Muhler

Catalysis at Surfaces

—

DE GRUYTER

Authors
Prof. Dr. Wolfgang Grünert
Faculty of Chemistry & Biochemistry
Industrial Chemistry
Ruhr University Bochum
Universitätsstraße 150
44801 Bochum
Germany

Prof. Dr. Martin Muhler
Faculty of Chemistry & Biochemistry
Industrial Chemistry
Ruhr University Bochum
Universitätsstraße 150
44801 Bochum
Germany

Prof. Dr. Wolfgang Kleist
Department of Chemistry
RPTU Kaiserslautern-Landau
Erwin-Schrödinger-Straße 54
67663 Kaiserslautern
Germany

ISBN 978-3-11-063247-7
e-ISBN (PDF) 978-3-11-063248-4
e-ISBN (EPUB) 978-3-11-063254-5

Library of Congress Control Number: 2023931058

Bibliographic information published by the Deutsche Nationalbibliothek
The Deutsche Nationalbibliothek lists this publication in the Deutsche Nationalbibliografie;
detailed bibliographic data are available on the Internet at http://dnb.dnb.de.

© 2023 Walter de Gruyter GmbH, Berlin/Boston
Cover image: Maximilian Göckeler and Wolfgang Grünert
Typesetting: Integra Software Services Pvt. Ltd.
Printing and binding: CPI books GmbH, Leck

www.degruyter.com

Preface

Catalysis is at the heart of the chemical industry, which uses solid catalysts for the production of commodity chemicals and intermediates, and is likewise an essential technology in environmental protection. It is going to play a key role in the upcoming transition of economies towards fully sustainable supplies for energy and materials, where energy needs to be stored in chemical bonds, e.g., of hydrogen, ammonia, or methanol, and substitutes need to be found for materials that are currently produced from oil or gas deposits. Against this background, catalysis will not only keep its eminent importance at an interface between basic and applied research, but will rather enhance its impact on society in the future.

Catalysis research is a truly interdisciplinary field. It requires a profound knowledge on various aspects of chemistry and physics as provided in this advanced textbook, which covers also the emerging fields of photo- and electrocatalysis and supplies an interface to catalytic reaction engineering. All essential tools of catalysis research are introduced, ranging from the synthesis and modification of porous solids via the study of their reactivity and of their structure, using bulk- and surface-sensitive techniques, to currently applied theoretical methods. A close-up to important aspects of surface catalysis describes basic strategies for improving the activity and the stability of catalysts and their selectivity towards the desired reaction products. It summarizes established knowledge about mechanisms of catalytic reactions and the sites, on which they proceed, and describes some relevant controversies in the field. Review papers are cited to give the reader access to more details on these topics and to potentially relevant information, which had to stay beyond the limits of this book. Intending to guide readers with a basic knowledge of chemistry to the recent state of catalysis research, the book is recommended to M.Sc. and PhD students, but it may be interesting also for engineers dealing with applications of catalysis and to scientists from adjacent fields.

Looking back on years with the growing manuscript, we are deeply indebted to many people who supported us with information, recommendations and encouragement. We want, in particular, to thank colleagues for critically reading sections on characterization techniques and theory: Angelika Brückner (Rostock), Volker Staemmler, Ulrich Köhler (both Bochum), Wolfgang Bensch (Kiel), Michael Hunger (Stuttgart), Jörg Kärger (Leipzig), Thomas Lunkenbein (Berlin), and Volker Schünemann (Kaiserslautern). We are also grateful to Ms. Alina Ouissa, M. Sc., Dr. Wilma Busser, and Dr. Katrin Lotz for the many beautiful figures they contributed and to Dr. Maximilian Göckeler for a fruitful cooperation on the cover image.

Thirty years ago, Hans Niemantsverdriet's "Spectroscopy in Catalysis" pleased students and impressed academic teachers by its clarity and the laid-back tone in the explanation of scientific contexts. Decades of teaching have raised the desire in the present authors to create an entrance of similar convenience into the whole multifaceted area of catalysis at surfaces, and we are now grateful to de Gruyter for the chance

https://doi.org/10.1515/9783110632484-202

to present the results of our effort to the catalysis community. Whatever the opinion of our readers about our didactic skills may be, we hope our book helps them to a swift way into this fascinating field.

Bochum, June 2023

Wolfgang Grünert
Wolfgang Kleist
Martin Muhler

Contents

1 Introduction

1.1 Catalysis, reaction rate, and equilibrium

A catalyst is a material that changes the rate of a chemical reaction without appearing among its products. This is almost literally the definition that Wilhelm Ostwald gave to catalysis in 1901 [1]. We now usually understand the "change" as an increase.

Ostwald's definition contains some important statements:

- Catalysis is about reaction rates, not about thermodynamics. It does not change equilibria: the Gibbs free energies of the reaction products formed from the given reactants remain the same. Catalysis is sometimes said to make impossible reactions happen. This may come true if the problem is a zero reaction rate, but not if the Gibbs energy of a mixture of reactants increased during the reaction.
- As the catalyst appears neither among the reactants nor among the products, it is not affected by the reaction. This needs to be qualified in the light of present-day knowledge: catalysts remain unchanged only on the timescale of the reaction, but due to side reactions, they are gradually modified themselves. These side reactions, which mostly result in deactivation, are among the topics of this book (cf. Sections 1.4.2 and 4.4).
- As the reaction equilibrium (with an equilibrium constant, $K = k_+/k_-$) remains unchanged while the forward rate constant k_+ increases, the reverse rate constant k_- must increase as well. This describes the principle of microscopic reversibility: catalysts accelerate forward and reverse reactions to the same extent.

Increasing the reaction temperature is another way to accelerate the chemical reactions as described, for instance, by the Arrhenius equation

$$k = A \, exp(-E_A/RT) \tag{1.1}$$

where A is the pre-exponential factor and E_A is the activation energy. Alternatively, the Eyring equation (2.26) is often employed in studies targeting basic aspects of the reaction kinetics.

Catalysts decrease the activation energy of a chemical reaction by opening reaction paths that avoid the highest energy barrier (Figure 1.1a). This does not automatically mean a higher rate, because at very high temperatures, the reaction with the highest activation energy, i.e., the noncatalyzed reaction, should dominate for mathematical reasons (Figure 1.1b). However, for catalyst systems encountered in our lives, these temperatures are beyond relevant ranges. Therefore, in our experience, the lower activation energy is correlated with a higher reaction rate.

https://doi.org/10.1515/9783110632484-001

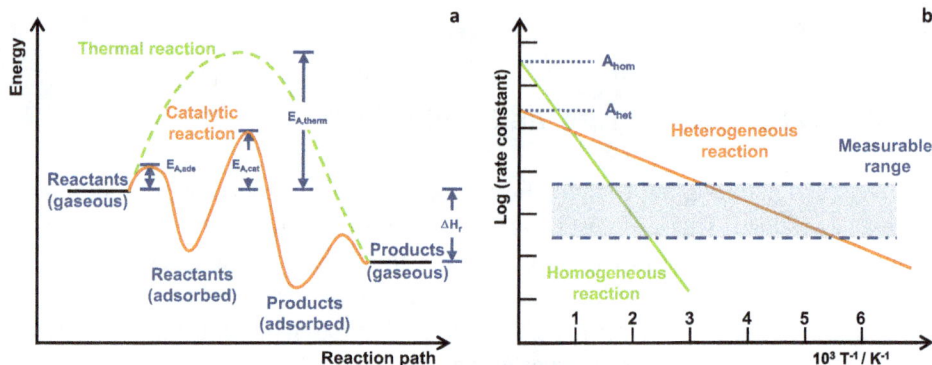

Figure 1.1: Catalysis as a kinetic phenomenon caused by a change of the reaction mechanism. (a) Reaction coordinate of a gas-phase reaction via a noncatalytic route or proceeding at a catalytic surface. (b) Rates and activation energies: due to higher activation energies, homogeneous reactions become faster than catalytic reactions, however, at temperatures beyond ranges of practical interest.

1.2 Catalyst and reactant phases

For the practical application of catalysis, it is a major difference if the catalyst is part of the same fluid[1] phase as the reactants, or if it interacts with the phase(s) containing the reactants via a phase boundary. These two options are referred to as homogeneous and heterogeneous catalysis, respectively. There is a third option – phase transfer catalysis, in which the catalyst changes between two liquid phases during the catalytic reaction cycle. Although scientifically very interesting, it is beyond the scope of this book. The reader may consult ref. [2] for more information.

The heterogeneous catalyst is usually a solid. The reaction proceeds at its catalytically active surfaces.[2] In the ideal heterogeneous system, there is no catalytic reaction in the fluid phase at all; therefore, mass transfer to and from the catalyst is an important part of the process (see Sections 1.4.3 and 2.6.2). In homogeneous catalysis, which proceeds in liquids with only a few exceptions, mass transfer affects the overall reaction rate only at extremely high catalytic reaction rates.

Catalysis at surfaces can also proceed under the influence of an electrical potential (electrocatalysis) or of light (photocatalysis). Both options might be also covered under the term "heterogeneous catalysis," which is, however, unusual. Instead, traditional heterogeneous catalysis is sometimes designated as "thermal catalysis" to differentiate it from photo- and electrocatalysis. In this book, all three versions are summarized as "surface catalysis."

---

1 A fluid phase is a liquid phase or a gaseous phase.
2 A liquid could be a heterogeneous catalyst only if there was absolutely no solubility of the reactants in it.

Electro- and photocatalysis are related to redox processes. These are separated into half reactions, which proceed on different electrodes or at different sites utilizing charge carriers generated by light absorption (cf. Sections 2.7 and 2.5.2, respectively). The overall rates of both half-reactions are identical, but their individual reaction events are not synchronous. Due to this separation, such reactions are not easily depicted on a single reaction coordinate, as in Figure 1.1a. The following simplified discussion may still provide initial insight into the nature of the phenomena.

On electrodes, the kinetic barriers in thermodynamically allowed reactions cause excess potentials (overpotentials), due to which observable reaction rates are obtained only at voltages exceeding the thermodynamic potential described by the Nernst equation (2.58). Catalytic electrodes open more efficient reaction mechanisms to the corresponding half reactions. Thus, they lower the overpotentials in a similar way as catalysts do with activation energies and with required reaction temperatures, under potential-free conditions (cf. Figure 1.1a). In photocatalysis, photon absorption causes an excitation of electrons into a higher state (or band, cf. Section 2.1). The redox reaction utilizes both the remaining holes and the excited electrons: their higher chemical potential opens reaction paths that are inaccessible in the dark. In the simple picture of Figure 1.1a, one might say that the faster reaction routes enabled by the catalyst are switched on by the absorption of the photon.

Biocatalysis is not easily ranked into a classification based on the phases involved. It deals with catalysis by enzymes, either in direct contact with the reactants or within living cells.

Homogeneous, heterogeneous, and biocatalysis have a number of similarities and common applications. Their nature, potential, and limitations can be illustrated by discussing some characteristic differences between them. In heterogeneous catalysis, the *concentration of "active sites"* (sites involved in the catalytic process, see Section 1.4.3) is generally low, because the interior of the solid catalyst is inaccessible to the reactants. Opposed to this, the active species can be anywhere in a homogeneous phase, i.e., their concentration can be high. In biocatalyis, it will be somewhat lower due to the larger size of biomolecules. Biocatalysts are known for their exceptional *chemo- and stereoselectivity* in reactions that may yield a variety of products. High selectivity can also be achieved with homogeneous and, sometimes, with heterogeneous catalysts. However, due to the molecular nature of the homogenous catalysts, the knowledge and predictive potential of science is better developed in this field than in heterogeneous catalysis, where catalysts are much more complex and less defined (see Section 1.5). It is therefore easier to optimize selectivity in the presence of homogeneous catalysts, although the great commercial success of heterogeneous catalysis shows that good selectivities can be achieved also with this type of catalyst.

Reaction conditions are most confined in biocatalysis: they are related to the conditions of life, namely, temperatures between 273 and 373 K and atmospheric pressure. Homogeneous catalysis is confined to temperatures that keep the liquid phase liquid (typically between 190 and 470 K), while there is no limit on pressure. Reactions catalyzed by solid surfaces are known for temperatures between 100 and 1,300 K, at pressures between high vacuum and hundreds of bars. Rating these conditions, one should avoid oversimplification. Low reaction temperatures are indeed desirable, but only if the reactions are exothermal. With endothermal reactions like dehydrogenations, acceptable conversions require high temperatures for thermodynamic reasons. With respect to pressure, atmospheric conditions are desirable for reactions in the liquid phase, but may be unfavorable for gas-phase reactions where they cause high plant volumes and low reaction rates. Compared with a plant operating at atmospheric pressure, the plant volume will be just 5% at 20 bar, and the reactant concentrations acting on the rate via the rate law will be larger by a factor of 20. Therefore, reaction pressures are a matter of an optimization balancing reaction rates with plant costs.

Separation of the reaction medium and the catalyst at the end of the reaction is trivial in heterogeneous systems, where the catalyst never becomes part of the reactant phase. It occurs spontaneously or may be achieved without damage to the catalyst activity using standard methods of gas–solid or liquid–solid separation technology. This advantage is a major reason for the predominance of heterogeneous catalysis in industry and environmental technology. In homogeneous systems, catalyst separation is much more complicated, and the choice of technology strongly depends on the properties of the reaction system. Continuous evaporation of the reaction product(s) or the use of biphasic reaction systems hosting catalyst and reactants in different phases are options opening up perspectives for industrial-scale application of homogeneous catalysts also. However, where these options fail, catalyst separation and circulation require great effort and often cause inacceptable losses or product contamination. Heterogenization of homogeneous catalysts, by attaching them to solid materials with high surface area, is therefore a permanent topic of technology-oriented research (cf. Section 2.3.3).

Separation of biocatalysts from the reaction media is less problematic. Due to the large sizes of enzymes, they can be removed from the products by ultrafiltration. Living cells may populate solid supporting media as in the trickle filter process used in sewage treatment, which may be considered as heterogenization of a biocatalyst.

1.3 Catalysis and daily life

Catalysis has always been a practical science giving mankind access to useful or desired substances. Owing to the close interrelation between early mankind and nature, the first catalytic process utilized on purpose was a natural (biocatalytic) one: fermentation was known already in old Sumer around 3000 BC. Its product, ethanol, was probably a desired rather than a useful one, although there were times and circumstances in history where the biocidal properties of ethanol made beer a less dangerous beverage than the available contaminated drinking water.

The identification of catalysts as compounds helping other substances to react with each other became possible only in the early years of modern science and technology.

In the Middle Ages, sulfuric acid was made by burning sulfur in the presence of water and saltpeter. SO_2 formed in an excess of O_2 was converted to the acid in a catalytic liquid-phase reaction, utilizing the oxygen transfer capabilities of the nitrogen oxides. In England, around 1750, after the replacement of the glass equipment by more robust lead-lined chambers, this became the first industrial-scale catalytic process – the lead chamber process. Shortly after 1780, conversion of starch to glucose, the esterification of acids with alcohols, and the decomposition of ethanol to ethene and water were described. All are acid-catalyzed – two of them in homogeneous phase and the third one over activated alumina. The first practical application of heterogeneous catalysis was the Döbereiner lighter (1823, Figure 1.2), in which a Pt sponge catalyzing water formation from H_2 and O_2 was heated by the reaction heat until it was able to light the hydrogen flame.

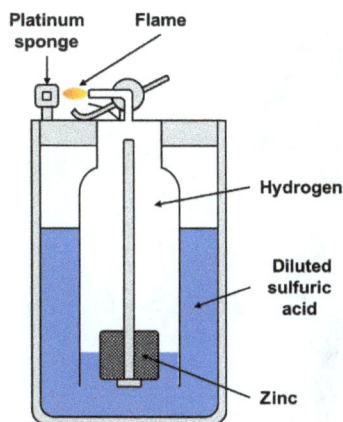

Figure 1.2: Döbereiner lighter, a schematic representation.

The underlying principle of catalysis was only gradually appreciated. In 1794, the Scottish chemist Elizabeth Fulhame reported on redox reactions of metal salts (being reduced to the metal) and metals (being oxidized), which occurred only in the presence of water. This water was regenerated at the end, i.e., it was involved as a catalyst [3]. In the beginning of the nineteenth century, many scientists noted the influence of hot metals or metal oxides on the decomposition of various substances. Summarizing such work, Berzelius concluded that there is yet another principle in chemistry beyond the well-known "affinity" – the "catalytic" (Greek for "loosening") force – a force that decomposes substances into subunits that would be recombined in a different way in the products. The insight that catalysts act only on reaction rates, which paved the way for the modern development of the field, is credited to Ostwald (see Section 1.1).

Ostwald's time saw an enormous upsurge of the chemical industry, driven also by discoveries related to heterogeneous catalysis. From the 1870s onward, a heterogeneously catalyzed process for SO_2 oxidation (the contact process) competed with the lead chamber process in sulfuric acid production, and replaced it gradually. In the new century, the Haber–Bosch process for NH_3 synthesis from atmospheric N_2 and the production of nitric acid from NH_3 by the Ostwald process provided an impressive example for the huge potential of technology in the hands of mankind and for the ever existing risks of its misuse. With these processes, production of nitrogen fertilizers could be greatly enhanced over the previous level, which was limited by the mining capacities of saltpeter deposits (e.g., in Chile). Starvation predicted from the developments of population and food production for the next decades was thus prevented. On the other hand, the accessibility of nitrates facilitated the production of explosives, which encouraged political leaders to include military options in their policies. Thus, nitrates from catalytically bound atmospheric N_2 were first used for ammunition in World War I during which millions of lives were lost.

Figure 1.3: Fritz Haber's setup for catalytic experiments under pressure, ca. 1908, as exhibited in the Deutsches Museum in Munich. Copyright JGvBerkel, public domain material.

Actually, the Haber–Bosch process combines two major achievements: the chemical fixation of nitrogen and the invention of a commercial-scale high-pressure technology (Haber's laboratory pressure equipment is shown in Figure 1.3). In the 1920s, BASF in Germany employed this high-pressure technology for the development of new catalytic processes, e.g., methanol synthesis and coal hydrogenation to liquid products. At the same time, Fischer and Tropsch developed the hydrogenation of CO to motor fuels into a technical process, which later played a great role in the preparations of Nazi Germany for the intended military expansion. In the late 1930s, the hydroformylation of alkenes to aldehydes using CO and H_2 was the first process catalyzed by a metal complex in the liquid phase to be developed to technical maturity. A plant was built in Germany, but not commissioned during World War II. While research in Germany was focused on the utilization of the abundant coal deposits, upgrading of oil fractions to improve their fuel quality was the focus in the United States. In the 1930s, catalytic cracking was invented by Houdry, and naphtha reforming was started using Mo-based catalysts, which were replaced by Pt-based catalysts in the 1940s.

After World War II, many older processes were modernized and revamped with new catalysts (e.g., methanol synthesis). A great number of new processes appeared in the industry, among them the catalytic polymerization of alkenes: first of ethene with silica-supported chromia (Phillips catalyst) and, shortly after, with combinations of group IVb compounds and Al alkyls (Ziegler–Natta catalysts, which are applicable for various alkenes), and after 1980, with the extremely active and selective metallocene catalysts. Hydrotreating of naphtha, and later of heavier oil fractions, the metathesis of alkenes, and the application of heterogeneous catalysts to reduce air pollution caused by cars (three-way catalyst, mid-1970s) and by power plants (selective catalytic reduction of nitrogen oxides in flue gases, 1980s) were introduced. For more detailed information, including the historical development also of biocatalysis, the reader is referred to ref. [4].

Obviously, catalysis is not at all a young science in its early, exciting stage. It is, however, extremely important for both chemical and environmental technology. For reasons to be discussed further, it goes on offering attractive opportunities for technological improvement and economic benefit in industry, and for the control of emissions from various sources.

"A small leak will sink a great ship." Catalysis shows nicely how this proverb can get a positive turn: a small amount of catalyst enables the conversion of huge quantities of reactants under milder conditions or with higher selectivity, thus providing savings in energy and feed consumption as well as cuts in waste production. Due to this promise, catalysis is employed in the chemical industry and refinery, wherever possible. Using and improving catalysts is a strategy that serves both economic and ecologic needs. Catalysis was identified as one of the 12 principles of green chemistry from early on [5].

Nowadays, more than 85% of all chemical products and almost all fuels used in the transportation sector pass at least one catalyzed step in their production [6, 7], ca. 90% thereof involving heterogeneous catalysts. About 20–30% of the world's gross domestic product (GDP) are created or influenced by processes involving catalysis. According to a recent study [8], the world market for catalysts had a volume of ≈$35.5 billion in 2020. The market volume of environmental catalysts has been largest over many years (≈37%), those of catalysts for refining, chemical processes, and polymerization were significantly smaller (26%, 22%, and 14%, respectively).

Due to the long service lives achieved with many catalysts, their contribution to the cost of products is small – on the order of 0.2% in chemical industry and 0.1% in refinery. Turned around, the value of products made by the catalysts priced ca. $22 billion/a (chemical processes, refinery, polymerization) should be up to three orders of magnitude higher. On the large scales of chemical production, small improvements in catalytic performance causing yields to increase by a few percent or the required temperatures to decrease by some degrees can result in very significant economic effects. This has always been a driving force behind the ongoing and even increasing efforts in global catalysis research. The other driving forces are the tightening of environmental regulations for emissions from various sources and the challenges of the upcoming transition of energy and raw material supply to sustainable conditions. In particular, the latter will also create new opportunities for electrocatalytic processes and for the application of photocatalysis. The perspective of catalysis in this development is discussed in Chapter 5 (see also refs. [6, 7]).

1.4 Basic concepts of surface catalysis

Catalysis research is about the improvement of catalyst performance, which is measured in terms of reaction rates, selectivity, and stability properties. In the following, basic concepts used in this research are introduced.

1.4.1 Reaction rate

For a homogeneous reaction, rate is defined as

$$r = \frac{1}{\nu_i} \frac{1}{V} \frac{dn_i}{dt} \tag{1.2}$$

where ν_i is the stoichiometric coefficient of reactant i and V is the reaction volume, which may change with conversion in a gas-phase reaction. If V remains constant, eq. (1.2) can be transformed into the well-known expression $\nu_i r = dc_i/dt$. The rate is often expressed in terms of conversion X, which is a basic concept of reaction engineering:

$$X = \frac{n_{0,A} - n_A}{n_{0,A}} = \frac{\dot{n}_{0,A} - \dot{n}_A}{\dot{n}_{0,A}} \tag{1.3}$$

where A is the key component in the reaction mixture (the one not in excess) and $\dot{n}$ designates a flow, e.g., in mol/h. The subscript "0" stands for initial in batch mode or for inlet in continuous reactor operation. The right-hand version relates to the flow regime typical for large-scale catalytic processes.

With $dX = -dn_A/n_{0A}$ resulting from eq. (1.3), eq. (1.2) can be expressed with the conversion X:

$$r = -\frac{1}{v_A}\frac{n_{0,A}}{V}\frac{dX}{dt} \tag{1.2a}$$

When a reactant i is converted over a catalyst, the change of its amount with time dn_i/dt should be related to the amount of catalyst, rather than to the reaction volume. A plausible quantity for reference would be the number of sites n_{site} supplied by the catalyst to support the reaction. The turnover frequency (TOF) obtained by this assignment

$$TOF = \frac{1}{v_i}\frac{1}{n_{site}}\frac{dn_i}{dt} \tag{1.4}$$

is most valuable from the viewpoint of science, but impractical for applied heterogeneous catalysis, where n_{site} is rarely known. Usually, the catalyst mass m_{cat} is used for reference instead, which results in the general definition of the reaction rate in heterogeneous catalysis

$$r_{cat} = \frac{1}{v_i}\frac{1}{m_{cat}}\frac{dn_i}{dt} \tag{1.5}^3$$

In homogeneous catalysis, in particular in catalysis with coordination compounds, rates are generally reported as TOF, which implies the assumption that all catalyst molecules contribute equally to this rate. Though sometimes found in literature, an analogous use of the TOF is rarely appropriate in heterogeneous catalysis, because the probability that all exposed atoms of the active element take part in the catalytic process is small in many cases. Turnover frequencies should be reported only if catalyst characterization provides reliable data on the percentage of the active element available for catalysis. Relating conversion rates to the total amount of the catalytically active element n_{cat} may still make sense, but the quantity should be given a different name, e.g., "normalized reaction rate" r_N

$$r_N = \frac{1}{v_i}\frac{1}{n_{cat}}\frac{dn_i}{dt} \tag{1.6}$$

In photocatalysis, rates are not proportional to the catalyst mass (see Section 3.2.2). Therefore, the quantum yield Φ is employed as a performance criterion, instead. It reports the probability that an absorbed photon induces a reaction event

---

3 Notably, eq. (1.5) is not useful for photocatalysis, where the tacit assumption that catalyst particles of identical properties make identical contributions to reactant conversion does not hold. For more details, see Section 3.2.2.

$$\Phi = \frac{dn_i/v_i}{dn_{ph,\,abs}} = \frac{dn_i/v_i\;m_{cat}\;dt}{dn_{ph,\,abs}/m_{cat}\;dt} = \frac{r}{J_{ph,\,abs}/m_{cat}} \tag{1.7}$$

where $n_{ph,abs}$ and $J_{ph,abs}$ are the number and the flux of the photons being absorbed. Obviously, the quantum yield depends on the energy range of the incident photons, which must be reported with experimental values of Φ. However, the quantum yield is also influenced by quantities that are not easily accessible, for instance, by the scattering properties of the photoabsorber. Therefore, the apparent quantum yield Φ_{app} is frequently reported instead, which relates the moles of product to the moles of the incident photons:

$$\Phi_{app} = \frac{r}{J_{ph,\,inc}/m_{cat}} \tag{1.7a}$$

In electrocatalyis, the reaction rate is detected by the current I associated with the reaction. The relation between I and the rate according to eq. 1.5 can be found by considering that each differential amount of converted reactant dn_i (normalized by the stoichiometric coefficient v_i) is related to a charge, $dQ = z\,F\,dn_i/v_i$, where z is the number of electrons exchanged per reaction event and F is the Faraday constant. After inserting $dn_i/v_i = dQ/zF$ into eq. (1.5), the current becomes

$$I = dQ/dt = m_{ecat}\,r\,z\,F$$

To comply with a rate expression, the current is related to a quantity expressing the amount of (electro)catalyst, e.g., its mass m_{ecat}:

$$I/m_{ecat} = r\,z\,F \tag{1.8}$$

When the content of the active element M (e.g., Pt) in the electrode is known, the current is usually related to its mass instead of the total mass

$$I_M = r\,z\,F\,m_{ecat}/m_M \tag{1.9}$$

Measured at a fixed reference potential, both quantities are referred to as mass-related activity (or mass activity, for the relation between rate and activity see Section 1.4.2) Alternatively the rate can be related to the electrochemically active surface area (ECSA) to obtain the specific activity

$$I_{sp} = r\,z\,F/ECSA \tag{1.10}$$

While well accessible for traditional Pt-based electrocatalysts via standard methods (cf. Section 3.4.12), the ECSA is difficult to derive for unconventional electrode materials, in which even the content of the active phase may be difficult to assess. Therefore, definitions (1.8)–(1.10) are all in use in electrocatalysis.

1.4.2 Activity, selectivity, and stability

Activity, selectivity, and stability are concepts describing key aspects of catalyst performance: all of them should be maximized at the same time. As "activity" is often used in an unreflected way nowadays, an effort to reshape this concept is made in the following.

Comparison of performance in a target reaction is a basic step in catalysis research. This is typically done by either measuring conversions (or reaction rates), which a set of samples achieves under specified conditions (p, T, reactant feed rates), or by determining temperatures at which such a set of samples achieves a desired rate or conversion. The "light-off temperature," T_{50}, which is the temperature required for 50% conversion, is a typical quantity derived for the latter type of comparison.

The concept of catalyst *activity* used all over the world of heterogeneous catalysis originates from such a comparison on the basis of one measurement per sample. It ranks catalysts: activity is higher when a higher conversion is achieved, or a lower temperature is required for a desired conversion *under the specified conditions*. Sometimes, however, the term "activity" is used without keeping to its original meaning: it is then just mixed up with "rate." In response to this, there has been effort to completely turn down its use in the scientific literature. This seems unrealistic to us. We propose instead to try and confine the term "activity" again to those situations where it is most suitable, as will be described in the following discussion.

Activity is be a property of a catalyst in a particular state (fresh, after treatment, after stress). It is always a rate *under specified conditions*, never a rate alone. Rate changes observed upon the variation of reaction conditions are completely due to the reaction kinetics: via the rate law in the case of concentration changes, and via the Arrhenius law (1.1) in the case of temperature variation. In the ideal case, the catalyst retains its state: when concentration or temperature changes are reverted, the catalyst again achieves the rate used to identify its activity earlier. If not, its state (and, therefore, its activity) has changed. It may, for instance, have been deactivated.

In this context, it is completely wrong to claim a higher activity when the conversion grows as a result of an increase in temperature or reactant concentration. Such changes are due to reaction kinetics, they do not indicate a new state of the catalyst. A high low-temperature activity, often targeted in literature, is a round circle: a catalyst achieving high rates already at low temperature is *very active* because other catalysts would achieve much lower rates under these conditions.

The activity concept dates back to a time where the catalyst surface was thought to be static: a stage on which the reaction theater plays. We know today that the surface region of solid catalysts responds to the chemical potential of the adjacent fluid phase (see, for instance, Sections 2.2.2, 2.5.1, and 3.5.3). Composition and morphology may change depending on the temperature and the gas-phase composition, but the rate of these changes depends on solid-state mobility and reaction conditions, and may be very slow. However, this new insight does not compromise the activity concept because the effect of feed composition and temperature variation can still be described by kinetic rate laws even if these implicitly reflect changes in surface properties. As long as the changes are reversible, the catalyst returns to the initial rate after re-establishing the initial conditions, i.e., it may be ranked by reporting an activity.

The problem is rather that these adaptation processes may differ between the catalysts to be compared, and that there may even be differences in the catalytic reaction mechanisms. Expressed more traditionally, the kinetic rate laws and/or activation energies may differ between catalysts to be studied. When, for instance, a reaction proceeds on two catalysts with different activation energies, their ranking may vary with temperature (cf. Figure 1.1b). Likewise, when reaction orders for reactants are different, a catalyst may outperform competitors at one feed composition but fail at another one. In this case, catalysts cannot be ranked by measuring a single rate: the activity concept fails, irrespective of the level of insight into the surface processes.

Box 1.1: Using "activity" consistently

- Use "activity" only for comparing the performance of catalysts: either of different catalysts or of a catalyst after different treatments.

- Avoid any indication that "activity" might change with temperature or even with reactant concentration; it is always the rate that changes upon variation of these conditions.

- Check if the content to be described is better covered by "activity" or by "rate" (or related terms, e.g., conversion). In doubt, resort to alternative expressions as "reactivity" or "performance" ("low-temperature performance" instead of "low-temperature activity")

In this book, we will use the term "activity" according to the rules presented in Box 1.1, which the reader may want to consider for her/his own scientific work. The activity concept is best suited to compare chemically similar catalysts or a catalyst in different states. Its failure should remind scientists that ranking catalysts by just one experiment bears the risk of overlooking good candidates.

The same statement holds for the ranking of electrocatalysts according to the observed lowering of the overpotential or of photocatalysts, based on the observed apparent quantum yield, which depends on the applied experimental conditions.

When the reactants can form several products, a successful catalyst development requires achieving both high rate and selectivity towards the desired product. *Selectivity* is a general concept not specific for catalysis. It reports the extent to which a desired product P_i is formed in a complex reaction system

$$S_{P_i} = \frac{\dot{n}_{P_i} - \dot{n}_{0,P_i}}{\nu_{P_i}} \frac{|\nu_A|}{\dot{n}_{0,A} - \dot{n}_A} \tag{1.11}$$

where A is the key reactant used for defining conversion (cf. eq. (1.3)).

Hydrogenation of CO

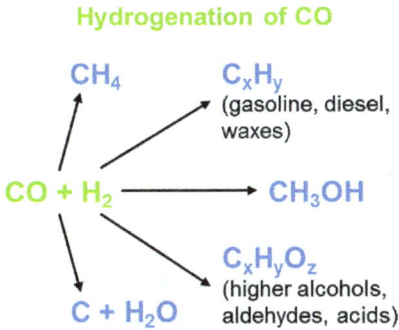

Figure 1.4: Different reaction paths available for syngas, depending on the catalyst and on conditions.

As an industrially important example, reaction products accessible from synthesis gas (a hydrogen-containing gas mixture, here with CO) are shown in Figure 1.4. The sum of selectivities of all products at a time is 1. As selectivity changes with time (or residence time) in consecutive reactions, a meaningful comparison of selectivities can be made only at identical (or similar) conversions. The yield of a product, Y_{P_i}, is

$$Y_{P_i} = \frac{\dot{n}_{P_i} - \dot{n}_{0,P_i}}{\dot{n}_{0,A}} \frac{|v_A|}{v_{P_i}} = S_{P_i} X \tag{1.12}$$

The third key catalyst property is stability. As mentioned in Section 1.1, catalysts change on a timescale, slower than that of the catalytic reaction. Except for the initial interaction between the catalyst and the feed, during which the activity may increase (cf. Section 3.1.5), catalysts always deactivate (see Section 4.4). Timescales encountered in deactivation range from seconds to years. In homogeneous catalysis, stability is expressed as the turnover number, which indicates how many moles of the reactant can be converted over a mole of catalyst before the rate becomes unacceptably small.

Fast deactivation does not necessarily exclude a catalyst from industrial application because, often, the catalytic behavior can be almost completely restored by a simple regeneration procedure not just once, but many times. Thus, catalyst stability may be targeted in two ways: either directly by extending the timescale on which the catalyst deteriorates or indirectly, if a fast deactivation cannot be avoided, by delaying the deterioration of the catalytic behavior, which the catalyst suffers in multiple regeneration steps, i.e. by increasing its stability toward regeneration.

1.4.3 Catalyst and catalytic process

In surface catalysis, the conversion of reactants to products occurs on a solid surface (4 in Figure 1.5). Before this can happen, the reactants must be adsorbed. Subsequently,

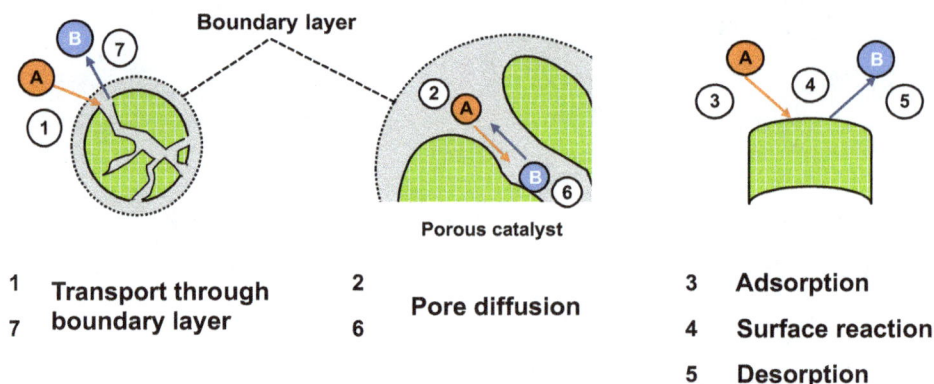

Figure 1.5: Elementary steps of a reaction catalyzed by a porous catalyst.

the products must be desorbed (3 and 5 in Figure 1.5). Adsorption designates the phenomenon that components of fluid phases are bound at solid surfaces. Under certain conditions, it can be written as a chemical reaction in which the "free" component A reacts with a free surface site (*) to the adsorbed state, A_{ads}, which is referred to as an adsorbate (cf. chemisorption, Section 2.4.1):

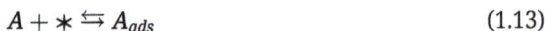

$$A + * \leftrightarrows A_{ads} \tag{1.13}$$

Desorption is the reverse process.

However, before A can stick to the surface, it must be close enough to it. In a fluid phase moving with an average velocity v parallel to a solid surface, a boundary layer forms, in which the local velocities decrease toward the surface, down to zero in the molecular layer that is in contact with the surface. On its way to the surface (1 in Figure 1.5), A will be transported by a mixed mechanism, which may comprise both convection and diffusion far from the surface, but proceeds exclusively by diffusion close to it ("film diffusion"). If the catalyst is porous, pore diffusion (2 in Figure 1.5) is an additional mass transfer step that has to be taken in to account. The scheme of elementary steps is completed by the reverse transfer of the products through the pores and the boundary layer (6 and 7 in Figure 1.5).

Often, one of these reaction steps is significantly slower than the remaining ones, and thus determines both the rate law and the net rate of the process. It is very important to identify this rate-determining step (rds) because it must be the target of any strategy for further catalyst development. It makes not much sense, for instance, to find out how to accelerate the surface reaction, if the rate is limited by film diffusion. Likewise, film diffusion, charge-transfer processes, and the bulk conductivity may be rate-determining in electrocatalysis. In photocatalysis, the recombination rate of the generated charge carriers in the bulk or at the surface of the photoabsorber and the subsequent charge transfer processes may limit the overall reaction rate.

In this book, the coverage of mass transfer phenomena is limited to the identification of their influence and on the consequences of such influence on the kinetic rate laws (Section 2.6.2).

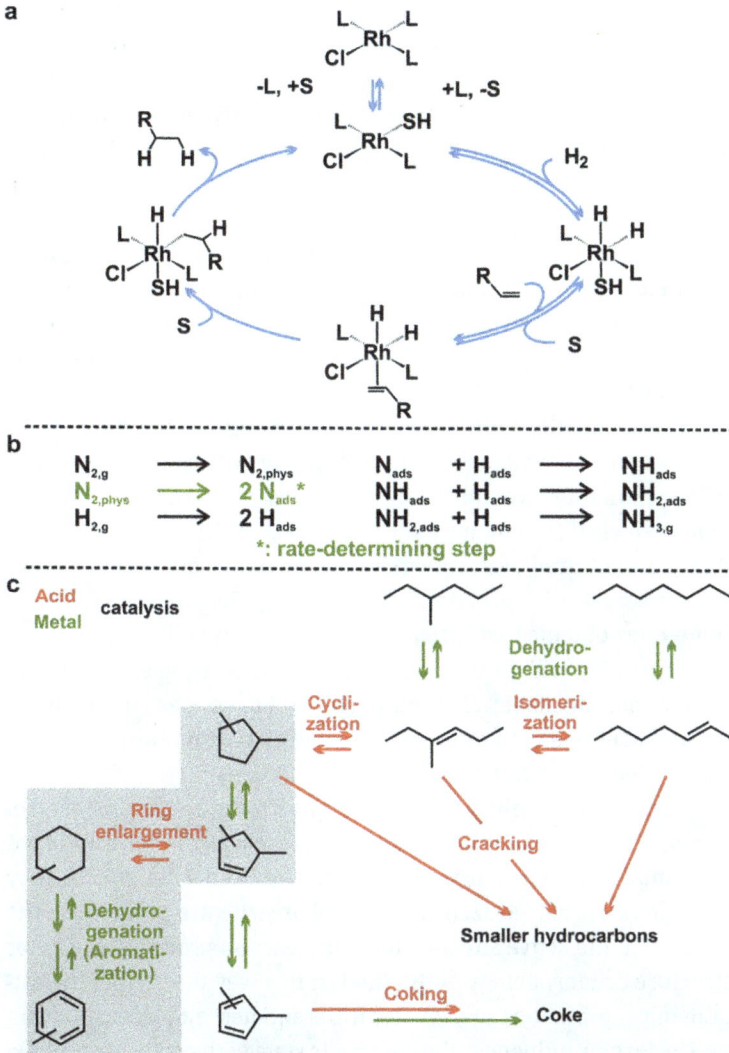

Figure 1.6: Examples of catalytic reaction mechanisms: (a) hydrogenation of an alkene at a homogeneous catalyst, (b) ammonia synthesis on a Fe surface (identification of N_2 dissociation as rate-determining step holds only for low reactant pressures), and (c) a reaction network for comparison: naphtha reforming over promoted Pt catalysts.

In catalyst research, the focus is on improvements in the sequence adsorption – surface reaction – desorption, with particular attention to the surface reaction. In coordination catalysis, a reactant is coordinated to the central atom, where it undergoes transformations before the product is released again (Figure 1.6a). A surface reaction could be written in a similar way, but here, the "central atom" is part of a solid surface. Instead of ligands, counter ions (often O^{2-}, OH^-) are bound to it with different binding types and forces. On a real surface, the active element may be present in different coordination geometries, and even oxidation states. Due to this complexity, there are only a few cases where proposals of complete catalytic cycles have been accepted as undisputed achievements by the scientific community. As an example, the mechanism of ammonia synthesis on Fe catalysts is shown in Figure 1.6b. It does not, however, specify the sites at which the steps proceed (see Section 4.2.3).

As a first step in the study of surface processes, researchers often try to identify the "active site." The active site is the *ensemble of surface atoms that participate in the rate- and/or selectivity determining step of the catalytic reaction mechanism*, but not yet with reactant(s) adsorbed. While this can be only a starting point in the elucidation of the surface processes, knowledge about the active site can already give valuable input for further catalyst improvement.

The next goal would be to identify a sequence of surface reaction steps, as exemplified in Figure 1.6b, and to find experimental evidence to prove or support its responsibility for the observed reaction rates. Notably, as different sites may catalyze a reaction with different rates and even via different mechanisms (cf. Section 4.5.6), it is important to make sure that the rates measured with research samples under maybe less severe conditions (e.g., at low pressures) or in special environments (e.g., in spectroscopic cells) are compatible with the rates obtained with real catalysts under a realistic (e.g., close to technical) reaction regime. We will, in the following, refer to a sequence of surface reaction steps as reaction mechanism, as shown in Figure 1.6b, whereas a scheme of (catalytic and/or noncatalytic) steps with intermediates desorbed from the surface (cf. Figure 1.6c) will be designated as *reaction sequence* or *network*.

In heterogeneous catalysis, the performance of the catalytically active element can be modified by the presence of additional components. These additives are referred to as promoters if they improve the performance. If they have adverse effects, they are called poisons. Promoters are categorized as chemical or structure promoters. The former interacts directly with the active site on which the reaction proceeds; the latter modifies the microstructure or morphology of the catalyst in a way that favors the catalytic performance. Chemical promoters are further differentiated into electronic and geometric promoters. The former influences the electronic state of the active site, while the latter modifies its structure. Modifying may mean destroying: such geometric promoters suppress undesired side reactions that require larger ensembles of atoms than the target reaction. Obviously, they are poisons for the undesired side reactions.

Work with promoters is a basic strategy in catalyst development (see Sections 4.2.4, 4.3.5, and 4.4.2). They are almost exclusively added during the synthesis of catalysts. In rare cases, they are provided or replenished with the reaction mixture. Poisons typically come with the feed, i.e., poisoning can be suppressed by appropriate feed conditioning. During catalyst synthesis, they can be added deliberately (as geometric promoters, see earlier), or they may get into the synthesis mixture unnoticed, when chemicals of insufficient purity are used. The latter may have caused one or the other contradiction between the experimental data obtained in different labs.

In some applications of catalysis, additives employed are referred to as co-catalysts. The term "co-catalyst" is not clearly defined. Dictionaries describe it as a substance, acting as a catalyst in tandem with another one. This covers additives playing very different roles: all kinds of promoters, but also activating reagents in polymerization, which convert the catalyst precursor into the active state, or ingredients of photocatalysts, which accelerate the electron transfer between the catalyst and the reactants, but depend on the previous photoexcitation. In this book, the term "co-catalyst" will be used only for the discussion of areas where it belongs to general terminology (polymerization and photocatalysis).

Catalytic reactions are performed in catalytic reactors, the design of which depends on the nature of the fluid phase. If the fluid is a liquid, the catalyst is often added as a dispersed powder. By agitating the liquid together with the powder, a slurry phase is created in this so-called slurry reactor. However, catalytic liquid-phase reactions can also be performed in a reactor, which is standard for catalytic gas-phase reactions – the fixed-bed reactor. A bed is created by filling solid particles into an empty tube. In the fixed bed, the bed is at rest and is passed by the gaseous and/or liquid reactants, typically downward. There are also moving beds where the solid catalyst is slowly removed at the bottom, and replaced at the top. When a bed of very small particles is exposed to a strong gas flow from below, it may get into a fluidization state, which is used for different forms of fluidized beds. For more details about these reactor types, see Section 2.8.

1.5 Surface catalysis: an art or a science?

Still, in the 1970s, when the author of this chapter made his acquaintance with catalysis, the statement "heterogeneous catalysis is an art; not a science" was quite popular. It summarized a practice in which the making of good catalysts involved many procedures, which were applied without really knowing the reasons for the improvements obtained. Reproducibility and scale-up of catalyst syntheses were often achieved only after tedious effort, and great differences in catalyst behavior between samples of identical structures (according to characterization data available) were not a rare case. Much of the know-how in preparation and handling of catalysts was proprietary and became available to the public only with large delay.

Actually, catalysis research had long gone beyond the initial trial-and-error strategies at that time. Invention and further optimization of many successful catalysts had been achieved on the basis of heuristic models, i.e. of simplified models of surface and surface processes, which were designed to reflect the known experimental facts without claiming the resolution of the molecular mechanisms.

Creating models of surface structure and surface processes for guidance of further research activity, which is sometimes referred to as knowledge-based approach, has remained a basic strategy in catalyst development. Models are, of course, much more refined nowadays. Together with contributions of surface science and theory, which will be outlined below, it has, meanwhile, resulted in a molecular-level understanding of surfaces and surface processes for quite a number of reaction systems. However, serious challenges remain all over the field of heterogeneous catalysis and, all the more, in electrocatalysis and photocatalysis. Moreover, while theory has achieved predictive power with respect to the behavior of catalytic sites for quite some catalyst types, the quantity of such sites available under reaction conditions when a catalyst was made according to a specific preparation protocol, is still far from being predictable.

If the predictive power of theoretical models indicates a grown-up science, surface catalysis is just coming of age. However, there are still catalyst types and reaction conditions for which the actual surface state is difficult to establish, and there are undisclosed reaction mechanisms and controversies about the active sites for important reactions. In catalyst preparation, reproducibility can still be a problem, e.g., when raw materials with different impurity levels are used or if complex preparation steps, dominated by nucleation and growth kinetics, such as precipitation, are crucial for the result. Still, there is a great deal of proprietary knowledge, related just to preparation and activation. Catalyst producers keep unauthorized people strictly off their recent products and allow academic researchers working with them only under tight confidentiality agreements.

Which are the reasons that a technologically important research field is only just about to come of age as a science after more than a century of effort by generations of scientists, after several contributions were awarded with Nobel Prizes (1908 – Wilhelm Ostwald, 1912 – Paul Sabatier, 1918 – Fritz Haber, 1932 – Irving Langmuir, 1963 – Karl Ziegler/Giulio Natta, 2007 – Gerhard Ertl)?

"The volume of the solids was made by god, their surface by the devil." This sentence by Wolfgang Pauli points out one of the reasons: characterization and theoretical description is more difficult for surfaces than for solids, because even ideal surfaces break the symmetry of a structure. The citation comes from a time when structure analysis of solids by diffraction methods was well developed, while the modern tools of surface analysis were not yet available. Meanwhile, surface science provides ample information about the structure, even of complex surfaces and about their interactions with adsorbates.

However, catalysis deals with real materials: metastable features like curvature or surface defects may be more relevant for it than equilibrated structures. Active sites or particles of the active element may be within the pore system of a support (cf. Section 2.3.1), where they are not well accessible for some characterization methods. They may be modified by interactions with the support, with promoters or poisons. Finally, as mentioned above, not all exposed sites of a catalytically active element may take part in the catalytic process, and often, it is only a minority that catalyzes the reaction of interest.

Likewise, among the adsorbates on a catalyst surface, the majority may be spectators under reaction conditions while the intermediates in a catalytic reaction sequence may be difficult to detect due to much shorter lifetimes.

Beyond this complexity of real catalytic materials, it is the abundance of reaction options in chemistry that presents ever new challenges to researchers. There is hardly any chemical reaction that could not be accelerated by catalysis. Likewise, the abundance of synthesis options has resulted in many new catalyst types. Over the past 50 years, chemists have discovered new types of porous materials almost every decade, from (synthetic) zeolites via zeotypes, (pillared) layered silicates, ordered mesoporous materials to metal-organic framework structures, and there is little reason to assume that this development may soon come to an end. New techniques to mix components of a solid on the molecular level were found; new approaches to finely distribute components on porous supports or on electrodes were described, up to an exact control of the particle size in the nm range (monodispersity). Some of this will be described later, in particular in Section 3.1. All the new materials have been and are being examined for their potential as catalysts or catalyst components for various reactions, which always includes a thorough characterization required to understand tendencies and rankings observed.

The recent development of the field might be described by pointing out three tendencies. Molecular-level understanding and predictability is being achieved for ever more complex catalytic reactions. Below this level, there is plenty of successful effort to understand details of surface structures and mechanisms for other catalytic processes, which are too complex to derive and validate models on the molecular level at this point of time. This stage is important, nevertheless, because it creates the basis for future studies of the fundamentals and contributes data relevant for the practical use of the catalyst type considered. Third, there is an ever extending exploratory work on new reactions and materials. Such work cannot be performed with the full arsenal of methods employed for the in-depth studies just mentioned. Its quality is definitely increasing as well, but the choice of experimental capacities used for it is determined by a strong focus on application.

The present molecular insight into many surface catalytic processes is due to the achievements of a second strategy in catalysis research, which has complemented and supported work with the knowledge-based approach since the early 1970s: the study of idealized models of catalytic surfaces and sites in other fields of chemistry, in particular, physical chemistry (surface science) and theory. There have also been impulses from molecular catalysis due to the analogy of active site models and coordination compounds. In recent years, work with quasimolecular structures, in which one or more oxygen atoms of the support are ligands in the coordination sphere around the active element (surface organometallic chemistry, cf. Section 3.1.3.4), has contributed valuable insight into reaction mechanisms and potential structures of active sites.

Surface science deals with structural and reactivity phenomena at the interface of solids with the vacuum or with fluid phases. Its close relation to catalysis was realized early on when the focus was still on interfaces between highly idealized, thermodynamically stable surfaces and vacuum. Researchers soon noticed that reactions between adsorbates or between adsorbates and the residual gas phase can also be studied under vacuum, and that metastable (highly indexed, see Section 2.2.1) surfaces exhibit particular catalytic effects [9]. The ongoing development brought a great wealth of insight in catalytic phenomena, e.g., the elucidation of the mechanism of ammonia synthesis over Fe [10] or the influence of subsurface hydrogen in selective hydrogenation over Pd [11], to name only a few.

Catalysis has enormously benefited from surface science, although the relation between both areas is complicated by two gaps: the pressure gap and the materials gap. The pressure gap is caused by the extreme difference in pressures under which experiments are performed. Almost all characterization techniques applied in surface science require ultrahigh vacuum (UHV), i.e., pressures below 10^{-8} mbar. Only in the 1990s did powerful nonlinear spectroscopic techniques become available that allow observing adsorbate signals from surfaces of only a few cm^2 in the presence of gas-phase reactants (see Section 3.4.8.2). On the other hand, catalytic reactions are performed at atmospheric pressure or above, up to 300 bar. As a result, catalytic reactions almost always proceed at higher surface coverages than in the reaction experiments of surface science. This changes the availability of sites free for the surface reaction; it may change the rate-determining step of the reaction mechanism (see example in Section 2.5.1) or perturb the surface exposed (see examples in Sections 2.5.1, 4.2.4, or 4.9.2). At the latter point, the pressure gap superimposes the materials gap, which relates to the fact that real catalysts are always more complex than the models for which the methods of surface science can provide meaningful information.

The problem of these gaps has been realized early on, and effort to narrow them (from both sides) or even bridge them has often provided new impetus to research. Working with moderately simplified materials, scientists in applied catalysis try and specify the sites responsible for the reaction of interest more reliably and under conditions extending to lower pressures. Surface scientists create more complex models within the limitations imposed by their characterization techniques. The potential of these methods has greatly extended over the years, often based on developments initiated in the surface-science labs. Figure 1.7a illustrates the evolution of models on the surface science side. Starting from low-indexed single-crystal metal facets, it arrived at nm-sized metal particles deposited on ordered oxide layers (e.g., Al_2O_3) grown on single-crystal facets of alloys (e.g., NiAl). The oxide layer is only a few atomic layers thick to retain the metallic conductivity required for methodical reasons [12]. The most recent step is the work with monodisperse metal nanoparticles of controlled size [13], which are now also deposited on glassy carbon substrates for electrocatalytic applications [14].

Theory is the second discipline contributing to the progress of catalysis by model studies. Theoretical treatment is, however, possible only for finite molecules or for infinite periodic structures built by repetition of a finite structural unit – both descrip-

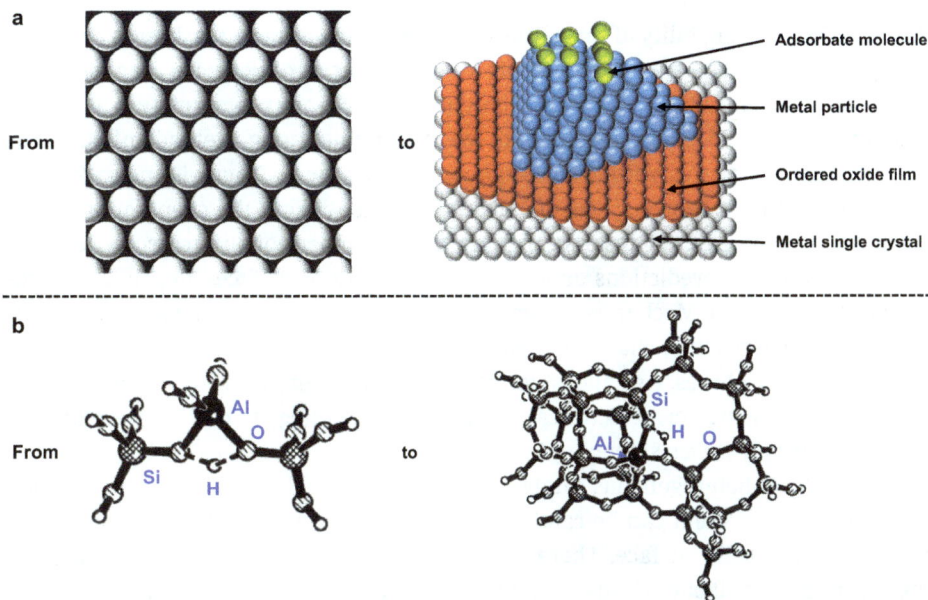

Figure 1.7: Development of model complexity in studies of surface science (a) and theory (b): (a) from stable single-crystal facets to models for oxide-supported nanoparticles; (b) from a minimum structure required to capture properties of zeolite Brønsted sites to extended models that allow studying effects of the framework geometry on the site and the influence of van der Waals interactions between reactant and zeolite surface on adsorption and catalysis; (b) adapted with permission from ref. [15]. Copyright 2001 American Chemical Society.

tions do not comply with catalytic surfaces. Theoreticians solved this problem by defining in the catalyst surface pseudomolecules that contain the assumed active site (cluster approach) or by working with artificial periodic structures, the repetitive unit of which contains the solid-fluid (mostly solid–gas) interface (see Section 3.5). These models can be treated with wave function-based methods or by using density-functional theory (DFT). The latter, though less accurate, can manage models of larger size and containing heavier atoms and is, therefore, applied predominantly.

The disparity between these models and the reality is, on one hand, due to limitations in the size of clusters or repetitive units imposed by computer capacities, for the cluster approach also by the problem that discontinuities at the boundary between cluster and surroundings are difficult to avoid completely. On the other hand, energies resulting from these calculations refer to 0 K. In addition, stability of adsorbates should be assessed by their Gibbs free energy rather than by the adsorption energy of one molecule.

In recent decades, great progress has been made to narrow these gaps. In the cluster approach, the size of the pseudomolecules was increased (an example for clusters used to evaluate acidic properties of zeolites is shown in Figure 1.7b), and methods to account for the influence of the chemical environment on clusters were developed (Section

3.5.2). In DFT, the eligibility of functionals for treating the different tasks was improved using benchmark data from experiments (e.g., frequencies of IR modes, energetic data like adsorption enthalpies and activation energies) or from wave function-based calculations. DFT-based energy evaluations were combined with methods of statistical thermodynamics to derive thermodynamic state functions. There is, nowadays, no serious research on basic aspects of surface catalysis without the cooperation of theoreticians. Theory can, for instance, help in the assignment of spectroscopic features. It can give realistic predictions on adsorbate coverages of surfaces, depending on the gas-phase conditions. Mechanistic discussions are aided by comparing energy barriers for the available reaction channels. Meanwhile, theoreticians compare activities of elements (or mixed phases, e.g., alloys) for certain reactions or electrode processes in the search for promising new (electro)catalysts. This new strategy in catalyst development is referred to as *"in silico* screening."

Modeling of heterogeneous catalysts by related molecular catalysts is most appropriate for reactions that can be catalyzed by an element either in the environment of ligands or on a solid surface. There are a few cases where this takes place, e.g., for the metathesis of alkenes or alkene polymerization reactions (cf. Section 4.5.2.3). The analogy between homogenous and heterogeneous catalysis is less stringent for acid catalysis, where molecular catalysts operate in the liquid phase, while the most important processes catalyzed by solid acids proceed in the gas phase where solvation effects are missing. As mentioned in Section 1.4, the analogy between the coordination compounds and the active sites of heterogeneous catalysts is obstructed by the different nature of the ligands. In addition, influences from the surroundings (e.g., electrical fields) are not covered by molecular models. Despite these disparities, both reaction mechanisms in coordination catalysis and stoichiometric reactions on ligands of coordination compounds have often served as a reservoir for options on hypothetic mechanistic steps in surface catalysis. Moreover, the above-mentioned incompatibilities are, meanwhile, being addressed in surface organometallic chemistry (cf. Section 3.1.3.4). Such overlap in the scientific interests of researchers in both fields (which also includes the effort in heterogenization of molecular catalysts, cf. Section 2.3.3), has been the basis for an ongoing symposium series devoted to the mutual exchange between the homogeneous and the heterogeneous catalysis communities, which are rather different in many other aspects of their scientific work. While this series was started in the 1970s, discussion between the communities has now intensified on a broad scale.

Despite the undoubted success of the knowledge- and model-based approaches to surface catalysis, they were challenged by the radically different approach of "combinatorial catalysis" in the 1990s. Inspired by the combinatorial methods in the synthesis of biomolecules and drugs, this approach promised an extreme acceleration of the trial-and-error catalyst screening by automatization and parallelization of synthesis and testing, combined with an automated evaluation of the test results (see Chapter 5). As the combinatorial aspect (creation of individuals from a finite set of building blocks by shuffling their sequence in the molecular structure) is alien to the synthesis and testing of solids, the approach is, meanwhile, referred to as high-throughput screening (HTS). Initial claims that HTS would antiquate traditional catalyst research have been attenuated, and searches to be performed with HTS are designed using the full level of insight provided by traditional catalyst research. While industrial catalyst development uses both "knowledge-based" and HTS strategies, HTS technology is less utilized in academic research due to its significant cost and the stronger focus of the academic groups on fundamental problems.

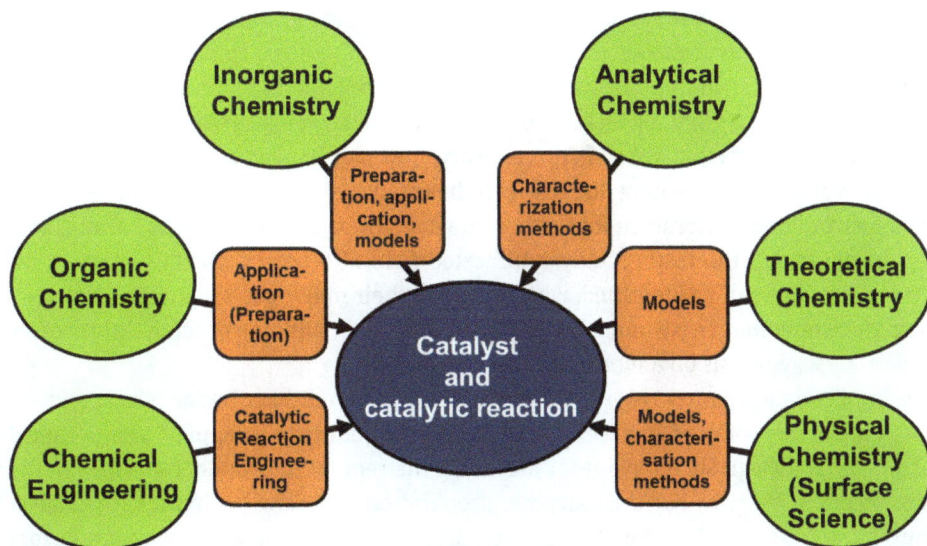

Figure 1.8: Surface catalysis as an interdisciplinary field of research.

1.6 Catalyst research: an overview

Previous sections have shown catalysis as a highly interdisciplinary field (see Figure 1.8). Catalysts, including model catalysts used in research, are mostly inorganic matter. Their preparation routes largely belong to the field of inorganic chemistry although syntheses with the catalytic elements ligated by organic ligands draw increasing attention, and hybrid materials combining organic and inorganic building blocks are extensively studied. While some of the largest processes with heterogeneous catalysts (and the most important applications of photo- and electrocatalysis) belong to inorganic chemistry as well, most reactions that utilize solid catalysts belong to organic chemistry. In the characterization of catalysts, many instrumental techniques of analytical chemistry are employed, often in special adaptations to the needs of catalyst research. Usually, appropriate discussion of the resulting data is impossible without a profound knowledge of the physical processes applied for the measurements, which adds physics and physical chemistry to the cooperating sciences. The enormous impact of model studies in physical chemistry (surface science) and theoretical chemistry on work with real catalysts has been outlined in Section 1.5.

Last but not the least, heterogeneous catalysts are applied in commercial-scale reactors to convert large product flows or in catalytic converters to detoxify various flue-gas streams. For this purpose, chemical engineers deal with catalysts using their specific methods to describe reaction rates, including the influence of mass and heat transfer, to model the interplay between hydrodynamics, reaction and transport rates in catalytic reactors, to find optimum reactor sizes for the requested product capacities, to design plants (plant models) where these reactors (and their models) are combined with steps of feed conditioning and product separation, and to optimize such plants. It goes without saying that all these steps are subject of ongoing research and improvement. Most of the methods used in this area are, however, beyond the scope of this book. For some of them still dealing directly with the catalyst, the reader will find brief introductions and useful literature for further information below (see Sections 2.6 and 2.8).

In the highly interesting and multifaceted field of research on surface catalysts, three directions can be identified, which differ in their major goals: application of catalysts, development of new/improved catalysts, and understanding of catalytic processes and catalyst synthesis on a molecular basis (Figure 1.9).

Application-oriented research intends to create conditions under which a given catalyst can be produced in necessary amounts and quality to be applied in a chemical plant, and to design such a plant for achieving the requested product capacities in the most economical way. For this purpose, preparation of a new (or improved) catalyst must be scaled up to allow for making kilograms or tons of the material without the loss of catalytic performance. Material data like densities or mechanical strength are to be determined, and reaction rates are to be measured under a wide range of (stationary) reaction conditions. Kinetic models capable of describing the observed catalytic behavior are to be developed and compared with respect to their performance. As the kinetic models become subroutines in larger packages employed for modeling reactors and even plants, focus is on the manageability of the model rather than on compatibility with the latest insight about the underlying reaction mechanism. Reactor models are established by combining the kinetic models with hydrodynamic models of the catalytic reactors.

The remaining two directions – catalyst development and molecular understanding of the catalytic phenomena – are strongly interwoven. Both directions benefit from the effort to identify active sites and the reaction mechanisms: on the one hand, its results may serve to establish more efficient heuristic strategies in the development of new catalysts; on the other hand, it can raise new ideas about more realistic catalyst models or relevant reaction steps to be studied with the methods of surface science and theory. Likewise, the search for relations between preparation conditions and the resulting surface structures is useful for the work on both the creation of high-performance catalytic materials and the elucidation of molecular mechanisms operating

Three major orientations of catalysis research

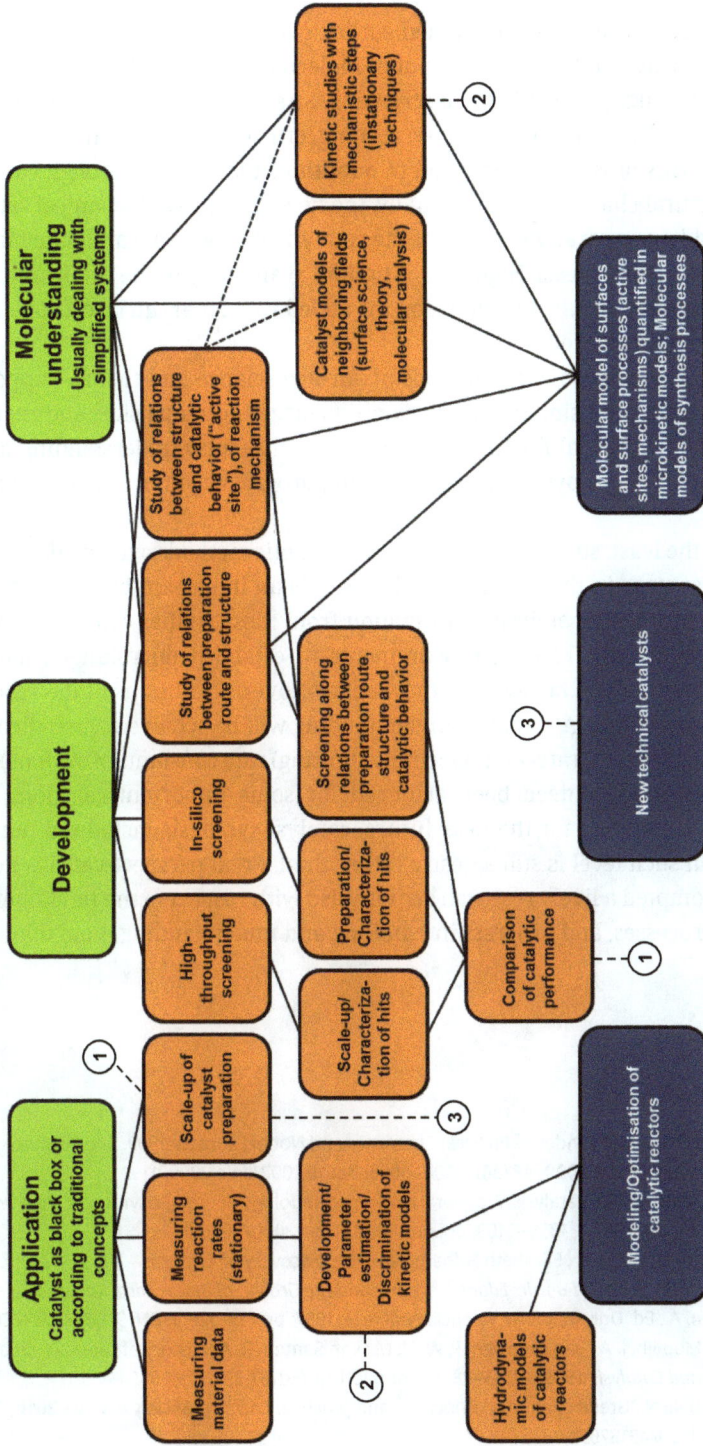

Figure 1.9: Fields of catalysis research and their interactions in the development of new catalytic materials and applications.

during the preparation steps. As mentioned earlier, catalyst development by heuristic strategies or, where available, with the input of molecular-level insight competes nowadays with high-throughput and *in silico* screening. Strategies can also be combined in a larger effort. Actually, for hits of *in silico* screening, efficient preparation methods are still to be found. Hits of HTS are prepared in quantities that allow thorough catalytic testing and structural characterization. Finally, the new or improved technical catalyst will be identified in comparative studies of the catalytic behavior of candidate materials, which include stability tests. High cost or limited availability of the catalyst components may influence the result, depending on the added value that can be created in the production process considered.

Work on catalyst models with methods of surface science and theory, supported by fundamental studies on simplified real catalysts targeting active sites and reaction mechanisms, remains crucial for further progress in molecular understanding of the catalytic phenomena. It is now increasingly driving progress even in the development of real catalysts.

Last but not the least, studies with instationary kinetic methods, which allow measuring rates of fast steps in the reaction mechanism, is an important branch of fundamental research in surface catalysis. Information from these studies can be combined with data from stationary kinetic measurements to establish microkinetic models, which describe the catalytic behavior with rate laws covering all steps of the reaction mechanism (cf. Section 4.6). A complete understanding will be achieved when theory is able to predict the relevant (rate- and selectivity-limiting) rate constants of such models.

While this level has indeed been achieved for some important reactions, and further progress is pending for the near future, catalyst synthesis includes a number of steps for which such level is still a desire rather than a real perspective. This deficit has, however, prompted a lively research activity also with respect to the fundamentals of preparation processes, and progress in handling and understanding them may soon be expected.

References

[1]   Ertl, G., Wilhelm Ostwald: Founder of Physical Chemistry and Nobel Laureate 1909. *Angewandte Chemie International Edition* **2009**, *48* (36), 6600–6606. doi:10.1002/anie.200901193

[2]   Makosza, M., Phase-transfer catalysis. A general green methodology in organic synthesis. *Pure and Applied Chemistry* **2000**, *72* (7), 1399–1403. doi.org/10.1351/pac200072071399

[3]   Laidler, K.; Cornish-Bowden, A., Elizabeth Fulhame and the Discovery of Catalysis – 100 Years before Buchner. In *New Beer in an Old Bottle: Eduard Buchner and the Growth of Biochemical Knowledge*, Cornish-Bowden, A., Ed. Universitat de València: València, 1997; pp 123–126. ISBN: 9788437033280

[4]   Kieboom, A. P.; Moulijn, J. A.; van Leeuwen, P. W. N. M.; van Santen, R. A., History of catalysis. *Studies in Surface Science and Catalysis* **1999**, *123*, 3–28. doi.org/10.1016/S0167-2991(99)80004-4

[5]   Anastas, P.; Eghbali, N., Green chemistry: Principles and practice. *Chemical Society Reviews* **2010**, *39* (1), 301–312. doi 10.1039/b918763b

[6]   Perathoner, S.; Centi, G.; Gross, S.; Hensen, E. J. M., Science and Technology Roadmap on Catalysis for Europe, https://www.euchems.eu/wp-content/uploads/2016/07/160729-Science-and-Technology-Roadmap-on-Catalysis-for-Europe-2016.pdf (accessed Feb. 18, 2023).

[7]   Catalysis – A Key Technology for Sustainable Economic Growth (Roadmap for Catalysis Research in Germany). http://gecats.org/gecats_media/Urbanczyk/Katalyse_Roadmap_2010_engl_final.pdf; https://dechema.de/dechema_media/Downloads/Publikationen/Katalyse_Roadmap_2022_ezl.pdf (both accessed April 11, 2023).

[8]   Catalyst Market by Type (Zeolites, Metals, Chemical Compounds, Enzymes, and Organometallic Materials), Process (Recycling, Regeneration, and Rejuvenation), and Application (Petroleum Refining, Chemical Synthesis, Polymer Catalysis, and Environmental): Global Opportunity Analysis and Industry Forecast, 2021–2030; https://www.alliedmarketresearch.com/catalysts-market (as of April 2023).

[9]   Somorjai, G. A.; Blakely, D. W., Mechanism of catalysis of hydrocarbon reactions by platinum surfaces. *Nature* **1975**, *258*, 580.

[10] Ertl, G., Surface science and catalysis-studies on the mechanism of ammonia synthesis: The P. H. Emmett award address. *Catalysis Reviews* **1980**, *21* (2), 201–223. doi 10.1080/03602458008067533

[11] Doyle, A. M.; Shaikhutdinov, S. K.; Jackson, S. D.; Freund, H. J., Hydrogenation on metal surfaces: Why are nanoparticles more active than single crystals? *Angewandte Chemie-International Edition* **2003**, *42* (42), 5240–5243. doi 10.1002/anie.200352124

[12] Freund, H. J.; Bäumer, M.; Kuhlenbeck, H., Catalysis and surface science: What do we learn from studies of oxide-supported cluster model systems? *Advances in Catalysis*, **2000**, *45*, 333–384. doi 10.1016/S0360-0564(02)45017-1

[13] Mostafa, S.; Behafarid, F.; Croy, J. R.; Ono, L. K.; Li, L.; Yang, J. C.; Frenkel, A. I.; Roldan Cuenya, B., Shape-dependent catalytic properties of Pt nanoparticles. *Journal of the American Chemical Society* **2010**, *132* (44), 15714–15719. doi 10.1021/ja106679z

[14] Reske, R.; Mistry, H.; Behafarid, F.; Roldan Cuenya, B.; Strasser, P., Particle size effects in the catalytic electroreduction of CO_2 on Cu nanoparticles. *Journal of the American Chemical Society* **2014**, *136* (19), 6978–6986. doi 10.1021/ja500328k

[15] Sierka, M.; Sauer, J., Proton Mobility in Chabazite, Faujasite, and ZSM-5 Zeolite Catalysts. Comparison Based on ab Initio Calculations. *The Journal of Physical Chemistry B* **2001**, *105* (8), 1603–1613. doi 10.1021/jp004081x

2 Surface catalysis: the scene and the play

This chapter summarizes basic knowledge on solids and their surfaces, on the elementary steps of catalytic reactions in the absence and in the presence of an electric potential, on rate laws used in application-oriented research on surface catalysis, and on catalytic reactors. While much of these traditional contents depict the catalyst as a scene in which the catalytic reaction is performed, the dynamic nature of the surface region, which is exemplified at many places throughout this book, should be kept in mind: other than in the theatre, the chemical scene can be strongly upset by the actors appearing in the magnificent catalytic play.

The first sections deal with topics that the reader should be familiar with from other courses or textbooks. They are briefly repeated with special reference to their application in catalysis, in order to provide a sound basis for the following chapters about tools of catalyst research and for the more profound discussion of interactions and mechanisms occurring on surfaces.

2.1 Solids

Although the catalytic cycle takes place at the surface, the solid exposing this surface is of interest as well. Its spatial and electronic structure influence the catalytic process via the surface.

Solids employed in surface catalysis may be metals, oxides, sulfides, and sometimes, chlorides or other compounds. They may be amorphous or crystalline, pure or mixed phases. In the crystals of mixed metals (alloys), the distribution of components may be random or ordered, i.e., forming a superstructure. Ordered alloys are, however, rare, because they require equilibration at high temperatures, which is usually avoided in catalyst preparation. In mixed (i.e., ternary and higher) oxides, the distribution of cations may be random or ordered, as well. Cation superstructures are characteristic of oxide compounds like perovskites, spinels, or ferrites, some of which have received much attention in the search for new redox catalysts. Due to the existence of complex and aggregated anions and cations,[1] the diversity of structures is enormous in this area. For more detail, the reader is referred to textbooks of inorganic chemistry.

In surface catalysis, much attention is paid to the amorphous state, because amorphous materials often exhibit high surface areas and metastable surface structures, which may possess interesting reactivities. Under thermal stress, the amorphous material may be partly or fully converted to a crystalline phase. Metal particles and many

---

[1] Anions – e.g., $(SO_4)^{2-}$, $(Mo_7O_{24})^{6-}$, $(PW_{12}O_{40})^{3-}$, up to charged networks like framework anions in clays, zeolites, layered silicates, etc., cations – e.g., NH_4^+, $(VO)^{2+}$, but also cationic layers in layered double hydroxides like hydrotalcites.

https://doi.org/10.1515/9783110632484-002

other materials, among them, phases with very high porosity like zeolites or metal-organic frameworks (MOFs), are crystalline.[2] When a crystal extends over the whole volume of a particle, this particle is monocrystalline. However, particles often consist of several crystals of the same structure, but erratically wedged together forming irregular grain boundaries. Such particles are polycrystalline (cf. Figure 3.72 below). Amorphous particles containing crystalline sections are semicrystalline. In complex materials, polycrystalline particles may contain segments that feature different crystal structures.

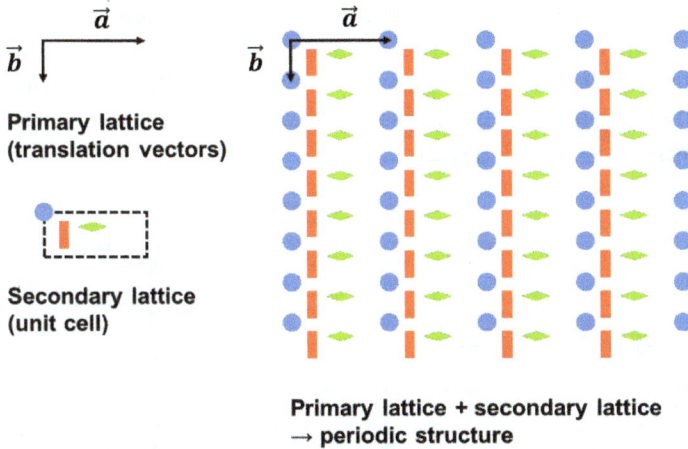

Figure 2.1: Two-dimensional lattice characterized by translational vectors $\vec{a}$ and $\vec{b}$ (primary lattice) and a non-primitive unit cell (secondary lattice). The lattice results from shifting the unit cell according to the pattern defined by the primary lattice.

Crystalline matter can be described as a translational lattice, in which the space is filled by the periodic repetition of a basic arrangement of atoms referred to as unit cell. The unit cell is shifted along three translation vectors, $\vec{a}$, $\vec{b}$, and $\vec{c}$, so that the position vectors of all its repetitions can be expressed by:

$$\vec{r} = n_a\vec{a} + n_b\vec{b} + n_c\vec{c} \tag{2.1}$$

where n_a, n_b, and n_c can be any natural number or zero. The operation is illustrated by a two-dimensional lattice in Figure 2.1. The basis of translation vectors describing the periodicity of the structure is referred to as primary lattice and the unit cell as secondary lattice. When the unit cell contains only one atom, the lattice is called primitive.

---

2 There is an amorphous state also of metals (for more information, see ref. [1]), but these materials are as yet rare and hardly relevant for catalysis.

Depending on the relations between the translation vectors, six crystal systems can be distinguished. Among them, there are three systems with $\vec{a}$, $\vec{b}$, and $\vec{c}$ at right angles, which differ in the relations between their lengths: all vectors of identical lengths in the cubic system, one of them deviating in the tetragonal system, and three different lengths in the rhombohedral system. In the remaining systems (hexagonal, monoclinic, and triclinic), some of the angles between translation vectors deviate from 90°. Most of the crystal systems are further differentiated when additional positions are occupied within the geometrical space spanned by the translation vectors, e.g., in its center or on its facets. This results in the 14 Bravais lattices, which can be found in any textbook of inorganic chemistry. It should be noted, however, that only the primitive versions of the lattices are usually depicted. A stronger populated unit cell will define as many identical sublattices as there are atoms in it. Their spatial arrangement is defined by the arrangement of the atoms in the unit cell (Figure 2.2).

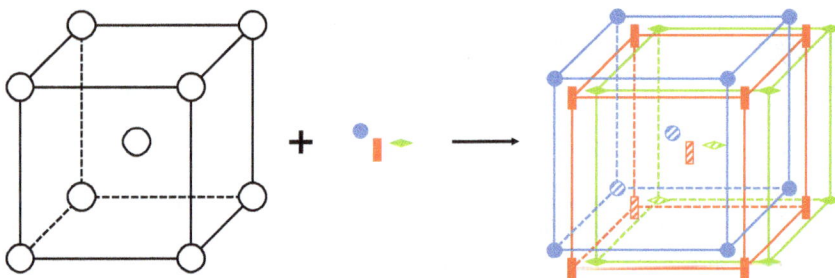

Figure 2.2: Non-primitive bcc lattice. Three sublattices originate from three atoms in the unit cell.

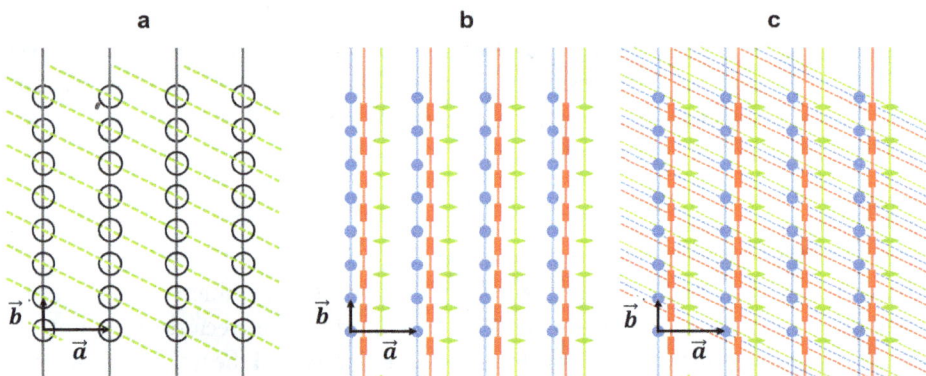

Figure 2.3: Lattice planes in primitive and non-primitive two-dimensional lattices: (a) primitive lattice with (100) and (110) planes (grey and green, respectively); (b, c) non-primitive lattice with (100) and (110) sub-planes, respectively.

The lattice plane is a crucial concept for the description of crystal structures. Lattice planes are sets of infinite numbers of parallel planes that intersect with the atom positions in the crystal. A set of lattice planes relating to a particular atom in a unit cell must cover all atoms of this kind in the crystal, i.e., in a primitive lattice, it covers all available atoms (Figure 2.3a). Lattice planes are labeled by Miller indices, which report the reciprocal intercepts of the planes with the axes next to the origin. For the dashed green lattice planes in Figure 2.3a, the intercepts are 1, 1, and infinite (for an assumed $\vec{c}$ perpendicular to $\vec{a}$ and $\vec{b}$). The corresponding set of Miller indices is (110). The intercepts of the grey planes are 1, ∞, and ∞, the Miller indices are (100). The number of possible sets of lattice planes is unlimited. Each one defines a direction in space. The distance between the members of a set depends, in a characteristic manner, on the direction of the planes and on the primary lattice, i.e., on lengths and directions of the translation vectors. Distances are largest between planes characterized by the lowest indices.

If the unit cell contains more atoms, these atoms define subsets of lattice planes, which have the same direction and spacing as the original ones (Figure 2.3b and c). For each lattice plane, there are as many subplanes as atoms in the unit cell. The distances between the planes of different subsets change with the orientation of the planes in a way that is determined by the atom arrangement in the unit cell, i.e., by the secondary lattice (cf. Figure 2.3b and c).

Solids are stabilized by the lattice energy which results from the bonding forces between their constituents. These forces may be covalent or ionic. Often, both electrostatic and covalent effects contribute to the bonding forces. Molecular crystals, in which periodic arrays of molecules are kept together by weak van der Waals or dipole forces, are unsuitable for catalysis due to their low stability.

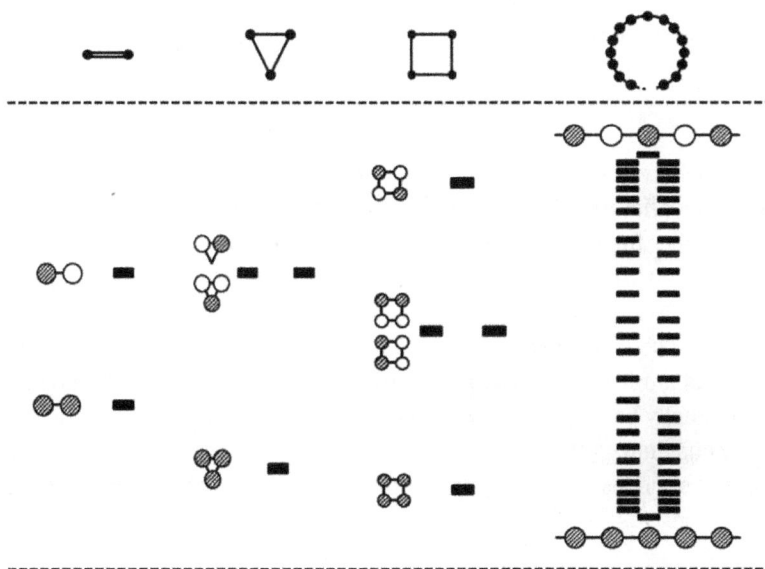

Figure 2.4: Energy levels in cyclic molecules formed by atoms via the interaction of s-orbitals. The sign of the orbitals is designated by their hatching. Adapted from ref. [2] with permission from Wiley-VCH Verlag GmbH & Co. KGaA, Weinheim, Germany.

In amorphous structures, covalent bonds are localized and involve only a few partners. In crystals, the periodicity of the structure results in the formation of energy bands. They consist of bonding orbitals with extremely narrow energy spacing that ideally extend over the whole crystal. The presence of bands results from the fact that a covalent bond involving identical contributions from n atoms forms n molecular orbitals[3]: 2 orbitals (bonding/antibonding) result from 2 atoms, 3 orbitals (bonding/nonbonding/antibonding) from 3 atoms, etc. Figure 2.4 shows this for s orbitals arranged in cyclic systems. A large cyclic system with non-perceptible curvature produces a spread of individual molecular orbitals with very small energy differences between them: this is already close to the case of a macroscopic lattice with myriads of atoms, though only a one-dimensional one.

As in the structures with only a few members, the lowest orbital of the band is most bonding, the highest one most antibonding. In between, the contribution to bond stability varies, correspondingly. In Figure 2.4, the sign of the atomic wave function contributing to the molecular orbital is designated by the hatching: the signs are the same at all atoms in the most bonding orbital and completely alternating in the most antibonding one, which has nodes between all atoms. Within the band, the number of nodes between atoms changes correspondingly.

In physics, the wave functions Ψ_k of the band orbitals (Bloch waves) are written as:

$$\Psi_k = \sum_n a_n \chi_n \tag{2.2}$$

where the χ_n are the basis functions at the lattice points, which are identical except for the position of their origins, and

$$a_n = exp(i\,n\,k\,a) \tag{2.3}$$

where a is the distance between the atoms of the one-dimensional lattice

The complex exponential function in a_n, which is related to the sine and the cosine of the argument,[4] describes the influence of nodes, which increases with the value of the parameter k. k has the dimension of a reciprocal length. At $k = 0$, $a_n = 1$ for all n, i.e., there are no nodes at all. At $k = \pi/a$, $a_n = (-1)^n$, i.e., the sign of the basis function alternates with n, resulting in nodes between all atoms. Due to the oscillating nature of the (co)sine, an increase in k beyond π/a brings no new information. Therefore, k is considered only between 0 and π/a (actually between $-\pi/a$ and $+\pi/a$, or $|k| \leq \pi/a$). Between $k = 0$ and $k = \pi/a$, the a_n are mostly imaginary, as wave functions are, anyway. They can be thought of as describing the number of nodes increasing with k in the way of a one-dimensional wave with decreasing wave length.

---

3 The following discussion can be found in more detail in refs. [2] and [3].
4 cf. the Euler formula: $exp(i\,n\,k\,a) = \cos(n\,k\,a) + i \sin(n\,k\,a)$; for $k = \pi/a$: $exp(i\,n\,\pi) = \cos(n\,\pi) + i \sin(n\,\pi) = (-1)^n$.

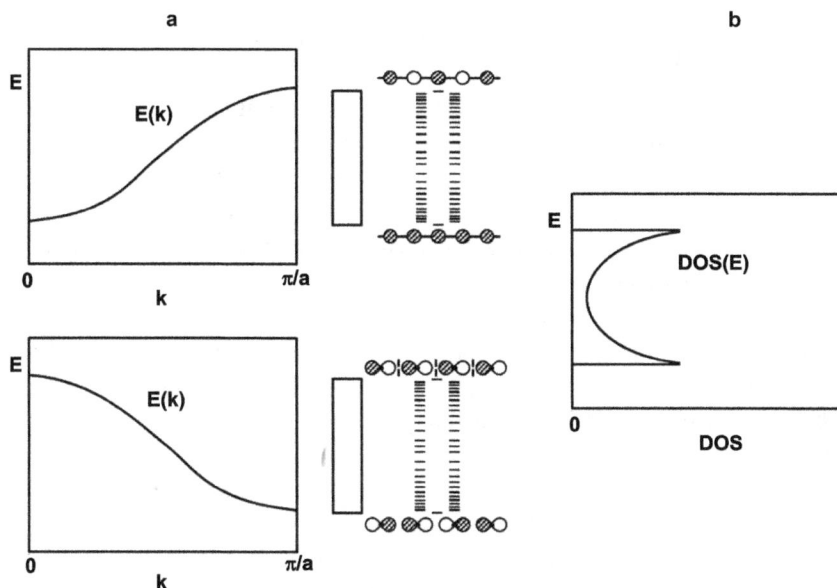

Figure 2.5: Band structure, symmetry of basis functions, and density of states. Adapted from ref. [2] with permission from Wiley-VCH Verlag GmbH & Co. KGaA, Weinheim, Germany.

When the energies of these levels are plotted vs. their differentiating parameter, k, a curve as shown in Figure 2.5a (upper diagram) results. The curve is not continuous, as the number of levels is finite though large. The width of the band depends on the degree of overlap between the involved atomic orbitals. Obviously, extended orbitals are formed only by the valence levels, and the smaller low-lying levels remain localized. In transition metals, the d bands are narrow, while the next s bands are wide.

The course of the energy dispersion curve is also influenced by the symmetry of the orbitals involved. Figure 2.5a shows also a band developed from p orbitals directed along the "crystal" extension. Obviously, the $k = 0$ combination is most antibonding here, and alternating signs ($k = \pi/a$) result in the most binding state. The resulting band structure mirrors the one obtained from the s orbitals.

Crystals are rarely one-dimensional. When a second direction is added (cf. Figure 2.1), nodes can occur in both directions. The k parameter, which describes the abundance of nodes, must now be able to differentiate the directions. It becomes two-dimensional as well, i.e. a vector: a wave vector. $\vec{k}$ describes waves in the contributions of basis functions at the various atom positions to the total wave function Ψ_k (eq. (2.2)), and these waves can now proceed in any direction of the two-dimensional lattice. $\vec{k}$ can be any sum of its components $\vec{k}_a$ and $\vec{k}_b$ along the unit vectors $\vec{a}$ and $\vec{b}$, which have the ranges $|k_a| \leq \pi/a$ and $|k_b| \leq \pi/b$. Obviously, the range that $\vec{k}$ can take is a rectangle, or a parallelogram for skew translation vectors. It is called the Brillouin zone. In three-dimensional space, it becomes a three-dimensional body, with extensions reciprocal to the lengths of the lattice vectors.

The discussion of three-dimensional band structures even of ideal lattices is beyond the scope of this book. There is, however, a convenient way to summarize such information: the density of states (DOS, Figure 2.5b). Dropping the information on spatial extension and symmetry properties of the basis functions, it reports how many states, n_{St}, there are in an interval between E and $E + dE$: $DOS(E) = dn_{St}(E)/dE$. In the physicochemical literature dealing with catalysts, DOS information is sometimes discussed. In this book, however, bands will be usually depicted as simple boxes indicating their energy extension, as shown right from the diagrams in Figure 2.5a. Seeing these boxes, the reader should, however, bear in mind some conclusions of this section:

- In each band, the lowest levels are bonding and the highest levels are antibonding. The bands will be filled with electrons, according to their availability. Only partly filled bands contribute to bonding; in full bands, the antibonding contributions cancel out the bonding ones.
- Insertion or withdrawal of electrons to or from a partly filled band affects the bonds between the atoms. The effect depends on the degree of band filling: bonds are destabilized by electron withdrawal at <50% band filling, but by electron insertion above 50%. Such electron shifts can occur during adsorption of molecules on metals. Related to a bulk band, the effect of this charge transfer would be negligible, but the electrons (or holes) remain near the surface attracted by the charge on the adsorbate. Destabilization due to such charge transfer may result in surface reconstruction phenomena (see Section 2.2.1).
- The extension of bands on the energy axis increases with the size (and, hence, the overlap) of the basis functions.
- Bands have symmetry properties depending on the shape of the basis functions. This is most pronounced in d bands and plays a major role in the bonding between adsorbates and surfaces.

Crystalline solids with a partly filled band are metals. The highest level occupied at 0 K is called the Fermi level (Figure 2.6a). Above 0 K, there is a thermal energy distribution of the electrons: some of them populating higher levels are missing at lower levels (Fermi-Dirac distribution, indicated in Figure 2.6a). The availability of free levels closely above the Fermi level is a prerequisite for metallic conductivity, because the kinetic energy of moving electrons cannot be accommodated in a full band.

Electrons can leave the metal after absorbing a sufficient amount of energy, e.g., from electromagnetic radiation or by collision with other electrons.[5] The energy level at which the electron is no longer influenced by the solid, but is still at rest (i.e., has no kinetic energy), is the vacuum level (Figure 2.6a). The difference between Fermi level

---

5 Only ionization into the vacuum will be considered here to avoid complications from interactions with gas molecules.

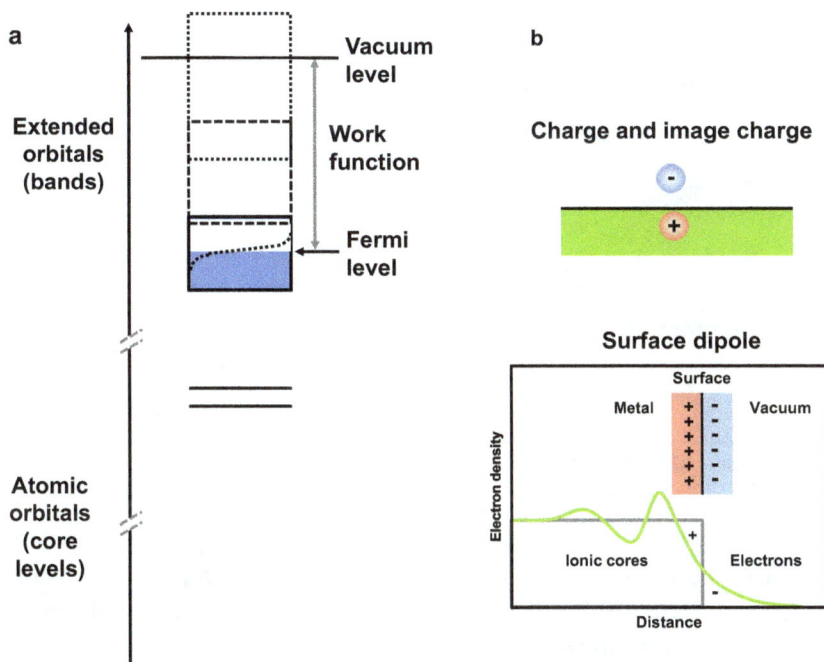

Figure 2.6: Electronic properties of metals: (a) core levels and extended states (bands), work function; (b) contributions to the work function: the image charge and the surface dipole (illustrated for the model metal jellium). (b) Adapted from ref. [4].

and vacuum level is the work function, φ (sometimes designated as W), which is a solid-state analogy of the ionization potential.

There are two major contributions to the work function of metals: those of the image charge and of the surface dipole. When an electron is close to the surface, it repels nearby electrons in the metal, so that the charge of the atomic cores is not completely screened anymore (Figure 2.6b). The attraction between the resulting image charge and the outgoing electron, which must be overcome during ionization, contributes to the work function, but it is not sensitive to the nature of the metal.

The surface dipole results from the fact that wave functions are not cut off at the metal surface, but extend into the vacuum. Due to these "evanescent waves," there are electrons outside the metal with a nonzero probability: they can be, for instance, tapped to produce electron beams for electron microscopy (cf. Section 3.4.11.1). The electrons outside are missing inside, which results in a surface dipole layer with an outward negative end (Figure 2.6b). The size of this dipole, which opposes the emission of electrons, depends on the nature of the metal. It can also be influenced by components deposited on the surface (e.g., promoters, cf. Section 4.2.4).

For understanding catalytic phenomena observed with transition metals, it is useful to know the trends of d-band properties and work functions across the 3d period of the periodic table. They are highlighted in Figure 2.7. Towards the right side of the period,

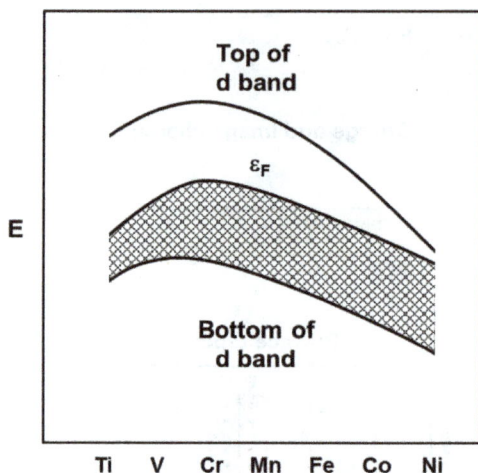

Figure 2.7: Energies and filling degrees of the d-band in first-row transition metals. Reproduced from ref. [2] with permission from Wiley-VCH Verlag GmbH & Co. KGaA, Weinheim, Germany.

the d-electrons become more stable due to the increasing incompletely screened nuclear charge. All features of the d-band are shifted down: its upper and lower edges, its center (not shown), and the Fermi level. Due to the more compact basis functions, the band becomes narrower, and it becomes obviously more filled. As the vacuum level remains constant (not shown), the decreasing Fermi level results in an increasing work function.

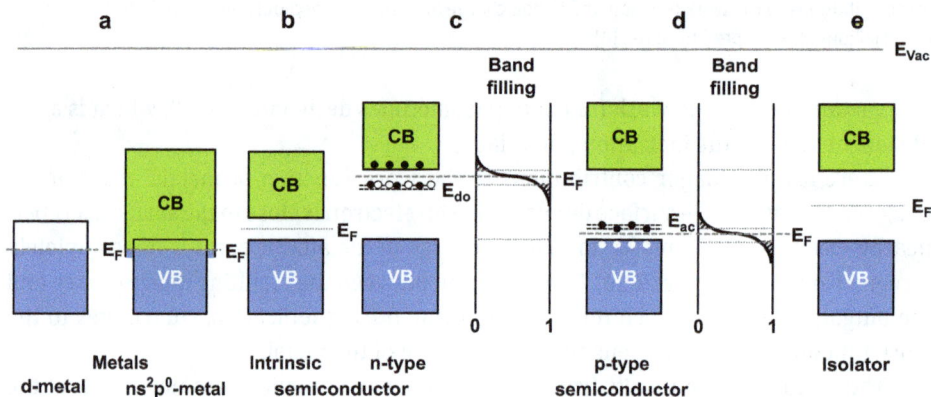

Figure 2.8: Mechanisms of electronic conductivity in solids: (a) metallic conductivity; (b) generation of charge carriers by thermal energy: intrinsic semiconductors; (c, d) generation of charge carriers by existing (or deliberately created) donator or acceptor levels (n- or p-type conductivity, respectively, in extrinsic semiconductors); and (e) isolator.

When the highest occupied band is full, there should be no electronic conductivity. However, depending on the energy gap (bandgap) between the filled and the next empty band, which are referred to as valence and conduction bands, respectively, transport of electric charge may be still possible. When the full valence band is superimposed by the conduction band, even metallic conductivity results: in alkaline earth metals, for instance, the s band is full, but the kinetic energy of flowing electrons can be accommodated by their transition into the superimposed p band (Figure 2.8a). If the bandgap is small enough to allow thermal electron excitation from the valence into the conduction band, the material is an intrinsic semiconductor (Figure 2.8b). If the bandgap is too large for this, the material is an insulator (Figure 2.8e). In both cases, the Fermi level is not well-defined. It is pinned to electrons trapped in isolated structural defects in the bandgap.

Semiconducting properties are strongly modified when there are structural defects in appropriate energy ranges. Donor levels shortly below the conduction band edge allow electron excitation into the conduction band at lower temperatures than in intrinsic semiconductors (n-type conductivity, Figure 2.8c). Likewise, acceptor levels slightly above the valence band edge trap thermally excited electrons from the latter, and conduction will occur by the displacement of holes (p-type conductivity, Figure 2.8d). The Fermi level is between the highest donor level and the conduction band in n-type semiconductors, and between the lowest acceptor level and the valence band in p-type semiconductors.

In semiconductor production, such defects are created on purpose by doping extremely pure isolators or semiconductors with altervalent elements. In oxidation catalysis, use of mixed oxides, which per se exhibit p- and/or n-type conductivity, is quite common. In such oxides, ion conductivity may occur as an alternative mechanism of charge transport. Both cations and anions may be charge carriers. Charge transport by oxygen anions is of particular importance for oxidation catalysis. The process requires the presence of anion (oxygen) vacancies in the structure. In an electric field, O^{2-} ions fill these vacancies while diffusing towards the anode, where they are oxidized. Vacancies are shifted towards the cathode, where they bind new oxygen, which is reduced there.

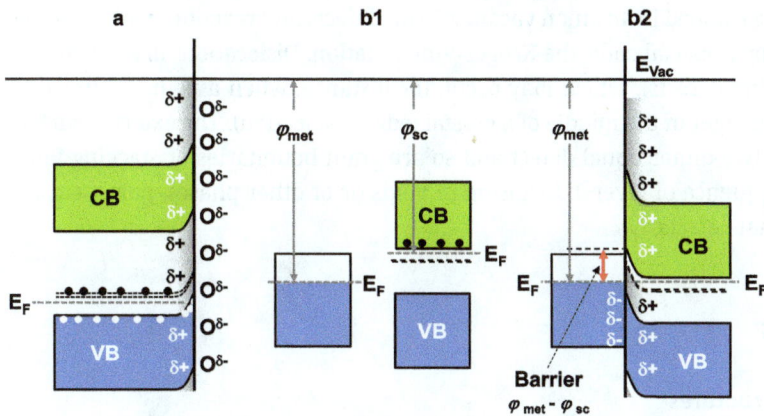

Figure 2.9: Band bending at semiconductor surfaces: (a) by adsorption of negatively charged species; (b1, b2) by contact with a metal (Schottky barrier); b1 shows semiconductor and metal without contact, b2 - with contact.

Near semiconductor surfaces, the energy of electrons can be influenced by the electric field of adjacent charges. As opposed to metals, where the field of external charges is shielded by the image charges forming right in the surface layer (Figure 2.6b), the field operates over a larger distance in nonmetals. Electron levels are changed by such fields (band bending): they are shifted upwards by a negative charge at the surface, which destabilizes electrons in the adjacent region. Electrons tend to withdraw towards the bulk, which results in an electron depletion zone near the surface.

Negative surface charges can arise from electronegative adsorbates (e.g., adsorbed oxygen, Figure 2.9a), or from a contact to a metal (Figure 2.9b). In both cases, the electrons causing the band bending originate from the semiconductor. In the adsorption example, adsorbed oxygen extracts them. In a metal-semiconductor junction, electron transport occurs if the Fermi level of the metal is below occupied semiconductor levels: the latter loses electrons to the metal until the Fermi levels are aligned (Figure 2.9b). The positive space charge left behind forms a dipole surface layer with the image charges in the metal. The resulting energy barrier between Fermi level and the lowest empty semiconductor level is called a Schottky barrier. It prevents the "lost" electrons from returning even under a moderate voltage bias with the positive end at the semiconductor, because this displaces more electrons from the depletion zone, which increases the barrier. Only a bias with reverse orientation causes a current to flow: the Schottky barrier is a rectifier. Schottky barriers occur in many catalysts that contain small metal particles deposited on semiconducting materials.

As indicated above, crystal defects can be relevant for surface catalysis, even if located in the bulk of the crystal. These defects are differentiated according to the dimension of their extension. Point defects are zero-dimensional. They include vacancies, interstitial atoms, which occupy positions unoccupied in an ideal crystal, and antisites – lattice points occupied by a wrong element. There are also color sites, in which one or two electrons are trapped in an anion vacancy. Point defects and reactions with them can be represented by a special code, the Kröger-Vink notation. Dislocations are one-dimensional defects (line defects), which may occur, for instance, when a plane occupied by atoms terminates right in the middle of a crystal (edge dislocation). The external surface of a crystal is a two-dimensional defect and so are grain boundaries or stacking faults (errors in the sequence of layers). Inclusion of voids or of other phases, pores, etc. are three-dimensional defects.

2.2 Surfaces

2.2.1 Surface structures

Surfaces are discontinuities in solids where atoms miss part of the neighbors they would have in the bulk, together with the stabilization by bonding to them. Due to the higher energy of surface atoms, solids tend to minimize their external surface area per volume and to preferentially expose surfaces with dense packing of surface atoms. Thermodynamics obviously favors large particles with stable surfaces, which are of little use for catalysis. Catalysis relies on metastable states of matter that exhibit higher reactivity: on small particles, which expose more surface area per volume, and on highly exposed sites, e.g., edges in stepped surfaces or kinks in these edges, where the atoms are more reactive than those exposed in the terrace faces.

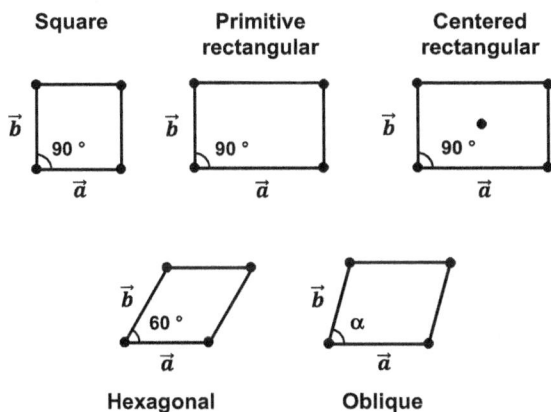

Figure 2.10: Two-dimensional Bravais lattices.

Surface structures may be periodical or disordered. A crystal usually exposes ordered surfaces, while an amorphous solid exposes disordered surfaces. Periodicity in two dimensions may be categorized in five basic patterns: the two-dimensional Bravais lattices (Figure 2.10). In three of them, the translation vectors, $\vec{a}$ and $\vec{b}$ are at a right angle. They are of equal length in the square lattice and of different lengths in the rectangular lattice, which exists in a primitive and a centered version. In the hexagonal lattice, $\vec{a}$ and $\vec{b}$ are of equal lengths and at an angle of 60°, while there are no specifications on the length of the vectors or on the angle between them in the oblique lattice.

When a monocrystal is cleaved, surfaces of different structure, roughness, and atom density are exposed. In Figure 2.11 and Figure 2.12, this is exemplified with the fcc and the bcc lattices cleaved along the (100), (110), and (111) lattice planes. Cutting an fcc crystal along (100) results in a square surface lattice, the translation vectors of which are at an angle of 45° to those describing the periodicity of the same plane in the crystal (Figure 2.11a). Truncation along (110) exposes a rectangular two-dimensional lattice with a longer translation vector in the z direction. Below the first layer, a second one, which includes the atoms in the face centers below the shaded plane in Figure 2.11b, is also visible. Cutting along (111) produces a hexagonal structure with the densest packing of surface atoms (Figure 2.11c) and, correspondingly, the highest stability.

The same cuts produce rather different surface structures when applied to the bcc lattice (Figure 2.12): The (100) plane is square as well, but with longer translation

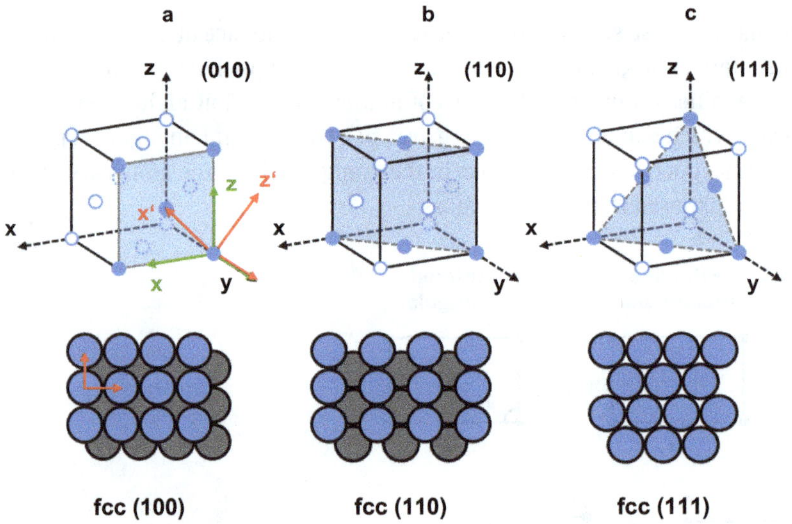

Figure 2.11: Relations between structures of (unreconstructed) surfaces and bulk structures: cutting an fcc crystal.

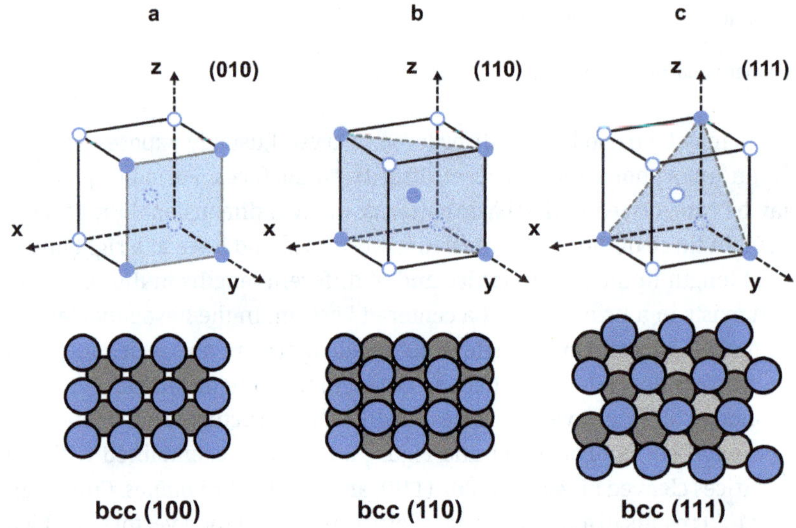

Figure 2.12: Relations between structures of (unreconstructed) surfaces and bulk structures: cutting a bcc crystal.

vectors. Below, the second layer including the center atom in Figure 2.12a is visible. Truncation along (110) exposes a centered rectangular surface structure (Figure 2.12b). The cut along (111) is hexagonal like its counterpart from the fcc lattice (Figure 2.11), but with much longer translation vectors: the central atom in Figure 2.12c belongs already to the next layer. The (111) plane of bcc exposes even the second layer below the outermost one. It is the roughest and the least stable surface structure among those described in Figure 2.11 and Figure 2.12.

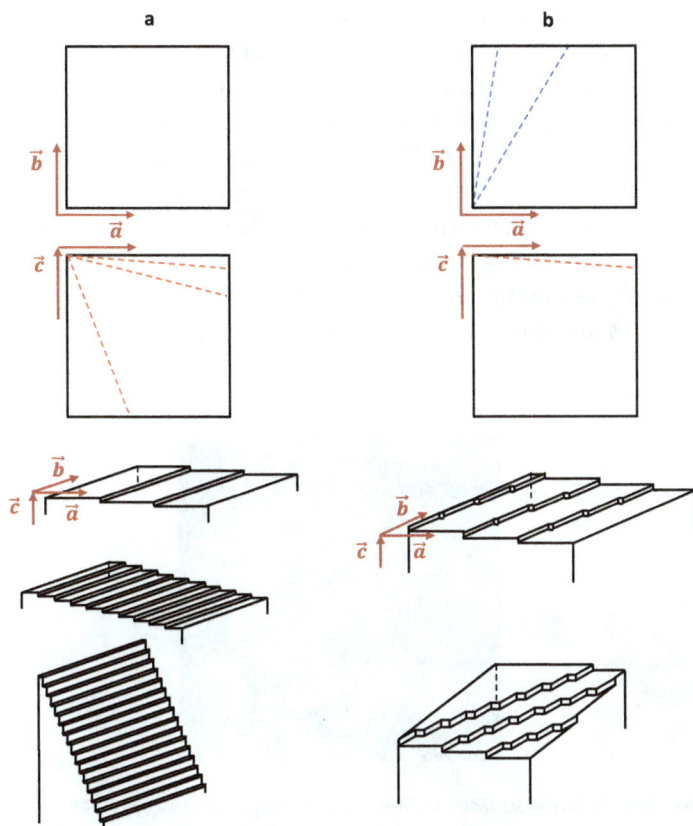

Figure 2.13: Terraces, steps, and kinks resulting from cuts of crystals at different angles to low-index planes.

Although stable metal surfaces characterized by low Miller indices can exhibit significant catalytic activity, they do not contain highly exposed atoms, which are required for reactions involving the scission of stable bonds in substrate molecules. Such atoms can be found in highly indexed surfaces, which are obtained when the crystal is cleaved along planes deviating from those spanned by only two of the three-dimensional translation vectors. Figure 2.13 exemplifies this for a cubic crystal. In Figure 2.13a, the crystal is first cut in the plane defined by $\vec{a}$ and $\vec{b}$. This exposes the (001) facet (not shown), irrespective at which position along the third direction $\vec{c}$ the cut is set. The cuts shown in Figure 2.13a run along planes spanned by $\vec{b}$ and a linear combination of $\vec{a}$ and $\vec{c}$, which inclines the exposed surface relative to (001). This inclination is realized by terrace structures, the properties of which depend on the angle relative to the (001) plane. At low angles, extended (001) facets are exposed, interrupted by rare monoatomic steps. With increasing angles, a growing number of steps decreases the extension of the (001) facets, and the steps uncover larger sections of the (100) lattice plane. At high angles, (100)-oriented facets are interrupted by (001)-oriented steps.

In these surfaces, the terrace edges are straight lines along the $\vec{b}$ direction. They become interrupted by kinks when the second vector defining the cutting plane is no longer constrained to $\vec{b}$ but is a linear combination of $\vec{b}$ and $\vec{a}$, as exemplified in Figure 2.13b. Similar to the terraces at small angles to (001), the kinks are monoatomic and rare at small angles to $\vec{b}$. They become more abundant and extended into the $\vec{a}$ direction at stronger deviations from $\vec{b}$. If the crystal is cut at large deviations from both (001) and $\vec{b}$, the resulting surfaces may contain (111)-oriented facets because of the higher stability of the (111) surface. Generally, the crystal exposes its most stable surface structures in as many extended facets as possible.

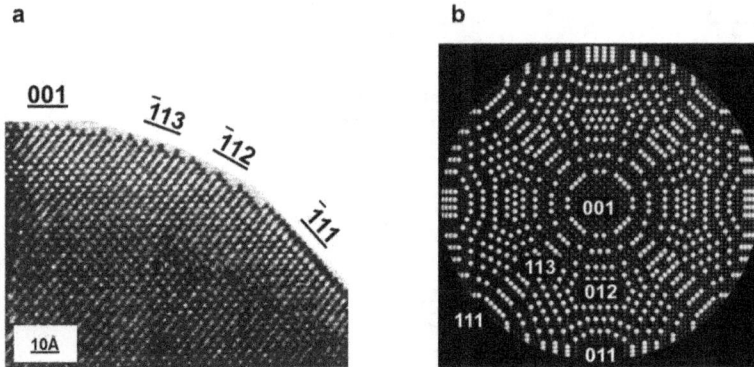

Figure 2.14: Exposure of stable facets in curved surfaces: (a) Transmission electron microscopy (TEM) image of a gold crystallite, (b) ball model of a (001)-oriented Pd tip with various facets exposed around. (a) reproduced with permission from ref. [5], copyright 1988 American Chemical Society, (b) reproduced from ref. [6] with permission from the Royal Society of Chemistry.

The latter statement also describes the way in which nature realizes curvature on particle surfaces. Figure 2.14a shows an electron micrograph of a gold particle exhibiting a pronounced curvature. From the right side with $(\bar{1}11)$[6] exposed, the $(\bar{1}11)$ terraces become shorter to the left and interrupted by steps, until (001) is exposed exclusively. In Figure 2.14b, a ball model of a metal tip pointing towards the reader with its (001) facet is depicted. Around this central plateau, smaller (001) facets bordered by terrace atoms (in lighter tone) are visible. They become narrower until other low-index facets are reached, e.g., (012) and (113). Further down the tip, (111) and (011) appear. Facets oriented in the other directions have related indices (not shown, e.g., $(\bar{1}11)$ instead of (111), or (101) and (110) instead of (011)).

---

6 $\bar{x}$ in Miller index designates −x.

fcc(110)

a

| Unreconstructed | With missing-row reconstruction |

b

Missing-row reconstruction on a gold crystal

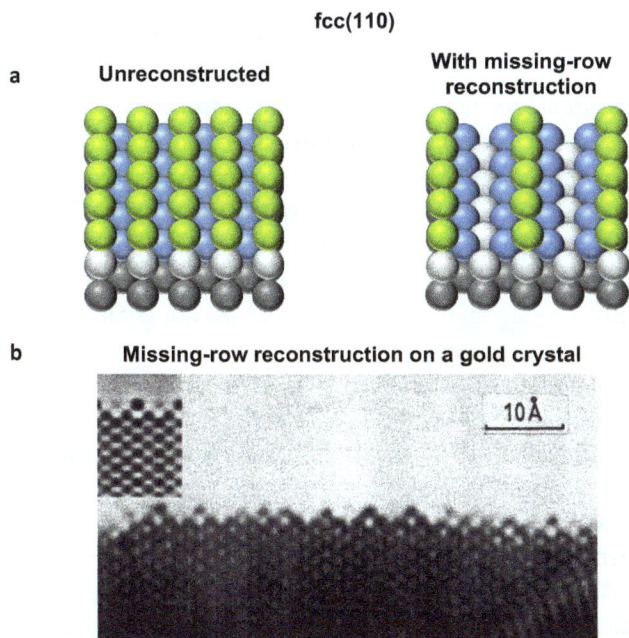

Figure 2.15: Reconstruction of metal surfaces: missing-row reconstruction of the fcc (110) facet. (a) Ball model with atoms in different layers differentiated by color; (b) TEM image showing the missing-row reconstruction of an Au (110) facet with structural imperfections; inset – result of image modeling, cf. Section 3.4.11.1. Reproduced with permission from ref. [5]. Copyright 1988 American Chemical Society.

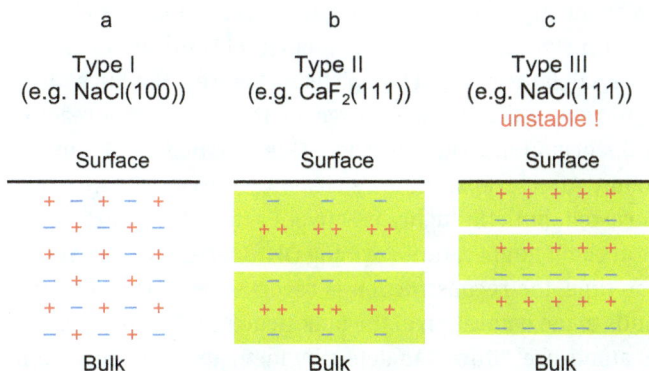

a	b	c
Type I (e.g. NaCl(100))	Type II (e.g. CaF_2(111))	Type III (e.g. NaCl(111)) unstable !

Figure 2.16: Stability of facets exposed by ionic compounds: the Tasker rules.

The surface structures discussed so far result from truncating crystals in certain directions. However, even low-indexed facets exposed by such a gedanken experiment are not necessarily stable. Different, more stable atom arrangements may be realized and are indeed detected, e.g., by surface diffraction (Low-energy electron diffraction, LEED, see Section 3.4.3.4). They are referred to as reconstructed surfaces or surface reconstructions. The so-called missing-row reconstruction frequently observed for the (110) facet of fcc metals is depicted in Figure 2.15a and shown for a real Au(110) facet in Figure 2.15b. Surface reconstructions can also be induced by the adsorption of species that destabilize the bonds between metal atoms, e.g., by inserting electron density into a largely filled conduction band (cf. Section 2.1).

In compounds, the presence of more than one atom type results in polarity and electrostatic interactions. This introduces an additional level of complexity when surface structures are to be derived by truncating crystals along low-indexed lattice planes. For the sake of simplicity, we will refer to the components as cations and anions in the following, even if there are covalent contributions in their bonds.

In polar compounds, the outermost layer of ideal crystal surfaces may expose cations or anions, or both. Ion arrangements normal to the surface result in dipole moments, which give rise to stability conditions for facets known as Tasker's rules [7]. According to Tasker, surfaces are unstable when the crystal structure results in a nonzero dipole moment along the surface normal.[7]

Surfaces can be categorized into three classes. In Tasker Type I surfaces, the ionic charge is balanced within the surface and within each layer below (e.g., (001) of rock salt-type lattices, Figure 2.16a). Any dipole along the surface normal is balanced by an adjacent opposite dipole. Tasker type II comprises charged layers, which can, however, be combined into sets that allow neutralization of the dipole moments. In the fluoride-terminated (111) surface of CaF_2, for instance, Ca^{2+} ion layers are sandwiched by F^- anion layers, which results in charge-balanced CaF_2 trilayers (Figure 2.16b).

While Tasker types I and II surfaces are stable, there is no combination of the charged layers that annihilates the dipole moment along the surface normal of Tasker type III surfaces. This holds, for instance, for the Ca-terminated (111) surface of CaF_2 or for the (111) surface of the rock salt lattice, which is depicted in Figure 2.16c. Such facets either reconstruct in order to reduce the surface charge, or they are highly reactive and stabilize by chemical transformations. The hexagonal O-terminated (0001) surface of ZnO combines both options. After preparation by cleaving a crystal in ultrahigh vacuum (UHV), it is reconstructed, but still highly reactive. Even in UHV, surface O^{2-} ions react with water traces at room temperature, forming OH^- species to decrease the negative charge. At the same time, the reconstruction is reverted: the OH-terminated ZnO(0001) surface corresponds to the truncation of the bulk structure. This process was long overlooked, because H atoms are difficult to detect by methods typically used in surface science (for the full story, see ref. [8]).

---

7 The surface normal is a vector orthogonal to a surface. The surface normal of the (xyz) lattice plane is designated as [xyz].

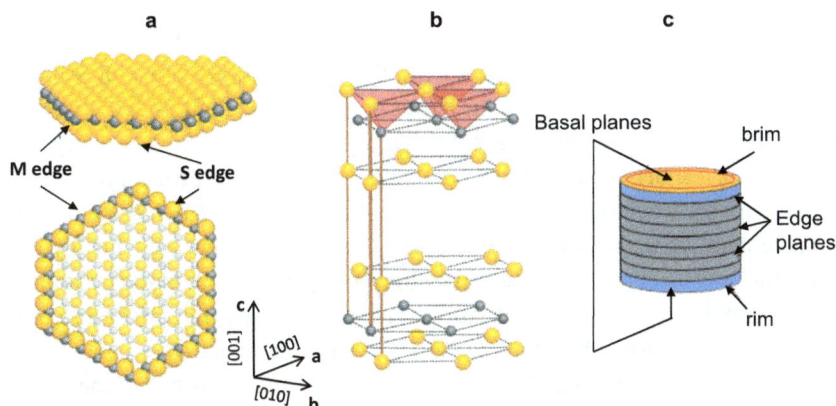

Figure 2.17: Anisotropy of structure and reactivity in MoS_2: (a) MoS_2 slabs, (b) stacking of slabs in a (2H-) MoS_2 crystal, (c) MoS_2 surface features relevant for catalysis.

Structural anisotropy between different facets, as discussed above for metals, is even more pronounced for polar compounds. This holds, in particular, when bulk structures consist of weakly interacting layers. MoS_2 crystals are stacks of slabs consisting of an Mo layer sandwiched between two sulfide layers (Figure 2.17a, b). The interactions between adjacent sulfide layers are weak. As a consequence, the slabs can be easily moved relative to each other, and MoS_2 can be used as a lubricant. Bulk MoS_2 crystals can be delaminated by a number of procedures to produce nanostacks that comprise only a few slabs, or even to monoslab structures (exfoliation). The hexagonal top and bottom sulfide layers of the slabs, the so-called basal planes, are stable and unreactive. The facets perpendicular to them expose Mo and/or S, depending on their orientation (Figure 2.17a). The abundance of exposed Mo and the nature of exposed sulfur (sulfide, disulfide, SH groups) depend on gas-phase composition and temperature [9]. Although details are still controversial, catalytic activity in almost all reactions catalyzed by MoS_2 has been attributed to sites on these edge planes, on the rim of MoS_2 stacks, or on basal-plane sites adjacent to the rim (brim sites, Figure 2.17c).

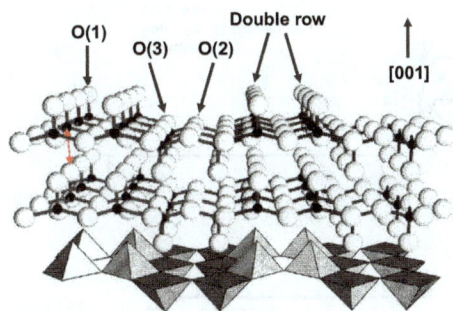

Figure 2.18: Anisotropy in oxide structures: V_2O_5. Adapted from ref. [10].

Figures 2.18 and 2.19 show examples of layered oxide structures that are relevant for heterogeneous catalysis. In V_2O_5 (Figure 2.18), the coordination sphere of V has a strongly distorted octahedral geometry: four bonds within the (001) plane are of nearly equal length, while those along the surface normal (i.e., the [001] vector) are very different: at each V ion, a short vanadyl (V=O) bond and a very long, i.e., weak interaction (red arrow in Figure 2.18) extend into opposite directions. This creates two-dimensional layers of square pyramids, as shown in the figure. They can be kept together by the weak forces of the long V-O bond because of the alternating orientation of the pyramids.

In the preferentially exposed (001) facet, three types of surface oxygen can be differentiated: vanadyl oxygen (O(1)) and bridging oxygen species coordinated to more than one V ions, either to two or to three of them (O(2) and O(3), respectively). While half of the V ions is shielded by vanadyl oxygen in the (001) facet, the remaining V ions are exposed. Any cut perpendicular to the (001) facet exposes V and O species in very different geometries and abundances. Vanadyl oxygen will never stick out of these surfaces. Where oxygen bridges to missing V ions are disrupted, either highly exposed V ions or charged surface oxygen groups (most likely saturated by H, i.e., surface OH groups) may be exposed.

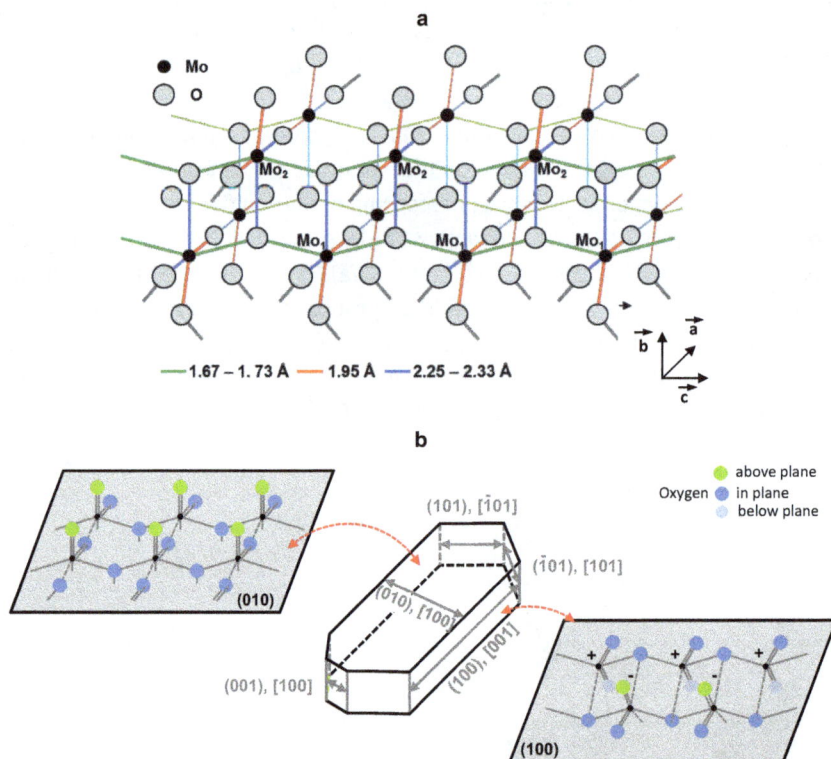

Figure 2.19: Anisotropy in oxide structures: MoO_3: (a) coordinations and bond lengths, adapted from ref. [11] with permission from the Royal Society of Chemistry (b) typical morphology and structure of facets exposed, adapted from ref. [12] with permission from Elsevier.

The layered structure of MoO_3 is depicted in Figure 2.19a. MoO_3 has two short Mo-O (molybdenyl) bonds and two long ones extending into opposite directions (highlighted red and blue, respectively, in the figure). The distorted octahedral coordination is completed by two more (green) bonds of intermediate, approximately equal lengths. In the coordinates used in Figure 2.19a, one of the Mo=O bonds sticks out of the (010) facet along $\vec{b}$, while the opposite long Mo-O bond represents a weak interaction with the next layer. The second Mo=O bond extends along $\vec{a}$ within the (010) plane, alternating with a weak Mo-O interaction. The remaining Mo-O create Mo-O-Mo chains along $\vec{c}$. In the preferentially exposed (010) facet, Mo is completely shielded by molybdenyl oxygen, as shown in Figure 2.19b. The figure also gives an example for a surface perpendicular to (010): in the (100) facet, rows of Mo=O groups alternate with rows of exposed Mo left behind after rupture of the long Mo-O bond. In reality, the high polarity of this surface will be attenuated by reactions with the surrounding atmosphere, resulting, for instance, in the formation of OH groups.

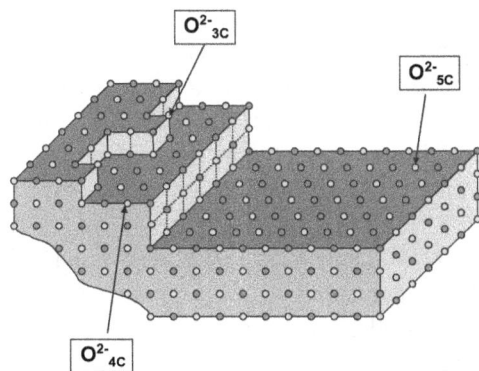

Figure 2.20: Heterogeneity of coordinations in terraced and kinked surface regions of polar compounds.

Similar to metal surfaces, surfaces of polar compounds may contain defects, terraced regions, and kinks in terraces. This adds a similar aspect of complexity as in the case of metals: atoms at terrace edges and kinks are less stabilized by bonds to their neighbors and, therefore, are more reactive. In the MgO surface seen in Figure 2.20 (rock salt lattice), the O^{2-} ions within terrace planes are stabilized by five instead of six Mg^{2+} cations, while those at terrace edges are stabilized by four and those in kinks by only three Mg^{2+} ions. The basicity of the oxygen anions grows with decreasing stabilization by adjacent cations. Likewise, the nucleophilicity of surface oxygen in redox-active oxides increases with the degree of its exposure. Upon heating in inert atmosphere, highly exposed oxygen exhibits the highest tendency to desorb as O_2, leaving the neighboring cations reduced (autoreduction). Analogously, the Lewis acidity of Mg^{2+} ions varies with their position in the exposed surface: it is highest, where least negative charge is available in the vicinity, i.e., in the kinks. Last but not the least, the ability of highly exposed cations to coordinate more than one adsorbate species is an additional reason for their extraordinary reactivity.

When crystals or amorphous particles consist of more than one component, the chemical composition of the surface becomes an additional variable, e.g., in alloys or ternary oxides. The ordering or even the composition of an alloy is often not preserved in the outermost surface layer. When alloyed metals have very different surface energies, the one with the lower surface energy will be preferentially, or even exclusively, exposed. While this has been known for long, discussion on catalysis with crystalline mixed oxides has been dominated by surface models derived from the truncation of bulk structures until recently. Using a spectroscopic method exclusively probing the outermost surface layer (Low-energy ion scattering, LEIS, cf. Section 3.4.6.3), it has been demonstrated, meanwhile, that the surface of such materials (e.g., of molybdates or vanadates) may also preferentially or exclusively expose one of the components — in the example — surface Mo or V oxide species [13].

2.2.2 Surface chemistry

Already in the previous section, it has been pointed out that surfaces may be stabilized by chemical reactions with components of the surrounding fluid and that species terminating a surface can change depending on the composition of this fluid and temperature. This is a special case of a general principle: solid surfaces tend to adapt to the chemical potential of the surrounding fluid, although the rates of this process may be very different in different systems.

When these rates are low, the surface termination of solids may be determined by the processes of their preparation. Oxide catalysts are often obtained by calcination of oxide hydrates or hydroxides previously precipitated from aqueous solutions. Depending on calcination conditions, their surfaces contain O^{2-} and hydroxyl groups in different ratios and may also expose the cation. Surfaces of oxides made by high-temperature processes (e.g., fumed silica by flame hydrolysis of $SiCl_4$, cf. eq. (2.5) below) may have different surface compositions. The surface chemistry of most solid oxides is confined to O^{2-} and hydroxyl groups, whereas the presence of peroxide has been reported only in rare cases.

O^{2-} ions and OH groups terminating oxide surfaces are not all equivalent, because they may be coordinated to a different number of cations. Figure 2.21 exemplifies this with one of the possible terminations of a fully hydroxylated γ-Al_2O_3 surface. In γ-Al_2O_3, Al^{3+} cations are located in octahedral and tetrahedral voids of an fcc close-packed lattice of O^{2-} ions. Figure 2.21a shows an A layer of this packing with the Al^{3+} ions on it: either in hollows formed by three adjacent O^{2-} ions (three-fold hollow sites, cf. Figure 2.49) or on top of an O^{2-} ion. When the next B layer is added, the coordination sphere around these Al^{3+} cations can become a tetrahedron (T_d) or an octahedron (O_h), as demonstrated in Figure 2.21b. There are two ways to the tetrahedral coordination: an Al^{3+} ion in a hollow A site may be topped by an O^{2-} ion, or an Al^{3+} ion on top of an O^{2-} ion of the

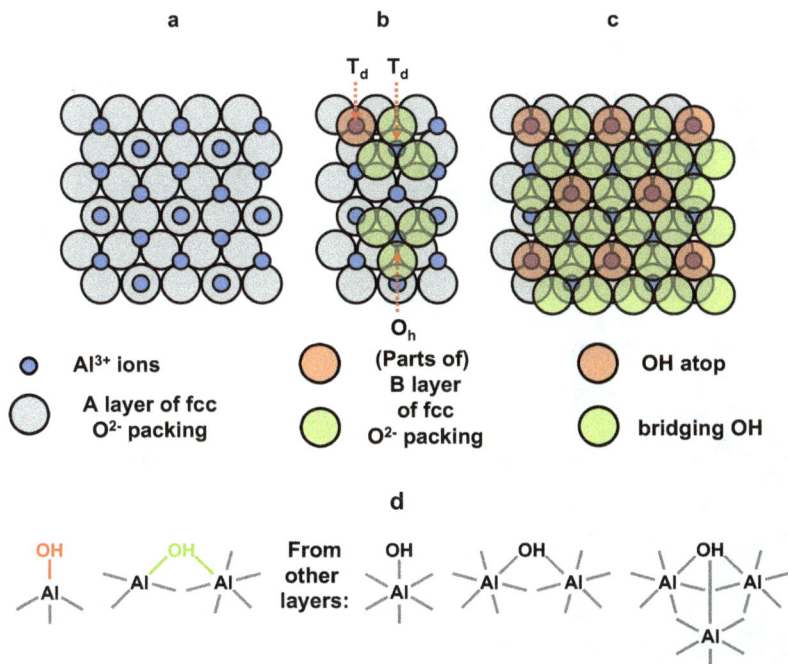

Figure 2.21: Inequivalence of OH groups on the surface of γ-Al$_2$O$_3$: (a) a layer of an fcc close packing of O^{2-} ions, with Al^{3+} cations; (b) individual O^{2-} ions of B layer creating O$_h$ or T$_d$ coordinations around Al^{3+}; (c) complete B layer capping the solid towards the atmosphere (i.e., OH-groups), atop and bridging OH groups differentiated by color; and (d) summary: types of OH groups derived from (c) and from other layers.

A layer is covered by a hollow site in the B layer. The octahedral coordination results when Al^{3+} ions are in threefold hollow sites of both layers (Figure 2.21b).

When the B layer is at the external surface and consists of OH$^-$ instead of O^{2-}, these OH groups may be on top of an Al^{3+} ion or bridging two of them (Figure 2.21c and d). The former (highlighted in red) belong to Al^{3+} in T$_d$ coordination and the latter (green) bridge between Al^{3+} in O$_h$ and T$_d$ coordinations. Thus, when the external OH layer is a B layer, two OH groups can be differentiated. By full analysis of the system, three more OH group configurations were identified [14] (Figure 2.21d). When these OH groups are isolated, i.e., not interacting with neighboring OH groups via H bridges, they can be differentiated by their signals in IR and ^{1}H-NMR spectroscopy.

Surface OH groups coordinated to 1, 2, 3, or more cations have been categorized as type I, II, III, or IV, respectively. With respect to their chemical environment, OH groups can also be differentiated as isolated, vicinal, and geminal hydroxyls, as illustrated with OH groups present on silica surfaces in Figure 2.22. When such a surface is heated, a siloxane bridge can be formed by condensation of two vicinal silanol groups.

Dehydroxylation proceeds when the oxides are heated in a dry gas or in vacuum. It gives access to the cations of surfaces that were fully hydroxylated earlier. While water is easily formed from vicinal OH groups, isolated OH groups can be stable up to high

Isolated   Geminal   Vicinal   Siloxane
Silanols   Silanols   Silanols

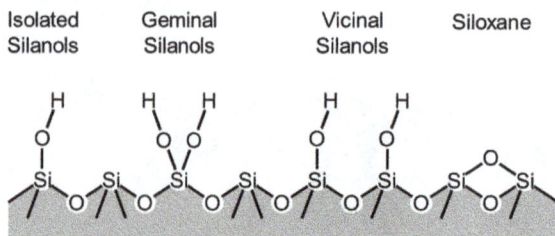

Figure 2.22: Surface OH groups on silica surfaces.

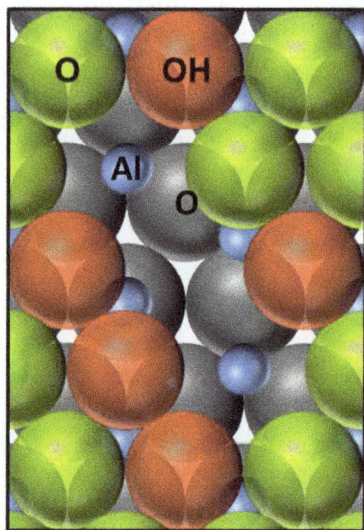

Figure 2.23: Ball model of a partially dehydroxylated γ-alumina surface.

temperatures. When OH groups and oxide ions below are nonequivalent, as in the case of γ-alumina, dehydroxylation results in several different environments around the exposed cations, as exemplified in Figure 2.23. The coordination sphere of Al^{3+}, which may offer up to three vacancies, may be partly blocked by OH^- and/or O^{2-}, and where three vacancies are free, adsorbate molecules may experience interactions with different neighboring OH groups or O^{2-} ions. OH-group densities of typical supports and their development with increasing temperature are discussed in Section 2.3.2.

Oxide surfaces exhibit both redox and acid-base properties. Acidity and basicity are related to the electronegativity of the cations involved. Exposed cations are Lewis acid sites; their strengths increase with their ionic potentials. They can polarize adsorbed water to become acidic. Mixed oxides often exhibit a higher acidity than any of the component oxides, because their structure contains the minority elements in an untypical coordination. Oxide ions coordinated to very weak Lewis acids, e.g., alkali or alkaline earth ions, exhibit a basic character.

$$R-OH_2^+ \rightleftarrows R-OH + H^+ \rightleftarrows R-O^- + 2\,H^+ \qquad (2.4)$$

When solid surfaces are immersed in water, they form a double layer, the polarity of which is critically influenced by the acid-base properties of the solid. The surface polarity of oxides in water is determined by the state of the OH groups terminating the lattice. They are protonated in strongly acidic medium and deprotonated in basic medium (eq. (2.4)). The pH value at which the surface is electrically neutral is called the isoelectric point (IEP).[8] It is lower for acidic oxides (e.g., ≈2 for SiO_2) than for basic oxides (12 for MgO), while amphoteric oxides are intermediate (Al_2O_3: 7–9, TiO_2: 6–7). The IEP is a critical parameter for preparation routes that involve the deposition of a component on an oxide surface from an aqueous solution like ion exchange or impregnation (see Section 3.1.3).

Redox properties of oxide surfaces generally correlate with the galvanic series, but the reactivity of sites may be strongly modulated by the degree of their exposure (see above) and by interactions with other surface species: adsorbates, promoters, and poisons.

By analogy to oxide materials, the surface chemistry of sulfide catalysts is dominated by exposed cations, sulfide ions, and sulfhydryl (–SH) groups, although disulfide groups are also discussed [10, 15]. This limitation results from the almost exclusive use of sulfide catalysts under high H_2 pressures. Oxygen-containing surface species are sometimes detected in such catalysts, but rather as residuals from the preparation process or as a result of experimental flaws. There are, however, reports claiming that the catalytic activity of some sulfide catalysts actually results from carbide species formed from feed components under reaction conditions [16].

The surface chemistry of high surface area carbon materials used in catalysis is very rich and can be experimentally tailored to some extent. The various types of carbons are highlighted in Section 2.3.2.6. Figure 2.24 exemplifies typical O-containing groups, which can be generated on the surfaces of such materials, e.g., by oxidation in hot HNO_3 or in HNO_3 vapors [17]. In addition to those shown in the figure, minority sites on aliphatic structures may be present. The thermal stability of these functional groups is very different. In particular, the acidic carboxyl groups, which can be used to adsorb precursor species for other active components, e.g., metals, are decomposed already at rather low temperatures.

Nitrogen-containing surface groups can be obtained by carbidization of N-containing precursors, e.g., polyacrylo-nitrile, or by high-temperature treatment of O-functionalized surfaces in NH_3. The groups obtained via these routes depend on temperatures employed. Nitriles, lactam, imide, and amino groups, for instance, can be found below

---

8 Sometimes also referred to as "point of zero charge" (PZC). Strictly, the PZC describes a surface not exposing charges at all, while the IEP designates surfaces with positive and negative charges, but none in excess. As the acid strength on real surfaces sites is distributed, the situation in catalysts is better described by the IEP concept.

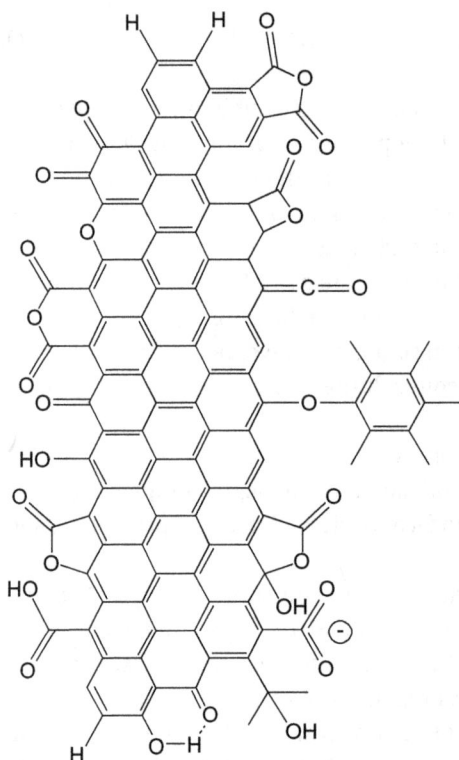

Figure 2.24: Structure model of an oxygen-functionalized carbon material showing typical O-containing surface groups. Reproduced from ref. [18] with permission of John Wiley &Sons.

570 K, pyridine groups up to 770 K, and pyrrolic and quaternary nitrogen at higher temperatures [19]. Other opportunities for surface functionalization of carbon materials, e.g., with S-containing groups or halide [18], have been less applied in heterogeneous catalysis so far, although sulfonation (in this case of polymers) is a standard procedure in the production of cation exchange resins, which are important catalysts for acid-catalyzed reactions in the aqueous phase.

2.3 Particle architecture in catalytic materials

2.3.1 Size, porosity, and structural heterogeneity of catalyst particles

Solid matter exposes more surface area the finer it is dispersed. The surface-to-volume ratio of a sphere with a diameter d is $(A/V)_{sph} = 6/d$.[9] Therefore, a fine distribution is a typical feature of catalytic materials, where particle sizes of 10 nm are often considered unacceptably large. However, free particles of such size cannot be easily managed. In

---

9 easily derived from the relation between the external surface area $A_{sph} = \pi d^2$ and the volume $V_{sph} = \pi d^3/6$.

particular, straightforward catalyst separation from the fluid phase, which is a major technological advantage of heterogeneous catalysis, is impossible with such particle dimensions. Obviously, manageability is another important objective of catalyst design, and tradeoffs with particle size must be found.

To expose expensive catalytic materials in a highly dispersed state, they are often deposited on another (cheaper) material, the support. Materials exposed are typically metals, oxides, or sulfides, but they may be also liquids, ionic liquids, or molecular catalysts tethered to the support surface (see Section 2.3.3). Catalysts completely made of the catalytic component, maybe with additives, are referred to as bulk catalysts, those having the catalytic component dispersed on a support as supported catalysts. The Fe catalyst for NH_3 synthesis,[10] solid acids (zeolites) employed for Friedel–Crafts alkylation of benzene with ethene or propene (eq. (4.9)), polyoxometalates useful for many acid-base and redox reactions [20], or the multicomponent molybdate-based mixed-oxide catalysts for the ammoxidation of propene (eq. (4.8)) are examples of bulk catalysts. There may be specific internal structural relations between components of bulk catalysts, as discussed with reference to Figure 4.21 for ammoxidation catalysts, or with reference to Figure 4.18 for Cu/ZnO-Al_2O_3 catalysts for methanol synthesis. Carbon- or MgO-supported Ru for ammonia synthesis, Pt-M particles (M = Re, Sn, or Ir) supported on Cl-promoted γ-Al_2O_3 for naphtha reforming, alumina-supported Co-Mo sulfides for the hydrodesulfurization of light oil fractions, and V-W/TiO_2 catalysts for the selective catalytic reduction (SCR) of NO by NH_3 (eq. (3.136)) are typical examples of supported catalysts. The concept includes materials with uneven distribution of the supported particles over the support (cf. Figure 3.17). Bulk catalysts, e.g., solid acids may be employed as supports in supported catalysts.

Box 2.1: IUPAC classification of pore sizes (d_p – pore width)

	d_p	$\leq$ 2 nm	micropores
2 nm	< d_p	$\leq$ 50 nm	mesopores
	d_p	> 50 nm	macropores

To achieve high accessibility of the catalytic components, catalyst particles are usually porous. The choice of particle architecture depends critically on the diffusion rates of reactants and products in the pores. Pore sizes can vary in wide ranges. Pores are therefore differentiated into micro-, meso-, and macropores as shown in Box 2.1.

As diffusion rates are very small in the liquid phase, catalytic liquid-phase reactions proceed predominantly at the external particle surfaces (cf. Section 2.6.2). Catalyst particles used in liquid-phase catalysis are therefore, often, small. They are suspended in the reaction mixture, i.e., reactions are performed in slurry reactors (cf. Section 2.8). Large production capacities can, however, be better achieved by passing the reactant(s) (often including a gas phase, e.g., H_2) through a fixed bed of catalyst particles

---

10 For more details on processes listed here – see Appendix A1.

(see Section 2.8). In such fixed beds, particle sizes must be above certain limits to avoid excessive pressure drops. In such catalysts, the catalytic component is therefore concentrated at the external surface (egg-shell distribution, Figure 3.17).

In the gas phase, diffusion is several orders of magnitude faster than in the liquid phase. Therefore, rates of catalytic gas-phase reactions are often determined by the kinetics of the surface processes, and a high porosity helps exposing the catalytic component, as much as possible. However, when the surface processes are very fast, reaction rates can be limited by the diffusion along the pores (cf. Figure 1.5). As in liquid-phase catalysis, the problem can be solved by depositing the active component only in the outer region of catalyst particles.

The pore system may serve additional purposes. In catalysts for processing heavy feedstock, for instance, deactivation may be delayed by trapping poisons and coke precursors in the periphery of the pellets, while the active sites are in the center (egg-yolk distribution, cf. Figure 3.17). In such catalysts, the pores at the periphery should be wide to minimize interference of mass transfer by accumulating deposits. In other reaction systems, pores of suitable sizes allow directing reactions via steric influences (shape selectivity, cf. Section 4.3.2).

When the target product is a reactive intermediate in a consecutive reaction, long diffusion paths must be avoided, because multiple additional surface contacts may cause its further conversion. Catalysts for the selective oxidation of hydrocarbons are therefore designed with low porosity to avoid overoxidation of the product. For the same purpose, oxidation catalysts are sometimes deposited in thin layers on ceramic spheres or rings, thus ensuring short diffusion lengths at low pressure drop and sufficient mechanical stability. The latter is important to prevent breakdown of catalyst pellets under the weight of tall fixed beds (cf. Section 4.4.1.4).

For applications that involve a challenging heat management (highly exothermal reactions) or a fast reversible catalyst deactivation, fluidized beds (see Section 2.8) are often employed. This technology requires small and highly attrition-resistant particles.

The origin of porosity in solid particles may be different, as depicted in Figure 2.25. Open porosity designates pore systems remaining between grains aggregated in the irregular stacking of polycrystalline particles (Figure 2.25a). At temperatures near the melting point, such pores can be filled by solid-state diffusion (sintering). As primary particles in polycrystalline catalyst pellets are often very small, gradual loss of porosity may occur even at temperatures well below this range. When pores/voids are defined by the crystal structure of materials (Figure 2.25b), the pore system is very regular. In solids with structure-inherent porosity, e.g., zeolites, metal-organic frameworks (cf. Section 2.3.2.4) as also salts of heteropoly compounds [20], pore sizes and shapes can be described in great detail, and the pore system can be reproduced with great reliability. In recent decades, templated pore systems (Figure 2.25c) have received much attention. They are formed when interstices between templating elements are filled by a (usually amorphous) material, e.g., silica or carbon, and the template is subsequently removed, e.g., by combustion or dissolution. The resulting pore systems can also be highly regular: such materials diffract electromagnetic waves although their pore walls are amorphous.

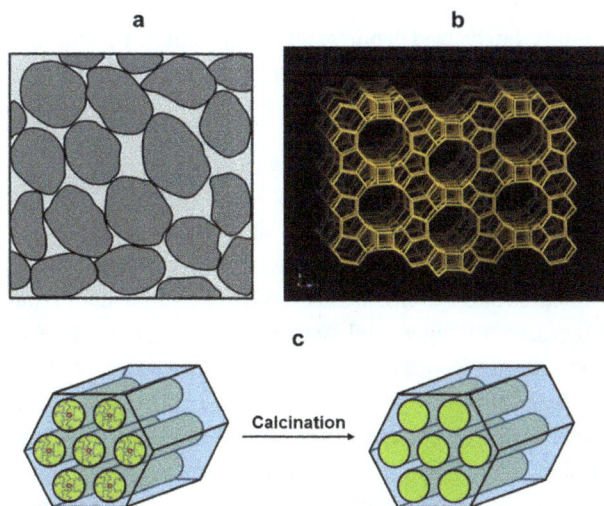

Figure 2.25: Origin of porosity in solids: (a) intergranular space; (b) structure-inherent porosity; and (c) templated pore systems. (b) from ref. [21] with permission from the International Zeolite Association (IZA).

While the properties of intergranular and templated pore systems can be tailored within certain limits by variation of the preparation conditions, such a degree of freedom is missing in the case of structure-inherent porosity. Depending on the preparation conditions, the crystalline phase may be formed more or less perfectly, in the form of larger or smaller particles, phase-pure or mixed with other phases, or not at all. The pore system is, however, always the same, because it is defined by the crystal structure. In reality, variations may occur due to a smaller or larger abundance of structural defects, which are, however, usually undesired. In some zeolites, however, pore dimensions may be fine-tuned by an exchange of structural charge-balancing cations.

Structure-inherent porosity is typically microporosity, which causes mass transfer problems in many applications. Therefore, methods have been devised for creating transport porosity within the microporous crystals (hierarchical pore systems). In zeolites (cf. Sections 2.3.2.3 and 3.1.1.2), this may be achieved by various approaches, e.g., by synthesizing the crystals around carbonaceous templates, which are subsequently removed by combustion or by leaching siliceous parts of (Si-rich) zeolite crystals (for more details, see ref. [22]). In these mesoporous zeolites, the micropore system remains intact, except for an enrichment of structural defects at the mesopore surfaces.

2.3.2 Supports for heterogeneous catalysts

Silica, alumina, amorphous and crystalline alumosilicates, titania, zirconia, ceria, and carbon materials are widely used as catalyst supports. Metal-organic frameworks (MOFs) have emerged more recently. Many of these supports are catalysts on their own. This chapter describes materials routinely purchased from external providers by catalysis labs, while other supports like ordered mesoporous materials (cf. Figure 2.25, Figure 3.4), which are typically prepared in-house, are discussed in Section 3.1.

2.3.2.1 Silica (SiO$_2$)

Below 846 K, the stable form of SiO$_2$ is α-quartz. The metastable products — silica gel, fumed silica, and kieselguhr — are employed for different purposes in catalysis.

Silica gels are formed when silicates like water glass are acidified and form sols that solidify in a sol-gel transition. Processes during sol-gel syntheses and opportunities to influence the resulting pore systems are discussed in Section 3.1.1.3. The porosity of the primary gel is modified by acid washing treatments and, in particular, by the conditions during the drying step. In xerogels made by conventional drying, stress by capillary forces inflicts losses, while the full porosity is retained in aerogels obtained by super-critical drying. For economic reasons, silica gels used in catalysis are mostly xerogels, although aerogels are also available on the market. Typical BET surface areas[11] of silica (xero)gels are between 200 and 600 m^2/g; average pore widths are between 2 and 6 nm.

$$SiCl_4 + 2\,H_2 + O_2 \rightarrow SiO_2 + 4\,HCl \qquad (2.5)$$

Pyrogenic silicas, which are made by hydrolysis of SiCl$_4$ in a H$_2$-fueled flame (eq. (2.5)), are referred to as aerosils or fumed silicas. The process produces 10–30 nm-sized primary particles fused into branched, chainlike structures, which aggregate into microspheres of 10–100 μm size. The BET surface area is strongly influenced by the conditions in the flame, it ranges between 100 and 400 m^2/g. Kieselguhr is a natural product (see below).

With an IEP around 2, silica is an acidic support (cf. Section 2.2.2), but with a low density of acid sites. After activation in vacuum at 450–470 K, the surface OH group density was found in a narrow range slightly below 5 OH/nm^2 for 100 SiO$_2$ samples of different origin, including fumed silica previously equilibrated in water at room temperature [23]. Upon evacuation at 673 K, the density decreased by almost 50%. For commercial fumed silicas, lower OH group densities of 2–3 nm^{-2} were observed, supposedly due to posttreatment steps for removal of residual Cl [24]. Due to their acidity, the OH groups are largely deprotonated in aqueous media. This interferes with most aqueous routes to well-dispersed supported oxide species, which offer the active element in anionic form (Section 3.1.3.1). On the other hand, silica has been widely used to disperse active elements by anchoring/grafting techniques (Section 3.1.3.4).

In catalysts working under serious transport limitations, active components may be supported on kieselguhr (synonymous with diatomaceous earth or diatomite), which is a natural product consisting of the siliceous cell walls of diatom algae. Their sizes are distributed in a wide range, but most frequently between 50 and 100 μm. The shells are

---

11 cf. Sections 2.4.3 and 3.4.1.3.

Figure 2.26: Scanning electron micrographs of diatomite (kieselguhr). Reproduced from ref. [25] with permission from Elsevier.

permeated by regular grids of 1–3 μm pores (Figure 2.26), which results in BET surface areas between 30 and 50 m^2/g, accessible mostly via macropores. Diatomite is often used to support liquid phases, because it exposes the liquid in lamellae across the shell apertures, instead of hiding it in long narrow pores (cf. Section 2.3.3).

Ordered mesoporous materials (OMM) are siliceous supports widely employed in catalysis research, which are, however, typically prepared in-house. For more information see Section 3.1.1.1.

2.3.2.2 Alumina (Al$_2$O$_3$)

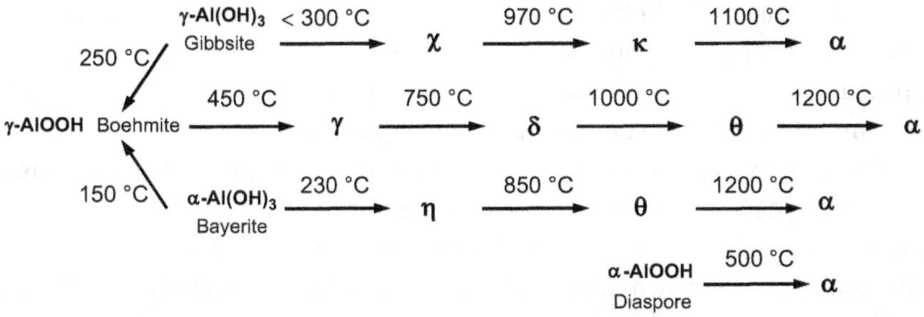

Figure 2.27: Transition aluminas accessible by the dehydration of aluminum hydroxides. Reproduced with permission from ref. [26]. Copyright 2002 American Chemical Society.

The aluminas employed in catalysis are metastable transition phases in the transformation of the hydroxides, bayerite or gibbsite[12] (α- or γ-Al(OH)$_3$, respectively) to the stable corundum (α-Al$_2$O$_3$), which is a bulk chemical used, for instance, for the production of ceramics and of aluminum (Figure 2.27). Boehmite (γ-AlOOH) is the starting point to obtain the most popular γ-Al$_2$O$_3$ by dehydration. It can be produced from both gibbsite

---

12 Gibbsite is also designated as hydrargillite.

and bayerite. The conversion from bayerite to boehmite requires hydrothermal conditions. Dehydration of bayerite under atmospheric conditions and slightly higher temperatures results in η-Al_2O_3. The high-temperature forms, δ and θ, receive less attention because of lower surface areas and reactivities.

γ- and η-Al_2O_3 are crystalline, though with highly defective structures. They have been described as defective spinels. In a spinel $Me^{2+}Me^{3+}_2O_4$, an fcc packing of oxygen anions hosts two cation sublattices: that of the bivalent cation Me^{2+} populates one out of eight tetrahedral voids, while that of the trivalent Me^{3+} – every second octahedral void.[13] As this cannot be neutral when Me^{2+} is replaced by a trivalent cation (Al^{3+}), it was assumed that some cation sites (one out of nine) are vacant in γ-Al_2O_3, although symmetry at the vacant positions remained controversial. Meanwhile, it has been proposed that 25% of Al^{3+} cations reside in tetrahedral voids, but without clear relation to the Me^{2+} sublattice positions of the spinel structure [27].

γ-Al_2O_3 is available with a wide range of BET surface areas (50–250 m^2/g) and average pore widths between 40 and 5 nm. Data on η-Al_2O_3 are less abundant, but surface areas of 250 m^2/g at average pore sizes of $\approx$5 nm have been reported, as well.

Due to the amphoteric nature of aluminas, γ-Al_2O_3 has an IEP near 7 (cf. Section 2.2.2). Its surface OH-group density by far exceeds that of silica — it is $\approx$16 OH/nm^2 at ambient and still 3–5 OH/nm^2 at 770 K. Further loss of OH groups is very gradual and accompanied by losses of BET surface area above 870 K. The transition into the stable α-alumina phase, which proceeds above 1,070 K depending on the experimental conditions, is again accompanied by a significant release of water, probably from isolated OH groups.

As discussed in Section 2.2.2, OH groups on γ-Al_2O_3 surfaces differ in their bonds to Al^{3+} cations, and therefore, also in their reactivity and IR signature (cf. Section 3.4.8.4). None of them is, however, sufficiently acidic to protonate pyridine. On the other hand, Al^{3+} ions exposed at surface oxygen vacancies are strong Lewis acids. Where γ-Al_2O_3 is employed as an acidic catalyst component, it acts either via its Lewis sites or is promoted with chlorine or even fluorine. Due to its neutral isoelectric point, it is well suited to disperse active oxide species via aqueous routes (Section 3.1.3.1).

α-Al_2O_3 is used to support active species in processes where high porosity of the catalyst is detrimental, e.g., selective oxidations (cf. Section 4.3.3). Ethylene oxide synthesis over Ag/α-Al_2O_3 catalysts is the most prominent example. α-Al_2O_3 can have BET surface areas up to 20 m^2/g [28], but for ethene oxidation, low surface areas of $\approx$1 m^2/g are most favorable. In the corundum structure, Al^{3+} populates two-thirds of the octahedral voids of a hexagonal oxygen anion packing. The surface OH group density at ambient conditions is $\approx$6 OH/nm^2.

α-Al_2O_3 may find new applications in catalysis after it has been obtained with surface areas around (140 m^2/g) via a mechanochemical synthesis, recently [29].

---

13 According to these concepts, η-Al_2O_3 is a defective inverse spinel, in which the sublattice hosted in tetrahedral voids is populated by Me^{3+}, while that in octahedral voids is shared by Me^{2+} and Me^{3+}.

2.3.2.3 Alumosilicates/zeolites

Alumosilicates are used in catalysis as solid acids, but also as supports for redox components. They may be amorphous or crystalline, which gives rise to drastic differences in properties and performance. This section is focused on zeolites, which are crystalline alumosilicates. Some basic information on amorphous alumosilicates follows at the end, and the preparation of zeolites is discussed in Section 3.1.1.2.

Combining excellent versatility with high thermal and chemical stability under a wide range of reaction conditions, zeolites have become important ingredients of technical catalysts in very different fields, both in acid-base and in redox reactions. Although several new classes of porous materials have been disclosed meanwhile, commercial success has so far remained on the side of ongoing innovation in zeolite catalysis.

According to the classical definition by D. W. Breck, zeolites are "crystalline, hydrated alumosilicates based on a three-dimensional framework that contains cations and water. The cations are mobile and undergo ion exchange. The water can be removed reversibly." Zeolites occur as natural minerals and as synthesized substances. Their name originates from the observation that the minerals release steam upon heating (Greek for "boiling stones," Cronsted 1756). Although natural zeolites are also examined for their catalytic properties, zeolites in commercial catalysts are exclusively synthetic.

In zeolite structures, Si or Al atoms tetrahedrally coordinated by O atoms combine into three-dimensional frameworks of different topologies (see Figures 2.28 through 2.32). Due to the charge imbalance caused by the replacement of Si^{4+} by Al^{3+}, the framework is a macro-anion, the charge of which increases with growing Al content. This charge is balanced by cations (after synthesis, typically, Na^+ or K^+). The general formula of zeolites can be therefore written as $M_{x/n}[(AlO_2)_x(SiO_2)_y] \cdot w\ H_2O$, where TO_2 (T = Al or Si) designates the basic tetrahedral structural units[14] and n the charge of the charge-balancing cation.

In dehydrated zeolites, the cations are located at positions determined by the spatial electrostatic field of the framework, where they can be also detected by XRD. They are part of the crystal structure but not of the framework. In hydrated form, their interaction with the framework is strongly attenuated, and they can be easily replaced by other cations or by protons (the latter requiring a detour in most cases, see Section 3.1.3.1). The exchangeability of zeolite cations is one of various sources of the amazing versatility of zeolites in catalysis: the introduction of protons turns them into solid acids; the introduction of redox-active cations converts them into redox catalysts.

In zeolites, the extra-framework space that accommodates the cations forms highly regular systems of voids within the framework. This pore system may consist of pores or of cages. The latter are voids connected to the rest of the pore system by smaller apertures, while the former are straight or meandering channels extending into one direction without significant variation of the cross section. Diffusion rates in these

---

14 As each O is coordinated by two T atoms, $TO_{4/2}$ represents a tetrahedron.

pore systems depend on the size of cage or pore entrances. Their widths are typically 0.3–0.5 nm, 0.5–0.6 nm, or 0.6–0.7 nm, depending on the structural motif forming the aperture (narrow, medium, or wide pore zeolites, respectively). The small pore sizes, which are completely in the micropore range, explain the molecular sieving properties of zeolites, which are employed in various processes of adsorption technology. In catalysis, the molecular sieving effect is utilized to achieve shape selectivity (see Section 4.3.2).

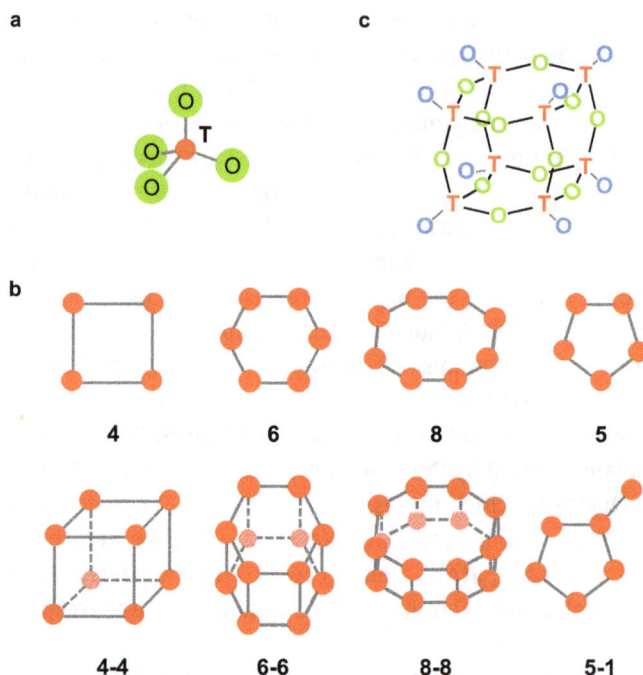

Figure 2.28: Elements of zeolite structures: (a) the primary building unit, T = Si or Al; (b) examples of secondary building units (SBUs, only T atoms are shown); and (c) complete double 4-ring, O atoms within the SBU and bridging to adjacent T atoms are differentiated by color.

Zeolite frameworks are assembled by combining characteristic structural elements that can be classified on three levels. The $TO_{4/2}$ tetrahedron as the ever-repeating primary building unit (Figure 2.28a) forms various secondary building units (SBUs), some examples of which are shown in Figure 2.28b. In these structural representations, only the T-atoms forming the structure are shown, and the bridging O atoms are omitted. The SBUs include rings and double rings containing different numbers of T-atoms and a characteristic five-ring motif extended by a single T atom — the pentasil unit. To remind of the presence of oxygen between the T atoms, a complete double four-ring element is depicted in Figure 2.28c. The structural elements defining the pore entrances in small-, medium- and large-pore zeolites are 8-, 10-, and 12-rings, respectively.

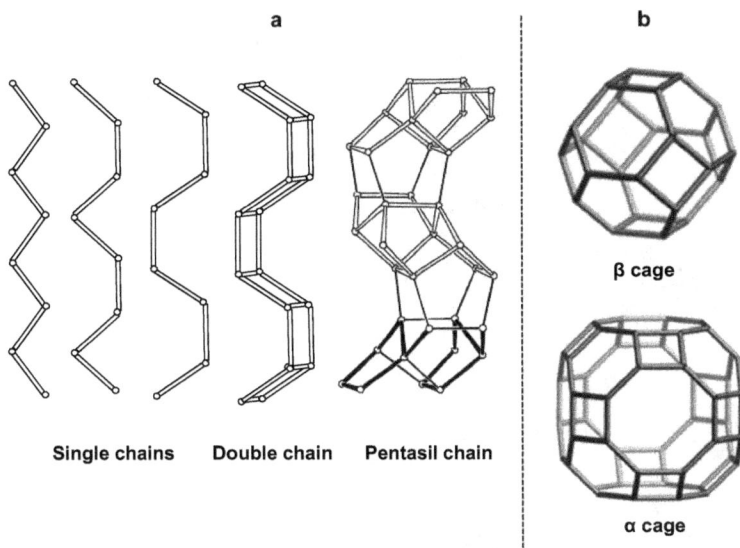

Figure 2.29: Elements of zeolite structures: tertiary building units (examples): (a) chains forming pore-shaped porosity: three types of single chains, a double chain, a pentasil chain, with the basic element composed of two pentasil units highlighted; (b) cages forming cage-type porosity – β cage, α cage. Adapted from ref. [21].

While the SBUs describe local structural details, the tertiary building units (TBUs) are the basic elements in the translational patterns of the zeolite structure. In zeolite structures with pore-shaped voids, they are typically chains and double chains extending into one direction. Examples are shown in Figure 2.29a. The pentasil chain presented there contains a basic element made of two pentasil units drawn in bold in the figure, which is repeated along the chain with 180° rotations. Cages are the TBUs of cage-type pore systems. In Figure 2.29b, the sodalite cage (or β cage), which forms most prominent zeolite structures, is depicted together with the α cage. The cage facets are formed by SBUs as well: those of the sodalite cage are 6-rings and 4-rings, while the α cage contains also 8-rings.

Three-dimensional structures are created from these TBUs by repeating the chains into the other two directions, unchanged or mirrored, unshifted or shifted, or by connecting cages by appropriate SBUs. Figure 2.30a shows how the pentasil chain extending along the c-axis is combined with its mirror image along the b-axis to obtain 10-ring apertures. Subtle differences in the stacking of these two-dimensional layers in the third dimension (a extension in Figure 2.30b) result in two different structure types: MFI[15] (e.g., zeolite ZSM-5) and MEL (e.g., zeolite ZSM-11).

---

15 According to a convention of the International Zeolite Association, zeolite structure types are denoted by codes consisting of three letters. The structure types can be realized with different chemical compositions. MFI, for instance, exists as an alumosilicate with a wide range of Si/Al ratios (ZSM-5), as a material with exclusively Si on T positions (silicalite-1), as a titanosilicate (TS-1), as gallo and ferro silicates, as a borosilicate (boralite), etc.

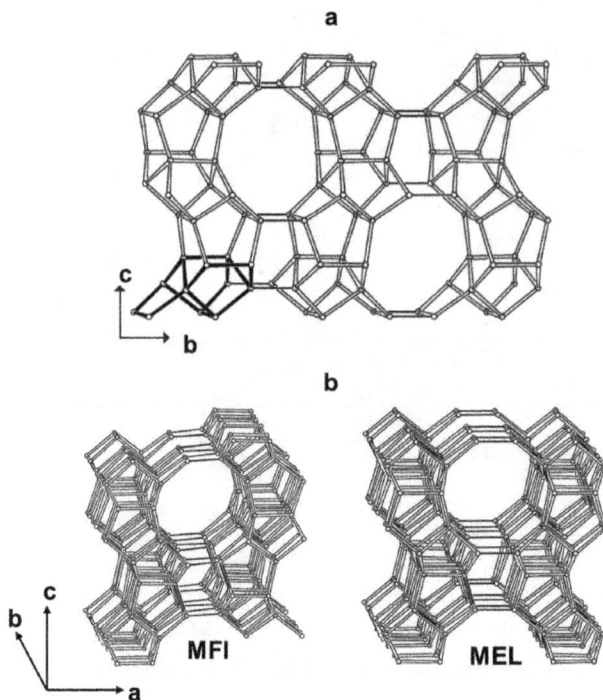

Figure 2.30: Examples for the combination of TBUs into zeolite structures (pore-type porosity):
(a) combination of pentasil chains into a two-dimensional layer (b/c plane); (b) stacking
two-dimensional layers in the third dimension (a/c plane, cf. coordinate system provided); different
structure types may be obtained via different stacking modes. Reproduced from ref. [21] with permission
from the IZA.

Cages can be combined with each other directly or indirectly via their facets. In Figure 2.31a, β cages are combined with each other via double 4-rings inserted between their 4-ring windows. In the resulting LTA structure type (e.g., zeolite A), this results in the formation of an α cage centered between eight β cages, which is accessible via an 8-ring. If the β (or sodalite) cages are linked directly, i.e., via their 4-rings instead of double 4-rings, the SOD (sodalite) structure emerges, from which the name of the cages is derived. Its cages are not accessible for most molecules, rendering it irrelevant for catalysis. In the FAU structure type (Figure 2.31b, faujasite, e.g., zeolites X or Y), the β cages are combined via double 6-rings, which creates both a larger interior cage (the supercage) and larger cage entrances. Figure 2.31b shows also the EMT structure type resulting from only minor layer stacking differences to FAU, which has, however, not yet attracted much attention in catalysis research.

As of 2022, 246 different structure types have been registered in substances that obey an extended zeolite definition, which specifies zeolites just according to their low framework density (< 20–22 T atoms per 1,000 Å^2, the limit depending on the smallest

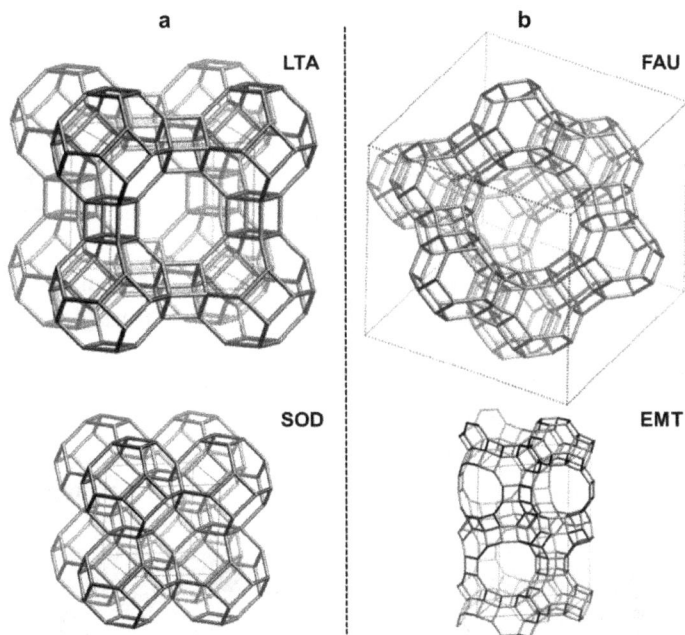

Figure 2.31: Examples for the combination of TBUs into zeolite structures (pore-type porosity): (a) combination of sodalite cages via 4-rings: with double 4-ring into LTA, with simple 4-ring into SOD, (b) combination of sodalite cages via double 6-rings with different stacking modes into FAU or into EMT. Reproduced from ref. [21] with permission from the IZA.

ring in the structure [30]). An up-to-date compilation of these structure types can be found on the website of the structure commission of the International Zeolite Association IZA (www.iza-structure.org). Despite great effort, it has been impossible to extend zeolite pore sizes beyond 1.2 nm, and the zeolites with the largest pore sizes are often difficult to access and/or too instable to have a perspective in catalysis.

Among the alumosilicate zeolites, there are structures compatible with only limited ranges of the Si/Al ratio (see below). For stability reasons, neighboring T-atoms can never be both Al (Loewenstein rule), which excludes zeolites with Si/Al < 1. Generally, zeolites with higher Al content are less stable: both hydrothermal stress (steam + heat) and interaction with acid can extract Al from the framework (dealumination). Notably, this does not necessarily result in a breakdown of the structure; rather, a stabilization of the framework can be achieved with appropriate dealumination procedures (cf. Section 3.1.4).

The revised zeolite definition allows covering a range of related materials that have also become important for zeolite catalysis. Thus, redox activity can be introduced into zeolites not only by ion exchange with redox cations, but also by partial substitution of their T-atoms with redox elements.[16] This is accomplished mostly during synthesis, but protocols for postsynthetic substitution have been reported as well. The materials obtained from such effort have been referred to as zeotypes. The introduction of Ti into an all-Si material of MFI type is a prominent example. The resulting TS-1 catalyst (cf. Table 2.1) is, nowadays, used in several large-scale oxidation processes with H_2O_2. ALPOs are porous aluminum phosphates realizing the same structure types as alumosilicate zeolites. They are relevant for adsorption rather than for catalysis, because due to their Al/P ratio of 1, they have no exchange capacity that might allow the development of acid-base or redox activity. These opportunities can be restored by replacing part of the Al in AlPOs by Si: due to the charge imbalance between Si and Al, the resulting silico-aluminum phosphates (SAPOs) again offer ion exchange capacity. Acidity and redox elements can thus be introduced into both SAPOs and alumosilicate zeolites, but even if frameworks are of the same structure type, the different framework composition may result in differences in the properties of the resulting catalysts.

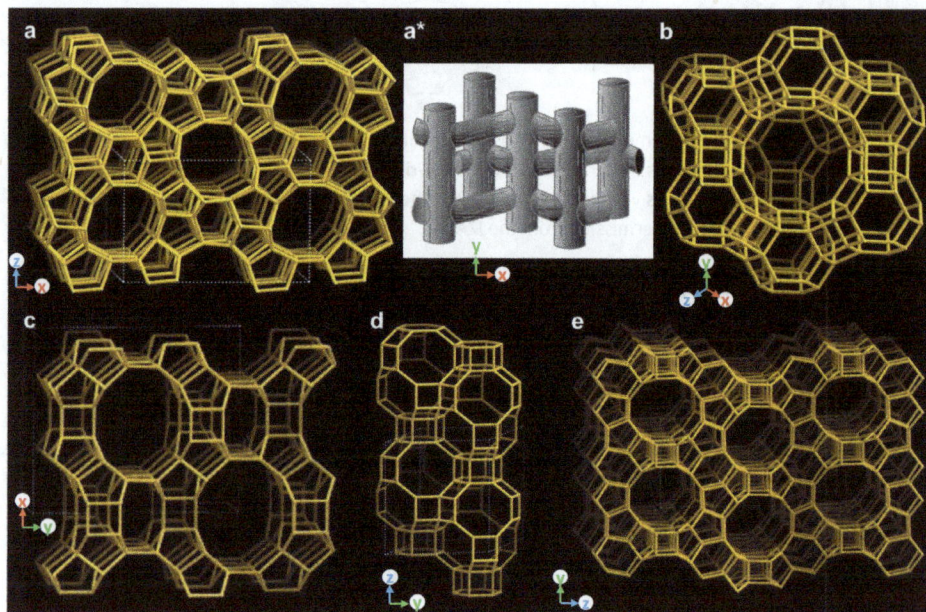

Figure 2.32: Structures of zeolites used in large-scale chemical processes: (a) ZSM-5 or TS-1 (MFI) (a* visualizes pore system); (b) zeolite Y or X (FAU); (c) mordenite (MOR); (d) chabazite, SSZ-13, SAPO-34 (CHA); and (e) Beta (*BEA). Adapted from ref. [21] with permission from the International Zeolite Association (IZA).

---

16 Substitutions are usually limited to a few percent of the T-atoms and do not change the structure. They are therefore referred to as "isomorphous" substitutions.

Table 2.1: Basic information on zeolites applied in large-scale processes of refinery, chemical industry, and environmental technology. For more information on applications, see Appendix A1.

Zeolite	Structure	Si/Al	Pore system	Application
A	LTA	1–2.5[1]	α cages accessible via 8-ring windows (0.41 nm, tunable by cations)	Detergents, adsorption, traditionally not in catalysis
X	FAU	1–1.5	Supercages (width – 1.12 nm) accessible via 12-ring windows (0.74 nm)	Adsorption, not in catalysis
Y	FAU	≥2.5		Hydrocarbon cracking (FCC)
ZSM-5	MFI	>12	2 channel types: straight (0.54 × 0.56 nm) and zigzag (0.51 × 0.55 nm); diffusion in 3rd dimension via channel intersections	Hydrocarbon cracking (FCC), Ethylbenzene synthesis (Amm)oxidations with H_2O_2, e.g., cyclohexane to hydroxylamine, propene to propene oxide
TS-1	MFI	no Al, ≈1 wt% Ti		
Mordenite	MOR	4–8	1-d channels (0.67 × 0.70 nm), with characteristic side pockets	Bifunctional catalysis, e.g., light naphtha isomerization
SSZ-13	CHA	4–25	Cages (width – 0.74 nm) accessible via 8-ring windows (0.37 nm)	Cu-SSZ-13, Cu-SAPO-34 in diesel exhaust treatment
Beta	*BEA[2]	>8	3 intersecting channels, two of them 0.76 × 0.67 nm, one of diameter 0.55 nm	Cumene synthesis

[1] until recently. Meanwhile, LTA is accessible even with high Si/Al ratios [31], which opens opportunities for applications in catalysis.
[2] the asterisk indicates that the lattice has characteristic imperfections

In Figure 2.32 and Table 2.1, a few general characteristics of important zeolites are summarized. Two of them (A and X) have not been applied in catalysis so far, but are mentioned nevertheless, due to their great role in other fields, e.g., as adsorbents (A, X) or detergent additives (A). Shape selectivity can be achieved with zeolites of ≅0.5 nm pore size (e.g., ZSM-5), which differentiate, for instance, between n- and iso-alkanes (cf. Section 4.3.2). In Friedel–Crafts alkylations of benzene with alkenes (cf. eq. (4.9)), H-ZSM-5 was suitable for the synthesis of ethyl benzene, but not of cumene, because its shape selectivity favored the undesired n-propyl benzene.

The acidity of zeolites originates from silanol groups that interact with an adjacent Lewis-acidic framework Al^{3+} site (Scheme 2.1a, see also Figure 1.7b). Unlike exchanged cations that bind to the framework by electrostatic forces, the proton is attached to the framework by a covalent bond. Due to the proximity between silanol oxygen and Al site, the anionic charge can be effectively stabilized in the deprotonated state. This results in high acid strengths. Although acid strengths depend on details of the bonding situation, their distribution in zeolites is discrete because of the rigidity of the crystalline

a.

b.

c.

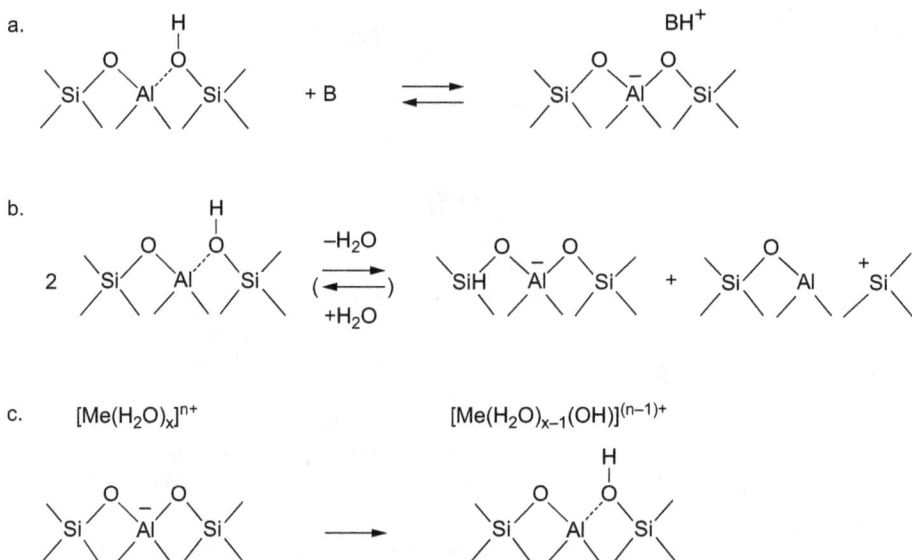

Scheme 2.1: Acid sites in alumosilicates: (a) deprotonation of a Brønsted site by a base, (b) interconversion between Brønsted and Lewis sites, and (c) formation of Brønsted sites via the Hirschler-Plank mechanism.

structure: there is only one or a few types of Brønsted site in each zeolite, as long as the Al content is low. In Al-rich zeolites, interactions between Brønsted sites mitigate their acidity and wash out the discrete distributions.

When zeolites are dehydrated under severe conditions, Brønsted sites can be converted into Lewis sites. The process is usually described by the equation shown in Scheme 2.1b. It may result in even stronger exposure of the Al site, but usually not in its detachment: dealumination of zeolites happens by hydrolysis rather than by dehydration. Lewis sites are always defect sites in zeolites. They can, however, be healed under suitable rehydration conditions.

Brønsted and Lewis sites can be formed also by multivalent cations introduced into zeolites, e.g., by La^{3+} or Ce^{3+} (Scheme 2.1c). At moderate dehydration temperatures, a Si-(OH)-Al site is formed by deprotonation of a water molecule from the coordination sphere of these ions (Hirschler-Plank mechanism). At lower temperatures, excess water still present interferes with the process; at higher temperatures, the site is converted into a Lewis site by complete dehydration (e.g., $(LaO)^+$). The Brønsted site is easily restored by rehydration. Rare-earth exchanged zeolites are widely applied in the cracking of heavy oil fractions (FCC process, cf. Appendix A1).

Despite the outstanding advantages of zeolites, amorphous alumosilicates have their specific applications in catalysis, e.g., for cracking of large molecules in FCC catalysts. They can be made quite inexpensively by co-precipitation, but more sophisticated preparations via sol-gel routes or even flame hydrolysis (cf. Section 3.1) have also been reported. They can be obtained with BET surface areas up to 600 m^2/g, but are also available in macroporous versions with much smaller surfaces.

Amorphous alumosilicates are much more flexible with respect to structure and properties than zeolites. As there is no Loewenstein rule for them, they may form alumina-like domains containing Al also in octahedral coordination. In precipitated alumosilicates, such heterogenization was reported to happen at (nominal) Al$_2$O$_3$ contents >25 wt-%. As in zeolites, the Brønsted acidity of amorphous alumosilicates arises from the interaction of Al sites with adjacent silanol groups (Scheme 2.1a). Due to their higher structural flexibility, the distances between Al and the silanol oxygen are, however, larger and less defined. Therefore, the strengths of acid sites in amorphous alumosilicates and related materials (e.g., in Al-modified OMMs, cf. Section 3.1.1.1) are lower than in zeolites by an order of magnitude, and their distributions are less discrete.

2.3.2.4 Metal-organic frameworks

Metal-organic frameworks (MOFs), also referred to as porous coordination polymers (PCPs) [32, 33], are a rather new class of microporous materials with structure-inherent porosity, which bears great promise for giving access to new catalysts and supports. Simple, mostly linear, coordination polymers have been known for decades, but first reports on two- or three-dimensional coordination polymers with a porous, crystalline structure appeared only in the 1990s. This initiated an enormous research effort, which was pioneered by the groups of Férey, Kitagawa and Yaghi, resulting in numerous new compounds.

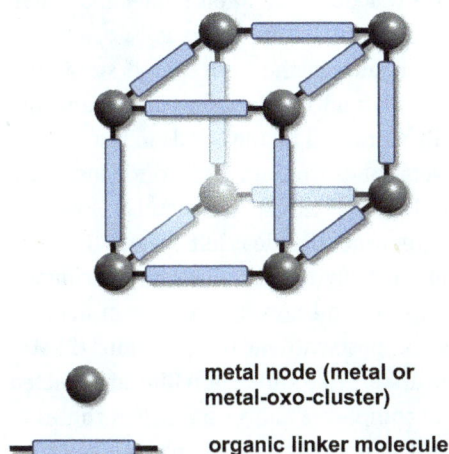

●   **metal node (metal or metal-oxo-cluster)**

▬▭▬   **organic linker molecule**

Figure 2.33: Construction principle of MOFs; the metal nodes are also referred to as SBUs.

MOFs are no organometallic compounds as their name might imply: they do not contain metal-carbon bonds. They are inorganic-organic hybrid materials constructed from metal ions (or metal-oxo-clusters, often referred to as metal nodes or secondary building units, SBU) and linkers. The linkers are bridging organic ligands with at least two functional groups capable of coordinating with the metal nodes (Figure 2.33). Aromatic dicarboxylates, diphosphonates, or disulfonates are typical examples, but also pyridine moieties or other N-containing groups, which lead to weaker interactions with the metal nodes. Like in zeolites, these basic components form larger, characteristic structural motifs, which are periodically repeated in the frameworks. The structure of these motifs and, consequently, the overall framework is determined by size and geometry of the organic linkers and by the characteristic coordination geometry of the metal nodes.

Only a few of the almost infinite number of possible MOF structures are, however, considered promising for use in catalysis, while the vast majority fails for a variety of reasons. The key properties for MOF-based catalysts are: (i) sufficient chemical and thermal stability, (ii) the presence of catalytically active sites, either free coordination sites at the metal nodes or functional groups extending from the organic linkers, and (iii) an accessible pore system allowing diffusion of substrate and product molecules within the framework [34].

The first criterion, by itself, rejects a large number of structures. Many frameworks with weakly coordinating linkers decompose even at room temperature or slightly above, in air or in moisture. For liquid-phase reactions, thermal stability at ≈470 K would be sufficient. However, chemical stability at this temperature is not less important: extreme pH values during the reaction or the presence of strongly coordinating functional groups in solvents, reactants, or products might result in leaching of the active component or even in the dissolution of the whole framework. Stability at 470 K is extraordinary in the realm of MOFs. As temperatures in catalytic gas-phase reactions are often higher than that, MOFs can be considered only for a limited range of gas-phase applications.

Many MOFs do not contain suitable catalytic sites in the as-prepared state (2nd criterion). Typical linkers offer just van der Waals and dipole interactions, and the coordination sites of the metal nodes are usually saturated by the bonds to the linkers. Dissociation of these bonds to give way for reactant adsorption required for measurable catalytic rates might well result in the dissolution of the MOF structure. However, there are also MOFs in which a residual vacancy per metal node is left even in the ideal structure, and linkers that contain functionalities of catalytic relevance at their aromatic backbones, e.g., carboxylic or sulfonic acid groups. In addition, side groups of the aromatic linkers can be converted into active sites via "postsynthetic modification" (PSM).

In PSM strategies, simple substituents at the linkers like amine or halide are reacted with organic molecules or even transition metal complexes along established routes to form new covalent bonds at these side groups. The approach allows placing complex and sterically demanding moieties in the pores, and in most cases, it is the only way to achieve this, because size or reactivity of these groups often prevent the formation of the desired MOF when they are present already during its synthesis. PSM reactions allow, for instance, generating chirality for enantioselective reactions or tuning the acid-base properties of the material by introducing a variety of organic functionalities.

PSM reactions can also be used for immobilizing well-defined transition metal complexes or ions at chelating side groups.

The third requirement, the accessibility of the pore structure, is easier to comply with, because the porosity exhibited by MOFs is structure-inherent. The majority of MOF materials are microporous, but some mesoporous structures have been also reported. To allow substrate molecules accessing the pores, pore sizes must be larger than a threshold value of around 0.4 nm, and pore blocking must be avoided. Pores might be blocked, for instance, by residual linker molecules that are not part of the lattice or by strongly interacting solvent molecules that cannot easily be replaced by the substrate. Pore stripping may be achieved by thermal or chemical treatments.

The design of "isoreticular" MOFs represents a straightforward approach to control the key properties just discussed. Isoreticular materials have structures with identical framework topology. Their linkers are molecules with closely related structures, but either of different sizes or with different functional side groups. The achievements of reticular strategies have been highlighted in ref. [35].

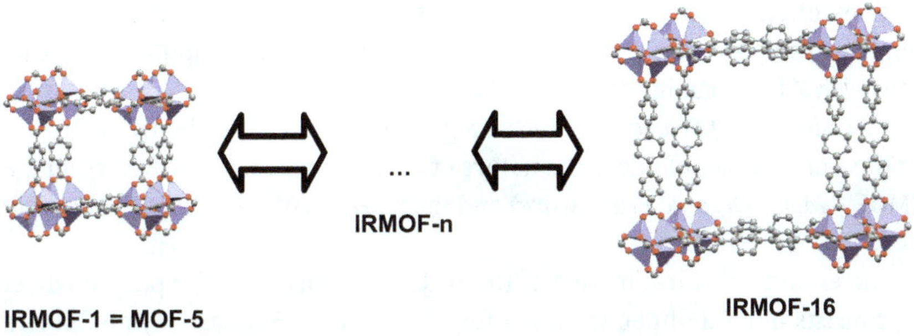

IRMOF-n

IRMOF-1 = MOF-5

IRMOF-16

Figure 2.34: Isoreticular metal-organic framework structures: the IRMOF series; the figure shows two out of 16 realized structures. Colors: purple polyhedral – Zn, red spheres – O, grey spheres - C, H omitted; Drawn with Mercury 1.0 (ref. [37]) using the database of the Cambridge Crystallographic Data Center (CCCD).

The most famous example for isoreticular MOFs is the so-called IRMOF series based on the structure of MOF-5, which has been reported by the group of Yaghi [36]. In the cubic structure of MOF-5 (or IRMOF-1), the metal nodes (SBUs) are Zn_4O tetrahedra, which are interconnected by benzene-1,4-dicarboxylate (BDC) linkers via the edges of the tetrahedral units (Figure 2.34). Isoreticular materials are obtained if the same SBUs are interconnected by other, structurally related linker molecules instead of BDC. When, for example, functionalized linkers like 2-bromobenzene-1,4-dicarboxylate (BrBDC) or 2-aminobenzene-1,4-dicarboxylate (ABDC) are employed for the isoreticular synthesis, the resulting crystalline structures (IRMOF-2 and IRMOF-3, respectively) have the same topology as IRMOF-1 and almost identical lattice constants.

In other cases, the SBUs typical of the IRMOF series are interconnected by longer linkers: biphenyl–4,4'–dicarboxylate (BPDC, IRMOF–10) or p-terphenyl–4,4''–dicarboxylate (TPBDC, IRMOF–16, cf. Figure 2.34), respectively. The pores in IRMOF–10 (1.54 nm) and in IRMOF–16 (1.91 nm) are significantly wider than those of IRMOF–1 (1.12 nm). It should be noted, however, that the use of longer linker molecules does not necessarily result in increased micropore volume, since interpenetration phenomena may occur if the linkers become too large. Interpenetration refers to the formation of a second lattice of the same topology occupying the void space of the first lattice, i.e., the existence of two interpenetrated lattices. As a consequence, the effective pore diameters and the available space within the pores decrease.

Isoreticular series may also comprise mixed-linker or multivariate MOFs, in which two or more linker molecules are distributed within a single framework. Likewise, mixed-metal MOFs, which contain two or more different types of metal nodes at equivalent lattice positions of the structure have also been reported. Obviously, such flexibility within a fixed structure type is of particular interest for applications in catalysis.

As mentioned above, the range of MOFs suitable for catalytic applications is relatively narrow. Some of the most promising and best-studied candidates, namely HKUST-1, MIL-53, MIL-101, UiO-67, and ZIF-8, are presented in Figure 2.35. Unfortunately, the nomenclature for MOFs is not as systematic as that for zeolites. New structures are counted in series, often named after the group or institution reporting them for the first time. Prominent examples are the "MIL-X" series from the Férey group (*Matériaux de l'Institut Lavoisier*, X is the running number of the structure), the "HKUST-X" series from the *Hong Kong University of Science and Technology*, and the "UiO-X" series from the *Universitetet i Oslo*. Some names allude to the nature of the materials, e.g., "MOF-X" from the Yaghi lab (MOF – *Metal-Organic Framework*) and the "ZIF-X" series (*Zeolitic Imidazolate Frameworks*).

The characteristic structural motive of the HKUST-1 structure is the paddle-wheel unit, which consists of a Cu dimer, to which four benzene-1,3-5-tricarboxylate linkers are coordinated (Figure 2.35a, left side). In the solvated state, one solvent molecule (at ambient typically water) is coordinated to each Cu site at the axial position of the paddle wheel. These solvent molecules can be desorbed by heating and/or evacuating to obtain coordinatively unsaturated metal sites (cus), which can be used for catalysis. The HKUST-1 structure contains three types of spherical pores with diameters of 0.5, 1.1, and 1.35 nm.

The MIL-53 structure contains one-dimensional zigzag chains of $[-M^{3+}-(OH)-]$ units, which are connected to four neighboring chains by terephthalate linkers (Figure 2.35b). All metal centers are octahedrally coordinated by six O atoms, four of which belong to carboxylate groups, and the remaining two to bridging OH groups. The resulting framework structure contains one-dimensional diamond-shaped pores with a diameter of 0.85 nm. Since the metal centers are fully coordinated, they cannot be used as active sites. However, catalytic sites can be introduced by attaching organic functionalities or immobilizing metal complexes via side groups of the linkers.

The metal centers of the MIL-101 structure are arranged in trimeric $(M^{3+})_3O$ units (Figure 2.35c). Six terephthalate linkers and three solvent molecules are coordinated to each of these units, forming the characteristic structural motif. The framework structure

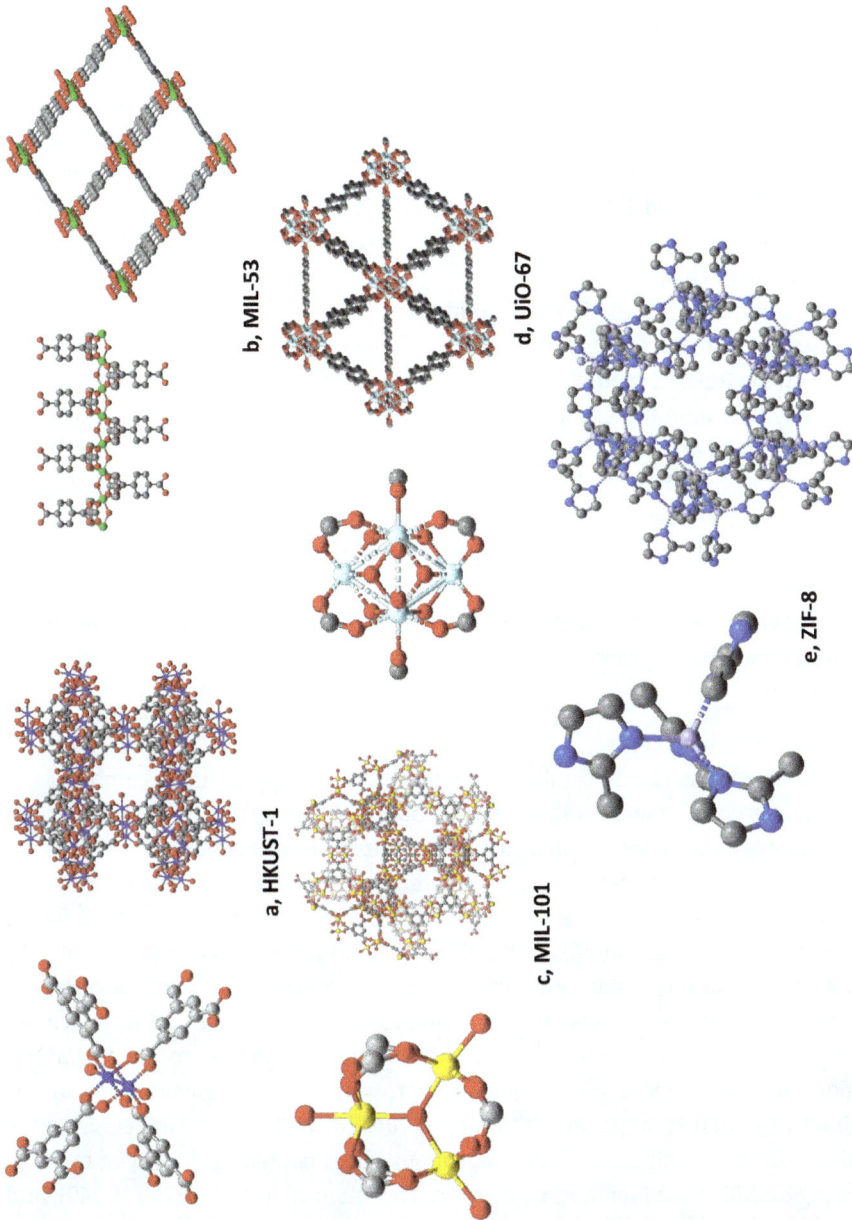

Figure 2.35: Examples of MOF materials studied for applications in catalysis: left – basic structural elements; right – integration of these elements into the structure; (a) HKUST-1, (b) MIL-53, (c) MIL-101, (d) UiO-67, (e) ZIF-8. Drawn with Mercury 1.0 (ref. [37]) using the database of the Cambridge Crystallographic Data Center (CCDC).

contains three different pores with diameters of 0.6, 3.0, and 3.8 nm. Catalytic sites can be obtained by desorption of the solvent molecules (cf. HKUST–1) or by PSM approaches.

The SBUs of UiO-67 are $[Zr_6O_4(OH)_4]^{12-}$ clusters, which are connected to 12 neighboring clusters by 1,1'–biphenyl–4,4'–dicarboxylate linkers (Figure 2.35d). The resulting framework structure contains larger octahedral cages (2.3 nm) and smaller tetrahedral cages (1.15 nm) connected by triangular pore windows (0.8 nm). Catalytically active sites can be functional groups at the linkers or Lewis acid sites resulting from linker defects at the Zr-oxo-clusters.

ZIF-8 belongs to the group of zeolitic imidazolate frameworks, a subgroup of MOFs, which feature structures closely related to those of zeolites. ZIF-8 combines the metal node Zn^{2+} with the linker 2-methylimidazolate (mim). Each metal center is tetrahedrally coordinated by four mim linkers, and the Zn-mim-Zn angle resembles the Si-O-Si angle in zeolites closely (Figure 2.35e). The $Zn(mim)_4$ group is, therefore, a structural analogue to the TO_4 primary building unit in zeolites (Figure 2.28a). In ZIF-8, the resulting framework structure features a sodalite net topology (cf. Figure 2.31) with a cage size of 1.16 nm and an aperture of 0.34 nm. The small pore size suggests that early promising results with this material were achieved just by the catalytic effect of the external surface. As in the case of zeolites, this has resulted in the development of several approaches to obtain ZIF-8 with a hierarchical pore system.

2.3.2.5 Titania, zirconia, ceria

TiO_2 plays an enormous role in catalysis both as a catalyst support and as a photocatalyst. Although rutile is its most stable modification in the temperature range relevant for catalysis, pure anatase can be obtained in the industrial sulfate process,[17] where the nature of the final product can be controlled by the calcination conditions [38]. This has been assigned to a particle-size dependence of the free energy, according to which anatase becomes the stable modification below a critical particle size of ca. 14 nm [39].

For most applications in catalysis, anatase is the preferred modification, although the TiO_2 exhibiting the most favorable performance in many photocatalytic processes (P25® from Degussa, now Evonik) is actually a mixture of anatase, rutile, and amorphous material, due to its origin from flame pyrolysis. There is considerable uncertainty about its actual composition, although a 70/30 or 80/20 ratio between anatase and rutile has been often reported. Its amorphous phase content is, however, not negligible: using an XRD-based technique, its share was quantified as 14% in ref. [40], and its removal resulted in significant gain in photodegradation activity. Another source revealed spatial heterogeneity in P25® composition, specifying the amorphous content between 0 and 13%, with anatase/rutile ratios between 4.9 and 5.7 [41].

---

17 The chloride process (via oxidation of $TiCl_4$) produces an anatase-rutile mixture (ref 38), which is converted to pure rutile in downstream calcination.

Above the critical particle size, the anatase-rutile transition is kinetically controlled and, therefore, subject to influences of a great many parameters like particle size and shape, surface area, atmosphere, heating rate, and purity. The first rutile characteristics can be usually detected after exposure to temperatures >870 K (see also Section 3.4.8.6). The transition is strongly affected by the presence of dopants. Alkali ions, for instance, and also surface vanadium oxide species are known to accelerate it, while silica, alumina [38], and also tungsten oxide species [42] exert a stabilizing effect. The reasons are not always clear, but it was reported that an increased concentration of O vacancies caused, for instance, by dopants or reducing conditions accelerates the phase transition, while saturation by surface OH groups delays it.

Depending on its thermal history, anatase can have very different BET surface areas. Products of more than 300 m^2/g are on the market. After exposure to temperatures typical of catalytic applications, e.g., 700 K, they stabilize at around 100 m^2/g, while the BET surface area of P25® is ≈ 50 m^2/g. OH-group densities of 5–6 nm^{-2} have been reported for pyrogenic and sol-gel derived products of predominant or exclusive anatase structure [24]. The IEP of TiO_2 is near-neutral (Section 2.2.2); its surface exposes both slightly acidic and slightly basic OH groups. Titania engages in interactions with supported species not only via these reactive OH groups, but also via effects caused by its non-negligible redox activity (SMSI, see Section 2.3.3).

Although zirconia has found only limited applications in catalysis so far, it should not be disregarded when new challenges are to be met, because it offers high thermal stability, an amphoteric character, and an abundance of reactive surface OH groups. ZrO_2 was employed to support Cu in catalysts for the synthesis of methanol, for the conversion of methanol with water to produce H_2 (methanol steam reforming) [43], for the hydrogenation of ethyl acetate [44], and for hydrodeoxygenation reactions. Other uses include Co-based Fischer–Tropsch catalysts, Co-Mo sulfide catalysts for hydrodesulfurization, and Ni catalysts for methane activation. ZrO_2 can be been also applied in its Y-stabilized version.

Among the three ZrO_2 polymorphs (monoclinic, tetragonal, cubic), only two are relevant for catalysis. Being the stable form up to 1,443 K, m-ZrO_2 is easily prepared via various routes [45]. t-ZrO_2 is formed below 1,443 K in a number of syntheses, but mostly mixed with m-ZrO_2, the amount of which tends to increase during thermal conditioning. However, pure high surface-area t-ZrO_2 is meanwhile also accessible, e.g., via a solvothermal route, which gives access to both m- and t-ZrO_2 with BET surface areas of ≥ 100 m^2/g [46].

Thermal history has a crucial influence on BET surface area and surface OH group density also in the case of zirconias. Although 200 m^2/g were obtained after exposure to 670 K with some m-ZrO_2 preparations, surface areas well below this have been reported in most studies, even after milder thermal stress [45]. m-ZrO_2 has an OH–group density of 12–14 nm^{-2} at ambient, but still ≈ 6 nm^{-2} after calcination at 1,173 K. Its IEP is between 5 and 6.

Some ZrO_2-based mixed oxides can be stabilized in the high-temperature ZrO_2 structures at lower temperature. The best known version is Y-stabilized ZrO_2 (YSZ), the structure of which can be tetragonal at low Y-content, but converges to cubic at >10 mol-% Y_2O_3 [47]. Doping Zr^{IV} oxide with Y^{III} cations sizably increases oxygen anion conductivity. YSZ is therefore widely used as membrane material in high-temperature fuel cells, but also as a support material for electrode catalysts in them. Zr oxide has also been included into other mixed-oxide phases, with quite different intentions. Deposition on SiO_2 or co-precipitation from Si and Zr sources results in supports exhibiting similar reactivity as ZrO_2 at much higher BET surface areas. On the contrary, in Ce-Zr mixed oxides, Zr is rather used to stabilize and activate the CeO_2 structure (see below).

ZrO_2 is also a catalyst on its own. It catalyzes the hydrogenation of alkenes, the dehydration of alcohols, and the carbonylation of methanol to dimetylcarbonate. Modified with H_2SO_4 or WO_3, it was the focus of a great research activity on its "superacid" properties, allowing, for instance, alkane isomerization under very mild conditions (for more detail, see ref. [48]).

The influence of the polymorph nature on the catalytic properties is not well understood. Methanol synthesis activity was up to an order of magnitude higher when Cu was supported on m-ZrO_2 [49]; alkane isomerization was faster on sulfated t-ZrO_2 [50].

The use of CeO_2 and of its mixed phases with other metal oxides, e.g., ZrO_2, for supporting metal particles or oxide species usually targets the attractive redox properties of the resulting systems (see Section 2.3.3). CeO_2 is rather special for its low OH-group densities. On its (111) facet, OH groups were detected only when Ce^{3+} obtained after a reductive treatment interacted with water: it reoxidized these cations to Ce^{4+} around 600 K [51]. Accordingly, extremely low OH-group densities between 0.3 and 0.7 nm^{-2} were found, for instance, on CeO_2 nanoparticles synthesized in supercritical water [52]. Despite missing OH functionalities, CeO_2-based materials offer other interactions that allow achieving high dispersion of metal particles or oxide species (see Section 2.3.3). BET surface areas of precipitated CeO_2 or mixed-oxide products are often between 50 and 100 m^2/g, but there is also CeO_2 with >150 m^2/g on the market. After deposition of Au on such a support, BET surface areas of >200 m^2/g were measured after calcination at 700 K, and still 25 m^2/g after calcination at 1,173 K [53].

Due to the Ce^{IV}/Ce^{III} redox couple, Ce-containing materials exhibit, by far, the highest redox activity among the supports introduced in this section. Their spectacular cooperation effects with metal particles will be discussed in the background of the behavior of the unmodified supports, as illustrated in Figure 2.36. It shows TPR profiles[18] of CeO_2 samples of different origins[19] and BET surface areas, and of a Ce-La mixed oxide.

---

[18] Temperature-programmed reduction (TPR) is a standard method for studying the redox behavior of catalysts, in which H_2 consumption is measured while the sample is heated in dilute H_2 along a linear temperature program (cf. Section 3.4.2.5)

[19] A and B are commercial samples; C through E were precipitated from ultra-dilute aqueous solutions (ref. [54]). It should be noted that their profiles in Figure 2.36 end with an isothermal period at 1,073 K. Comparison of the profiles below this temperature reveals, however, that the contribution of the low-temperature peak is higher in C than in B.

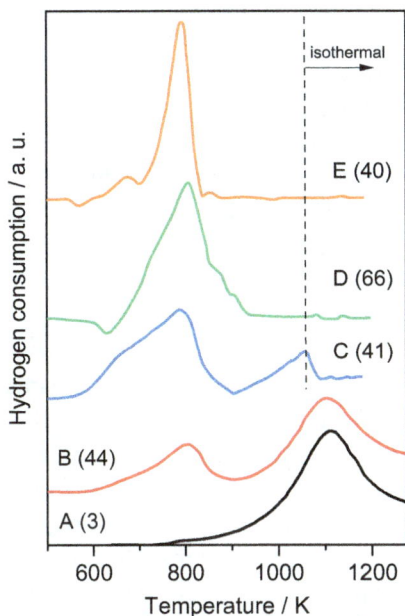

Figure 2.36: Temperature-programmed reduction of CeO$_2$ (A, B, C) and a CeO$_2$-based mixed oxide (CeO$_2$/La$_2$O$_3$, D, E); numbers in parentheses report the BET surface areas in m^2/g.

They are dominated by two peaks around 800 K and 1100 K, which can be ascribed to the reduction of Ce^{4+} in the outermost surface layer and in the bulk, respectively [55]. Accordingly, the low-temperature peak is insignificant in the profile of sample A (3 m^2/g). Its different contribution to the profiles of samples B and C, both with ≈40 m^2/g, suggests the existence of an additional influence on the redox behavior: apparently, an abundance of structural defects allows the reduction proceeding into deeper layers. Indeed, when La was substituted on cation sites of the CeO$_2$ lattice (sample D, CeO$_2$/La$_2$O$_3$ = 4), Ce^{4+} was completely reduced below 900 K [54]. The high-temperature peak was not restored even when the BET surface area of the material was decreased to 40 m^2/g by calcination at 1,073 K instead of 773 K (sample E).

The enhanced participation of Ce^{4+} in redox processes is a general trend in Ce-M mixed oxides, particularly in Ce$_x$Zr$_{1-x}$O$_2$ materials used as an oxygen storage material in three-way catalysis. At the same time, Ce-Zr mixed oxides outperform CeO$_2$ with respect to thermal stability. More details about Ce-based oxides relevant for catalysis can be found in refs. [55, 56].

2.3.2.6 Carbon materials

In this section, carbons, which are extensively used in surface catalysis as supports for (electro)catalysts and as electrode materials, are discussed following ref. [57]. Carbons may be amorphous (active carbon, carbon black) or exhibit long-range order as graphite, carbon nanotubes, graphene, diamond, and templated carbons. In the latter, the order relates only to the pore system.

Carbons can be prepared with enormous porosities and with different pore-size distributions. While they are a priori hydrophobic, their surface chemistry can be varied in a wide range by functionalization with O- or N-containing species, the surface density of which can be further modified by thermal treatments (cf. Section 2.2.2). Due to their relatively weak interactions with deposited species, complete reduction of metal precursors, which is often difficult with base metals on oxide supports, is easily achieved on carbon supports. Carbons are stable in almost all liquid environments, in inert atmospheres also at high temperatures. In oxidizing or hydrogenating environments, they can be, however, gasified. On the other hand, combustion of the support is an easy way to recover valuable carbon-supported metals from deactivated catalysts. While some carbon materials are very pure, active carbons may contain contaminants (ashes) in varying amounts, due to their provenance from natural products. Likewise, carbon nanotubes usually contain residues of the catalysts used in the growth process (cf. Section 3.1.1.5).

The term **"active carbon"** designates amorphous carbon materials with very high porosity and internal surface area, which offer large capacities for the adsorption of molecules from fluid phases. They are made starting from carbonaceous raw materials like coal, wood, coconut shells, residues from oil distillation, or from agriculture, which are subjected to a procedure consisting of carbonization and activation stages.

In the carbonization stage, the raw material is heated to temperatures up to 1,123 K under anaerobic conditions, which eliminates water, heteroatoms, aliphatic structures, and dust, and contracts and consolidates the carbon structure. Porosity is enhanced in an activation stage, subsequent or parallel to carbonization. For subsequent activation, the intermediate is subjected to a thermal treatment with steam, CO_2, or air, or mixtures thereof, at temperatures between 1073 and 1,473 K, which gasifies more labile parts of the material and creates porosity also by the release of extra gas molecules (cf. eqs. (2.6, 2.7)).

The activation can be catalyzed or inhibited by inorganic ashes present in the raw material. For the parallel process, also referred to as chemical activation, such catalysts (or activating agents, e.g., H_3PO_4, HNO_3, or $ZnCl_2$) are impregnated into the raw material. During the thermal treatment, these agents favor dehydration, influence pyrolytic processes, for instance, by inhibiting tar formation and improving thus the carbon yield, and decrease the temperatures required for achieving the desired porosity. Although the properties of activated carbons depend on quite a number of parameters [57], the porosity can be better controlled by chemical activation. The final products must be, however, washed to remove residues of the activating agents.

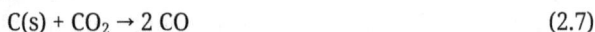

$$C(s) + 2\,H_2O \rightarrow CO_2 + 2\,H_2 \tag{2.6}$$

$$C(s) + CO_2 \rightarrow 2\,CO \tag{2.7}$$

Typical internal surface areas of active carbons are between 700 and 1,200 m^2/g, which suggests a large weight of microporosity in the pore-size distribution. The proportions between micro-, meso- and macropores depend, for instance, on the choice of the raw materials. Active carbon is initially granular, but it can be ground to powders and processed with binders to any shape.

Graphite is rarely used in catalysis due to its low surface area and its inertness towards functionalization. Both surface area and reactivity can, however, be expanded by intercalation-based methods and by high-energy ball milling.

Glassy carbon (or vitreous carbon) is a microcrystalline, but macro isotropic carbon material, which does not crystallize to graphite, even at 3270 K. It is prepared by pyrolyzing polymeric materials like PVC, cellulose, or phenolic resins. Though being highly porous, it is impermeable for gases, because its voids are neither accessible nor interconnected. Being hard and brittle like glass, it stands out for its excellent thermal and chemical stability and its high, isotropic conductivity for electricity and heat flow. It can be also produced with high surface areas by carburizing foamed polymer resins (reticulated vitreous carbon). Glassy carbon is widely used as an electrode material, on which electrocatalysts (e.g., Pt/C) are deposited, but it has been also employed to support active metal or oxide components.

Carbon blacks are soot-like materials, which are produced by incomplete combustion or by pyrolysis of hydrocarbons under conditions that ensure its high carbon content. It contains typically >97% C, the remainder being H, S, and O, and low amounts of ashes and solvent extractable fractions (both <1%). Carbon black is available from different processes. Furnace black, which is made by injection of heavy fractions from oil distillation or other refinery processes into gas-fueled flames, caters to 95% of the market. Acetylene black obtained in a similar way from acetylene stands out for its high purity. Thermal blacks are pyrolytic blacks obtained in a cyclic process, in which reactors are sequentially heated by burning fuel, charged with the hydrocarbons to be pyrolyzed to C and H_2, and discharged. Thermal black made from natural gas is highly pure, as well.

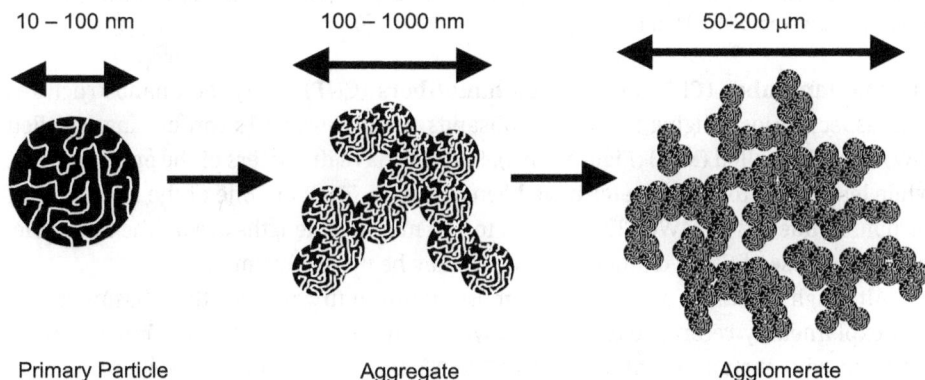

Figure 2.37: Microstructure of carbon-black particles.

The microstructure of carbon black is highlighted in Figure 2.37. Primary particles form aggregates similar to branched chains (more spherical in thermal black). These agglomerate to μm-sized units, which may be processed to larger granules. The microstructure indicated in Figure 2.37 allows a wide range of BET surface areas, which extends up to 1,500 m^2/g for furnace blacks (ca. 80 m^2/g for acetylene blacks), while the surface area of thermal blacks remains below 50 m^2/g due to a larger size of the primary particles (100–500 nm).

Due to their good electrical conductivity, carbon blacks with smaller primary particle sizes are popular supports for electrocatalysts, e.g., in fuel cells. Conductivity can be significantly enhanced when the carbon is partially graphitized by treatment at 2,770–3,270 K.

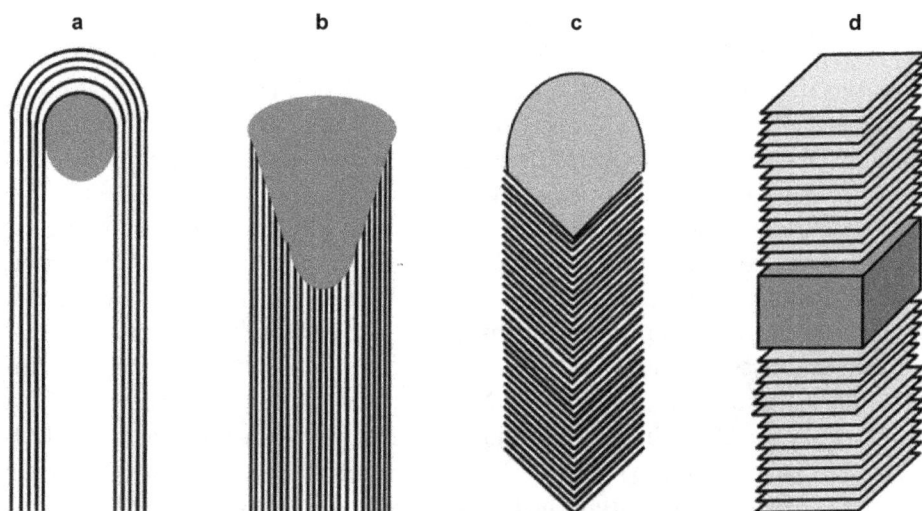

Figure 2.38: Carbon nanotubes and nanofibers: (a) multi-walled CNT; (b) r-CNF (ribbon); (c) h-CNF (herringbone); and (d) p-CNF (platelet). The grey areas designate metal particles remaining from the growth process. Reproduced from ref. [57].

Carbon nanotubes (CNT) and **carbon nanofibers (CNF)** are carbon nanostructures with aspect ratios ranging from ten thousands to millions. CNTs can be single-walled (SW) or multi-walled (MW) (Figure 2.38a). SWCNTs have diameters of the order of 1 nm, while lengths up to half a meter have been achieved. They are one of the allotropes of carbon. Diameters of MWCNTs range up to 100 nm; their lengths are on the mm scale, similar to those of CNFs, diameters of which may be up to 500 nm.

Although MWCNTs have attracted most attention in catalysis, their basics can be best explained by reference to SWCNTs, which can be considered as graphene layers wrapped with certain orientations. Any SWCNT can be imagined to be sliced parallel to its axis, unrolled, and flattened on a plane (Figure 2.39). Any atom A exactly on the cutting line will miss on the other side of the flattened tube surface. The roll-up vector

$\vec{R}$ connecting these points A_1 and A_2 is normal to the cutting line, while its length corresponds to the tube circumference. Using a base of two vectors $\vec{u}$ and $\vec{v}$, which connect A_1 with the next atoms of the same orientation of sp^2 bonds (Figure 2.39), $\vec{R}$ can be expressed as

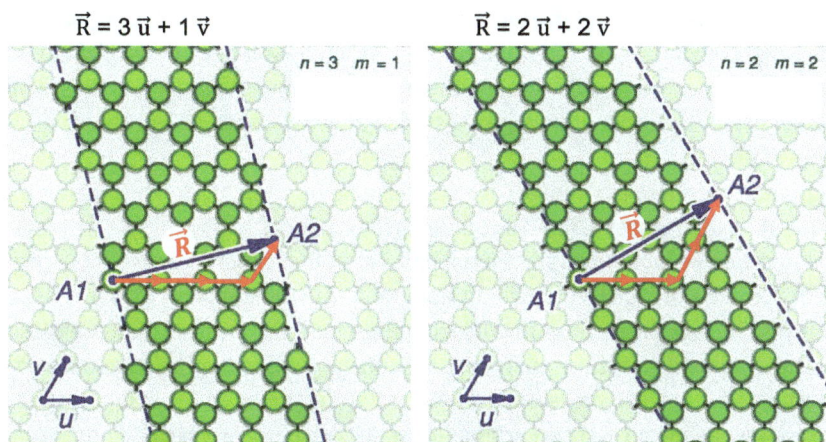

$$\vec{R} = 3\,\vec{u} + 1\,\vec{v} \qquad\qquad \vec{R} = 2\,\vec{u} + 2\,\vec{v}$$

Figure 2.39: SWCNTs by rolling a graphene sheet; tubes with different properties are obtained by rolling into different directions (two examples). Note that there are two options for the orientation of the sp^2 bonds around C atoms. The corresponding atoms are distinguished by slightly different tones of green. Adapted from a graphics published under https://en.wikipedia.org/wiki/File:Nanotube_strip_%2B03_%2B01. pdf#filelinks (accessed January 20, 2023).

$$\vec{R} = n\,\vec{u} + m\,\vec{v} \qquad\qquad (2.8)$$

The n, m pair completely characterizes the structure of SWCNTs, giving access also to its circumferences and to the angle between axis and base vectors. In Figure 2.39, this is demonstrated with two examples ((n,m) = (3,1) and (2,2)). SWCNTs with n = m are metallic conductors. Most other orientations produce semiconductors, some with very small bandgaps.

MWCNTs consist of many SWCNTs arranged like elements of a telescope tube (Figure 2.38a), with a distance of 0.34 nm between successive graphene layers, as in graphite. They usually exhibit metallic conductivity, because at least one of the SWCNTs stacked in it will be a metallic conductor for reasons of statistical probability [57].

MWCNTs are prepared via a catalytic process, during which the catalyst particles remain in the tubes (Section 3.1.1.5, Figure 3.10). Procedures for getting rid of these contaminants, which, at the same time, open the interior space for access by the fluid phase, have been reported in literature. Procedures for depositing active species on CNTs must handle the hydrophobicity of their surfaces. While they can be wetted by alcohols (ethanol, isopropanol), their wettability by water can be improved by oxidizing treatments (cf. Section 2.2.2). The problem is nicely illustrated by the comparison of different preparation routes of Co-CNT catalysts for Fischer–Tropsch synthesis in ref. [58].

CNTs combine high surface areas with complete accessibility of the sites: porosity in the macroscopic granules consisting of highly entangled tubes is in the macro- and mesopore range. Together with their high electrical conductivity, this qualifies them, in particular, for the use in liquid media and for electrocatalysis. The interior space of MWCNTs has attracted particular attention because of significant deviations of catalytic effects from those observed in comparable unconstrained situations [59]. As these are caused by the curvature in the aromatic π-systems, they are confined to tubes with a few nm inner diameter.

Although many of these advantages are offered also by CNFs, their use in catalysis has been limited, so far. Figures 2.38b-d illustrate three types of CNFs with different relations between graphene layers and fiber axis: parallel in the ribbon type (r-CNTs), perpendicular in the platelet type (p-CNFs), and oblique in the herringbone type (h-CNFs). Surfaces of p- and h-CNFs differ dramatically from those of CNTs: while the latter expose exclusively the graphene structure, surfaces of the former resemble the edge plane of graphite crystals, i.e., they are more reactive. CNFs can be modified by activation processes that remove the most reactive carbon atoms, increase surface area and porosity, and populate the surface with functional groups.

Templated carbons are obtained by the nanocasting (or hard templating) approach, in which a siliceous pore system (e.g., of an ordered mesoporous material like MCM-41, cf. Section 3.1.1.1) is filled by a carbon precursor like furfuryl alcohol, phenol/formaldehyde, or sucrose, which is then carbonized by thermal treatment. The carbons obtained after dissolution of the siliceous mold have highly regular pore systems and can exhibit BET surface areas up to 2,000 m^2/g. The method gives access to materials that are of great interest for research but are not easily scaled up.

The disclosure of the two-dimensional carbon allotrope **graphene** has inspired a great effort to use this material in surface catalysis. For practical reasons, it must be integrated into three-dimensional structures for this purpose, which allow easy access to the graphene surfaces. Approaches to solve this problem and examples for applications, in particular, for electrocatalysis and photocatalysis, are reported in ref. [60].

2.3.3 Support and supported species

Although the intention for using supports was originally just to disperse expensive active components, there are other important roles that supports can play in the catalytic process. They can take part in the reaction with the help of own catalytic sites, they can confer mechanical stability to a catalyst where this would not be achievable with the active component alone, and they can influence the structure of the supported components and their behavior during catalysis and under thermal stress. All this is discussed in the present section.

Depending on the nature of supported species and support, the interactions between them may be very different. They are strongest when **oxides** are **supported on oxide surfaces**. Therefore, structures of supported oxide phases differ dramatically from those of the respective bulk oxides, except for very high loadings of the active component. The differences result from the tendency of systems to minimize their surface energy. Due to the strong interactions offered by most oxide support surfaces, supported oxides tend to form monolayer surface oxide structures under dry oxidizing conditions, particularly at low coverages. The structures grow into the third dimension only when metal oxide loading exceeds the theoretical monolayer capacity.[20] The actual speciation depends, however, on the chemical potential in the gas phase (e.g., the water content, the presence of reductants) and on sample history [62].

Under ambient conditions, hydrophilic supports adsorb multilayers of water. Surface oxide species dissolve in this water film, forming structures similar to those found in unconstrained aqueous solutions at pH values that are primarily determined by the acid-base properties of the support, i.e., by its IEP (cf. Section 2.2.2). In an aqueous film on MgO, for instance, vanadium mostly forms orthovanadate (VO_4), on the less basic alumina – metavanadate $(VO_3)_n$. On the slightly acidic supports, ZrO_2 and TiO_2, higher oligomers become more important, e.g., decavanadate, which is predominant on SiO_2 [62]. When such catalyst is dried, these species are adsorbed on the surface by electrostatic interactions. Upon calcination, surface oxide species are formed by condensation of OH groups from surface and adsorbates (cf. Section 3.1.3.1), which decomposes layered oligomer moieties into two-dimensional oxide structures.

Typical structures detected on surfaces of oxide supports are shown in Figure 2.40. At low coverages, isolated metal cations are coordinated by terminal oxygen ligands and bonded to the surface via bridging oxygens. The number of terminal oxygens was elucidated by Raman spectroscopy with ^{18}O-equilibrated samples [62]. With increasing metal oxide loading, oligomeric structures that contain oxygen bridging between metal ions appear and become predominant (Figure 2.40). In well-prepared catalysts, three-dimensional oxide clusters appear only above the theoretical monolayer capacity, when the support is completely covered with surface metal oxide species [63].[21]

---

20 From Raman spectroscopic studies, theoretical monolayer capacities were reported to be between 4 and 5 atoms/nm^2 for Mo or W on Al_2O_3, TiO_2, and ZrO_2, between 5 and 6 for Nb, and even between 7 and 8 atoms/nm^2 for V, on the same supports. Surface capacities on SiO_2 were found to be more than an order of magnitude lower, much more than expected from the smaller OH group density on this support, probably due to low reactivity of the majority of these groups [61].

21 This does not exclude the existence of catalysts, in which three-dimensional oxide species and exposed support surface coexist: although monolayer dispersion of supported oxides may be achieved also by thermal spreading phenomena in many systems (cf. Section 3.1.3.7), conditions or duration of thermal processing may have been insufficient to completely disperse oxide aggregates deposited in preceding steps.

Tetrahedral
mono-oxo, e.g. VO$_4$
(see below)

Tetrahedral
dioxo, e.g. MoO$_4$

Tetrahedral
trioxo, e.g. ReO$_4$

Isolated surface
VO$_4$ monomer
species

Polymeric surface
VO$_4$ species

Surface vanadium oxide
monolayer and V$_2$O$_5$ NPs

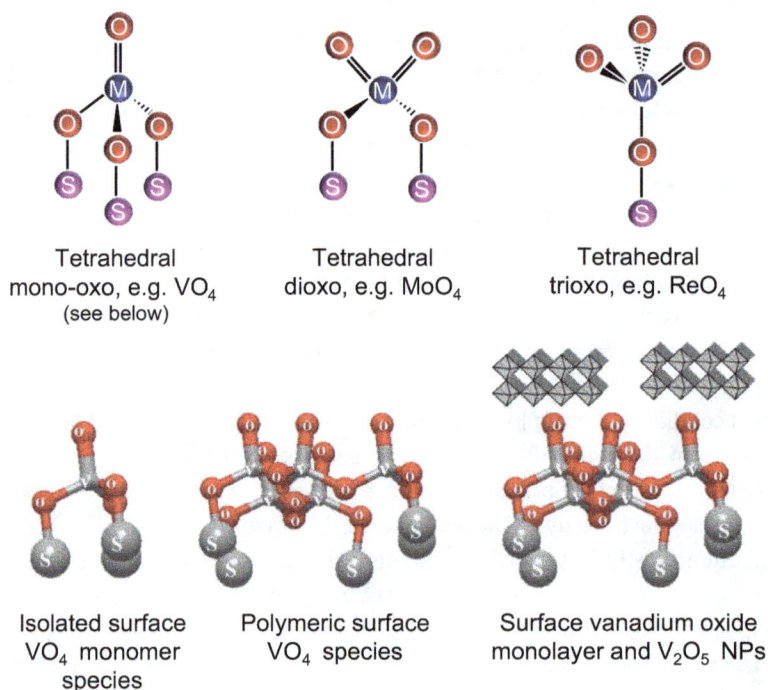

Figure 2.40: Surface oxide species in supported oxide catalysts. Upper row – different forms of monomeric species, reproduced from ref. [62] with permission from Springer Nature. Lower row – species observed in supported V oxide catalysts, reproduced from ref. [64] with permission from Elsevier.

The coordination around the cations in Figure 2.40 is tetrahedral. Higher coordinations have been reported for some elements in isolated sites and within oligomeric surface species.

Structural perturbation and electronic influences of the support cause dramatic changes in the reactivity of the exposed oxide species, e.g., towards reduction by H$_2$. Dispersion of W oxide species on alumina, for instance, dramatically stabilizes the W(VI) oxidation state. Below the theoretical monolayer capacity, W(VI) reduction starts several hundred degrees higher in WO$_3$/Al$_2$O$_3$ than in unsupported WO$_3$ (cf. Figure 3.59). Opposed to this, the reduction onset of TiO$_2$-supported W(VI) oxide species is downshifted relative to WO$_3$ [65], although the surface tungsten oxide species are dispersed in a similar way as on alumina. Figure 3.59 shows also that oligomeric surface oxide species are more reactive than isolated sites, and that oxide clusters formed at coverages slightly above the monolayer capacity may be more reactive than bulk oxides, apparently due to their defective structures. The actual differences, however, depend strongly on the type of reactivity addressed by the fluid-phase composition and the reaction conditions. Some information on this point can be found in ref. [62].

When surface oxide species are reduced, oxygen bridges between cation M and support (cf. Figure 2.40) are broken. The attenuated interaction between support and supported species can result in some clustering, which is reverted upon reoxidation. According to studies summarized in ref. [66], it is this oxygen bridging to the support, by which the catalyst participates in many redox reactions, rather than the oxygen in M-O-M bridges or the terminal oxygen. This explains a parallel influence of the support on reaction rates achieved with supported V oxide species in reactions like methanol oxidation, SO_2 oxidation, selective propene oxidation, or selective reduction of NO by NH_3. The support property behind this influence is the cation electronegativity, which is illustrated by a correlation between these electronegativities and rates of propene oxidation to acrolein over supported V oxide catalysts [67].

When the external surface areas of catalysts shrink as a result of thermal stress, the fate of the system depends on the miscibility of its components. In many cases, there are no stable mixed phases containing both supported and support cations. With decreasing BET surface area, the supported phase rearranges according to its growing coverage, and begins to segregate as a bulk oxide when the monolayer capacity is exceeded. In other cases, supported cations diffuse into supports at high temperatures forming solid solution phases (Cr_2O_3/Al_2O_3) or even compounds (V_2O_5/CeO_2). When, however, small catalyst particles aggregate into bigger ones, supported species may end up in the interior, despite mutual immiscibility with the support; they will then remain at the grain boundaries.

Interactions between **metal particles** and **oxide supports** are less intense, and their nature depends on the electronic structures of interacting materials. Most supports applied in heterogeneous catalysis are either semiconductors (TiO_2, CeO_2) or insulators. As mentioned in Section 2.1, contact between metals and semiconductors results in the formation of space charges, e.g., in a Schottky barrier, when $\varphi_{met} > \varphi_{sc}$ (Figure 2.9).[22]

In addition to this long-range effect, there are physical and chemical interactions confined to the interface, which create interface states in the bandgap. If very abundant, the interface states can even determine the barrier height instead of metal and semiconductor work functions. In catalysts, the relations between metals and insulator surfaces are usually dominated by this "Fermi level pinning," while metal-semiconductor contacts are mostly Schottky contacts [68].

Right at the interface, adhesion energies may be generated by local charge redistribution, by population of metal-induced gap states, or by chemical bonds [68]. Electrostatic contributions arise from interactions of ion charges with their image charges in the metal (cf. Section 2.1, Figure 2.6b). In the metal d-band, the population of orbitals may change near the surface under the influence of the ionic charges. In wide bandgap insulators, dispersion forces may arise from mutual polarization of support and metal particles. Metal-induced bandgap states have been proposed to result from interactions of evanescent waves of the metal (cf. Figure 2.6b) with oxide states of the semiconductor, which

---

22 In the opposite case, e.g., with alkali metals, the metal donates electrons, and the space charge zone is enriched by electrons rather than depleted. In this case, which is less relevant for catalysis, the contact is ohmic.

produces a smooth density of bandgap states within several Å even from ideal interfaces. Chemical bonds between metal atoms and surface oxygen or supported cations may be ionic or (polar) covalent. In the latter case, bond polarity depends on the electronegativity of bond partners. Ionic bonds require a previous oxidation of metal atoms.

Chemical interactions at semiconductor-metal contacts causing space charges are much more intense and spatially extended. The electrical fields set up by the differing work functions (cf. Figure 2.9b) favor mass transport: while the outward negative charge at a Schottky contact favors cation diffusion, the outward positive charge at an ohmic contact (footnote 22) facilitates O^{2-} diffusion (for a more detailed discussion of the underlying Cabrera-Mott theory, see ref. [68]). Various reactions can occur around the interface: a metal with a low work function may be oxidized by the semiconductor (eq. (2.9a), where $||$ denotes the interface),

$$Me(1) \ || \ Me(2)O_x \rightarrow Me(1)O_y \ || \ Me(2)O_{x-y} \qquad (2.9a)$$

and the resulting metal cations may form a mixed phase with the support, which provides an additional driving force,

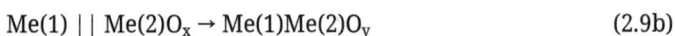

$$Me(1) \ || \ Me(2)O_x \rightarrow Me(1)Me(2)O_y \qquad (2.9b)$$

A metal with higher work function may be sandwiched by a layer of slightly reduced support species (decoration or encapsulation).

$$Me(1) \ || \ Me(2)O_x \rightarrow Me(2)O_{x-\delta} \ || \ Me(1) \ || \ Me(2)O_x \qquad (2.9c)$$

Under severely reducing conditions, support cations may be reduced and form alloys with the supported oxide

$$Me(1) \ || \ Me(2)O_x \rightarrow Me(1)Me(2)_y \ || \ Me(2)O_z \qquad (2.9d)$$

The extent of oxide formation according to (2.9a) depends on the work function difference between metal and support. Large differences as in alkali or earth alkaline metals result in multilayers of Me(1) oxides already at room temperature, while the oxidation of early transition metals (4 eV < φ_{met} < 5 eV) is much more confined to the interface. Notably, the oxidation is slowed down when donor levels are introduced into the semiconductor, e.g., bulk Ti^{3+} defects into TiO_2. They raise the semiconductor Fermi level to slightly below the conduction band edge, which decreases the positive charge developing on the metal, i.e., the driving force for O^{2-} diffusion.

Noble metals with $\varphi_{met} \geq 5$ eV being part of Schottky junctions can be covered by layers of defective support species upon severe reduction. The support cations (e.g., interstitial Ti^{4+} in TiO_2) diffuse towards the metal in the field of the negative charges trapped by the metal (Figure 2.9b). n-Donation (Ti^{3+}) enhances the driving force. Actually, due to relatively small differences in work functions, encapsulation proceeds to an observable extent only with partially reduced supports.

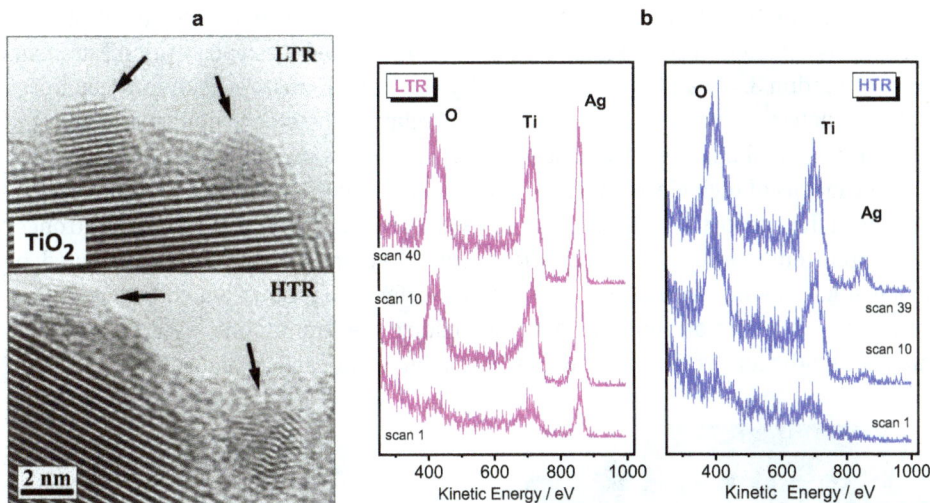

Figure 2.41: Strong metal-support interaction in an Ag/TiO$_2$ catalyst; the catalyst was reduced in dilute H$_2$ at 473 K (Low-temperature reduction, LTR) or at 773 K (High-temperature reduction, HTR). Encapsulation after HTR detected by high-resolution TEM (a) and low-energy ion scattering (LEIS) (b). Reproduced with permission from refs. [69] (a) and [70] (b). Copyright 1999 (a) and 2004 (b), American Chemical Society.

The phenomenon was first observed with (real) Pt/TiO$_2$ catalysts in 1978 [71] and has been referred to as strong metal-support interaction (SMSI) since that time. Due to the encapsulation, the noble metal becomes increasingly unavailable for catalysis with growing reduction severity, but as long as it is accessible, it may exhibit changes in reaction selectivities due to intensified and modified support influences. The SMSI state can be completely reversed by oxidation of the catalyst.

Reference [68] reports SMSI phenomena also for ideal single-crystal support surfaces. However, some differences between the idealized cases and real catalysts have remained unexplained. While Au and Ag were found to be unavailable for SMSI on ideal rutile facets (in agreement with model predictions [68]), SMSI has been clearly observed with real anatase-supported Au and Ag. In Figure 2.41, this is illustrated for a 7 wt% Ag/TiO$_2$ catalyst reduced in H$_2$ at 473 K (LTR) or at 773 K (HTR) by high-resolution TEM and LEIS. The micrographs show the Ag particles wetted, but not covered by amorphous material after LTR, while they may be buried in it after HTR (Figure 2.41a). The completely surface sensitive, but abrasive LEIS method (cf. Section 3.4.6.3) revealed that Ag exposure was actually negligible after HTR, and that the cover consisted of Ti and O atoms (Figure 2.41b).

The spatial relation between metal atoms and surface cations or anions at metal-support interfaces has been dealt with in theoretical studies. Below a critical value of the metal electronegativity (according to ref. [68] ca. 1.9), the metal atoms tend to interact with O^{2-} ions. For more noble metals, the arrangement depends on a number of influences. On γ-Al_2O_3, for instance, atoms in very small Pt nanoparticles were shown to interact with both Al and O. This results in a raft-like (instead of cuboctahedral) morphology as detected in early studies with EXAFS (Extended X-ray absorption fine structure, cf. section 3.4.4) and later confirmed by electron microscopy (summarized in ref. [72]). On defect-free parts of surfaces, epitaxy relations between support and adjacent layer of the metal particle may exist. The interactions are usually specified as weak.

For metals of high electronegativity, the actual situation is, however, dominated by defects in the support surface, such as steps, oxygen vacancies, etc. Due to the stronger interaction energies generated by these sites, they are preferentially populated by metal atoms and nucleate particle formation. Surface OH groups, which do not interact strongly with atoms of noble metals, may oxidize the first atomic layer of less noble metals, and the ions remain sandwiched between metal particle and support.

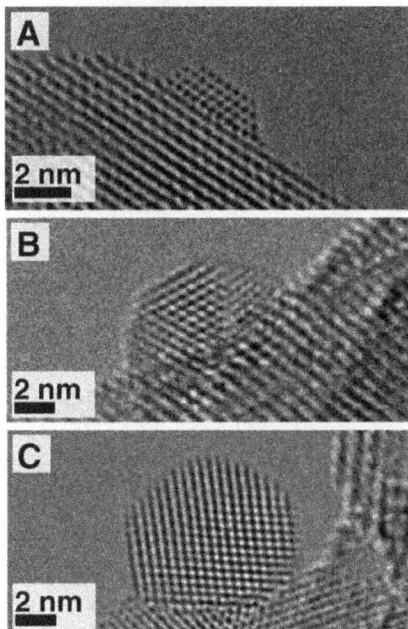

Figure 2.42: HRTEM micrographs of Pd nanoparticles on CeO_2. From ref. [73]. Reprinted with permission from AAAS.

Due to their higher surface energies, metals usually assume three-dimensional shapes when growing larger on oxide surfaces. In thermal equilibrium, particle shapes are obtained, which minimize the surface energy, as can be modeled by the Wulff construction (cf. ref. [74]). Some shapes found for Pd particles supported on CeO_2 are shown in Figure 2.42.

The models discussed so far do not cover an important aspect of metal-support interactions: the activation of the adjacent support surface by the metal particle. Already in the 1990s, noble metal particles supported on (or promoted by) CeO_2 were found to be more active in redox reactions with CO (CO oxidation or water-gas shift, eq. (2.10)) than if deposited on irreducible supports [55].

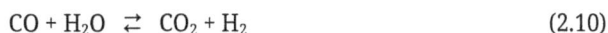

$$CO + H_2O \ \rightleftarrows \ CO_2 + H_2 \tag{2.10}$$

The effect was ascribed to a particular redox activity of support oxygen around the metal particles. The "active" oxygen is supposed to arise from the perimeter of the metal particles, the contact line common to metal, support, and fluid phase.

More recently, this assignment was confirmed by a study, in which CO oxidation rates over monodisperse CeO_2-supported Pt, Pd, and Ni particles (three different sizes; for Pd, see Figure 2.42) were shown to follow a single correlation with the perimeter length, irrespective of the nature of the metal [73]. Even in the CO oxidation over Au/TiO_2 around room temperature, the oxygen introduced into the CO_2 was shown to originate from the TiO_2 support, most likely from the Au-TiO_2 perimeter region [75] (see also Section 4.5.6).

While formally in agreement with eq. (2.9d), spectacular reports on silicide formation in SiO_2-supported metal catalysts well below 900 K suggest an activating effect of metals even on the very stable SiO_2. In their LEIS study with a Cu/SiO_2 model catalyst, van den Oetelaar et al. observed a decreasing Cu/Si intensity ratio during mere evacuation at 773 K. At 900 K, the Cu had completely disappeared, but was uncovered by the abrasive technique immediately below a Si layer shielding it from the vacuum [76]. The process could be reverted by reoxidation of the catalyst, as known for the SMSI effect. The identity of the intermediate Cu silicide was confirmed by Auger electron spectroscopy (cf. Section 3.4.5).

Metal nanoparticles in zeolite cavities are influenced by the electric field set up by the zeolite constituents. In the particles, which consist only of a few atoms, valence orbital energies respond to changes of the electron density at adjacent framework oxygen, which can be manipulated, for instance, by replacing cations by protons that form a covalent bond with the O atoms. Under these influences, reaction rates of some hydrocarbon reactions (neopentane hydrogenolysis, but also decalin hydrogenation) were found to vary by orders of magnitude, which was important for the development of sulfur-tolerant hydrocracking catalysts [77, 78]. The properties of Pt clusters were influenced by the electron density of support oxygen also on supports with open porosity [78], where, however, particle sizes required to identify the effects are not easily stabilized.

In traditional catalysts for the hydrotreatment of oil fractions, the active components (Mo or W **sulfides**, promoted by Co or Ni) are supported on **Al_2O_3**. Its major purpose is to disperse MoS_2 or WS_2 in nanoslabs, the edges of which are decorated by atoms of the promoter (cf. Figure 4.15). In the preparation, reactions resulting in this state compete with processes damaging the targeted catalyst structure.

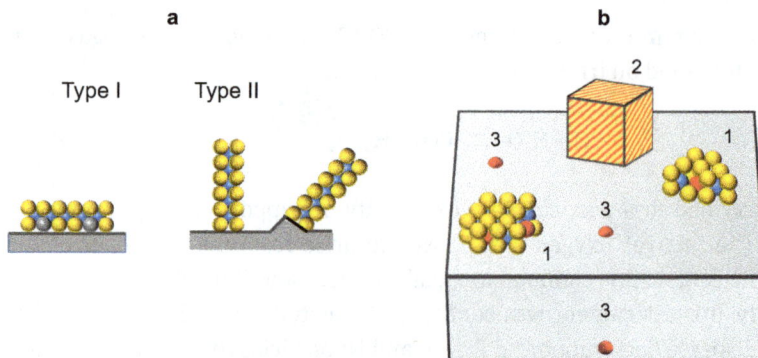

Figure 2.43: Relations between sulfide species and support in supported sulfide catalysts for the hydrotreatment of oil fractions: (a) type I- and type-II-MoS$_2$; (b) productive and parasitic promotor species in a Co-MoS$_2$/Al$_2$O$_3$ catalyst: 1 – CoMoS phase, 2 – Co$_9$S$_8$, 3 – Co ions in spinel sites of Al$_2$O$_3$. b adapted from ref. [79]. Color code of atoms, red – Co, blue - Mo, yellow – S.

After the impregnation of the support with precursors for Mo (or W) and the promoter, the conditions during the subsequent thermal treatment are most critical in limiting the formation of Mo-O-Al bridges and the diffusion of Co or Ni into the alumina lattice. During the sulfidation (in the lab with H$_2$S/H$_2$ mixtures, in plants with labile S compounds dissolved in the feed), terminal oxygen of surface Mo oxide species is exchanged with S at low temperatures (<400 K), where supported Co(Ni) oxide species may be already sulfided [80]. Mo-O-Mo bridges react at somewhat higher temperatures, while conditions required to break the Mo-O-Al bridges tend to cause sintering of the primary MoS$_2$ monoslabs to stacks. The intrinsic activity of sites on (promoted) MoS$_2$ slabs has been found to be significantly lower when the slabs are still linked to the support via residual Mo-O-Al bridges (type I-MoS$_2$). In the more active type II-MoS$_2$, the slabs often protrude at some angle from the support surface (Figure 2.43a).

Excessive thermal stress prior to sulfidation will result in a moderately active catalyst containing type-I CoMoS (NiMoS) structures with some parasitic promoter species (bulk sulfide, Co(Ni) in support, Figure 2.43b). Decades of research have resulted in effective procedures for minimizing losses of promoter to spinel and bulk sulfide structures and maximizing yields of highly active type-II CoMoS (NiMoS) moieties with low stacking degree. Delaying the sulfidation of the promoter metal until the MoS$_2$ (WS$_2$) structure is formed was essential for this success. It was achieved by binding the metals with complexones added to the impregnation solution. Modifying the Al$_2$O$_3$ surface by additives (phosphate or boron) is another option to influence the sulfide/support interaction. Sintering of the emerging sulfide phase can be also suppressed when the exothermal sulfidation process is performed in the liquid phase. For more detail, the reader may consult ref. [80].

Supported liquid-phase (SLP) **catalysts** are a catalyst type right at the border between homogeneous and heterogeneous catalysis. For stability reasons, the vapor pressure of the supported liquid or its solubility in a surrounding phase must be extremely low.

In the classical systems, molten salt mixtures or phosphoric acid are supported. The role of the support is again to create a large interface between the supported liquid and the surrounding fluid. Due to transport resistances in filled pores, micro- and mesoporosity makes no sense in SLP catalysts. The most suitable materials feature thin walls with micrometer apertures that span lamellae of the supported liquid, similar to structure packings in distillation columns. Therefore, diatomite (Figure 2.26) is often used for SLP catalysts.

In SLP catalysis, observed reaction rates are influenced by the mass transfer through the gas–liquid interface and by the catalytic reaction rates in the liquid. The situation resembles, to some extent, the reaction under pore-diffusion limitations (Section 2.6.2). Depending on the relation between rates of diffusion in the liquid and of the reaction, the reaction may proceed in the whole liquid volume (slow reaction) or only at the gas–liquid interface (fast reaction). For more information, the reader is referred to textbooks of reaction engineering or to ref. [81].

The K-promoted V_2O_5 catalyst for SO_2 oxidation to SO_3 is the most important traditional SLP catalyst. During the reaction, a melt containing K_2SO_4, $K_2S_2O_7$, but also vanadyl sulfates is formed from the precursors, V_2O_5 and K_2SO_4, and the feed [82]. Below 700 K, the melt solidifies, and the activity suddenly drops by orders of magnitude. The "solid phosphoric acid" (SPA) catalyst may have played even a significant role in twentieth century history. Invented in the U.S. well before World War II, it was applied in refinery processes, e.g., the oligomerization of short alkenes or the alkylation of isobutane [81]. Some historians list the superiority of the resulting gasolines over the Fischer–Tropsch-derived fuels (mostly n-alkanes) used by the German army among the reasons for the fast advance of the U.S. army after landing in Normandy in 1944.

The SLP concept is one of the basic approaches for the heterogenization of molecular catalysts.[23] Although such heterogenization constrains the active species to a small part of the reactor volume (the solid surface) and tends to introduce transport limitations, it has been an important research objective over decades, in order to combine the advantages of molecular catalysis (cf. Section 1.2) with facile phase separation in heterogeneous systems [83]. In the SLP approach, catalysts need to be adapted to the limited choice of solvents suitable for this technology, which is, however, less challenging than requirements for the direct fixation of catalysts on the support surface (see below). SLP catalysts were developed mostly for reactions best suited for coordinative catalysis, e.g., hydroformylation of propene, but also of unsaturated alcohols, and selective (or asymmetric) hydrogenations. Solvents employed include ethylene glycol, molten triphenylphosphane, for reactions in hydrophobic liquids even water [83]. Although stability over hundreds of hours, as reported for some systems, is extraordinary for homogeneous catalysts, the SLP concept has not yet been commercialized to a significant extent.

---

23 Alternatively, the catalyst can be bound to the support by covalent or non-covalent bonds (see below) or dissolved in a liquid phase immiscible with the one providing the reactants and withdrawing the products. This multiphase technology is established in several large-scale applications.

With the advent of the ionic liquids (ILs) [84], the potential of the SLP concept has been enormously extended and diversified. The high polarity and negligible volatility of ILs allows numerous applications with molecular catalysts in both liquid- and gas-phase reactions. Thin IL films can be used to modify active sites of traditional heterogeneous catalysts. These new concepts have been referred to as SILP (supported ionic liquid phase) and SCILL (supported catalyst with ionic liquid layer), respectively. The SILP concept includes cases where the activity is provided not by a dissolved catalyst, but by the IL itself, e.g., by Brønsted or Lewis acidity of cations or anions.

IL layers are mostly made by impregnation. They interact with the surface via electrostatic forces, hydrogen bonds, and capillary forces, but chemical anchoring via linkers between cation or anion and the surface is also possible. Their thickness depends on the IL loading, but very thin layers are possible: data in literature range between 0.5 and 3 nm. When loading and pore system of the support are well-optimized, a high extension of the liquid-gas interface and a nearly complete utilization of the IL volume for catalysis can be achieved. Such economical use of the rather expensive ILs has been a strong stimulus for the development of SILP and SCILL catalysts.

Due to the minor layer heights, the ILs are significantly influenced by both the support surface and the adjacent fluid phase. The differences from bulk ILs include melting point depressions, deterioration of thermal stability, and enrichment or depletion of cations or anions at the interfaces. By modifying species of interest (cations, anions, dissolved catalysts) with suitable chemical groups, their concentration at the IL-fluid interface can be enhanced. Due to their high polarity, bulk ILs exhibit significant internal order, which is affected in the thin supported layers, by perturbations originating at the interfaces. Details, however, depend strongly on the system [83].

ILs are powerful solvents for polar species; their miscibility properties can be widely tuned. However, for application in SILP environments, ligands in established catalysts may be required to be replaced by ion-tagged analogues to allow for high active site concentrations [84]. Continuous operation of SILP catalysts was demonstrated for many reactions from the classical domain of homogeneous catalysis, e.g., hydroformylations, (asymmetric) hydrogenations, methanol carbonylation, as also water-gas shift and acid-catalyzed Friedel–Crafts alkylations [84].

In SCILL, the solubility properties of the IL change the chemical potential of reactants and products at the active sites. The IL may also block sites that are active for side reactions or modify their behavior by interacting with them. In the selective hydrogenation of dienes or alkynes (cf. Section 4.3.4), coating Pd-based catalysts with ILs was found to allow for higher alkene yields, though at decreased activities [84]. The effect was ascribed to the IL blocking the active sites for re-adsorption of the alkene, but not for the stronger interacting diene or alkyne reactants. The resulting benefit has been used in a first industrial process based on an IL-modified catalyst [85]. For SILP catalysts, pilot plant work was reported [86].

As mentioned above, surfaces of oxide supports can also be employed for direct **heterogenization of molecular catalysts** (transition metal complexes), without solvation by a solvent layer. While the risk of losing activity by such fixation is obvious, it should be noted that stability (for continuous processes) or reusability (for batch processes) are targets no less important than activity.

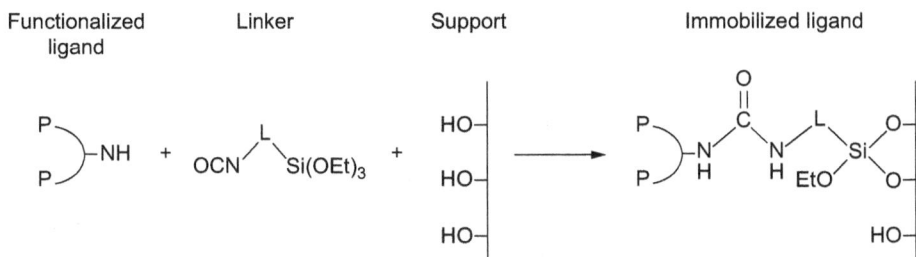

Scheme 2.2: Synthesis of a link between a complex ligand and a support surface.

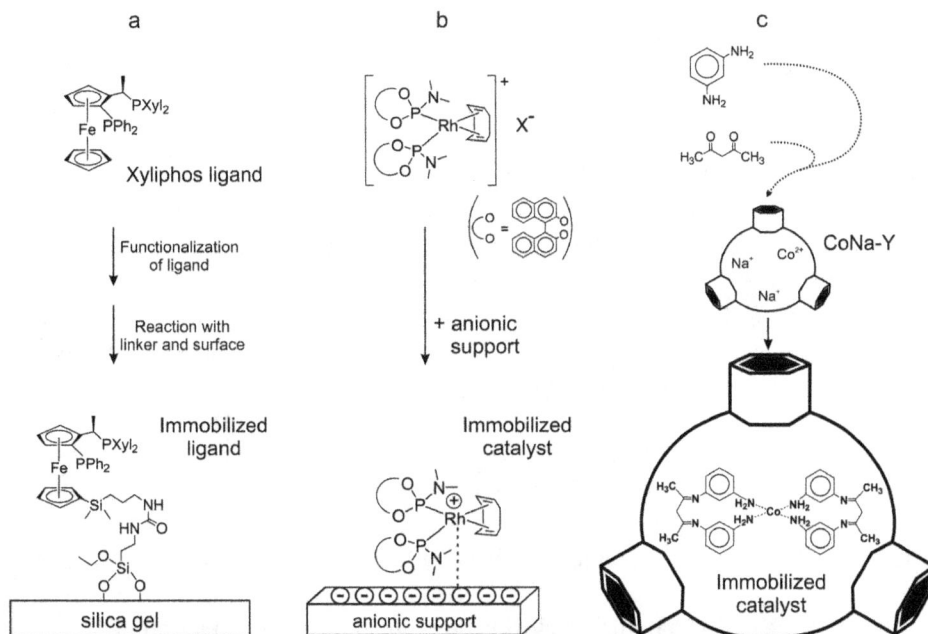

Figure 2.44: Approaches for the immobilization of transition metal complexes on solid supports: (a) by covalent bonds; (b) by electrostatic forces; and (c) by steric constraints (ship-in-bottle catalysts). a and b adapted from ref. [87], c adapted from ref. [88].

Various strategies are available for anchoring transition metal complexes at the surfaces of inorganic supports or of polymers. They may be differentiated according to the nature of the bond between support and complex – covalent, ionic, or by steric constraints. The covalent link between complex and surface is established via one of its ligands rather than via its central atom. Typically, linker groups are placed between complex and surface to minimize adverse interactions between them. Scheme 2.2 highlights a typical procedure in which the linker is attached to the SiO_2 surface via ethoxy silyl groups, while a reaction between isocyanate and an −NH group extends the linker by a ligand, which is ready to bind a metal precursor added in the next step. In Figure 2.44a, a xyliphos ligand is shown attached to a silica surface. Together with Ru (or Ir) and additional ligands, it becomes a silica-supported catalyst for asymmetric hydrogenation.

Complex ions can be exchanged into ion exchange resins (Figure 2.44b), which saves a lot of synthesis effort required for covalent ligand fixation. On the other hand, the method exposes the central atom to the anionic charge of the exchange sites and to influences of adjacent non-exchanged sites, because exchange degrees achieved are usually low. Heterogenization via electrostatic forces can be stable only in reactions, in which the central atom is never discharged along the catalytic cycle, because this would initiate leaching.

In cage-type zeolites, transition metal complexes can be synthesized by assembling suitable ligands around cations previously exchanged into the cavities. As the resulting complexes are too large to escape through the apertures (Figure 2.44c), such systems are referred to as ship-in-bottle catalysts. Similar work has been done with MOFs more recently. While the steric constraints prevent leaching very effectively, application of the microporous catalysts in the liquid phase is penalized by mass transfer limitations. The intrinsic activity of encaged complexes was sometimes found to exceed that of the unconstrained ones, probably due to distortions of complex geometry imposed by the zeolite cage.

Summarizing decades of research effort, ref. [87] stated that matching the activity of the homogeneous version with the heterogenized one seems to be the exception rather than the rule. Improvement by heterogenization was achieved only in rare cases. In addition, the performance of catalysts heterogenized via covalent or ionic interactions was very often affected by leaching. Reasons for the limited success of this effort were instructively discussed for some important reactions (metathesis, C-C couplings, and hydrogenation). Stability was disclosed as the more urgent problem with the pertinent catalysts, which is solved by heterogenization only in very special cases.

It should be noted that this conclusion does not hold for polymerization catalysts, because these are not reused (cf. Sections 2.3.4 and 4.5.2.3). On the contrary, polyolefins made with metallocene catalysts achieved significant market penetration only after the (homogeneous) metallocenes had been attached to supports, because this allowed applying the catalysts in existing polymerization reactors. Problems encountered during the heterogenization of metallocene polymerization catalysts are discussed in ref. [89].

2.3.4 Alternative catalyst architectures

Catalyst types beyond the traditional bulk and supported versions were developed to combine multiple catalytic functions in one catalyst or to suppress adverse interactions between ingredients of multicomponent catalysts. Approaches range from simple layer structures in commercial three-way catalysts to sophisticated encapsulation strategies.

Layered structures may be flat (layered washcoats in monoliths, cf. Section 2.8.2) or a shell around a core. Three-way catalysts are complex systems containing different noble metals and additives required to simultaneously oxidize CO and hydrocarbons and reduce NO under oscillating feed conditions (cf. Appendix A1). To prevent adverse interactions and to encourage favorable interactions between components (cf. Sections 4.4.1.3 and 4.4.2), the catalyst is broken down into different subsystems, which are deposited sequentially on the monolith walls. The resulting layered structure may be topped by a porous non-reactive layer keeping away particulate contaminants (cf. Section 4.4.2).

Core-shell structures may be devised to prevent sintering of the core material by intercepting contact to nearby particles, or to combine the catalytic reactivities of core and shell material. NH_3 decomposition to N_2 and H_2, for instance, which is part of a potential ammonia-based hydrogen storage technology, can be catalyzed by simple iron oxides, though at temperatures that induce sintering (800–1,000 K). Coating α-Fe_2O_3 particles with a thin layer of porous SiO_2 was shown to keep their decomposition activity stable even at 1,023 K [90]. Cu/ZnO-Al_2O_3 catalyst particles coated with a H-ZSM-5 layer are an example for the combination of reactivities [91]. In a syngas feed, methanol is formed on the core catalyst and converted to the clean fuel dimethyl ether, while passing the zeolite shell.

As mentioned in Section 2.3.1, egg yolk versions of supported catalysts, with the active sites in the particle center, allow trapping poisons in the external particle regions. In catalysts for fluid catalytic cracking (FCC), this is better achieved by integrating μm-sized grains of the zeolite catalyst into a macroporous matrix, which offers much space for storing deposits, but also contains a number of other ingredients essential for good catalyst performance. The FCC catalyst, a real piece of catalytic artwork, is discussed in some detail in Section 4.4.2 (Figure 4.25).

As a special case, the fate of solid alkene polymerization catalysts during the reaction will be briefly illustrated. Starting as supported catalysts that expose the active site precursors on their pore walls, they keep this state only during initialization and the first chain growth steps. Soon, however, the intergranular spaces of polycrystalline particles can no longer accommodate the growing polymer. Particles are fragmented, while polymerization goes on at the sites attached to support debris, which becomes surrounded by a growing polymer shell. Fragmentation occurs on a timescale of seconds to (a few) minutes, while chain growth goes on for hours and produces polymer particles with dimensions 1–2 orders of magnitude larger than the catalyst particles [92]. As a result, the content of catalyst residues in the grown polyolefin is very small, below legal limits even for toxic catalytic elements as Cr, which are dilute already in the initial catalyst.

Layer-by-layer (OnionSkin)

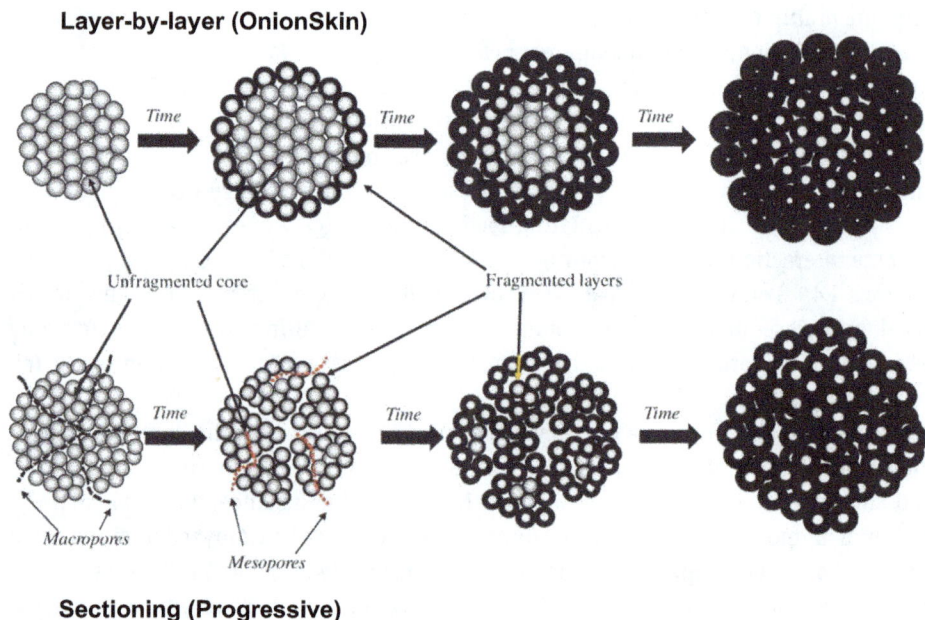

Sectioning (Progressive)

Figure 2.45: Two models for the fragmentation of polycrystalline catalyst particles during alkene polymerization. Reproduced from ref. [92] with permission from Wiley-VCH Verlag GmbH & Co. KGaA, Weinheim, Germany.

To give a flavor of these processes, Figure 2.45 illustrates two basic fragmentation schemes that can be used to model resulting particle morphologies. They are versions of a multigrain model, which treats the catalyst as an isotropic agglomerate of active micrograins. If the porosity is low, polymerization will first proceed at the outermost layer only, until its active micrograins become detached from the layer below, giving the monomer full access to the latter and "igniting" the micrograins there. Polymerization gradually proceeds towards the center, and rates and properties depend on the radial position in the particle (layer-by-layer model). At high particle porosity, polymerization and fragmentation start everywhere in the particle, first at the walls of macropores and then progressing to regions with smaller pores. For more detail about relations between particle fragmentation, morphology, and polymerization kinetics, the reader is referred to ref. [92].

2.4 Adsorption

2.4.1 Adsorption and desorption: phenomena and classification

Solid surfaces in contact with a gas or a liquid attract components of this fluid until equilibrium is established between concentrations at the surface and in the fluid. This general phenomenon is called adsorption, the reversed process desorption. Its participants (free and adsorbed species of i and the solid) are referred to as adsorptive, adsorbate, and adsorbent, respectively. Beyond being elementary steps of surface catalysis (cf. Section 1.4.3), adsorption and desorption are also the basis of an important separation technology in chemical engineering. In this book, they will be, however, treated only with respect to their implications for catalysis.

Adsorption is always exothermal. Even if it is split into two species, the adsorptive loses degrees of freedom upon becoming an adsorbate. As a result, the adsorption entropy, ΔS_{ads} is negative, and for the Gibbs free adsorption energy (2.11) to become negative, ΔH_{ads} must be <0, as well.

$$\Delta G_{ads} = \Delta H_{ads} - T \Delta S_{ads} \qquad (2.11)$$

For catalysis, the knowledge of both the adsorption equilibrium and the rates of adsorption/desorption is of great interest. The rates become relevant when adsorption or desorption are the slowest steps in the catalytic reaction mechanism and limit the overall reaction rate (cf. Section 1.4.3). Adsorption-based techniques for the characterization of surface properties (cf. Section 3.4.1) rather rely on established equilibria. Adsorbed species are often investigated by spectroscopic techniques to elucidate the interactions between solid and adsorbate with respect to catalytic reaction mechanisms.

Adsorption processes are described by relating adsorbed quantities of the adsorptive i per gram of adsorbent ($n_{i,ads}$ or V_{ads}) to adsorptive concentrations, c_i (or partial pressures p_i for the gas phase) and temperatures, T. In characterization studies, the target quantity is usually the monolayer capacity $n_{i,mono}$. The ratio $n_{i,ads}/n_{i,mono}$ is the adsorbate coverage θ_i, as long as it is ≤ 1. If it is >1, it can be used to assess the average height of the adsorbate layer.

Kinetic or equilibrium data are most conveniently presented in two dimensions, keeping one of the three quantities involved constant. Below, only the adsorption isotherm $n_{i,ads} = f(p_i)_T$ (or $f(c_i)_T$) will be used, which is the most popular representation of equilibrium data, but there are also adsorption isobars $n_{i,ads} = f(T)_{pi}$, adsorption isochors $p_i = f(T)_{n_{i,ads}}$, and adsorption isosters $p_i = f(T)_{\theta_i}$. The following discussion is focused on the adsorption from the gas phase, which has been studied in considerable detail. For some general information on liquid-phase adsorption, see Section 2.4.4.

Adsorption processes are classified according to either the nature of adsorbate-surface interactions or to the shape of the equilibrium curve, i.e., the adsorption isotherm. The former differentiates between chemisorption and physisorption. Chemisorption is based on chemical bonds to the surface, which confines it to a monolayer, whereas physisorption involves the formation of multilayers up to the condensation of the adsorptive on the surface. The classification according to the shape of the adsorption isotherm is phenomenological and includes chemisorption and physisorption, as well as capillary condensation of the adsorptive in the pores (see Section 3.4.1).

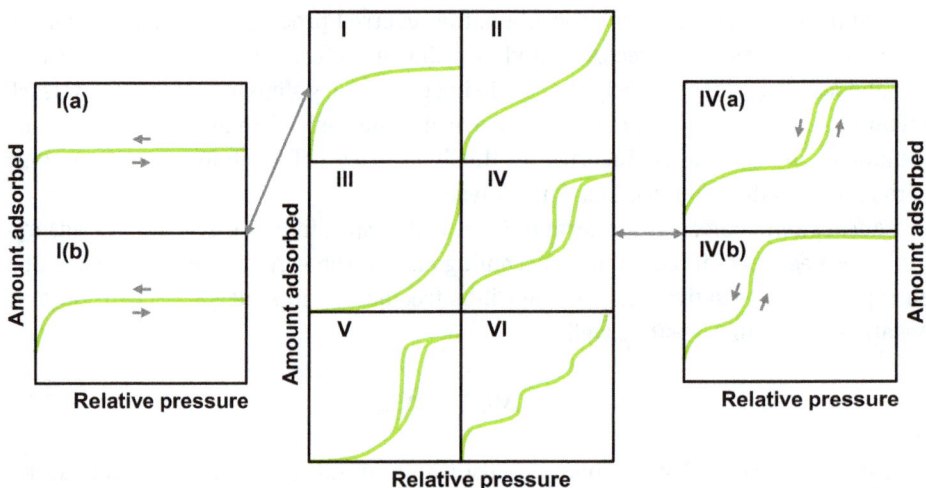

Figure 2.46: The IUPAC classification of adsorption isotherms; the previous classification shown in the center was extended by sub-types (to the left and to the right) recently. The relative pressure, which ranges from 0 to 1 (not shown), relates the pressure p_i of the adsorptive i to its vapor pressure $p_{0,i}$ at the measurement temperature. On a flat surface, the adsorptive condenses at $p_i/p_{0,i} = 1$.

Figure 2.46 shows the traditional IUPAC classification of adsorption isotherms obtained by adding a sixth type to an older classification by Brunauer (types I to V). It has been updated recently by differentiating types I and IV into two sub-types [93] as indicated in the figure. The type I isotherm is obviously the only one, which is able to describe the monolayer nature of chemisorption. It can, however, also be observed for microporous samples with low external surface area: micropores are filled with liquid already at low relative pressures (see Section 3.4.1), while the small external surface causes only minor additional adsorption. The subdivisions, Ia and Ib differentiate between filling of ultramicropores (<0.7 nm wide) and of supermicropores (0.7–2 nm wide) [93, 94].

Type II and type III isotherms originate from adsorption on nonporous materials. Type II describes cases where the interactions between surface and adsorbate are stronger than those within the adsorbate layer. Point B was proposed to indicate the monolayer capacity, but its identification is ambiguous without appropriate data treatment. Type III is found in systems where the interactions among the adsorbate species are stronger than the (weak) interactions between adsorbates and surface: the adsorbate layer reluctantly forming at low p/p_0 becomes increasingly stable at higher cover-ages. Water adsorption on hydrophobic surfaces is a typical example for such systems.

Type IV and Type V isotherms describe multilayer adsorption on (meso)porous materials. The enhanced uptake at intermediate p/p_0 is due to capillary condensation in the pores, because the equilibrium vapor pressure over the curved menisci in the pores is smaller than that over the flat liquid surface (p_0). For reasons discussed in Section 3.4.1, condensation and evaporation pressures at a temperature may differ depending on whether the adsorptive pressure was increased or decreased during the measurement. Therefore, the isotherm may or may not expose a hysteresis, which differentiates types IVa and b. Like in the case of types II and III, the difference between type IV and V arises from the different relations between adsorbate-adsorbate and adsorbate-surface interactions. Type VI isotherms are rare. They were observed for nonpolar spherical adsorptives (Ar, Kr) on extremely homogeneous surfaces, e.g., graphitized carbon [95].

2.4.2 Physisorption and chemisorption

As mentioned above, physisorption and chemisorption differ according to the nature of interactions between surface and adsorbate: physisorption originates from weak unspecific forces resulting from van der Waals or dipole interactions, whereas chemisorption involves chemical bonds between surface atoms and the adsorbate. Even if these bonds are weak, they are specific for the kind of surface atoms involved.

Specific or unspecific interactions: all other differences between chemisorption and physisorption result from this primary one. Chemisorption is limited to a *mono-layer*, because only the solid offers the specific interactions. Physisorption can occur on any kind of surface, be it a bare solid surface, a physisorbed or even a chemisorbed layer. Therefore, the adsorbate accumulates in *multilayers* up to the liquefaction of the adsorptive at $p/p_0 = 1$. In chemisorption, the maximum *layer density* is determined by the surface structure of the adsorbent because of the chemical bonds between surface atoms and adsorbates: there are well-defined chemisorption sites. Chemisorption is therefore considered *localized*, even though adsorbates may move between sites (see below). Physisorbed layers ideally arrange without reference to the surface structure below, because variations of the interaction energy across the surface are small, compared to thermal energy. Therefore, physisorption is *delocalized*. Layer densities are higher in physisorption than in chemisorption and converge towards the density of the liquid adsorptive with increasing layer thickness. Finally, the difference in *adsorption enthalpies*, which is often itemized first to distinguish both types of adsorption, arises from the different nature of surface-adsorbate interactions, as well. These enthalpies are up to 20 kJ/mol for physisorption and of the order of a chemical reaction enthalpy for chemisorption, often >100 kJ/mol, though sometimes significantly smaller. Only interactions in chemisorption are strong enough to activate adsorbate molecules for becoming involved in catalytic surface reactions.

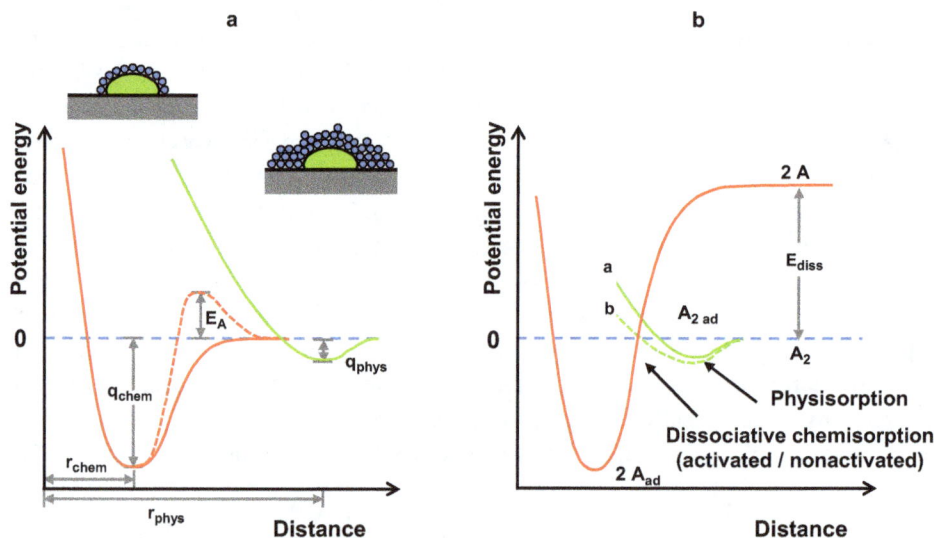

Figure 2.47: Physisorption and chemisorption: potentials experienced by species moving orthogonal to the surface: (a) molecular adsorption, activated (dashed) and nonactivated; (b) dissociative chemisorption of a physisorbed A_2 molecule, nonactivated (dashed) or activated.

The interaction potential experienced by an adsorbate that is approaching a surface from the gas phase is depicted in Figure 2.47a. In physisorption, the potential curve is shallow. Therefore, the equilibrium adsorbate position is more distant from the surface in physisorption than in chemisorption. Physisorption may precede chemisorption. The transition between the two states may be activated (dashed line in Figure 2.47a) or nonactivated. Without any activation energy, the process is extremely fast. When the molecule is physisorbed initially, it is in thermal equilibrium with the surface, and only adsorbates with energy at the upper end of the thermal (Maxwell-Boltzmann) energy distribution are able to pass the activation barrier. Chemisorption can, however, also occur without initial physisorption. The energy required to pass an existing activation barrier must be then provided by the kinetic energy of the adsorptive.

Figure 2.47a describes molecular chemisorption, which may modify bonding interactions in the molecule, but does not result in its fragmentation. A potential diagram typical of dissociative chemisorption is depicted in Figure 2.47b. At the intersection of the curves describing the interactions between the surface and the molecule or its fragments, the molecule is split up to access a more stable state: the adsorption of the fragments. Upon desorption, the fragments recombine, because they would become highly unstable at larger distances from the surface. Dissociative chemisorption may be activated or nonactivated as well (cf. Figure 2.47b). H_2 is dissociatively adsorbed on many metal surfaces, nonactivated on noble metals like Pt, but activated on some other metals like Cu. The dissociative adsorption of N_2 on Fe surfaces is activated, as well.

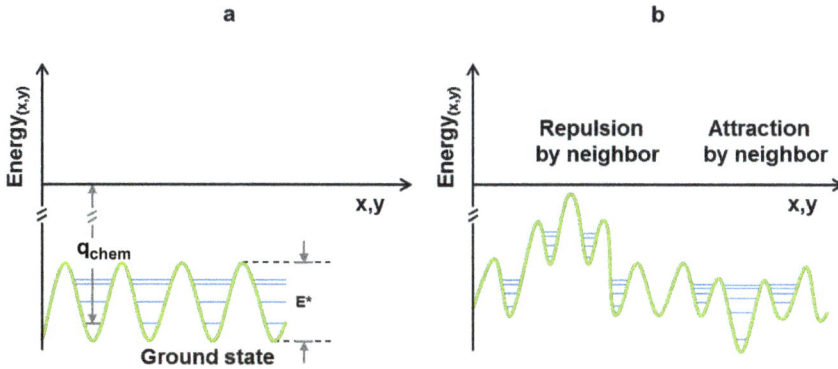

Figure 2.48: Chemisorption: potentials experienced by a species moving across a surface at a constant distance r_{chem} (cf. Fig. 2.47) without neighbors (a) or influenced by neighbors (b). Adapted from ref. [96].

Figure 2.48a depicts a chemisorption potential landscape extending parallel to the surface at a fixed distance. As mentioned above, the chemical bond between surface and adsorbate favors a particular position of the adsorbate relative to the surface atom(s) forming the adsorption site. The remaining positions are disfavored, which results in a pronounced corrugation of the profile with barriers between the adsorption sites. The relation between barrier height and thermal energy decides whether adsorbates stick to their sites at a temperature or hop between them. Figure 2.48b exemplifies the effect of co-adsorbates, which depends on the nature of interactions between partners: if they are attractive, the adjacent sites are stabilized, while repulsive interactions destabilize adjacent sites. In this case, adsorbates maximize the distances between their positions. When their coverage increases, this can be best achieved by forming regular adsorption patterns on the surface. Attractive interactions, on the contrary, result in the formation of dense islands.

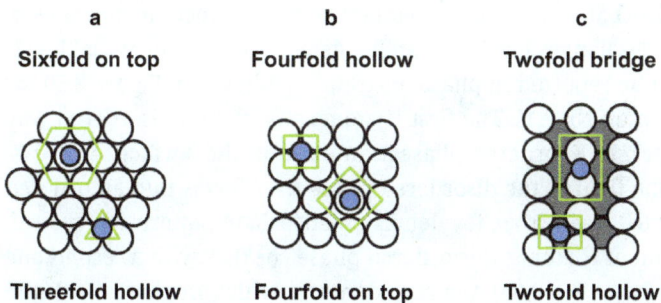

Figure 2:49: Nomenclature of adsorption sites on different surface lattices: (a) hexagonal, (b) square, (c) rectangular.

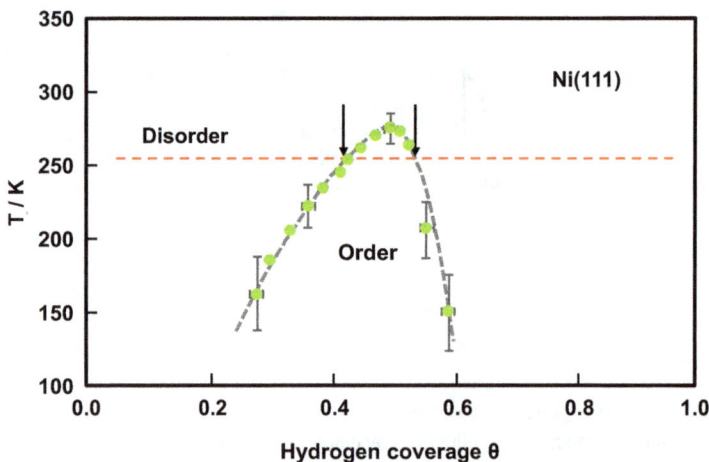

Figure 2.50: Phase diagram for the adsorption of hydrogen on Ni(111); for a discussion of red line and arrows, see text. Adapted from ref. [97], with permission from AIP publishing.

Adsorption sites are classified according to the number of surface atoms coordinating to the adsorbate (Figure 2.49). If it is one, the site is on top (or atop); if it is two, the site is bridging, and if the adsorbate interacts with three of more atoms, the site is hollow. Sites are further differentiated according to the rotational symmetry of the coordination sphere around. In Figure 2.49, the symmetries of the sites are sixfold and threefold (a), fourfold (b), and twofold (c). The lower adsorbate in c is also in a twofold site despite having four neighbors at identical distances, because the arrangement needs a 180° rotation to reestablish its shape.

On ideal surfaces, adsorbates may form regions, in which their density (local coverage degree) and their interrelation (ordered/disordered) are homogeneous, e.g., islands of adsorbates with attractive interactions, domains with ordered arrangements of repulsively interacting adsorbates. They are two-dimensional thermodynamic phases, referred to as surface phases. Ranges for their stability depending on state variables (T, adsorbate coverage θ) can be reported in phase diagrams, as shown in Figure 2.50 for the adsorption of hydrogen on Ni(111). The first H atoms adsorb in a disordered way at any temperature. Below 270 K, ordered phases can exist on the surface. At 250 K, for instance (red line in the figure), the disordered phase transforms into an ordered one at H coverage θ near 0.4 (see arrow, for details on adsorbate pattern etc. see ref. [97]). Upon further addition of H atoms, this ordered phase coexists with a second one of higher density (see arrow at $\theta \approx 0.53$), the abundance of which increases with the average H coverage θ. When θ reaches the value characteristic of the second phase, the one with $\theta_{loc} \approx 0.4$ has completely disappeared. Additional H atoms destabilize, however, also the ordered phase with $\theta_{loc} \approx 0.53$, and the whole surface will be again covered by H atoms in random arrangement, though at higher density.

In most cases, the order in the adsorbed phases is influenced by the arrangement of sites in the surface beneath, but not necessarily identical: the adsorbates may form superstructures. Superstructures differ from those of the adsorbing surfaces in their translation vectors, and the differences are used to identify them. The most universal system is the matrix notation, which takes advantage of the fact that any two-dimensional vector may be expressed as a linear combination of two other ones. Indexing the vectors of the superstructure vectors with s, they are obtained by eq. (2.12) i.e., the matrix m unambiguously relates superstructure and surface structure.

$$\vec{a_s} = m_{11}\,\vec{a} + m_{12}\,\vec{b}$$
$$\vec{b_s} = m_{21}\,\vec{a} + m_{22}\,\vec{b} \qquad m = \begin{pmatrix} m_{11} & m_{12} \\ m_{21} & m_{22} \end{pmatrix} \qquad (2.12)$$

a	b	c
Hexagonal surfaces, e.g. fcc(111), hcp(0001)	**Square surfaces,** e.g. fcc(100), bcc(100)	**Square surfaces,** e.g. fcc(100), bcc(100)

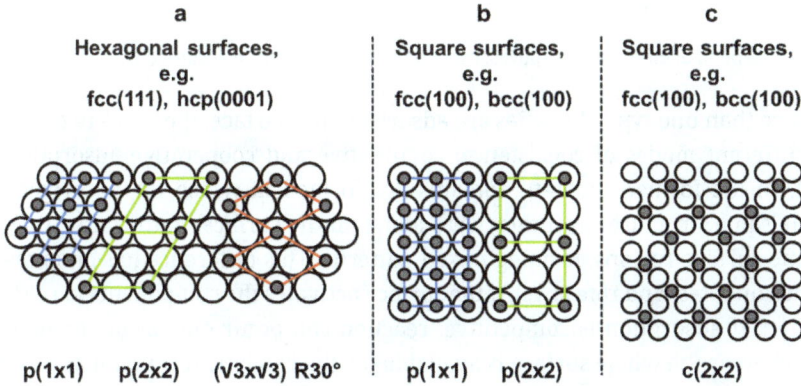

p(1x1) p(2x2) (√3x√3) R30°	p(1x1) p(2x2)	c(2x2)

Figure 2.51: Examples for adsorbate superstructures and their identification by Wood's notation. Adapted from ref. [4].

The more popular Wood notation reports by which factor the superstructure vectors are extended over those of the surface structure, and if and how they are rotated (Figure 2.51). In a (√3 x √3)R30° superstructure, for instance, both translation vectors are elongated over those of the surface by √3 and rotated by 30°. In more dilute patterns (e.g., (2 x 2)), a central position is available for an additional adsorbate. These patterns are therefore differentiated by p and c, the latter with the central position occupied. The identity of surface and adsorbate periodicity is encoded as p(1 × 1).

It should be kept in mind that both notations describe the superlattice by reference to the surface. Figure 2.51 shows nicely that a p(2 × 2) superlattice looks quite different on a hexagonal and on a square surface. The notations refer only to periodicity, but not to the type of sites occupied: the square c(2 × 2) superlattice in Figure 2.51c would receive the same designation if the species were adsorbed atop, instead of in fourfold hollow sites.

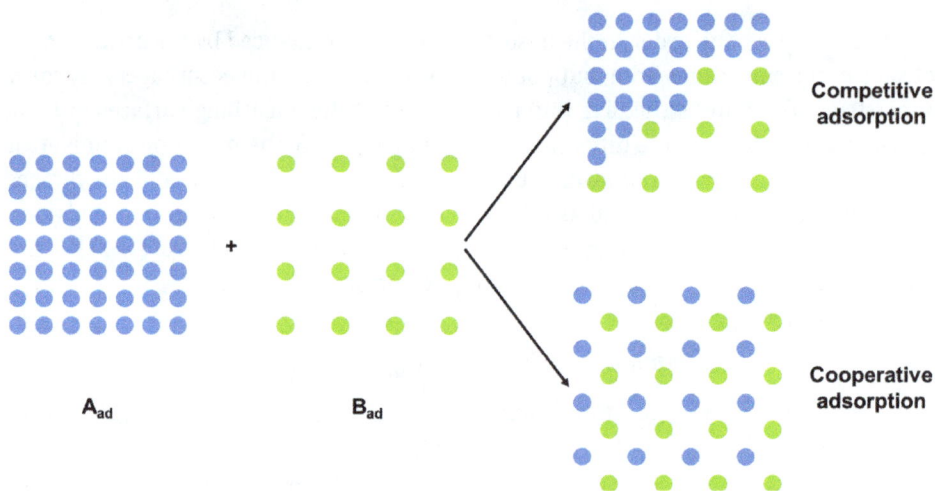

Figure 2.52: Co-adsorption of different adsorptives: cooperative and competitive adsorption.

When more than one type of species are adsorbed on a surface, there are two fundamentally different modes of coexistence: competitive and cooperative adsorption (Figure 2.52). In the former, separate domains are formed in which the adsorbates arrange in the same structures as if they were alone on the surface; in the latter, they share the same surface regions, forming mixed patterns. This difference has an enormous influence on rates (and rate laws), if the surface activates the components to react with each other. If adsorption is competitive, reaction can occur only at the domain boundaries, whereas the whole surface is available for the catalytic reaction if adsorption is cooperative.

2.4.3 Description of equilibria in gas-phase adsorption

Box 2.2: Assumptions required to derive the Langmuir and the BET isotherm equations.

Langmuir model	BET model
(a) all adsorption sites equivalent	(a) first layer: $\Delta H_{ads} \neq f(\theta)$, higher layers: $\Delta H_{ads} = \Delta H_{LV}$
(b) no interactions between adsorbates	(b) number of layers $\rightarrow \infty$
(c) monolayer adsorption	(c) evaporation equilibrium influenced by solid only in first layer

---

The classical models to describe adsorption isotherms are the Langmuir model for chemisorption and the Brunauer-Emmett-Teller (BET) model for physisorption. The basic assumptions for these models are compared in Box 2.2. The Langmuir isotherm can be derived using a kinetic approach, because the rates of adsorption and desorption are equal in equilibrium:

$$k_{ads}\, p_i\, (1 - \theta_i) = k_{des}\, \theta_i \tag{2.13}$$

where k_{ads} *and* k_{des} are the rate constants of adsorption and desorption. This can be easily rearranged into:

$$\theta_i = \frac{n_{i,ads}}{n_{i,mono}} = \frac{K\, p_i}{1 + K\, p_i} \tag{2.14}$$

with the adsorption constant $K = k_{ads}/k_{des}$. The surface coverage θ_i is proportional to the adsorptive pressure at low p_i, but converges towards 1 at high p_i. An isotherm shape consistent with eq. (2.14) is depicted in Figure 2.46 (Type I isotherm). In dissociative adsorption, the rate laws used in eq. (2.13) are second order: adsorption in the coverage of free sites $(1-\theta_i)$, desorption in the adsorbate coverage θ_i, and the Langmuir isotherm becomes:

$$\theta_i = \frac{\sqrt{K\, p_i}}{1 + \sqrt{K\, p_i}}$$

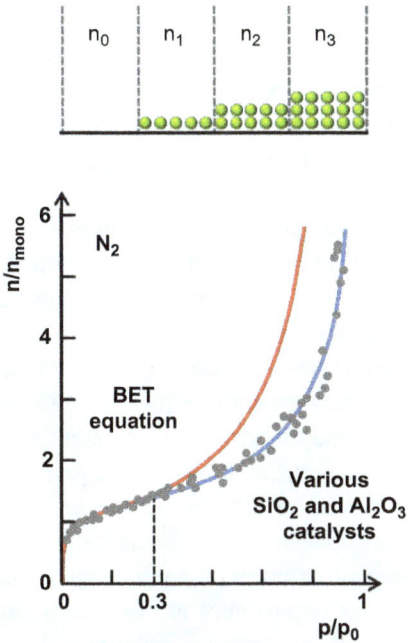

Figure 2.53: The BET isotherm: illustration of the packing model for the adsorbate layer used in the derivation of the BET equation (cf. Appendix A2) and comparison of experimental data (N_2 on nonporous silicas and aluminas) with predictions of the BET equation parametrized at $p/p_0 > 0.3$. Diagram reproduced from ref. [95] with permission from Elsevier.

For deriving the BET isotherm, energetic homogeneity of the surface is assumed as well (Box 2.2), but now, a second and further layers are formed. Again, equilibrium is described by equating the rates of adsorption and desorption, but now for all sections of equal adsorbate height (counted by j), see eqs. (2.15). The model includes *j = 0*, i.e., sections of bare surface coexisting with sections covered by 1, 2, and more stacked atomic layers (cf. Figure 2.53), which suggests that the interactions between adsorbate and surface are stronger, but of the same order of magnitude as those among adsorbate species.

$$1^{st} \text{ layer: } k_{ads,1}\, p\, \theta_0 = k_{des,1}\, \theta_1 \qquad (2.15a)$$

$$j^{th} \text{ layer: } k_{ads,j}\, p\, \theta_{j-1} = k_{des,j}\, \theta_j \qquad (2.15b)$$

The derivation, which can be tracked in Appendix A2, results in a description of the adsorbed amount (here mol/g adsorbent) depending on the pressure of the adsorptive, p (index i dropped).

$$n_{ads} = n_{mono}\, c\, \frac{p/p_0}{\left(1 - p/p_0\right)\left(1 + p/p_0(c-1)\right)} \qquad (2.16)$$

where n_{mono} is the monolayer capacity (for c, see below).

The typical shape of the BET isotherm is shown in Figure 2.53. Its steep initial increase, which decays with growing p/p_0, originates from the preference of surface-adsorbate over adsorbate-adsorbate interactions, which is reflected in the c parameter of the equation (see Appendix A2). A high value of c (in the figure – c ≈ 150) results in a knee shape of the initial region. Smaller values change the shape of this region towards that of a type III isotherm. At higher values of p/p_0, the influence of the c parameter is smaller, because this region is dominated by adsorbate-adsorbate interactions.

While the isotherms presented so far are most popular, they are not necessarily the most suitable ones. Usually, there is no invariability of the adsorption enthalpy with changing surface coverage as required for the Langmuir model. According to calorimetric studies, a majority of sites with little variation of the adsorption enthalpy is often accompanied by a significant minority of more strongly interacting sites. The merit of the Langmuir model for describing a catalytic process on such a surface depends on the role of these sites. If the more reactive minority is rapidly poisoned and the steady-state reaction proceeds on the majority sites, the Langmuir model is sufficient, but if the reaction requires the higher reactivity of the minority sites (the other ones being just spectators), it fails.

Adsorption isotherms based on other relations between adsorption enthalpy and coverage can be used to describe chemisorption steps in catalysis on nonhomogeneous surfaces. They differ in the treatment of the coverage dependence of the adsorption enthalpy. The Freundlich isotherm (2.17) is based on a logarithmic dependence and the Temkin isotherm (2.18) on a linear decrease with increasing coverage.

	$\theta_i = f(p_i)$	$\Delta H_{ads} = f(\theta_i)$	
Freundlich	$\theta_i = a\, p_i^{1/n_F}$	$\Delta H_{ads} = -\Delta H_{ads,av} \ln \theta_i$	(2.17)
Temkin	$\theta_i = K_1 \ln K_2\, p_i$	$\Delta H_{ads} = \Delta H_{ads,max}\,(1 - \alpha\, \theta_i)$	(2.18)

In both models, the number of sites is constant and the adsorption stoichiometry does not change with coverage. The Freundlich exponent is given as $1/n_F$, because it has a physical meaning: $n_F = -\Delta H_{ads,av}/RT$, where $\Delta H_{ads,av}$ is the average adsorption enthalpy. Notably, both isotherms cannot describe important features of chemisorption: they are not confined with respect to the monolayer coverage, and the Temkin isotherm predicts negative coverages at low pressures. However, the Freundlich isotherm can be useful at low coverages, while the Temkin isotherm was, for instance, successfully applied to create a microkinetic model describing the interaction between CO and Cu surfaces that included calorimetrically derived relations between adsorption enthalpies and surface coverages [98].

All these isotherms can be converted into linear forms that relate a quantity involving the adsorbed amount with another one that includes the adsorptive pressure (cf. linearized BET isotherm, eq. (3.34)). These linear forms can be used to check which isotherm is most suitable for the given experimental data.

As in chemisorption, experimental physisorption data cannot be completely represented by the most popular (i.e., the BET) isotherm, even if systems obviously violating the underlying assumptions are excluded (e.g., those with type III isotherms). In Figure 2.53, adsorption capacities measured with different nonporous silicas and aluminas are plotted and compared with the BET equation parametrized to optimally reproduce the low-pressure region. The BET equation typically holds only for $p/p_0 \leq 0.3$, Above, it fails for a number of reasons, one of it being the unrealistic assumption that the adsorbed phase contains parts with layer thickness up to infinity at $p/p_0 < 1$. Indeed, Brunauer, Emmett, and Teller also proposed isotherm equations based on more realistic assumptions (cf. ref. [95]). As these equations contain additional parameters that need to be fitted, the easy-to-handle eq. (2.16) has remained most popular.

The similarity of the N_2 adsorption data for different solids (cf. Figure 2.53) may suggest that a universal adsorption isotherm can be found when plotting data of various systems in reduced coordinates (such as n_{ads}/n_{mono} and p/p_0). Notably, this expectation failed. Although interactions between adsorbates and surface atoms are weak, they can differ sufficiently between different materials to cause differences in isotherm shapes, which are beyond experimental errors and tolerable deviations. Generalization was more successful when empirical standard isotherms were established for specified combinations of adsorptive and adsorbent, e.g., for N_2 on oxides, on graphitized surfaces, or for Xe on the same materials. The t-plots employed to identify different types of porosity in solid samples (see Section 3.4.1.1) are procedures that originated from these efforts.

2.4.4 Description of equilibria in liquid-phase adsorption

The investigation of adsorption from the liquid phase is much more demanding due to the presence of the solvent. An electrochemical double layer is formed when a solid is immersed into a liquid (electrolyte), typically an aqueous solution. Solids usually have a surface charge, either by the ions exposed in the external layer or by being connected to a potential. Due to Coulomb interactions, ions of the opposite charge are attracted from the electrolyte to shield the first layer electrostatically. During adsorption, the solvation shell of the adsorptive has to be partly removed, and the adsorptive has to cross the electrochemical double layer. Correspondingly, solvation occurs during desorption. The adsorption equilibria are not only a function of temperature and concentration, but also of the pH and the ionic strength. Furthermore, the solvent may compete for the adsorption sites. Despite the much higher complexity of the field, the same isotherms as those used in gas-phase adsorption are commonly applied to report equilibrium data in liquid-phase adsorption. For a more detailed treatment, the reader is referred to ref. [99].

2.5 Surface reactions

2.5.1 Thermal catalysis

The surface reaction is the step that rearranges bonds between atoms involved, which transforms reactants into intermediates or products. In simple reactions, it is the only step between adsorption and desorption. In the reaction coordinate shown in Figure 1.1a, it is the step with the highest activation energy. According to the Arrhenius equation (1.1), it is therefore rate-determining, because the number of sites, which influences the rate via the pre-exponential factor, cannot be higher for the reaction than for the adsorption step. As mentioned before, adsorption or desorption may, however, be rate-determining in other cases. Examples for this will be given below: NH_3 synthesis over Fe catalysts under realistic pressures (cf. Figure 4.16), or the dehydrogenation of cyclohexane over Pt catalysts in Section 2.6.1.

The simple reaction coordinate in Figure 1.1a neglects that different reaction channels with comparable activation energies may be available to activated reactants, i.e., different products may be formed in the surface reaction. Such competition may occur already during adsorption, e.g., between molecular and dissociative adsorption of O_2. Very often, adsorption of reactants and desorption of products are linked not by just one, but by a series of consecutive surface reaction steps. Each of them may, in principle, branch the series and open routes to different reaction products. In the following, surface reactions are, however, discussed under the assumption that they are the only or the rate-determining step (rds) between adsorption and desorption. The other, fast steps are rather difficult to address in research (cf. Section 4.6). Often, the rds determines selectivity as well, but rate- and selectivity- determining steps of a catalytic reaction are not necessarily identical.

When reactants exchange electrons with each other during the surface reaction, the solid is classified as a redox catalyst. On acid-base catalysts, the surface reaction is dominated by acid-base interactions. Catalytic reaction mechanisms may involve consecutive redox and acid-base reaction steps that proceed at adjacent redox and acid-base sites. Catalysts offering this opportunity are referred to as bifunctional. In this rather traditional classification, catalysts for some alkene reactions like metathesis or polymerization, which are initiated by the formation of a bond between a Lewis acidic transition metal cation and a C atom are, however, not categorized as acid-base catalysts.

In the surface reaction, a reactant activated by chemisorption may interact with another adsorbate, with a gas-phase reactant, with a component of the catalyst surface (e.g., oxygen), or it may dissociate into fragments. Some of these options are parts of classical reaction mechanisms proposed in the early days of catalysis, which are nowadays named after scientists with outstanding merits for the development of the field.

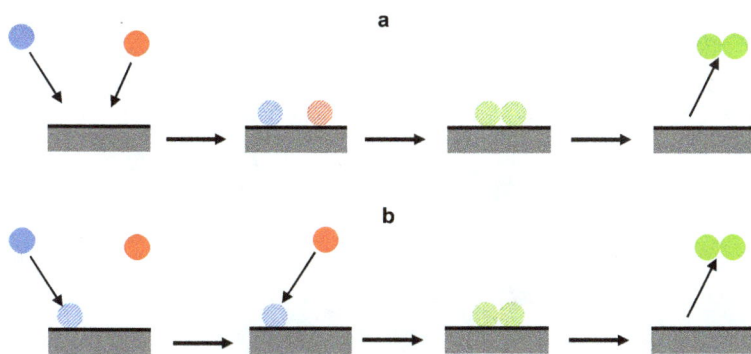

Figure 2.54: Classical mechanisms of bimolecular surface reactions: (a) Langmuir-Hinshelwood mechanism; (b) Eley-Rideal mechanism.

Surface reactions may be monomolecular and bimolecular. During the monomolecular reaction, a reactant previously activated by molecular chemisorption dissociates. This should be distinguished from dissociative adsorption (Figure 2.47b), where the reactant was either physisorbed or not adsorbed at all, initially. Both steps usually initiate a series of surface reactions in which the adsorbed fragments are converted to stable molecules that can desorb from the surface.

Two of the classical mechanisms involve bimolecular surface reaction steps. In the Langmuir-Hinshelwood (LH) mechanism, two reactants adsorb on the same type of sites and react with each other[24] (Figure 2.54a); in the Eley-Rideal (ER) mechanism, an adsorbed reactant reacts with a molecule from the gas phase (Figure 2.54b). In the LH mechanism, the competition of reactants for a limited number of sites has specific con-

---

[24] To be differentiated from the option that reactants adsorbed on different sites (i.e., not competing for space on the surface) react with each other, which does not bear a name.

sequences for the reaction rate. It is highest when both reactants populate the surface with equal abundance, and it decreases to zero when one of them occupies all sites, preventing access of the other one. This occurs, when the product $K_i p_i$ in eq. (2.14) or (2.29a) grows large, either with a large adsorption constant, K_i or a high partial pressure, p_i of (only) one reactant, or with both together. In reactions following the ER mechanism, there is no rate maximum upon variation of partial pressures. Instead, the rate tends towards a limiting value when the coverage of the adsorbing reactant approaches 1.

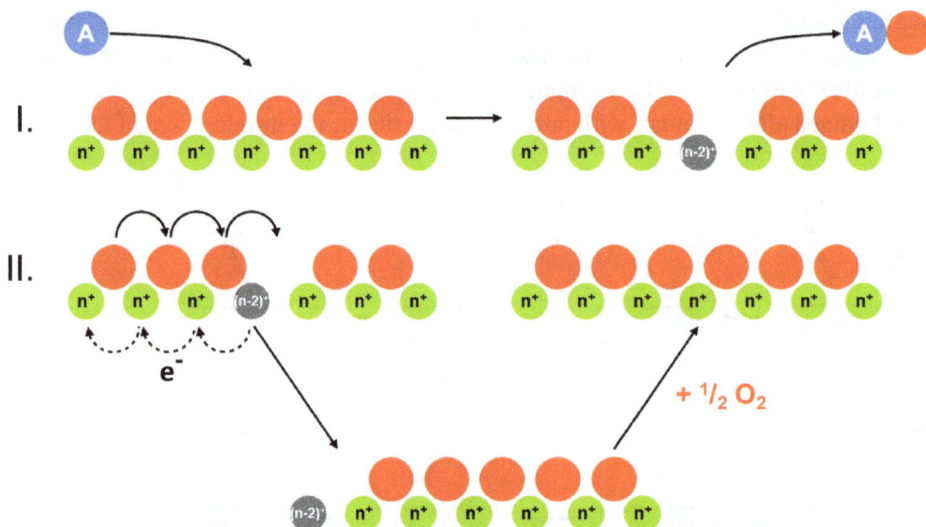

Figure 2.55: The Mars – van Krevelen mechanism describing the oxidation of reactants (A) by lattice oxygen. In two asynchronous steps, the surface becomes reduced by oxidation of the reactants (I) and re-oxidized by reduction of O_2 (II). As sites for step I are usually not very efficient for step II, the steps proceed on different sites, which includes O^{2-} diffusion and electronic conduction between the involved sites into the mechanism.

Reactions of adsorbates with a component of the catalyst surface change the latter. As part of a catalytic mechanism, such a reaction is therefore always combined with a process that restores the previous state of the surface. The Mars-van Krevelen mechanism of oxidation reactions (Figure 2.55) runs along these lines. In the first step, the reactant is oxidized by lattice oxygen, which leaves the surface in a reduced state. In the second step, the catalyst is reoxidized by gas-phase oxygen. Usually, this reoxidation occurs at a different site, which is better suited for activating O_2 by supplying to it four electrons and accommodating the resulting two O^{2-} anions than the vacancy left behind by the oxidation of the reactant. Transport of electrons from these vacancies to the oxygen activation sites and reverse transport of oxide anions from the oxygen activation sites to the vacancies is, therefore, a crucial part of the reaction mechanism. This transport is not confined to the external surface. Therefore, electronic and anion conductivity are important properties for catalysts that operate according to the Mars-van Krevelen mechanism (cf. Section 4.3.3).

Although the redox cations change their oxidation state between a lower and a higher level with each reaction event, they predominantly adopt one of them, depending on the rates of catalyst reduction and reoxidation. If the catalyst is reduced slowly, it becomes quickly reoxidized, i.e., if reactant oxidation is rate limiting, the surface is largely in the oxidized state. If reoxidation is sluggish, the re-oxidized sites will soon find a reactant to take the oxygen, and the cations will be in the reduced state, most of the time.

$$\tfrac{1}{2} O_2 \;+\; Cat \;\longrightarrow\; O^* \;+\; Cat'$$

$$O^* \;+\; A \;\longrightarrow\; AO_{ads}$$

$$AO_{ads} \;+\; Cat' \;\longrightarrow\; AO \;+\; Cat$$

Scheme 2.3: Oxidation of a reactant with adsorbed oxygen species. Their nature and mechanisms of activation remain unspecified.

Alternatively, reactants may be oxidized by gas-phase oxygen, which has been previously activated by the surface (see Scheme 2.3), but not stabilized as lattice oxygen by the electrical fields of surrounding cations. Scheme 2.3 does not specify the mechanism of O_2 activation or the nature of the activated oxygen species. These species are often very reactive, which results in high rates, but low selectivities (see Section 4.3.3, electrophilic oxygen).

The Horiuti-Polanyi mechanism (Figure 2.56a), which was proposed already in 1934, refers to alkene hydrogenation, another important application of heterogeneous catalysis. The mechanism comprises the adsorption of both H_2 and the alkene, the latter forming two σ bonds to metal atoms (di-σ complex). The crucial step is the addition of one H atom resulting in a half-hydrogenated state, in which the C-C bond being hydrogenated has already acquired the rotational degree of freedom typical of a single bond. The reaction is completed by the addition of a second H atom and the desorption of the alkane, which vacates the sites for the next cycle.

It should be noted that most of these mechanisms were proposed at times when even the existence of surface-bound intermediates was a bold hypothesis, and yet they have been substantiated by rich experimental evidence, meanwhile. The rate maximum upon variation of reactant partial pressures predicted by the LH mechanism has been found in many cases. The ER mechanism is rather rare, and it was believed for some time that it describes a limiting case rather than a real process. Meanwhile, there are rather convincing reports showing its relevance. The Mars-van Krevelen mechanism predicting reactant oxidation by lattice oxygen can be identified by pulsing small amounts of the reactant mixed with $^{18}O_2$ instead of $^{16}O_2$ over the catalyst. If the reaction follows this mechanism, the first products released from the surface will be without the

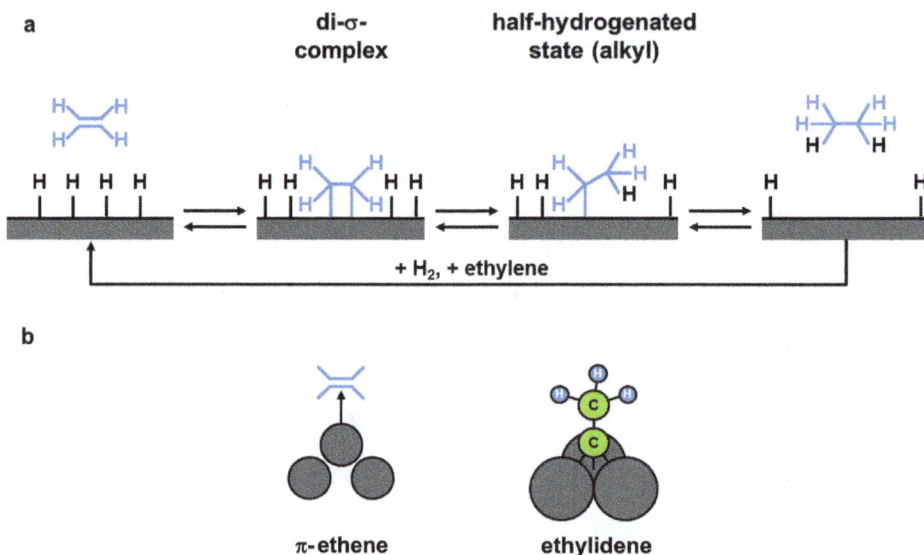

Figure 2.56: The Horiuti-Polanyi mechanism of alkene hydrogenation: (a) original form; (b) alternative forms of adsorbed ethene detected in surface science studies, ethylidene being a spectator, and π-ethene an intermediate.

label (cf. TAP reactor, Section 3.2.1). This has been found for many important selective hydrocarbon oxidations in the gas phase.

The stepwise character of the Horiuti-Polanyi mechanism was supported by the observation that hydrogenation is often accompanied by (non-equilibrated) HD exchange and by cis-trans isomerization. However, attempts in surface science to detect the ethene adsorbate preceding the half-hydrogenated state on the metal surface failed, until the advent of techniques that allow vibrational spectroscopy of adsorbates on cm^2-sized samples at mbar pressures and in atmospheres containing a large excess of the corresponding adsorptive (sum frequency generation, SFG, see Section 3.4.8.2).

Under traditional UHV conditions, π-bound ethene (Figure 2.56b) was detected below 50 K on Pt(111). At higher temperatures, it was converted into di-σ-ethene (Figure 2.56a), which, however, dehydrogenated to an ethylidene species (Figure 2.56b) above 240 K [100]. The latter were observed under similar conditions also on Rh(111). When ethene pre-adsorbed on (111) facets at low temperature (with or without H_2 co-adsorbed) was heated, ethane was formed in a temperature range in which only ethylidene was detected on the surface. However, ethylidene was known to be a spectator: its hydrogenation had been shown to be six orders of magnitude slower than that of ethene in independent work.

Only under realistic conditions ($\approx$145 mbar H_2, 45 mbar ethene), additional adsorbates were found to coexist with ethylidene in the relevant temperature range: π-ethene, and di-σ-ethene. The abundance of π-ethene, which binds to a single surface atom, could be correlated with the hydrogenation rate. The remaining species, which appear to compete for three-fold hollow sites, are only spectators [100]. Therefore, the Horiu-ti-Polanyi mechanism was largely confirmed, however, with the di-σ-complex replaced by π-bound ethene, which escaped observation in UHV due to its low abundance. In addition, it was found that hydrogen has two options to be adsorbed: on the same site as ethene, i.e., competitive, and on a noncompetitive site. Therefore, the kinetics of hydrogenation reactions do not comply with the LH mechanism under all conditions.

The work of Somorjai and coworkers showed the (modified) Horiuti-Polanyi mechanism to operate in a wide range of reactant pressures, i.e., it was not strongly affected by the pressure gap on the metals studied. As mentioned in Section 1.5, this does not hold for all catalytic processes. At higher pressures, LH-type reaction steps may be prevented by one reactant blocking the sites; the surface may be covered by deposits or modified by reactions with reactants. The reaction may still be observed if the surface offers other types of sites (usually with lower activity), which may operate different mechanisms and may require different conditions.

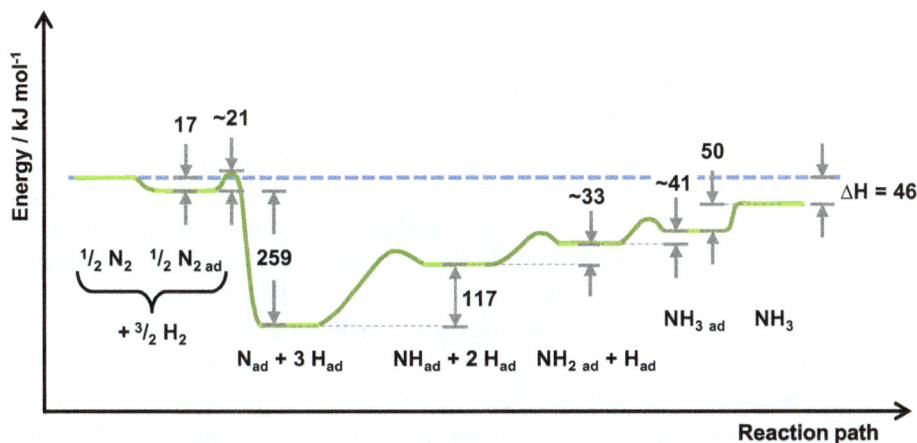

Figure 2.57: Potential energy diagram of NH$_3$ synthesis over Fe catalysts for low reactant pressures. The energies are averaged over three most important crystal facets. Adapted from ref. [101].

$$N_2 + 3 H_2 \rightleftarrows 2 NH_3 \tag{2.19}$$

Ammonia synthesis (eq. (2.19)) over Fe catalysts is an example for an intermediate situation between the extremes just described. The reaction mechanism identified in surface science studies by G. Ertl and coworkers is highlighted in a reaction coordinate in Figure 2.57 (see also Figure 1.6b)). The rds is the dissociation of N$_2$ starting from a physisorbed state. It should be noted that the activation energy of this step is very low: its small reaction rate is not due to a high barrier, but due to a low abundance of the active sites (for more details, see Section 4.2.3). The resulting adsorbed N atoms are then hydrogenated by adsorbed H atoms in a sequence of LH-type steps via NH$_{ads}$ and NH$_{2,ads}$ to NH$_{3,ads}$, before NH$_3$ can be desorbed. The mechanism is valid also for real reaction conditions. However, NH_3 adsorbs strongly on Fe surfaces, and substantial ammonia partial pressures severely retard NH_3 formation. Under these conditions, the Fe surface is almost completely covered with adsorbed NH_3 and NH_x intermediates, which changes the rds. It is then in the sequence comprising NH_x hydrogenation and the desorption of ammonia, which is required to vacate sites for N_2 dissociation.

Mechanisms involving the reaction of intermediates with (lattice) oxygen of the catalyst are not confined to oxide catalysts. This case can be nicely discussed using the example of CO oxidation over noble metal surfaces. Being simple with respect to reactants and products, this reaction was studied in great detail in traditional surface science, which has been extended to realistic pressures more recently [102]. At low pressures of both reactants, CO oxidation is a textbook example for a reaction following the LH mechanism. Both reactants adsorb on the metal surface, O_2 adsorption being dissociative, and CO_{ads} combines with O_{ads} to CO_2.

However, already under these idealized conditions, there are differences between metal surfaces and adsorbate-induced changes of surface structures. Thus, co-adsorption of CO_{ads} and O_{ads} was found to be cooperative (see Figure 2.52) on Rh(111) [103], but competitive on Pd(111) and Pd(100) [102]. On a clean Pt(110) surface, the missing-row reconstruction (cf. Figure 2.15a) is reverted in the presence of CO, because CO is adsorbed more strongly on the (less stable) unreconstructed facet. As O_2 adsorption is favored on this facet as well, it exhibits a high rate of CO oxidation. In a stationary flow of a CO/O_2 mixture, the accelerated reaction causes a depletion of CO in the gas phase. As a result, Pt(110) returns to its less active reconstructed version. Now, the CO concentration increases again, the Pt(110) reconstruction is reverted and the cycle is repeated, i.e., the reaction rate exhibits stable temporal oscillations [103]. Such oscillations can be observed with many (mostly exothermal) reactions, and their rates can exhibit spatiotemporal patterns even on ideal surfaces. The mechanism just described is the most elementary one. Often heat and mass transfer processes are also involved in these phenomena. In the present case, for instance, the rate peak could be additionally enhanced by local hotspots due to non-dissipated reaction heat.

Increasing the O_2 pressure opens additional options for the change of the catalyst structure: the surface may be oxidized. Various types of O-modified surface structures have been detected on different facets of Pt, Pd, and Ru [102]: adsorbate-induced reconstructions, surface oxides that are confined to the outermost layer of the metal, epitaxial three-dimensional oxide layers, and bulk oxide without structural relation to the underlying metal. Each metal facet has its own chemistry. Often, O-modified surface structures were detected under conditions allowing a strongly increased CO oxidation rates, which suggests a high activity of these structures. High CO oxidation activity was indeed detected with some (but not all) facets of stable noble metal oxides, but their suitability as models for the extremely thin surface oxide phases present under reaction conditions is under discussion. Therefore, a reliable assignment of phases active under particular conditions is still missing. There are also cases (in particular with Pt), where surface oxides seem to be less active than the bare metal surface [102].

Surface oxides also provide new options for reaction mechanisms of CO oxidation. The Mars-van Krevelen mechanism was identified for some cases [102], where CO_{ads} reacts with lattice oxygen bridging two metal atoms, which is then replaced by dissociative adsorption of O_2. Opposed to this, an ER-type mechanism, in which lattice oxygen reacts with gas-phase CO was discussed for some other cases. In a computational study, it was shown that the dissociative adsorption of O_2 is unfavorable on $PtO_2(110)$. Instead, the lowest energy barriers were found in a mechanism in which CO_{ads} reacts with O_2 adsorbed in an oxygen vacancy to $CO_2 + O_{lat}$. The resulting (bridging) lattice O atom easily oxidizes a second CO atom, which restores the vacancy [104].

The classical mechanisms can be easily used to derive kinetic rate laws under plausible assumptions (cf. Section 2.6.1). Even though they describe the catalytic processes well under certain reaction conditions, molecular surface processes are more complex. The effort to describe them taking in account all mechanistic steps, ideally also changes of the surface structure with variations of the reaction conditions, is referred to as microkinetic modeling. In addition, the range of reaction conditions may include regimes where transport processes influence the observed reaction rates. How these problems are treated in the description of catalytic reaction kinetics is described in sections 2.6 and 4.6.

2.5.2 Photocatalysis

In photocatalysis, chemical reactions are initiated by electromagnetic radiation in the presence of a light-absorbing substrate, which takes part in the chemical transformation of the reactants: the photocatalyst. In heterogeneous photocatalysis, the substrate is a semiconductor (SC), such as TiO_2 or ZnO. Light absorption creates electron—hole pairs (excitons) in it by excitation of electrons from the valence band to the conduction band (1 in Figure 2.58):

$$h\nu \, [+ \, SC(s)] \rightarrow [SC(s) +] \, e^- + h^+ \qquad (2.20)$$

The electron-hole pairs diffuse from the bulk to the surface where they can undergo oxidation and reduction reactions with reactants in the fluid phase (3 and 4 in Figure 2.58). Reactions with acceptor molecules A and donor molecules D create the primary redox products A^- and D^+, respectively, which are usually short-lived free radicals. These react further, forming the final products, which desorb from the surface.

The product quantum yield (cf. eq. (1.7)) describes the efficiency of the overall process. It is determined by three factors: (1) the efficiency of the electron-hole pair formation, (2) the transfer of charge carriers to the surface, and (3) the formation of primary redox products A^- and D^+. The overall efficiency is mostly affected by the undesired recombination of electrons and holes, directly after charge carrier generation in

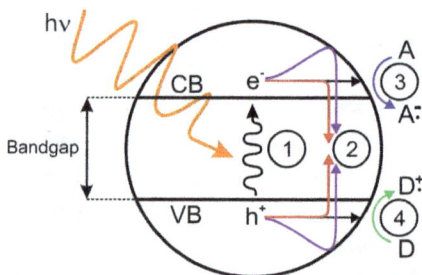

Figure 2.58: Elementary steps of a photocatalytic reaction.

the bulk, in secondary interactions with trapped electrons or holes on the way to the surface or at the surface (2 in Figure 2.58).

The formation of the primary redox products is a surface process, for which most photoabsorbers are not well suited. It may be dramatically accelerated by depositing co-catalysts (for reduction and/or oxidation) on their surfaces. The co-catalysts also favor charge separation by trapping electrons (in the case of metals by forming Schottky barriers, cf. Figure 2.9b) or holes. Pt, Ag, or Rh nanoparticles are typical reduction co-catalysts; metal oxide nanoparticles like RuO_2, CrO_x and IrO_2 are typical oxidation co-catalysts. More details on this topic can be found in Section 4.8.

In many photocatalytic reactions, e.g., in processes useful for water or air purification, the formation of the primary redox products is followed by a series of consecutive steps, which do not necessarily involve electrons and holes provided by the catalyst. Intermediates may even leave the surface and convert reactants in the surrounding liquid phase. In the oxidation of water contaminants, for instance, the excited photocatalyst (usually TiO_2) converts H_2O and O_2 to reactive intermediates such as OH radicals or superoxide radicals. The latter can be further converted to H_2O_2 in a series of reaction steps.

$$H_2O + h^+ \rightarrow H^+ + OH^{\cdot} \tag{2.21}$$

$$O_2 + e^- \rightarrow O_2^{-\,\cdot} \rightarrow \ldots \rightarrow H_2O_2 \tag{2.22}$$

These active species decompose contaminants like dyes or volatile organic compounds (VOCs) in solution. Even impurities in ppb level can be completely removed, resulting in full mineralization, i.e., the formation of CO_2 and water.

Whereas the mineralization of VOCs in the presence of O_2 is an exergonic reaction, photocatalysis also allows performing endergonic reactions ($\Delta G > 0$), for which absorbed light is used as energy supply. Overall water splitting (eq. (2.23)) is a challenging reaction that requires the transfer of four electrons to oxidize two OH^- ions to O_2 (oxygen evolution reaction, OER) and to reduce four H^+ to two H_2 molecules (hydrogen evolution reaction, HER). Intense research on these processes aims at the sustainable generation of H_2 from H_2O by sunlight (for more details, see Section 4.8).

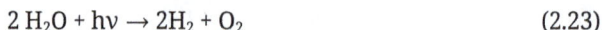

$$2\,H_2O + h\nu \rightarrow 2H_2 + O_2 \tag{2.23}$$

Likewise, the photocatalytic reduction of the greenhouse gas CO_2 by H_2O releasing O_2 (see, for instance, eq. (2.24)) aims at the synthesis of carbon-based fuels like methanol.

$$CO_2 + 2\ H_2O + h\nu \rightarrow CH_3OH + 3/2\ O_2 \tag{2.24}$$

This multistep reaction, which requires the transfer of six electrons, is often referred to as artificial photosynthesis. Numerous parallel reactions can result in various products such as CO or CH_4, rendering selectivity control a demanding task. Despite many reports on these reactions, their quantum yields and reaction rates are still too low for a practical perspective.

Research in the past decade has shown that approaches relying on a simple screening of materials will be insufficient for improving photocatalytic processes to a level, which allows realizing the large practical potential that they offer. A combined improvement of structured absorber and co-catalyst materials resulting in complex catalyst architectures will be essential to expand the field of applications, which also requires advanced reaction engineering solutions to improve light absorption and mass transfer.

2.6 Reaction mechanisms and reaction kinetics in thermal catalysis

2.6.1 Reaction kinetics of surface processes

The kinetics of a chemical reaction describes the change of reactant and product concentrations over time at a given set of conditions (temperature, concentrations/partial pressures). For this purpose, a suitable rate definition is equalized with the rate law of the reaction to obtain a differential equation, the solution to which gives the desired information. For a second-order homogeneous reaction A + B → C, for instance, with B in excess and no change in reaction volume (A – key component, $\nu_A = -1$), starting from eq. (1.2) yields:

$$r = -\frac{dc_A}{dt} = k\ c_A\ c_B \tag{2.25}$$

where the temperature dependence of the rate constant is described by the Arrhenius equation (1.1) or the Eyring equation

$$k = \frac{k_B T}{h}\ exp\left(\Delta S^{\ddagger}/R\right) exp\left(-\Delta H^{\ddagger}/RT\right) \tag{2.26}$$

In eq. (2.26), $\Delta H^{\ddagger}$ is the activation enthalpy, $\Delta S^{\ddagger}$ the activation entropy, and k_B the Boltzmann constant.

In a reaction network, the main reaction is accompanied by side reactions, which consume reactants or products. The individual reactions of the network are described by combining the rate definition with a suitable rate law for each relevant reaction as in eq. (2.25). The solution of the resulting system of differential equations allows modeling the behavior of the reaction network over time at the given reaction temperature. For more details on the reaction kinetics in homogeneous systems, the reader is referred to textbooks on chemical kinetics [105, 106].

In surface catalysis, surface species react with each other or with reactants from the surrounding fluid phase. Therefore, rate laws of surface reactions combine surface concentrations with each other or with fluid-phase concentrations. With the rate definition related to the catalyst mass (eq. (1.5)), eq. (2.25) might be transformed into:

$$r_{LH} = -\frac{1}{m_{cat}} \frac{dn_A}{dt} = k \, c_{s,A} \, c_{s,B} \qquad r_{ER} = -\frac{1}{m_{cat}} \frac{dn_A}{dt} = k \, c_{s,A} \, p_B$$

(2.27a (LH))

(2.27b (ER))

Equations (2.27a) and (2.27b) describe bimolecular surface reactions of LH- or ER-type, respectively (cf. Section 2.5.1). The surface concentrations are related to the monolayer capacity, $n_{i,mono}$ (cf. Section 2.4.3) via $c_{s,i} = n_{i,mono} \theta_i$. In the rate laws, $n_{i,mono}$ (or $n_{i,mono}{}^2$) are included into the rate constants. For LH- and ER-type bimolecular surface reactions, this results in

$$r_{LH} = -\frac{1}{m_{cat}} \frac{dn_A}{dt} = k \, \theta_A \, \theta_B \qquad r_{ER} = -\frac{1}{m_{cat}} \frac{dn_A}{dt} = k \, \theta_A \, p_B$$

(2.28a (LH))

(2.28b (ER))

Equations (2.28a) and (2.28b) describe the reaction behavior only if the surface reaction is the slowest among the elementary steps (cf. Figure 1.5). In addition, a possible reverse surface reaction is neglected. As mentioned earlier, rates of transport steps and of adsorption/desorption may influence the kinetics of the reaction as well: we return to these cases below. When the rate is determined by the surface reaction, adsorption equilibria can be assumed as established, and surface coverages can be described by adsorption isotherms.

In the practice of reaction kinetics, these considerations are taken into account to a different extent. In early engineering work, simple power rate laws:

$$r = k \, \Pi_i \, c_i^{n_i},$$

for which eq. (2.25) is an example (with $n_A = n_B = 1$), were employed, fully neglecting the heterogeneous structure of the reaction system, and fitted to experimental data. Reaction orders n_i were used as fit parameters; non-natural and negative numbers were accepted as fit results. While this can be appropriate in some situations, such models can be valid only in very narrow ranges of reaction conditions.

On the other hand, microkinetic modeling, which takes into account all steps of the reaction mechanism, is challenging (cf. Section 4.6) and often, cannot be achieved within limited times and resources, even nowadays. A closer inspection of the catalytic process results, however, in a number of approximate rate laws, which can be derived using plausible assumptions.

The existence of an rds is the most successful of these assumptions. It results in the Hougen-Watson rate laws (see below), some of which are, however, rather complicated. If the Bodenstein principle is applicable, it may give access to simple rate expressions. Its fundamental requirement, $d\theta_i/dt = 0$ for all reactive intermediates i cannot hold, of course, at very low or very high conversions. Where it holds, there is no accumulation or depletion of intermediates: rates of mechanistic steps are all identical and equal to the overall reaction rate. When the spectroscopic characterization of the working catalyst suggests coverage of the surface by one particular intermediate, this may provide an additional condition that allows deriving simplified rate laws in combination with the Bodenstein principle (masi concept, see below).

Rate laws derived according to these principles, which are sometimes referred to as "global" kinetic rate laws, are often very successful. However, a successful fit of the experimental data to a specific rate law is not conclusive proof of the underlying reaction mechanism. On the other hand, the failure of a rate law in fitting the experimental data definitely disproves the assumed mechanism.

Returning to the case of rate-determining surface reaction (e.g., bimolecular reactions (2.28a and b)), the coverages of A and B may be expressed by equations describing the established adsorption equilibria. Despite its limitations, the Langmuir isotherm (2.14) is usually employed for this purpose, however, in a modified version, which covers the competition of different species (A, B, but also of reaction products) for the same type of adsorption site. For only two competing species, A and B, the modified version can be easily derived by applying the kinetic approach (2.13) to both of them:

$$k_{ads,A}\, p_A\, (1 - \theta_A - \theta_B) = k_{des,A}\, \theta_A, \text{ and } k_{ads,B}\, p_B\, (1 - \theta_A - \theta_B) = k_{des,B}\, \theta_B$$

Box 2.3: How to derive the Langmuir isotherm for competitive adsorption

$$K_A\, p_A\, (1 - \theta_A - \theta_B) = \theta_A \text{ and } K_B\, p_B\, (1 - \theta_A - \theta_B) = \theta_B$$

$$\theta_B = \frac{K_B\, p_B\, (1 - \theta_A)}{1 + K_B\, p_B}$$

$$K_A\, p_A\, (1 - \theta_A) - \frac{K_A\, p_A\, K_B\, p_B\, (1 - \theta_A)}{1 + K_B\, p_B} = \theta_A$$

multiply and cancel . . .

$$K_A\, p_A - K_A\, p_A\, \theta_A = \theta_A + K_B\, p_B\, \theta_A$$

$$\theta_A = \frac{K_A\, p_A}{1 + K_A\, p_A + K_B\, p_B}$$

After introducing the adsorption constants of A and B ($K_i = k_{i,ads}/k_{i,des}$), one of the coverages, θ_i, may be isolated and substituted into the other equation. Some rearrangements (cf. Box 2.3) result in eq. (2.29a), which resembles the Langmuir isotherm for a single adsorbate, j: the competing adsorbate just causes an additional term, $K_j\,p_j$ in the denominator. A generalized Langmuir isotherm for competing adsorbates is given in (2.29b).

$$\theta_A = \frac{K_A\ p_A}{1 + K_A\ p_A + K_B\ p_B} \qquad \theta_B = \frac{K_B\ p_B}{1 + K_A\ p_A + K_B\ p_B} \qquad (2.29a)$$

$$\theta_A = \frac{K_A\ p_A}{1 + \sum_i K_i\ p_i} \qquad (2.29b)$$

With eq. (2.29a), the rate laws of bimolecular reactions proceeding according to the LH or ER mechanisms ((2.28a) or (2.28b), respectively) become

$$r_{LH} = \frac{k\ K_A\ p_A\ K_B\ p_B}{(1 + K_A\ p_A + K_B\ p_B)^2} \qquad r_{ER} = \frac{k\ K_A\ p_A\ p_B}{1 + K_A\ p_A} \qquad \begin{array}{l}(2.30a\ (LH))\\[6pt](2.30b\ (ER))\end{array}$$

These rate laws are very different from power rate law expressions, but tend toward such expressions under certain circumstances. For instance, if both p_A and p_B are low or both adsorption constants are small (i.e., $K_A\,p_A + K_B\,p_B \ll 1$), both pass over to second-order rate laws ($r = k'\,p_A\,p_B$). Equation (2.30a) is a function peaking at intermediate p_A or p_B. This can be deduced (for variation of p_A) by applying the condition $(\partial r_{LH}/\partial p_A)_{p_B} = 0$, or indirectly by considering the equation at high $K_A\,p_A$ ($\gg 1 + K_B\,p_B$). It then becomes $r_{LH} = k''\,p_B/p_A$, i.e., the rate vanishes at high partial pressures of the varied component, as indicated in Section 2.5.1. For the ER mechanism (eq. (2.30b)), p_A cancels under these conditions, and the rate becomes first order in the non-adsorbing component, B.

Equations (2.30) have been derived under the condition that no reaction products are adsorbed by the catalyst surface. In reality, this condition fails frequently. The pressure of the adsorbing product(s) then appear(s) in the denominator similar to the reactant pressures (cf. (2.29b)), which results in decreasing rates at high product concentrations. This phenomenon is referred to as product inhibition. It is a consequence of the limited number of adsorption sites on the surface.

In a full discussion, adsorption, surface reaction, and desorption are all included as reversible steps. For a first-order reaction of A to product P, this can be written as:

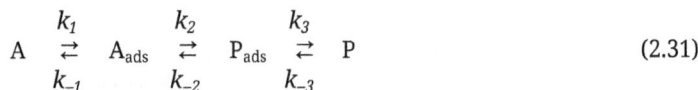

$$\mathrm{A} \underset{k_{-1}}{\overset{k_1}{\rightleftarrows}} \mathrm{A_{ads}} \underset{k_{-2}}{\overset{k_2}{\rightleftarrows}} \mathrm{P_{ads}} \underset{k_{-3}}{\overset{k_3}{\rightleftarrows}} \mathrm{P} \qquad (2.31)$$

where the rate constants of adsorption (desorption) of A and P are designated as k_1 (k_{-1}) and k_{-3} (k_3), respectively. The rate constants of forward and reverse surface reaction are

k_2 and k_{-2}. The net reaction rate is always the difference between rates of forward and reverse reactions:

$$r_{ads,A} = r_1 - r_{-1} = k_1\, p_A\, (1 - \theta_A - \theta_P) - k_{-1}\, \theta_A \tag{2.32a}$$

$$r_{surf} = r_2 - r_{-2} = k_2\, \theta_A - k_{-2}\, \theta_P \tag{2.32b}$$

$$r_{des,P} = r_3 - r_{-3} = k_3\, \theta_P - k_{-3}\, p_P\, (1 - \theta_A - \theta_P) \tag{2.32c}$$

We will now assume the adsorption of A to be rate-limiting without neglecting the desorption of A_{ads}. Surface reaction and desorption of P are then in quasi-equilibrium, i.e., forward and reverse rates are fast and identical ($r_{surf} = r_{des,P} = 0$), readily processing the molecules delivered by the adsorption ($r_{ads,A} \neq 0$). These conditions allow isolating θ_A and θ_P, which can be inserted in eq. (2.32a), as shown in some detail in Appendix A 2.2. In the derivation, the adsorption constants, K_A and K_P are used, with $K_A = k_1/k_{-1}$, but $K_P = k_{-3}/k_3$. Together with the equilibrium constant of the surface reaction $K_2 = k_2/k_{-2}$, these constants are related to the equilibrium constant of the whole reaction:

$$K = p_P/p_A = \frac{\theta_A}{p_A\, \theta_{free}} \frac{\theta_P\, p_P\, \theta_{free}}{\theta_A} \frac{}{\theta_P} = K_A\, K_2\, K_P^{-1}$$

After rearrangements shown in A 2.2, the rate law becomes:

$$r = \frac{k_1\, (p_A - p_P/K\ )}{1 + K_P\, p_P + K_A\, p_P/K} \tag{2.33}$$

With other reaction orders and assumptions about the rate-determining step, a great number of rate equations was derived: the Hougen-Watson (HW) rate laws (cf. Table 2.2). They have the general form:

$$r = \frac{(kinetic\ term)\ (potential\ term)}{(adsorption\ term)^{exponent}} \tag{2.34}$$

Obviously, the kinetic term is just k_1 in (2.33), but $k\, K_A\, K_B$ and $k\, K_A$ in (2.30a) and (2.30b), respectively (cf. Table 2.2a). The potential (or driving force) term includes the back reaction; it is $p_A - (p_p/K)$ in (2.33), but $p_A\, p_B - 0$ in both versions of (2.30), because the back reaction was neglected (Table 2.2b). In the HW equations, the adsorption term has the general form, $1 + \Sigma_i K_i\, p_i$, where summands are to be modified depending on the underlying mechanism (cf. Table 2.2c). For the first-order reaction with rate-determining adsorption of A, $K_A\, p_A$ is replaced by $K_A p_p/K$. The exponent can be found in Table 2.2d: it is 1 for rate-determining non-dissociative adsorption (eq. (2.33)), and 2 for rate-determining bimolecular surface reactions (eq. (2.30)).

Table 2.2: Structure of tables with terms for Hougen-Watson rate laws; the full tables can be found in refs. [107, 108].

a) Potential terms

Rate determining ...	Reactions			...
	$A \rightleftarrows P$	$A \rightleftarrows P + S$	$A + B \rightleftarrows P + S$	
Adsorption of A	$p_A - p_P/K$	$p_A - p_P p_s/K$	$p_A - p_P p_s/Kp_B$	
Adsorption of B	–	–	$p_B - p_P p_s/Kp_A$	
Desorption of P	$p_A - p_P/K$	$p_A/p_s - p_P/K$	$p_A p_B/p_s - p_P/K$	
Surface reaction	$p_A - p_P/K$	$p_A - p_P p_s/K$	$p_A p_B - p_P p_s/K$	
...				

b) Kinetic terms: rate-determining adsorption/desorption ...

... of A	... of B	... of P	... of A (dissociative)*	...
k_A	k_B	$k_P K$	$(s/2)k_A$	

*s – number of fragments

Surface reaction rate-determining

Condition	Reactions			...
	$A \rightleftarrows P$	$A \rightleftarrows P + S$	$A + B \rightleftarrows P + S$	
No dissoc. of A*	$k_s K_A$	$k_s K_A$	$k_s K_A K_B$	
A dissociates	$sk_s K_A$	$sk_s K_A$	$s(s-1)k_s K_A K_B$	
B not adsorbed	$k_s K_A$	$k_s K_A$	$k_s K_A$	
...				

*k_s corresponds to k in (2.28a) and to k_2 in (2.32b)

c) In adsorption term $1 + K_A p_A + K_B p_B + K_S p_S$ replace "X" by "Y" (see table below); no changes for rate-determining surface reaction

X for rate determining ...	Y for reactions			...
	$A \rightleftarrows P$	$A \rightleftarrows P + S$	$A + B \rightleftarrows P + S$	
Adsorption of A: $K_A p_A$	$K_A p_P/K$	$K_A p_P p_s/K$	$K_A p_P p_s/Kp_B$	
Adsorption of B: $K_B p_B$	–	–	$K_B p_P p_s/Kp_A$	
Desorption of P: $K_P p_P$	$K K_p p_A$	$K K_p p_A/p_s$	$K K_p p_A p_B/p_s$	
Diss. adsorption, etc.	–	–	–	

d) Exponents n: adsorption/desorption rate-determining
Non-dissociative adsorption: n = 1, dissociative adsorption: n = 2
Surface reaction rate-determining

Condition	Reactions			...
	$A \rightleftarrows P$	$A \rightleftarrows P + S$	$A + B \rightleftarrows P + S$	
No dissociation of A	1	2	2	
A dissociates	2	2	3	
A dissociates, B not adsorbed	2	2	2	
...	...	...	...	

The use of the most abundant surface intermediate (masi) for deducing rate laws is less universal than the assumption of a rate-determining step. It may be supported by characterization data, although the most abundant surface species is not necessarily a reaction intermediate – it might be just a spectator. Where it is applicable, the masi approach changes the balance of the surface sites into $\theta_{free} \approx 1 - \theta_{masi}$. For instance, in a reaction mechanism leading to the final product via three practically irreversible surface steps (e.g., the consecutive dehydrogenation of cyclohexane without intermediate desorption), the adsorbed product, D was observed to be most abundant (* – surface site):

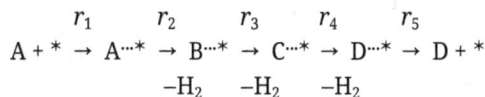

$$
\begin{array}{ccccccccc}
 & r_1 & & r_2 & & r_3 & & r_4 & & r_5 \\
A + * & \rightarrow & A{\cdots}* & \rightarrow & B{\cdots}* & \rightarrow & C{\cdots}* & \rightarrow & D{\cdots}* & \rightarrow & D + * \\
 & & & & -H_2 & & -H_2 & & -H_2 &
\end{array}
$$

With the Bodenstein principle, $r_1 = r_5 = r_{tot}$, and therefore, $r_{tot} = k_1 p_A (1 - \theta_D) = k_5 \theta_D$. With $\theta_D = r_{tot}/k_5$, this becomes $r_{tot} = k_1 p_A (1 - r_{tot}/k_5)$, Finally, r_{tot} can be isolated: $r_{tot} = k_1 p_A/(1 + K p_A)$.

Note that $K = (k_1/k_5)$ is not an equilibrium constant in this case, but just a ratio of rate constants.

2.6.2 Influences of transport limitations on reaction rates

As indicated in Figure 1.5, mass transfer from the flowing fluid to the catalyst particle and vice versa (film diffusion) and diffusion within porous particles are elementary steps of heterogeneous catalysis as well, and may become rate-limiting when the remaining steps are all significantly faster. Slow heat transfer may induce temperature differences between catalyst particles and fluid and even temperature gradients within the particles, which influence the observed reaction rates.

External mass and heat transfer can be targeted by designing reactors with suitable flow patterns; pellets causing internal transport limitations may require optimizing their sizes, shapes, and porosities to maximize the catalyst surface area accessible for reactants with given diffusivities. Such engineering work, which also includes modeling reaction regimes dominated by transport limitations and their transition to regimes with kinetic control, is beyond the scope of this book. For studying these aspects, the reader is referred to refs. [105–110]. Chemists should, however, definitely know how to identify transport limitations in experiments targeting surface processes. They should understand their consequences for the catalytic process and the strategies to reduce their influence. To achieve this, mass transfer will be treated in the following on an elementary quantitative level, while phenomena related to heat transfer limitations will be discussed qualitatively.

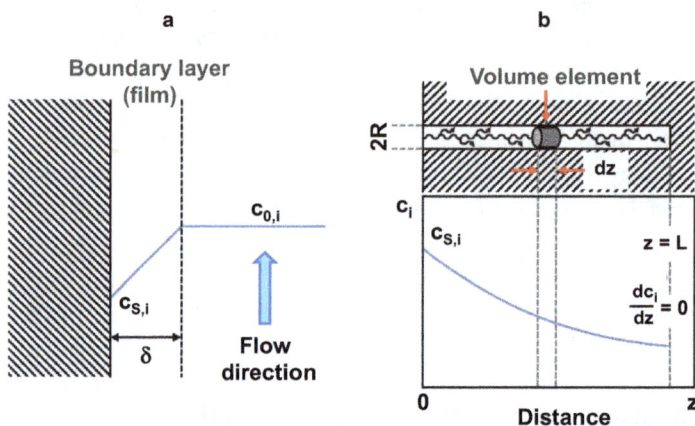

Figure 2.59: Film diffusion and pore diffusion: (a) concentration profile near a catalytic surface causing a reaction rate comparable with the diffusion rate; (b) concentration profile along a cylindrical pore with catalytic pore walls causing a reaction rate comparable with the diffusion rate.

Reactant transport towards a catalytic wall (a pellet surface) is perpendicular to the flow direction of the fluid (Figure 2.59a). In turbulent flow, most of the distance is covered by convective transport with negligible flow resistance. The resistance arises mostly from crossing a boundary layer close to the surface, where the fluid is laminar or (at the surface) stagnant. Since transport through this boundary layer is dominated by diffusion, it is often referred to as film diffusion. In laminar flow, the boundary layer is much thicker. In narrow spaces, e.g., monolith channels, it extends over the whole cross section of the flow.

Mass transfer is driven by concentration differences. Due to the catalytic reaction, the surface concentration $c_{S,i}$ of the reactant is smaller than its core concentration, $c_{0,i}$. In a simplified model (Figure 2.59a), c_i decays linearly across the thickness, δ of the boundary layer. In the steady state, the reactant flow by transport $(dn_i/dt)_{Diff}$ equals the amount of reactant converted per time $(-(dn_i/dt)_{React}$, because the reaction causes the concentration to decrease). For a surface section S, the former may be expressed by Fick's 1st law as $-S\,D_i\,(c_{S,i} - c_{0,i})/\delta$, where D_i is the diffusion coefficient. Using a surface-related reaction rate, r_S,[25] the reactant consumption can be expressed as $-v_i\,S\,r_S$. In the resulting equation:

$$-v_i\,r_s = -D_i\,(c_{S,i} - c_{0,i})/\delta,$$

different rate laws can be inserted. For a first-order surface reaction, insertion of $r_s = k_s\,c_{S,i}$ allows isolating $c_{S,i}$:

$$c_{S,i} = \beta\,c_{0,i}/(k_s + \beta).$$

$\beta = D_i/\delta$ is a mass transfer coefficient — actually a rate constant of mass transfer. The measured rate, r_{eff} may now be also expressed by a rate law containing only the observable concentration $c_{0,i}$:

$$r_{S,eff} = k_S\,c_{S,i} = k_{S,eff}\,c_{0,i}.$$

Inserting $c_{S,i}$ yields:

$$r_{S,eff} = \frac{k_S\,\beta}{k_S + \beta}\,c_{0,i} \quad \text{or} \quad r_{S,eff} = \frac{1}{1/\beta + 1/k_S}\,c_{0,i} \qquad (2.35a)$$

Equation (2.35a) gives access to the effective rate constant $k_{S,eff}$. Its relation to the rate constants of mass transfer and surface reaction is, however, better expressed via its reciprocal form:

$$\frac{1}{k_{S,eff}} = \frac{1}{\beta} + \frac{1}{k_S} \qquad (2.36a)$$

A reciprocal rate constant specifies a resistance. Equation (2.36a) shows that film diffusion and surface reaction are processes in series. Their resistances add up to the total resistance, $1/k_{S,eff}$.

In the derivation above, the external catalyst surface area, S is the cross section of the diffusional flow. In catalysis, the mass-related rate definition (1.5) is, however, more popular. Equations (2.35a) and (2.36a) can be easily rearranged to include this definition. Obviously, $(dn_i/dt)_{React}$ can be expressed by both mass- and surface-related rates:

---

25 r_S is obtained by inserting S instead of m_{cat} into eq. (1.5).

$r\, m_{cat} = r_S\, S$, and, with rate law inserted, $m_{cat} k\, c_{S,i}^n = S\, k_S\, c_{S,i}^n$.

$c_{S,i}^n$ can be cancelled, because concentration profiles do not depend on the rate definition. Therefore, the effective mass-related rate, r_{eff} is (cf. eq. (2.35a)):

$$r_{eff} = r_{S,eff}\, \frac{S}{m_{cat}} = \frac{S}{m_{cat}}\, \frac{1}{1/\beta + S/k_S\, m_{cat}}\, c_{0,i}$$

$$r_{eff} = \frac{1}{m_{cat}/S\, \beta + 1/k_S}\, c_{0,i} \tag{2.35b}$$

and the effective rate constant results from:

$$\frac{1}{k_{eff}} = \frac{m_{cat}}{S\, \beta} + \frac{1}{k_S} \tag{2.36b}$$

These relations can be elegantly summarized by using the second Damköhler modulus, Da_{II}. For a reaction of nth order, it is defined as:

$$Da_{II} = \frac{k_S\, c_{0,i}^{n-1}}{\beta}\, \frac{V_{cat}}{S} \tag{2.37}$$

i.e., $Da_{II} = V_{cat}\, k_S/S\, \beta$ for a first-order reaction. The second Damköhler modulus compares the maximum kinetic reaction rate (unaffected by diffusion) with the maximum diffusion rate (i.e., with $c_{S,i} = 0$). It refers, however, to a homogeneous model with rates related to the catalyst volume, V_{cat}, instead of the mass, m_{cat}. With a modified definition adapting Da_{II} to mass-related catalytic rates ($Da_{II}^* = m_{cat}\, k_S/S\, \beta$), equations (2.35b) and (2.36b) can be easily transformed into:

$$r_{eff} = \frac{k_S}{Da_{II}^* + 1}\, c_{0,i} \tag{2.35c}$$

and

$$k_{eff} = \frac{k_S}{Da_{II}^* + 1} \tag{2.36c}$$

The equations for the original homogeneous model are identical, just with Da_{II} instead of Da_{II}^*.

Film diffusion keeps the observed rate, r_{eff} below the one at which the catalyst would operate without mass transfer influence (r). The loss may be reported by an effectiveness factor, η_{ext}, which compares rates with and without film diffusion limitation:

$$\eta_{ext} = r_{eff}/r \tag{2.38}$$

With eq. (2.35b) or (2.35c) and $r = k_S\, c_{0,i}$, η_{ext} becomes:

$$\eta_{ext} = \frac{1}{m_{cat}\, k_{s/s}\, \beta + 1} = \frac{1}{Da_{II}^{*} + 1} \tag{2.39}$$

Rate limitations by mass transfer reveal themselves by very low (apparent) activation energies of the reactions studied. The rate equation of diffusive mass transport is Fick's first law (see above); its rate constant is $\beta = D/\delta$, and the kinetic order is 1, because the surface concentration, $c_{S,i}$ becomes zero when diffusion is very slow. Activation energies of transport processes in fluid phases are only a few kJ/mol, far below those of chemical reactions. When the activation energy of a catalytic reaction decreases with growing temperature tending towards 10 kJ/mol, transition to the film diffusion regime is likely. At the same time, the reaction order of the key reactant should tend towards 1, which is certainly less specific.

In research, it is important to make sure that catalytic rate measurements (see Section 3.2) are not corrupted by external mass transfer influences, if possible with only a few measurements. This can be achieved by comparing runs, in which the same catalyst/feed ratio is realized with different feed velocities. A higher feed velocity enhances the influence of the turbulent flow pattern, which results in decreased film thickness, δ and improved mass transfer. When the conversion does not depend on the feed velocity in this series, limitations by film diffusion can be excluded. Experiments can be made in a flow reactor or in a recycle reactor. In the former, the catalyst mass is varied, and the feed velocity is scaled correspondingly. In a recycle reactor, it is sufficient to vary the intensity of internal recycling.

Almost all chemical reactions involve release or consumption of heat, which needs to be drained from or provided to the catalytic surface. Analogous to mass transfer, reaction heat is transported perpendicular to the fluid and against resistances mostly residing in the laminar boundary layer adjacent to the surface (cf. Figure 2.59a). In a simple picture, heat transfer can be described by the Fourier equation:

$$\frac{dQ}{dt} = -\lambda\, S\, \frac{T_S - T_0}{\delta} \tag{2.40}$$

where T_S and T_0 are the temperatures at the surface and in the fluid stream, λ is the heat conductivity in the boundary layer, and λ/δ can be lumped into a heat-transfer coefficient, α. Equation (2.40) implies that temperatures at the surface and in the flowing fluid stream are never identical, except for thermoneutral reactions. If heat effects are small or the transfer coefficient, α is large (e.g., due to a low value of δ), the temperature difference can be negligible.

On the other hand, slow heat transfer causes measurable temperature differences. In an endothermal reaction, the surface becomes colder; in an exothermal reaction it will be hotter than the fluid stream, and catalytic rates will be lower or higher, correspondingly, than in the isothermal case. A higher reaction rate means that the effectiveness factor, η_{ext} exceeds 1. Although things are more complicated in reality, because rates enhanced by delayed heat transfer may run into mass transfer limitations, $\eta_{ext} > 1$ has been observed often.

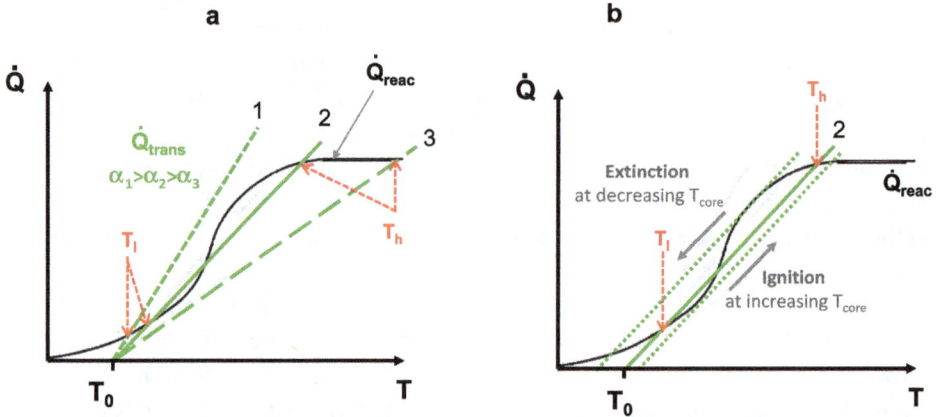

Figure 2.60: Heat transfer through a boundary layer: ignition and extinction: (a) bi-stable regimes of an exothermal reaction at a catalytic wall: rates of heat release and heat transfer compared for different heat transfer coefficients. (b) Hysteresis of reaction regimes: transition from the lower to the higher stationary point upon increasing the core temperature T_0 (ignition), and vice versa (extinction) upon decreasing T_0.

In the stationary state, the heat flow $\dot{Q}_{reac}$ originating from the surface reaction must be completely drained or provided by transport ($\dot{Q}_{trans}$, cf. eq. (2.40)). In Figure 2.60a, the temperature dependences of $\dot{Q}_{reac}$ and $\dot{Q}_{trans}$ are plotted for an exothermal reaction. For $\dot{Q}_{trans}$, three straight lines starting at the core temperature, T_0, represent cases with different heat transfer coefficients, α. $\dot{Q}_{reac}$ exhibits the typical S-shape with a strong increase at intermediate temperatures followed by an asymptotic behavior near full conversion. Stationary operation is indicated by the intersection points of $\dot{Q}_{reac}$ and $\dot{Q}_{trans}$ vs. T. When heat transfer is fast (α_1), this point is at a low surface temperature (T_l). When the heat transfer coefficient is low (α_3), a large driving force is required: the stationary point is at a high temperature (T_h) and, thus, at nearly full conversion, where $\dot{Q}_{trans}$ can catch up with the nearly invariant $\dot{Q}_{reac}$. At intermediate heat transfer coefficient (α_2), there are three intersections between the heat flow curves (see also Figure 2.60b). Again, T_l and T_h indicate stationary states, while the system is unstable at the intermediate point.

The actual surface regime depends on system history — if the system is heated from below T_0 or cooled from a high temperature (Figure 2.60b). Due to the weak temperature dependence of hydrodynamic parameters, heating or cooling of the flowing fluid can be represented by parallel shifts of the $\dot{Q}_{trans}$ line. When T_0 has been approached from below, the surface operates at T_l. Upon further increase of the core temperature, the surface temperature T_S moves along the $\dot{Q}_{reac}$ curve, until the two lower intersection points of the heat flow curves merge into one, which is the osculation point of the $\dot{Q}_{trans}$ tangent. At this moment, T_S increases irresistibly and stabilizes only on the upper branch of the $\dot{Q}_{reac}$ curve. This switching phenomenon is called ignition. When the flow is now cooled, the surface temperature remains on the upper branch, even when the core temperature falls below T_0. Only when the two upper intersection points of the heat flow curves merge into the osculation point of the respective $\dot{Q}_{trans}$ tangent, T_S returns to the lower branch of the $\dot{Q}_{reac}$ curve by extinction.

Bistable behavior with ignition/extinction phenomena and hysteresis of regimes is not possible for endothermal reactions. For exothermal reactions, it is not a rare phenomenon. In the example of Figure 2.60, it occurs in different ranges of the core temperature T_0, which depend on the heat transfer coefficient. $\dot{Q}_{trans}$ for α_1 has only one intersection point with $\dot{Q}_{reac}$ at the given core temperature (Figure 2.60a), but when T_0 is increased, there is a (narrow) temperature range with three intersections, also for α_1. Likewise, a (very broad) temperature range with bistable behavior exists for α_3, below the given T_0.

Ignition/extinction phenomena are typical for highly exothermal reactions with significant activation energies, e.g., total oxidations in not very dilute feed streams. As exemplified in Figure 2.60, their treatment is based on balancing heat flows ($\dot{Q}_{trans} = \dot{Q}_{reac}$). In the present example, the balance space was just the surface of a catalyst particle. Real balance spaces can be large: sections of reactor tubes or whole (well-mixed) reactors, where ignition/extinction phenomena may be observed in hotspots or within the entire reactor. More detail on these features can be found in textbooks on reaction engineering [105–110].

Film diffusion and pore diffusion differ fundamentally in their sequence of steps: while they are strictly consecutive in film diffusion (Figure 2.59a), reaction may occur already along the diffusion path in pore diffusion (Figure 2.59b). This results in specific differences in their mathematic description, despite characteristic analogies. Pore diffusion is treated below, based on the simple cylindrical pore model. Results from the more realistic, somewhat more complicated porous sphere model will be communicated without proof. Possible simultaneous film diffusion limitations are neglected in these simple approaches. For more information, the reader is referred to textbooks of reaction engineering [105–110].

When pore diffusion and surface reaction proceed at comparable rates, the reactant concentration, c_i decreases along the pore length coordinate, z, as shown in Figure 2.59b. The condition $(dc_i/dz) = 0$ at the pore bottom results from an oversimplification of the figure: the model is meant to represent a pore of a length $2L$, which is open at both ends. At $z > L$, the concentration follows the mirrored profile of $z < L$.

Due to the continuous variation of c_i along the pore, a space for balancing species engaged in diffusion and reaction can have only a differential extension along z. It is a cylinder with top surfaces of πR^2 and a jacket surface of $2\pi R\,dz$, with R being the pore radius (Figure 2.59b). Species of i that entered at z but did not make it to $z + dz$ were obviously reacted at the pore wall. Using Fick's first law for describing diffusion, a surface-related definition of the reaction rate and assuming a first-order rate law for the surface reaction, this yields:

$$-\pi R^2 D \left(\frac{dc_i}{dz}\right)_z + \pi R^2 D \left(\frac{dc_i}{dz}\right)_{z+dz} = 2\pi R\,dz\,k_S\,c_i$$

which can be easily converted into:

$$\frac{\left(\frac{dc_i}{dz}\right)_{z+dz} - \left(\frac{dc_i}{dz}\right)_z}{dz} - \frac{2\,k_S\,c_i}{R\,D} = 0 \quad \text{and} \quad \left(\frac{d^2c_i}{dz^2}\right) - \frac{2\,k_S\,c_i}{R\,D} = 0$$

To remove quantities containing units from this differential equation, dimensionless (reduced) coordinates are introduced: $Z = z/L$, and $C_i = c_i/c_{S,i}$, where $c_{S,i}$ is the reactant concentration at the external surface. By equating $k_S\,dS = k_S\,2\pi R\,dz$ with $k\,dV = k\,\pi R^2\,dz$, the surface-related rate is converted into a volume-related rate (homogeneous model). Inserting these relations into the differential equation and multiplying by $L^2/c_{S,i}$ yields:

$$\left(\frac{d^2C_i}{dZ^2}\right) - L^2 \frac{k}{D} C_i = 0.$$

The expression preceding C_i is the square of the Thiele modulus

$$\Phi_o = L\,\sqrt{k/D} \tag{2.41}$$

With this definition, the differential equation describing the cylindrical pore model becomes:

$$\left(\frac{d^2C_i}{dZ^2}\right) - \Phi_0^2\,C_i = 0 \tag{2.42}$$

Any function solving this differential equation must fulfill the boundary conditions $C_i = 1$ at $Z = 0$ and $dC_i/dZ = 0$ at $Z = 1$.

The rather straightforward solution of (2.42), which can be tracked in Appendix A 2.3, results in:

$$C_i = \frac{\cosh(1-Z)\Phi_0}{\cosh \Phi_0} \tag{2.43}$$

For the porous sphere model, spherical coordinates are used. Instead of considering diffusion in empty pores, an effective diffusivity, D_{eff} is attributed to the porous material. It depends on properties of the pore system according to:

$$D_{eff} = D \frac{\epsilon}{\tau} \tag{2.44}$$

In eq. (2.44), ε is the void fraction of the pellet, the ratio between its pore volume V_p and its total volume V:

$$\varepsilon = \frac{V_p}{V} \tag{2.45}$$

τ is the tortuosity factor, which accounts for the fact that direct connections between points in space are not always available in pore systems and detours must be conceded. Typical values for ε are around 0.5, while τ can be as large as 10. With a reduced radius coordinate $R = r_{rad}/r_{sph}$[26] starting at the center of the sphere and a first-order rate law, the differential equation (analogous to eq. (2.42)) reads:

$$\left(\frac{d^2 C_i}{dR^2}\right) + \frac{2}{R}\left(\frac{dC_i}{dR}\right) - \Phi_0^2\, C_i = 0 \tag{2.46}$$

With a Thiele modulus of

$$\Phi_o = r_{sph} \sqrt{k/D_{eff}} \tag{2.47}$$

and boundary conditions of $C_i = 1$ at R = 1, and $dC_i/dR = 0$ at R = 0, the solution is:

$$C_i = \frac{\sinh R\, \Phi_0}{R \sinh \Phi_0} \tag{2.48}$$

In Figure 2.61a, the dependence of the reduced concentration, C_i on the reduced pore depth coordinate, Z and the Thiele modulus, Φ_0 is depicted for the cylindrical pore model (eq. (2.43)). Results for the porous sphere model, with R proceeding inward out, are almost indistinguishable. The decay of C_i along the pore (or towards the pellet center) is small when $\Phi_0 < 1$. Beyond $\Phi_0 = 1$, the reactant becomes depleted in the pore, and for

---

26 In the following, the variable radius coordinate will be designated as r_{rad} to avoid confusion with the reaction rate r. r_{sph} is the pellet radius.

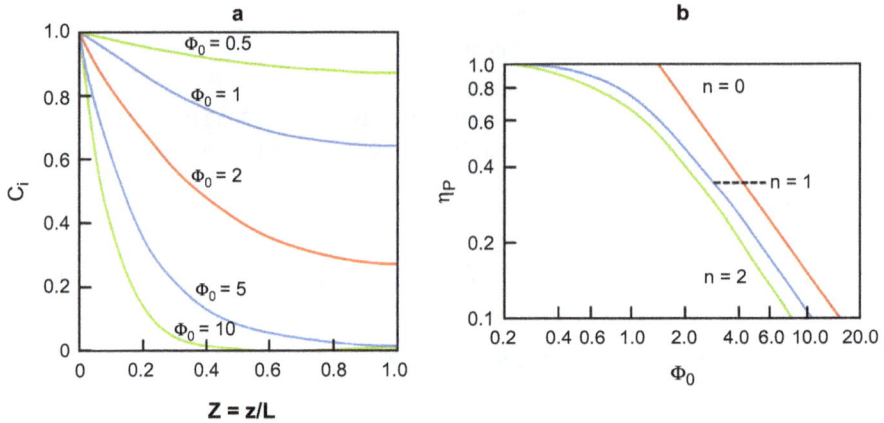

Figure 2.61: Influence of pore diffusion on internal concentration profiles and effectiveness factors: (a) intra-pore profiles for different Thiele moduli, first-order reaction, cylindrical pore model; (b) dependence of the effectiveness factor on the Thiele modulus for different reaction orders, spherical model.

$\Phi_0 > 5$, it no longer reaches the bottom of the pore (or the pellet center). At $\Phi_0 > 10$, the reaction proceeds largely near the mouth of the pore. Expensive catalyst components deposited in the particle interior are wasted under these conditions.

For the interpretation of the Thiele modulus, it is instructive to look at its squared form:

$$\Phi_0^2 = r_{sph}^2 \frac{k}{D_{eff}} \left(= \frac{r_{sph} \, k \, c_{S,i}}{D_{eff} \, c_{S,i}/r_{sph}} \right)$$

The inspection of numerator and denominator shows that the Thiele modulus compares maximum rates of competing processes: the maximum surface reaction rate (with $c_{S,i}$ across the whole pellet) and the maximum diffusion rate (at a driving force of $(c_{S,i} - 0)/r_{sph}$). Φ_0^2 is completely analogous to the second Damköhler modulus (2.37), in which V/S in the numerator has the dimension of a (characteristic) length, as well.

The Thiele modulus is actually defined for any reaction order n, and it can be generalized for arbitrary pellet geometries:

$$\Phi_0 = L_c \sqrt{k \, c_{S,i}^{n-1} / D_{eff}} \tag{2.49}$$

where L_c is a characteristic length (e.g., the sphere radius).

An effectiveness factor, η defined by analogy to eq. (2.38) can be used to quantify the retardation of reactions also by pore diffusion:

$$\eta_p = r_{eff}/r \tag{2.50}.$$

According to a derivation outlined in Appendix A2, the relation between η_P and the Thiele modulus for a first-order reaction is:

$$\eta_P = \frac{3}{\Phi_0} \left(\frac{1}{\tanh \Phi_0} - \frac{1}{\Phi_0} \right) \tag{2.51}.$$

In Figure 2.61b, the dependence of the effectiveness factor on the Thiele modulus is illustrated for reaction orders of 0, 1, and 2 in logarithmic coordinates. The curves are for spherical pellets, but the influence of the particle shape is minor. The impact of pore diffusion on the reaction rate increases with the reaction order. It is sizeable, but low at $\Phi_0 < 1$, whereas it can be dramatic at high values of the Thiele modulus.

The definition of Φ_0 (2.47) implies, therefore, that diffusion limitations occur at large pellet sizes, low effective diffusivities, and high surface reaction rates. The latter should be definitely kept in mind: a low effectiveness factor does not indicate a low activity, but rather a high potential for improvement, because reaction rates are actually high. At $\eta_P = 0.1$, for instance, a tenfold rate could be achieved by making the whole internal surface available for catalysis. Decreasing particle sizes and engineering pore systems for larger void fractions and, in particular, smaller tortuosities, are promising strategies. It should be kept in mind that this applies to gas-phase reactions. Liquid-phase reactions of relevant rates are always subject to severe pore diffusion limitations because of the low diffusivities in liquids. According to a study of van der Pol et al. [111], microporosity reduces the diffusion length that allows participation of pore surfaces in liquid-phase reactions to the order of micrometers.[27]

According to Figure 2.61b, the course of all η_P vs. Φ_0 curves is parallel at $\Phi_0 > 5$. In the first-order model (2.51), the expressions in parentheses cancel in this range, and η_P tends towards $3/\Phi_0$. Generalizing this into

$$\eta_P \cong \text{const.}/\Phi_0 \tag{2.52}$$

for $\Phi_0 > 5$, and expressing the Thiele modulus by eq. (2.49), the effective reaction rate may be written as:

$$r_{eff} = r \, \eta_P = k \, c_{S,i}^n \, \text{const.}/\Phi_0 = k \, c_{S,i}^n \, \text{const.} \, \sqrt{D_{eff}} / L_c \, \sqrt{k \, c_{S,i}^{n-1}}$$

$$r_{eff} = \text{const.}' \, \sqrt{D_{eff} \, k} \, \sqrt{c_{S,i}^{n+1}} \tag{2.53}$$

---

27 Applying the Weisz modulus (see (2.56) below), the authors assessed the impact of pore diffusion in TS-1 zeolite (cf. Section 2.3.2.3) during the hydroxylation of phenol (in acetone) to dihydroxobenzenes by H_2O_2. The effectiveness factor, η_p, which was ≈ 0.8 at crystallite sizes of ≈ 0.3 mm, decreased to < 0.01 for 5 mm crystallites.

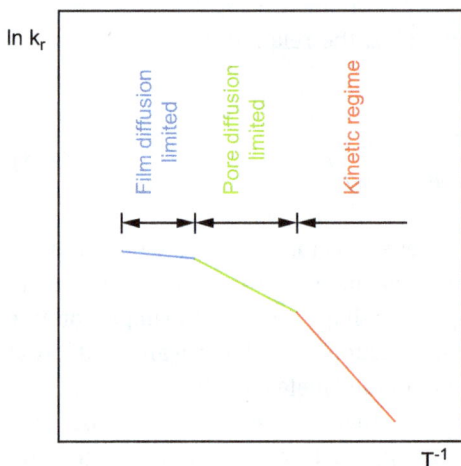

Figure 2.62: Influence of pore diffusion and of film diffusion on the temperature dependence of measured (i.e., effective) rate constants: an Arrhenius plot.

Equation (2.53) has the form of a rate law: the order is (n+1)/2 and the rate constant is proportional to $(D_{eff}\,k)^{0.5}$. This shows that both reaction order and activation energy change during the transition to the pore-diffusion limited regime. Only the first-order reaction stays first-order, a reaction order of zero becomes 0.5, and a second-order reaction tends towards n = 1.5. The changes of the (apparent) activation energy are more sizeable. According to the Arrhenius equation (1.1),

$$\sqrt{D_{eff}\,k} = const. \sqrt{exp^{-(E_A + E_{Diff})/RT}}.$$

As E_{Diff} is very small, the apparent activation energy under pore-diffusion limitation is half the activation energy of the surface reaction:

$$E_{A,eff} \cong 0.5\,E_A \tag{2.54}$$

The temperature dependence of measured rate constants during transition from the kinetic regime via the pore-diffusion to the film-diffusion limited regime is highlighted in Figure 2.62.

As discussed above for film diffusion, it is imperative to check for pore diffusion limitations, when preparing for rate measurements with an unknown catalyst. The standard approach is to conduct experiments at the same catalyst/feed ratio, but with the catalyst crushed and sieved to have it in different particle sizes. When the results are identical, pore diffusion limitation can be excluded: it would be revealed by increased conversions over smaller particles.[28]

---

28 under the condition of nonnegative reaction order.

For pore diffusion, there is, however, also a convenient method to assess the diffusion influence by a single experiment. The Thiele modulus is not suited for this purpose, because it contains the unknown surface rate constant. Instead, the Weisz modulus,

$$\Phi_W = \eta_P\,\Phi_0^2 \tag{2.55}$$

can be determined from quantities accessible by measurement or well-founded estimates. Using eq. (2.49), Φ_W becomes:

$$\Phi_W = \frac{r_{eff}}{r}\,\Phi_0^2 = \frac{r_{eff}}{k\,c_{S,i}^n}\,\frac{L_c^2\,k\,c_{S,i}^{n-1}}{D_{eff}}$$

$$\Phi_W = \frac{r_{eff}L_c^2}{D_{eff}\,c_{S,i}} \tag{2.56}$$

For estimating D_{eff} in eq. (2.56), diffusion coefficients can be obtained by a correlation reported in ref. [112]. Together with $\varepsilon \approx 0.5$, a tortuosity factor τ of 2 is usually assumed. When $\Phi_{WP} \ll 1$, pore diffusion influences can be safely excluded, at $\Phi_{WP} \gg 1$, they are very likely.

Similar as in the external boundary layer, heat transfer can be also limited by resistances in the pore system. Due to the resulting intra-particle temperature gradients, the center of the particle becomes hotter or colder than the external surface. For highly exothermal reactions, hotspots in the particle center combined with high activation energy of the reaction can again result in effectiveness factors >1. Moreover, mass transfer limitations also have consequences for the product selectivity in consecutive or parallel reaction schemes. These topics are beyond the scope of this book and can be looked up in textbooks of reaction engineering [105–110].

2.7 Electrochemistry and electrocatalysis

The separation of redox reactions into two half-reactions proceeding at different electrodes offers two attractive opportunities: to enforce thermodynamically "impossible" (uphill) reactions by applying electrical energy and to convert chemical energy stored in thermodynamically "possible" (downhill) reactions into electrical energy. Electrolysis cells and galvanic cells are the technical devices for realizing these opportunities. Both can benefit from catalysts operating on their electrodes — among the galvanic cells, in particular, the fuel cell. Electrolytic water splitting into H_2 and O_2 and water formation from these reactants in various types of electrolyzers and fuel cells, respectively, are examples for applications that illustrate the high relevance and increasing importance of electrocatalysis. They offer the promise to help replacing a fossil-fuels-based energy system by a hydrogen-based one.

Electrolysis can be used to split water into H_2 and O_2 using excess green energy. Unlike electricity, these gases can be stored at scales required to compensate for energy shortage in national grids, due to lack of sunlight for photovoltaics or of wind for wind turbines. The required energy is then recovered by converting H_2 and O_2 to water in fuel cells. In the following, water electrolysis will be used to exemplify some basic principles of electrochemistry and electrocatalysis.

To drive an uphill redox reaction by electricity, its (positive) Gibbs free energy, ΔG^0, must be balanced by adding an electric potential E of at least

$$E = -\Delta G^0/z\,F \tag{2.57}$$

where z is the number of electrons transferred and F is the Faraday constant. This "reversible" cell potential, E, is distributed between anode and cathode. At equilibrium, the corresponding half-reactions and their reverse reactions proceed with equal rates or currents, i.e., there is no net conversion. The currents at the reversible potential are called "exchange currents." Depending on electrode materials and surface properties, they are the material contribution to electrochemical reaction rates.

The reversible potentials at the two electrodes vary with the concentrations (or thermodynamic activities a) of reactants i in their oxidized and reduced forms, according to the Nernst equation:

$$E = E_0 + \frac{RT}{zF}\ln\frac{a_{i,ox}}{a_{i,red}} \tag{2.58}$$

where E_0 is the standard potential for the half-reaction proceeding at the electrode. In the galvanic series, the standard potentials of all possible half-reactions are ranked relative to a reference, the hydrogen evolution reaction (HER, eq. (2.59)), the potential of which was defined as 0 V. Water electrolysis (2.61) combines the HER (2.59) at the cathode with the oxygen evolution reaction (OER, (2.60)) at the anode.

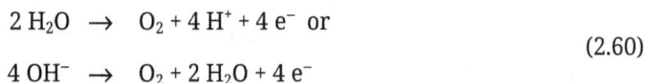

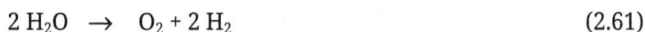

$$\begin{aligned} 2\,H^+ + 2\,e^- &\rightarrow H_2 \text{ or} \\ 4\,H_2O + 4\,e^- &\rightarrow 2\,H_2 + 4\,OH^- \end{aligned} \tag{2.59}$$

$$\begin{aligned} 2\,H_2O &\rightarrow O_2 + 4\,H^+ + 4\,e^- \text{ or} \\ 4\,OH^- &\rightarrow O_2 + 2\,H_2O + 4\,e^- \end{aligned} \tag{2.60}$$

$$2\,H_2O \rightarrow O_2 + 2\,H_2 \tag{2.61}$$

For the electrolytic hydrogen production, alkaline electrolysis is an established technology. Highly concentrated KOH solutions (25–30 wt%) provide sufficient conductivity to minimize the resistance of the electrolyte. At pH 14, the standard potentials for reactions

(2.59) and (2.60) are −0.83 V and +0.4 V, respectively, which adds up to the reversible potential of water splitting of 1.23 V (ΔG^0 = 286 kJ mol^{-1} [113, 114]).

However, due to various reasons, the cell voltages required for splitting water into H_2 and O_2 exceed the reversible potential determined by thermodynamics (eq. (2.62)). Such overpotentials may be due to ohmic resistances R in electric lines, contacts, or in the electrolyte. Other contributions, designated by η in eq. (2.62), arise at the two electrodes, e.g., by gas bubble formation and, most importantly, by kinetic barriers in the redox processes proceeding there. In electrochemistry, kinetic barriers can be overcome by enhancing the driving force, i.e., the potential. The relation between rate and potential will be discussed below, as well as limitation of rates by mass or charge transfer at high potentials.

$$E_{cell} = \sum_{An,Cath} E_{rev} + \sum_{An,Cath} \eta + \sum I \cdot R \qquad (2.62)$$

In water splitting, the kinetic barriers are most significant in the anodic OER, because of the sluggish reaction mechanism of the four-electron transfer required for oxidizing two O^{2-} to O_2 (often abbreviated as "water oxidation"). With anode materials considered to be noncatalytic, e.g., glassy carbon, an overpotential of ca. 1.2 V was required for a current density of 10 mA cm^{-2} in 1 M NaOH. Carbon corrosion observed in this potential range contradicts the assumed inertness [115] and, correspondingly, even higher values have been found on more stable boron-doped diamond electrodes [116].

$*$ = metal site

$$* + OH^- \xrightarrow{\Delta G_1} OH + e^-$$

$$* OH + OH^- \xrightarrow{\Delta G_2} * O + H_2O + e^-$$

$$* O + OH^- \xrightarrow{\Delta G_3} * OOH + e^-$$

$$* OOH + OH^- \xrightarrow{\Delta G_4} * + O_2 + H_2O + e^-$$

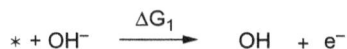

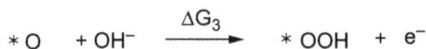

Scheme 2.4: Elementary steps of the oxygen evolution reaction (OER) in an alkaline electrolyte ($*$ – metal site).

These high overpotentials are prohibitive for the utilization of water electrolysis on a large commercial scale. Actually, anode materials in technical electrolyzers are catalytic already, nowadays. Under the reaction conditions, even noble-metal surfaces such as Ir or Ru are oxidized, forming IrO_2 and RuO_2 surface layers. In alkaline media, mixed transition metal oxide catalysts based on Ni, Fe, Mn, or Co are commonly used as OER electrocatalysts. Under these conditions, the reaction proceeds, most likely, according to the reaction mechanism shown in Scheme 2.4, where each step is a one-electron transfer [117]. Formation of O_2 from two adjacent *O groups, which seems an easier way to clear the reaction sites than the detour via the hydroperoxide, was found to be prevented by a high kinetic reaction barrier [118]. It is, therefore, not part of the mechanism shown in Scheme 2.4. More information about OER catalysts and surface processes during the reaction can be found in Section 4.9.

Frustrating work on difficult problems sometimes results in the development of interesting fallback solutions. In water splitting, oxidizing a reactant other than water (i.e., avoiding the OER) is an example for such an approach. Such replacement reactants are referred to as sacrificial agents. In electrochemical H_2 generation from water, the sacrificial agent should be oxidized at lower potentials than H_2O or OH^- and by a less demanding mechanism (i.e., with lower overpotentials) in order to decrease the overall potential required for the process. The sacrifice is not necessarily useless: by employing alcohols, e.g., polyols accessible from renewable resources, valuable products may be obtained in the oxidation step. However, such a fallback approach, which may be useful as an intermediate stage, can never absolve one from the pain of finding a full solution to the problem. In the present case, the demand for valuable products from sacrificial agents is rather unlikely to scale with the H_2 quantities produced for an energy grid, even if the setup is employed only for smoothing fluctuations in green electricity production.

For assessing the performance of electrocatalysts, it is necessary to find out how they influence the relation between the driving force (the kinetic overpotential $\eta = E - E_{rev}$, in V) and the rate (the current density j, e.g., in mA/cm^2). In the region around the reversible potential, this is described by the Butler-Volmer equation (2.63):

$$j = j_{fwd} - j_{rev} = j_0 \left(\exp\left[\alpha_a\, z\, F\, \eta/RT \right] - \exp\left[-\alpha_c\, z\, F\, \eta/RT \right] \right) \tag{2.63}$$

where j_0 is the exchange current density, and α_a and α_c are charge transfer coefficients at anode and cathode. Often, α_c is assumed to be $\alpha_c = 1 - \alpha_a$. The total current is the sum of anodic and cathodic currents (actually the difference due to differing directions). Around equilibrium, both currents are nearly or totally balanced, while one of them dominates at higher potentials. The Butler-Volmer equation implies that the rate of an electrochemical reaction depends exponentially on the (over)potential, just as the thermal reaction rate depends on temperature (cf. Arrhenius equation (1.1)).

At higher overpotentials (> ±120 mV), where the reactions can be considered irreversible, the relation between overpotential and current density is usually expressed by the Tafel equation (2.64), which exists in a cathodic and an anodic form.

$$\eta = a + b \log j \tag{2.64}$$

Combined with eq. (2.63), cathodic or anodic currents can therefore be expressed as in eq. (2.65) (in the anodic form):

$$\eta = -\left(\frac{2.303\ R\ T}{a_a\ z\ F}\right)\log\ j_0 + \left(\frac{2.303\ R\ T}{a_a\ z\ F}\right)\log\ j \tag{2.65}$$

Equation (2.65) relates the overpotential not only to the observed current density j, but also to the exchange current density j_0. By plotting η vs. $\log j$, the Tafel parameters, a and b, can be extracted. The exchange current density, j_0 is then determined from the intercept as a measure of the catalytic activity. The Tafel slope, b gives information about the number of electrons, z that are transferred in the rds. Its role in electrocatalytic mechanisms is discussed in Section 4.9.

Electrocatalysis improves the efficiency of electrochemical processes. In the effort to improve electrocatalysts, the first aim is to design active sites with higher intrinsic activity, thus decreasing the kinetic barriers for the reaction [119]. As a result, the electron transfer overpotential is lowered, which can differ for the anodic and cathodic reaction, as accounted for by the transfer coefficients. The second goal is to increase the number of these active sites in order to achieve high current densities [119]. Both routes of catalyst improvement ideally increase j_0 of a given reaction. However, this is only valid until diffusion of the reactants through the boundary layer at the electrode starts limiting the rate.

Mass transport and charge transfer are coupled processes, but which of them becomes the rds depends on many variables, such as reactant concentrations, the mode of transport (convection by stirring, electrode rotation, etc.), and the electrochemical conditions. In general, for small deviations from the equilibrium potential, mass transport is fast, so charge transfer is limiting. However, increasing the potential further, reactants are converted and depleted at a rate that mass transfer cannot supply enough reactants at the electrode surface. In this case, mass transport becomes rate-determining. For methods to identify the rds, the reader is referred to textbooks of electrochemistry [120].

2.8 Reactors for surface-catalytic reactions

2.8.1 Ideal and real reactors

At the beginning of Chapter 2, the solid and its surface were compared with a (dynamic) scene for the magnificent spectacle of catalysis. In a more technical way of thinking, the catalytic conversion of reactants needs still another scene to be realized: the catalytic reactor. This section introduces reactors for using surface-catalytic reactions in industry, and environmental protection. The use of reactors for measuring catalytic reaction rates in laboratories will be addressed in Section 3.2.

A reactor is a container in which a chemical reaction proceeds – be it a test tube, an Erlenmeyer flask, a monolith in the exhaust duct of a car, a set of thousands of parallel tubes in a multitubular reactor, or a set of hundreds of pairs of electrodes separated by membranes in an electrolyzer. Reactors are differentiated according to various aspects, among which the mode of operation and the number of coexisting phases are the most fundamental ones.

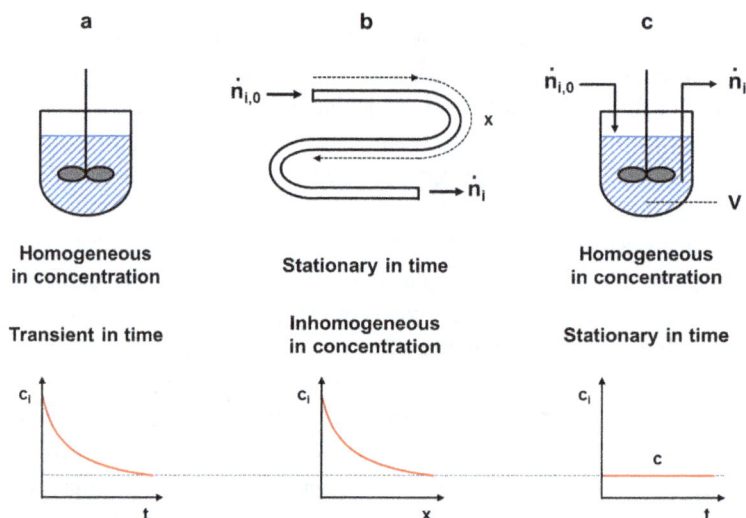

Figure 2.63: Ideal reactor types, differentiated according to the dependence of concentration on time and position in them: (a) well-mixed tank reactor (batch reactor); (b) continuously stirred tank reactor (CSTR); and (c) plug flow reactor (PFR).

Reactors can be operated in continuous and discontinuous modes. Discontinuous ("batch") reactors are filled and kept under specified reaction conditions for a specified time, after which products are recovered. This mode is typical for exploratory laboratory work or small-scale production with liquid-phase reactions. The idealized form of the batch reactor is the well-mixed tank reactor (Figure 2.63a), in which the concentrations are identical and change the same way at all points within the reactor volume. Continuous operation means that reactants are introduced to the reactor and products are withdrawn from it in a way that prevents accumulation or depletion of the reactor content.

In continuous operation, the degree to which a volume element added at a point of time is mixed with elements having entered earlier (backmixing) has a major influence on the output achieved from a reaction under given conditions. The continuously stirred tank reactor (CSTR, Figure 2.63c) and the plug flow reactor (PFR, Figure 2.63b) are idealized reactor types representing the extremes. In the CSTR, all new volume elements are mixed with the reactor content instantly (at least at a rate far above any reaction rate). It is homogeneous, with the same concentration level everywhere, which includes the withdrawn product flow. In the PFR, there is no backmixing, per definition. This can be approximated only in a tubular reactor.

It is important to differentiate batch reactor and CSTR, which are both internally mixed, but actually very different in their influence on conversions and selectivities. There is, rather, a real analogy between batch reactor and PFR: in both, reactant and product concentrations vary over a coordinate, which is the reaction time in the batch reactor, and the length in the PFR (Figure 2.63). In the CSTR, concentrations are always at the final level, i.e., low for the reactant(s). In reactors achieving the same conversion of the reactant, its concentration in the CSTR is lower than at any point in the PFR, except for the outlet. As reactants usually have a positive reaction order in rate laws, rates in the CSTR are always lower than in the PFR (or the batch reactor).

On the other hand, the homogeneous nature of the CSTR facilitates the determination of reaction rates: for a set of conditions, there is only one rate in the CSTR, while rates change over time in the batch reactor and over length in the PFR. For balance reasons, the difference between the flows of component i entering and leaving the CSTR, $\dot{n}_{0,i}$ and $\dot{n}_i$, respectively, is caused by its conversion in the reaction. Using definitions of the reaction rate according to (1.2) or (1.5), this yields:

$$\dot{n}_{0,i} - \dot{n}_i = v_i V r \tag{2.66a}$$

for a homogeneous reaction, and

$$\dot{n}_{0,i} - \dot{n}_i = v_i \, m_{cat} \, r \tag{2.66b}$$

for a catalytic reaction.

With respect to the effect of flow patterns on reaction kinetics, it is also important to note that backmixing favors the conversion of intermediate products in consecutive reactions and, likewise, reactions of lower kinetic order in competitive schemes. The rationale behind this is that the CSTR permanently operates at the target reactant conversion. In a consecutive reaction, this mixture also contains the intermediate product(s) formed at this conversion degree: they are available for the next reaction all the time, unlike in plug flow kinetics, where they are formed and disappear along the reactor tube. As reactant concentrations are always low in the CSTR, reactions of high kinetic reaction order are penalized most seriously. For more details, the reader is referred to textbooks of reaction engineering [105–110].

Regarding the differentiation between reactors according to phase inventory, devices for monophasic mixtures such as the stirred tank reactor and the tubular reactor seem to be beyond the scope of this book. Filled with a slurry (tank reactor) or with a catalyst bed (tubular reactor), they reappear, however, among the multiphase reactors to be dealt with next. In reactors containing solid phases, the crucial topics are the distribution of phases in space and the insertion or withdrawal of heat. Insertion of photons is the key aspect in photocatalytic reactors and insertion and withdrawal of electrons via electrodes in electrochemical devices.

2.8.2 Reactors for thermocatalytic processes

To achieve high productivity of the catalytic process, the exposed catalytic surface area per m^3 of reactor space should be maximized. For fast reactions, mass and/or heat transfer limitations should be suppressed. This calls for high relative velocities between fluid and solid phases.

The technical solution depends strongly on the situation to be handled and on reaction rates provided by the available catalysts. If these are relatively low, the reactor space may be densely packed with catalyst granules, which can be fully utilized because of the absence of transport limitations. The dimensions of such fixed beds (or packed beds) depend on the requirements of heat management (see below). With increasing catalytic reaction rate, the reaction zone in the catalyst particles withdraws to the external surface (cf. 2.6.2) and film diffusion limitations may come into play. Decrease of pellet sizes and increase of linear gas velocities to fix these problems may cause extensive pressure drops. In such situations, reactors operating with fine particles or thin layers like fluid beds or monoliths, which offer low flow resistances may be preferable. For avoiding undesired consecutive steps in some extremely fast reactions, special ultrashort contact time assemblies are available.

In catalytic reactors, the fluid phase may pass through a catalyst bed, which is at rest or moves slowly by gravity; it may stir a bed of fine particles into motion; or it may just pass over a bed of fine particles. The latter option exists almost exclusively for gas-phase processes. Figure 2.64 shows the most important reactors of this type. Rotary kilns (Figure 2.64a) are hardly ever used for catalytic reactions, but they are a good choice for thermal treatments during catalyst synthesis as an alternative to flowing hot air above powders in pans or crucibles (cf. Section 3.1.4).

In monoliths (Figure 2.64b), the gas phase passes through narrow parallel channels, the walls of which are coated with a catalyst layer of a few μm thickness (the washcoat, cf. Section 3.1.2). Pressure drop in the channels is small. Walls can be made of ceramics (cordierite) or thin metal foils (Figure 2.65a). There are also cases, where the whole monolith body is made of a catalyst material cheap enough to tolerate that it is utilized only to a small extent.

In washcoats, pore diffusion problems are negligible due to the short diffusion lengths. On the other hand, the small channel dimensions prevent a transition of the gas flow into the turbulent regime. In the laminar flow pattern along the channels, mass transfer towards the washcoat proceeds almost exclusively by diffusion, which can significantly affect net reaction rates (Section 2.6.2). There is quite some research effort to replace the parallel-channel scheme of monoliths by more complicated structures that allow introducing turbulence into the flow pattern, e.g., ceramic foams or metal sheet packings.

Figure 2.65b shows a wall-flow monolith, in which channels are blocked from the exit or entrance sides alternatingly. Gas entering from one side is forced to cross over to a neighboring channel through the porous channel wall. Such monoliths are used to filter soot particles out of diesel exhaust, and their walls are usually furnished with

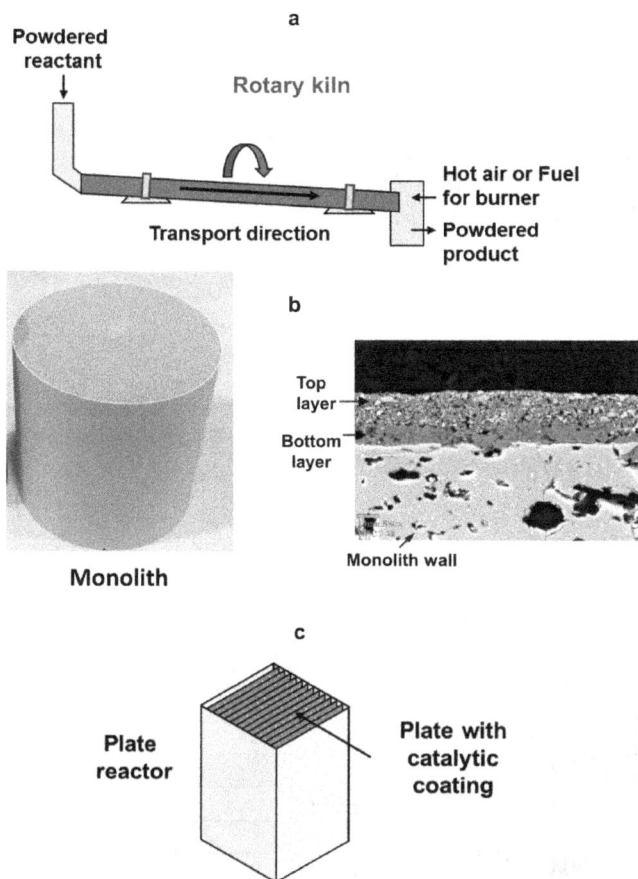

Figure 2.64: Reactors with gas-phase passing over catalyst layers: (a) rotary kiln, schematic; (b) catalytic monolith, left: overview (courtesy Dr. V. S. Narkhede, Süd-Chemie India), right: scanning electron micrograph of channel wall with double-layer washcoat (reproduced from ref. [121] with permission from Elsevier); and (c) plate reactor, schematic.

catalytic materials that help burning the accumulated soot in regeneration phases (cf. Appendix A1).

Plate catalysts (Figure 2.64c) have competed with monoliths in the abatement of nitrogen oxides from power plant flue gases for decades. More recently, the plate geometry has been adapted in microreactor systems that allow handling highly exothermal reactions, because the slits can be charged with reactants and a coolant in an alternating sequence. Both plate assemblies and monoliths are also being examined for use in liquid-phase or even triple-phase catalytic reactions.

Figure 2.66 presents three versions of fixed beds in gas-phase reactions. Only the simple container shown in Figure 2.66a is actually referred to as fixed–bed reactor. Due to the large distance between its center and the walls, where heat could be exchanged, it is a quasi-adiabatic reactor. Fixed-bed reactors are used for reactions with low reac-

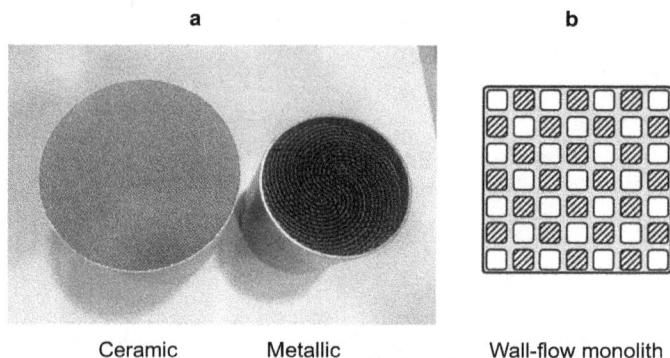

Figure 2.65: Different types of monoliths: (a) ceramic and metal monolith (courtesy Dr. V. S. Narkhede, Süd-Chemie India). (b) Wall-flow monolith with channels blocked at the exit or the entrance alternatingly. Channel walls are porous: the wall-flow monolith is a filter device.

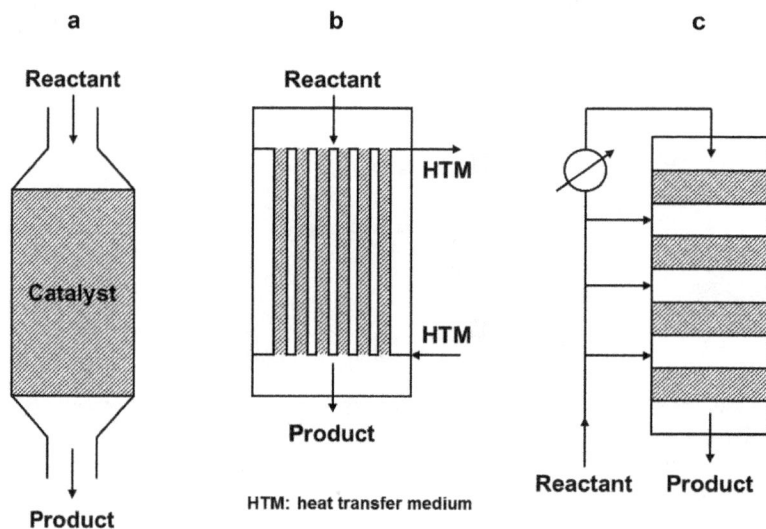

Figure 2.66: Reactors with packed beds: (a) fixed-bed reactor; (b) multitubular reactor; and (c) staged reactor, with intermediate quenching.

tion enthalpies, e.g., isomerizations, or for conversions of very dilute streams. Removal of S-containing contaminants from light oil fractions by treatment with H_2 (hydrodesulfurization, HDS) is an example for such application, because the feed contains the contaminants only in small concentrations. There are also cases where reactions are performed under adiabatic conditions, and heat management is postponed to subsequent equipment.

Opposed to this, multitubular reactors (Figure 2.66b) are designed for maximum heat transfer performance in highly exothermal or endothermal reactions. They are applied, in particular, if there is a risk of thermal reactor instability (runaway[29]) or if reactants or products decompose at high temperatures. They may be also used to achieve very high reaction temperatures. To maximize the heat transfer surface area per catalyst volume, the catalyst is distributed over many parallel tubes of a few cm width (down to 2 cm), which are cooled by pressurized water or molten salts or fired with burners in the space in between the tubes. Multitubular reactors may contain up to 30,000 tubes. They offer tight control on the temperature, in particular with exothermal reactions, but also, on the residence time by suppressing backmixing in the gas phase. This is most welcome in selective hydrocarbon oxidations, where the target products are intermediates vulnerable towards overoxidation, and justifies the considerable investments.

The staged reactor (Figure 2.66c) combines adiabatic catalyst layers with zones in which reaction heat is recovered or provided. Several options are available for such heat management, which are only briefly mentioned here. Figure 2.66c highlights the quenching technology, in which cold reactant streams added between the layers make up for temperature increases due to released reaction heat. This approach has long been applied in reactors for the synthesis of methanol or of ammonia. Alternatively, reaction heat may be recovered by heat exchangers. Placing them outside the reactor as, for instance, in plants for SO_2 oxidation, is a traditional and cheap option for this technology. It is, however, not applicable for plants operating at high pressures, where heat exchangers are often placed between catalyst layers, nowadays, and catalyst layers may be arranged radially instead of axially, as shown in Figure 2.66c.

Strongly endothermal reactions as naphtha reforming (cf. Appendix A1) can be performed in a series of reactors reminiscent of the staged-reactor scheme. Each reactor is preceded by a furnace that overheats the feed, thus providing the reaction heat required in it. Older naphtha reforming plants used four fixed beds in series, the last of which was periodically regenerated and then reinserted at the front end. In more recent plants, the reactors are placed one on top of the other, and the catalysts moves down the tower gradually. The gas flow has, however, remained the same: it is overheated before entering the reactor and is directed to reheating furnaces after each section, except for the last one.

$$4\,NH_3 + 5\,O_2 \rightarrow 4\,NO + 6\,H_2O \tag{2.67}$$

$$2\,CH_4 + 2\,NH_3 + 3\,O_2 \rightarrow 2\,HCN + 6\,H_2O \tag{2.68}$$

$$2\,CH_3OH + O_2 \rightarrow 2\,CH_2O + 2\,H_2O \tag{2.69}$$

---

[29] Runaway occurs when the reaction heat of a strongly exothermal reaction cannot be withdrawn at the rate it is released. For details, see textbooks of reaction engineering.

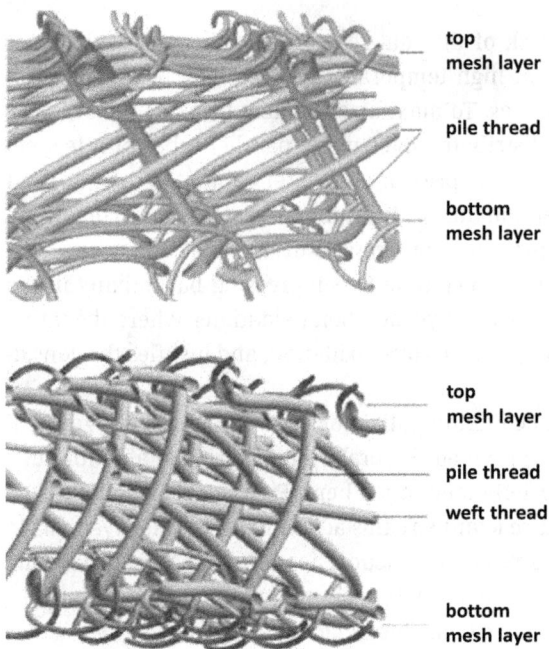

Figure 2.67: Structure of knitted catalyst gauzes (Multinit® gauzes); reproduced from ref. [122] with permission from Elsevier.

Ultrashort contact time catalysis is an option to save valuable products of oxidation processes from total oxidation (cf. Section 4.3.3). Oxidations are performed at very high temperatures that allow achieving high or even complete conversion at contact times of milliseconds, and products are quenched subsequently. The morphology of these metal catalysts depends on their mechanical properties. Pt and Pt-Rh alloys can be drawn into very fine wires of 50–100 µm diameter, which can be knitted into gauzes (Figure 2.67). Pt-Rh gauzes (with ca. 1,000 meshes /cm^2) are employed in the oxidation of NH_3 to NO in the Ostwald process for HNO_3 production (reaction (2.67)) and in the oxidative conversion of methane and ammonia to HCN (Andrussov process, reaction (2.68)). A similar process is available for the oxidation of methanol to formaldehyde (reaction (2.69)), but the silver catalyst employed cannot be processed into gauzes. It is exposed as a thin layer of silver granules instead. Millisecond contact times can also be realized with short monoliths coated with the catalytic metal.

There are two fundamentally different versions of catalysts in motion: the moving bed and fluidized beds in different subtypes. The moving bed is always part of a catalyst cycle tuned to known rates of catalyst deactivation. The deactivated catalyst is recovered at the bottom of the reactor (or of a tower of piled reactor sections, see above), after which it is regenerated and fed again at the top.

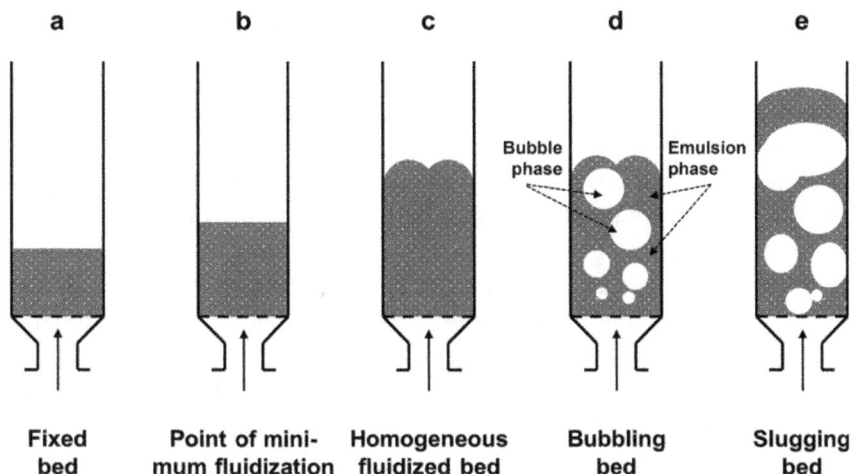

Figure 2.68: From the fixed bed to fluidized beds in different versions.

The fluidized bed (Figure 2.68) is a phenomenon that can be observed when beds consisting of fine particles of suitable materials (e.g., sand) are exposed to a gas flowing from below at gradually increasing velocities (Figure 2.68a). At a critical velocity, the bed expands somewhat and the particles start moving like boiling water (Figure 2.68b): due to its flow resistance, the bed is lifted by the gas, slightly expanding at the same time. The expansion decreases the flow resistance of the bed: it falls back in the initial state, and the process starts again. As a result of this instability, the particle bed behaves like a liquid, it is fluidized. It forms a meniscus parallel to the earth's surface. Objects made of light materials swim on it, while heavy objects sink. As there is a permanent mixing of the particles, the bed exhibits excellent heat transfer properties.

The flow pattern of the gas phase is complicated and depends on the state of the bed. The bed may be completely homogeneous (Figure 2.68c). However, in many cases, gas bubbles containing no solid particles are formed within the homogeneous regions, in which particles are suspended in the flowing gas (bubbling bed, with bubble phase and emulsion phase, cf. Figure 2.68d). While a more detailed treatment of such states is beyond the scope of this book, the statement that the degree of gas-phase backmixing in fluidized beds is significant, but far below complete will suffice for the discussions in it. For more detail, the reader is referred to ref. [123]. Despite nonzero backmixing, fluidized beds are also employed for large-scale selective hydrocarbon oxidation processes, e.g., for the synthesis of acrylonitrile.

All fluidized beds in Figure 2.68 are stationary. Even in the slugging fluid bed, where bubble diameters grow up to the reactor diameter (Figure 2.68e), only fines are blown out because the particles trickle down in a thin layer along the reactor wall. At even higher gas velocity, it is not only the porous bed, which is lifted by its flow resistance, but also the individual particles: they are blown out. This is the principle of the riser, into which streams of both reactant gas and catalyst particles are fed continuously. After a joint vertical flight and a turn by 90°, they are separated by a drastic increase of the flow cross section followed by a series of cyclones. In the FCC process, the riser is driven by mixing the feed at its boiling point with a stream of strongly overheated catalyst particles coming from the regenerator, which results in explosive feed evaporation. The regenerator is a stationary fluid bed, where the coke is burnt off to recover activity and overheat the catalyst.

For catalysts in moving beds, attrition resistance adds to the specifications on mechanical stability requested from heterogeneous catalysts. Conditions are most stressful in stationary fluid beds and in the riser. This holds not only for the catalyst but also for the equipment, e.g., immersed cooling coils or the 90° turn at the top of a riser. They are exposed to a sandblasting regime over months, which puts narrow limits on the choice of suitable materials. Moreover, not all powders are suited for fluidization. The Geldart classification reports which materials can be fluidized when finely divided and which type of fluidized bed results [124]. Despite these limitations, fluidized beds play an enormous role in catalytic technology.

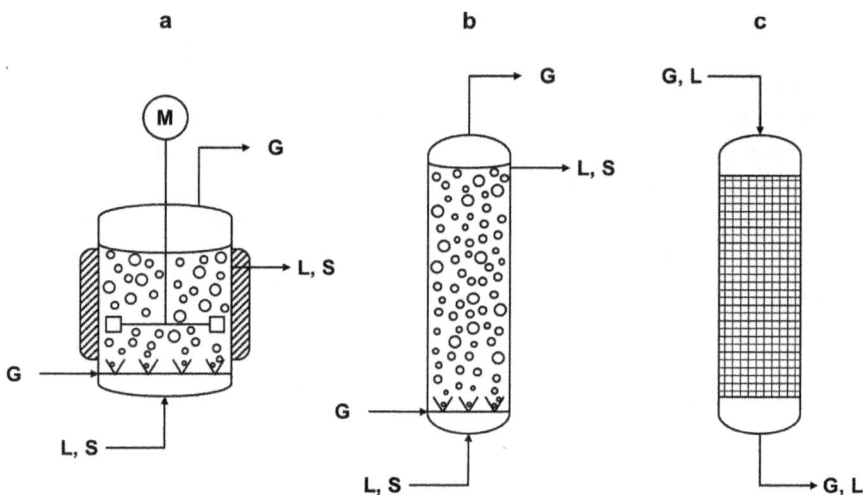

Figure 2.69: Reactors for catalysis in the liquid phase and for three-phase catalysis: (a) (continuously) stirred tank reactor; (b) bubble column; and (c) trickle bed reactor.

In solid–liquid systems, driving forces for phase separation are much weaker than in solid–gas systems. Therefore, the choice of reactor options for heterogeneous catalysis in the liquid phase (or for three-phase catalysis) is more limited. The most important reactor types are shown in Figure 2.69. Tank reactor and bubble column (Figure 2.69a and b) are used for processing liquids in catalyst slurries. In the bubble column, the gas is used for agitating the remaining phases. It is, therefore, suitable only for three-phase reactions, e.g., hydrogenations with H_2. The gas is introduced by distributors, which may be static like perforated or porous plates or dynamic (nozzles). The flow pattern in the column can be easily modified. When, for instance, the upward flow driven by the bubbles is confined to the center by an open tubular installation, the liquid flows down along the walls, and a jet loop reactor results. The bubble column shown in Figure 2.69b can be also converted into a column cascade by inserting additional perforated plates, where the gas bubbles, which grow by coalescence on their way up, are collected and redistributed.

The tank reactor (Figure 2.69a) is also eligible for reactions without a gaseous component. When a gas is involved, it can be fed by a gas distributor. However, there are also stirrers with hollow shafts and apertures or nozzles for gas distribution.

In tank and bubble column reactors, there is practically no direct contact between the gas and the catalyst surface: the gaseous reactant participates in the reaction via the liquid. This puts serious limitations on reaction rates in systems with low solubility of the gaseous reactant. When, however, a trickle bed (Figure 2.69c) is properly run, the solid catalyst is not completely wetted by the liquid reactant. This leaves space for gas–solid interactions and, most significantly, for contact lines between all three coexisting phases, where active sites can be accessed by reactants from both the liquid and the gas.

The trickle bed is a fixed bed fed with liquid from the top and with gas from the top (Figure 2.69c) or from the bottom. No energy is required for mixing, and phase separation is trivial. On the other hand, the theoretical description of trickle beds is very intricate. The flow patterns of the liquid phase, which may form films, droplets, and rivulets, are complicated and influenced by a large number of relevant variables like liquid/gas velocities, flow ratios between phases, and liquid holdup. The reactor performance may be strongly affected by maldistribution of the liquid phase over the bed cross section, up to channeling. All this is difficult to capture in models, which are, therefore, not yet on a level of maturity that would allow scale-up from small-size immediately to commercial-scale equipment, as is nowadays performed with some other reactor types. Despite this problem, the opportunity to obtain triple-phase contact and a relatively simple technical realization have resulted in trickle-bed applications on very large scales, e.g., in the hydrotreatment (e.g., HDS, hydrodenitrogenation (HDN)) of heavy oil fractions.

2.8.3 Reactors for photocatalytic processes

Photocatalytic reactions are mostly performed in the liquid phase using slurry reactors with cylindrical light sources such as Xe or Hg lamps, usually in batch or semi-batch operation (Figure 2.70a), and sometimes in flow reactors (Figure 2.70b). As an alternative, immobilized photocatalyst films may be applied, which are also suitable for continuous operation. In liquids, the use of porous photocatalyst films may be, however, complicated by diffusion limitations (cf. Section 2.6.2). The solubility of gaseous reac-

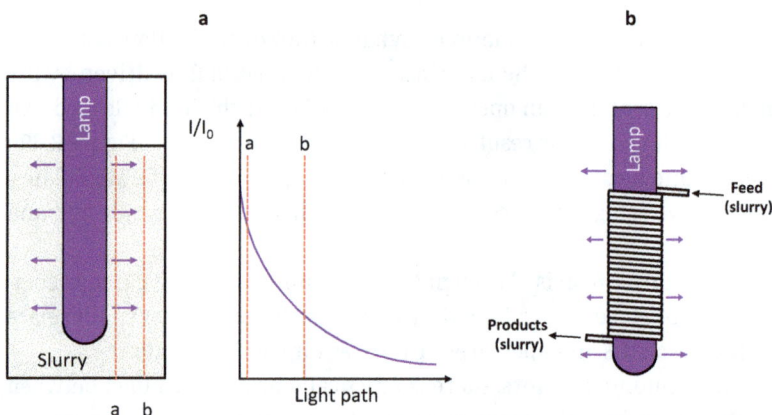

Figure 2.70: Effect of photoreactor geometry on the light path and the resulting light attenuation in the reactor: (a) batch reactor; (b) flow reactor.

tants and their transport to the catalyst surface may also be limiting factors, because photoreactors are usually not constructed for high pressure. An overview of reactor types for continuous and batch operation can be found in refs. [125, 126].

The necessity to shine light into the reactor has several implications. Photon absorption naturally limits the penetration depth of photons in the liquid reactant, but in concentrated slurries, the catalyst itself aggravates the problem by shadowing. In real photoreactors, nonuniform light distribution may be a major constraint, leading to locally perturbed reaction rates that affect overall conversions. Light attenuation in the reactor can create dark zones, which may favor the back reaction of reversible reactions. These problems may be minimized by shortening the light path in the reactor, e.g., by using microreactors, which is, however, payed for by a loss of radiative energy, resulting in decreased quantum yields.

Designing photoreactors is a compromise between efficiency and productivity. Productivity increases with increasing amount of absorbed photons, since the rate of the initial electron-hole formation depends on the local volumetric rate of photon absorption. However, using high light intensities leads to increased losses of radiative energy, which lowers the efficiency. The efficiency of the overall process also comprises the efficiency of external light sources. Traditional UV light sources such as Hg lamps have low internal quantum yields due to heat generation. The emerging LED technology is a promising alternative, as the efficiency of LEDs is growing steadily, whereas their price continues to decrease.

For scaling up photocatalytic processes, the numbering up of individual reactor units is often mentioned as the best choice, because the described characteristics of photon transport render a simple enlargement of photoreactors difficult.

2.8.4 Reactors for electrochemical conversions

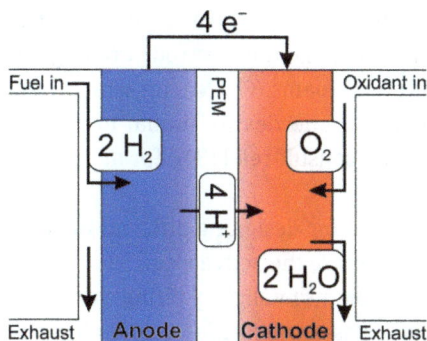

Figure 2.71: Schematics of a proton-exchange membrane fuel cell (PEMFC); not to scale, thickness of layers are in the μm range.

Electrochemical methods directly convert electrical energy into chemical energy, or vice versa. Major applications are electrolysis such as the chlor-alkali electrolysis, and fuel cells, e.g., proton-exchange membrane fuel cells (PEMFC).[30] The principle of a PEMFC is briefly depicted in Figure 2.71. At the anode, the fuel (H_2; in other devices, methanol) is oxidized producing protons and electrons. The electrons generate an electric current through the external circuit and reduce O_2 in the cathode reaction, which results in water formation. The protons needed for water formation have to diffuse from the anode to the cathode through the proton-exchange membrane (PEM), which should be as thin as possible to lower its resistance. Very thin membranes raise problems of mechanical stability and pinhole formation, which results in crossover of H_2 to the cathode. In a typical membrane-electrode assembly (MEA) at the heart of a PEM fuel cell, thicknesses of electrodes and of the membrane are in the μm, rather than in the mm range. The process can be reverted by applying a suitable potential at the electrodes, which converts the PEMFC into a PEM water electrolyzer.

There are also other types of fuel cells, which are described in refs [127, 128].

The cell voltage, E_{cell}, is a crucial factor determining the efficiency of electrochemical cells. According to eq. (2.62), it deviates from the reversible value, $\sum_{An,Cath} E_{rev}$ predicted by thermodynamics; it is higher in the case of electrolysis and lower in case of fuel cells. The deviations are due to overpotentials $\sum_{An,Cath} \eta$ at both cathode and anode arising from kinetic limitations that call for better electrocatalysts, from mass transfer limitations, and from ohmic resistances. Mass transfer can be accelerated by creating turbulent flow patterns in the electrolyte adjacent to the electrode, the ohmic resistances, $\sum R$, can be lowered by using short distances between the electrodes, by large electrode surfaces, and by increasing the conductivity of the electrolyte, which can be achieved by adding salts and working at elevated temperature.

---

30 Electrochemical cells for research work are discussed in Section 3.2.3.

Three-dimensional electrodes, which are familiar from batteries, can also improve the performance of electrolyzers by increasing the activity per geometrical surface. Due to the critical role of transport limitations, optimization of porosity and a hydrodynamic regime resulting in forced convective flow through the electrode are of great importance. 3d-electrodes are typically made of metal foams (Co, Ni), on which catalytic materials may be deposited. Their hydrodynamic properties are similar to those of fixed beds [129]. For a recent review, the reader may consult ref. [130], for an impression of ongoing research – ref. [131].

In electrolyzers, gas evolution is a great challenge for designing efficient devices, because gas bubbles in contact with the electrode decrease the active electrode area. In common devices, electrodes are vertical, but improvements by slightly tilting electrodes are discussed in literature. Reactions that consume gases proceed almost exclusively at the three-phase boundary between gas, electrolyte, and catalyst/electrode. Gas diffusion electrodes hosting both electrolyte and gas in their pore system allow establishing and maximizing these phase contact areas, and hence the catalytic efficiency. They require precise pressure control to avoid flooding.

The previous discussions relate to design and operation of individual electrochemical reactors (cells), which provide very limited production capacities. Upscaling to industrially relevant capacities is achieved by numbering up individual cells to large stacks rather than by enlarging the cells.

References

[1]   Wang, W. H.; Dong, C.; Shek, C. H., Bulk metallic glasses. *Materials Science & Engineering R-Reports* **2004**, *44* (2–3), 45–89. doi 10.1016/j.mser.2004.03.001

[2]   Hoffmann, R., *Solids and Surfaces: A Chemist's View on Bonding in Extended Structures*. Verlag Chemie: Weinheim, 1988. ISBN: 9780471187103

[3]   Hoffmann, R., A chemical and theoretical way to look at bonding on surfaces. *Reviews of Modern Physics* **1988**, *60* (3), 601–628. doi 10.1103/RevModPhys.60.601

[4]   Niemantsverdriet, J. W., *Spectroscopy in Catalysis*. 1st ed.; VCH: Weinheim, 1995. ISBN: 9783527287260

[5]   Marks, L. D.; Smith, D. J., Atomic imaging of particle surfaces. *ACS Symposium Series* **1985**, *288*, 341–350. doi.org/10.1016/S0081-1947(08)60135-6

[6]   Barroo, C.; Lambeets, S. V.; Devred, F.; Chau, T. D.; Kruse, N.; De Decker, Y.; de Bocarme, T. V., Hydrogenation of NO and NO_2 over palladium and platinum nanocrystallites: Case studies using field emission techniques. *New Journal of Chemistry* **2014**, *38* (5), 2090–2097. doi 10.1039/c3nj01505j

[7]   Tasker, P. W., Stability of ionic-crystal surfaces. *Journal of Physics C-Solid State Physics* **1979**, *12* (22), 4977–4984. doi 10.1088/0022-3719/12/22/036

[8]   Wöll, C., The chemistry and physics of zinc oxide surfaces. *Progress in Surface Science* **2007**, *82* (2), 55–120. doi.org/10.1016/j.progsurf.2006.12.002

[9]   Prodhomme, P.-Y.; Raybaud, P.; Toulhoat, H., Free-energy profiles along reduction pathways of MoS_2 M-edge and S-edge by dihydrogen: A first-principles study. *Journal of Catalysis* **2011**, *280* (2), 178–195. doi 10.1016/j.jcat.2011.03.017

[10]   Kämper, A.; Hahndorf, I.; Baerns, M., A molecular mechanics study of the adsorption of ethane and propane on $V_2O_5(001)$ surfaces with oxygen vacancies. *Topics in Catalysis* **2000**, 11–12, 77–84. https://doi.org/10.1023/A:1027239612464

[11]   Volta, J. C.; Forissier, M.; Theobald, F.; Pham, T. P., Dependence of selectivity on surface-structure of MoO_3 Catalysts. *Faraday Discuss* **1981**, *72*, 225–233. doi 10.1039/dc9817200225.

[12]   Abon, M.; Massardier, J.; Mingot, B.; Volta, J. C.; Floquet, N.; Bertrand, O., New unsupported [100]-oriented MoO_3 catalysts II. Catalytic properties in propylene oxidation. *Journal of Catalysis* **1992**, *134* (2), 542–548. doi.org/10.1016/0021-9517(92)90341-E

[13]   Merzlikin, S. V.; Tolkachev, N. N.; Briand, L. E.; Strunskus, T.; Wöll, C.; Wachs, I. E.; Grünert, W., Anomalous surface compositions of stoichiometric mixed oxide compounds. *Angewandte Chemie International Edition* **2010**, *49* (43), 8037–8041. doi 10.1002/anie.201001804

[14]   Ratnasamy, P.; Knözinger, H., Catalytic aluminas: Surface models and characterization of surface sites. *Catalysis Reviews* **1978**, *17* (1), 31–70. doi 10.1080/03602457808080878

[15]   Lauritsen, J. V.; Bollinger, M. V.; Laegsgaard, E.; Jacobsen, K. W.; Nørskov, J. K.; Clausen, B. S.; Topsøe, H.; Besenbacher, F., Atomic-scale insight into structure and morphology changes of MoS_2 nanoclusters in hydrotreating catalysts. *Journal of Catalysis* **2004**, *221* (2), 510–522. doi 10.1016/j.jcat.2003.09.015

[16]   Kelty, S. P.; Berhault, G.; Chianelli, R. R., The role of carbon in catalytically stabilized transition metal sulfides. *Applied* Catalysis A: General **2007**, *322*, 9–15. doi 10.1016/j.apcata.2007.01.017

[17]   Xia, W.; Jin, C.; Kundu, S.; Muhler, M., A highly efficient gas-phase route for the oxygen functionalization of carbon nanotubes based on nitric acid vapor. *Carbon* **2009**, *47* (3), 919–922. doi 10.1016/j.carbon.2008.12.026

[18]   Stein, A.; Wang, Z.; Fierke, M. A., Functionalization of porous carbon materials with designed pore architecture. *Advanced Materials* **2009**, *21* (3), 265–293. doi 10.1002/adma.200801492

[19]   Kundu, S.; Xia, W.; Busser, W.; Becker, M.; Schmidt, D. A.; Havenith, M.; Muhler, M., The formation of nitrogen-containing functional groups on carbon nanotube surfaces: A quantitative XPS and TPD study. *Physical Chemistry Chemical Physics* **2010**, *12* (17), 4351–4359. doi 10.1039/b923651a

[20]   Enferadi-Kerenkan, A.; Do, T.-O.; Kaliaguine, S., Heterogeneous catalysis by tungsten-based heteropoly compounds. *Catalysis Science & Technology* **2018**, *8* (9), 2257–2284. doi 10.1039/C8CY00281A

[21]   Baerlocher, C.; McCusker, L. B. Database of Zeolite Structures. http://www.iza-structure.org/databases/.

[22]   Möller, K.; Bein, T., Mesoporosity – A new dimension for zeolites. *Chemical Society Reviews* **2013**, *42* (9), 3689–3707. doi 10.1039/C8CY00281A

[23]   Zhuravlev, L. T., The surface chemistry of amorphous silica. Zhuravlev model. *Colloids and Surfaces A – Physicochemical and Engineering Aspects* **2000**, *173* (1–3), 1–38. doi 10.1016/s0927-7757(00)00556-2

[24]   Mueller, R.; Kammler, H. K.; Wegner, K.; Pratsinis, S. E., OH surface density of SiO_2 and TiO_2 by thermogravimetric analysis. *Langmuir* **2003**, *19* (1), 160–165. doi 10.1021/la025785w

[25]   Sun, Z. M.; Yang, X. P.; Zhang, G. X.; Zheng, S. L.; Frost, R. L., A novel method for purification of low grade diatomite powders in centrifugal fields. *International Journal of Mineral Processing* **2013**, *125*, 18–26. doi 10.1016/j.minpro.2013.09.005

[26]   Digne, M.; Sautet, P.; Raybaud, P.; Toulhoat, H.; Artacho, E., Structure and stability of aluminum hydroxides: A theoretical study. *The Journal of Physical Chemistry B* **2002**, *106* (20), 5155–5162. doi.org/10.1006/jcat.2002.3741

[27]   Krokidis, X.; Raybaud, P.; Gobichon, A.-E.; Rebours, B.; Euzen, P.; Toulhoat, H., Theoretical study of the dehydration process of boehmite to γ-alumina. *The Journal of Physical Chemistry B* **2001**, *105* (22), 5121–5130. doi.org/10.1006/jcat.2002.3741

[28]   van den Reijen, J. E.; Versluis, W. C.; Kanungo, S.; d'Angelo, M. F.; de Jong, K. P.; de Jongh, P. E., From qualitative to quantitative understanding of support effects on the selectivity in silver catalyzed ethylene epoxidation. *Catalysis Today* **2019**, *338*, 31–39. doi.org/10.1016/j.cattod.2019.04.049

[29] Amrute, A. P.; Łodziana, Z.; Schreyer, H.; Weidenthaler, C.; Schüth, F., High-surface-area corundum by mechanochemically induced phase transformation of boehmite. Science **2019**, *366* (6464), 485–489. doi 10.1126/science.aaw9377

[30] Baerlocher, C.; McCusker, L. B.; Olson, D. H., *Atlas of Zeolite Structure Types*. Elsevier: Amsterdam, 2007. ISBN: 9780444530646

[31] Jo, D.; Ryu, T.; Park, G. T.; Kim, P. S.; Kim, C. H.; Nam, I.-S.; Hong, S. B., Synthesis of High-Silica LTA and UFI Zeolites and NH_3–SCR Performance of Their Copper-Exchanged Form. *ACS Catalysis* **2016**, *6* (4), 2443–2447. doi 10.1021/acscatal.6b00489

[32] Furukawa, H.; Cordova, K. E.; O'Keeffe, M.; Yaghi, O. M., The chemistry and applications of metal-organic frameworks. *Science* **2013**, *341* (6149), 1230444. doi 10.1126/science.1230444

[33] Ryu, U.; Jee, S.; Rao, P. C.; Shin, J.; Ko, C.; Yoon, M.; Park, K. S.; Choi, K. M., Recent advances in process engineering and upcoming applications of metal–organic frameworks. *Coordination Chemistry Reviews* **2021**, *426*, 213544. doi.org/10.1016/j.ccr.2020.213544

[34] Dhakshinamoorthy, A.; Li, Z.; Garcia, H., Catalysis and photocatalysis by metal organic frameworks. *Chemical Society Reviews* **2018**, *47* (22), 8134–8172. doi 10.1039/C8CS00256H

[35] Freund, R.; Canossa, S.; Cohen, S. M.; Yan, W.; Deng, H.; Guillerm, V.; Eddaoudi, M.; Madden, D. G.; Fairen-Jimenez, D.; Lyu, H.; Macreadie, L. K.; Ji, Z.; Zhang, Y.; Wang, B.; Haase, F.; Wöll, C.; Zaremba, O.; Andreo, J.; Wuttke, S.; Diercks, C. S., 25 Years of Reticular Chemistry. *Angewandte Chemie International Edition* **2021**, *60* (45), 23946–23974. doi.org/10.1002/anie.202101644

[36] Eddaoudi, M.; Kim, J.; Rosi, N.; Vodak, D.; Wachter, J.; Keeffe, M.; Yaghi, O. M., Systematic design of pore size and functionality in isoreticular MOFs and their application in methane storage. *Science* **2002**, *295* (5554), 469. doi 10.1126/science.10672

[37] *Mercury* 1.0; https://www.ccdc.cam.ac.uk/solutions/csd-core/components/mercury/ (accessed Dec 4, 2022): Cambridge, 2022.

[38] Hanaor, D. A. H.; Sorrell, C. C., Review of the anatase to rutile phase transformation. *Journal of Materials Science* **2011**, *46* (4), 855–874. doi 10.1126/science.10672

[39] Zhang, H.; Banfield, J. F., Thermodynamic analysis of phase stability of nanocrystalline titania. *Journal of Materials Chemistry* **1998**, *8* (9), 2073–2076. doi 10.1126/science.10672

[40] Lebedev, V. A.; Kozlov, D. A.; Kolesnik, I. V.; Poluboyarinov, A. S.; Becerikli, A. E.; Grünert, W.; Garshev, A. V., The amorphous phase in titania and its influence on photocatalytic properties. *Applied Catalysis B: Environmental* **2016**, *195*, 39–47. doi.org/10.1016/j.apcatb.2016.05.010

[41] Ohtani, B.; Prieto-Mahaney, O. O.; Li, D.; Abe, R., What is Degussa (Evonik) P25? Crystalline composition analysis, reconstruction from isolated pure particles and photocatalytic activity test. *Journal of Photochemistry and Photobiology A – Chemistry* **2010**, *216* (2–3), 179–182. doi 10.1016/j.jphotochem.2010.07.024

[42] Kompio, P. G. W. A.; Brückner, A.; Hipler, F.; Auer, G.; Löffler, E.; Grünert, W., A new view on the relations between tungsten and vanadium in V_2O_5-WO_3/TiO_2 catalysts for the selective reduction of NO with NH_3. *Journal of Catalysis* **2012**, *286*, 237–247. doi 10.1016/j.jcat.2011.11.008

[43] Scotti, N.; Bossola, F.; Zaccheria, F.; Ravasio, N., Copper–Zirconia catalysts: Powerful multifunctional catalytic tools to approach sustainable processes. *Catalysts* **2020**, *10* (2), 168. doi 10.3390/catal10020168

[44] Schittkowski, J.; Tölle, K.; Anke, S.; Stürmer, S.; Muhler, M., On the bifunctional nature of Cu/ZrO_2 catalysts applied in the hydrogenation of ethyl acetate. *Journal of Catalysis* **2017**, *352*, 120–129. doi 10.3390/catal10020168

[45] Kouva, S.; Honkala, K.; Lefferts, L.; Kanervo, J., Review: Monoclinic zirconia, its surface sites and their interaction with carbon monoxide. *Catalysis Science & Technology* **2015**, *5* (7), 3473–3490. doi 10.1039/C5CY00330J

[46]   Li, W.; Huang, H.; Li, H.; Zhang, W.; Liu, H., Facile synthesis of pure monoclinic and tetragonal zirconia nanoparticles and their phase effects on the behavior of supported molybdena catalysts for methanol-selective oxidation. *Langmuir* **2008**, *24* (15), 8358–8366. doi 10.1021/la800370r

[47]   Krogstad, J. A.; Lepple, M.; Gao, Y.; Lipkin, D. M.; Levi, C. G., Effect of yttria content on the zirconia unit cell parameters. *Journal of the American Ceramic Society* **2011**, *94* (12), 4548–4555. doi.org/10.1111/j.1551-2916.2011.04862.x

[48]   Yan, G. X.; Wang, A.; Wachs, I. E.; Baltrusaitis, J., Critical review on the active site structure of sulfated zirconia catalysts and prospects in fuel production. *Applied Catalysis A: General* **2019**, *572*, 210–225. doi.org/10.1016/j.apcata.2018.12.012

[49]   Rhodes, M. D.; Bell, A. T., The effects of zirconia morphology on methanol synthesis from CO and H_2 over Cu/ZrO_2 catalysts: Part I. Steady-state studies. *Journal of Catalysis* **2005**, *233* (1), 198–209. doi.org/10.1016/j.jcat.2005.04.026

[50]   Stichert, W.; Schüth, F.; Kuba, S.; Knözinger, H., Monoclinic and tetragonal high surface area sulfated zirconias in butane isomerization: CO adsorption and catalytic results. *Journal of Catalysis* **2001**, *198* (2), 277–285. doi 10.1006/jcat.2000.3151

[51]   Kundakovic, L.; Mullins, D. R.; Overbury, S. H., Adsorption and reaction of H_2O and CO on oxidized and reduced Rh/CeOx(111) surfaces. *Surface Science* **2000**, *457* (1), 51–62. doi.org/10.1016/S0039-6028(00)00332-0

[52]   Seo, J.; Moon, J.; Kim, J. H.; Lee, K.; Hwang, J.; Yoon, H.; Yi, D. K.; Paik, U., Role of the oxidation state of cerium on the ceria surfaces for silicate adsorption. *Applied Surface Science* **2016**, *389*, 311–315. doi.org/10.1016/j.apsusc.2016.06.193

[53]   Karpenko, A.; Leppelt, R.; Plzak, V.; Cai, J.; Chuvilin, A.; Schumacher, B.; Kaiser, U.; Behm, R. J., Influence of the catalyst surface area on the activity and stability of Au/CeO_2 catalysts for the low-temperature water gas shift reaction. *Topics in Catalysis* **2007**, *44* (1), 183–198. doi 10.1007/s11244-007-0292-x

[54]   Katta, L.; Thrimurthulu, G.; Reddy, B. M.; Muhler, M.; Grünert, W., Structural characteristics and catalytic performance of alumina-supported nanosized ceria-lanthana solid solutions. *Catalysis Science & Technology* **2011**, *1* (9), 1645–1652. doi 10.1039/c2cy00449f

[55]   Trovarelli, A., Catalytic properties of ceria and CeO_2-containing materials. *Catalysis Reviews – Science and Engineering* **1996**, *38* (4), 439–520. doi 10.1080/01614949608006464

[56]   Montini, T.; Melchionna, M.; Monai, M.; Fornasiero, P., Fundamentals and catalytic applications of CeO_2-based materials. *Chemical Reviews* **2016**, *116* (10), 5987–6041. doi 10.1021/acs.chemrev.5b00603

[57]   Serp, P.; Machado, B. F., Carbon (nano)materials for catalysis. In *Nanostructured Carbon Materials for Catalysis*. Royal Society of Chemistry: Cambridge, 2015. doi.org/10.1039/9781782622567

[58]   Eschemann, T. O.; Lamme, W. S.; Manchester, R. L.; Parmentier, T. E.; Cognigni, A.; Rønning, M.; de Jong, K. P., Effect of support surface treatment on the synthesis, structure, and performance of Co/CNT Fischer–Tropsch catalysts. *Journal of Catalysis* **2015**, *328*, 130–138. doi.org/10.1016/j.jcat.2014.12.010

[59]   Pan, X. L.; Bao, X. H., The effects of confinement inside carbon nanotubes on catalysis. *Accounts of Chemical Research* **2011**, *44* (8), 553–562. doi 10.1021/ar100160t

[60]   Qiu, B. C.; Xing, M. Y.; Zhang, J. L., Recent advances in three-dimensional graphene based materials for catalysis applications. *Chemical Society Reviews* **2018**, *47* (6), 2165–2216. doi 10.1039/c7cs00904f

[61]   Wachs, I. E., Raman and IR studies of surface metal oxide species on oxide supports: Supported metal oxide catalysts. *Catalysis Today* **1996**, *27* (3–4), 437–455. doi 10.1016/0920-5861(95)00203-0

[62]   Strunk, J.; Banares, M. A.; Wachs, I. E., Vibrational spectroscopy of oxide overlayers. *Topics in Catalysis* **2017**, *60* (19–20), 1577–1617. doi 10.1007/s11244-017-0841-x

[63]   Briand, L. E.; Tkachenko, O. P.; Guraya, M.; Gao, X.; Wachs, I. E.; Grünert, W., Surface-analytical studies of supported vanadium oxide monolayer catalysts. *The Journal of Physical Chemistry B* **2004**, *108*, 4823–4830. doi 10.1021/jp037675j

[64]   Kim, T.; Wachs, I. E., CH_3OH oxidation over well-defined supported V_2O_5/Al_2O_3 catalysts: Influence of vanadium oxide loading and surface vanadium–oxygen functionalities. *Journal of Catalysis* **2008**, *255* (2), 197–205. doi.org/10.1016/j.jcat.2008.02.007

[65]   Engweiler, J.; Harf, J.; Baiker, A., WO_x/TiO_2 catalysts prepared by grafting of tungsten alkoxides: Morphological properties and catalytic behavior in the selective reduction of NO by NH_3. *Journal of Catalysis* **1996**, *159* (2), 259–269. doi.org/10.1006/jcat.1996.0087

[66]   Wachs, I. E., Recent conceptual advances in the catalysis science of mixed metal oxide catalytic materials. *Catalysis Today* **2005**, *100* (1–2), 79–94. doi 10.1016/j.cattod.2004.12.019

[67]   Zhao, C. L.; Wachs, I. E., Selective oxidation of propylene over model supported V_2O_5 catalysts: Influence of surface vanadia coverage and oxide support. *Journal of Catalysis* **2008**, *257* (1), 181–189. doi 10.1016/j.jcat.2008.04.022

[68]   Fu, Q.; Wagner, T., Interaction of nanostructured metal overlayers with oxide surfaces. *Surface Science Reports* **2007**, *62* (11), 431–498. doi.org/10.1016/j.surfrep.2007.07.001

[69]   Claus, P.; Hofmeister, H., Electron microscopy and catalytic study of silver catalysts: Structure sensitivity of the hydrogenation of crotonaldehyde. *The Journal of Physical Chemistry B* **1999**, *103*, 2766–2775. doi 10.1021/jp983857f

[70]   Grünert, W.; Brückner, A.; Hofmeister, H.; Claus, P., Structural properties of Ag/TiO_2 catalysts for acrolein hydrogenation. *The Journal of Physical Chemistry B* **2004**, *108* (18), 5709–5717. doi 10.1021/jp049855e

[71]   Tauster, S. J.; Fung, S. C.; Garten, R. L., Strong metal-support interactions - Group 8 noble metals supported on TiO_2. *Journal of the American Chemical Society* **1978**, *100* (1), 170–175.

[72]   Batista, A. T. F.; Baaziz, W.; Taleb, A.-L.; Chaniot, J.; Moreaud, M.; Legens, C.; Aguilar-Tapia, A.; Proux, O.; Hazemann, J.-L.; Diehl, F.; Chizallet, C.; Gay, A.-S.; Ersen, O.; Raybaud, P., Atomic scale insight into the formation, size, and location of platinum nanoparticles supported on γ-alumina. *ACS Catalysis* **2020**, *10* (7), 4193–4204. doi 10.1021/acscatal.0c00042

[73]   Cargnello, M.; Doan-Nguyen, V. V. T.; Gordon, T. R.; Diaz, R. E.; Stach, E. A.; Gorte, R. J.; Fornasiero, P.; Murray, C. B., Control of metal nanocrystal size reveals metal-support interface role for ceria catalysts. *Science* **2013**, *341* (6147), 771. doi 0.1126/science.1240148

[74]   Chorkendorff, I.; Niemantsverdriet, J. W., *Concepts of modern catalysis and kinetics*. 2 ed.; VCH: Weinheim, 2007.

[75]   Widmann, D.; Behm, R. J., Activation of molecular oxygen and the nature of the active oxygen species for CO oxidation on oxide supported Au catalysts. *Accounts of Chemical Research* **2014**, *47* (3), 740–749. doi 10.1021/ar400203e

[76]   van den Oetelaar, L. C. A.; Partridge, A.; Toussaint, S. L. G.; Flipse, C. F. J.; Brongersma, H. H., A surface science study of model catalysts. 2. Metal–support interactions in Cu/SiO_2 model catalysts. *The Journal of Physical Chemistry B* **1998**, *102* (47), 9541–9549. doi 10.1021/jp9829997

[77]   Miller, J. T.; Koningsberger, D. C., The origin of sulfur tolerance in supported platinum catalysts: The relationship between structural and catalytic properties in acidic and alkaline Pt/LTL. *Journal of Catalysis* **1996**, *162* (2), 209–219. doi.org/10.1002/cctc.201801027

[78]   Ramaker, D. E.; de Graaf, J.; van Veen, J. A. R.; Koningsberger, D. C., Nature of the metal-support interaction in supported Pt catalysts: Shift in Pt valence orbital energy and charge rearrangement. *Journal of Catalysis* **2001**, *203* (1), 7–17. doi 10.1006/jcat.2001.3299

[79]   Topsøe, H., The role of CoMoS type structures in hydrotreating catalysts. *Applied Catalysis A*: General **2007**, *322* (1), 3–8. doi 10.1016/j.apcata.2007.01.002

[80]   Bensch, W., Hydrotreating: Removal of sulfur from crude oil fractions with sulfide catalysts. In *Comprehensive Inorganic Chemistry II*. 2nd ed.; Poeppelmeier, J.; Reedijk, K., Eds. Elsevier: Amsterdam, 2013; pp 287–321. dx.doi.org/10.1016/B978-0-08-097774-4.00723-3

[81]   Villadsen, J.; Livbjerg, H., Supported liquid-phase catalysts. *Catalysis Reviews-Science and Engineering* **1978**, *17* (2), 203–272. doi 10.1080/03602457808080882

[82] Chinchen, G. C.; Davies, P.; Sampson, R. J., The historical development of catalytic oxidation processes. In Anderson, J. R.; Boudart, M., Eds., *Catalyis: Science and Technology*, Vol. 8. Springer: Berlin, Heidelberg, 1987; pp 1–67. ISBN: 9783642932786.

[83] Franciò, G.; Hintermair, U.; Leitner, W., Unlocking the potential of supported liquid phase catalysts with supercritical fluids: Low temperature continuous flow catalysis with integrated product separation. *Philosophical Transactions of the Royal Society A* **2015**, *373* (2057), 20150005. doi:10.1098/rsta.2015.0005

[84] Steinrück, H.-P.; Wasserscheid, P., Ionic liquids in catalysis. *Catalysis Letters* **2015**, *145* (1), 380–397. doi 10.1007/s10562-014-1435-x

[85] Cokoja, M.; Adams, D. S.; Cooper, D., Ultra-selective acetylene hydrogenation catalyst developed to boost profitability in ethylene production. *The Catalyst Review* **2018**, *31* (11), 8–9.

[86] Franke, W.; Hahn, H., Ein Katalysator, der an seine Grenzen geht. *Element (Evonik Science Newsletter)* **2015**, *51*, 18–23.

[87] Hübner, S.; de Vries, J. G.; Farina, V., Why does industry not use immobilized transition metal complexes as catalysts? *Advanced Synthesis and Catalysis* **2016**, *358* (1), 3–25. doi.org/10.1002/adsc.201500846

[88] Ebadi Amooghin, A.; Sanaeepur, H.; Omidkhah, M.; Kargari, A., "Ship-in-a-bottle", a new synthesis strategy for preparing novel hybrid host–guest nanocomposites for highly selective membrane gas separation. *Journal of Materials Chemistry A* **2018**, *6* (4), 1751–1771. doi 10.1039/C7TA08081F

[89] Kristen, M. O., Supported metallocene catalysts with MAO and boron activators. *Topics in Catalysis* **1999**, *7* (1), 89–95. doi 10.1039/C7TA08081F

[90] Feyen, M.; Weidenthaler, C.; Güttel, R.; Schlichte, K.; Holle, U.; Lu, A.-H.; Schüth, F., High-temperature stable, iron-based core–shell catalysts for ammonia decomposition. *Chemistry – A European Journal* **2011**, *17* (2), 598–605. doi.org/10.1002/chem.201001827

[91] Yang, G. H.; Tsubaki, N.; Shamoto, J.; Yoneyama, Y.; Zhang, Y., Confinement effect and synergistic function of H-ZSM-5/Cu-ZnO-Al$_2$O$_3$ capsule catalyst for one-step controlled synthesis. *Journal of the American Chemical Society* **2010**, *132* (23), 8129–8136. doi 10.1021/ja101882a

[92] Alizadeh, A.; McKenna, T. F. L., Particle growth during the polymerization of olefins on supported catalysts. Part 2: Current experimental understanding and modeling progresses on particle fragmentation, growth, and morphology development. *Macromolecular Reaction Engineering* **2018**, *12* (1), Art. Nr. 1700027. doi.org/10.1002/mren.201700027

[93] Thommes, M.; Kaneko, K.; Neimark Alexander, V.; Olivier James, P.; Rodriguez-Reinoso, F.; Rouquerol, J.; Sing Kenneth, S. W., Physisorption of gases, with special reference to the evaluation of surface area and pore size distribution (IUPAC Technical Report). In *Pure and Applied Chemistry*, 2015; Vol. 87, p 1051. doi 10.1515/pac-2014-1117

[94] Cychosz, K. A.; Guillet-Nicolas, R.; García-Martínez, J.; Thommes, M., Recent advances in the textural characterization of hierarchically structured nanoporous materials. *Chemical Society Reviews* **2017**, *46* (2), 389–414. doi 10.1039/c6cs00391e

[95] Gregg, S. J.; Sing, K. S. W., *Adsorption, Surface Area and Porosity*. 2nd ed.; Academic Press: London, 1982. ISBN: 0123009561

[96] Christmann, K., *Introduction to Surface Physical Chemistry*. Springer: Berlin, Heidelberg, 1991; ISBN: 9783662080092 p 288.

[97] Christmann, K.; Behm, R. J.; Ertl, G.; Vanhove, M. A.; Weinberg, W. H., Chemisorption geometry of hydrogen on Ni(111) - Order and disorder. *Journal of Chemical Physics* **1979**, *70* (9), 4168–4184. doi 10.1063/1.438041

[98] Naumann d'Alnoncourt, R.; Bergmann, M.; Strunk, J.; Löffler, E.; Hinrichsen, O.; Muhler, M., The coverage-dependent adsorption of carbon monoxide on hydrogen-reduced copper catalysts: the combined application of microcalorimetry, temperature-programmed desorption and FTIR spectroscopy. *Thermochimica Acta* **2005**, *434* (1), 132–139. doi.org/10.1016/j.tca.2005.01.045

[99]   Builes, S.; Sandler, S. I.; Xiong, R., Isosteric heats of gas and liquid adsorption. *Langmuir* **2013**, *29* (33), 10416–10422. doi 10.1021/la401035p

[100]   Cremer, P. S.; Somorjai, G. A., Surface science and catalysis of ethylene hydrogenation. *Journal of the Chemical Society, Faraday Transactions* **1995**, *91* (20), 3671–3677. doi 10.1039/ft9959103671

[101]   Ertl, G., Surface science and catalysis – Studies on the mechanism of ammonia synthesis: The P. H. Emmett award address. *Catalysis Reviews* **1980**, *21* (2), 201–223. doi 10.1080/03602458008067533

[102]   van Spronsen, M. A.; Frenken, J. W. M.; Groot, I. M. N., Surface science under reaction conditions: CO oxidation on Pt and Pd model catalysts. *Chemical Society Reviews* **2017**, *46* (14), 4347–4374. doi 10.1039/c7cs00045f

[103]   Ertl, G. Nobel Prize Lecture 2007. https://www.nobelprize.org/uploads/2018/06/ertl_lecture.pdf (accessed Feb 21, 2023).

[104]   Gong, X. Q.; Raval, R.; Hu, P., General insight into CO oxidation: A density functional theory study of the reaction mechanism on platinum oxides. *Physical Review Letters* **2004**, *93* (10). 106–104. doi 10.1103/PhysRevLett.93.106104

[105]   Emig, G.; Klemm, E.; Hungenberg, K.-D., *Chemische Reaktionstechnik*. 6th ed.; Springer: Berlin-Heidelberg, 2017.

[106]   Scott Fogler, H., *Elements of Chemical Reaction Engineering*. 6th ed.; Pearson: Boston: 2020. ISBN: 9780135486221

[107]   Baerns, M.; Behr, A.; Brehm, A.; Gmeling, J.; Hofmann, H.; Onken, U.; Renken, A., *Technische Chemie*. Wiley-VCH: Weinheim, 2006. ISBN: 9783527310005

[108]   Froment, G. F.; Bischoff, K. B.; De Wilde, J., *Chemical Reactor Analysis and Design*. 3rd ed.; John Wiley & Sohns: New York, Chichester, 2010. ISBN: 9780470565414.

[109]   Levenspiel, O., *Chemical Reactor Engineering*. 3rd ed.; John Wiley & Sons: New York, 1998. ISBN: 9780471254249

[110]   Murzin, D. Y., *Engineering Catalysis*. de Gruyter: Berlin, 2020. doi 10.1515/9783110614435

[111]   van der Pol, A. J. H. P.; Verduyn, A. J.; van Hooff, J. H. C., Why are some titanium silicalite-1 samples active and others not? *Applied Catalysis A: General* **1992**, *92* (2), 113–130. doi.org/10.1016/09 26-860X(92)80310-9

[112]   Hirschfelder, J. O.; Curtiss, C. F.; Bird, R. B., *Molecular Theory of Gases*. John Wiley & Sons: New York, 1954. doi 10.1002/pol.1955.120178311

[113]   Hunter, B. M.; Gray, H. B.; Müller, A. M., Earth-abundant heterogeneous water oxidation catalysts. *Chemical Reviews* **2016**, *116* (22), 14120–14136. doi 10.1021/acs.chemrev.6b00398

[114]   Roger, I.; Shipman, M. A.; Symes, M. D., Earth-abundant catalysts for electrochemical and photoelectrochemical water splitting. *Nature Reviews Chemistry* **2017**, *1* (1), 0003. doi 10.1038/s41570-016-0003

[115]   Yi, Y.; Weinberg, G.; Prenzel, M.; Greiner, M.; Heumann, S.; Becker, S.; Schlögl, R., Electrochemical corrosion of a glassy carbon electrode. *Catalysis Today* **2017**, *295*, 32–40. doi.org/10.1016/j. cattod.2017.07.013

[116]   Sillanpää, M.; Shestakova, M., Chapter 3 - Emerging and combined electrochemical methods. In *Electrochemical Water Treatment Methods*, Butterworth-Heinemann: Oxford 2017; pp 131–225. doi.org/10.1016/B978-0-12-811462-9.00003-7

[117]   Dau, H.; Limberg, C.; Reier, T.; Risch, M.; Roggan, S.; Strasser, P., The mechanism of water oxidation: From electrolysis via homogeneous to biological catalysis. *ChemCatChem* **2010**, *2* (7), 724–761. doi 10.1002/cctc.201000126

[118]   Mehandru, S. P., Oxygen evolution on a $SrFeO_3$ anode. *Journal of the Electrochemical Society* **1989**, *136* (1), 158. doi 10.1149/1.2096577

[119]   Seh, Z. W.; Kibsgaard, J.; Dickens, C. F.; Chorkendorff, I.; Nørskov, J. K.; Jaramillo, T. F., Combining theory and experiment in electrocatalysis: Insights into materials design. *Science* **2017**, *355* (6321), eaad4998. doi 10.1126/science.aad4998

[120] Bard, A. J.; Faulkner, L. R., *Electrochemical Methods: Fundamentals and Applications*. 2nd ed.; John Wiley & Sons Ltd: New York 2000. ISBN: 9780471043720.

[121] Heck, R. M.; Farrauto, R. J., Automobile exhaust catalysts. *Applied Catalysis A – General* **2001**, *221* (1–2), 443–457. doi 10.1016/s0926-860x(01)00818-3

[122] Pérez-Ramírez, J.; Kapteijn, F.; Schöffel, K.; Moulijn, J. A., Formation and control of N_2O in nitric acid production: Where do we stand today? *Applied Catalysis B: Environmental* **2003**, *44* (2), 117–151. doi.org/10.1016/S0926-3373(03)00026-2

[123] Kunii, D.; Levenspiel, O., *Fluidization Engineering*. Butterworth-Heinemann: Newton MA, USA, 1991.

[124] Geldart, D., Types of gas fluidization. *Powder Technology* **1973**, *7* (5), 285–292. doi.org/10.1016/0032-5910(73)80037-3

[125] Noel, T., *Photochemical Processes in Continuous Flow-Reactors*. World Scientific: London 2017. doi 10.1142/q0065

[126] de Lasa, H.; Serrano, B.; Salaices, M., *Photocatalytic Reaction Engineering*. Springer-Verlag: New York, 2005. ISBN: 9780387275918

[127] Kreysa, G.; Ota, G.; Savinell, R. F., *Encyclopedia of Applied Electrochemistry*. Springer-Verlag: New York, 2014. ISBN: 9781441969965

[128] Wendt, H.; Kreysa, G., *Electrochemical Engineering*. Springer-Verlag: Berlin-Heidelberg, 1999. ISBN: 9783662038512

[129] Langlois, S.; Coeuret, F., Flow-through and flow-by porous electrodes of nickel foam. II. Diffusion-convective mass transfer between the electrolyte and the foam. *Journal of Applied Electrochemistry* **1989**, *19* (1), 51–60. doi 10.1007/BF01039389

[130] Xu, J.; Ma, Y.; Xuan, C.; Ma, C.; Wang, J., Three-dimensional electrodes for oxygen electrocatalysis. *ChemElectroChem* **2022**, *9* (2), e202101522. doi.org/10.1002/celc.202101522

[131] Aijaz, A.; Masa, J.; Rösler, C.; Xia, W.; Weide, P.; Fischer, R. A.; Schuhmann, W.; Muhler, M., Metal–organic framework derived carbon nanotube grafted cobalt/carbon polyhedra grown on nickel foam: An efficient 3D electrode for full water splitting. *ChemElectroChem* **2017**, *4* (1), 188–193. doi.org/10.1002/celc.202101522

3 Tools of catalysis research

A chapter under this heading may be rightly expected to present information on how the performance of catalysts is assessed, how structural features that may correlate with the catalytic properties are elucidated, and how a data basis required to engineer the environment for applying catalysts in large-scale reactors is created.

There is, however, no catalysis research without catalyst preparation. Therefore, we categorize catalyst preparation as a tool of catalysis research as well. It is actually both a tool and a topic of catalysis research, where studies may target a better understanding and the skillful application of new reactions and synthesis principles discovered, e.g., in inorganic chemistry, for making catalysts. This is actually not much different from other tools of catalysis research, the application of which is usually based on earlier work for creating environments in which the technique (e.g., a spectroscopic method) supplies information most relevant for understanding catalysts and catalytic processes.

3.1 Catalyst preparation

The preparation of a heterogeneous catalyst usually involves four major stages: (1) the synthesis of a solid precursor, which includes its separation from other (e.g. liquid) phases and a drying step, (2) the shaping of the catalyst precursor, (3) thermal conditioning steps, and (4) the activation of the catalyst. While (1) and (4) are naturally the outset and the end of the sequence, there is quite some variability in between. Thermal treatments can be applied to intermediate products. When the solid is a support, the active component is deposited on it in an additional stage, again followed by thermal conditioning. Catalysts manufactured for technical processes are shipped in a form that can be handled without any danger. Irrespective of an academic or industrial environment, the activation of the catalyst is usually performed in the same reactor in which the targeted catalytic process is going to take place.

The sequence of the intermediate stages in the preparation of a particular catalyst depends on a number of aspects, which are beyond the scope of this book. In the following, the four major stages will be discussed without reference to a particular preparation sequence.

3.1.1 Formation of solids from fluid phases

The liquid phase is the most convenient starting point for catalyst preparation because it allows an easy homogenization and mixing of the ingredients. Preparation routes that exclusively involve solid-state reactions (e.g. NH_3 synthesis catalysts via sintering of the required oxide components, or mechanochemical syntheses) are rare. Some preparation

https://doi.org/10.1515/9783110632484-003

techniques bypass the liquid state by oxidizing or pyrolyzing gaseous compounds in the gas phase or even in flames (see Sections 3.1.1.5 and 3.1.1.6).

Liquids containing the ingredients for catalytic solids may be obtained by the simple dissolution of solid raw materials, by peptization of powders into colloidal dispersions, by the (partial) dissolution of gels, and even by the fusion of solid ingredients. Solids may be recovered from the liquids by precipitation, by crystallization, by sol-gel transitions, by the solidification of melts, or by combustion.

While precipitation is based on a phase transition, i.e., a physical process, gelation is caused by a chemical reaction – the formation of macromolecules from smaller units by condensation reactions. Crystallization may proceed via both pathways. Molecular or ionic crystals are formed by the ordered deposition of molecules or ion pairs, crystallization of ordered macrostructures, e.g. zeolites, involves condensation steps between the dissolved oligomeric precursor structures (Section 3.1.1.2). While crystallization provides well-ordered particles directly from the liquid, emphasis in precipitation is rather on small sizes and narrow size distributions of the recovered solid particles. They may be amorphous, semicrystalline, or even crystalline. Well-ordered crystalline products may be obtained from precipitates in subsequent thermal stages.

Phase formation from the liquid determines important properties of the solid product, which can be modified to a certain extent, but not completely changed in the following steps. Such properties are the size and the size distribution of the solid particles and the relation between components in mixed products (i.e., well-mixed or segregated).

3.1.1.1 Precipitation and coprecipitation

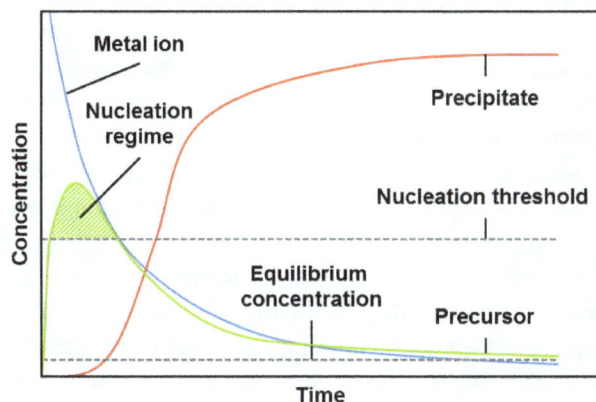

Figure 3.1: La Mer model of a precipitation process. A precipitating agent is added in excess to a solution of a cation at t = 0. In the absence of extrinsic sites that could aid nucleation, the solid phase is not formed unless the concentration of the colloidal precursor of the particles exceeds the nucleation threshold. Adapted from ref. [1].

While the start of a precipitation can be clearly identified (see below), its end is less defined because the precipitate can change significantly in subsequent aging processes. Following practical considerations, we treat aging as a part of the preparation step, which extends from the mixing of the components to the separation of the solid product from the liquid phase.

When there is a contact between a solid substance and its solution, dissolved species precipitate or crystallize when their concentration exceeds a threshold defined by thermodynamics (e.g., the solubility product if the solid dissociates into ions). In the absence of a solid phase, such supersaturated solution may, however, be metastable, because creating an interface between a new solid phase and the solution requires a driving force, a certain degree of supersaturation. As long as this is not available, the species to be precipitated remain dissolved as precursors of the solid phase. The highest precursor concentration accessible in a supersaturated solution is the nucleation threshold, where solid phase formation becomes inevitable (see La Mer plot, Figure 3.1).

While the nucleation rate is low at the nucleation threshold, it increases with a further growing degree of supersaturation. This provides opportunities to tailor the properties of the resulting solid phase: the higher the nucleation rate at a given feed amount (or feed rate in a continuous process), the smaller the particles that will be formed. Small particles, which are typically preferred in catalysis, can be obtained by "burst nucleation," a sudden strong increase in supersaturation, which produces many nuclei in a short time and causes a drastic decrease of the precursor concentration. Ideally, there will be no more nucleation beyond this "burst": all particles have the same age and similar sizes. If, however, high structural perfection is targeted (e.g., in crystallization), nucleation may be circumvented by adding well-ordered crystals (seed crystals) to the supersaturated solution and keeping concentrations below the nucleation threshold.

In the preparation of mixed solid phases, the homogeneous distribution of components is difficult to achieve if the solubility of the two precipitates is very different. In this situation, core-shell particles with shells rich in the better soluble component covering cores deficient in it are often obtained. A similar problem arises in sol-gel transitions, with precursors exhibiting very different condensation tendencies. The problem may be alleviated by choosing different precipitating agents (or leaving groups for sol-gel transitions) or by binding the components in mixed complexes before these are converted into the targeted mixed phases.

As a process involving the formation of a new phase via a nucleation step, precipitation requires an extremely careful control of all process variables to avoid reproducibility problems. In supersaturated solutions, precipitation may be initiated already below the nucleation threshold by items that could function as seed for nucleation (dust particles suspended in the solution, eddies at fast moving agitators, rough portions of vessel walls), or by concentration inhomogeneities. Ideally, each new droplet of the precipitation agent is mixed with the liquid at a rate exceeding the nucleation rate by orders of magnitude. In reality, nucleation in a diffusion zone around the droplets competes with mixing and convection driven by the agitator in a complex way.

Concentration inhomogeneity may be avoided by generating the precipitating agent *in situ*, e.g., OH⁻ ions by the thermal decomposition of dissolved urea:

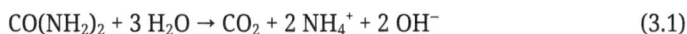

$$CO(NH_2)_2 + 3\ H_2O \rightarrow CO_2 + 2\ NH_4^+ + 2\ OH^- \tag{3.1}$$

For the use on a technical scale, such a reaction should be very fast to allow precipitation of large quantities within a short time. This condition is not fully met for reaction (3.1) even near the boiling point of water. Nonetheless, the reaction is applied in a popular method to deposit active components on support surfaces in homogeneous distribution (urea deposition/precipitation, Section 3.1.3.3).

Precipitates often contain large amounts of fines that cannot be recovered from the liquid phase by standard methods such as filtration. Therefore, an aging period is mandatory even if the synthesis aims at catalysts or supports in powdered form. Aging causes larger particles to grow at the expense of smaller ones (Ostwald ripening). This may be accompanied by chemical transformations, in particular, in multicomponent systems. Thus, the composition and structure of the solid phases deposited on the filters may substantially differ from those of the initial precipitate (see below).

Figure 3.2 shows the typical arrangements used to precipitate solids for catalyst preparation. Compared to the continuous mode (Figure 3.2c, d), the batch mode (Figure 3.2a, b) is less demanding with respect to process control. Arrangements, in which one component is added dropwise to the second one (e.g., precipitating agent A into a metal salt solution B, cf. Figure 3.2a), are the simplest, but by no means the best ones. They suffer strongly from concentration inhomogeneities as described above, in particular initially, when the first droplets of A plunge into a concentrated reagent B. During batch precipitation, the average concentration of B in the solution decreases and that of A increases. Proceeding to a high excess of A, B can be utilized very completely. However, as solubility products are small, there is usually no need for this: the process may rather be stopped when the amount of added A has reached the required stoichiometric ratio to B.

Simultaneous addition of both A and B to the solvent in the right stoichiometry (Figure 3.2b) requires precise control of their feed rates, but removes most of the problems with respect to concentration inhomogeneity. This is not because such inhomogeneities are avoided, but they have no consequences because both concentrations increase from a very low level. The solubility product is exceeded with the reactant concentrations at the stoichiometric ratio. The end of the process is not dictated by one of the reagents running out. Instead, it is rather the capacity of the equipment to handle the emerging slurry that limits the process. Due to these differences to the batch process, the arrangement with simultaneous addition of both reactants is referred to as "semi-batch".

Precipitation can be also performed in a continuous operation, either in tubular reactors or in continuously stirred reactors (Figure 3.2c, d). Both are derived from idealized reactor types – the plug flow reactor (PFR) and the continuously stirred tank

a

Precipitating agent

b

Precipitating agent   Metal solution

c

Precipitating agent

Mixing chamber

Tubular reactor

Metal solution

d

Precipitating agent   Metal solution

To separation

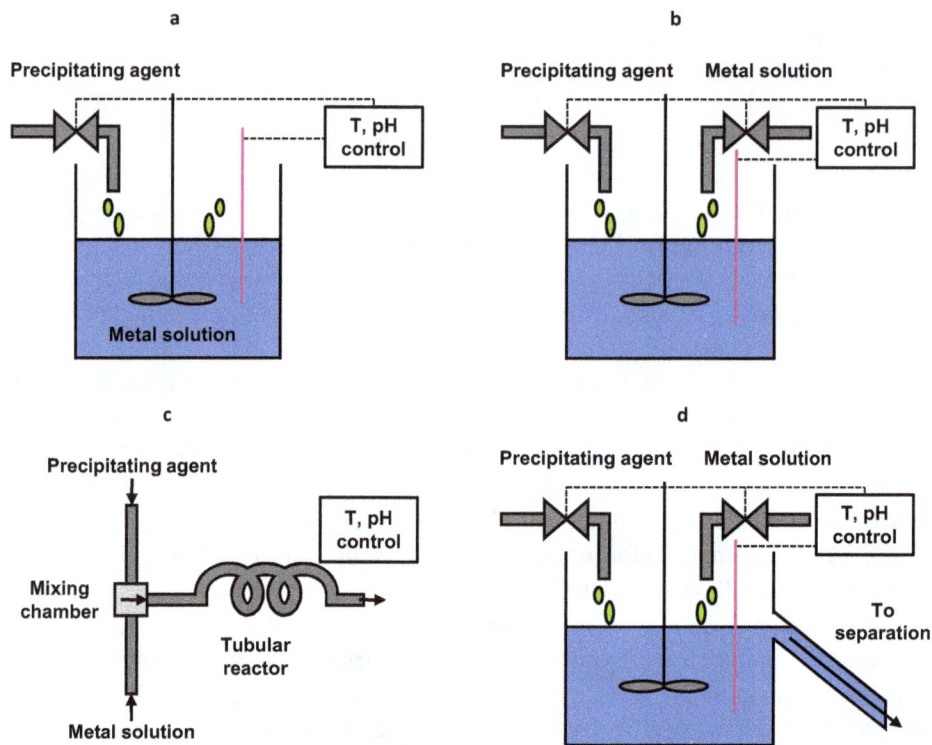

Figure 3.2: Reaction engineering options for the precipitation of a solid material: (a) batch mode: precipitation agent added to metal solution; (b) semi-batch mode: both metal salt and precipitating agent added simultaneously; (c) continuous mode, flow reactor; (d) continuous mode, stirred tank reactor. (a, b, d) Adapted from ref. [1].

reactor (CSTR, see Section 2.8.1). In continuous operation, reproducibility of the product properties is improved by a tighter control of the reaction conditions, but flexibility is limited. Therefore, the precipitation process should be thoroughly explored prior to designing a continuous scheme.

In the tubular reactor (Figure 3.2c), reactants A and B are charged to a mixing chamber at a specified ratio, which determines the degree of supersaturation to be reached in the following nucleation. Precipitation and aging proceed in a long tube (or coil), directly after the mixing chamber. Static mixers may be additionally installed in the tube.

As outlined in Section 2.8.1, there is only one mixture composition in (ideal) continuously stirred reactors, which corresponds to a high reactant conversion in the reaction performed. This shows that the continuously stirred reactor (Figure 3.2d) is not suited to concentrate a clear solution of A and B for inducing nucleation, as shown in Figure 3.1: it can work only with the existing crystals. It is therefore used for seed crystallizations where seeds are continuously charged and grow in the reactor. In stirred reactors, the species entering the vessel remain in it for very different times: there is a broad residence time distribution even in the ideal CSTR (see textbooks of chemical reaction engineering, e. g. refs. [2, 3]). This applies also to the seed crystals and causes a broad distribution of crystal sizes at the exit. In the tubular reactor, all species have (almost) the same residence time. Consequently, crystal size distributions are narrower.

The preparation of $Cu/ZnO-Al_2O_3$ catalysts for methanol synthesis is a typical example of the application of (co)precipitation in catalyst manufacturing. Intimate mixing of Cu^{2+} and Zn^{2+} ions cannot be achieved when the components are precipitated as hydroxides. Instead, solutions of Na_2CO_3 and of Cu and Zn nitrates in appropriate ratios (typically 70/30) are simultaneously added to water as shown in Figure 3.2b. One of the feeds contains Al nitrate to achieve an Al content of a few wt.-%. Depending on the pH value during precipitation, the Cu/Zn ratio in the primary mixed hydroxycarbonate phases (e.g., rosasite $(Cu_x, Zn_{1-x})_2(OH)_2(CO_3)$, auricalcite $(Cu_x, Zn_{1-x})_5 (OH)_6(CO_3)_2$) may vary, and an undesired phase containing Na^+ and Zn^{2+}, instead of Cu^{2+} and Zn^{2+}, may be formed. The latter dissolves during aging, allowing more Zn to be incorporated into the (Cu, Zn) phases. After thorough washing to remove Na^+ and CO_3^{2-} ions, the materials are mildly calcined (see Section 3.1.4). Careful reduction of such precursor materials results in Cu metal particles of a few nm size in close interaction with a ZnO phase of similar size, which is kept highly defective by Al^{3+} cations on lattice positions while its sintering is delayed by Al^{3+} cations decorating the surfaces [4] (cf. Section 4.2.3).

Ni/Al_2O_3 catalysts for steam reforming of methane, Fe-based Fischer–Tropsch catalysts, and alumina supports for a variety of supported catalysts are other solids prepared by precipitation. Notably, Ni/Al_2O_3 is a supported catalyst made by reduction of a coprecipitate, which is a procedure different from the typical sequential route via preparation of the support and its subsequent loading with the active components (cf. Section 3.1.3). For the preparation of Al_2O_3, aluminate solutions are acidified. Depending on the precipitation conditions, either the metastable α-$Al(OH)_3$ (bayerite) or γ-$Al(OH)_3$ (gibbsite) are formed. Their conversion into the transition aluminas used in catalysis has been discussed in Section 2.3.2.2.

Precipitation is usually performed to fix cations in a solid (oxide) phase. The anions are washed out or, if bound in the precipitate, eliminated in the subsequent thermal steps. Therefore, anions such as carbonate, acetate, oxalate, or sometimes nitrate are preferred, which do not leave residues upon decomposition.

In the synthesis of ordered mesoporous materials (OMM), precipitation occurs in the presence of ordered micellar phases. In aqueous systems containing surfactants, the simple globular micelles, which appear just above the critical micellar concentration, are replaced by ordered liquid-crystalline phases at high surfactant concentrations. They contain the surfactant molecules in very different spatial arrangements. The surfactants may, for instance, form hexagonal arrays of cylindrical rods with the residual water in the interstitial space (Figure 3.3).

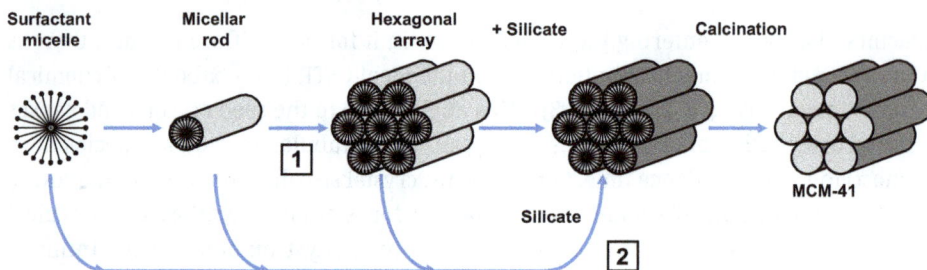

Figure 3.3: Mechanisms for the precipitation of ordered mesoporous materials (OMMs), (1) liquid-crystal-initiated route; (2) silicate-initiated route; surfactant and Si source might be, for instance, cetyltriammonium bromide (CTAB) and tetraethlyorthosilicate (TEOS), respectively. Reproduced with permission from ref. [5]. Copyright 1992 American Chemical Society.

When the water contains a precursor for silica formation, the precursor molecules around the templating rods may condense to a solid phase. A siliceous replica of the micellar phase is formed, the pores of which are finally vacated by the combustion of the surfactant. This "liquid-phase templated mechanism" (route 1 in Figure 3.3) was proposed for the formation of OMMs initially. However, conditions for transitions between the involved liquid crystalline phases were observed to be significantly influenced by the silica precursor, which seems to be actively involved in the creation of the spatial patterns (route 2 in Figure 3.3) [5]. Apparently, the relevance of the two mechanisms depends on the process conditions.

Despite early unnoticed patents in this field and similar work in Japan [6], the popularity of OMMs dates back to the work of Mobil researchers published in 1992 [5]. The synthesis was later modified by using tri-block copolymer templates, which are also capable of forming liquid crystalline phases in water. As the order of the liquid crystals is imprinted into the silica, OMMs cause diffraction signals although their pore walls are amorphous: diffraction is caused by the regularity of contrast differences between the pore walls and the empty pores. The translational vectors of the "pore patterns" (Figure 3.4) are several nanometers long. Therefore, the reflections appear at very low diffraction angles (cf. Section 3.4.3.3).

Figure 3.4 shows the pore geometries of popular OMMs. The pore systems of MCM-41 and SBA-15 (MCM for Mobile composition of matter, SBA for Santa Barbara amorphous material) consist of a hexagonal array of parallel pores (see Figure 3.4). MCM-48 contains two three-dimensional cubic pore systems, which are not interconnected (Figure 3.4). The sizes of the extremely uniform pores can be tuned in ranges of 3–4 nm (MCM-41, MCM-48) or 3–12 nm (SBA-15) by variation of the synthesis conditions, and further by introducing additives to the surfactants. Pore walls are dense in MCM-41 and MCM-48. Those of SBA-15 are thicker and contain micropores, which can be influenced by post-synthetic treatments. To achieve acidic or redox properties, Al^{3+} or redox-active cations can be introduced into the walls in small quantities. Alternatively, the pores may be loaded with catalytically active components.

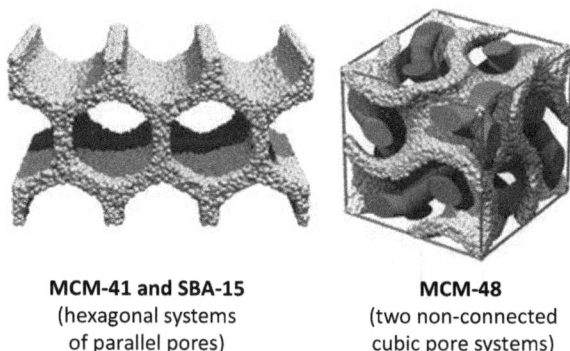

MCM-41 and SBA-15
(hexagonal systems
of parallel pores)

MCM-48
(two non-connected
cubic pore systems)

Figure 3.4: Pore-systems of well-known OMMs. Reproduced from ref. [7] with permission from the Royal Society of Chemistry.

OMMs are relatively expensive due to the loss of the template during synthesis. Despite extensive research effort, they have not yet found large-scale applications as active components or supports in heterogeneous catalysis. On the other hand, their pore sizes can be determined independently by X-ray diffraction (XRD) and electron microscopy, which gives access to the distances between their pore centers and the thickness of their pore walls. OMMs are, therefore, important benchmark materials in pore size analysis. Their use has resulted in great progress of the field (see Section 3.4.1).

3.1.1.2 Crystallization: synthesis of zeolites and MOFs

Crystallization is usually applied to obtain the desired product in the form of large, well-ordered particles, which are not likely to bear promise in catalysis. It is therefore rarely used in catalyst production – except for cases where the crystals form a highly porous and reactive phase. Zeolites are the typical example for this kind of crystalline porous solid. Hence, this section deals with the preparation of zeolites (Section 2.3.2.3), but some information about the preparation of metal-organic frameworks (2.3.2.4) is given as well.

Box 3.1: Recipe for a template-free zeolite synthesis, from ref. [8].
For the (template-free) synthesis of 56 g of zeolite Na-MOR, 19 g of NaOH are dissolved in 40 g of water. After dissolving 14.3 g of $NaAlO_2$ in this liquid, 645 g of water are added, before 98.2 g of silica powder are stirred in the resulting solution for 30 min. For crystallization, the mixture is kept in an autoclave at 170 °C for 24 h.

Zeolites are synthesized, starting from gels that contain the ingredients in appropriate quantities. The gels include a Si source (e.g., silica sol, water glass, SiO_2, or Si alkoxides), an Al source (e.g., gibbsite, Al salts, or sodium aluminate), a mineralizing agent (OH^-, sometimes F^-), which provides the required amount of alkali cations M^+, a structure-directing agent (SDA), and water. Their composition is specified by reporting the molar ratio between the ingredients after normalizing the Si, Al, and M content to SiO_2, Al_2O_3,

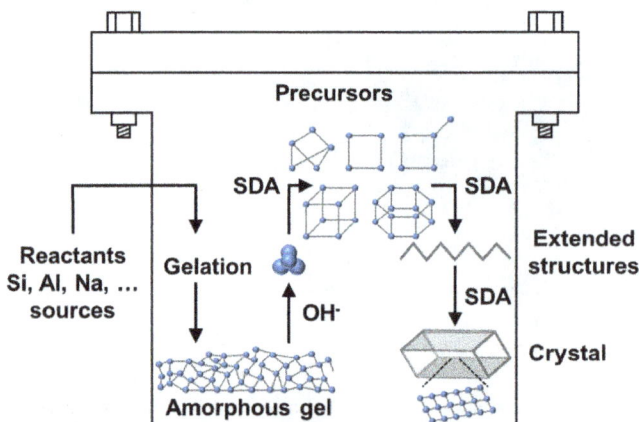

Figure 3.5: Basic steps of zeolite synthesis: gelation, dissolution of the gel, formation and growth of precursor structures, and crystallization. Adapted from ref. [9].

and M_2O, respectively. The gel is allowed to age. This equilibrates the solid gel phase with dissolved (alumo)silicate oligomers, which form geometric patterns under the influence of the SDA (Figure 3.5). The mixture is then exposed to a new set of conditions, which enhances gel dissolution and facilitates supersaturation, nucleation, and crystal formation. As an example, a typical recipe for a zeolite synthesis is shown in Box 3.1.

Zeolite synthesis requires careful control over a number of parameters. Most zeolites form only in specific ranges of the Si/Al ratio, and the amount of alkali needs to be within certain limits as well. These requirements result in ranges of gel compositions, which must be met to obtain a specific zeolite structure. Such "synthesis fields" may be reported in triangular diagrams with SiO_2, Al_2O_3, and M_2O on the sides (Figure 3.6a), which may be stacked to include a fourth parameter (e.g. the temperature) [10]. The typical alkali contents result in pH values of the gel between 9 and 13. Within the synthesis fields, variation of the alkali content can influence both rates and yields of crystallization, and even the Si/Al ratio in the product and its crystal morphology.

Temperature and time are relevant synthesis parameters as well. Crystallization time is important because zeolites are metastable phases, which may be rearranged into more stable products when the process is not terminated in due time. The undesired consecutive product may be another zeolite as exemplified in Figure 3.6b, or a denser crystalline phase.

Structure-directing agents (SDA) are cations that favor the formation of typical structural motifs from the oligomeric species, equilibrated with the aging gel (Figure 3.5). Most SDAs are alkyl ammonium cations, which are referred to as templates, but alkali cations can exert similar effects. In the hydration sphere around alkyl ammonium cations, the O atoms of water molecules are arranged in a similar way as in the zeolite structure. When H_2O is replaced by silicate or aluminate species (the latter neutralizing the cationic charge of the template), structural elements of the zeolite are built up like a cage or a

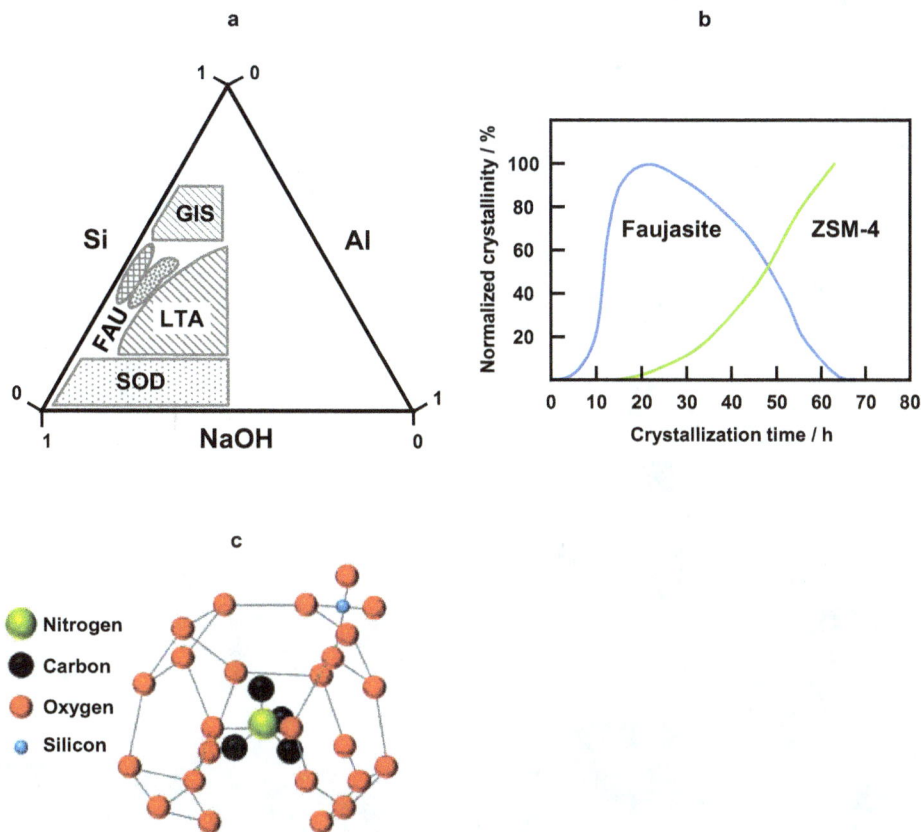

Figure 3.6: Relevant features of zeolite synthesis: (a) synthesis field for a specific set of conditions; GIS – gismondine polymorphs, SOD – sodalite; with Na silicate as Si source (diagrams strongly depend on the nature of Si source [10]). Adapted from ref. [11] with permission from Elsevier, (b) Consecutive crystallization of different solid structures during zeolite synthesis. (c) Example for the relation between SDA and alumosilicate structure (only one Si atom of framework is shown) in an as-prepared zeolite. (b, c) Adapted from ref. [9] with permission from Wiley-VCH Verlag GmbH & Co. KGaA, Weinheim, Germany.

channel section (Figure 3.6c). The template remains trapped in the cavity and is removed at the end by combustion.

Zeolites can also be obtained by seed crystallization. As the seed favors its structure over other ones, which might crystallize in its absence, seeds are also categorized as SDAs. Some zeolites are accessible by template-free syntheses (cf. Box 3.1), apparently because the structure-directing effect of the alkali cations is sufficient in these cases.

As mentioned in Section 2.3.2.4, MOFs are inorganic-organic hybrid materials, in which metal nodes (or SBUs) are connected by bi- or polytopic organic ligands, the linkers. Typical metal precursors are simple salts, like nitrates, acetates, hydroxides, or halides. Linkers are in most cases derived from aromatic carboxylic acids, like benzene-1,4-dicarboxylic acid, benzene-1,3,5-tricarboxylic acid, or 1,1′-biphenyl-4,4′-dicar-

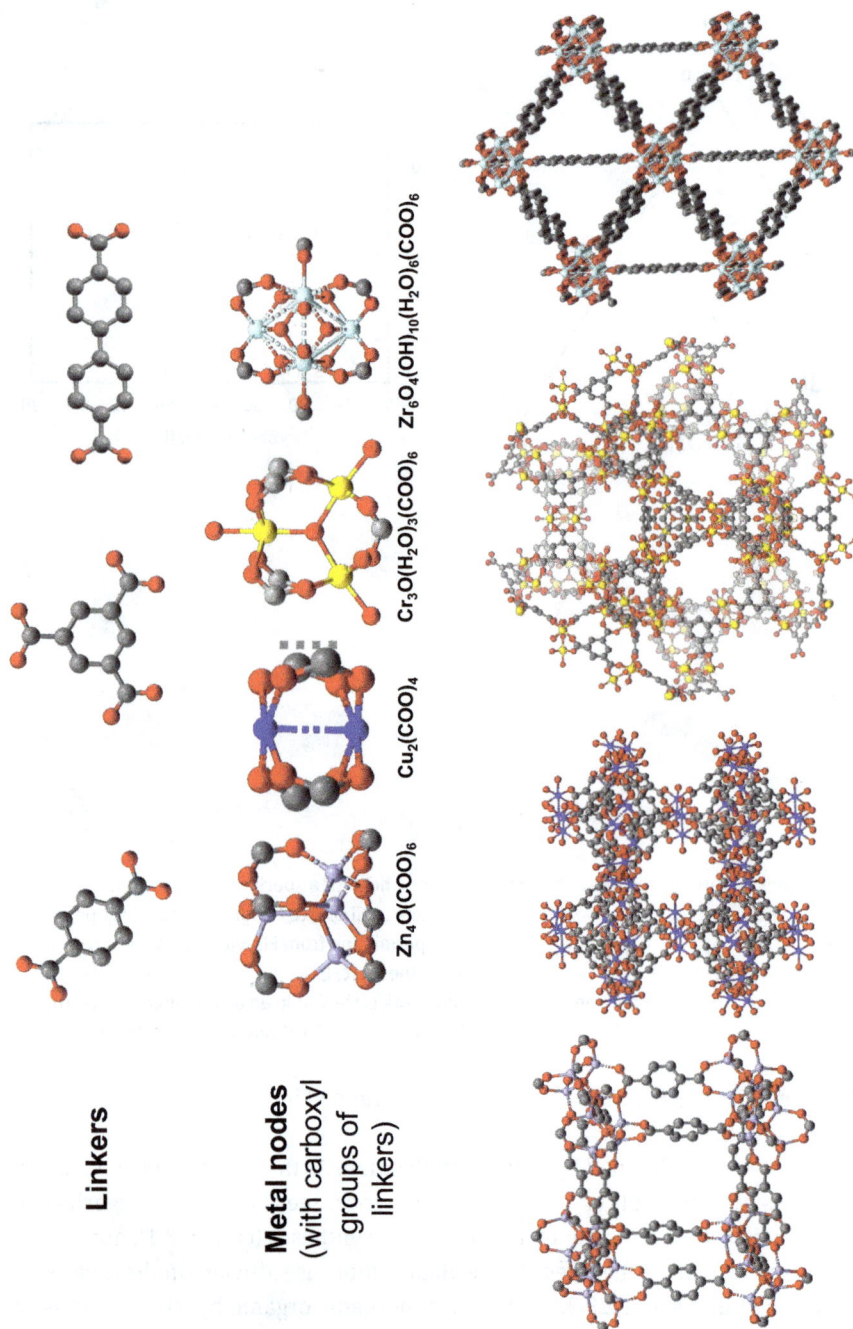

Figure 3.7: Examples of organic linker molecules (top) for SBUs, with coordination of linker carboxyl groups indicated (center), and for MOFs containing these SBUs (bottom; from the left - MOF-5, HKUST-1, MIL-101, UiO-67). Drawn with Mercury 1.0 (ref. [12]) using the database of the Cambridge Crystallographic Data Center (CCDC).

boxylic acid, see Figure 3.7. Phosphonates or sulfonates, which are also suitable, have received less attention. In many MOF syntheses, weakly coordinating monodentate functionalities (e.g., 4,4′-bipyridine) are applied, but the resulting frameworks are not sufficiently stable for use in catalysis.

MOF syntheses are often performed in sealed autoclaves, employing organic solvents like N,N-dimethylformamide (DMF), alcohols and/or water under autogenous pressure (solvothermal conditions). Starting from a clear solution, the SBUs are formed during the synthesis process (Figure 3.7, center). Their topology depends on the nature of the metal (oxidation state, coordination geometry, etc.), the structure of the linkers, and the synthesis conditions. Under suitable reaction conditions, well-defined, highly porous, and crystalline two- or three-dimensional networks are crystallized. Generalization of a rich synthesis experience, with reference to the MOF construction principles (cf. Figure 2.33), has made this kind of synthesis rather predictable [13]. In contrast to zeolite synthesis, organic templates are not employed, but monotopic ligands (e.g., benzoic acid) may be added to the reaction mixture as "modulators" in order to control the formation and stabilization of the desired SBUs.

Beyond this classical approach, MOFs can also be prepared in microwave-assisted protocols, by electrochemical, mechanochemical, and sonochemical methods [14]. With regard to catalysis, there is much attention for multi-component (or multi-variate) MOFs, which contain different metals and/or more than one type of organic linkers in a single framework, because they might give access to bifunctional catalysts with unprecedented properties [15].

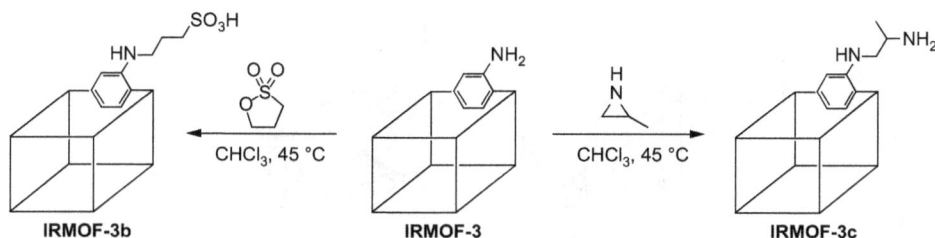

Figure 3.8: Post-synthetic modification of IRMOF-3: introduction of acidic or basic sites. Reproduced from ref. [16] with permission from the Royal Society of Chemistry.

Post-synthetic modification (PSM) comprises strategies targeting the modification of linkers or manipulations at free coordination sites of the SBU, e.g., for removing solvent molecules to create vacancies useful for catalysis. Covalent modification of the linker allows employing the whole arsenal of organic synthesis to create suitable active sites in the pores. As an example, Figure 3.8 shows how the site structure of IRMOF-3, which contains a basic nitrogen, can be modified at this group by suitable reactions [16].

3.1.1.3 Sol-gel transition

Gelation is another phenomenon giving access to solid oxide materials, e.g., for supports, mixed-oxides, or supported catalysts [17, 18]. It occurs when a suitable precursor, which is dissolved in a solvent, is destabilized to start condensation processes, finally resulting in a gel. A gel is a mixed material consisting of a solid and a fluid phase, which are both continuous: from any starting point, it is possible to travel to any other point in the same phase, without entering another phase.[1] On the way to the gel state, the system first becomes a sol, a clear solution of colloidal moieties, which are intermediates in the ongoing condensation processes. When the emerging solid structures become continuous, the mixture is no longer fluid. The concomitant transition, from a gradual to a very steep increase of the viscosity, marks the gel point, for which no other definition, e.g., on a thermodynamic basis, is available [19]. Although the gel point is the most conspicuous feature of the sol-gel transition, it is not its end, because precursor conversion is not yet complete. The ongoing reaction is superimposed by beginning aging processes (see below).

Gels may be obtained, starting from different precursors. For the synthesis of catalytic materials, Si or metal alkoxides like TEOS ($Si(OEt)_4$), $Al(O^iPr)_3$, or $Ti(O^nBu)_4$ are most popular. They are typically dissolved in the base alcohol of the alkoxide. However, solutions containing anions with metals in high oxidation states like silicate or vanadate can be also activated for the sol-gel transition. Alternatively, the sol state may be accessed from solid precursors in powdered form. This process is referred to as peptization. Sol-gel processes via routes other than hydrolytic [18] are beyond the scope of this book.

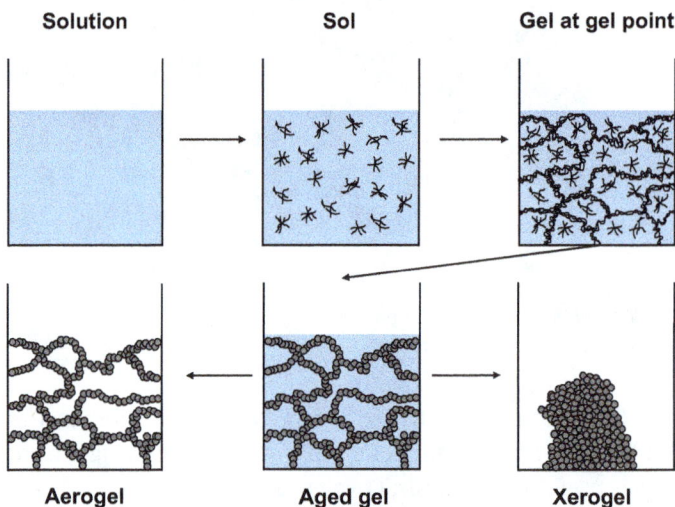

Figure 3.9: Schematic representation of the sol-gel process. Adapted from ref. [18], including information from ref. [19].

---

1 This differentiates sol-gel processes from the usually undesired flocculation: flocculates are not continuous.

Hydrolysis $\qquad$ $(RO)_n-M-OR + H_2O \rightarrow (RO)_n-M-OH + ROH$ $\qquad$ (3.2a)

Neutralization $\qquad$ $[SiO_4]^{4-} + n\,H_3O^+ \rightarrow [H_nSiO_4]^{(4-n)-} + n\,H_2O$ $\qquad$ (3.2b)

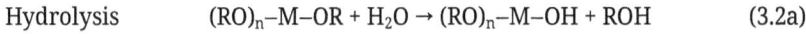

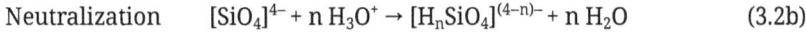

In the sol-gel process, the dissolved precursors are activated by a reaction that produces a hydroxyl (eq. (3.2)) as a favorable leaving group for the subsequent condensation steps (3.3). In syntheses with alkoxides, the OH group is obtained by a controlled addition of water (hydrolysis (3.2a)). In the production of silica gel, activation proceeds by partial neutralization of a Na silicate solution (3.2b).

During the condensation, a water or an alcohol molecule is released when bonds are formed between the alkoxide groups (3.3a). Analogously, water or an OH⁻ anion are split off during the condensation of silicate anions (3.3b).

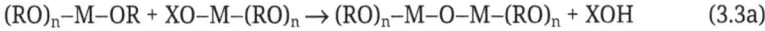

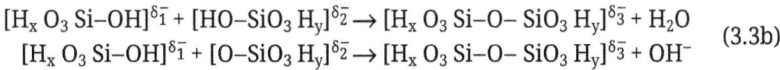

$$(RO)_n-M-OR + XO-M-(RO)_n \rightarrow (RO)_n-M-O-M-(RO)_n + XOH \qquad (3.3a)$$

$$(X = H, R)$$

$$[H_x O_3 Si-OH]^{\delta\bar{1}} + [HO-SiO_3 H_y]^{\delta\bar{2}} \rightarrow [H_x O_3 Si-O- SiO_3 H_y]^{\delta\bar{3}} + H_2O$$
$$[H_x O_3 Si-OH]^{\delta\bar{1}} + [O-SiO_3 H_y]^{\delta\bar{2}} \rightarrow [H_x O_3 Si-O- SiO_3 H_y]^{\delta\bar{3}} + OH^- \qquad (3.3b)$$

Condensation (3.3) can be catalyzed by either acids or bases.

The intermediate stage of the clear sol (cf. Figure 3.9) can be processed into coatings, fibers, droplets of various sizes, etc., often in procedures, which simultaneously accelerate the condensation reactions, e.g., by increasing the temperature or shifting the pH value. The porous solid body obtained from the gelation of a sol in a vessel is referred to as a monolith and the time required to access the gel point as the gelation time. Gels entrapping water as the liquid phase are called hydrogels, while the term lyogel denotes a gel with a different liquid phase.

The structures of colloids forming in the sol and growing into the gel depend on various parameters, among which the ratio between the rates of hydrolysis and condensation is important. Hydrolysis rates tend to increase with the charge on the metal ion involved, which is determined not only by the nature of the metal, but also by the nature of the alkoxide group. The ratio is also influenced by the temperature and the water/alkoxide ratio in the reaction mixture.

The influence of the pH value is complex. In the gelation of Si alkoxides, hydrolysis rates exceed condensation rates in an acidic environment, whereas condensation becomes faster in basic medium. Together with mechanistic differences (for more detail on the role of electrophilic or nucleophilic reagents, see ref. [18]), this results in the growth of largely linear, weakly branched chains under acidic conditions. In the gel state, the solid phase is formed by entangled, weakly crosslinked threads capable of filling space to a large extent. Such gels tend to exhibit microporosity and relatively low pore volumes. A stronger branching tendency at high pH values leads to the formation and growth of three-dimensional clusters, which are crosslinked in the gel and form mesopore systems with high pore volumes.

This versatility with respect to the pore system is most pronounced in the synthesis of silica gel by the acidification of silicate solutions. At pH 7–10, far from the isoelectric point, coalescence of clusters is prevented by their negative surface charge, and colloids can grow up to hundreds of nanometers sizes in the sol. When the repulsion between the colloids is attenuated by neutralization or by the addition of salts [18], the coalescence of primary particles into gel networks can be triggered. By varying the residence time in the basic environment,[2] the properties of the pore system can be adjusted in wide ranges.

As mentioned above, the condensation of oligomers is not completed at the gel point (Figure 3.9), but continues in the subsequent aging phase where it is superimposed by stabilization processes, tending to decrease the excessive solid–liquid interface just formed (analogous to Ostwald ripening). When the gel has reached its target properties (Figure 3.9), it is washed to remove the unreacted precursor species.

In the subsequent drying step, the solvent molecules are removed from the pores. This can be achieved either by conventional methods like solvent evaporation or by a supercritical drying procedure. During solvent evaporation, strong capillary forces act on the pores and cause a (partial) collapse of the porous architecture. The resulting xerogels (Figure 3.9) exhibit relatively high densities and mechanical stabilities, while BET surface areas may still be in the range of hundreds of m^2/g. In contrast, supercritical drying generates aerogels with extremely low particle densities (Figure 3.9). For this purpose, the solvent molecules in the pores are first exchanged by supercritical CO_2. The subsequent release of the gaseous CO_2 under mild conditions avoids capillary forces and the loss of porosity.

The final step of the sol-gel process is a thermal treatment, which is necessary to transfer the xerogels or aerogels into thermally stable oxide materials that can be utilized for various applications.

Beyond binary oxides like Al_2O_3, TiO_2 or SiO_2, which are frequently used as supports in heterogeneous catalysis, multimetallic composites are also accessible by sol-gel routes. In this case, the rates of hydrolysis and condensation need to be controlled for more than one metal species, which makes the situation far more complicated. Precursors for the metals of interest may have different electronic properties (e.g., electrophilicities) and exhibit strongly different hydrolysis rates. As this favors homocondensation (M_1–O–M_1) at the expense of heterocondensation (M_1–O–M_2), gels with non-uniform distribution (or even segregation) of components are obtained. Various options to handle this problem are outlined in ref. [18]. As an example, a chelation of the more reactive cations with polydentate organic ligands like carboxylates or amines can be applied to

---

2 Sol droplets may, for instance, fall through an oily layer for the growth of primary particles before entering the acidic aqueous layer below for gelation.

obtain more homogeneous materials, since both the hydrolysis and condensation rates can be influenced by varying the pH value and/or the ligand/metal ratio.

The amorphous mixed oxides obtained by gelation are often the end point of the synthesis. However, such materials are at the same time attractive starting points for the preparation of crystalline compounds like spinels, perovskites, or hydrotalcites. Traditionally, such compounds are made by calcining compressed mixtures of powdered binary oxides. In the gels, cations are already mixed on the atomic level.[3] Therefore, the desired crystal structure can be achieved at much lower thermal load than from oxide powders, which results in products with smaller particle sizes and higher reactivities.

Catalytic components, e.g., transition metal ions or complexes, organic molecules, or even enzymes can also be directly incorporated into gels by adding them as dopants during the synthesis. The species of interest can be, for instance, inserted into the primary colloidal clusters by co-condensation with the gel precursor molecules. Alternatively, molecular catalysts or pre-formed nanoparticles may be adsorbed at the surface of primary particles or entrapped in the voids between them during gelation and desolvation steps. Such strategies allow also tailoring the polarity of surfaces or modifying them with organic molecules.

In conclusion, sol-gel processes offer rich opportunities for the synthesis of porous metal oxides and supported catalysts, with a high degree of control over the properties of the pore system and the distribution of components on the atomic level. It should be noted, however, that alkoxide precursors are rather expensive and scale-up to an industrial level may be not straightforward.

3.1.1.4 Solidification of melts

Catalysts or catalyst precursors made via solidification of melts are rare in catalyst synthesis because the high temperatures required for melting favor the formation of dense phases. The precursor for ammonia synthesis catalysts is a prominent example for this route: it is obtained by melting oxides of Fe and of the promoter elements in an electric arc. After crushing the melt, the precursor is reduced in the reaction mixture, which also induces the porosity that allows reactants to diffuse into the interior of the initially dense particles. (cf. discussion of Figure 4.17; as mentioned above, the precursor can also be made by a sintering process).

Raney catalysts are derived from melts as well. Raney nickel is made by dissolving nickel in molten aluminum and quenching the resulting alloy. During the quenching step, promoters may be added. The precursor is shipped as a fine powder, which must be activated by concentrated NaOH. This leaching step removes most of the Al and

---

3 Mixtures at an atomic level may be alternatively achieved by the citrate route (Pechini process, liquid mix process) where the cations are chelated or coordinated in mixed complexes in the liquid phase.

creates pores in the initially dense particles. Raney Ni is the best-known catalyst of this type, which is applied for reduction processes, both in the lab and on industrial scales. Raney Co and Raney Cu extend the choice of catalysts from this family.

3.1.1.5 Solids from the gas phase

In modern material technology, synthesis of thin films from gaseous precursors via physical or chemical vapor deposition (PVD, CVD) plays an enormous role. While PVD is less relevant for catalysis due to its inability to access the interior pores of particles, catalyst synthesis and modification by CVD and related methods is an active field of catalysis research. This section deals with chemical vapor synthesis (CVS), which is closely related to CVD. In CVS, the solids are created from the gas phase. Deposition technologies, which modify an existing surface, follow in Sections 3.1.3.4 and 3.1.3.6.

As in CVD, in CVS, a precursor with sufficient thermal stability and vapor pressure is transported into a reactor in a carrier gas flow. The reactor is empty and is kept at a high temperature. The precursor molecules undergo thermal decomposition or react with carrier gas components such as O_2, which results in homogeneous nucleation, particle growth, and aggregation. Due to efficient fluidization, no solid films are formed on the reactor walls. Particles that are still in the aerosol state may be sintered or calcined in downstream reactors. Finally, they are collected in a suitable filtering device.

Starting from diethyl zinc, for instance, well-defined ZnO nanoparticles can be synthesized in the aerosol state in an O_2-containing carrier gas. From alcoholate precursors, oxides can be obtained even in inert gas, e.g., SiO_2 from TEOS ($Si(OC_2H_5)_4$), TiO_2 from $Ti(OC_3H_7)_4$, and Al_2O_3 from $Al(OC_4H_9)_3$ [20]. The properties of the particles can be varied by adding O_2, which increases the water concentration during nucleation.

The synthesis of multi-walled carbon nanotubes (MWCNTs, cf. Section 2.3.2.6) from gaseous carbon sources is another example of CVS. The CNT growth reaction requires an efficient growth catalyst. Suitable catalysts are based on supported Fe, Co, or Ni metal particles, which are able to break C-H and C-C bonds in hydrocarbons such as methane (e.g., Ni in steam reforming, cf. Section 4.2.2). Various hydrocarbons can be used as carbon sources, as also alcohols and syngas. Growth conditions and the resulting yields depend on the reactivity of the carbon source. Temperatures up to 1,250 K are required to obtain CNTs from methane, whereas ethylene reacts between 870 K and 970 K. Doping of CNTs with heteroatoms (e.g., O or N), which is usually performed by post-synthetic modification,[4] can also be achieved by reacting carbon sources containing the element of interest, e.g., acetonitrile.

During the growth process, C atoms that are accumulated on the surface or dissolved in the bulk of the catalyst particles are segregated in the form of graphene sheets, which results in the growth of CNTs, with diameters similar to the particle diameters. Depending on the strength of particle-support interactions, the tubes either grow above

---

4 cf. Section 2.2.2.

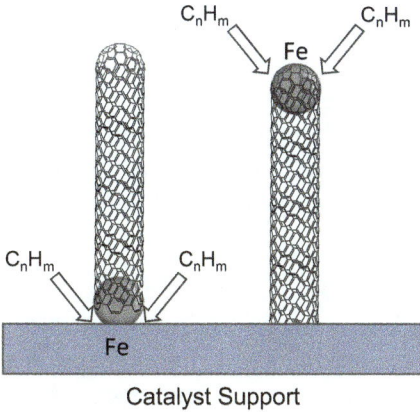

Figure 3.10: MWCNTS growing via base growth or tip growth.

the supported particles (base growth) or below. In the latter case, the tubes are attached to the support, while the particles sit at their tips (tip growth, cf. Figure 3.10). To suppress deactivation of the metal particles by encapsulation with amorphous carbon, H_2 is usually added to the carbon source. Thus, yields of up to 200 g of CNTs per g catalyst can be achieved. They are determined by the initial growth rate and the mean lifetime of the active sites [21]. The latter can be increased by adding promoters such as Mo or Mn oxides.

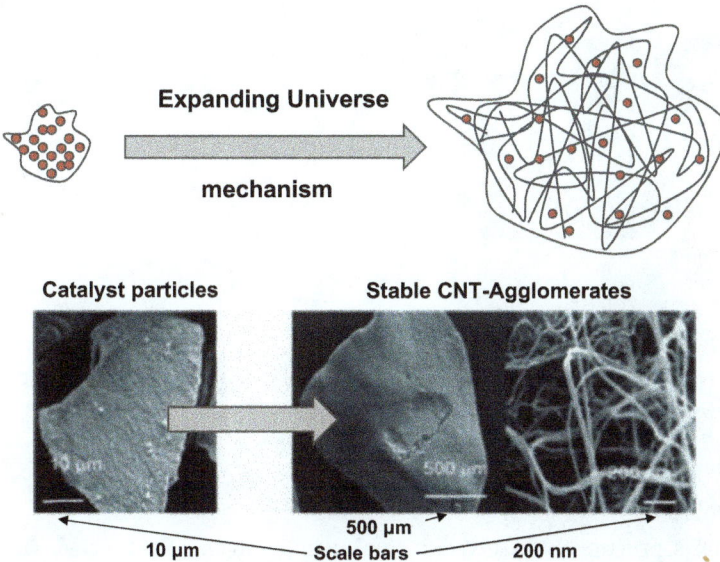

Figure 3.11: Large-scale synthesis of MWCNTs: the expanding universe mechanism. Adapted from ref. [22] with permission from Wiley-VCH Verlag GmbH & Co. KGaA, Weinheim, Germany.

CNTs can be produced on commercial scales using fluidized-bed reactors ([23], cf. Section 2.8.2), which allow continuous replacement of the catalyst and recovery of the product consisting of granules of strongly entangled CNTs (Figure 3.11). Like in polymerization (Section 2.3.4), the catalyst particles are fragmented under the mechanical stress caused by the growing CNTs, but they remain connected by the agglomerated tubes, which is essential for the fluidization. The growth mode has been termed "expanding universe mechanism" [22]). In some catalytic applications, especially in electrocatalysis, the residual growth catalyst can catalyze side reactions and must be removed by washing procedures.

3.1.1.6 Flame-based synthesis routes

When an evaporated precursor is fed into a flame, it can react by hydrolysis or by combustion. In flame hydrolysis, a volatile precursor is reacted in a H_2/air flame to obtain oxide products with very small primary particle sizes and large surface areas. Flame hydrolysis is applied on commercial scales to synthesize Cl-free SiO_2 and TiO_2 nanoparticles from $SiCl_4$ and $TiCl_4$ precursors, respectively (cf. eq. (2.5)). The properties of these pyrogenic oxides, in particular of the titania product P25®, have been discussed in some detail in Sections 2.3.2.1 and 2.3.2.5.

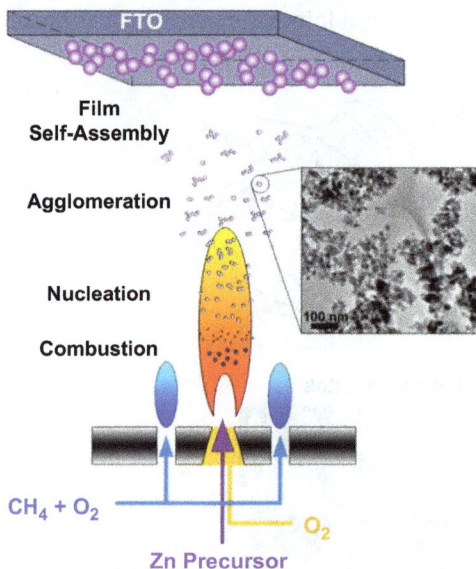

Figure 3.12: Spray flame pyrolysis of Zn naphthenate for the synthesis of ZnO nanoparticles. Reproduced from ref. [24] with permission from Elsevier. FTO - fluorine doped tin oxide.

In flame spray pyrolysis, precursors dissolved in combustible solvents are reacted. The liquid is fed into the center of a two-phase nozzle (Figure 3.12), where it is dispersed into an aerosol by O_2 operating as the dispersion gas. The combustion of the precursor solution in the resulting spray flame is stabilized by a surrounding premixed methane flame called the pilot flame. The process requires a precise control of all the burner parameters.

The flame spray method allows the large-scale synthesis of mixed metal oxides, such as spinels or perovskites, with relatively high surface areas. Many other catalytic materials, e.g., supported oxide and even noble metal catalysts are accessible as well. For more detail, the reader is referred to ref. [25].

3.1.2 Shaping of catalyst pellets

In many large-scale processes, the reactor is filled with catalyst pellets, which stay at rest while being passed by the fluid reactant(s) (cf. Section 2.8.2). The pellets in these fixed beds may have various forms such as spheres, cylinders, or hollow cylinders, with typical sizes of up to several millimeters in order to keep the pressure drop low. Reactors with catalysts in vigorous motion as in fluid beds or in slurries work with particles in the sub-mm range, but even these small particles sometimes consist of several smaller-sized primary particles. Therefore, compacting powdered catalytic components to mechanically stable pellets or particles is a procedure applied to almost all catalysts employed in the technical processes.

Although sol-gel processes may result in amorphous monophasic blocks that fill vessels of any size (cf. Figure 3.9), mm-sized catalyst pellets are usually polycrystalline. They are held together by physical and/or chemical adhesion forces.

Physical adhesion is based on dispersive or dipole forces or on hydrogen bonding, but it also includes frictional effects like the mutual fixation of rough interfaces. In compressed particles, form locking, e.g., when crystallites of various sizes, shapes, and orientations are wedged together, may support stability. Chemical adhesion via bonds between adjacent crystallites is characteristic of ceramic materials.[5] In the temperature range accessed in catalyst preparation, chemical adhesion is often established by oxygen bridges formed at the contact points when OH groups from the two surfaces that are in touch, condense. The effect can be greatly intensified when mixtures for pellet formation are enhanced by adding binders, e.g., colloidal suspensions of silica or alumosilicate, which offer an abundance of OH groups and can fill the interstices between grains. During thermal processing, they bind to these grains and cure themselves by dehydratization. Binders can also form their own (macro)pore systems that give access to the bound particles.

In Figure 3.13, typical unit operations for shaping catalysts into pellets or small particles are illustrated. Extrusion (Figure 3.13a) is a technique widely used in the ceramics industry. An extruder processes a slurry with a low water content, which is further densified while being pressed toward the narrow orifice by the screw. The cross-section of the orifice determines the size and cross-section of the thread leaving the extruder. Figure 3.13a shows the cross-sections that provide cylinders and hollow cylinders after

---

5 In ceramics, interstices are filled by diffusing material (sintering) or by the melting of constituent phases at the firing temperature (wetting).

a

Drive Raw material Screw Die Product

c

I   II   III   I

b

Feed
(small particles or
hydrogel droplets)

Rotary
pan

Product
spheres

d

Process air   Feed   Spray
machine   Cyclone

Drying chamber

Product

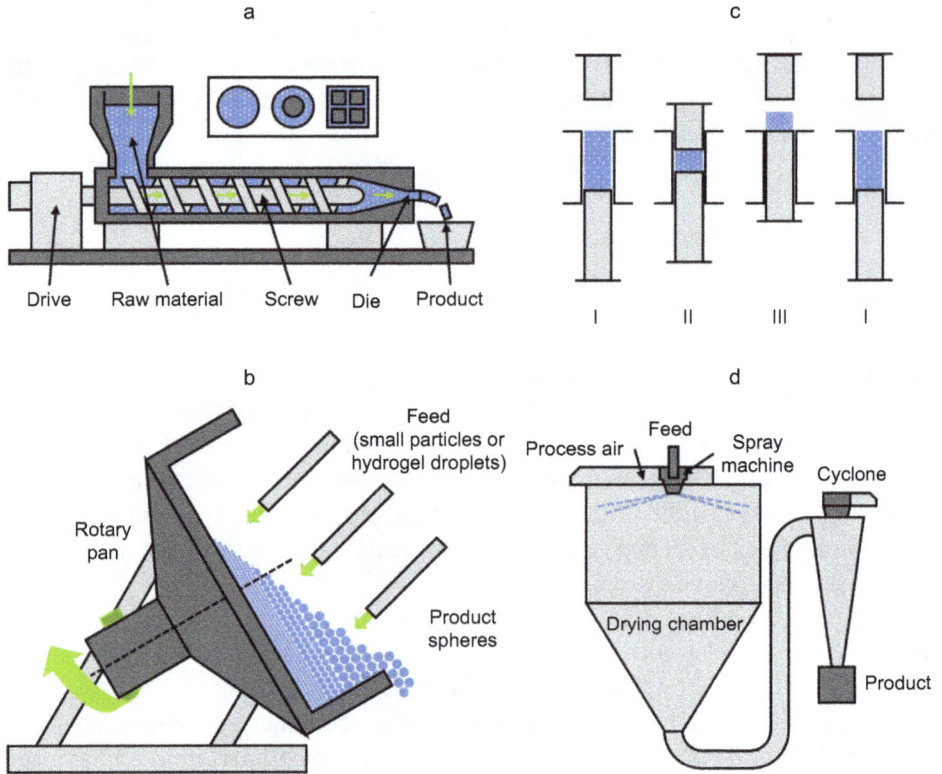

Figure 3.13: Shaping of catalyst pellets: (a) extruder; (b) rotating dish granulator; (c) pelletizer; (d) spray drying (co-current mode).

chopping. The third pattern would result in an extrudate containing four parallel channels, with square cross-sections. It symbolizes the extrusion of a green body for ceramic monoliths (cf. Section 2.8.2).

Figure 3.13b shows a rotating dish granulator, which is also used for fertilizer granulation. While a powder (e.g., alumina) and a granulating liquid (typically water) are fed continuously, granules form according to the snowball effect and are recovered from the dish, in more sophisticated versions even differentiated according to their sizes. In the mechanical pelletizer (Figure 3.13c), powder is compacted by brute force. The figure shows the three phases of the operation for one press mold, out of dozens working in parallel. Phase I starts with both pistons retracted; the lower one still sealing the mold from below and the upper one giving room to devices that fill the powder into the molds or later recover the finished tablets. When the molds are filled with powder, both pistons hit simultaneously in order to achieve comparable material density on both sides of the pellet (II). Subsequently, both pistons move upward: the upper one into its previous position and the lower one ejects the pellet for being recovered (III). It is then retracted into its initial position to close the cycle.

Small particles may be produced by spray granulation in the gas phase, where a stream of liquid (a solution or a slurry of the catalyst precursor, or a melt) is sprayed into a stream of hot air (Figure 3.13d). When the air flows in countercurrent to the liquid, a fluidized bed may form (fluid bed granulation). The residence time of the solid particles in it can be varied, which results in a variation of the pellet sizes in the product. Sub-millimeter particle sizes, as required for catalysis in fluidized-bed or in slurry processes, are, however, also accessible by spray granulation into a co-current gas stream as shown in Figure 3.13d, where excessive aggregation is prevented by the very short residence times.

The shaping processes introduced in Figure 3.13 result in self-supporting granules of different sizes. In some other cases, however, catalysts are applied in the form of coatings, stabilized by a supporting surface below, e.g., in washcoats on monolith walls (Figure 2.64b) or as coatings on plate catalysts (Figure 2.64c). As mentioned in Section 2.3.1, thin layers of oxidation catalysts are sometimes coated on ceramic bodies to suppress the consecutive oxidation of the valuable product.

Coating is achieved by contacting the surfaces with a slurry containing very small catalyst particles and appropriate organic or inorganic binders, followed by drying and thermal conditioning. The contact is often accomplished by simply dipping the supporting structure into the slurry and retracting it slowly (dip coating). For washcoating, monoliths are likewise immersed in the catalyst-containing slurry. After retracting them, the remaining liquid menisci are blown out of the channels. The steps may be repeated before the deposited layer is dried and calcined.

3.1.3 Deposition of active components on the internal surface of porous supports

In supported catalysts, active component(s) and promoter(s) should be distributed over the support surface in high dispersion, either in uniform concentration or according to specified concentration profiles within the pellets. Large-scale procedures should avoid excessive cost and environmental risks (e.g., toxic waste). Usually, the liquid phase is employed to transport the component(s) to be deposited to their target locations in the particles, but gas-phase transport and even solid-state transport phenomena are utilized as well.

The following sections, which deal with the deposition of active components on supports, cover techniques used on industrial scales as well as methods of potential promise, which have not yet found broad application. Impregnation is by far the simplest and most widely employed start into the preparation of supported catalysts. However, as interactions and mechanisms operating during the deposition process are better utilized in the more tedious equilibrium adsorption method, our discussion begins at this end.

When interactions between the support and the deposited species are weak, the latter may be re-distributed while the solvent is evaporated from the pores. Such processes, which are specific for confinement in pore systems, are therefore discussed in the present section, and not in section 3.1.4 that deals with thermal treatments.

3.1.3.1 Equilibrium adsorption and ion exchange

During equilibrium adsorption, the support is allowed to equilibrate with an excess volume of an aqueous solution of the component to be deposited. The supernatant liquid is then removed, and the support is washed thoroughly with water to get rid of all weakly interacting species. Finally, the product is dried and calcined. In research-oriented work, ion exchange is usually performed with powdered supports to avoid delays due to slow liquid-phase diffusion within the pores.

The primary interaction between the dissolved species and the support surface is electrostatic. In this respect, the method is similar to ion exchange. It should be recalled that oxide surfaces charge positively or negatively in aqueous media, depending on the relation between their isoelectric point (IEP) and the pH value (Section 2.2.2). Only ions of opposite charge are attracted by the surface. If they do not contain the element of interest, the procedure is unlikely to result in a good catalyst.

Processes occurring during equilibrium adsorption on a support of open porosity, including subsequent drying and calcination, are illustrated in Figure 3.14 [26]. A negatively charged oxide surface, with a protonated base BH$^+$ enriched in the adjacent water layer, is depicted as the initial state (I). The complex cations ML$_4$$^{2+}$, surrounded by diffuse water shells enriched with the counter ion A$^-$, are still in the bulk liquid phase. In II, the ML$_4$$^{2+}$ complex has approached the surface; the charge of which repels the anions A$^-$. The surface oxygen may now adsorb the complex cation by two mechanisms: it may enter its hydrate shell, forming an ion pair with it (outer sphere complex, or ion exchange, III), or it may replace a ligand L and form a coordinative bond to the metal cation (inner sphere complex, IV). The surface participates in this interaction as a polyanion or a reactant, respectively. Both types of interactions have been described in literature (cf. ref. [27]). The adsorbed water molecules may survive mild drying, but will be removed, together with the ligand L, during calcination (V). The last two steps describe how the support integrates the supported element M into its lattice, which requires that M can form an oxide phase miscible with the support. This is very often not the case, although mixed phases from support and supported elements have been observed after thermal stress in some systems, e.g., V^{4+} diffusion into TiO$_2$ or (Al,Cr)$_2$O$_3$-mixed phases formed in Cr$_2$O$_3$/Al$_2$O$_3$ catalysts.

In equilibrium adsorption, the pH value does not only determine the charge of the support surface, but it also influences the structure of the dissolved oxo anions. The consequences for the ion-exchange preparation of supported molybdenum oxide catalysts are illustrated in Figure 3.15. After dissolution of the precursor ammonium heptamolybdate (AHM, (NH$_4$)$_6$Mo$_7$O$_{24}$), the Mo oxo anions are monomeric (molybdate, MoO$_4$$^{2-}$) in the basic range. Around neutral pH, MoO$_4$$^{2-}$ ions are in equilibrium with heptamolybdate;

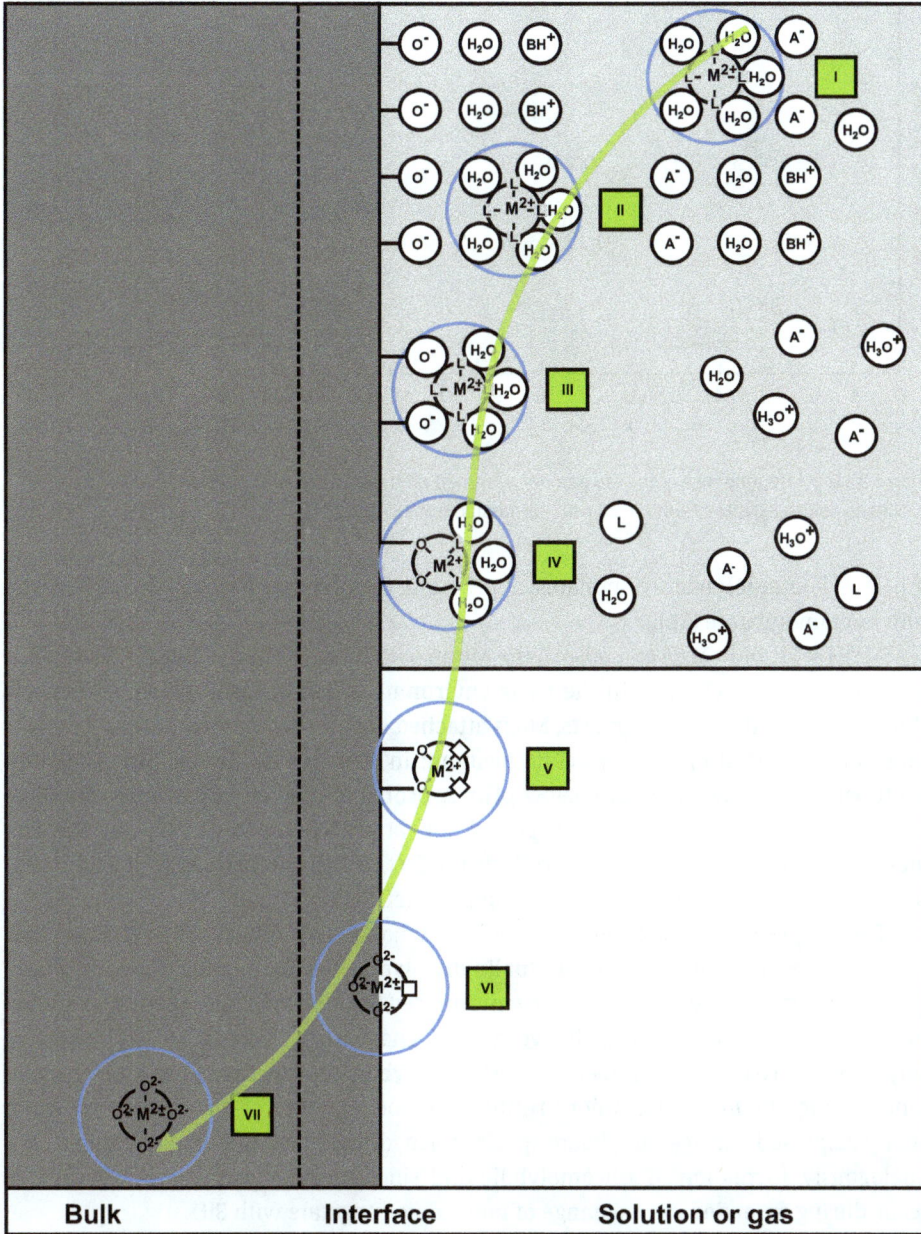

Figure 3.14: Steps of equilibrium adsorption at an oxide surface. Adapted from ref. [26] with permission from Elsevier.

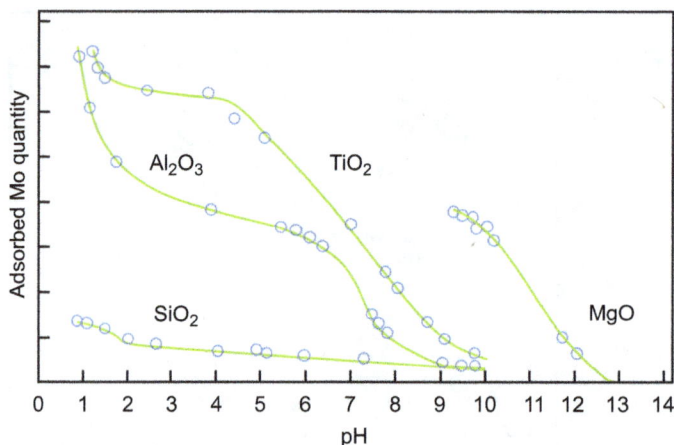

Figure 3.15: pH Dependence of the Mo amount adsorbed on the surfaces of supports with different isoelectric points. Adapted from ref. 28 with permission from Elsevier.

at pH < 6, heptamolybdate dominates; at pH = 1, octamolybdate $((Mo_8O_{26})^{4-})$ may coexist with heptamolybdate [28].

At pH $\approx$ 11, only MgO can adsorb the Mo species; because of its high IEP, its surface remains positively charged in the basic environment. Mo is adsorbed as monomeric $MoO_4{}^{2-}$ ions. On the other supports, Mo is attached both as heptamolybdate and molybdate ion, in neutral or slightly acidic medium; in stronger acidic medium, predominantly as heptamolybdate; and below pH = 2, probably in even larger units. For SiO_2, electrostatic adsorption would be expected only at pH < 2 due to its very low IEP. The small size of the step in Mo loading reflects the fact that the areal density of acidic sites is lower on SiO_2 than on the remaining supports (cf. Section 2.2.2).[6]

The simple picture of deposition via ion exchange is complicated by the fact that in some cases, the support may gradually dissolve in the interacting aqueous phase. This dissolution can be favored by reagents offered for ion exchange. Thus, at extended contact times for ion exchange between AHM and γ-Al_2O_3, $[Al(OH)_6Mo_6O_{18}]^{3-}$ heteropolyanions were found to compete with heptamolybdate for surface sites. After thermal conditioning, the heteropolyanions resulted in MoO_3 patches of low dispersion, while highly dispersed surface molybdate species formed from heptamolybdate anions [29]. Analogously, formation of silicomolybdic and silicotungstic species was reported to occur during extended ion exchange of molybdate/tungstate with SiO_2.

---

6 The gradual increase of the Mo loading on SiO_2 between pH 10 and 3 seems to contradict the electrostatic adsorption model. It should be, however, taken into account that the strength of the acid sites (like any other property related to interaction energies) is distributed around the average reflected by the IEP. Obviously, there are sites that can be already protonated when the pH is still above the IEP, as can also be seen in the other curves. Regarding SiO_2, the wide pH range over which the adsorption of Mo extends below the IEP suggests a very broad distribution of the acid site strengths.

While the surface charge of oxides with open porosity depends on the pH value, there are other solids where the surface charge is defined by their chemical nature, e.g., cation or anion exchange resins, zeolites and related materials like SAPOs, clays, or layered double hydroxides. In the following, ion exchange into zeolites, which is used to load these materials with active components even in large-scale applications, is discussed in some detail. Due to the chemical nature of zeolites (see Section 2.3.2.3), the zeolite framework always performs as a polyanion.

$$M^+ + ZO^-Na^+ \rightleftarrows Na^+ + ZO^-M^+ \qquad (3.4)$$

The introduction of monovalent cations into zeolites, as described in eq. (3.4), is an equilibrium reaction. As a consequence, a complete exchange of the initial Na$^+$ ions by a different cation, M$^+$, requires multiple repeats with excess concentrations of M$^+$, which can cause waste problems if M is expensive or environmentally harmful. In the zeolite cavities, the cations are still hydrated after mild drying. After calcination, they take positions that minimize their potential energy in the electrostatic field of the framework. They can migrate in electrical fields, both in the hydrated state or, at high temperatures, in dry atmosphere. Zeolites are, therefore, cation conductors.

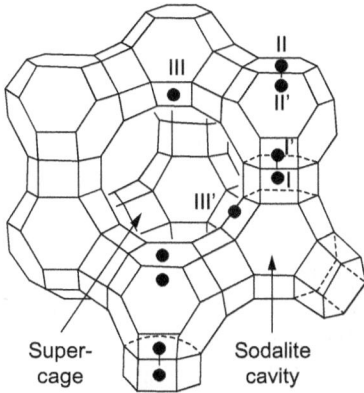

Figure 3.16: Cation sites in faujasite (FAU). Adapted from ref. [30].

For several types of zeolites, the preferred location of monovalent cations after dehydration was determined by structural investigations. Figure 3.16 shows sites of cations in faujasite. They can be located in the double six-rings that connect the β cages (SI) within the β cages in front of the different six-ring windows (SI', SII') and in the supercage in front of six-rings (SII) and four-rings (SIII). Only those in the supercage can be accessed by molecules larger than H$_2$. With increasing cation concentration, SII is populated first, followed by SI and SI' (only one of them if they are related to the same six-ring), and SIII.

The exchange of Na$^+$ by protons is of particular importance because H-forms of zeolites can be excellent acid catalysts. Some of them are used in industrial processes (see Appendix A1). Direct exchange with (dilute) acids according to eq. (3.4) is feasible only with Si-rich frameworks, e.g., of Na-ZSM-5 (Si/Al ≥ 15). At higher Al content, the acid sites become more unstable towards the removal of the adjacent Al ions (dealumination). Therefore, the H form of Al-rich zeolites is obtained using a detour via the NH_4-form, which is converted into the H-form by calcination (eqs. (3.5) and (3.6), with ZO^- symbolizing a cation exchange site):

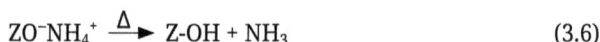

$$NH_4^+ + ZO^-Na^+ \rightleftarrows Na^+ + ZO^-NH_4^+ \tag{3.5}$$

$$ZO^-NH_4^+ \xrightarrow{\Delta} Z\text{-}OH + NH_3 \tag{3.6}$$

Opposed to the larger cations, protons are bound to the lattice covalently, forming a bridging OH group between a Si and an Al framework site as shown in Figure 1.7.

Exchange of cations with higher charge results in a more complex situation, in particular in Si-rich zeolites, because the neutralization of the cation charge would now require more than one framework Al site. Several options for charge neutralization in such situations are known.

Box 3.2: Special sites/configurations of polyvalent transition-metal ions in zeolites; dimer structure adapted from ref. [31] with permission from Elsevier.

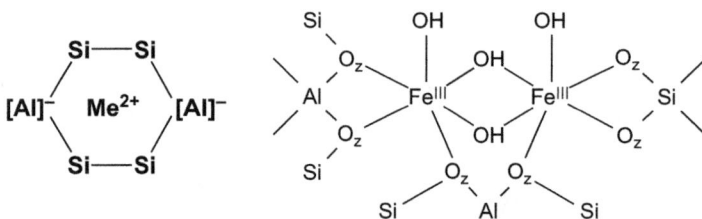

In Al-rich zeolites, the charge of polyvalent cations can be neutralized by more than one framework Al site because of the small distances between them. Even in zeolites with Si/Al ratios up to 15, framework Al sites can be close enough to jointly stabilize a bivalent cation, e.g., Fe^{2+} or Co^{2+} (see Box 3.2). As there will be no third negative charge around, further oxidation of cations in such sites is much more difficult than in other environments.

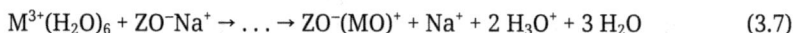

$$M^{3+}(H_2O)_6 + ZO^-Na^+ \rightarrow \ldots \rightarrow ZO^-(MO)^+ + Na^+ + 2\,H_3O^+ + 3\,H_2O \tag{3.7}$$

If there are not enough framework Al sites at short distance, the cation charge can also be neutralized by extra-framework oxygen. Equation (3.7) shows how a trivalent cation can be stabilized by a single framework charge. However, extra-framework oxygen might also stabilize oligomeric oxide structures in the cavities: dimers, trimers, etc. (cf. example of a dimer in Box 3.2). Indeed, some zeolites were shown to accommodate significantly larger amounts of Cu^{2+} or Fe^{3+} than expected based on their exchange capacity

(*viz.* Al content). In the case of Cu, the presence of massive oxide particles in well-prepared "overexchanged" samples could be excluded spectroscopically [32]. Cu trimers in
MOR have been thoroughly studied as sites for the oxidation of methane to methanol [33].
Fe zeolites are much more complicated, but the presence of binary structures and even
of higher oligomers at high Fe contents could be proven in a recent Mössbauer study [34].

Binary or higher oligomers are, however, prone to clustering under severe conditions, in particular, in the presence of moisture. This clustering may be accompanied
by transport toward the external zeolite surface. When ZSM-5, initially containing Fe
ideally dispersed in the form of $[FeCl_2]^+$ species, was subjected to a sequence of hydrolysis, drying, and calcination, some of the iron ended up as oxide crystal on the external
ZSM-5 surface and some below, clustered in disordered oxide structures, most likely
mixed with debris of the locally damaged zeolite (cf. [35]). Some bivalent cations, e.g.,
Co^{2+}, cannot be stabilized in zeolites at all except by two nearby framework Al sites as
shown in Box 3.2 [36].

It should be noted that there is also a solid-state route for ion exchange (cf. Section
3.1.3.7).

3.1.3.2 Impregnation

In impregnation, the support is contacted with a solution containing the element of
interest for some time. It is then dried and conditioned by calcination. Obviously, the
method is less time-consuming than equilibrium adsorption or ion exchange, and it
produces less waste. Its major versions are dry and wet impregnation.

In dry (or incipient wetness) impregnation, which is often employed with powdered supports, the volume of the impregnation solution is adapted to the pore volume
of the support, which can be determined by performing the procedure with pure water.
The support never becomes wet, because each additional drop is immediately sucked
up due to the capillary forces in the particles. When pore filling is indicated by increasing adhesion between the particles, the procedure is stopped and the dissolved catalyst
precursor is deposited in the pores by the evaporation of the solvent. If the deposited
amount is short of the target loading, the procedure is repeated.

In wet impregnation, the volume of the impregnation solution exceeds the pore
volume, and the active component finally impregnated onto the support also originates
from the supernatant liquid. Notably, upon evaporation of the supernatant liquid, the
precursor is exclusively deposited on the support surface, which provides plenty of
nuclei for particle formation. One might expect this to cause a strong enrichment of
the active component near the external pellet surface, but in practice, the process is
more complex and its outcome depends on the properties of the pore system and on the
solvent evaporation rate. Surface enrichment may be mitigated by diffusion from the
liquid-gas interface, where the precursor is permanently enriched by solvent evaporation toward the liquid–solid interface, where it is depleted by precipitation. Drying may
result in further re-distribution as outlined below.

In impregnation, the time for interaction between the dissolved species and the surface is limited. The pores are filled by convective flow quite rapidly, but ions interacting with the surface are delayed, and go on moving by diffusion once there is no more convection. Thus, adsorption by electrostatic interactions takes place to some extent, but when the solvent is evaporated, the remaining ions are precipitated on the surface. The precipitate is not necessarily identical with the compound dissolved in the impregnation liquid. For example, a cation introduced as a nitrate might precipitate as a hydroxide, a hydroxynitrate, or as a nitrate, depending on solubilities. As there is no washing in impregnation, the active element may be present in various structures before calcination. The role of the support in the process is as ill-defined as the speciation of the element introduced. It is a macromolecular counter ion or a reactant for cations attached to the surface via outer or inner sphere complex formation; it provides nucleation sites for those ions precipitating during solvent evaporation.

Inhomogeneous distribution of the active component is a major problem in impregnation, in particular with shaped pellets. When powders are impregnated, its effects are mitigated when pellets are formed from the powders.

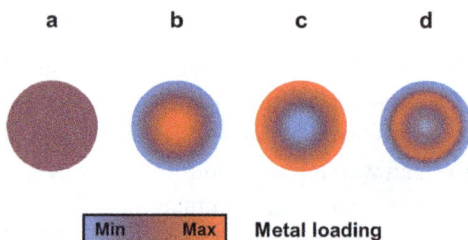

Figure 3.17: Concentration profiles of the active component in a spherical catalyst pellet; (c) is referred to as egg-shell profile, (b) as egg-yolk profile, and (a) is a homogeneous distribution.

If the interactions between the surface sites and the deposited species are strong, the latter are trapped near the outer pellet surface. If the resulting egg-shell profile (Figure 3.17c) is undesired, a more homogeneous deposition can be achieved by adding a species that competes for the surface sites without being adverse to catalyst activity.

The competition between $PtCl_6^{2-}$ and Cl^- ions for adsorption sites on γ-Al_2O_3 is a classic example for this situation. When alumina is impregnated with (dilute) H_2PtCl_6, platinum is deposited as a thin layer near the external surface, but in the presence of chloride in suitable concentration, it becomes well distributed over the pellet. At the same time, Cl^- anions acidify the alumina surface, which is desired for some important applications of Pt/γ-Al_2O_3 catalysts. More recently, the related Pd/γ-Al_2O_3 system was studied in considerable detail. In ref. [37], it has been described, how all the concentration profiles shown in Figure 3.17 can be obtained intentionally in this system.

While concentration profiles induced by strong interactions between the deposits and the surface withstand solvent evaporation, weakly interacting deposits can be significantly re-distributed during this step. To understand problems and countermeasures, it is useful to consider the three major stages of the evaporation process

[27]. After the initial heating phase, in which the evaporation rate increases with the temperature, most of the liquid is evaporated at a constant rate in a pseudo-stationary state. During this stage, the liquid in the pores is still continuous, and evaporation occurs via menisci at or near the external surface, to which the liquid is transported at the same rate by capillary forces. At a certain point, the drying front moves into the pellet, which decreases its surface as also the drying rate. The residual liquid in the pores becomes discontinuous, which prevents further significant transport of the dissolved material.

During the pseudo-stationary phase, two opposite mass flows compete in the pores, and their relation determines the resulting concentration profiles. The convective flow mentioned above enriches even weakly interacting precursors near the external surface. At the same time, this enrichment sets up a concentration gradient, which induces back-diffusion of the precursor into the interior. Slow evaporation allows for more back-diffusion and results in more homogeneous deposit concentration profiles. There is even the chance to change an initial eggshell profile toward a more homogeneous distribution, if the deposit can be dissolved at the higher evaporation temperature. Initially homogeneous profiles can be stabilized when the convective flow is impeded by a higher viscosity of the solvent, which can be achieved by adding organic molecules like ethylene glycol to the impregnation solution.

For a glimpse on the full complexity of the processes, we recall that the intraporous deposition of the precursor is mostly a precipitation. Its result depends on the rate at which the nucleation threshold is exceeded. During slow evaporation, nucleation rates are low, which favors large particles that are well-distributed in the pellet. With increasing evaporation rate, particle sizes tend to decrease. When, however, back-diffusion can no longer compete with convection, precipitation becomes confined to the external surface region, which again favors larger particle sizes. Reference [27] provides a number of instructive examples for these phenomena.

Impregnation is a standard technique for depositing active components on CNTs also. As their surfaces are hydrophobic after synthesis, they need to be modified by polar groups. Some methods for this have been mentioned in Section 2.2.2. Although such treatments create groups capable of ion exchange on the CNT surfaces, (e.g., carboxylate), the amounts deposited may exceed the ion exchange capacity by far. As the surfaces of the interwoven CNTs form a very open porosity (cf. Section 2.3.2.6), mass transport during solvent evaporation, as discussed above for narrow pores, is negligible. This does not hold, however, for the deposition of active components in the interior space of CNTs. How to handle this challenge can be found in ref. [38].

3.1.3.3 Deposition-precipitation

When precipitation is performed in the presence of a solid, e.g., a support, the process is called deposition-precipitation. Similar to wet impregnation, described in Section 3.1.3.2, the support lowers the energy barrier for nucleation, thereby favoring precipitation onto its surfaces over nucleation in the liquid phase. Still, the latter can occur, if the nucleation threshold is locally exceeded due to the inefficient distribution of the added precipitating agent.

This problem can be avoided when the precipitating agent is gradually released by a chemical reaction from a co-dissolved precursor, e.g., urea (urea deposition precipitation, cf. eq. (3.1)). The method allows depositing large amounts of active component on weakly interacting supports with good dispersion of the active metals (e.g., 50% Ni/carbon nanotubes, 60% Pt/carbon black, cf. ref. [27]).

Problems can arise when the support starts dissolving during the extended interaction with the fairly basic medium. As a result, the active element can be precipitated as a compound, which is difficult to reduce, e.g., Ni on SiO_2 as phyllosilicate instead of $Ni(OH)_2$. High NH_3 concentration can also result in re-dissolution of the precipitate by complexation. For this reason, several alternative versions of homogeneous deposition-precipitation were developed, e.g., the release of OH⁻ by thermal decomposition of nitrite ions

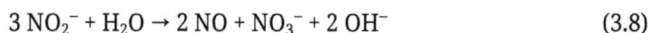

$$3 NO_2^- + H_2O \rightarrow 2 NO + NO_3^- + 2 OH^- \tag{3.8}$$

A special version of deposition-precipitation can be used to prepare highly dispersed gold nanoparticles on various supports (cf. ref. [27]).

3.1.3.4 Anchoring, grafting, and surface organometallic chemistry

While the adsorption of dissolved precursor species relies on their electrostatic interactions with the support sites derived from surface OH groups, anchoring strategies use OH groups to fix the catalytic element via covalent bonds. Alternatively, hydrogen chemisorbed on metal particles can be employed to attach precursors to the metal surface. Precursors eligible for such anchoring approach must have a high affinity to react with the reactive surface groups in a well-defined way. With respect to surface OH groups, this is the case with many organometallic compounds and metal alkoxides, as also with metal carbonyls and some metal halides:

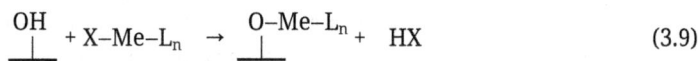

$$\underset{|}{OH} + X–Me–L_n \ \rightarrow \ \underset{|}{O–Me–L_n} + \ HX \tag{3.9}$$

(X = Cl, OR, R). Anchoring via surface hydrogen is usually performed with metal alkyls, e.g.:

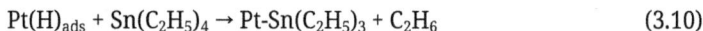

$$Pt(H)_{ads} + Sn(C_2H_5)_4 \rightarrow Pt\text{-}Sn(C_2H_5)_3 + C_2H_6 \tag{3.10}$$

Anchoring is the doorstep to two techniques, which are different in both goals and methods: grafting (with the related field of Surface organometallic chemistry, SOMC) and Atomic layer deposition (ALD). While grafting and SOMC require the anchored species to be well isolated, ALD targets a complete coverage of the surface with a mono-atomic layer of precursor species in each step. Grafting or SOMC means performing chemistry with well-defined surface groups, while ALD aims at preparing metal (or oxide) particles at both high surface density and dispersion.

Box 3.3: Instruction for grafting vanadium oxide species on supports, from ref. [39]. In this paper, surface structures obtained after depositing various vanadium amounts on the supports by this technique and by wet impregnation are compared.

The support (γ-Al$_2$O$_3$ or fumed SiO$_2$), which had been previously calcined at 823 K, was evacuated at 393 K in a flask to remove any moisture adsorbed during the short exposure to the ambient. After re-filling the flask with Ar, the powder was suspended in toluene, and vanadium(V) oxytriisopropoxide (VO(Oi-Pr)$_3$) was added as the reagent. After 4 h, the supernatant toluene was removed by a canula, and the powder was washed three times with toluene. The material was then dried and conditioned by a procedure consisting of a slow ramp to 573 K in N$_2$, an isothermal period, followed by a switch to air and a ramp to 773 K for final calcination.

To achieve the site isolation required for grafting/SOMC, the support is initially heated in flowing gas or in vacuum to remove the adsorbed water and to tune the surface OH-group density. The precursor can react with the surface from the gas phase (see Section 3.1.3.6) or from a solution in an aprotic solvent. To remove any unreacted precursor, the material is subsequently purged with inert gas or washed with the pure solvent. Ligands of the attached species can then be modified by suitable reactions, e.g., by oxidation or by ligand exchange. The preparation technique as a whole is referred to as grafting (for an example see Box 3.3). In its subdomain SOMC, precursors and the resulting surface structures are organometallic species.

Anchoring molecular compounds of metals on surfaces and chemical manipulation of the attached species were pioneered by Ermakov in Russia [40] and by Iwasawa in Japan [41] in the 1970s. In a typical study, Mo was attached to SiO$_2$ or Al$_2$O$_3$ using Mo(C$_3$H$_5$)$_4$ (Scheme 3.1a). The anchored Mo(C$_3$H$_5$)$_2$ complex was reduced to a surface-bound Mo^{II} species, which was confirmed by the release of 2 mol of C$_3$ hydrocarbons per mol Mo. The Mo^{II} species, which is not easy to obtain in homogeneous systems, was oxidized stepwise to surface-bound Mo^{IV} and Mo^{VI}, and it was possible to revert the sequence by reductions in H$_2$. Using the precursor Mo$_2$(η^3-C$_3$H$_5$)$_4$, dimeric Mo oxo structures were synthesized on SiO$_2$. As with monomeric surface species, the ligands were stripped using H$_2$, and redox reactions resulting in structures depicted in Scheme 3.1b could be performed. From EXAFS studies, the Mo-Mo distance in the Mo^{II} and Mo^{IV} surface dimers was determined to be between 2.53 and 2.80 Å, longer than in the initial Mo dimer complex (2.183 Å). On different SiO$_2$ supports, Mo^{IV} and Mo^{V} surface dimer species were observed, with and without bridging the oxygen between the Mo atoms (Scheme 3.1b).

Model catalysts prepared along similar routes were applied in various test reactions [40, 41]. SiO$_2$-bound Mo^{II}, both monomeric and dimeric, was highly active in the hydrogenation of ethene, and of 1,3-butadiene to butenes. While surface-bound Mo^{IV}

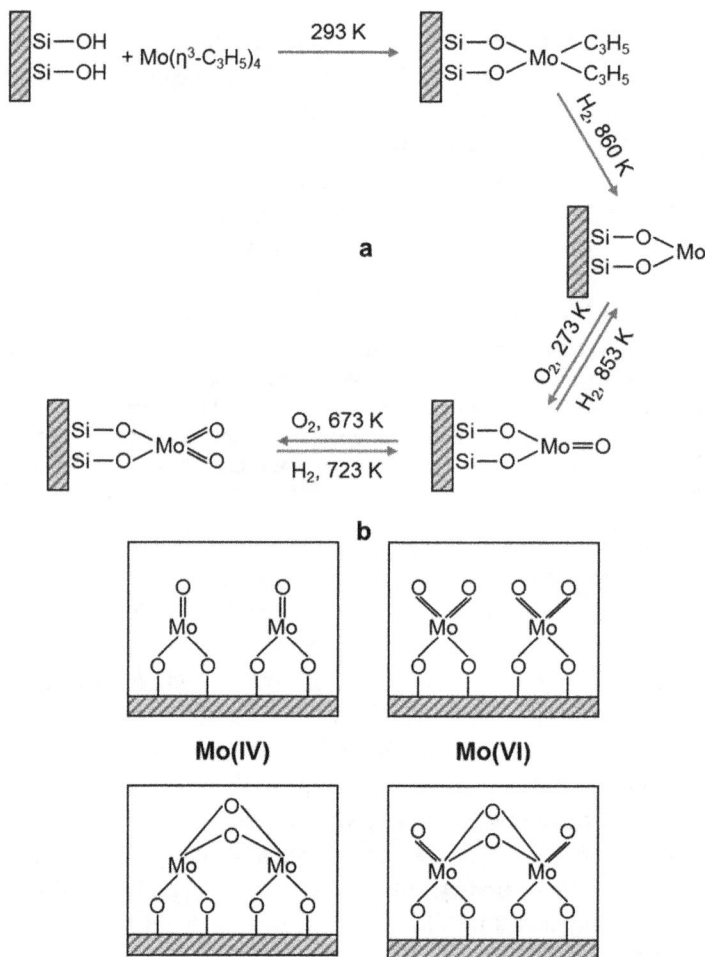

Scheme 3.1: Anchoring and grafting of Mo species on SiO$_2$: (a) anchoring Mo allyl complexes to SiO$_2$, with subsequent conversion to Mo species of different Mo oxidation states; (b) binary Mo surface species produced from a binuclear Mo allyl complex. Reproduced from ref. [41] with permission from Elsevier.

species were the best precursors for propene metathesis catalysts, Mo^{VI} oxo species were inactive[7]. Surface-bound Mo^{VI} species were highly efficient in the selective oxidation of propene. Attractive selectivities to acrolein were, however, obtained only with Mo^{VI} dimers bridged by μ-oxygen atoms.

---

7 It was found later that Mo^{VI} and W^{VI} oxo species can also be precursors of highly active metathesis sites when properly pretreated [42, 43].

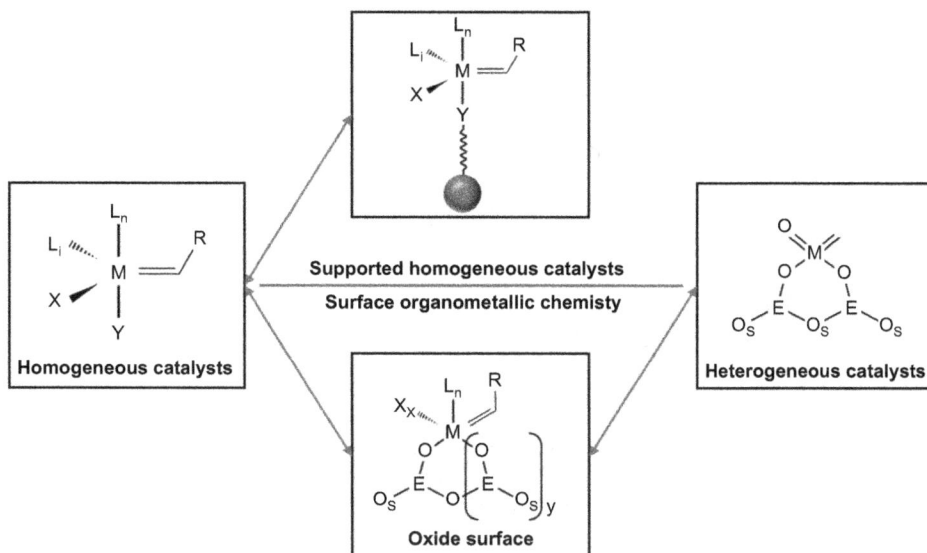

Figure 3.18: Surface organometallic chemistry and heterogenization of homogeneous catalysts: different strategies at the interface between homogeneous and heterogeneous catalysis. Adapted from ref. [44].

While the importance of such studies for understanding active sites on surfaces is obvious, application of the organometallic toolbox to anchored surface species (i. e., SOMC) has extended the potential of grafting methods tremendously. SOMC should be differentiated from the apparently similar strategy of supporting homogeneous catalysts on solid surfaces (cf. Section 2.3.3). The difference is illustrated in Figure 3.18. While the active element is attached directly to the support surface in SOMC, catalytic complexes are preferentially fixed to the surface via flexible linkers, which shield them from the often adverse chemical influence of the support.

SOMC was started with the intention to model active sites and the mechanistic steps proceeding on the surfaces of real catalysts. The role of the support as a rigid ligand is therefore a subject of study rather than an undesired feature. Meanwhile, the development has gone far beyond the original goals. Groups working on SOMC have found many well-controlled routes to promising catalysts (see, e.g., ref. [45]), but have also shown previously unknown reactions to proceed on heterogeneous catalysts, e.g., alkane metathesis.

Alkane metathesis (Scheme 3.2) is catalyzed by surface Ta or W hydride species ($[Ta]_s$-H, $[W]_s$-H). Scheme 3.2 exemplifies, how $[Ta]_s$-H can be obtained by anchoring a Ta(V) carbene complex to the silica surface and hydrogenating the product. Metathesis of alkenes and of alkynes has been known for long, and while alkenes of low polarity can also be reacted over heterogeneous catalysts (cf. Section 4.5.2.3), the field is a stronghold of homogeneous catalysis. Alkane metathesis, however, was reported in the late 1990s [46] and is known only in heterogeneous catalysis to the best of our knowledge.

$$\text{e.g., } [Ta]_s\text{- H}$$

$$2\ CH_3\text{-}CH_3 \rightleftharpoons CH_4 + CH_3\text{-}CH_2\text{-}CH_3$$

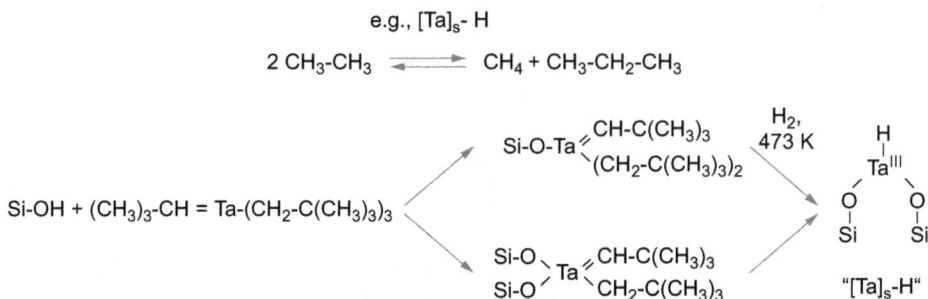

Scheme 3.2: Surface organometallic chemistry (SOMC): synthesis of surface Ta hydride species (lower part), which catalyze the metathesis of alkanes (top line: ethane metathesis). Reproduced from ref. [47]. Copyright 1996 American Chemical Society.

Different from the early mechanistic proposal in ref. [46], alkane metathesis was later found to proceed via surface Ta or W carbene species, similar as alkene metathesis (cf. Figure 4.37). The carbenes are formed from the alkanes by H abstraction, and the product carbenes are hydrogenated finally. More detail can be found in ref. [48], where additional examples for the extraordinary reactivity of $[Ta]_s$-H and $[W]_s$-H are reported.

As mentioned above, anchoring can also be used to attach active elements (or promotor species) to surfaces of supported metal particles. An early example was shown in eq. (3.10) above. By saturating an alkyl ligand, the H atom enables a direct (and exclusive) deposition of Sn alkyls on the Pt surface. As the stability of these alkyls is low, early work was focused on the modification of supported Pt by surface alloying with Sn [49]. Later on, it was shown that residual surface alkyl species, preserved by a controlled decomposition route, can create remarkable selectivity effects in the hydrogenation of unsaturated alcohols run in hydrophilic media (e.g., alcohol/water mixtures) [50].

While SOMC has revealed routes to remarkable catalytic materials, many of these routes are difficult to scale-up and/or rely on expensive reagents. Although commercial success was limited by these aspects, results obtained by SOMC often disclose opportunities of improvement of known catalyst systems or indicate likely steps in reaction mechanisms.

As ALD is closely related to CVD, it will be discussed in Section 3.1.3.6.

3.1.3.5 Deposition of colloids and melt infiltration

In the techniques discussed so far, precursors are deposited on surfaces (or attached to them) and converted to the desired chemical form in subsequent steps, e.g., to metal particles. Colloidal chemistry provides an alternative route: the deposition of ready-made colloidal metal or oxide nanoparticles. The colloids are first synthesized from a homogeneous solution of the catalyst precursor, which additionally contains capping agents that prevent coalescence and excessive growth of the particles. For their deposition on the support, a conventional technique is used, typically wet impregnation.

While colloidal dispersion can be achieved for various kinds of solids, routes for catalyst synthesis mostly involve metal colloids. They can be obtained via three major approaches, which have been reviewed in ref. [51]: by chemical reduction, by electrochemical synthesis, and by thermal decomposition of organometallic compounds. In the chemical reduction route, burst nucleation (cf. Section 3.1.1.1) is achieved by injecting a reductant ($NaBH_4$, H_2, hydrazine, alcohols) into a solution of the precursor compound (e.g., $HAuCl_4$, H_2PtCl_6, metal salts). Surfactants, soluble polymers, oleic acid, or long-chain phosphins are typical stabilizers present in the solution. In the polyol method, polyalcohols like ethylene glycol, are reductants and stabilizers at the same time. In electrochemical synthesis, the metal content of the colloids originates from a sacrificial anode. The dissolved cations are discharged at the cathode and nucleate into nanoparticles before they are trapped by the stabilizer, typically, tetraalkyl ammonium ions. Colloid synthesis by thermal decomposition is performed by injecting an organometallic precursor into a hot solution of the stabilizer (hot injection).

The quality of colloid-derived catalysts does not only depend on the properties of the nanoparticles, but also on their spatial distribution on the support. As the concentration of the colloidal solution needs to be kept small to confine the risk of coalescence, wet impregnation is the preferred deposition method. Due to the low diffusivity of the colloids, the time required to achieve homogeneous distribution across a pellet is much longer than in impregnations of low-molecular precursors. As the interactions between the stabilizing shell and the support are often weak, colloids can be re-distributed toward the external surface during solvent evaporation (cf. Section 3.1.3.2). Inhomogeneous distribution of colloids on supports, in particular egg-shell profiles were often reported [27].

When electrostatic interactions dominate the relation between the colloid shell and the support, the behavior of the deposits critically depends on the pH value (cf. Section 3.1.3.1). The colloidal shell may have its own IEP, different from that of the support. In this situation, a stable deposition will be achieved only in the narrow range of pH values, which gives rise to opposite charging of the colloids and the support surface.

After impregnation, the particle surfaces are still shielded from the surrounding fluid by the stabilizing shell. Although colloids may catalyze reactions even through this shell, which is apparently not always dense, its removal is standard in the colloidal deposition route. It can be achieved by thermal decomposition in inert gas, with noble metals by calcination, or by extraction. Thermal methods bear the risk of particle growth; extraction may cause re-distribution and even loss of particles. The effects of such post-treatment are not yet sufficiently studied. Ref. [27] cites a case where an Au/TiO_2 catalyst made of Au colloids of initially 3 nm size and stabilized by poly(vinylalcohol) (PVA) was five times more active in CO oxidation after extraction of just 20% of the PVA than after total PVA removal by calcination, although the particle size after extraction was only slightly lower than after calcination (4.8 nm *vs.* 6.1 nm).

Melt infiltration [27] takes advantage of the fact that some precursors, e.g., acetyl acetonates or crystal-water containing nitrates, melt at low temperatures. Support and precursor are just physically mixed and heated above the melting temperature of the latter, which results in pore filling due to the capillary forces. The method is similar to impregnation, but does not require a solvent. Due to the higher melt viscosities, longer periods of time are required for the distribution of the precursor over a catalyst pellet than in impregnation. Under suitable conditions, melt infiltration allows achieving very high loadings of the support: according to ref. [27], the pore system of SBA-15 could be completely filled with amorphous Co nitrate species by infiltrating a $Co(NO_3)_2 \cdot 6\ H_2O$ precursor.

Melt infiltration should be differentiated from solid-solid methods (Section 3.1.3.7), which operate far below the melting point of the precursor. For more information about the technique, the reader may consult ref. [27].

3.1.3.6 Deposition of precursors from the gas phase

Chemical vapor deposition (CVD) allows synthesizing nanoparticles or thin films on catalyst supports from the gas phase. It requires precursors with sufficient thermal stabilities and suitable vapor pressures, which are charged to the support, either admixed to a flowing inert gas or under vacuum. Metal halides, hydrides, alcoholates, or metal-organic compounds are typical precursors used for CVD. In the latter case, the technique is designated as metal-organic CVD (MOCVD). The primary interactions between the precursors and the surfaces are not always well studied, but are in many cases likely to resemble eq. (3.9).

The complex allyl cyclopentadienyl palladium (Pd(allyl)Cp), for instance, sublimes at 323 K and adsorbs readily on O-functionalized CNTs at 353 K. During adsorption, functional groups of the support replace organic ligands of the complex. Pd nanoparticles with a narrow size distribution between 2 and 4 nm were formed during the subsequent reduction in H_2 at the same temperature [52]. The creation of Cu-Zn interactions by exposing a Cu/Al_2O_3 catalyst to $Zn(C_2H_5)_2$ is another instructive example of MOCVD, which allowed emphasizing the importance of Cu-Zn interactions for the methanol synthesis reaction [53]. Combining MOCVD with CVS (cf. Section 3.1.1.5), a supported metal catalyst can be made completely from gaseous precursors. In the proof of principle, Pd was deposited on CVS-prepared SiO_2 and TiO_2 particles in a sequence of a CVS and a MOCVD step, the latter using the Pd(allyl)Cp precursor (see above). The narrow Pd particle size distributions that were obtained allowed deriving realistic turnover frequencies and structure-activity correlations in ethylene hydrogenation [54].

In Atomic layer deposition (ALD), anchoring reactions similar to eq. (3.9) are performed with surfaces fully covered by reactive OH groups. As these are offered only by the support, the reaction is self-limiting. In a stepwise procedure, residual ligands of the precursor that forms the first layer are hydrolyzed to obtain anchoring groups for a second layer. Highly uniform, conformal, and dense films were, for instance, achieved by exposing a hydroxylated substrate to trimethylaluminum (Al(CH$_3$)$_3$, TMA) under vacuum at elevated temperatures. The reaction between TMA and the substrate OH groups resulted in a monolayer of Al-methyl species. After evacuating residual TMA, a hydroxylated Al$_2$O$_3$ surface was obtained from this layer by admitting water vapor. In the example cited, the chamber was evacuated again, and the four-step ALD cycle was repeated many times, which resulted in precise thickness control [55]. Homogeneous layers can be grown in μm dimensions. Notably, ALD can deposit layers also in pores of the substrate.

With metals, layers may break up into nanoparticles, which allows applying ALD for the synthesis of supported metal catalysts. Only a small number of ALD cycles is required to obtain highly dispersed noble metal particles. In ref. [27], Pd particles up to 2.9 nm size were reported to grow uniformly distributed on an Al$_2$O$_3$-coated silica surface when up to 25 cycles were performed with the precursor Pd(hfac)$_2$ (hfac – hexafluoro acetylacetonate). Particles of just 1.1 nm size were accessible with a lower number of cycles. By using two different precursors, well-mixed PtRu particles of 1–2 nm size could be synthesized on Al$_2$O$_3$.

Stabilization of nm-sized metal particles by the deposition of alumina patches on them is another opportunity for ALD in catalysis [27]. For this purpose, some layers of Al may be deposited as trimethylaluminum, both on metal and support, according to the protocol mentioned above. Alumina is then formed by a final calcination. It was observed that the metal remained accessible over many cycles. Noble metal particles covered by alumina patches did not grow under thermal loads that caused severe sintering of unprotected particles.

While the scientific interest in techniques involving precursor deposition from the gas phase is plausible, availability, prices, in particular of organometallic precursors applied in MOCVD, and safety requirements are important aspects that may impede the application of such catalysts on large scales. However, even their preparation in the lab requires attention to some principles of reaction engineering to avoid pitfalls.

Thus, fixed-bed reactors are inappropriate for CVD preparations because of the concentration gradients along their axis. Instead, fluidized-bed reactors can be applied for all stages of the synthesis – like drying of the support, deposition, conditioning (e.g., by hydrogenation), and calcination. They provide efficient mixing of the fluidized support particles, and establish an isothermal regime (cf. Section 2.8.2). Saturation of the carrier gas with liquid precursors can be achieved in leak-tight fully metal-sealed stainless-steel bubblers. Solid precursors can be evaporated using furnaces or laser heating.

Due to the application of ALD in semiconductor production, fully automated setups with fluidized beds are commercially available, which provide pulse valves with minimized dead volumes, and enable precise pressure and temperature control. To facilitate diffusion of the precursor in porous supports, these setups usually operate below atmospheric pressure. They allow applying the complex pulse profiles consisting of flushing, deposition, and reaction steps, in a reproducible way, many times. The opportunities offered by ALD have raised hopes that catalyst preparation can be performed with similar controls over product properties as the production of semiconductor devices, one day. Research is under way along these lines, but results have not yet been communicated.

3.1.3.7 Solid-state methods

Similar to melt infiltration, procedures in solid-state methods are very simple: support and precursor are mixed (with zeolites, the precursor may be impregnated onto the external surface of the crystals), and heated. Temperatures involved are, however, well below the melting temperature of the precursors. The preparation of MoO_3/Al_2O_3 or MoO_3/TiO_2 catalysts by heating the supports mixed with MoO_3 to 723–773 K is a well-known example, likewise the "solid-state ion exchange" (SSIE) of cations into zeolites by heating their H- (or NH_4–) form mixed with the corresponding chlorides (cf. (eq. (3.11), for a monovalent chloride and the H-form of a zeolite Z). The spreading of MoO_3 over Al_2O_3, initially reported in China [56], was studied in much detail by Knözinger et al. in the 1980ies (see, e.g., ref. [57]). SSIE is nowadays used to introduce cations into zeolites in several technical applications.

$$MCl + Z–OH \rightarrow HCl + ZO^-M^+ \tag{3.11}$$

It is not easy to decide the form in which the active element spreads out across the surfaces in these preparations. The low sublimation points of some metal chlorides may suggest transport through the gas phase. MoO_3 can form a rather volatile oxyhydroxide $MoO_2(OH)_2$ in the presence of moisture, which causes Mo losses in some industrial processes (cf. Section 4.4.1.2). However, SSIE also proceeds swiftly with alkali chlorides [58], and even Cu_2O and CuO were observed to fuel SSIE of Cu^{2+} into H-ZSM and NH_4-Y zeolites, though at higher temperatures than CuCl and $CuCl_2$ [59]. In [57] it was shown that MoO_3 spreads over the Al_2O_3 surface under completely dry conditions, but does not bind to the support via oxygen bridges. Surface molybdates were observed only during spreading in a water-containing atmosphere.

The mechanism of dry spreading is apparently based on surface diffusion. Crystals of a non-volatile source decompose when the interface energies between the dispersed species and the support surface exceed the cohesion energy between the layers in the source crystals. As long as the pore system of the support is void of dispersed species, there is a gradient driving their diffusion.

When precursors of significant volatility are used, gas-phase transport, paralleling surface diffusion, may enhance transport rates.

3.1.4 Thermal conditioning

Thermal treatments serve to stabilize the dried catalysts and to bring them reproducibly into a state in which they can be easily shipped and filled into reactors for the subsequent activation step. Before dealing with aspects related to the choice of treatment conditions, a general rule should be noted: a catalyst should have been exposed to a temperature of some K above the range specified for its operation regime. When this is skipped, catalyst performance can be affected by the omitted stabilization processes proceeding in the background.

Besides the temperature, the composition and flow rate of the gas phase are important process parameters. Calcination (thermal treatment in air) is the cheapest version. It may be performed in stagnant air or in flowing air, in forced flow through a catalyst bed, or in air flowing across vessels with powdered material. The O_2 content of the gas phase may be decreased by blending air with nitrogen, or by applying pure inert gas. Sometimes, the gas flow may contain additional components (e.g., NO, see below). Treatments in reducing atmospheres are not considered in Section 3.1.4: they produce highly active surfaces and are part of activation protocols.

Heating dried products of catalyst synthesis may induce many different processes. Water is released from various sources; other solvents are evaporated if applied in the preparation. Depending on the ingredients, other gases may arise from precursor decomposition. In the presence of O_2, organic ingredients are oxidized. Components of the solid phase may be oxidized as well, depending on solid-state mobility, either only at the surface or in ranges extending into the bulk. Annealing in inert gas can cause (auto) reduction of labile surface species.

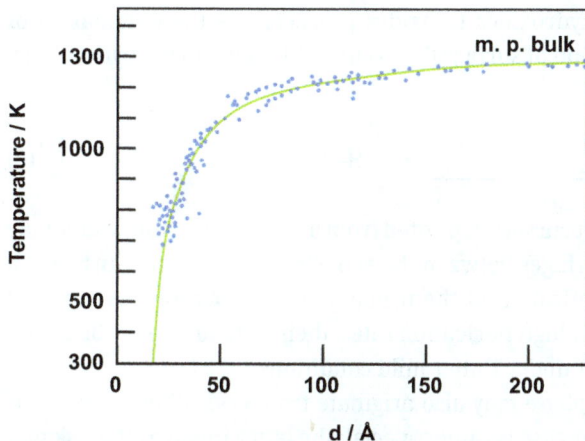

Figure 3.19: Dependence of the melting point (m. p.) of gold on particle size. Adapted with permission from ref. [60]. Copyright 1976 American Physical Society.

In addition, amorphous precursors may crystallize, defects in primary structures may be healed, phase transitions may be initiated, spreading of precursor components may be induced (cf. Section 3.1.3.7), and particle sizes may start increasing (cf. Section 4.4.1.3). These processes involve diffusion within the solid phases or at interfaces between grains, which gain momentum at higher temperatures. Bulk diffusion, i.e. (inter)change of atom positions within a crystal, can be observed above the Tammann temperature; healing of surface defects starts at the Hüttig temperature. As a rule, Tammann and Hüttig temperatures are related to the melting temperature T_M of solid crystals, according to

$$T_{Hüt} \approx 0.3\, T_M \quad T_{Tamm} \approx 0.5\, T_M \text{ (all temperatures in K)} \tag{3.12}$$

When applying these rules, it should be, however, kept in mind that melting temperatures depend on the particle sizes: as an example, Figure 3.19 shows the particle-size dependence of the melting temperature of gold [60].

The growth of particles is usually referred to as "sintering" in catalysis literature. This deviates somewhat from the general use of this term, which specifies a process, in which a powder is densified to an agglomerate with gradually decreasing porosity, and finally to a bulk body by solid diffusion processes at temperatures shortly below the melting point. On the other hand, spreading of oxide particles by surface diffusion (cf. Section 3.1.3.7) may decrease oxide aggregates left on the surface from a suboptimal preparation route. In systems allowing such spreading processes, the coverage of monolayer surface oxide species may improve during thermal treatments, although it is not granted that the spreading rates are sufficient to iron out all blobs produced by a poor deposition procedure.

Dehydroxylation is the most universal source of water evolving during thermal treatment. Conversion of (oxy)hydroxides to oxides proceeds via the elimination of structural OH groups. Condensation of surface OH groups (3.13) lowers the hydrophilicity of the surface.

$$\underset{|\quad\;|}{OH\; OH} \;\to\; \underset{\diagup\diagdown}{O} \;+\; H_2O \tag{3.13}$$

When supports are covered by precursors deposited from aqueous solutions, water may be released by forming oxygen bridges between the active element and the surface. All these processes result in a densification of the primary solid structures. As the solid phases have often been created at high nucleation rates, their structures may be rather labile, and stabilization may start under rather mild conditions.

Water appearing in the gas phase may also originate from crystalline hydrates or from the pores of gels. The former case is rather special; the latter involves the risk that the delicate pore system of the gel is damaged (cf. Section 3.1.1.3). Therefore, water is sometimes replaced by other solvents before drying gels.

Depending on preparation details, other components than water may be released during thermal treatment, e.g., alternative solvents or gases like NO, NO_2 (from nitrate decomposition), CO_2 or even CO. Carbon oxides result from combustion or decomposition of anions, like carbonate or oxalate, and from the combustion of organic material, e.g., of templates. In some preparations, organic material is trapped on purpose in solid particles formed, e.g., by precipitation. This allows creating a secondary pore system in the particles by combustion of the organics. Total oxidation to CO_2 and H_2O may be expected if the catalyst contains a redox active component. If not, e.g., in template removal from zeolites or OMMs, the effluent may contain significant quantities of CO. While most processes occurring during thermal treatment result in a densification of structures and a loss of external surface area, the clearance of pore systems by combustion of templates or sacrificial organics or by evaporation of solvents are examples for the opposite.

Processes during thermal treatment may be endothermal or exothermal, and their heat effects may cause significant deviations of the sample temperature from the nominal one. Evaporation of solvents, dehydrations, and dehydroxylations are endothermal. The same holds for most reactions of precursor decomposition, except for those involving oxidations. Combustion of organic components is obviously strongly exothermal. It is often started with low oxygen content in the gas phase to avoid hotspots in the sample. Phase transitions during temperature increase are, in principle, endothermal. In precursors for catalysts, however, the new phase often consists of larger particles with smaller external surface areas, and due to the release of surface energy, the net heat effect may be exothermal. This holds, in particular, for the breakdown of metastable supports into stable phases, e.g., γ-Al_2O_3 into α-Al_2O_3. The sizeable exothermal effects of these transitions arise not only from the gain of surface energy, but also from the formation of more ordered and more stable structures. Likewise, crystallization of amorphous phases causes exothermal effects.

Calcination may cause an exposure of catalysts or precursors to a water-containing gas phase at high temperatures. Such hydrothermal stress may induce undesired structural changes in the material being calcined. The presence of water, even in low concentrations, favors mobility phenomena by a variety of mechanisms, some of which are not yet well understood. It may result in the premature loss of surface area by sintering, in excessive growth of particles, in premature stabilization of metastable structures, and in re-distribution or even the loss of the active component.

As mentioned in Section 2.3.2.3, hydrothermal stress causes the dealumination of H- (or NH_4-) zeolites: the detachment of the framework Al sites from their framework. Except at very high Al contents, this does not result in structural collapse. Instead, the defective, more siliceous zeolite framework remains intact on the whole due to concomitant healing processes, and resists better the loss of residual Al. The remaining acid sites are stronger than those in the Al-rich starting materials. As the dealuminated material has survived (and will in future survive) hydrothermal conditions, it is called "stabilized." Zeolites, stabilized by steaming, are used in important technical processes, e.g., FCC (cf. Appendix A1).

Dealumination of zeolites by steaming is special, because it inflicts a limited structural damage to improve a property more important than structural perfection: stability. In most other cases, the water content needs to be kept very low, which can be achieved by raising temperatures slowly, keeping gas flows high, and arranging forced flow through the material to be treated (calcined). Fine powders (e.g., zeolites), which cannot be calcined in fixed beds, should be exposed to the flowing gas in shallow pans, instead of high crucibles (shallow-bed conditions, as opposed to deep-bed conditions in stagnant air for steaming). While rotary kilns are another favorable option in this situation, damage by evolving moisture can be suppressed most efficiently in fluidized beds (cf. Section 2.8.2).

Attention paid to the risks of thermal treatments was sometimes limited in the past, but it has meanwhile been appreciated how badly inappropriate treatment can spoil promising approaches. Calcination of precursors for methanol synthesis catalysts is an example for the benefit of tight control on process conditions. As the reduction of Cu^{2+} in the mixed (Cu,Zn)hydroxycarbonates formed in the initial coprecipitation (cf. Section 3.1.1.1) would require high temperatures, the precipitates are calcined to obtain the easy-to-reduce CuO. However, the dispersion of the oxide, and thus of the metal resulting from subsequent reduction, peaks at calcination temperatures as low as 573–623 K, where the product contains small CuO particles intermixed with finely divided zinc hydroxycarbonate residues [61]. At higher calcination temperatures, these residues are also converted to ZnO, which starts to aggregate. This removes a significant obstacle to sintering of both CuO during calcination and of Cu during the final reduction.

As mentioned in Section 2.3.3, calcination of MoO_3/Al_2O_3 precursors for hydrodesulfurization catalysts is another example where excessive severity may result in failure. The appropriate conditions result from a tradeoff: while calcination keeps the surface Mo oxide species in the precursor well dispersed, the reluctance of the Mo-O-Al bridges to yield to sulfidation may result in sintering of the target sulfide slabs. In Section 2.3.3, other strategies to optimize the interactions between the Mo species (and promoters) and the support are discussed as well.

Recent work with nitrates impregnated onto SiO_2 (actually SBA-15) by K. P. de Jong's group [27] resulted in an unusual calcination strategy involving gas-phase NO, which prevents the formation of O_2 during the decomposition of nitrate ions:

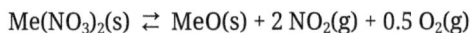

$$Me(NO_3)_2(s) \rightleftarrows MeO(s) + 2\ NO_2(g) + 0.5\ O_2(g)$$

$$Me(NO_3)_2(s) + NO \rightarrow MeO(s) + 3\ NO_2(g)$$

For Ni and some other elements, the decomposition of nitrates (or actually hydroxynitrates, which may form during drying) produced big particles when the decomposition products – O_2 or NO_2 – were present in the gas phase, while dilute NO resulted in highly dispersed oxide particles. Particle sizes obtained in pure inert gas were intermediate. For Ni, NO was found to favor the formation of NiO nuclei at low temperatures by scavenging the adsorbed oxygen radicals originating from the nitrate ions. As growth rates are low under these conditions and NiO catalyzes nitrate decomposition as well, the number of NiO nuclei became very large and the particle size small. The deterioration in the absence of NO can be explained by assuming that the combination of oxygen radicals to O_2 is rate-determining for the (uncatalyzed) nitrate decomposition: the reaction then starts at higher temperatures where the higher growth rates result in larger particles. With NO_2 and/or O_2 in the gas-phase, nitrate decomposition is further delayed for equilibrium reasons, with even stronger adverse effects for the NiO particle size.

3.1.5 Activation

During activation, the catalyst precursors are transformed into the active state to achieve their full catalytic potential. Surfaces become highly reactive, often pyrophoric, and cannot be exposed to the ambient atmosphere any more. Therefore, the catalytic reactor is the place where catalysts should be activated. However, to avoid facing downstream equipment with problematic product streams, which might contain large loads of water or H_2S, catalysts are sometimes activated (hydrogenated, sulfided) in special equipment and subsequently passivated by contact with traces of O_2 under very mild conditions. After installation in the catalytic reactor, re-activation by contact with the feed causes less trouble with the undesired effluent components. In HDS technology, this approach even competes with a procedure where the fully activated catalyst is sealed in containers and installed in commercial-scale reactors under protective gas.

Activation is most straightforward with acid catalysts, which require only a moderate heat treatment to desorb water adsorbed on their active sites during handling at ambient conditions. The conversion of methanol to hydrocarbons over acidic zeolites is special in this respect, because it proceeds via an adsorbate pool, which is permanently fed by (catalytic) incorporation of reactant molecules and releases product molecules by cracking steps (cf. Sections 4.3.2 and 4.5.5). The buildup of these pools causes an initial transient, even with a perfectly activated catalyst. Basic catalysts may require considerable thermal stress to get rid of carbonates formed with atmospheric CO_2.

Activation procedures and phenomena during the initial interaction with the feed are most variable in redox catalysis, where the catalytic elements are usually in higher oxidation states after calcination than during steady-state operation. There are cases where even the active phase develops from a precursor phase only on stream, e.g., the formation of vanadyl pyrophosphate ($(VO)_2P_2O_7$), which catalyzes the selective oxidation of n-butane to maleic anhydride, from $(VO)HPO_4$ precursors. It may last for hours or for days, depending on conditions [62]. In the stationary operation of Fischer–Tropsch synthesis, adsorbed CO and hydrogen coexist on the catalyst surface with C_xH_y adsorbates of different chain lengths, which are generated by splitting CO into C_1 species and water, followed by surface polymerization (cf. Section 4.5.3). In the initial transients with the Fe- or Co-based catalysts, processes like precursor reduction (by CO and H_2), carbidization of the metals (for Fe), their reoxidation by H_2O becoming abundant with increasing rates of CO dissociation, and buildup of the adsorbate phase superimpose in a complex way.

Catalysts for (de)hydrogenation reactions (including the hydrogenation of N_2 or CO) are initially reduced in H_2 or in H_2-containing feeds. As suggested in Section 3.1.4, water released during these reductions can damage the dispersion of the emerging reduced phase. In work with 0.5 wt.-% Pt/γ-Al$_2$O$_3$ catalysts, for instance, the author of this chapter observed that hydrogen chemisorption capacities measured to assess the Pt particle size (cf. Section 3.4.1.6) decreased by 15–20%, when samples previously calcined at 773 K were briefly exposed to air for determining their weights prior to sample reduction, as compared to a procedure where such exposure was avoided: the moisture trapped by the dry support surface (ca. 200 m^2/g) was sufficient to inflict a measurable damage on the Pt dispersion, resulting from the final reduction. Therefore, catalyst reduction is performed under strict control of gas-phase water contents even in huge reactors in industry. In a typical procedure, the temperature is raised using a slow ramp until some K below the onset of reduction. Its further increase is then controlled in such a way that the water content of the effluent remains below a specified threshold at all times.

HDS catalysts are activated by exposing them to H_2 mixed with a sulfur compound at increasing temperatures. On the lab scale, H_2S/H_2 mixtures are employed for this purpose, in industry – labile compounds, like dimethylsulfide or thiophene, are dissolved in the process feed. Passivated catalysts (see above) may be, for instance, impregnated with a hydrocarbon mixture enriched with such sulfur compounds. In the reactor, oxygen is removed from the passivated layer by H_2, which recovers some of the HDS activity. S atoms trapped from the impregnated compounds heal defects imposed on the sulfide structure during passivation so that the catalysts gradually achieve their nominal activity.

Catalysts experience their first contact with the feed, for which they are designed, either right at the start of the activation procedure or after some preliminary step like reduction or sulfidation. Their response to the full feed depends on the nature of the precursor or the pre-activated form, on the reactivity, which these offer during their conversion to the stationary surface state, on the feed composition, and on the reaction conditions. Strong deviations between the initial and stationary activity and selectivity are typical for redox catalysts, although an initial decay of the activity to a stationary level, caused by deactivation of the most active sites, can also be observed with solid acids. In redox catalysis, the initial activity may exceed or be short of the stationary level.

The initial transients, in which oxidation catalysts adapt to their stationary oxidation state, are sometimes referred to as catalyst formation processes. They may be confined to the atomic layers adjacent to the external surface or comprise bulk phases like the conversion of (VO)HPO$_4$ to (VO)$_2$P$_2$O$_7$ mentioned above. In catalysts for selective oxidations, the oxygen removed initially is more reactive than the lattice oxygen available during the stationary operation along the Mars-van Krevelen mechanism (cf. Section 2.5.1). This does not necessarily result in an excessive release of H_2O and CO_2, because H and C atoms of the first reactant molecules may be used to build up a characteristic coverage of surface groups like OH, carboxylate, or carbonate. The further course of the transient depends on the stability of the target product toward overoxidation. If it is vulnerable, total oxidation dominates initially, and selectivity increases with decreasing average oxidation state of the active element. If it is stable, the initial total oxidation may turn into a phase with enhanced formation of the target product, the yields of which decrease toward the stationary phase. In such a case, a change of the gas-phase composition to allow a higher stationary oxidation state of the active site may be promising.

Supported metal nanoparticles may start with a better performance than achieved in the steady state, because the high particle dispersions obtained during precursor reduction are usually not stable under reaction conditions. When the formation of the active surface requires the participation of reactants or intermediates (e.g., carbided surfaces or surfaces from adsorbate-induced reconstructions), stationary behavior will be achieved via increasing performance.

The surface of nm-sized metal particles always contains highly exposed atoms, which exhibit enhanced reactivities (cf. Sections 2.1, 4.2.2, and 4.2.3). In Pt-catalyzed hydrocarbon conversions (e.g., naphtha reforming, Appendix A1), which are always performed in the presence of H_2 to delay coke deposition on the catalyst, these "hyperactive" sites catalyze hydrogenolysis instead of dehydrogenation. The hydrocarbon feed may be converted to methane over many hours until these sites are gradually deactivated by carbon deposits. The problem is solved by charging sulfur-containing feed initially. Adsorbing on almost all surfaces, sulfur is a strong poison for most catalytic reactions (cf. Section 4.4.1.1). However, under the conditions of naphtha reforming, its adsorption is reversible on the (de)hydrogenation sites, while the hyperactive sites remain blocked. After such stabilization by the injection of S-containing feed, the performance gradually improves to the stationary level. This discussion does not apply to Pt-Sn-based reforming catalysts, which can be charged with the full feed immediately after reduction, because the hyperactive sites are poisoned by the tin component.

Knowledge about the initial transient catalyst behavior is important for finding suitable regimes during the startup of the catalytic reactors on commercial and lab scales. However, research about the nature of these phenomena can also help in understanding the catalytic surface processes because the transients end in the stationary surface conditions.

3.1.6 Preparation of catalytically active electrodes

Catalytic electrodes need to ensure a good electric contact and a strong adhesion between the catalyst and the electrode material. An even distribution of the catalyst powder over the electrode surface is also important. Electrodes used on the lab scale may be planar (e.g., glassy carbon) or porous (e.g., Ni foam). Catalysts are deposited on them from inks and glued to them by binders, which are stable in the electrolyte and withstand the applied potentials, e.g., by Nafion®. In the ink, the catalyst powder is suspended in a carrier liquid (e.g., a water/ethanol mixture), in which the binder is dissolved. For improving the electrical contact between the catalyst particles and the electrode, conductors such as carbon black can be added. The inks can be applied to planar electrodes by drop coating and to porous electrodes by spray coating. Preparation is completed by a drying step, which may start during coating when the electrode material is kept at an elevated temperature.

Achieving evenly coated electrodes with areas of up to the m^2 range is a major challenge for industrial applications. To prepare membrane-electrode assemblies (MEAs, cf. Section 2.8.4), techniques such as screen printing, calendaring, and hot pressing are applied to produce pinhole-free coatings with even thickness. In gas diffusion electrodes (GDEs), which are applied, for instance, in fuel cells and in chlor-alkali electrolysis with oxygen-depolarized cathodes (see Appendix A1), it is essential to avoid flooding. This is achieved by adding Teflon or Nafion® as a hydrophobic component. Gas diffusion electrodes are also promising for the electrochemical reduction of CO_2, which is currently a hot research topic (cf. Section 4.9.2). The preparation of such electrodes, in which the three-phase boundary between catalyst, electrolyte, and gas phase is optimized *vs.* various electric losses and transport resistances, mainly in the electrolyte, is a challenge for electrochemical engineering.

3.2 Rate measurements in surface catalysis

3.2.1 Thermal catalysis

Catalytic experiments are always rate measurements, even if the setup allows only establishing averaged rates. In simple screening using autoclaves, the degrees of conversion after a specified time are compared. This is based on the rate definition (1.2a), much alike to screening in the flow regime where conversions at a specified ratio between the catalyst weight "W" (in g) and the feed velocity F (in ml/min) are compared, because the weight/feed ratio W/F (in min g/ml) is proportional to the contact time at the catalyst. In a complex reaction system, the rates of the competitive steps can be determined directly from the conversion data, while rates of consecutive steps can be extracted only with computational effort. This section deals with the design of experiments suited to access unbiased rate data for the stationary state or rate information from transient measurements, which allows accessing rates of intermediate steps. For the sake of simplicity, we confine the treatment to a single catalytic reaction or a set of parallel reactions.

In any rate measurement, concentrations of reactants and products are measured before and after the feed has been converted over a catalyst. The catalyst is exposed in a reactor, the hydrodynamic characteristics of which have a profound influence on the result. The following treatment deals with the proper choice of a reactor for rate measurements. Basics and details of the various chromatographic, spectrometric, and even magnetic methods for concentration analysis are beyond the scope of this book.

It should be kept in mind that rate measurements must be preceded by tests on the absence of film and pore diffusion limitations under the given reaction conditions (cf. Section 2.6.2). This request is less stringent when just good and bad samples are to be differentiated in a preliminary screening, because a catalyst working in the diffusion-limited range will always outperform poor catalysts. However, for making choices between catalysts of comparable performance or for rate measurements required in engineering or science, tests to exclude mass transfer limitations are absolutely mandatory.

Reactors for catalytic rate measurements comply with the general classification of reactors (cf. Section 2.8.1): there are batch or flow reactors, and among the latter, there are reactors targeting either plug flow or full backmixing. In the flow regime, reaction rates can be studied under stationary or under transient conditions. Transient flow regimes are unusual in technical processes, which are much easier handled at steady states. In kinetic studies, transient methods are used to study reaction steps, which are at quasi-equilibrium in the stationary state (cf. Section 2.6.1). When the latter is suddenly disturbed, the response exposes these steps non-equilibrated, which allows measuring their rates under favorable conditions.

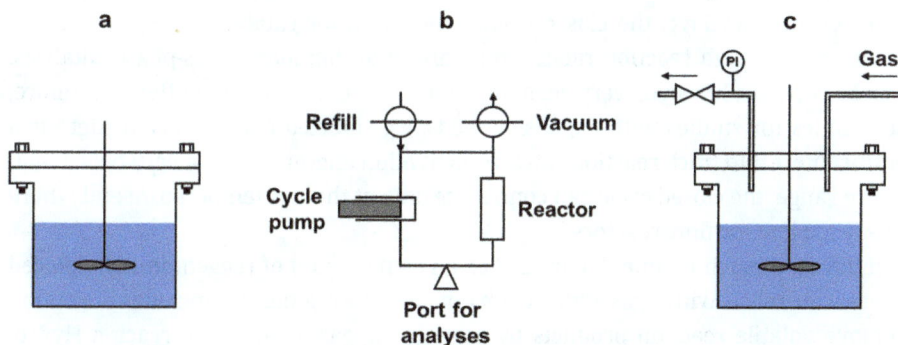

Figure 3.20: Reactors for measuring catalytic reaction rates: (a) batch mode: autoclave; (b) batch mode: closed cycle (loop); (c) semi-batch mode: autoclave with gas supply.

Batch reactors are typically applied for rapid screening of catalysts for liquid-phase reactions, in particular when elevated pressures are needed, because high-pressure equipment for continuous operation is much more expensive and intricate in handling than simple autoclaves (Figure 3.20a). Their obvious advantages of low cost and easy sample change come with a number of drawbacks, some of which can, however, be mitigated by special design features.

As the filled autoclave needs to be heated to the reaction temperature and cooled after a specified time, the duration of the run is not well defined. This can be improved by appliances allowing the injection of the catalyst or of a reactant at the reaction temperature. The end point needs less attention because rates are lower then. Product analysis is possible only when the autoclave is opened. There are, however, special designs that allow intermediate sampling. In the closed autoclave, the gas-phase composition cannot be influenced: it is completely defined by the initial pressure of the gaseous reactants, by their ongoing conversion, and by the temperature dependence of liquid-phase properties like vapor pressures and solubilities of the reactants and the products. All products of the main reaction and also of the side reactions remain in the liquid phase. If one of them is a catalyst poison, the results may be strongly biased. Both problems can at least be partly removed when the gas phase can be exchanged (see below), which is, however, already beyond the definition of a batch reactor. Finally, due to the adsorption of reactants or products on the catalyst, it may be difficult or even impossible to obtain good elemental balances in the analysis of these components. This holds, in particular, for catalysts comprising supports with very high specific surface area, e.g., active carbons, which can withdraw a significant percentage of reactants or products from the liquid phase.

The closed cycle (Figure 3.20b) is the gas-phase analog to the autoclave. At high circulation frequencies, the per-pass conversion over the catalyst is low. During a reaction, the gas phase composition is therefore almost identical in the whole cycle, with gradually growing conversion, as would be the case in a mixed reactor.

Unlike the autoclave, the closed cycle is not used for catalyst screening because it cannot handle high reaction rates, which are quite familiar in gas-phase catalysis. The closed cycle is, however, very economical in feed consumption. It offers, therefore, opportunities for studies with expensive feeds, e.g., labeled compounds. If there is a scientific interest to track reaction rates (or activation energies) over a very broad temperature range, the closed cycle can contribute data at the low-temperature end where rates are too low for flow reactors.

In the semi-batch regime (Figure 3.20c), a given amount of reagent initially placed in the reactor reacts with a gas added from outside. The regime also includes the option to remove volatile reaction products by passing the gas through the reactor. Hydrogenation reactions with controlled H_2 pressure are typical examples of the first case. These opportunities mitigate some limitations of the closed autoclave mentioned above. Removal of inhibiting products can be, however, achieved only if they are sufficiently volatile.

In the flow regime, the time available for the reaction is the residence time in the reactor. In homogeneous reaction systems, where the rate is related to the reactor volume V (eq. (1.2)), the hydrodynamic residence time is

$$\tau = V/\dot{V} \tag{3.14}$$

where $\dot{V}$ is the volumetric flow rate. As the reaction rate is related to m_{cat} in catalysis (eq. (1.5)), a modified residence time is used instead:

$$\tau_{mod} = m_{cat}/\dot{V} \; (= W/F) \tag{3.15}$$

with the unit [s g ml^{-1}]. τ_{mod} is often referred to as weight/feed ratio W/F. The dimensions of τ and τ_{mod} differ by the ratio between the catalyst mass and the feed volume (with the unit [g ml^{-1}]), which interconverts between the heterogeneous and homogeneous cases. It can also relate the units of rates (homogeneous – [mol l^{-1} s^{-1}], heterogeneous – [mol g^{-1} s^{-1}]) and of rate constants (e.g., first order: homogeneous – [s^{-1}], heterogeneous – [l g^{-1} s^{-1}]).

There are also alternative ways to communicate the relation between catalyst amount and feed velocity, e.g., the liquid hourly space velocity (LHSV) and the gas hourly space velocity (GHSV). They are defined as the ratio between $\dot{V}$ of the liquid or gaseous feed (the latter always given at STP[8]) and the catalyst *volume* V_{cat}. The dimension of both LHSV and GHSV is therefore a reciprocal of time (e.g., [h^{-1}]), which is sometimes referred to as "vvh" (volume per volume and hour) for gas-phase processes.

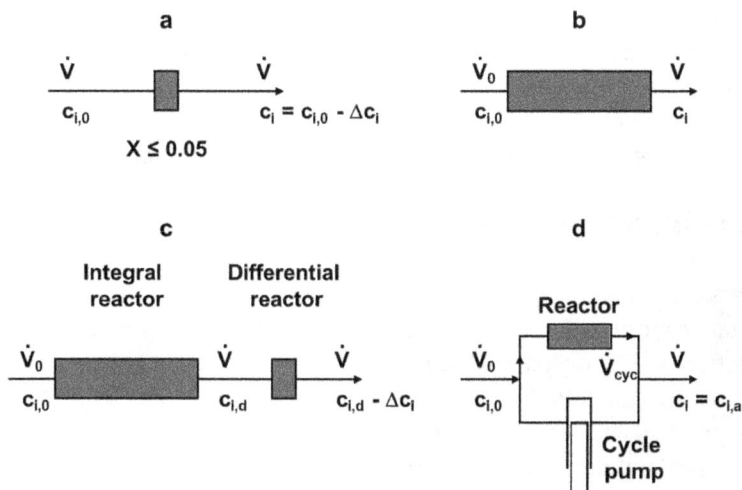

Figure 3.21: Flow reactors for rate measurements (I): (a) differential reactor; (b) integral reactor; (c) differential reactor with preceding differential reactor; (d) differential loop reactor with external cycle.

As mentioned in Section 2.8, the flow regime can be realized in tubular reactors (idealized – plug flow), where the reaction progresses along the reactor length, or in back-mixed reactors (idealized – CSTR). The latter include cycles with reactant feed and product withdrawal.

---

8 standard temperature and pressure, 273 K, 1 bar.

In tubular reactors (Figure 3.21a, b), concentrations, and with them the reaction rate, vary along the reactor, but product analysis is possible only at the end[9]. It represents, therefore the rates in averaged form. A rate constant can be deduced from such data only if the (correct!) rate law is integrated and the measured conversion/ residence-time pairs are compared with the conversion/time profile resulting from the model. If, however, conversion is confined to the differential range ($X \leq 0.05$), the rate according to eq. (1.5) can be measured directly:

$$r_{cat} \approx \frac{1}{v_i}\frac{1}{m_{cat}}\Delta \dot{n}_i = \frac{1}{v_i}\frac{\dot{V}}{m_{cat}}\Delta c_i \tag{3.16}$$

where $\dot{V}$ can be considered constant because of the limited conversion. This is referred to as the differential reactor approach whereas tubular reactors without constraints on conversion are integral reactors (Figure 3.21a, b, respectively).

The advantages of differential reactors (Figure 3.21a) are obvious: they allow measuring rates without the integration of a hypothetical rate law. By examining the response of rates to the variation of concentrations, kinetic orders may be established. As the rate is (nearly) constant along the reactor, there is no variation of the heat effect either. Therefore, isothermal conditions can be easily established. Due to the high flow rates $\dot{V}$ required to keep the conversions low, problems with film diffusion limitation (Section 2.6.2) are unlikely.

Unfortunately, these advantages come with serious drawbacks. The differential regime is very challenging for product analysis. This holds in particular for reactions in which the conversions cannot be determined via the reaction products because the differences in reactant concentrations ($c_{0,i}$ and $c_{0,i} - \Delta c_i$) are very small. For this reason, the limitation on differential conversion is sometimes heavily stretched in literature. Notably, rates measured in the differential reactor are always initial rates. Inhibiting effects of a reaction product, which are very familiar in catalysis, remain undetected. In equilibrium-limited reactions, the influence of the reverse reaction (cf. Hougen-Watson rate laws, Section 2.6.1) cannot be assessed.

Although the integral reactor (Figure 3.21b) mitigates challenges to product analysis and covers the whole range of conversions, the abandonment of the differential condition results in a number of problems, even beyond the requirement to integrate the rate law. Heat production or consumption, which is proportional to the reaction rate, varies along the reactor, and when heat effects are significant, isothermal conditions are difficult to achieve even with segmented heating or cooling schemes. The situation may be even more complicated when reactions with different heat effects form a consecutive reaction scheme. Temperature profiles may arise not only along the reactor

---

9 Custom-built devices with appliances to withdraw fluid samples along the reactor length are special equipment for reactor engineering research (ref. [63]).

(i.e., axially), but also on the radial coordinate, because heat transfer to or from the interior catalyst portions requires temperature gradients toward the reactor wall. Heat and mass transfer limitations may complete the list of potential challenges encountered in work with integral reactors.

Even if not all these challenges take action, extraction of constants for kinetic rate laws from the integral reactor data requires solving a multiparameter reactor model, which normally includes the temperature dependence of the rate constants due to the inevitable deviations from isothermicity. In chemical reaction engineering, such models are often established to demonstrate a grip on the processes in a reactor relevant for technical application rather than to establish physically significant rate parameters. Alternatively, kinetic models are established under less challenging conditions and inserted into reactor models with parameters kept fixed or allowed to be further optimized.

Placing a differential reactor downstream of an integral reactor (Figure 3.21c) removes some of the challenges just described. The integral reactor is required only to progress the reaction to levels of conversion where product inhibition or the influence of the back reaction become significant. If the analytics available copes with analyzing the differential conversions, this combination allows using the advantages of the differential reactor at any "integral" conversion.

Working with well-mixed ("gradientless") reactors is, however, the more universal approach to well-defined regimes for kinetic rate measurements. From the species balance in the CSTR (eq. (2.66b)), the rate of a catalytic reaction can be obtained by

$$v_i\, r = \frac{\dot{n}_{0,i} - \dot{n}_i}{m_{cat}} = \frac{\dot{V}_0 c_{0,i} - \dot{V} c_i}{m_{cat}} \tag{3.17}$$

As there is only one reactant concentration c_i in the CSTR (cf. Section 2.8), there is only one rate even at high conversions, which can be determined without the integration of the rate law. As there is only one rate of heat evolution or consumption, temperature profiles along the catalyst bed are minor, although the bed temperature may differ from the gas-phase temperature. By varying the W/F ratio and the feed composition (e.g., ratio between the reactants and the admixture of the reaction products), rate laws can be elucidated and data sets for the estimation of kinetic parameters can be collected.

In heterogeneous gas-phase catalysis, backmixing is either simulated by recycling or realized by employing an agitator. In the former case, a cycle containing the catalyst is included in the feed flow (Figure 3.21d). Conversions along the reactor in the loop become differential when the cycle flow $\dot{V}_{cyc}$ strongly exceeds the exit flow $\dot{V}$, which ensures that volume elements leaving the loop have been through many cycles. In each cycle, reactant molecules are converted according to the reaction rate supplied by the catalyst under the given conditions (T, p, composition in the loop), which makes an integral conversion from many differential per-pass conversions. To safely achieve differential conditions, the recycle ratio $\dot{V}_{cyc}/\dot{V}$ should approach 100.

a

b

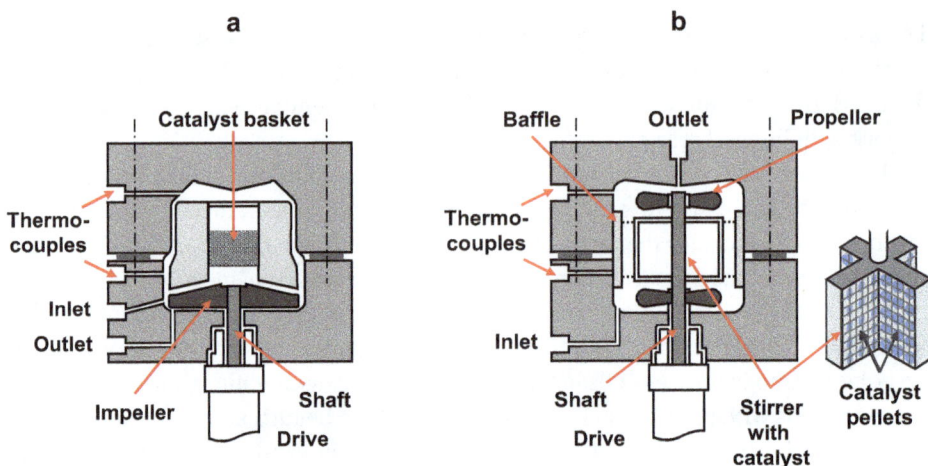

Figure 3.22: Flow reactors for rate measurements (II): (a) differential loop reactor with internal cycle; (b) spinning basket reactor.

There are differential loop reactors with external and internal recirculation (Figures 3.21d and 3.22a, respectively), which differentiates if the gas leaves the heated zone during the recycle or not. The external cycle (Figure 3.21d) is a rather unpretentious setup that is easy to build in-house. It allows the additional modification of the experiment, e.g., by condensing a reaction product from the cycle. In any case, as re-heating the recycled flow upstream of the reactor may require high temperatures at the heat-transfer surfaces, it should be made sure that these consist of catalytically inert material.

Internal recirculation is typically driven by a radial impeller, which forms a radial blower together with the housing, and a mold that frames the catalyst basket (Berty-type reactor, Figure 3.22a). The blower drives the cycle gas through a concentric draft between the housing and the internal mold. The gas is then redirected to the catalyst basket from which it is sucked by the impeller. While the fluid movement is directional in the Berty reactor, the spinning basket reactor (Carberry reactor, Figure 3.22b) is a true stirred tank. In this reactor, the gas phase is agitated by a stirrer that contains the catalyst pellets in its "blades." Despite the porosity of these "blades," the gas phase is partly taken into a rotary state, which results in a centrifugal movement. Baffles around the circumference disturb this movement and create turbulence, which supports the mixing process.

Both reactors were successfully employed in many kinetic studies, but the directed flow through the catalyst (Berty-type) achieves larger relative velocities between the fluid and the solid phase and is, therefore, less prone to problems with film diffusion limitations. Unlike the Carberry reactor, the Berty reactor allows measuring the reaction temperature directly in the catalyst bed. Both reactor types can also be used for studying catalytic reactions in the liquid phase and in three-phase systems.

The methods discussed so far deal with rate measurements under stationary conditions where fast elementary steps are in quasi-equilibrium. Fast reaction steps can be observed in transient regimes, when the reaction system is forced to cross over to another stationary state or when conditions are continuously changed. Such changes may be imposed on temperature or reactant concentrations: rates of the whole reaction or of elementary steps (e.g., reactant adsorption or desorption) may be tracked while the temperature is increased, pulse quantities of the reaction mixture may be injected into an inert gas flowing over the catalyst, or the composition of a stationary feed stream may be changed suddenly (step change). Most of these methods are used in catalysis research for other purposes as well, but some of them are exclusively tools for research on kinetics and mechanisms. In the following, methods based on concentration variations will be introduced, while temperature variation will be discussed, together with other temperature-programmed techniques in Section 3.4.2.

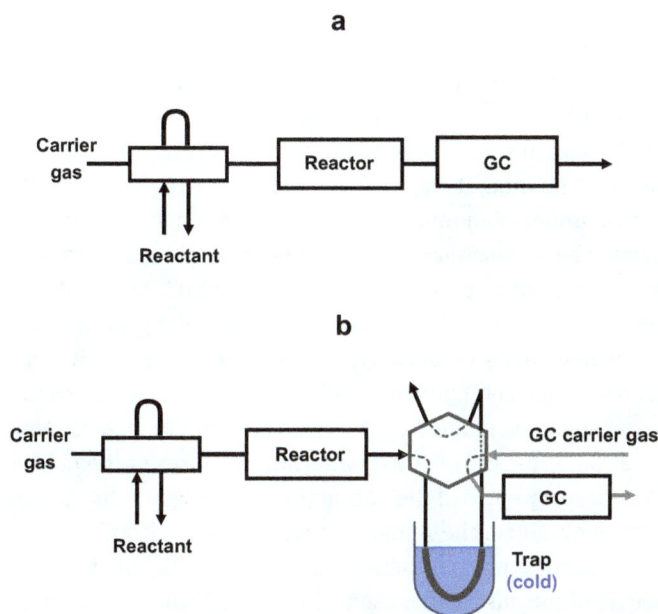

Figure 3.23: Setups for studying catalyst performance in the pulse mode: (a) with online analysis; (b) with off-line analysis.

In Figure 3.23, early setups for probing the reactivity of catalysts in the pulse mode are depicted. The version with online analysis was obtained by placing a small reactor between the injection port and the column of a commercial gas chromatograph (Figure 3.23a). Although the injected pulse is broadened while passing the catalyst, product separation on the column remains acceptable in many applications. The simplicity of the setup is, however, at the expense of flexibility: gas velocity and pressure must comply with the conditions required for chromatographic product separation. These limitations can be removed by collecting the product pulse in a cold trap (Figure 3.23b). After switching the six-port valve, quick heating of the trap injects the pulse onto the column in a narrow profile.

The pulse method is quick and cheap. Because both reactants and catalysts are required in only minor amounts, it is attractive for work with labeled compounds. Due to the low reactant quantities converted, there is a better chance to keep isothermicity than in a stationary regime. For the same reason, changes of the catalyst surface caused by each pulse are usually negligible. If there are fast changes, their course can be resolved just by analyzing a series of reactant pulses.

These advantages come, however, with serious drawbacks. The method probes catalysts in their initial state, which can be very far from their stabilized state (cf. Section 3.1.5). Reaction rates (conversions) that are obtained result from surface coverages very far from those under stationary conditions. Highly active sites present in small amounts may, for instance, contribute to these rates more than in the stationary regime, where a large quantity of less active sites hardly populated by the pulse may determine the rate. In addition, inhibition by reaction products is likely to be underestimated in the pulse mode because sites are refreshed by desorption between pulses.

For these reasons, the pulse technique is not used for catalyst screening any more. On the other hand, as a transient method, it should offer good opportunities to study reactions that are equilibrated under stationary conditions. Modeling pulse experiments at atmospheric pressure, where coverages may vary up and down to large extents in short times, is, however, very complex and not a promising approach for the determination of unknown parameters. Although there was some effort along these lines, pulsing into carrier-gas flows is now rarely employed for research on mechanistic steps and their rates. The application of this technique is rather limited to special purposes.

Such purposes may be mechanistic studies with labelled reactants or poisoning studies. In the latter case, the catalyst is charged with alternating pulses of the poison and of the reactant(s). With appropriate size of the poison pulses, relevant information about the site structure may be obtained: the number of sites may be titrated, and if there are sites of significantly different intrinsic activity, they may be distinguished by the loss of activity per molecule of the adsorbed poison. The method has been recently employed to identify the state of Zn in the active sites of $Cu/ZnO-Al_2O_3$ methanol synthesis catalysts [64]. The pulse technique may also be used to study rapidly changing catalyst states, e.g., the formation of stationary catalyst activity by the reduction of calcined transition metal oxides in reactant pulses.

The Temporal analysis of products (TAP) reactor (Figure 3.24) is an advanced version of the pulse reactor in which pulses are charged into vacuum, instead of being injected into a carrier gas. The concentration profile of the effluent flow is resolved on the time scale of fast mass-spectrometric analysis. Pulses are administered by high-speed valves, which provide pulse widths of ≤ 100 µs. Reactants may be pulsed premixed or separately with time delay using different valves (pump-probe experiments). Reactant amounts per pulse can be limited to quantities small enough to achieve predominant transport by Knudsen diffusion, which excludes gas-phase collisions between molecules. In oxidation catalysis, this is important to avoid corruption of results by gas-phase oxidation of reactants or products. Quantities of 10^{13}–10^{15} molecules/pulse routinely applied are well below the amount of active sites in the sample in most cases.

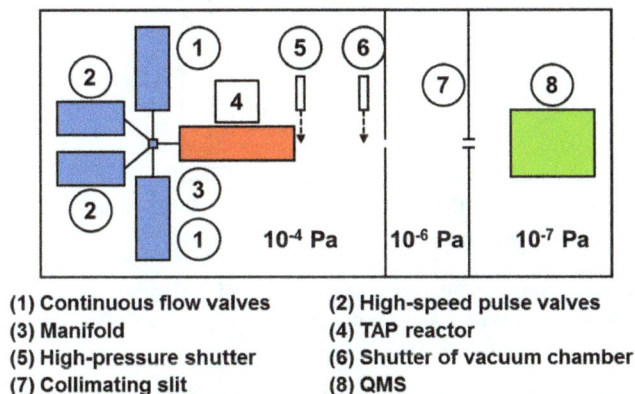

(1) Continuous flow valves   **(2) High-speed pulse valves**
(3) Manifold   **(4) TAP reactor**
(5) High-pressure shutter   **(6) Shutter of vacuum chamber**
(7) Collimating slit   **(8) QMS**

Figure 3.24: Temporal analysis of products (TAP): a pulse reactor under high vacuum.

While the base pressure in the reactor and the surrounding vacuum chamber is ≈10^{-4} Pa (Figure 3.24), the QMS works at 10^{-7} Pa. Therefore, the effluent flow is subjected to a differential pumping scheme operated by powerful vacuum pumps. Due to the very small pulse sizes, effluent quantities arriving at the QMS are low, even for this highly sensitive technique. Therefore, results are averaged over a series of pulses to obtain acceptable signal-to-noise ratios.

The TAP reactor is also fitted with two continuous-flow valves (Figure 3.24), which allow conditioning the catalyst or its precursor, prior to the pulse experiment, e.g., by calcination, by reduction, or by stabilization in the feed stream. To avoid overloading the pumps, the reactor is then closed by a high-pressure shutter, and the effluent leaves the reactor by an outlet not shown in the figure. The course of the conditioning stage can be monitored by a small effluent stream leaked into the vacuum system through a tiny aperture in the high-pressure shutter (not shown in Figure 3.24).

The TAP reactor was developed as a tool for exploring catalytic reaction networks. The vacuum condition and the absence of carrier gas molecules decreases the frequency of visits at the surface by the molecules passing the catalyst bed. This helps differentiating the primary from the consecutive reaction products, which leave the catalyst with time delay. Figure 3.25 shows an example where CH_4 and $^{16}O_2$ were pulsed simultaneously on a V_2O_5(1%)/SiO_2 catalyst at 900 K. The response suggests that methane is sequentially converted to formaldehyde, CO, and CO_2 without shortcuts: methane is not directly converted into carbon oxides, and CO_2 is not directly formed from CH_2O.

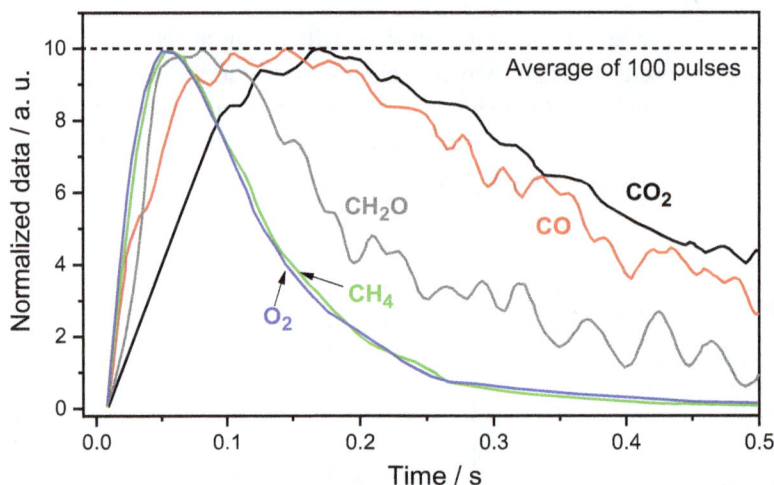

Figure 3.25: Identifying a reaction sequence with the TAP reactor. Adapted from ref. [65] with permission from Springer Nature. Pulse heights are normalized in the graphic.

However, TAP also allows identifying features of reaction mechanisms, e.g., unstable intermediates desorbed from surfaces. In studies of the oxidative coupling of methane, for instance (eq. (3.18)), methyl radicals were observed by the QMS in the effluent of CH_4/O_2 pulses charged onto MgO [66].

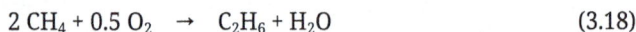

$$2\ CH_4 + 0.5\ O_2 \quad \rightarrow \quad C_2H_6 + H_2O \tag{3.18}$$

In oxidation catalysis, the TAP reactor can be favorably applied to elucidate whether a reactant is activated by oxygen from the gas-phase or from the lattice. A reaction product formed via the Mars-van Krevelen mechanism (Figure 2.55) is expected to contain only lattice oxygen (cf. Section 2.5.1). When the reaction is performed by pulsing the reactant with $^{18}O_2$, the product should not contain any ^{18}O, at least for the initial period of time.[10] Due to the small pulse sizes and the low contact times, experiments of this type are much more conclusive in the TAP reactor than with any arrangement at normal pressure. They may fail only in the unlikely case that oxygen exchange is much faster than the oxidation reaction (see footnote).

Pump-probe experiments are another powerful tool for the elucidation of reaction mechanisms, for instance of oxidation reactions. When a reactant is activated by adsorbed gas-phase oxygen (Scheme 2.3), no reaction products are formed when

---

10 The latter specification accounts not only for the fact that the surface takes up ^{18}O during reoxidation, but also for the possibility that there may be a parallel isotope exchange between the gas-phase $^{18}O_2$ and oxygen of the surface. In this case, the ^{18}O-labeled reaction product will appear in the effluent pulses with a rate that depends on the relation between the rates of isotope exchange and the oxidation reaction.

it is pulsed over the catalyst alone. When, however, the reactant pulse is preceded by an oxygen pulse, reaction products can be observed, provided that the delay between pulses is not too long. By varying this time interval, the life time of the activating oxygen species may be probed, and in favorable cases, even its decay kinetics may be studied. Upon reversal of the pulse sequence, the reaction product is formed during the oxygen pulse as long as the reactant is not completely desorbed from the surface.

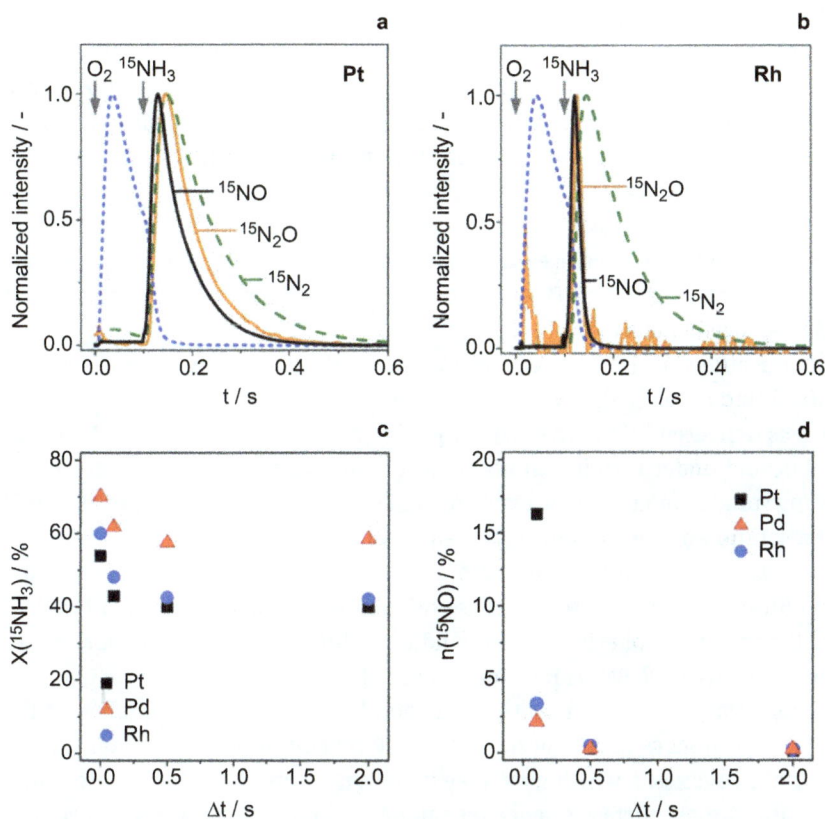

Figure 3.26: A TAP study on the reaction mechanism of NH_3 oxidation over noble metals: (a, b) pump-probe experiments with Pt and Rh; (c) dependence of ammonia conversion on the delay between O_2 and $^{15}NH_3$; (d) dependence of the NO yield on the delay between O_2 and $^{15}NH_3$. Reproduced from ref. [67] with permission from Elsevier. Pulse heights are normalized in (a) and (b).

Figure 3.26 summarizes a study in which pump-probe experiments were used to illustrate the dependence of product selectivities on the oxygen coverage on metal surfaces [67]. The work deals with NH_3 oxidation to NO (eq. (2.67)) over Pt/Rh gauzes, the first step of the Ostwald process. In Figure 3.26a and b, experiments are compared in which O_2 and NH_3 were pulsed in equal quantities over Pt and Rh wires, respectively, at

1,073 K.[11] While the response of O_2 always decreased sharply when NH_3 was added with a delay of 0.1 s, the responses of NO and N_2O were quite different: from Pt, all products were desorbed in broad signals, from Rh, NO and N_2O appeared as narrow spikes after a somewhat shorter time. Being the primary product, NO always desorbed earlier than N_2O and N_2. The slightly shorter delay of the NO peak over Rh indicates a higher activity of this metal, the narrower peak width results from a higher tendency to engage in side reactions, which is confirmed by the NO yields summarized in Figure 3.26d: a high NO yield is achieved only over Pt.

While these observations could also have been probably made by pulsing NH_3 and O_2 simultaneously, the promise of the pump-probe approach is illustrated by Figure 3.26c and d, where the effect of the delay between the pulses on NH_3 conversions and NO yields is shown. This effect is very characteristic: while the yield drops to almost zero at $\Delta t = 0.5$ s, the influence of the delay on NH_3 conversion is only moderate: NH_3 is apparently also converted at low oxygen coverage, but in side reactions rather than to NO. This agrees with the well-known importance of high O coverages for good NO selectivities. Although the TAP reactor works under vacuum, nearly full NO selectivity was achieved even with Rh when O_2 and NH_3 were pulsed together at a ratio of 10.

Other applications of the TAP reactor and an overview over the recent developments can be found in ref. [68].

The curves delivered by the TAP reactor give access to rate constants and adsorption constants, and under certain conditions, also to diffusivities in pore systems, but only via fitting suitable models to the experimental concentration responses. While this requires the solution of partial differential equations, which cover diffusion, adsorption, surface reaction, and desorption according to the assumed kinetic models, the vacuum condition and the intense gas-phase mixing by diffusion render the task less challenging than for atmospheric pressure conditions. Information about opportunities with existing software packages can be found in ref. [68].

Despite its great potential for mechanistic and kinetic work, one problem of the pulse reactor is even more pronounced for the TAP reactor: it operates under untypically low surface coverages, which may have a dramatic impact on surface processes. Like in the case of most surface-science techniques, there is a pressure gap between the conditions in the TAP and in the usual catalytic reactors. There are definitely cases where results of TAP studies can be directly applied to relevant catalytic conditions, but it needs to be sorted out if this is possible for each investigated system.

Unlike the TAP reactor, Steady state isotopic transient kinetic analysis (SSITKA) probes the catalyst surface in its stationary state, which is established in response to the chemical potential of the reaction mixture. SSITKA is, therefore, a most efficient tool for the work on reaction mechanisms and microkinetic modeling. A step change of

---

11 The experiments were performed with $^{15}NH_3$ to allow mass-spectrometric differentiation of N_2O and CO_2, the latter originating from surface contaminations of the wires.

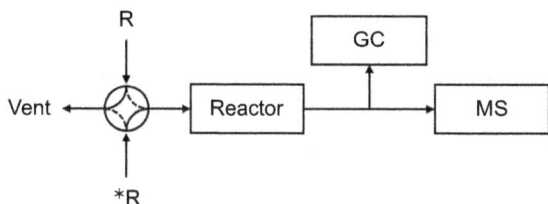

Figure 3.27: Basic scheme of Steady-state isotopic transient kinetic analysis (SSITKA).

the reactant concentration is applied to the catalyst without disturbing the stationary state, because the only feature changed in the step is the isotopic label on one reactant. Correspondingly, the response recorded shows the decay of the initial reactant label and the increasing concentrations of the labeled products.

Figure 3.27 shows a scheme of the setup: a four-port valve upstream of the reactor allows charging the catalyst alternately with two feed streams, which are identical except for the label on one component. The nature of the inert gas may also be switched: a mixture of 1% $^{14}NH_3$, 5% O_2 in He might be, for instance, switched to 1% $^{15}NH_3$, 5% O_2 in Ar. The transient response is analyzed by a mass spectrometer. In the experiment, care must be taken to separate pressure fluctuation effects caused by valve switching from signals due to changing concentrations. A gas chromatograph may be required to supply supplementary full analyses of the effluent if this is required for the interpretation of the QMS data. The assumption of an unchanged stationary state neglects possible kinetic isotope effects, which would cause differences in product composition, before and after the switch.

In the reactor, the unlabeled reactant is distributed between the gas phase, the active sites and, maybe, the spectator sites. On the active sites, the reactant is adsorbed in the form in which it enters the rate-determining step because all preceding steps are fast. The interpretation of the response depends on the nature of the mechanism. It is most straightforward if desorption of the unlabeled reactant from the sites can be neglected. In this case, the quantity of active sites n_{act} is directly counted by the amount of unlabeled reaction product recovered from the catalyst:

$$n_{act} = \int_0^\infty r_{(u)} dt = \frac{\dot{V}}{m_{cat}} \int_0^\infty c_{(u)} \, dt \tag{3.19}$$

where $r_{(u)}$ is the decaying formation rate of the unlabeled product ($r_{(u)} = \dot{V} c_{(u)}/m_{cat}$). As there is a finite quantity of reactants, the decay profile allows conclusions on the reaction rate: if the unlabeled reactant decays fast, the rate is high. The integral

$$\tau = \int_0^\infty \frac{r_{(u)}}{r_{stat}} dt = \frac{\dot{V}}{m_{cat} r_{stat}} \int_0^\infty c_{(u)} \, dt = \frac{n_{act}}{r_{stat}} \tag{3.20}$$

(r_{stat} – stationary reaction rate) is the average residence time of the reactant on the sites, which is the reciprocal of the rate constant in the case of a first-order reaction.

When atoms of an element are paired in a reaction (e.g., N_2 formed from NO and NH_3), the ratio between the unlabeled and the mixed-labeled products provides important insight into the mechanism. When desorption of (unchanged) reactants from the active sites can be neglected, reactants detected in the effluent originate from the gas phase or from spectator sites. Non-adsorbed reactants decay on a similar time scale as the inert gas; spectator sites usually release their adsorbates at clearly different rates, which allows identifying them.

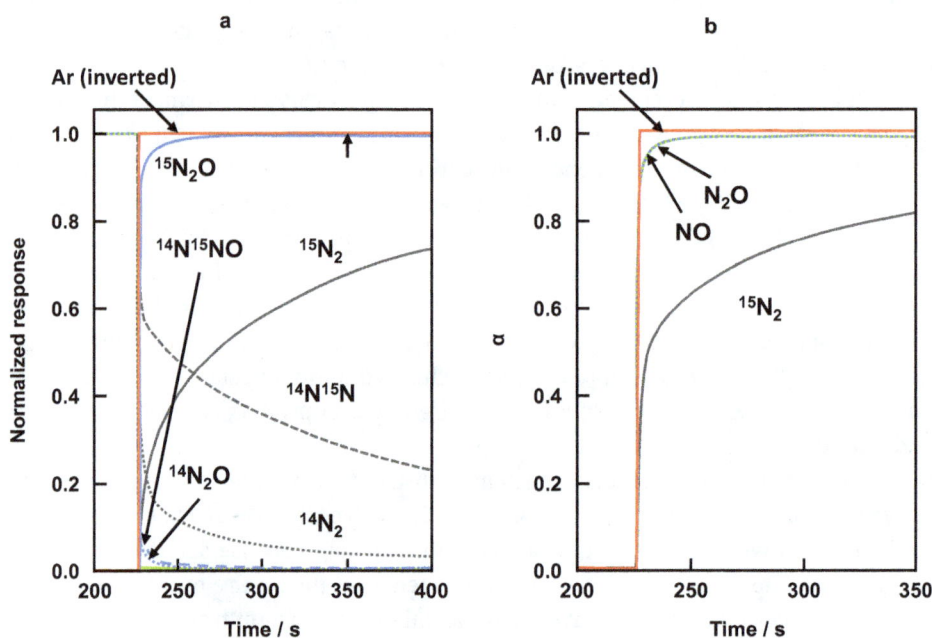

Figure 3.28: SSITKA-analysis of the reduction of NO by H_2 over a Pt catalyst: (a) response of product composition on the switch from ^{14}NO (in Ar) to ^{15}NO (in He); (b) response of fraction α of labeled molecules (NO, total for N_2O, ^{15}NO). Adapted with kind permission of Elsevier B. V. from ref. [69].

The example shown in Figure 3.28 deals with the reduction of NO by H_2 over a Pt/SiO_2 catalyst (3.21),

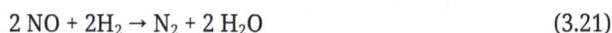

$$2\,NO + 2H_2 \rightarrow N_2 + 2\,H_2O \tag{3.21}$$

which is accompanied by N_2O formation. When ^{14}NO is switched to ^{15}NO at 333 K, $^{14}N_2O$ is almost immediately replaced by the mixed-labeled compound and by $^{15}N_2O$ (Figure 3.28a) . The decay of $^{14}N_2$ and the transients of $^{14}N^{15}N$ and $^{15}N_2$ are much more extended. As the products contain two N atoms, the fraction of labeled or unlabeled molecules (α, $1-\alpha$, respectively) of a species was used instead of concentrations to describe the transients. Figure 3.28b shows that the response of N_2O is hard to distinguish from the reactant switch, while the response of N_2 is much slower.

Using integrations analogous to (3.19) and (3.20) with ($1-\alpha$) instead of $c_{(u)}$, the coverages of N_2, N_2O, and, at higher temperatures, also of NH_3, and their surface residence times were evaluated for a range between 323 K and 460 K. As expected, τ_{N2} was more than 20-fold of τ_{N2O} at the lower end, but the relation was reverted at 400 K.

For more examples, including a discussion of equilibrium reactions and of reactions with kinetic isotope effects, see ref. [70].

3.2.2 Reactors for photocatalytic rate measurements

As mentioned in Section 1.4.1, photocatalytic reaction rates are not proportional to the catalyst mass. We justify this fundamental statement by looking at a monomolecular photochemical reaction.

$$A + h\nu \rightarrow P$$

The consumption of A depends on the number of absorbed photons $dn_{Ph,a}$ rather than on its concentration c_A. For irradiation with a wavelength λ, this can be written by reference to eq. (1.7) as

$$\frac{dc_A}{dt} = \Phi_\lambda \frac{dn_{Ph,a}}{V\,dt} = \Phi_\lambda\,I_{abs}\ (= k^* c_A) \tag{3.22}$$

where Φ_λ is the quantum yield at the wavelength λ and I_{abs} - the absorbed photon flux per volume. The right-hand part of (3.22) indicates that we are to examine the nature of a formal first-order rate constant k^*. Further, we use the absorbance $A = -ln\,(I_t/I_0)$, where I_t and I_0 are the transmitted and incident photon fluxes, respectively. It depends on the absorber concentration c_{abs}, the optical path length d, and the (natural) molar extinction coefficient ε' via the Lambert–Beer law $A = \varepsilon'\,c_{abs}\,d$ (cf. eqs. (3.100) and (3.101) below). As the absorbed photon flux I_{abs} is $I_0 - I_t$, it can be written as

$$I_{abs} = I_0\,(1 - exp(-A_\lambda))$$

and, confining A_λ to <0.05, which allows using $exp(-A_\lambda) \approx 1 - A_\lambda$,

$$I_{abs} = I_0\,A_\lambda = I_0\,\varepsilon'\,c_{abs}\,d$$

In a photocatalytic reaction, the absorber is the catalyst rather than the reactant. Relating the mass and the concentration of catalyst via a proportionality $c_{cat} = const \cdot m_{cat}$, eq. (3.22) becomes

$$\frac{dc_A}{dt} = \text{const} \, \Phi_\lambda \, I_0 \, \varepsilon' \, m_{cat} \, d \; (= k^* c_A) \tag{3.23}$$

or $k^* = \text{const} \, \Phi_\lambda \, I_0 \, \varepsilon' \, m_{cat} \, d/c_A$. It should be noted that this treatment is strongly simplified because absorption by a solid catalyst is better described by the diffuse reflection model discussed in Section 3.4.7.2 than by the Lambert–Beer law.

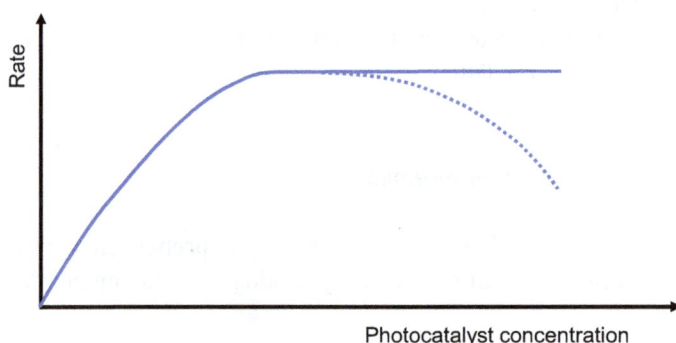

Figure 3.29: Dependence of photocatalytic reaction rates on catalyst concentration. Adapted with kind permission of the ACS from ref. [71].

As expected, the effective rate constant k^* deduced from eq. (3.23) contains the catalyst mass, but also quantities, which depend on m_{cat}. The optical path length d, for instance, decreases with growing catalyst mass because of shielding effects. This results in a non-linear increase of the reaction rate, which typically tends to a constant value at high catalyst concentrations (Figure 3.29). There are, however, even cases where rates decay beyond an optimum catalyst concentration due to the reduced penetration depth and increased backscattering of the incident light. Ranking catalyst performance without taking notice of these phenomena can result in serious mistakes. As suggested by eq. (3.23), k^* is further influenced by the concentration of the reactant (which may also absorb at the wavelength λ), and by ε', which describes an absorption process that is significantly influenced by particle morphology and arrangement due to the scattering influences (cf. Section 3.4.7.2). For more detail, the reader is referred to ref. [71].

To respond to this problem, it has been proposed to relate rates of reactant conversion to the incident intensity I_0 instead of I_{abs}, and to express performances as photonic efficiencies $\zeta(\lambda)$ or as apparent quantum yields [71, 72].[12] Related to the monomolecular reaction mentioned above, $\zeta(\lambda)$ is

$$\zeta(\lambda) = \frac{1}{I_0} \frac{dn_A}{dt} \tag{3.24}$$

---

12 Apparent quantum yields have been used in different definitions in literature, cf. refs. [71, 72].

Comparison of performance indicators based on I_0 require, however, that the fraction of light absorbed is the same in all reactors used for the measurements. This is highly questionable, in particular for traditional batch slurry reactors, where the light intensity is not constant in the whole reaction volume, but can differ strongly due to the absorption and scattering by the dispersed semiconductor particles.

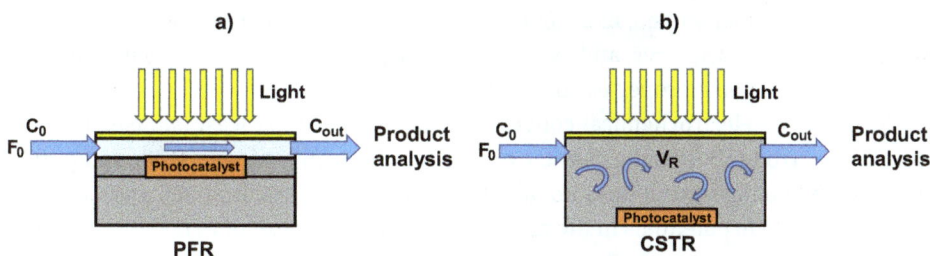

Figure 3.30: Flow schemes for measuring photocatalytic reaction rates in the gas phase: (a) targets at plug flow conditions; (b) at ideal backmixing. Adapted from ref. [73] with permission from De Gruyter.

For kinetic work, continuous-flow photomicroreactors, in which the catalyst is immobilized in a shallow bed, are therefore recommended [71]. Figure 3.30 illustrates this concept with two versions for gas-phase studies, which target plug flow or ideal backmixing. The one in Figure 3.30a may also be suited for liquid-phase studies, and it may be converted to a backmixed version by integrating it into a recycle scheme (cf. Section 3.2.1). These reactors circumvent many problems of traditional batch reactors due to a large wall area per unit reactor volume, fast mixing, and enhanced heat transfer. Due to their small dimensions, light attenuation in them is minor or even negligible, i.e., the local rate of photon absorption, which varies strongly over the reactor volume in traditional slurry reactors, is either constant or reproducibly and uniformly distributed in them. Therefore, the use of the incident instead of the absorbed intensity (cf. discussion of eq. (3.24)) is less likely to corrupt comparisons between catalysts.

Measuring at a well-defined wavelength instead of using wave-length ranges is a third requirement for a reliable characterization of the photocatalytic performance (together with working at the optimum catalyst concentration and with well-defined and minor spatial distribution of photon absorption rates). Monochromatic light can be obtained by using light emitting diodes (LEDs) or can be prepared from polychromatic light with cut-off filters. Additional water filters prevent the heating of the system by removing IR radiation emitted by the light source.

The use of a photomicroreactor for the kinetic study of a photocatalytic reaction is nicely illustrated in ref. [74]. A profound analysis of the problems related to modeling the kinetics of photocatalytic reactions is given in ref. [75].

3.2.3 Measuring electrochemical reaction rates

Electrocatalytic reactions and their kinetics can be very complex and are not always well understood. Their rates are primarily measured as currents. To assess the catalytic effect, the current is related to properties of the catalyst: its mass, or mass or surface area of the active component (cf. eqs. (1.8) through (1.10)). Measured currents depend on a number of phenomena, like competition between electrode kinetics and reactant diffusion, Ohmic resistances, and capacitive effects, which need to be specified or kept under control to make proper use of the data.

To drive the electrochemical conversion, each setup has at least two electrodes, a working electrode (WE) and a counter electrode (CE). Such a two-electrode setup indicates the cell voltage, but it does not allow relating the potentials at WE and CE to the standard electrode potentials in the electrochemical series, i.e. it does not give access to the overpotentials involved. Therefore, a currentless reference electrode (RE) is additionally immersed in the electrolyte. Its fixed, known standard potential, allows referring potentials measured at WE and CE to the standard hydrogen electrode. All methods for evaluation and characterization of electrocatalysts (incl. voltammetry, cf. Section 3.4.12) employ such a three-electrode setup. It is usually placed in a single vessel. Alternatively, the cell may consist of two compartments separated by a frit or an ion-exchange membrane. In such a H-cell, the WE is protected from influences by the CE, e.g., by the species dissolved from it.

Although technical processes may work in diffusion-limited regimes, it is essential for scientific purposes to measure the kinetics of the electrode processes. For this, the diffusion boundary layers at the electrode (cf. Section 2.6.2) need to be thin and well-controlled.

The rotating disk electrode (RDE, Figure 3.31a) is the standard tool for achieving such a well-defined regime. Subjected to centrifugal forces caused by the rotating electrode, the electrolyte in the boundary layer is flung away from the center and replaced by new electrolyte that is primed perpendicular to the electrode surface. This forced flow pattern strongly reduces the boundary layer thickness δ, which is inverse proportional to the square root of the angular frequency, ω. In the fully diffusional regime, the limiting current I_l no longer depends on the potential and is described by the Levich equation,

$$I_l = a \, z F A D^{2/3} \, v^{-1/6} \, C \, \omega^{-1/2} = B \, \omega^{-1/2} \tag{3.25}$$

where z is the number of electrons per elementary step, F the Faraday constant, D the diffusion coefficient of the reactant, C its concentration in the bulk phase, v the kinematic viscosity of the electrolyte, and a a conversion factor depending on the unit for ω (e.g., 0.201 for ω in rpm). B is sometimes referred to as the Levich constant. For the full

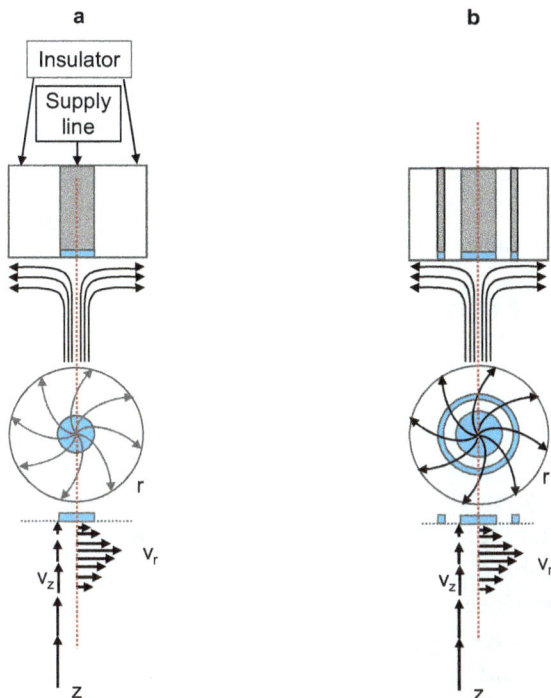

Figure 3.31: Rotating disk electrode (a) and rotating ring disk electrode (b) with flow patterns. Adapted from ref. [76] with permission from Elsevier.

potential range, the resistance of the setup is the sum of the kinetic and the mass-transport contributions, which results in the Koutecky-Levich equation (given for ω in rpm)

$$\frac{1}{I} = \frac{1}{I_K} + \frac{1}{0.201 \, z \, F \, A \, D^{2/3} \, \nu^{-1/6} \, C} \, \frac{1}{\omega^{1/2}} \tag{3.26}$$

In Figure 3.32, the evaluation of RDE data in a Koutecky-Levich plot for separating the kinetic from diffusion currents is illustrated. Polarization curves measured at different rpm (typically between 400 and several thousand) are shown in Figure 3.32a. Plotting j^{-1} vs. $\omega^{-1/2}$ at each potential provides a family of parallel straight lines, the intercepts of which with the ordinate give access to the respective kinetic currents (Figure 3.32b). These currents could be further analyzed via the Butler-Volmer or Tafel equations ((2.63), (2.64), respectively). The Levich constant B obtained from the gradient (compare eqs. (3.25) and (3.26)) can be used to determine the number z of electrons per elementary step, e.g., to check z values obtained from the more familiar Tafel plots. B is also accessible by analyzing the plateau currents with eq. (3.25).

The rotating ring disk electrode (RRDE) is a powerful extension of the RDE concept that offers remarkable opportunities both for mechanistic studies and for kinetic measurements. In an RRDE (Figure 3.31b), a second electrode is arranged as a concentric ring

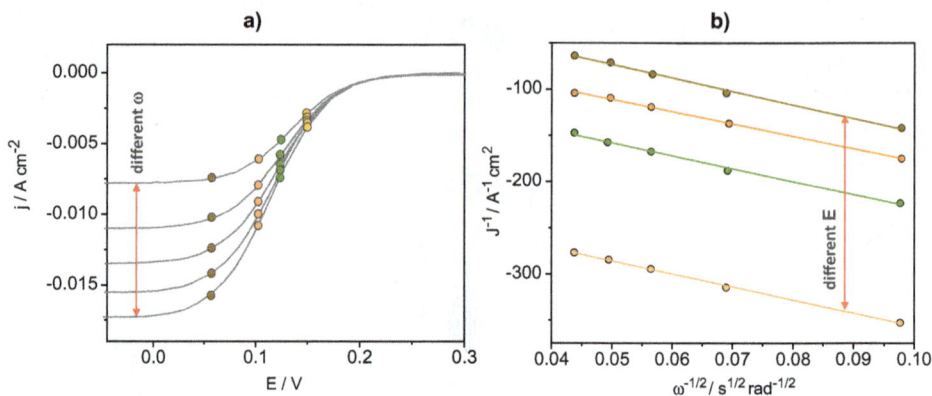

Figure 3.32: Koutecky-Levich analysis of an electrode process: (a) RDE data for different ω (with current density plotted instead of current); (b) linearized plot of equation (3.26).

around the central disc, separated from it by an insulating material and connected to a potentiostat, which needs to be capable of controlling four potentials simultaneously. Due to the laminar outward flow of the electrolyte, the species formed at the central disc can be reacted in consecutive steps, induced by the potential at the external ring. This electrode can also be employed to identify a consecutive product in a reaction mechanism and to determine its abundance by subjecting it to a potential range where the candidate species are known to react. In the oxygen reduction reaction (Section 4.9.1), it can be, for instance, used to detect the amount of the undesired byproduct H_2O_2 [77].

Due to the still considerable thickness of the diffusion layers, the RDE is suited for kinetic measurements only with relatively slow reactions. Higher reaction rates can be detected with microelectrodes. The thickness of the diffusion layers is strongly diminished at curved surfaces, e.g., of hemispherical electrodes, the radius r of which is well below the layer thickness $\delta = \sqrt{\pi D t}$ [78]. The "edge-effect" occurring at such sharp tips facilitates electrolyte exchange; the small electrode area results in decreased double-layer capacitances. As overall currents are much lower than with macroscopic electrodes (on the order of pA), the ohmic losses are very small. As a result, steady states of Faradaic processes are established very fast, potential scanning rates can be much higher than those used with macroscopic electrodes, and signal-noise ratios in steady state measurements are improved.

The characteristic length of the microelectrodes, their diameter, is typically <50 μm, but even carbon nanotubes of ca. 100 nm diameter have already been employed for this purpose. Microelectrodes can be used at very high current densities for fast transient processes, for sensing in flowing media, for experiments in media of poor conductivity, even without electrolyte, and in equipment targeting high spatial resolution, e.g., the Scanning electrochemical microscope (SECM, cf. Section 3.4.11.3). More information on the preparation, properties, and application of microelectrodes can be found in refs. [78, 79].

3.3 Application-oriented characterization of solid catalysts

Information on materials properties of solid catalysts are required for different purposes. In catalysis research, attention is almost exclusively paid to properties expected to be relevant for the understanding of the catalytic behavior. This section deals with properties that are important for the practical application of catalysts: crushing strengths, piled weights, particle size distributions, and particle densities, which are required for the design of packed beds and attrition resistance, which are relevant in moving catalyst beds. Research-oriented characterization follows in Section 3.4.

3.3.1 Density and porosity of particles and beds

Solid densities are determined in pycnometers, which allow measuring the fluid volume replaced by a known mass of the sample. The particle mass of porous solids may be related to the volume of the solid matrix V_s or, alternatively, to the total particle volume, which includes the pore volume ($V_s + V_p$). This results in different density definitions: the skeletal (or material) density ρ and the apparent density ρ_{app}:

$$\rho = {}^m\!/V_s \tag{3.27}$$

$$\rho_{app} = {}^m\!/(V_s + V_P) \tag{3.28}$$

The particle porosity ε (cf. eq. (2.45)) is therefore

$$\varepsilon = 1 - {}^{\rho_{app}}\!/\rho \tag{3.29}$$

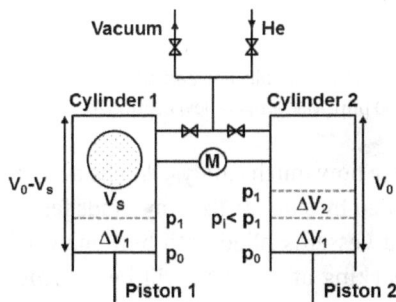

Right: $p_0 V_0 = p_1 (V_0 - \Delta V_1 - \Delta V_2)$
Left: $p_0 (V_0 - V_S) = p_1 (V_0 - \Delta V_1 - V_S)$

$\cdots\cdots\cdots$

$V_S = \Delta V_2 p_1 / (p_1 - p_0)$

Figure 3.33: Determining the skeletal density of catalysts with a He pycnometer.

Figure 3.33 shows the principle of a He pycnometer. Two cylinders of identical diameters, one of which contains the sample, are filled with He at the same pressure p_0. When the pistons are moved by the same volume difference ΔV_1 to compress the gas, a pressure p_2 is achieved in the sample chamber, which can be equalized in the other chamber only by moving its piston by an additional volume ΔV_2. From Boyle's law applied to both cylinders (see Figure 3.33), the sample volume V_s becomes

$$V_s = \frac{p_1}{p_1 - p_0} \Delta V_2 \tag{3.30}$$

Hg pycnometers are suited for measuring the apparent densities of porous materials because Hg does not enter the pores. They are, however, no more in use because this quantity is obtained as a result of porosity analysis in Hg porosimeters (cf. Section 3.4.1.5).

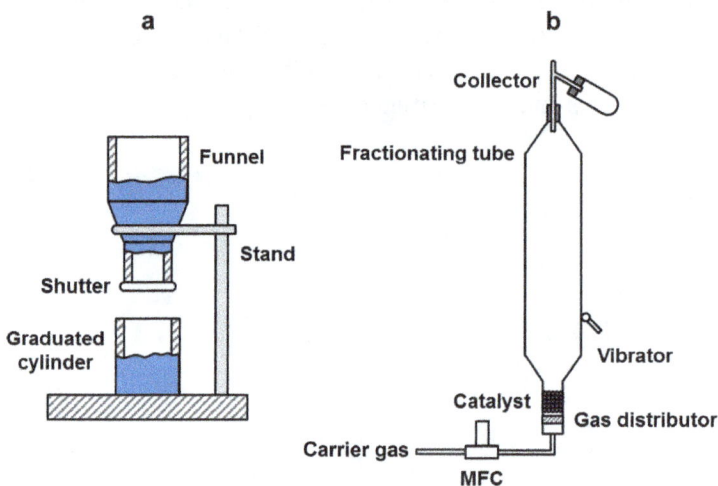

Figure 3.34: Methods for the determination of bulk densities (a) and for establishing particle size distributions of fines ((b), wind screening, sizes between 1 and 50 μm). MFC – mass flow controller.

Bulk densities of catalyst beds are needed to evaluate how much catalyst can be accommodated in a given reactor volume. They are obtained by standardized methods familiar from the technology of bulk materials. When a vessel is filled with bulk material just poured from a certain height, the achieved packing of particles will be far from dense. Such bulk (or poured) densities ρ_b are measured with devices in which batches of the bulk material (e.g., the catalyst) are allowed to suddenly fall into a graduated cylinder by retracting a shutter (Figure 3.34a). By vibrating or tapping the cylinder, the bed density will increase because particles rearrange. This results in the tapped (or compacted) density ρ_t, typically the 1.2–1.4-fold of ρ_b.

Bed densities in catalytic reactors are between ρ_b and ρ_t: catalysts are not poured into them in sudden flows, and reactors cannot be tapped or vibrated. The achieved packing also depends on the particle shape, and in narrow tubes, on the relation between the tube diameter and the particle size. Bed densities can be evaluated from empirical relations using ρ_b and ρ_t of the catalyst, the relation between the sizes of catalyst and tube, and shape parameters.

Although catalysts are routinely screened to obtain them in specified ranges of particle size, their full particle size distribution is rarely determined. Some methods used for such analysis are briefly reviewed here nevertheless because of their close relation to other relevant techniques and to typical catalytic reaction regimes like the fluidized bed.

Dry sieve analysis is a generalization of the screening process just mentioned. With sieves of different mesh piled into a column, which is rocked by a shaker, the size distribution can be established for wide ranges down to 150 μm, though with limited resolution. When the sieves are rinsed during shaking (wet sieve analysis), sizes down to 50 μm can be differentiated.

Below this range, particles can be classed by hydrodynamic techniques such as wind screening or sedimentation. Both methods are based on the comparison between gravity and flow resistance of the particles, which sink in a water column in sedimentation, or are suspended in an upward air flow in wind screening. Figure 3.34b shows the principle of vertical wind screening, which allows the fractionation of powders with very high resolution. With increasing gas velocity, fines are blown out of the catalyst bed. A fractionating tube allows larger particles that may have been raised by turbulences to sink back to the catalyst bed. Fines are collected at the top, where their mass can be related to the gas velocity.

In catalysis research, particle size distributions are typically determined for supported metal particles rather than for catalyst particles, mostly by electron microscopy (cf. Section 3.4.11). Other methods, like XRD or Small angle X-ray scattering (SAXS), supply information on particle size distributions implicitly, and its isolation is laborious and subject to uncertainties.

3.3.2 Stability toward mechanical stress

The mechanical stability of catalyst particles is studied along typical protocols of materials testing, where the strain causing failure is recorded for specimens of a representative sample. In Figure 3.35a, this is exemplified for the crushing strength of spherical pellets. To get data of the statistical significance, samples comprise 20–50 specimens.

The strain exerted by a flat piston is, however, not representative for the situation in a catalyst bed, where pellets have several contact points around their outer surfaces and load patterns change when one of the neighbors fails. Therefore, the particle stability is also tested in simulated catalyst beds, which are enclosed in a pressure vessel and

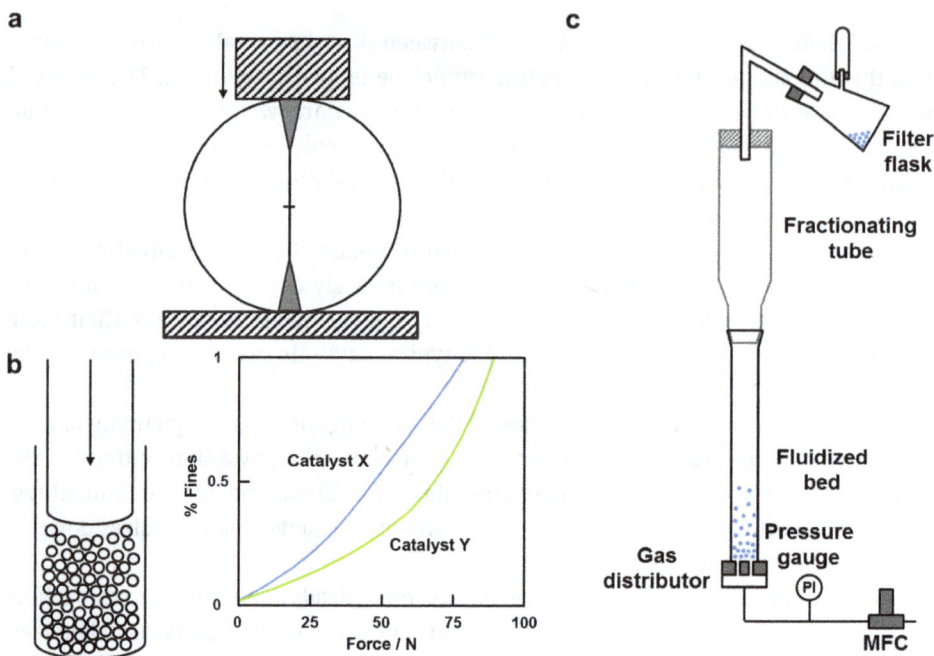

Figure 3.35: Methods for studying the mechanical stability of catalysts: (a) crushing strength of single pellets; (b) crushing strength of pellets in a particle bed; (c) attrition resistance of catalysts for fluidized beds.

subjected to preselected strain levels by a piston (Figure 3.35b). The quantity of fines can be plotted *vs* the increasing strain, as shown in the figure for two different catalysts.

Attrition resistance is crucial for catalysts to be used in fluidized beds. It is studied in setups that simulate the industrial application – typically a fluid bed fitted with a fractionation tube and a filter that recovers the fines (Figure 3.35c), similar to wind screening analysis (Figure 3.34b). Materials are ranked according to their rates of fines formation in typical hydrodynamic regimes.

More information about testing and improving the mechanical stability of catalysts can be found in ref. [80].

3.4 Research-oriented characterization of solid catalysts

3.4.1 Analysis of surface area and porosity

Adsorption covers accessible solid surfaces with adsorbate molecules, which suggests its suitability for measuring the size of these surfaces. This is easily accomplished for adsorption from the gas phase, which can be monitored by volumetry or gravimetry. Due to its unspecific nature, physisorption gives access to the total surface area of a solid, while chemisorption measures surfaces only of components offering suitable interactions (e.g., metal particles). When physisorption is used, the multilayer character of the adsorbate must be taken into account. As mentioned in Section 2.4.1, adsorption-based analytical techniques rely on established equilibria, which can be described by the Langmuir isotherm for chemisorption and by the BET isotherm for physisorption (eqs. (2.14) and (2.16), respectively).

In porous media, the buildup of physisorption multilayers is superimposed by capillary condensation because of the depression of equilibrium vapor pressures over curved menisci. As these capillary phenomena are widely used for the analysis of pore size distributions and pore shapes, they are treated next, prior to the discussion of the related characterization techniques.

3.4.1.1 Capillary phenomena in pores

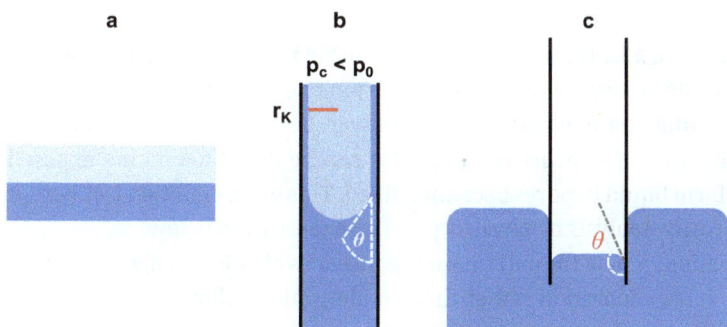

Figure 3.36: Capillary phenomena with wetting and non-wetting liquids: (a) flat meniscus; (b) filling of a capillary by a wetting liquid; and (c) reluctance of a non-wetting liquid, e.g., Hg, to enter a pore.

In porous media, condensation occurs at lower adsorptive pressures than over a flat surface, if energy can be gained by wetting the pore walls (Figure 3.36). In adsorption isotherms of type IV and V, this capillary condensation causes a steep increase of the adsorption capacity at intermediate p/p_0 (Figure 2.46), in type-I isotherms, it occurs at very low p/p_0. The analysis of pore size distributions is based on the dependence of the pore-filling pressure p_c ($< p_0$) on the pore size. Liquefied wetting adsorptives form a curved meniscus at the pore walls, with a contact angle $0° < \theta < 90°$ (Figure 3.36b). Opposed to this, the non-wetting liquid mercury, which forms contact angles $> 90°$ (Figure 3.36c), enters the pores only upon the application of an external pressure $p > p_0$. This is the basis of an alternative method of pore size characterization, which will be briefly discussed below (Hg porosimetry, Section 3.4.1.5).

In classical porosity analysis by physisorption, the relation between p/p_0 and the size of the pores being filled is described by the Kelvin equation

$$\ln {p_c}/{p_0} = -\frac{2}{r_K} \frac{V_M \sigma}{RT} \cos \theta \tag{3.31}$$

where V_M is the molar volume of the liquid adsorptive, σ – its surface tension, and r_K – the radius of the meniscus (Kelvin radius). The equation suggests a simple relation: smaller condensation pressures indicate smaller pores. While this is not wrong, it does not account for the fact that the Kelvin radius is different from the pore radius r_{Po} (Figure 3.36b), because the pore walls are covered by an adsorbate layer at any adsorptive pressure. The thickness t of this layer, therefore, also depends on p/p_0

$$r_{Po} = r_K(p/p_0) + t\,(p/p_0) \tag{3.32}$$

t can be expressed by

$$t = h\, {n_{ads}}/{n_{mono}} \tag{3.33}$$

where h is the height of an adsorbate monolayer, e.g., 0.354 nm for N_2. Upon decreasing p by an increment, the desorbed amount originates from two sources: from filled pores running empty and from the exposed adsorbate layer becoming thinner. Likewise, the species adsorbing upon a stepwise increase of p can be deposited on the exposed adsorbate layers or form liquid in pores becoming filled. This differentiation is inherent in all methods of porosity analysis by physisorption (see discussion below).

During condensation, a new (liquid) phase is formed, which is usually delayed by nucleation barriers. It is evaporation rather than condensation, which is controlled by thermodynamics. However, there are situations where even evaporation is delayed, e.g., when liquid adsorptive is entrapped in a space with a narrow exit. Obviously, evaporation can start only when p/p_0 has been decreased to the value related to the radius of this exit (p_c/p_0). Then, the whole (wider) space will be emptied spontaneously (percolation). When such an exit from a larger space is below a critical value (in the most popular adsorbate/adsorbent systems ≈ 4 nm), evaporation may proceed by yet another mechanism called cavitation. In this process, a gas bubble is formed in the entrapped liquid at a relative pressure p/p_0, still above p_c/p_0. It drives the evaporation through the narrow exit, where a liquid lamella remains after the cavitation process. While thermodynamic evaporation and percolation are related to properties of the pore system, cavitation is not.

Due to the different roles of thermodynamics and kinetics during the condensation and evaporation in pores, the adsorption isotherm may depend on the sense of pressure changes – decreasing or increasing. This results in the hystereses characteristic for isotherms of porous materials (types IVa and V, Figure 2.46), the shapes of which are related to the pore shapes dominating in the sample. There is a IUPAC classification for hysteresis types as well (Figure 3.37), which contained four types before it was updated to six isotherms by subdividing type H2 and adding type H5 recently [81].

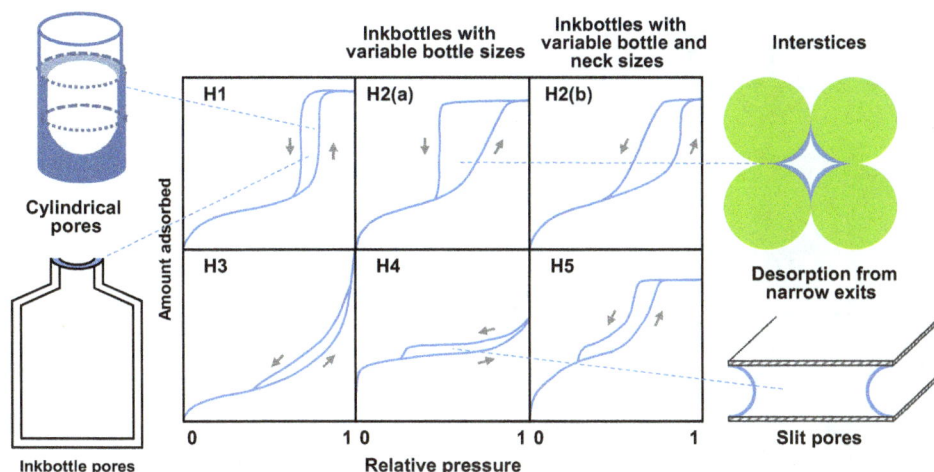

Figure 3.37: Types of hystereses in adsorption/desorption isotherms and the underlying pore shapes. Graphs adapted from ref. [81] with permission from de Gruyter.

A hysteresis of type H1 can result from open cylindrical pores where the meniscus is cylindrical during adsorption, but spherical during equilibrium desorption. The same shape can originate from inkbottle pores (Figure 3.37), provided that the wider bottle corpus supporting the meniscus during the adsorption is cylindrical. If its diameter varies or if there is a distribution of the bottle diameters, the hysteresis shifts to type H2. The vertical desorption branch describes the delayed evaporation through bottle-necks of identical size, but it may also result from cavitation. For information on how to differentiate these processes, see ref. [82]. If the bottleneck diameters are distributed, type H2b will be found. Condensation in irregular interstices between the particles and the evaporation thereof also result in type H2b.

Slit pores are special because there is no capillary condensation on their flat walls: slits with parallel walls are filled below $p/p_0 = 1$ only when their flat adsorbate layers come in touch (Figure 3.37). After condensation, however, the liquid film between the plates causes a curved meniscus during equilibrium desorption, and, therefore, a different desorption branch of the isotherm (type H4). Type H3 describes less idealized situations, where angles and distances between the slit walls are distributed, or non-rigid aggregates of plate-like materials. Type H5 arises from pore systems containing both open and partially blocked pores [81].

Characteristic deviations from a standard isotherm (cf. Section 2.4.3), which are caused by porosity, can be used to identify the shape of the predominant pores. For this purpose, theoretical layer heights t_{st} are calculated from equation (3.33), where the index "st" is used to indicate that a standard isotherm has been used to predict $n_{ads} = f(p/p_0)$. When t_{exp}, calculated from eq. (3.33) with the experimental adsorption data, is plotted vs. t_{st} (Figure 3.38), a straight line at $t_{exp} = t_{st}$ indicates the absence of porosity. An upward deviation suggests capillary condensation in mesopores or interstices. Slit

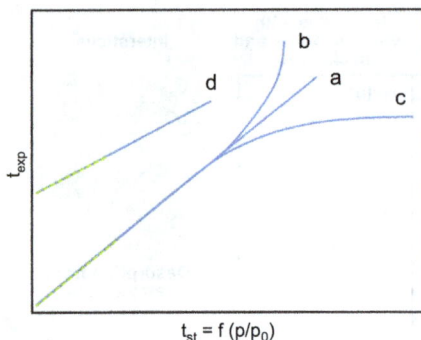

t_st = f (p/p_0)

a - Nonporous solid
b - Capillary condensation in
     cylindrical pores, inkbottle pores,
     interstices
c - Slit pores
d - Micropores

Figure 3.38: Identifying the nature of pore systems via the t-plot.

pores result in a decreasing slope, because when slits between parallel plates are filled, their surface is no longer available for adsorption. Microporosity causes t plots with an intercept at $t_{st} = 0$, which can be explained by pore filling at very low p/p_0.

3.4.1.2 Measuring adsorption data

Adsorption capacities are determined by volumetric or gravimetric techniques. For chemisorption capacities, a pulse method is also popular because it can be easily integrated into setups used to measure kinetic data in gas-phase catalysis.

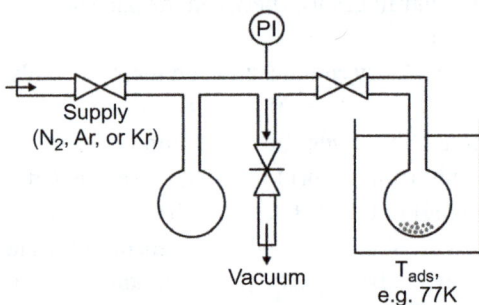

Figure 3.39: Volumetric method for measuring adsorption isotherms.

The principle of the static volumetric method is depicted in Figure 3.39. The setup contains a storage volume and a sample flask. Their volumes and those of all connections in between are exactly known. After thermal treatment for degassing, etc., the sample flask is evacuated and cooled to the measurement temperature T_{ads}. For each data point, the storage volume is then filled with an amount n_0 of the adsorptive, which can be evaluated via the ideal gas law using the adsorptive pressure p_{st}, the temperature T, and the known volume. When storage volume and the sample flask are connected, the pressure decreases for two reasons: the adsorptive expands into the flask (1) and adsorbs on the sample surface at T_{ads} (2). The adsorbed amount is the difference between the quantity of the adsorptive offered (n_0) and the quantity remaining in the gas-phase after equilibrium has been established ($n_{gas}(p_e)$). The latter is again accessible via the ideal gas law using the equilibrium pressure p_e, the volumes, and the related temperature(s): the flask at T_{ads}, the remaining volumes at T. After this step, sample flask and storage volume are again separated and the latter is filled at a higher pressure (or evacuated if the desorption branch is recorded). Then, the next data point can be obtained by connecting these parts again. The empty volume of the flask containing the sample, which is required to calculate $n_{gas}(p_e)$, is initially determined by a blank run with an inert gas such as He.

The measured adsorption capacity n_{ads} is the difference between the two large quantities n_0 and $n_{gas}(p_e)$, which is a constellation prone to producing experimental error. In the present case, however, both quantities can be determined with sufficient accuracy. The most critical point is the effect of temperature variation between T_{ads} and the (room) temperature above the sample flask (Figure 3.39) on $n_{gas}(p_e)$. From the equipment side, it is addressed by using small pipe diameters and ensuring constant levels of the cooling medium. The effect can be further minimized by keeping pressures low during the measurement.[13] For this reason, samples with small surface areas are investigated using adsorptives that exhibit low vapor pressures at the measurement temperatures (Kr, Xe, see Table 3.1).

Table 3.1: Relevant properties of adsorptives used for the characterization of surface areas and porosities by physisorption: T_{ads} – preferred adsorption temperature; p_0 – vapor pressure of liquid adsorptive at T_{ads}; a_m – surface area per physisorbed molecule or atom.

Molecule	T_{ads} K	p_0 at T_{ads} Pa	a_m nm^2/molecule
N_2	77	1.013×10^5	0.16
Ar	77	2.78×10^4	0.15
Kr	90	2.74×10^3	0.20
Xe	90	8.75	0.23

---

13 At low pressures, the amounts n_0 offered per a step are smaller. Correspondingly, the percentage of n_0 adsorbed by the sample is larger than at higher pressures. The enhanced measurement effect decreases the influence of errors.

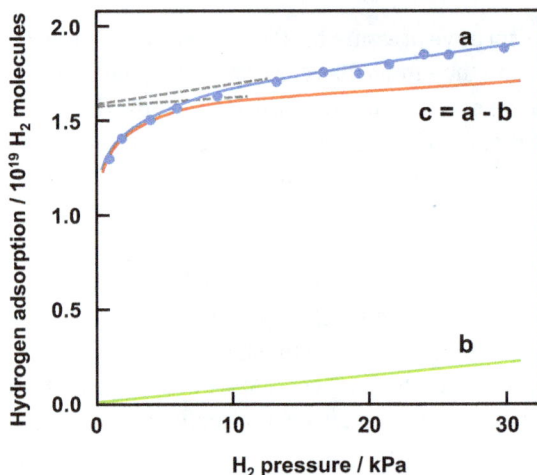

Figure 3.40: Correction of chemisorption data from static volumetric techniques: (a) measured data, with correction by back-extrapolation; (b) second adsorption run after evacuation of sample, subsequent to (a); and (c) correction according to bracketing approach.

Chemisorption is often employed to characterize supported metal catalysts. When such data are measured using the static volumetric technique, corrections are needed to differentiate strong from weak chemisorption, from physisorption on the support, and from spillover (migration of species activated by the metal onto the support, cf. Section 4.5.4). Due to these phenomena, the adsorbed amount may not exhibit the expected asymptotic behavior, but may continue to rise with increasing adsorptive pressure, as shown from a measurement of H_2 chemisorption on Pt/SiO_2 in Figure 3.40 (data points). The simplest correction would be back-extrapolation of the increasing tendency to zero pressure (a in Figure 3.40). Results of this procedure are often considered as the monolayer capacity.

In the alternative "bracketing" approach, the catalyst is evacuated after the first adsorption measurement, which removes weakly adsorbed species, including spilt-over H atoms, which are desorbed after returning to the metal particle. Repeating the measurement with the sample still containing the strongly adsorbed species (b) discloses the amount of weak adsorbates, which can be subtracted from the original result to obtain the corrected data (c). This method provides the amount of strongly chemisorbed species, which is, however, not necessarily identical with the monolayer capacity in chemisorption (cf. Section 2.4.2).

While the static volumetric technique is rather universal, the pulse technique outlined in Figure 3.41 is suitable only for strong and nonactivated chemisorption, which proceeds very fast. The setup depicted in Figure 3.41a is adapted for the adsorption of H_2 pulses from a carrier gas by a metal catalyst. The catalyst or its precursor is initially reduced in H_2, which is fed via the four-way valve. By switching this valve, chemisorbed hydrogen is then stripped off the catalyst surface by feeding the carrier gas at the reduction temperature for some time. After cooling the catalyst to the measurement temper-

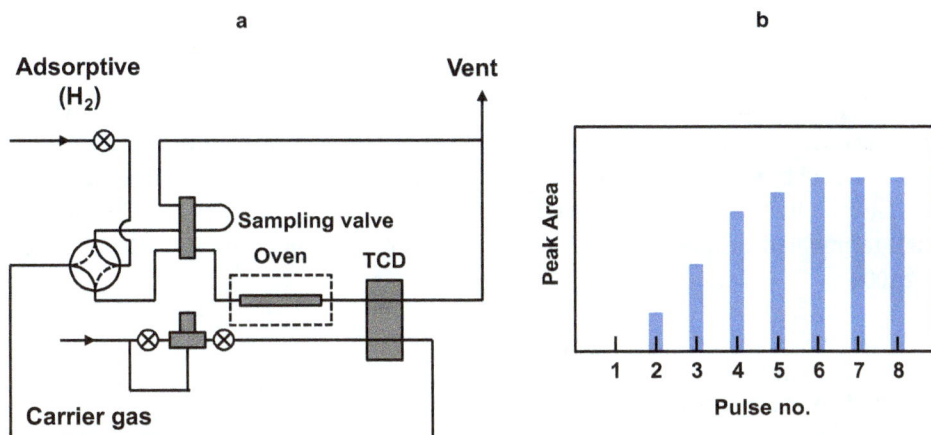

Figure 3.41: Pulse technique for the determination of chemisorption capacities: (a) schematics of setup (TCD – thermal conductivity detector); (b) development of TCD signal.

ature, it is charged with H_2 pulses using the sampling valve, and the effluent is analyzed by the thermal conductivity detector (TCD). The sample adsorbs H_2 from the first pulses until it is saturated, as exemplified in Figure 3.41b. Using the last unchanged pulses for calibrating the sensitivity of the TCD, the adsorbed amount can be easily evaluated from the peak intensity that is missing in the first pulses.

The method is not as easy as it seems because it requires an extremely pure carrier gas to avoid a detrimental impact of oxygen on the small catalyst sample during stripping and between the pulses. If adsorption is not strong enough, the adsorbate bleeds out between the pulses, resulting in baseline drifts and reduced accuracy. However, if properly applied, the method allows a reliable comparison among catalysts, and its results are well correlated with those of the more demanding stationary method applied in the bracketing mode.

3.4.1.3 Specific surface area: the BET method

From the BET equation (2.16), the monolayer capacity can be extracted using its linearized form

$$\frac{p}{n_{ads}\,(p_0-p)} = \frac{1}{n_{mono}c} + \frac{c-1}{n_{mono}c}\frac{p}{p_0} \tag{3.34}$$

A plot of $p/n_{ads}\,(p_0-p)$ vs. p/p_0 in the range $p/p_0 \leq 0.3$ can be fitted with a linear relation, in which the intercept is $1/n_{mono}c$ and the slope is $(c-1)/n_{mono}c$. Some examples are shown in Figure 3.42. From the slope and the intercept, both n_{mono} and c are easily calculated. The BET surface area is obtained by combining the monolayer capacity with the area a_M occupied by a molecule or atom in the physisorbed layer:

$$S_{BET} = n_{mono}a_m \qquad (3.35)$$

In Table 3.1, the values for a_m are collected for frequently applied adsorptives, together with measurement temperatures T_{ads} and vapor pressures at these temperatures.

It should be kept in mind that the analysis makes sense only if the data comply with the basic assumptions made to derive the BET equation (Box 2.2). For N_2 physisorption on O-terminated surfaces, this is, for instance, accepted when the c parameter obtained is ≥ 100.

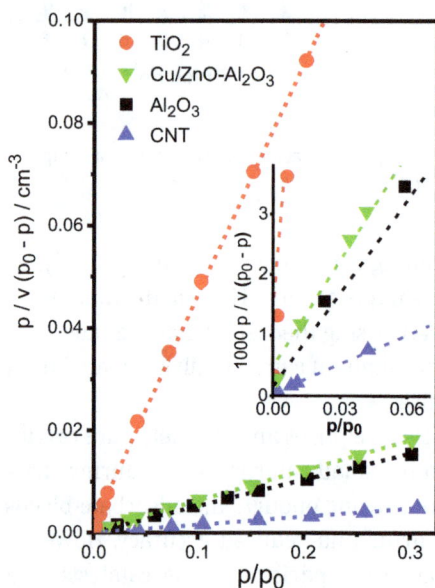

Figure 3.42: Linearized BET plots for N_2 on different adsorbents; inset – low p/p_0 range magnified to show the intercepts. Resulting BET surface areas (in m^2/g): TiO_2 – 9.7;, $Cu/ZnO-Al_2O_3$ – 72.1; Al_2O_3 – 86.6, CNT – 274.

The small intercepts of the curves in Figure 3.42 (see inset) result from these high values of c. They allow estimating the surface areas with a simplified method by establishing the equilibrium only once and connecting the data point obtained with the origin (one-point method). While these estimates may be acceptable for some purposes, the use of the full adsorption isotherm significantly improves the results in terms of accuracy and reproducibility.

As long as there is no microporosity, BET surface areas of >100 m^2/g are typically determined with standard deviations of $\pm$ 1%$_{rel}$ in modern commercial equipment. For low surface areas, e.g., of the order of 5 m^2/g, reproducibility is smaller, differing strongly among available instruments. Microporosity results in capillary phenomena at $p/p_0 \ll 0.3$, which strongly obscures the linearized BET-plot according to eq. (3.34). Therefore, results of the linearization analysis obtained from samples containing micropores cannot be considered as real surface areas although they can be used to assess the accessibility of the pore system.

3.4.1.4 Porosity analysis by physisorption

As mentioned above, the capillary phenomena can be employed to analyze the pore size distribution in porous materials. For materials exhibiting type IV isotherms, the Barrett-Joyner-Halenda (BJH) method is traditionally applied for this purpose. Its ingredients are (i) the Kelvin equation, in its version for cylindrical pores (3.31), for relating the radius of the meniscus r_K to p/p_0, (ii) a relation between p/p_0 and the layer height t (defined by eq. (3.33)), which is derived from a standard isotherm (cf. Sections 2.4.3 and 3.4.1.1), and (iii) – eq. (3.32) to evaluate the pore radius r_{Po}. When a liquid volume dV_k is condensed or evaporated in a pore with a Kelvin radius r_K, the related pore volume dV_{po} is

$$dV_{Po} = \left({r_{Po}}/{r_K} \right)^2 dV_K,$$

and the surface area in these pores is

$$dS_{Po} = 2 \, {dV_{Po}}/{r_{Po}}.$$

As mentioned above, a gas volume $\Delta V_{ads,g}$ adsorbed by increasing the adsorptive pressure, e.g., from p_1 to p_2, is partly condensed in pores of a Kelvin radius $r_K(p_2/p_0)$. The remainder ends up on the free surface, increasing the adsorbate layer on it from $t(p_1/p_0)$ to $t(p_2/p_0)$. Obviously, however, this contribution of the free surface changes as pores are filled or emptied by varying p/p_0. Therefore, the pore size distribution must be calculated stepwise. The BET surface area deduced from a range of low pressures, in the absence of condensation, may be a starting point for S. As capillary condensation takes place, the areas in the filled pores are subtracted, until only the external surface of the remains at $p/p_0 \to 1$. From there, the desorption branch may be evaluated.

The calculation results in distributions of pore volumes or intra-pore surface areas among the pores of different sizes (diameters!) in the sample. They can be shown as differential or as cumulated distributions, as exemplified in Figure 3.43. The differential distribution tells which volume (Figure 3.43a) or surface area is available in pores of the size d_{po} ($\pm dd_{po}$, e.g., ± 0.5 nm). Cumulative distributions look different. The example in Figure 3.43b reports what percentage of the total pore volume is in pores *smaller than* d_{po}. As expected, it increases strongly at ≈ 4 nm, but not even to 60 vol–%, which reflects that large pores contribute more to the pore volume than small pores with the same abundance. The distribution in Figure 3.43c shows which surface area in pores is *wider than* d_{po}. It changes more strongly at the most frequent pore size, because the influence of the diameter is smaller on the pore surface than on the pore volume.

The BJH method has been recommended for analyzing pore sizes between 3 and 20 nm. At the upper limit, the changes in the curvature of the meniscus upon variation of pore size become very small; hence, the resolution on the pore size coordinate becomes unacceptable. At the lower limit, the Kelvin equation (3.31), which contains only properties of the adsorptive, fails because a stronger influence of the adsorbent on

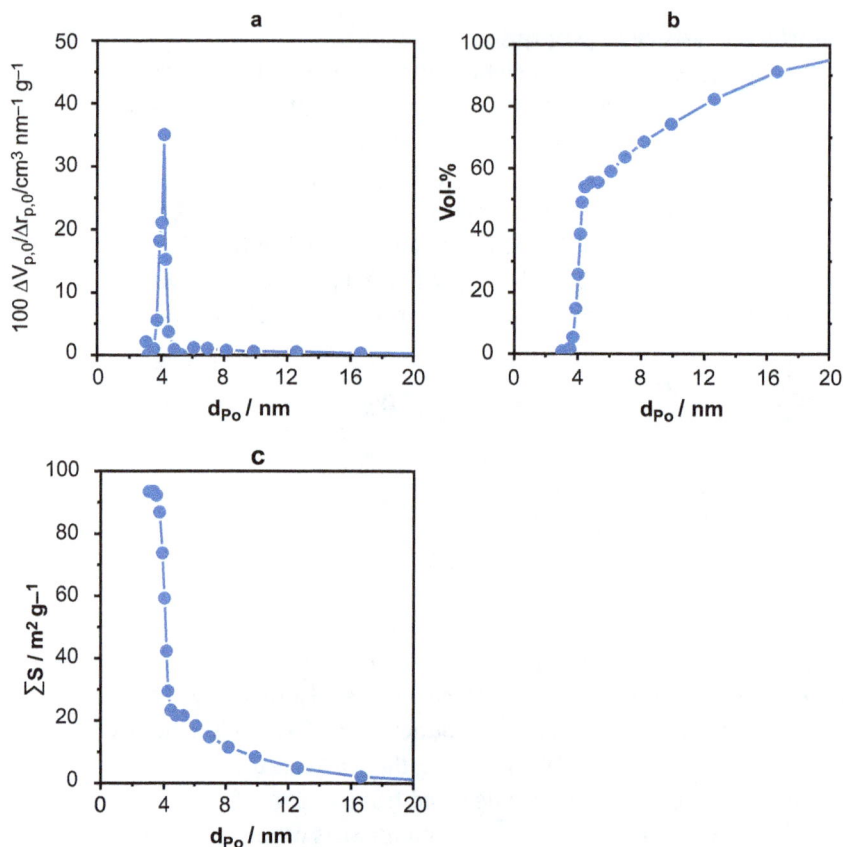

Figure 3.43: Examples of pore size distributions in different formats: (a) differential distribution of pore volume; (b) cumulative distribution of pore volume in pores smaller than d_{Po}; (c) cumulative distribution of surface area in pores wider than d_{Po}.

the adsorbate affects the fluid structure in narrow pores. As a result, thermodynamic properties of the condensed phase in confined spaces differ from those in the unconfined state [82]. Below 5 nm pore size, BJH results can be considered only as rough estimates. It has been shown recently that it underestimates the real pore sizes by 20–30%, even below 10 nm ([81, 82], see also Figure 3.44b).

For pore sizes <3 nm, a number of models were developed that describe the dependence of pore filling on p/p_0 on the basis of statistical thermodynamics. Meanwhile, an approach that uses potential data evaluated by nonlocal DFT (NLDFT) is preferred, which also allows evaluating the delayed condensation due to metastable states of the pore fluid. Testing the models with OMMs exhibiting extremely homogeneous pore systems (e.g., MCM-41, see Section 3.1.1.1) proved the superiority of the DFT-based approaches. They allow successful data treatment not only for micropores, but also over the entire range of mesopore sizes [81, 82], although they require input regarding the nature of the predominating pores, and there may be problems with the interaction parameters for some materials.

N_2 physisorption, in combination with BJH remains, however, a useful method for the routine analysis of mesopores. For the investigation of micropore systems, Ar is preferred over N_2 because of the slow adsorption kinetics of the latter in very small pores.

3.4.1.5 Porosity Analysis by mercury intrusion

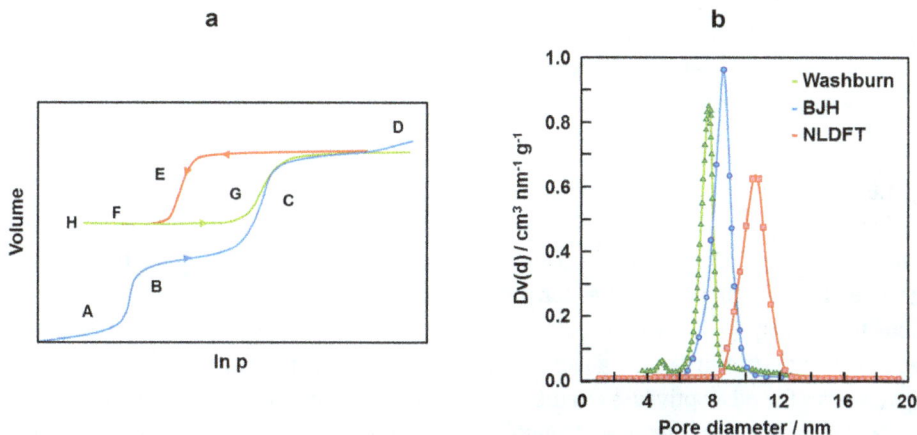

Figure 3.44: Pore size analysis by mercury penetration, compared with physisorption techniques: (a) dependence of the Hg volume on pressure during a Hg intrusion measurement (for explanation of A through H see text); (b) comparison of pore sizes distributions obtained for a siliceous OMM (KIT-6) by Hg intrusion and by N_2 physisorption, treated with the BJH or the NLDFT methods. Adapted from ref. [82] with permission from the Royal Society of Chemistry.

Mercury porosimetry, in which the pressure dependence of Hg intrusion into pores (cf. Figure 3.36c) is measured, is the standard method for the analysis of macropores, but its range extends down to the smaller mesopore range. Hg is injected under vacuum into a cell containing the porous sample. When this cell is pressurized, the decreasing total volume of (Hg + sample) is recorded. The relation between the applied pressure and the pore size is described by the Washburn equation

$$p \, r_{Po} = -2 \, \sigma \cos \theta \tag{3.36}$$

with symbols as used for eq. (3.31). Pressures required for filling the smaller mesopores range up to 4,000 bar.

Figure 3.44a shows the curves typically obtained in a Hg porosimeter experiment. Initially, the powder is compacted (A), and Hg intrudes into voids between the grains (B). At higher pressures, B is superimposed by actual pore filling (C), which is sometimes followed by compression of the solid matrix (D). Upon pressure release, the curve usually takes a different course, which results in a hysteresis with extrusion (E) at different pressures than with intrusion. Often, the hysteresis is not closed because some Hg may be trapped in pores. When the pressure is raised in a second cycle, pore filling starts again (G), now without influence from other processes, and the next pressure release results in a closed hysteresis.

In Figure 3.44b, the pore size distributions of the OMM KIT-6 obtained by different treatment of gas adsorption data and by Hg porosimetry are compared. Although at the limits of its scope, Hg porosimetry gives similar results as the classical BJH treatment of adsorption data. Both underestimate, however, to some extent, the real pore size obtained by the NLDFT approach (see previous section).

The apparent density of the catalysts (eq. (3.28)) can be derived from the intrusion curves, applying corrections that cover phases A and B in Figure 3.44a.

3.4.1.6 Analysis of particle dispersion by chemisorption

While physisorption gives access to the total surface area of catalysts, the active surface area of supported metal particles can, in principle, be determined by chemisorption. However, for this purpose, values for the area required by one adsorbate species (analogous to a_m in eq. (3.35)) would be required. As the dimensions in the adsorbate layer depend on the structure of the adsorbing surface in chemisorption, there are no universal a_M values for adsorptives, as in the case of physisorption. In addition, chemisorption studies target the number n_{exp} of metal atoms exposed rather than the surface area, because the former is more relevant for catalysis. For assessing n_{exp}, one needs information about the adsorption stoichiometry: how many surface atoms are counted by one adsorbate species.

Such adsorption stoichiometries were established long ago from plausibility considerations combined with comparisons between results obtained with different adsorptives and with data on particle sizes established by different techniques. For hydrogen, a stoichiometry of 1 H/M is widely accepted, although deviations were sometimes claimed, e.g., for extremely disperse samples. For CO, the stoichiometry is somewhat below 1 CO/M (down to 0.7 CO/M) and depends on systems and conditions, because CO populates adsorption sites also that bridge two or even three metal atoms (cf. discussion of Figure 3.122 below). O_2 and N_2O are the other popular adsorptives, the O atoms of which count two metal atoms each.

H_2 can be used to study the active surface area of various metals, like Pt, Rh, Ni, or Fe. In the case of Pd, conditions resulting in hydride formation must be avoided. Therefore, CO is preferred, which is sometimes employed also for Pt. The interaction of N_2O with Cu surfaces is a reactive chemisorption, because N_2 is released upon the deposition of O atoms on the Cu sites. Volumetry is, therefore, not suitable for measuring N_2O chemisorption. Instead, the pulse technique, with detection of the unreacted N_2O or of evolving N_2, may be applied. In an alternative mode called Reactive frontal chromatography (RFC), a step front of N_2O is charged on the catalyst rather than pulses.

Adsorption	$2\,Pt + 1\,H_2 \rightarrow 2\,Pt\text{-}H$
Titration	$2\,Pt\text{-}H + O_2 \rightarrow Pt\text{-}O\text{-}Pt + H_2O$ (support)
Backtitration	$Pt\text{-}O\text{-}Pt + 2\,H_2 \rightarrow 2\,Pt\text{-}H + H_2O$ (support)

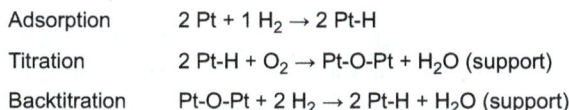

Scheme 3.3: H_2/O_2 titration of surface Pt atoms.

With H_2 and O_2, a titration procedure is available that increases the detectable effect per exposed metal atom. It is started with the adsorption of H_2 (Scheme 3.3). Subsequently, O_2 is offered under conditions that result in monolayer formation without bulk oxidation. The adsorbed H atoms are reacted to water, which spills over to the support, and O atoms adsorb on the metal. With most noble metals, a next run with H_2 results in water formation as well as H atoms adsorbing again. The stoichiometry is now $H_2 : Pt = 1 : 1$ instead of $1 : 2$ in the initial adsorption. Over some noble metals, this titration can be repeated many times with stable H_2 and O_2 consumptions. Where, however, strong O-M bonds are formed, back-titration of the adsorbed oxygen fails, which can be used, for instance, to differentiate the exposed metal atoms in alloy catalysts.

In chemisorption experiments with metal catalysts, the target quantity is the dispersion f, the ratio between the number of exposed metal atoms and the total number of metal atoms in the sample

$$f = {}^{n_{exp}}\!/_{n_{tot}} \tag{3.37}$$

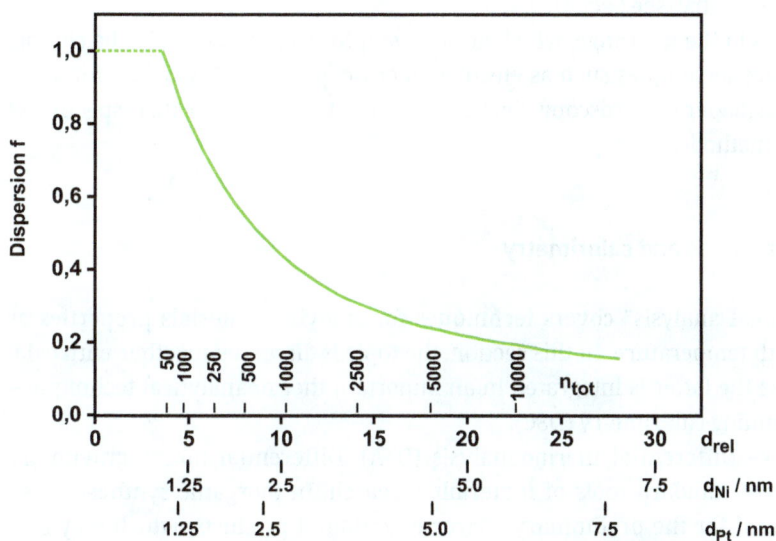

Figure 3.45: Relation between the size of the spherical particles and dispersion. For the use of the multiple abscissa see text. Adapted from ref. [83] with permission from Elsevier.

Dispersion is related to particle size, but the exact relation depends on the particle shape. Figure 3.45 summarizes this relationship for spherical particles. According to this graph, f remains 1 up to 50 atoms, which corresponds to a diameter of ≈1 nm for the metals mostly employed. Particles of 4–5 nm size, which contain ≈ 5,000 atoms, still expose a quarter of them. The multiple abscissa of the graph gives direct access to the number of atoms in spherical particles of Pt or Ni. For other metals, this graduation can be derived via the universal diameter d_{rel}, also shown at the abscissa, which represents the particle diameter divided by the nearest neighbor distance of any metal.

Notably, the model is meant only to provide an impression on how many atoms are contained and exposed in nm-sized metal particles. Atoms at the interface between the metal and the support are neglected (i.e., considered exposed). For very small particles, the spherical shape assumption is very likely to fail due to the support influence (cf. Section 2.3.3).

Chemisorption techniques are also applied to oxides or sulfides. Their adsorption sites are, however, usually saturated with (formally) anionic species (O^{2-}, OH^-, S^{2-}, SH^-). In addition, there is much more anisotropy in particle shapes and reactivity of facets in such materials than in metals (see Section 2.2.1). Adsorption sites may be created by removing oxide or sulfide species, and chemisorption tracks to what extent this happens by the given treatments. Typical examples are the adsorption of NO, O_2, or CO on sulfides, and of O_2 or H_2 on oxides. While such studies target the number of exposed sites, the nature of the adsorbates can be studied by vibrational spectroscopy (see Section 3.4.8), and their interaction strength with the surface sites by thermal desorption and calorimetry (see Section 3.4.2).

Particle sizes in the nm range, which are relevant for catalysis, can also be characterized by various techniques such as electron microscopy, XRD, EXAFS, and for some elements by Mössbauer spectroscopy (Sections 3.4.11, 3.4.3.3, 3.4.4, 3.4.10, respectively) or by magnetic methods.

3.4.2 Thermal analysis and calorimetry

The term "thermal analysis" covers techniques for studying materials properties as they change with temperature. In this section, the topic is discussed together with calorimetry, because the latter is integrated in an important thermoanalytical technique – Differential scanning calorimetry (DSC).

Methods like Differential thermoanalysis (DTA), Differential thermogravimetry (DTG), or DSC are standard tools of materials research. In inorganic synthesis, they are routinely used for the preliminary characterization of products, which may also be catalyst precursors. Observing the response of samples to a temperature increase has, however, also found applications specific to catalysis research. The effects targeted in these applications are the desorption of molecules previously adsorbed on the surface (Temperature-programmed desorption, TPD), the catalytic reactions between gas-phase components and the adsorbates (Temperature-programmed surface reaction, TPSR), the reduction or (re-)oxidation of the catalyst in different states (Temperature-programmed reduction, TPR, and oxidation, TPO), and, in a rather special version, the sulfidation of precursors for sulfide-based catalysts (Temperature-programmed sulfidation, TPS). In the following, the methods specific for catalysis research are discussed at some length, while the standard techniques, familiar from inorganic chemistry, are summarized very briefly. DSC will be treated together with the stationary version of calorimetry.

3.4.2.1 Standard methods: DTA and DTG

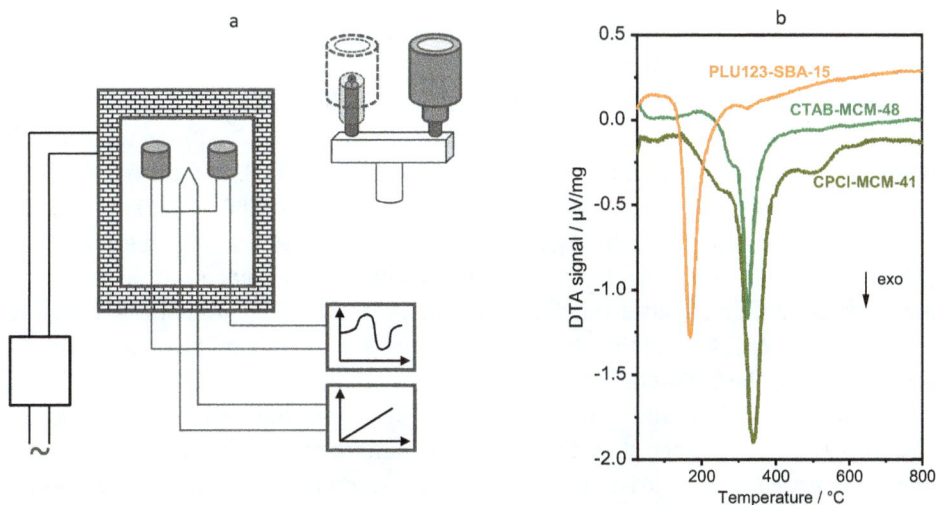

Figure 3.46: Differential thermoanalysis (DTA): (a) schematics of a DTA setup, with sample holder detailed; (b) DTA curves measured during template combustion from OMMs of different origin (see text). (b) Adapted from ref. [84] with permission from Elsevier.

In DTA, temperature differences between a sample and an inert reference, which arise from processes in the sample, are recorded during a specified temperature program, typically a linear ramp. The setup consists of a sample holder with thermocouples and containers for sample and reference, a furnace, a temperature programmer, and data-recording equipment (Figure 3.46a). The sample holder is a block made of metal or ceramics, which should allow for a good heat distribution at not too high heat conductivity. The furnace ensures a homogeneous spatial temperature distribution, at least around the sample holder. Often, the sample can be kept in a controlled atmosphere during the measurement.

In DTA, the temperature difference between the sample and the reference is plotted *vs.* the temperature or the time. As there are different conventions regarding the orientation of the exothermal and endothermal effects on the ordinate, this orientation should be indicated (Figure 3.46b). The reference should have similar thermal properties as the sample. Sample batches previously exposed to the full temperature ramp may be employed for this purpose.

Thermal effects encountered in samples are frequently endothermal, e.g., the release of crystal water, the dehydration of hydrophilic surfaces, of hydrated cations (e.g., in zeolites), the decomposition of precursors (e.g., carbonates), the reduction of precursors, or phase transitions. Notably, in materials relevant for catalysis, transitions from metastable to stable phases (e.g., γ-Al_2O_3 → α-Al_2O_3, zeolites → dense alumosilicates) often involve great gains in surface energy, which may turn the thermal effect into an exothermal one. Exothermal effects may also arise from the crystallization of amorphous material or from the oxidation of catalyst components or precursors (e.g., oxalates).

In Figure 3.46b, DTA curves recorded during template combustion from ordered mesoporous materials (OMMs, cf. Section 3.1.1.1) are reported. They belong to MCM-41, synthesized with the non-standard template cetyl pyridinium chloride (CPCl), to MCM-48 synthesized with cetyltrimethylammonium bromide (CTAB), and to SBA-15, which was templated with a nonionic surfactant (Pluronic 123). The curves obtained from the materials synthesized with the cationic surfactants (CPCl, CTAB) are similar, though differing in some details, e.g., in the temperature of the main exothermal effect (341 °C *vs.* 325 °C). This effect is preceded by a smaller exothermal effect below 300 °C and by endothermal effects around 200 °C or below 150 °C. The latter was assigned to water desorption [84]. Additional exothermal effects above the main signal are most pronounced for the CPCl-templated MCM-41.

On the basis of parallel DTG profiles with mass-spectrometric effluent analysis (see below) and of the previous own work, the authors concluded that the endothermal effect around 200 °C arises from the decomposition of the template molecules by Hofmann degradation [84]. In the following, beginning combustion causes net exothermicity, which still competes with expiring Hofmann degradation and endothermal cracking, and results in the exothermal signal around 325 °C (for an explanation of the shoulder see below). Due to oxygen deficiency in the pores, some of the long-chain alkenes are converted to coke in this region. Its oxidation gives rise to the high-temperature signals. The higher temperature of the main exothermal effect in CPCl-MCM-41 and its stronger high-temperature signals were assigned to a better stability of the CPCl toward halide elimination and Hofmann degradation.

Template removal from SBA-15 proceeds completely differently. The endothermal effect of water desorption is immediately followed by the main exothermal signal peaking well below 180 °C, but heat release extends up to rather high temperatures. The triblock copolymer is less stable than the cationic surfactants. Despite the lower temperature, the main peak results from a combined decomposition and combustion process, accompanied by coking. Coking was proposed to occur mainly in the micropores located in the mesopore walls of SBA-15, which explains the delayed combustion of these residues [84].

DTA is nowadays rarely offered as a standalone method. Instead, it is integrated into instruments for DSC (cf. Section 3.4.2.2) or DTG. For more details about the method, the reader is referred to refs. [85, 86].

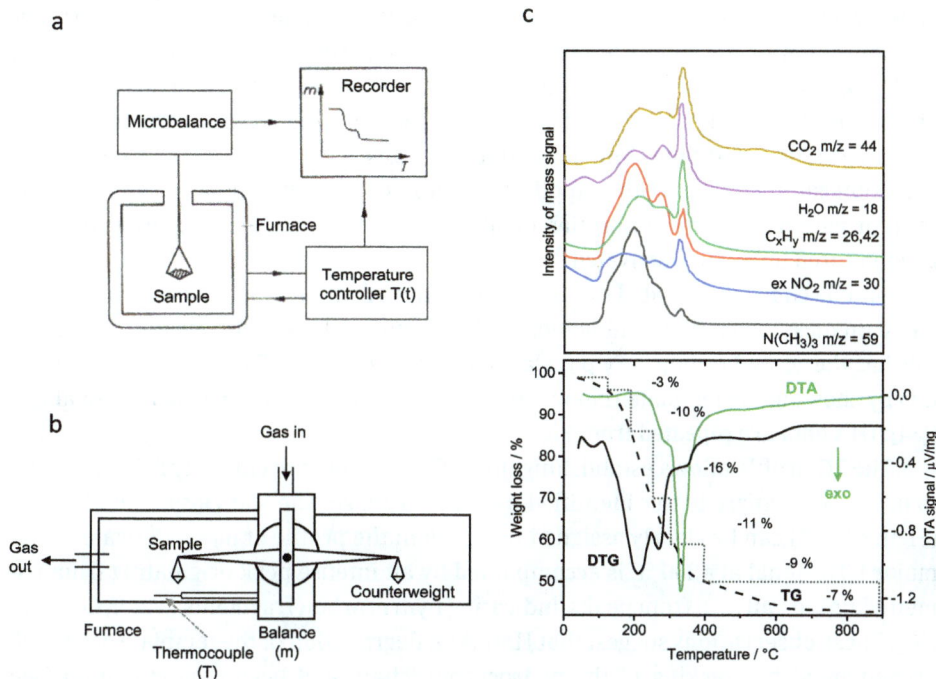

Figure 3.47: (Differential) Thermogravimetry: (a) schematics of a TG setup; (b) zero-point balance integrated into the gas flow of a G setup, horizontal gas flow (schematic); (c) combination of (D)TG with mass-spectrometric effluent analysis: TG and MS profiles measured during template combustion from MCM-48 synthesized with CTAB. (c) Adapted from ref. [84] with permission from Elsevier.

In thermogravimetry (TG), changes of sample weight, mostly losses, are recorded during a linear temperature ramp. To help differentiating processes that cause losses at similar temperatures, the profile of the weight changes is also differentiated (DTG). Like in DTA, the sample is heated in a furnace providing a homogeneous temperature distribution around the sample (Figure 3.47a). The microbalance at the heart of the analyzer is often a zero-point balance where the position of a beam carrying sample and countermass (Figure 3.47b) is kept constant by (electro)magnetic forces, which compensate for sample weight changes. Weight effects are measured by currents required for their compensation. There are also instruments with magnetic suspension of the sample, which allow separating its environment from the balance mechanism, or with completely different principles of mass determination (see below).

In catalysis research, (D)TG is often performed in flowing reactants, e.g., dilute O_2 or H_2. Figure 3.47b shows an arrangement with horizontal gas flow. Vertical gas flow allows for narrower sample compartments.

Thermogravimetry is often used in preliminary work, e.g., to avoid subjecting catalysts to unnecessary thermal stress. Temperatures sufficient for complete precursor decomposition or for complete removal of templates from as-synthesized zeolites or OMMs can be determined. Reduction of catalyst precursors as well as oxidation of catalysts can be studied, e.g., after previous reduction or after exposing them to a feed under specified conditions. The deposition of carbonaceous residues as well as their oxidation can be tracked. Other applications of DTG deal with the dehydration of surfaces or with the thermal desorption of adsorbates (cf. Section 3.4.2.2). Adsorption can be monitored only if it is activated.

Sample changes indicated by the microbalance are often difficult to explain merely based on mass losses. The combination of (D)TG with mass-spectrometric effluent analysis has therefore become very popular in catalysis research. Figure 3.47c exemplifies this by showing the template removal from the sample CTAB-MCM-48 discussed above ([84]; DTA data are repeated from Figure 3.46b).

The TG profile allows establishing mass changes (here exclusively losses) related to processes that are better identified by the DTG curve. The mass loss around 100 °C (Figure 3.47c) can be clearly assigned to a peak in the profile of m/e = 18 (water). The major DTG signal at ≈200 °C is accompanied by an intense peak originating from trimethylamine, but also from peaks, indicating hydrocarbon fragments, water, CO_2, and NO_2. These observations suggest that Hofmann degradation of the template occurred, accompanied by cracking of the hydrocarbon chain and beginning oxidation (see discussion of Figure 3.46b). The decrease of the signals of $(CH_3)_3N$ and of the cracking products indicates that endothermal reactions fade out above 200 °C, while oxidation goes on. As a result, the overall heat effect becomes exothermal (see DTA signal).

The major heat effect occurs at 325 °C and is accompanied by peaks in the desorption rates of all products. It results, presumably, from the combustion of intact molecules, while the species oxidized earlier were probably reactive intermediates of the degradation processes. Their expiration may be the reason for the shoulder preceding the main peak. Above 350 °C, carbon-rich material is oxidized: while CO_2 is released until high temperatures, the H_2O signal decays to nearly zero, shortly above the peak temperature. Notably, NO_2 is also released up to high temperatures, which shows that Hofmann degradation is not the only source for nitrogen release from CTAB. In oxygen-deficient regions of the sample, N can be apparently included in carbonaceous material, which is combusted above 350 °C.

In traditional microbalances, there is no forced gas flow through the sample: the gas passes the pan in which the gas phase above the catalyst layer is almost stagnant, i.e., transport of reactants or reaction products is dominated by diffusion. Therefore, the risk that rates of reactions with gas-phase components or of reactions releasing gaseous products are mass transfer limited is very high.

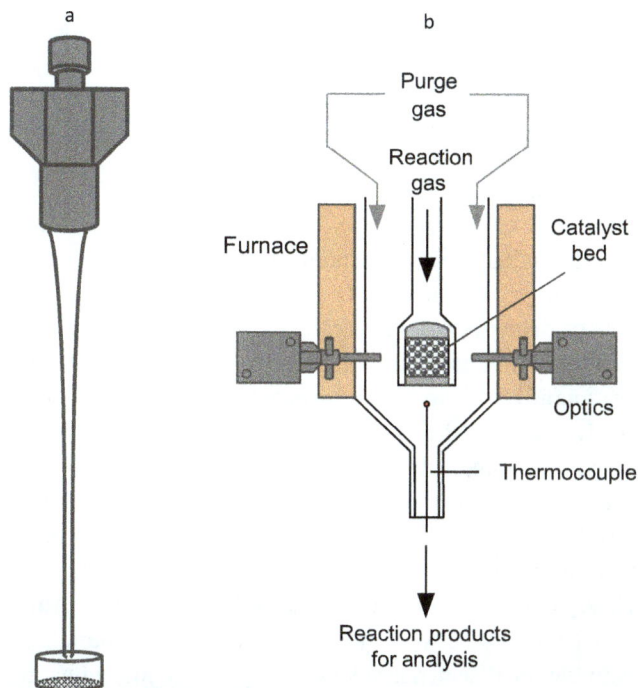

Figure 3:48: Tapered element oscillating microbalance (TEOM): (a) oscillating sample compartment; (b) oscillator integrated into a catalytic setup, with optical elements to measure its oscillation frequency.

These disadvantages are removed by the Tapered element oscillating microbalance, in which the mass is determined via the eigenfrequency of an oscillating sample compartment (TEOM, Figure 3.48). It was originally developed for the analysis of particulate matter in the ambient. In catalytic applications, the sample compartment replaces the original particle filter at the tip of a glass capillary (Figure 3.48a). The capillary is inserted into a furnace, the reactants are fed through the capillary, and a carrier gas flow flushes the reaction products to analysis by a QMS or a GC (Figure 3.48b). The TEOM allows correlating tendencies in catalytic reaction rates with changes in catalyst mass. Typical applications include studies on the relation between coke deposition and the degree of deactivation, on the reaction kinetics of the catalytic CNT growth (cf. Section 3.1.1.5), and on the relation between the buildup of the adsorbate layers and the reaction rates in reactions involving hydrocarbon pools (methanol-to-hydrocarbons, MTH,[14] cf. Sections 4.2.3 and 4.3.2).

---

14 "MTH" is used in this book as a generalizing designation for a reaction family that has spawned several industrial processes with better known names: MTG, MTO, and MTP – the conversion of **M**ethanol **T**o a highly aromatic **G**asoline, to **O**lefins, or to **P**ropylene.

Both, traditional microbalances and the TEOM, have also been used to study adsorption equilibria, however, in isothermal rather than in the temperature-programmed mode.

3.4.2.2 Calorimetry in stationary and dynamic modes

Calorimeters allow the direct measurement of heat effects related to changes in a material. Traditional static calorimetry operates at a nominal temperature, which is kept constant except for temperature excursions resulting from the heat effects to be evaluated. Differential scanning calorimetry combines calorimetry with thermal analysis: heat effects are measured while the nominal temperature is increased, usually proportional to time. It allows quantifying the heat effects, which are detected only qualitatively in DTA. In catalysis research, the processes studied by calorimetry are typically the adsorption/desorption of molecules, and sometimes, surface reactions with the solid acting as a catalyst or as a reactant. Instruments used for this purpose are referred to as adsorption and reaction calorimeters, respectively. In DSC, the focus is similar, but the scope of the applications is broader.

Calorimeters are categorized depending on how the heat effect is handled. **Adiabatic calorimeters** exclude any heat flow between the sample and the surroundings: Their temperatures must remain identical, which is achieved by permanently adjusting the temperature of the surroundings to that of the sample. The heat effect is determined from the power required for this on the basis of calibrations. Power-compensating calorimeters in which the sample temperature is kept constant by neutralizing the heat effect via a feedback loop right at the sample are sometimes classified as quasi-adiabatic. In **heat-flow calorimeters**, the heat is completely transferred to (or from) the surroundings via a well-defined thermal contact. "Complete" means that the sample temperature returns to its initial value. To achieve this, the heat capacity of the surroundings must be very high. During the heat flow, the sample temperature goes through a transient, which allows evaluating the heat effect on the basis of calibrations. There are also calorimeters in which the heat is distributed on (or originates from) both the sample and the surroundings. In such **isoperibolic calorimeters**, the surroundings are kept at a constant temperature, which differs from the sample temperature at the end of the run.

For studies on gas-phase catalysis, heat-flow calorimeters are most convenient because they can be combined with gas-dosing equipment (see Figure 3.50). DSCs can operate as heat-flow or as power compensation calorimeters. In gas–liquid systems, the distribution of the heat effect between the sample and the surroundings is difficult to predict; therefore, they are studied with isoperibolic calorimeters. All these instruments are differential calorimeters, which compare thermal responses of the sample with those of a reference not exhibiting the feature of interest.

In Tian-Calvet heat flow calorimeters (Figure 3.49), the thermal contact is combined with the temperature measurement: hundreds of thermowell wires span between the containers for sample and reference and the surrounding calorimetric block. Conduct-

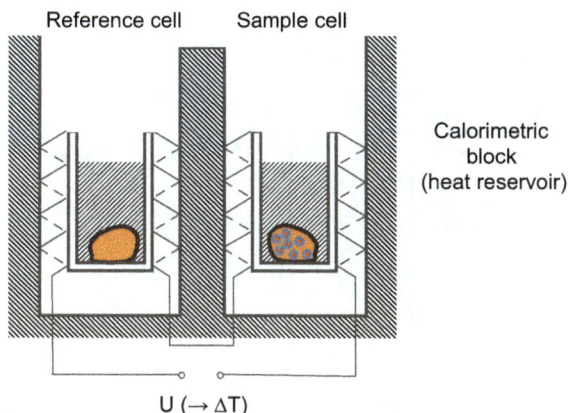

Figure 3.49: Schematic of a Tian-Calvet heat-flow calorimeter.

ing the heat flow, they are at the same time interconnected to form two thermopiles that allow very precise temperature measurements. A heat effect caused by the sample at a point of time may be neutralized by conduction to or from the heat reservoir or cause a change of the sample temperature T_S. The corresponding heat balance reads

$$-\frac{dQ}{dt} = \alpha\,S(T_S - T_R) + C\frac{d(T_S - T_R)}{dt}$$ (3.38)

where α is the heat transfer coefficient, S – the external surface area of the sample container, C – its heat capacity (including the sample), and T_R – the temperature of the heat reservoir. Equation (3.38), often expressed with different symbols, is referred to as the Tian-Calvet equation. Integrating over time gives access to the released heat (with T_S-T_R expressed as ΔT):

$$-\int_0^{Q(t)} dQ = \alpha\,S\int_0^t \Delta T\,dt + C\int_{\Delta T(0)}^{\Delta T(t)} d\Delta T$$

On the right-hand side, the second integral disappears because ΔT is zero at the start and at the end of each run. Thus, the heat effect is obtained by integration over the ΔT transient using a heat transfer parameter αS determined via calibration:

$$Q = -\alpha S \int_0^t \Delta T\,dt$$ (3.39)

Adsorption calorimetry relates the released heat to the quantity of adsorbed molecules. For this purpose, the heat-flow calorimeter is integrated into a static volumetric setup for adsorption measurements, as depicted in Figure 3.50a (see also Figure 3.39).

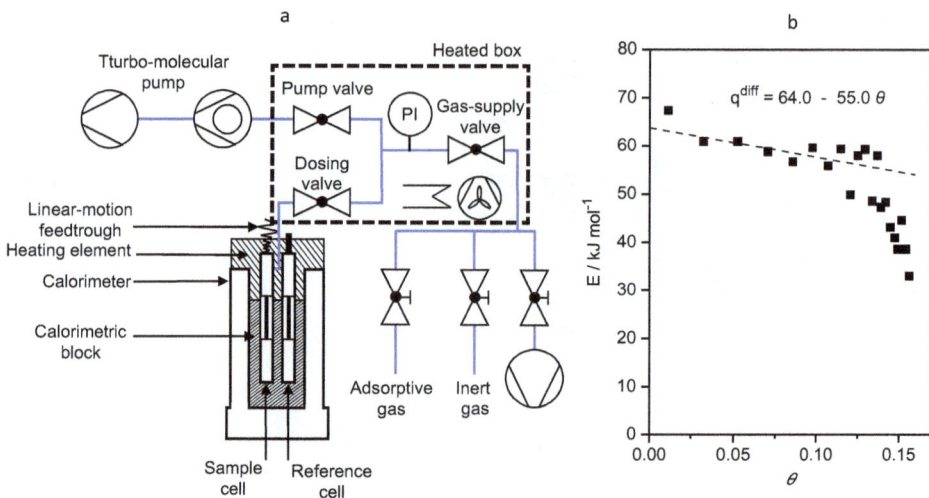

Figure 3.50: Combining volumetric adsorption measurements with heat-flow calorimetry: (a) schematics of the setup; (b) development of the differential adsorption heat q^{diff} with adsorbate coverage (CO on a Cu/ZnO-Al$_2$O$_3$ catalyst, q^{diff} given in kJ/mol). Adapted with permission from ref. [87]. Copyright 2006 American Chemical Society.

Figure 3.50b shows a typical result of such a study, reporting differential adsorption heats q^{diff} of CO on a catalyst containing Cu, ZnO, and Al$_2$O$_3$ in a molar ratio of 1 : 6 : 3.[15] q^{diff}, i.e. the heat measured for very small CO doses exhibits a linear decrease with the CO coverage θ:[16]; obviously, adsorption of CO on Cu surfaces should be described by the Temkin isotherm (2.18) rather than by the Langmuir isotherm.

Data accessible by combining volumetry and calorimetry include adsorption isotherms and rate constants of adsorption and desorption, which allow for crosschecking the adsorption constants determined at equilibrium. Differential molar adsorption entropies can be determined as well, because $dS_{ads} = dq_{ads}/T$ (cf. ref. [87]). The series of differential heats or entropies can be converted into integral quantities.

Adsorption calorimetry becomes challenging when the adsorbate reacts with the catalyst or with a preadsorbed species. When the reaction is slow, it may superimpose the adsorption kinetics and feign an ongoing adsorption, although equilibrium is already established. Simple irreversible reactions can be studied in the pulse mode, but experimental heats will be lumped from the heat effects of all elementary steps and will refer only to a small percentage of the active sites (cf. Section 3.2.1). The application of calorimetry in catalysis is, therefore, focused on the characterization of surface properties by probe molecules. Studies on the distribution of acid and basic site strengths are

---

15 A methanol synthesis catalyst, but with extremely low Cu content.
16 θ was related to the exposed Cu surface determined by RFC, cf. Section 3.4.1.6. It was not investigated in ref. [87] if there is a weaker chemisorption mode beyond $\theta = 0.15$.

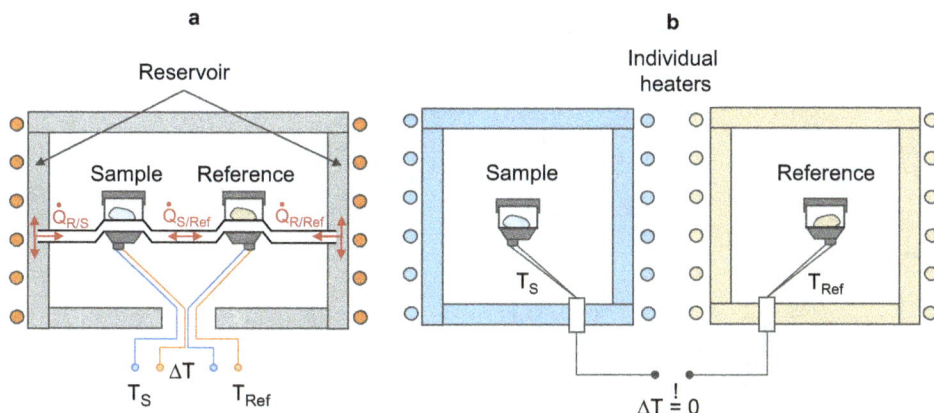

Figure 3.51: Differential scanning calorimetry: schematic of a heat flow DSC (a) and of a power compensation DSC.

discussed with many examples in refs. [88] and [89], combined with work performed with other thermoanalytical techniques. As shown in Figure 3.50 for Cu nanoparticles, the distribution of interaction strengths on metal catalysts can be investigated with probe molecules like CO.

Differential scanning calorimeters allow evaluating the heat effects over a wide range of conditions, though with a somewhat lower precision than static calorimeters. To facilitate heating according to the specified temperature profiles, their calorimetric blocks are less massive. In most heat-flow DSCs, sample and reference are in the same oven (Figure 3.51a). The sample is thermally connected to both the reservoir and the reference by a plate of high thermal conductivity. Although heat flows are less defined than in the Tian-Calvet calorimeter (Figure 3.49), accuracies are satisfactory for purposes of catalysis research. In power-compensating DSCs, the sample and the reference are enclosed by separate identical ovens (Figure 3.51b), and the difference of heating power required to keep both ovens on the same temperature profile is recorded. In both DSC versions shown in the figure, the oven(s) can be flushed with reactive gases or evacuated during the experiment.

Calorimetric studies of adsorption or catalysis at the solid–liquid interface are much more complex and are still in an infant state. Neither a heat flow between sample and calorimeter block nor a temperature measurement in the sample are feasible without interference by the solvent through which the adsorptive (or reactant) would be applied. Isoperibolic calorimetry is, therefore, the method of choice. The heats obtained cannot be assigned to a single interaction, e.g., to the interaction between adsorbate and surface, because the adsorbate also perturbs the interactions of the surface with the solvent, and the surface perturbs the interactions of the solvent with the adsorptive.

3.4.2.3 Temperature-programmed desorption (TPD) and adsorption (TPA)

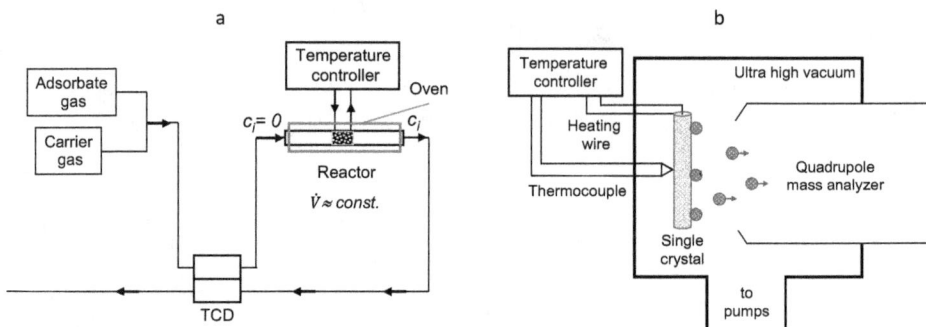

Figure 3.52: Temperature-programmed desorption (TPD) and Thermal desorption spectroscopy (TDS). Basic elements of experiments: (a) TPD; (b) TDS. Adapted from ref. 90.

In TPD, a catalyst sample previously loaded with an adsorbate is heated using a linear temperature ramp during which the desorption process is monitored. Figure 3.52 shows two versions of this method, named with slightly different designations. In temperature-programmed desorption, the adsorbate is desorbed into a flowing gas whereas in thermal desorption spectroscopy, it is desorbed into vacuum. The setup for TPD consists of equipment for gas dosing, a reactor with a temperature controller, and a concentration analyzer (Figure 3.52a). When heat conductivities of the desorbing adsorbate and the carrier gas differ sufficiently, the process can be sensitively monitored by a TCD. Alternatively, a mass spectrometer is employed, which can also detect competing desorption processes (e.g., if the catalyst was not adequately conditioned) or unexpected conversions of the adsorbate by the catalyst.

TDS belongs to surface science. The analyzer is a QMS, the sampling port of which is moved close to the surface under study, e.g., a single-crystal facet (Figure 3.52b). Temperature ramps employed are mostly between 2 and 30 K/s, two orders of magnitude faster than in TPD where the typical ramps are 5–10 K/min. In TDS, adsorption and start of temperature ramps are often at very low temperatures, e.g., 100 K. Due to the very fast temperature ramps and the low gas-phase concentrations in the vacuum, desorption is irreversible in TDS, while re-adsorption may broaden the desorption profiles in TPD.

TPD is yet another method to determine the adsorption (chemisorption) capacity of surfaces. In addition, it allows differentiating adsorption sites and determining their quantities and their adsorption strengths. H_2 and CO are typical probe molecules for metal surfaces. For acid sites, bases like NH_3 are applied. The analogous approach to study basic sites via desorption of weak acids like CO_2 is complicated by the availability of many reaction channels between the surface and the adsorbate (cf. Scheme 3.4 below). TPD is then accompanied by the decomposition of the different forms of carbonates and formates, and there is no guarantee that the surface returns to its original state after the experiment. Studies of basic surfaces by CO_2 adsorption are, therefore, in the focus of IR spectroscopy rather than of TPD.

Heating oxygen-containing catalytic materials in inert gas to detect labile oxygen species is also referred to as TPD in literature, although the desorbed O_2 is not adsorbed initially. The technique targets the structure of surfaces rather than their reactivity. Although it is very popular in the study of carbon-based materials, which exposes various O-containing functional groups (cf. Figure 2.24), it will not be discussed in more detail here (see, however, ref. [91]).

TPD of H_2 or CO from supported metal catalysts revealed, for instance, a broadening of the main signal with decreasing average particle size, which can be ascribed to the increased impact of highly exposed (edge/kink) sites. Their contribution to the signal can be enhanced by exposing the sample to a temperature near the desorption peak prior to the run, which removes most adsorbates from the terraces. In TDS, adsorbates on different facets of single crystals can be differentiated by their desorption temperatures. This can be used to analyze the exposure of facets in supported catalysts by the TPD of the same probe molecule, although the reliability of conclusions is somewhat affected by the lower resolution of desorption energies (temperatures) to be achieved in TPD. The support or promoter influences on adsorption properties can be studied by TPD as well.

TPA starts with the bare catalyst, which is heated in a gas flow containing the adsorptive, following a linear temperature ramp. It makes sense only for activated adsorption; nonactivated adsorption processes cannot be resolved. With increasing temperature, the catalyst withdraws the adsorptive from the flowing gas until it is saturated; later it desorbs the adsorbate again. In the TPA profile, the partial pressure p_i of the adsorptive initially decreases below that of the feed ($p_{0,i}$) Subsequently, it increases above it before finally returning to $p_{0,i}$.

TPA is more intricate than TPD with respect to both the experiment and the data treatment. Depending on the adsorption kinetics, the temperature needs to be started even below 270 K. This method is used in the advanced characterization of catalytic surfaces, e.g., for establishing adsorption kinetics to be used in microkinetic models (cf. Section 4.6). Typical TPA experiments are described in ref. [92], where the reader can also find how conclusions can be derived by combining them with TPD studies.

Figure 3.53 presents two typical examples for the application of TPD in catalysis. TPD of ammonia is a standard technique for the characterization of acid sites. In Figure 3.53a, it is applied to zeolite catalysts – three Fe-ZSM-5 samples and a H-ZSM-5 with Si/Al ≈ 14, which was the parent zeolite for two of the Fe-modified catalysts. The third one, Fe-Z(B), was made by CVD of $FeCl_3$ into a Si-rich H-ZSM-5 (Si/Al ≈ 40), in which much of the stoichiometric Al was not incorporated in the framework [93]. The ammonia was adsorbed at 373 K to avoid contributions of physisorbed NH_3.

In the TPD profiles, there are two major signals arising from adsorption sites of different strengths. The method gives no evidence on the nature of these sites, but it is known that the high-temperature signal is dominated by NH_3 from the strong Brønsted sites of zeolites. However, Lewis sites exposing the framework Al are very strong as well,

a

b

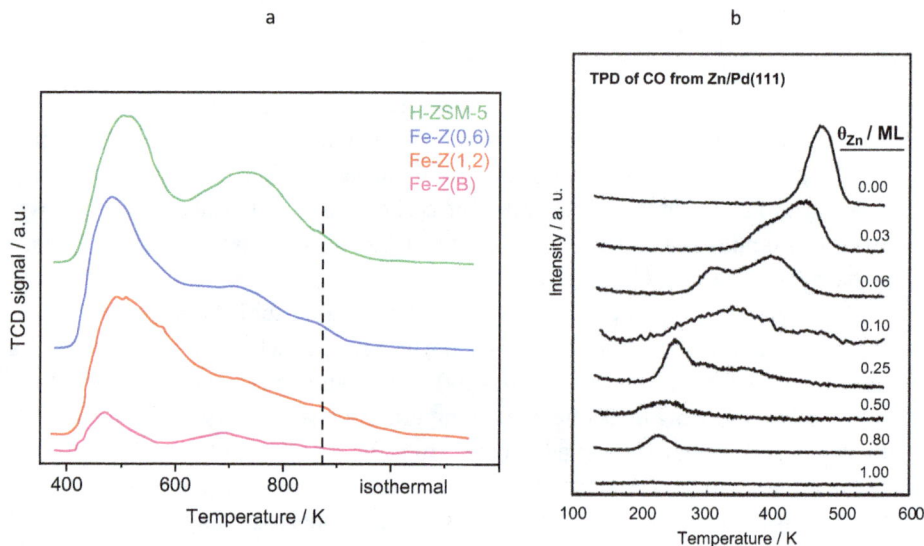

Figure 3.53: Examples for the application of TPD and TDS: (a) TPD of NH₃ from Fe-modified zeolites; (b) TDS of CO from Zn/Pd(111) alloy surfaces. Reproduced from refs. [93] (a) and [94] (b) with permission from Elsevier.

as long as the Al atoms are not detached. The low-temperature signal may be ascribed to various sites, e.g., to silanol groups at the external surface or in structural defects, and to Lewis-acidic extra-framework alumina. In the TPD profiles of Figure 3.53a, the decrease of the high-temperature signal with growing Fe content suggests an exchange of zeolite protons by the Fe cations. At the same time, the intensity increases slightly below 600 K, which may be ascribed to the Fe ions, either as Lewis sites or to Fe-OH groups. In Fe-Z(B), this signal was not observed at all. Spectroscopic studies showed the Fe phase being dominated by small oxide aggregates in this sample.

Figure 3.53b shows the thermal desorption spectra of CO from Pd(111), which was annealed with Zn atoms at increasing coverages of up to 1 monolayer (ML) to obtain surface alloys. From pure Pd, CO desorbs in a single peak at 460 K. A Zn surface coverage of 0.03 is sufficient to substantially perturb the profile, and at $\theta = 0.06$, the highest desorption signal has downshifted by 60 K. With increasing Zn coverage, the profile broadens drastically, while the amount of the desorbing CO decreases by just about 10% until $\theta = 0.25$. At $\theta \geq 0.5$, the adsorption capacity breaks down, although 50% of the Pd atoms are still exposed, while it is zero on a pure Zn surface [94].

The interpretation of these data was supported by a series of vibrational spectra (HREELS, see Section 3.4.8.2), which differentiate CO on threefold, bridge, and atop sites (cf. Section 2.4.2). On Pd(111), CO adsorption is most stable in threefold hollow ("Pd₃") sites. At the low CO coverages used, atop and bridge sites were not populated; therefore, only one signal appears in the TDS (Figure 3.53b). On Zn atoms, CO does not adsorb at all. Already below 0.1 ML, the deposition of Zn atoms has two major effects: Pd₃ sites are destroyed by Zn, leaving behind the bridged or atop sites for CO, and the intact Pd₃ sites are influenced by the adjacent Zn atoms. These modifications are examples of what is called "ensemble effect" and "ligand effect" in the catalysis with alloys (see Sections 4.2.2 and 4.3.5).

A model of the surface at 0.06 ML Zn showed the most likely numbers of Zn atoms adjacent to the Pd_3 sites to be 2, 1, or 0; obviously, there are still Pd_3 sites on this surface without direct contact to Zn. As the influence of Zn is weakest in these sites, the desorption signal at 400 K was ascribed to them. This results in the remarkable conclusion that even such indirect influence of Zn caused a 60 K shift of the desorption temperature of CO, from 460 K at 0 ML to 400 K at 0.06 ML. From the remaining, not always unambiguous assignments, we cite only one for the peak at 230 K dominating the profiles at $\theta \geq 0.5$. On the basis of both surface models and HREELS spectra, it was ascribed to desorption from Pd atop sites influenced by the adjacent Zn [94].

If signals are well separated, TPD may give access to surface coverages, to the kinetics (rate constants, activation energies) of desorption, and, via the equilibrium conditions, also of adsorption. Such information is important for the advanced characterization of catalytic surfaces. It is also useful for the development of microkinetic models for reactions, in which the adsorbate is a reactant. Therefore, some proven methods for extracting data of desorption or adsorption kinetics from TPD profiles are briefly discussed in the following.

It should be noted that a TPD profile gives direct access to the changes of the desorption rate r_{des} over time (or temperature). Considering desorption as back reaction of adsorption, r_{des} is

$$r_{des} = - \frac{1}{m_{cat}} \frac{dn_{ads}}{dt} = \frac{\dot{n}_{des}}{m_{cat}} \tag{3.40}$$

(cf. eq. (1.5), n_{ads} – amount of adsorbate on the surface). The amount (per second) desorbed into the carrier gas is removed by convection: $\dot{n}_{des} = \dot{n}_{conv}$. Assuming that the carrier gas velocity $\dot{V}$ is not significantly changed by the desorption, this results in

$$m_{cat}\, r_{des} = \dot{V}\, c_i, \text{ and}$$

$$r_{des} = \frac{\dot{V}}{m_{cat}}\, c_i \tag{3.41}$$

In the model, the reactor is considered well-mixed, which may be justified by small catalyst volumes and short catalyst layers.

In the following, r_{des}, according to eqs. (3.40) and (3.41) will be equalized with the rate laws of desorption, e.g.,

$$r_{des} = k_{des}\, \theta^n - k_{ads}\, c_i\, (1 - \theta)^p \tag{3.42}$$

(cf. eq. (2.13)), where n and p are the kinetic orders of desorption and adsorption, respectively. In simplified approaches, re-adsorption is neglected and the rate is described by the Polanyi-Wigner equation

$$r_{des} = v \, n_{mon}^n \, \theta^n \, exp\left(-E_{A,des}/RT\right)$$ (3.43)

in which the rate constant is split into the product of the frequency factor v [s^{-1}], the monolayer capacity n_{mon} [mol g^{-1}] as a part of the surface concentration term (cf. Section 2.6.1), and the temperature dependence, with the activation energy of desorption $E_{A,des}$ [kJ mol^{-1}].[17] Using eqs. (3.40) through (3.43), the basis for modeling the change of the adsorbate coverage θ with temperature T is easily derived.

Written with the monolayer capacity n_{mon}, eq. (3.40) reads

$$r_{des} = -n_{mon} \, \frac{d\theta}{dt}$$ (3.44)

Equating r_{des} from (3.41) and (3.42) allows isolating c_i

$$r_{des} = \frac{\dot{V}}{m_{cat}} \, c_i = k_{des}\theta^n - k_{ads} \, c_i \, (1-\theta)^p$$

$$c_i = \frac{m_{cat} \, k_{des} \, \theta^n}{\dot{V} + m_{cat} \, k_{ads} \, (1-\theta)^p}$$

This can be inserted into the equation obtained by equating r_{des} according to (3.44) and (3.41)

$$r_{des} = -n_{mon}\frac{d\theta}{dt} = \frac{\dot{V}}{m_{cat}}c_i = \frac{\dot{V}}{m_{cat}}\frac{m_{cat}k_{des}\theta^n}{\dot{V} + m_{cat}k_{ads}(1-\theta)^p}$$

Due to the linearity of the temperature profile $T = T_0 + \beta T$, the dependence on time can be easily transferred into a dependence on temperature ($dt = dT/\beta$), which results in

$$\frac{d\theta}{dT} = -\frac{\dot{V}}{\beta \, n_{mon}}\frac{k_{des} \, \theta^n}{\dot{V} + m_{cat} \, k_{ads} \, (1-\theta)^p}$$ (3.45)

Although this equation allows modeling all TPD profiles, we use it rather to discuss two limiting cases with respect to the rate of re-adsorption, which is either zero or very

---

17 In literature, the n_{mon} term is sometimes dropped, which results, however, in a disagreement of units on the two sides of eq. (3.43).

fast. The latter corresponds to an established adsorption equilibrium. With $k_{ads} = 0$, eq. (3.45) becomes very simple. In eq. (3.46), k_{des} is therefore split, by analogy, to the Polanyi-Wigner equation (3.43)

$$\frac{d\theta}{dT} = -\frac{k_{des}\theta^n}{\beta n_{mon}} = -\frac{\nu n_{mon}^{n-1}}{\beta}\theta^n exp\left(-\frac{E_{A,des}}{RT}\right) \qquad (3.46)$$

At established equilibrium, $m_{cat}\, k_{ads}\,(1-\theta)^p$ is much larger than $\dot{V}$. Using the same notation for the rate constants, the adsorption enthalpy appears in the temperature dependence because it is the difference between the activation energies involved

$$\frac{d\theta}{dT} = -\frac{\dot{V}}{\beta\, m_{cat}}\frac{\nu_{des}}{\nu_{ads}} n_{mon}^{n-p-1}\frac{\theta^n}{(1-\theta)^p}\, exp\left(-\frac{\Delta H_{ads}}{RT}\right) \qquad (3.47)$$

While eq. (3.46) allows deriving relatively simple methods for extracting the kinetic parameters from experimental data (see below), eq. (3.47) requires the application of numerical methods for fitting the model to experimental profiles.

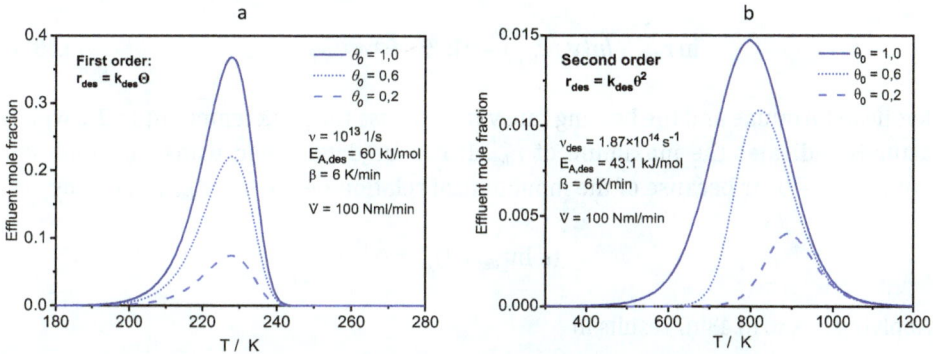

Figure 3.54: TPD profiles for first-order (a) and second-order desorption (b), re-adsorption excluded.

Data reduction is somewhat simplified by the limited choice of orders for desorption and adsorption: They are typically first or second order. In the absence of re-adsorption, the desorption order can even be derived from the profile shapes obtained in runs with different initial coverages.

In Figure 3.54, desorption profiles simulated on the basis of eq. (3.46) for first- and second-order desorption are compared.[18] The properties of the curve sets reflect rather general differences between the reactions of first and higher orders. In a first-

---

18 Notably, $E_{A,des}$ was kept constant in these simulations, as opposed to analogous simulations shown in ref. [95].

order reaction, there is no influence of the initial concentration on the dependence of conversion on time. Actually, the curves in Figure 3.54a differ only by a factor. Related to the profile for $\theta_0 = 1$, the factor is 0.6 for $\theta_0 = 0.6$, and 0.2 for $\theta_0 = 0.2$. Indeed, conversions at a temperature T (i.e., the ratios between integrals up to T and below the whole profile) are the same for all three curves. When the order is >1, the enhanced concentration dependence of the rate results in a stronger deceleration at low reactant concentrations. In Figure 3.54b, peak temperatures increase with decreasing θ_0; higher temperatures are required to achieve comparable conversions. Notably, in the second-order desorption, the profiles tend toward a common trend at high temperatures.

In the following, three proven methods for deriving the parameters of desorption kinetics are introduced. The **Redhead formulae** have been widely applied in surface science to relate TDS peak temperatures with $E_{A,des}$ by a single measurement. They involve, however, at least one estimated parameter. In TPD, **heating rate variation** is a standard method for the determination of $E_{A,des}$. When there is a series of high-quality TPD profiles with varying initial coverages θ_0, **complete analysis** also gives access to rate laws and frequency factors.

All three methods can be derived, starting from the Polanyi-Wigner equation (3.43) in its logarithmic form

$$\ln r_{des} = \ln\left(v\, n_{mon}^n\right) + n \ln \theta - E_{A,des}\big/RT \tag{3.43ln}$$

Redhead formulae and the heating rate variation use the peak temperature T_m, which actually indicates the maximum of r_{des}. It can be determined using the condition $(dr_{des}/dT)_{Tm} = 0$, or because of the monotonical relation between $\ln r_{des}$ and r_{des}, by

$$(d \ln r_{des}/dT)_{Tm} = 0$$

Applying this to (3.43ln) results in

$$(d \ln r_{des}/dT)_{Tm} = \frac{n}{\theta_m}\left(\frac{d\theta}{dT}\right)_{Tm} + E_{A,des}\big/RT_m^2 = 0$$

where θ_m is the coverage at the peak temperature. Therefore,

$$\left(\frac{d\theta}{dT}\right)_{Tm} = -E_{A,des}\theta_m\big/nRT_m^2$$

Combining the rate definition (3.44), the linear temperature profile, and the Polanyi-Wigner equation (3.43) results in another expression for $d\theta/dT$:

$$r_{des} = -n_{mon} \frac{d\theta}{dt} = -n_{mon} \beta \frac{d\theta}{dT}, \text{ and}$$

$$\frac{d\theta}{dT} = -\frac{v}{\beta} n_{mon}^{n-1} \theta^n \exp\left(-E_{A,des}/RT\right).$$

By equating the expressions for $d\theta/dT$,

$$-E_{A,des}\theta_m/nRT_m^2 = -\frac{v}{\beta} n_{mon}^{n-1} \theta_m^n \exp\left(-E_{A,des}/RT_m\right) \text{ is obtained, finally}$$

$$E_{A,des}/RT_m^2 = -\frac{n\,v}{\beta} n_{mon}^{n-1} \theta_m^{n-1} \exp\left(-E_{A,des}/RT_m\right) \tag{3.48}$$

From (3.48), the **Redhead formulae** are obtained by inserting n = 1 or 2 [96]:

First order

$$E_{A,des}/RT_m^2 = -\frac{v}{\beta} \exp\left(-E_{A,des}/RT_m\right) \tag{3.49}$$

Second order

$$E_{A,des}/RT_m^2 = -\frac{v}{\beta} n_{mon}\theta_0 \exp\left(-E_{A,des}/RT_m\right) \tag{3.50}$$

Equation (3.50) contains the approximation that the coverage at the maximum θ_m is ca. half the initial coverage θ_0.

Although $E_{A,des}$ cannot be isolated from eqs. (3.49) and (3.50), it can be evaluated for the first-order from one experimental peak temperature T_m, provided the frequency factor v is known, for which $\approx 10^{11}$ or 10^{13} s^{-1} have been considered good estimates. For using (3.50), θ_0 is also required, which can, however, be determined from the desorption curve.

By analyzing data with different values of v/β, Redhead found that $E_{A,des}$ is proportional to T_m in a wide range of this parameter [96]. This allows simplifying eq. (3.49) into

$$E_{A,des} = RT_m \left(\ln \frac{v\, T_m}{\beta} - 3.64 \right) \tag{3.51}$$

The ranges of β covered in Redhead's analysis ($\beta \geq 1$ K s^{-1}) suggest that (3.51) may not be valid for TPD into flowing gas. The Redhead formulae assume that $E_{A,des}$ does not depend on the adsorbate coverage θ, which is nowadays considered a drastic simplification.

The basic equation of the **heating rate variation** approach is obtained from eq. (3.48) via a few simple rearrangements:

$$\ln \frac{\beta}{T_m^2} = \ln \left(\frac{v\, n\, R\, n_{mon}^{n-1}}{E_{A,des}} \theta_m^{n-1} \right) - \frac{E_{A,des}}{RT_m} \tag{3.52}$$

In eq. (3.52), β is an experimental parameter and T_m is an experimental result; all other quantities except for θ_m are constant, though mostly not known. For n = 1, plots of

$ln(\beta/T_m^2)$ vs. $1/T_m$ are linear with the slope $-E_{A,des}/R$. This also holds for other orders, provided that there is no strong influence of β on θ_m.

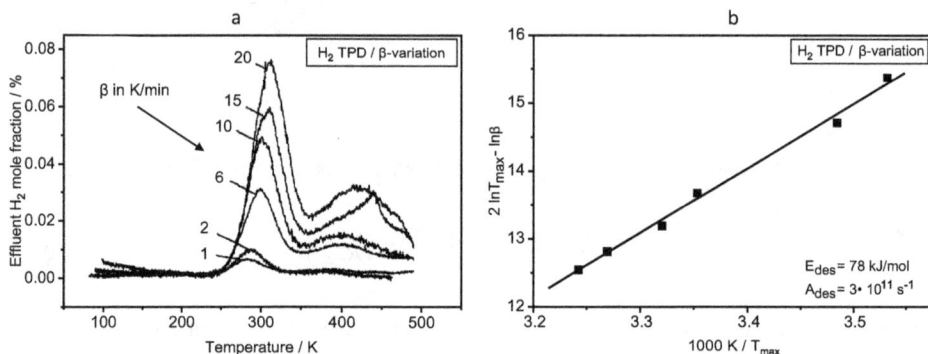

Figure 3.55: Desorption kinetics via heating rate variation: TPD of H_2 from a Cu/ZnO-Al$_2$O$_3$ catalyst: (a) TPD profiles measured with different heating rates β; (b) determination of the activation energy of desorption. Adapted from ref. [97] with permission from Springer Nature.

As an example, Figure 3.55a shows the TPD profiles of H_2 from a Cu/ZnO-Al$_2$O$_3$ methanol synthesis catalyst, recorded with heating rates β between 1 and 20 K min^{-1} [97]. Only the first peak around 300 K originates from H atoms adsorbed on Cu.[19] The shift of its peak temperature T_m with increasing β is obvious. At the same time, the signal intensities grow, because the same amount of adsorbate is desorbed into decreasing quantities of carrier gas. The plot of $ln(\beta/T_m^2)$ vs T_m^{-1} is indeed linear (Figure 3.55b), which suggests that the inherent assumption of constant $E_{A,des}$ is valid. From the slope, an activation energy of H_2 desorption from Cu interacting with ZnO of 76 ± 2 kJ mol^{-1} was obtained. It should be noted that this cannot be equated with the adsorption enthalpy ΔH_{ads}, because adsorption of H_2 on Cu is activated.

Later work showed significant disagreement in profile widths when the experimental curves were simulated on the basis of eq. (3.46) using the kinetic parameters obtained by heating rate variation (Figure 3.55b). The deviations were most drastic at high θ_0. Convincing agreement could be, however, achieved with very similar parameters by additionally introducing a dependence of $E_{A,des}$ on the coverage: $E_{A,des} = (75 - \theta^{2.6})$ kJ/mol [98]. Thus, heating rate variation supplied good estimates of the kinetic parameters, but failed to indicate a (moderate) variation of $E_{A,des}$.

Figure 3.56 shows a set of TPD profiles that can be used for **complete analysis**, obviously from a second-order desorption. The determination of desorption rates and coverages is demonstrated for an arbitrary temperature T_s. The curves show desorption rates, which are proportional to the measured gas-phase concentrations (eq. (3.41)). The

---

19 The broad signal above 350 K resulted from the interaction of Cu with impurity water in the carrier gas and could be eliminated in later work.

coverage θ is the ratio between the hatched area indicating the amount of adsorbate still on the surface at T_s and the total area of the respective profile. It is quite obvious that the coverages at a temperature T_s vary with the initial coverage θ_0 in the data of Figure 3.56.

For determining the kinetic parameters from the data, the logarithmic Polanyi-Wigner equation (3.43ln) is used. A plot of $\ln r_{des}$ vs. $\ln\theta$ at T = constant is linear and reveals the reaction order n as its slope. One may conveniently check if n varies with temperature. $E_{A,des}$ is obtained by plotting $\ln r_{des}$ vs. T^{-1} at constant coverage θ. The slope of these linear plots is $E_{A,des}/R$. With good primary data, this can be performed for any adsorbate coverage θ; thus, complete analysis allows detecting if $E_{A,des}$ depends on the coverage. From the intercepts (cf. eq. (3.43ln)), even the pre-exponential factor $\nu\, n_{mon}^{n}$ is accessible.

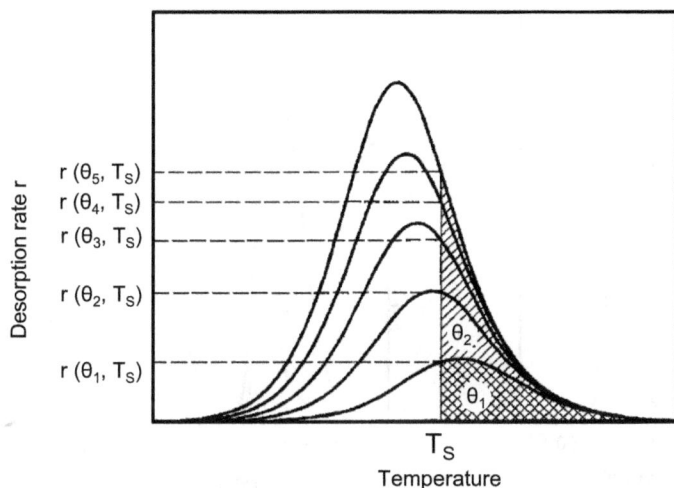

Figure 3.56: Complete analysis of TPD profiles. Reproduced from ref. [99] with permission from Taylor & Francis.

3.4.2.4 Temperature-programmed surface reaction (TPSR)

TPSR is closely related to TPD, but now the gas phase contains a component that is able to involve the adsorbate in a catalytic reaction, e.g., H_2. The effluent contains the adsorbate and the reaction product(s), which can be monitored only with a mass spectrometer. TPSR differentiates spectator sites, releasing the adsorbate unchanged, and reaction sites of different activities, which release the product at different temperatures. It supplies useful information also if the catalyst offers competitive reaction channels for the adsorbate.

The term "TPSR" was originally used only for the temperature-programmed reaction of a deposit, e.g., coke, with a gas-phase component like H_2 or O_2. The latter would be designated rather as a temperature-programmed oxidation today (TPO, see Section 3.4.2.5). The analogous experiment with adsorbates was referred to as temperature-programmed reaction (TPR). It is this type of experiment, which is referred to as TPSR in this book in order to avoid confusion with the acronym TPR that means tempera-ture-programmed reduction of catalyst precursors, see Section 3.4.2.5. On the contrary, the interaction of a catalyst with the full reaction mixture during a linear temperature program, which is sometimes used as a rapid protocol for testing catalyst activity, is not covered by the term "TPSR" at all; in TPSR, the supply of one reactant is limited by its adsorption capacity.

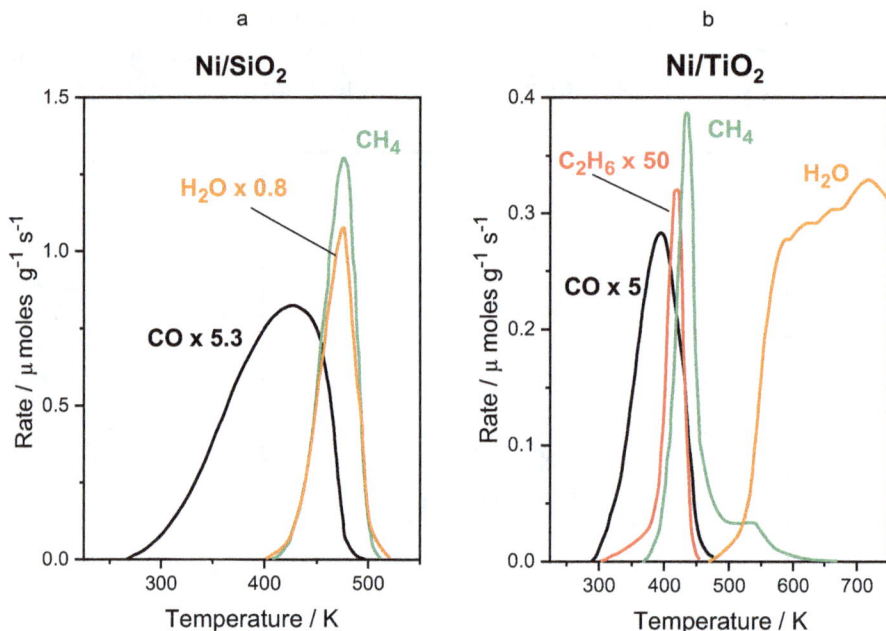

Figure 3.57: Temperature-programmed surface reaction (TPSR) of CO adsorbed on supported Ni catalysts, with H_2 from the gas-phase. Ni was supported on SiO_2 (a) or TiO_2 (b). Adapted from ref. [99] with permission from Taylor & Francis.

In Figure 3.57, TPSR profiles of adsorbed CO on supported Ni catalysts reacting with gas-phase H_2 are compared. In both experiments, CO starts desorbing intact from the spectator sites at low temperatures. From Ni/SiO_2 (Figure 3.57a), both reaction products CH_4 and H_2O are desorbed in equimolar ratio in a narrow peak at ≈470 K. From Ni/TiO_2 (Figure 3.57b), methane is desorbed below 450 K, while water appears strongly delayed from the hydrophilic support. It was not explained in ref. [99] why the H_2O signal from Ni/TiO_2 is clearly overstoichiometric; options may be dehydroxylation or reduction of TiO_2 around the Ni particles (cf. SMSI effect, Section 2.3.3).

The profiles allow differentiating three types of sites – spectator sites and two types of reactive sites. The spectator sites are less abundant than the reactive sites (by 30–40%), their adsorption energies (actually, $E_{A,des}$, but CO adsorption on Ni is not activated) are widely distributed, in particular on SiO_2. There are also remarkable differences with respect to the methanation sites. Closer inspection (cf. the different scales of the ordinate axis) shows that Ni is more active on Ti, but the number of sites is much lower. A large amount of less active sites available on SiO_2 is almost completely missing on TiO_2. The formation of ethane only on Ni/TiO_2 (Figure 3.57b) is another difference between the catalysts. Among CO hydrogenation catalysts, Ni is known to be very selective for methane (cf. Section 4.5.3). Therefore, the formation of ethane, though at low selectivity, under conditions that strongly favor hydrogenation of C_1 surface species over their combination to C_2 moieties (no CO, only H_2 in the gas phase) is quite special.

TPSR profiles can be used to extract kinetic information for the reaction in a similar way as the desorption of adsorbates from TPD profiles (Section 3.4.2.3). Activation energies and pre-exponential factors obtained from TPSR profiles refer to the rate-determining step of the surface reaction, which may be, of course, the desorption of the reaction products in some cases. TPSR can reflect various influences on a catalytic site: by promoters, by poisons, by supports, by specific treatments, etc. However, as in pulse-catalytic approaches, gas-phase composition and surface coverages during TPSR differ strongly from those in steady-state kinetic experiments using the complete feed. The merit of TPSR is the detection of active sites, their differentiation, and even quantitative description in catalytic systems. What contribution do these sites make to the rates achieved when the same catalyst operates in a technical reactor is, however, not always easy to forecast from these data.

3.4.2.5 Temperature-programmed reduction (TPR), oxidation (TPO), and sulfidation (TPS)

In this section, methods are introduced in which a feed component reacts with the catalyst or a catalyst precursor. The by far most popular method of this group is Temperature-programmed reduction (TPR), which probes the reducibility of the sample, usually in H_2, sometimes in CO, or using a reactant as reductant. Temperature-programmed oxidation (TPO) is not applied to catalyst precursors because these are typically conditioned by calcination. A TPO run can follow a TPR run to find out if the changes during reduction are reversible. TPO transfers the catalyst components to their highest oxidation states, which are good reference states. When there is only one reducible element, TPO can, therefore, be applied after a catalytic run to measure the average oxidation state of the catalyst in the reaction environment. It is also applied to catalysts deactivated by carbonaceous deposits because these can be differentiated by their affinity to oxygen. Temperature-programmed sulfidation (TPS) is a rather special technique dealing with the preparation of sulfide catalysts from oxide precursors. Sulfidation rates depend strongly on the structure of these precursors and on influences of supports and promotors on them (cf. Section 2.3.3). Therefore, TPS can contribute to the appropriate choice of precursor structures and treatment conditions for the optimization of sulfide catalysts.

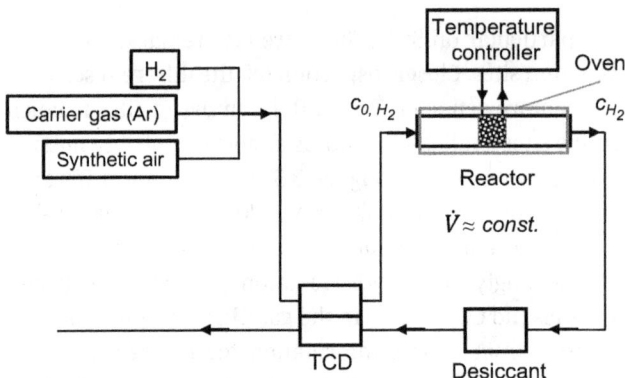

Figure 3.58: Temperature-programmed reduction (TPR): basic elements of the experiment. Adapted with kind permission of Wiley-VCH from ref. [90].

Figure 3.58 shows a typical **TPR** experiment. It is very simple and similar to TPD. The carrier gas is now a mixture with a pre-defined H_2 content, the deviations from which are monitored by a TCD or a QMS. As the reduction product water would affect the analysis by the TCD, it is trapped upstream of the detector.

To achieve good data quality, the relation between the amounts of reducible substance in the sample and of H_2 supplied by the gas must be within certain limits. In an excess of H_2, sample reduction causes only minor concentration changes, while an excess of the sample may cause temporary depletion of gas-phase H_2, which results in distortions of the TPR profile. Recommendations for an appropriate choice of parameters can be found in reference [100].

A TPR profile tells when a precursor is completely reduced, i.e., at which temperature unnecessary stress would be inflicted on the catalyst. From the H_2 consumption, average oxidation states can be assessed, provided the initial one is known (it is the highest possible in many cases). Unfortunately, this information is not very specific. If there is only one reducible element, the average may still represent the coexisting sites in different oxidation states; if there are more reducible elements, even the allocation of the reduction equivalents to them remains unclear. Although assignments based on chemical intuition may be useful, they are not necessarily safe, because reducibility in mixed samples may be affected by the interactions between the components. Co-reduction, where H atoms activated on metal particles obtained from the more reducible component spill over to the less reducible one and reduce it, sometimes even within the same TPR signal (cf. Section 4.5.4), is an example for such surprises. Thus, if the state of the sample during or after TPR is of interest, H_2 consumption data must be combined with complementary characterization, e.g., by XPS or XAFS.

Although TPR is not a high-level technique either experimentally or methodically, it provides important evidence about solid-state interactions in the mixed-oxide phases. Conclusions are mostly drawn on a fingerprint basis. Bulk oxides have specific TPR profiles, which should superimpose when the oxides are just mixed. Deviations from the expected pattern are therefore assigned to chemical interactions or solid-state reactions between the components. It should be noted that such an assignment neglects the phenomenon that TPR profiles of oxide powders can also vary with the particle size under otherwise identical conditions, but deviations from the physical mixture pattern are often so dramatic that the fingerprint approach is justified. On the other hand, while TPR indicates the interactions, it does not reveal their nature, which necessitates complementary structural investigations.

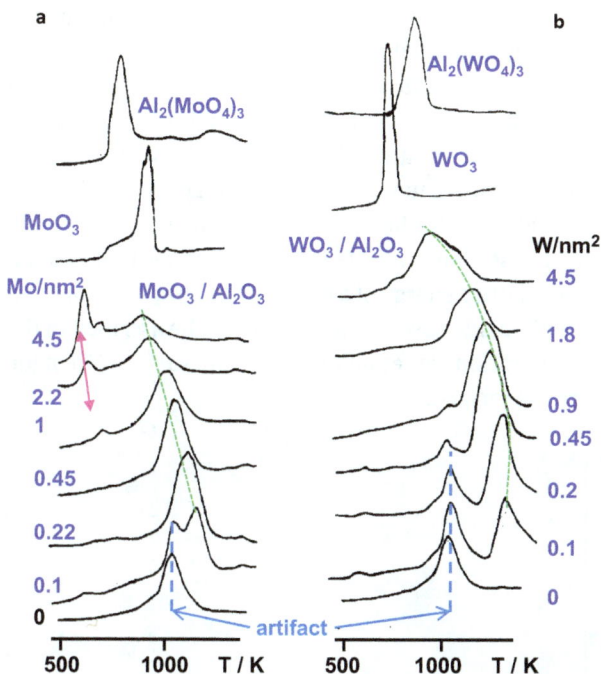

Figure 3.59: Interactions between metal oxides and γ-Al$_2$O$_3$, (a) MoO$_3$; (b) WO$_3$. Adapted with kind permission of Elsevier from ref. [101]. The feature at ≈1,000 K (pure Al$_2$O$_3$, low Mo (W) contents) was later identified as an artifact.

As an example, TPR profiles of W(VI) and Mo(VI), supported on γ-Al$_2$O$_3$ in different loadings, are compared with those of bulk compounds (oxides, molybdates) in Figure 3.59. In the experiments, sample sizes were adapted to have comparable amounts of reducible substance in all runs. W and Mo contents ranged from very low loadings to values just around the monolayer capacity, which is 4–4.5 W (Mo)/nm^2.

The difference between the TPR profiles of bulk and supported oxides is most striking for tungsten (Figure 3.59b). The bulk compounds caused single peaks centered at ca. 700 K (oxide) and 850 K (molybdate). Al$_2$O$_3$ is not reducible in the temperature range covered; the signal around 1,000 K was later assigned to an artifact. At low W(VI) content, temperatures above 1,300 K are required to reduce W(VI)!

Stabilization is a consequence of a structural effect already discussed in Section 2.3.3 – the formation of a two-dimensional surface oxide phase, distinctly different from $Al_2(WO_4)_3$. At low loadings, the surface W(VI) oxide species are isolated and most stabilized. With increasing loading, they start aggregating via oxygen bridges, which improves their reducibility, but only slightly above the theoretical monolayer capacity does reduction start at temperatures typical for bulk compounds. The conclusion drawn from this observation, according to which the W oxide layer does not grow into the third dimension until the Al_2O_3 surface is completely covered by the monolayer, was later confirmed by spectroscopy and surface analysis. Why the reduction of isolated Al_2O_3-supported W(VI) is so difficult, has remained unexplained so far.

The consequences of Mo(VI) interaction with alumina are less dramatic, but still significant (Figure 3.59a). The TPR profiles of MoO_3 and $Al_2(MoO_4)_3$ exhibit major peaks at ca. 925 and 780 K, respectively, while temperatures beyond 1,100 K are required to reduce 0.1 Mo(VI)/nm². However, at 1 Mo/nm² a new signal appears around 700 K. With increasing Mo loading, it enhances and shifts far below the reduction onset of the bulk compounds. The signal was assigned to second-layer Mo (VI) oxide species that is stabilized neither by the support nor by the lattice energy of a crystal. That they were present far before the monolayer of surface Mo oxide species was closed may be ascribed to the preparation route chosen. Other preparation routes allow closing the surface Mo oxide monolayer before additional layers are formed.

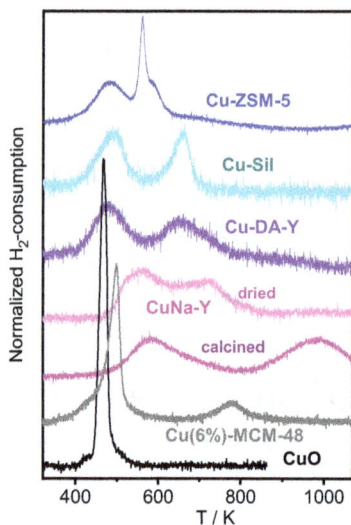

Figure 3.60: Reducibility of Cu(II) in CuO, in an OMM, and in zeolites studied by TPR; Sil – siliceous MFI, DA-Y – siliceous FAU.

In Figure 3.60, the influence of the zeolite matrix on the reduction of Cu^{II} is demonstrated by TPR profiles. Under normal conditions, bulk CuO is reduced in a single step. The intermediate Cu^{I} can be detected only at very small H_2 concentrations and low heating rates. The reaction is considered autocatalytic. Once some Cu metal has been formed, it activates hydrogen, which accelerates the reduction of the remaining oxide species. This is the background of the CuO profile in Figure 3.60, and likewise of the main peak of CuO in MCM-48, where a minor signal around 800 K was probably caused by Cu silicate species formed from amorphous parts of the sample.

In zeolites, the reduction always proceeds in two steps, but the separation between them is very different. It is small in Cu-ZSM-5 and in the dried CuNa-Y; it is wider in the siliceous forms of the zeolites, and it is extreme after the calcination of CuNa-Y. To indicate only some of various interesting features in these samples, we point out that the reduction of Cu^{2+} starts at higher temperatures in the polar Y than in zeolites with Si-rich framework, and that calcination results in an extreme stabilization of Cu cations in Y. The latter arises probably from very efficient cation stabilization in the narrow β cages, which are not populated during exchange due to the bulky hydration spheres. At higher temperatures, Cu ions migrate to these sites where they are very difficult to reduce to the metal.

The two peaks in the TPR profiles of Cu zeolites are usually explained by the sequence $Cu^{II} \rightarrow Cu^{I} \rightarrow Cu^{0}$. Therefore, only Cu^{I} should be expected between the rate maxima. In EXAFS studies under comparable conditions, it was, however, found that highly disperse nuclei of metallic copper are already formed slightly above the first maximum. In the minimum, they coexist with the major Cu^{I} component, and in some cases, even with residual Cu^{II}. Only in the calcined CuNa-Y, metal formation did not start until the onset of the second TPR reduction signal at 800 K [102, 103].

The differentiation of carbonaceous residues according to their susceptibility to oxidation is a most attractive opportunity for **TPO**. The nature of carbonaceous residues can be very different, from highly aromatic, even graphitic material (hard coke) to aliphatic and alicyclic molecules just trapped in micropores, e.g., of zeolites (soft coke, cf. Section 4.4.1.1). TPO helps in establishing conditions for the combustion of coke species and to find out which type of coke needs to be removed for regenerating most of the catalyst activity.

The use of TPO to determine average oxidation states of catalytic elements in the reaction atmosphere is more challenging, because after flushing the sample with inert gas, adsorbates or carbonaceous residues may remain. The oxygen consumed by them can be detected in the carbon oxides and in water. Their desorbed quantities (or even their desorption profiles) need to be measured along with the concentration profile of O_2. This allows evaluating the quantity of oxygen consumed by the catalyst, either as a total or differentiated along the temperature ramp. An application of this method is described in reference [104].

TPS has been instrumental in achieving insight into the processes proceeding when supported oxide precursors are converted into hydrotreating catalysts (cf. Sections 2.3.3 and 3.1.5). Sulfidation in H_2S/H_2 is performed using a linear temperature ramp, and the concentrations of effluent H_2, H_2S, and H_2O are monitored by mass spectrometry. The mutual relationship of the consumption and desorption events allows conclusions on surface reaction steps. When, for instance, an oxide is reduced prior to sulfidation, H_2 consumption parallels H_2O formation, while H_2S remains unaffected:

$$MoO_3 + H_2 \rightarrow MoO_2 + H_2O \tag{3.53a}$$

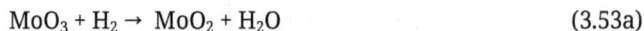

Alternatively, reduction may be preceded by ligand exchange:

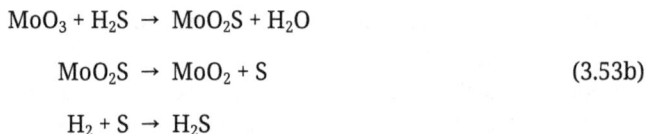

$$MoO_3 + H_2S \rightarrow MoO_2S + H_2O$$

$$MoO_2S \rightarrow MoO_2 + S \tag{3.53b}$$

$$H_2 + S \rightarrow H_2S$$

This route can be identified by H_2S consumption, followed by parallel H_2 consumption and H_2S formation.

Sequences and temperatures of events are influenced by the interactions of the precursors with the support and with promoters, which are strongly modified by the initial ligand exchange. For more details, the reader may consult reference [105].

3.4.3 Scattering and diffraction

3.4.3.1 Elastic and inelastic scattering

The characterization techniques to be introduced in the following are based on the interaction of electromagnetic or electron waves with matter. They can be well illustrated using the general concept of absorption and scattering as the primary modes of interaction. Electromagnetic waves offer photons of $E = h\nu$ for this interaction, while all energies up to the kinetic electron energy E_{el} are available from electron waves.

Interactions between the incoming wave and an electron in the sample may be elastic or inelastic. The most important mode of inelastic interaction occurs when the energy transferred allows an electron in the sample accessing an empty electron state with energy, E_{fin}, from its initial state with the energy E_{ini}. A photon can be absorbed when the difference, $\Delta E_{fin/ini}$, is equal to its energy $h\nu$, which transfers the system into an excited state:

$$h\nu = \Delta E_{fin/ini} \tag{3.54}$$

The effect can be measured by comparing the intensities of the exciting wave before and after the sample. Electrons exciting such transition are not absorbed, they suffer an energy loss instead: the incoming electron leaves the sample with a lower energy of

$$E_{el,fin} = E_{el,ini} - \Delta E_{fin/ini} \tag{3.55}$$

The effect is recorded by monitoring the decelerated outgoing electron(s) using energy-differentiating electron detectors (see Section 3.4.5).

Absorption also covers the photoelectric effect, where the photon energy exceeds the amount necessary to raise the sample electron to the vacuum level (cf. Section 2.1). The resulting photoelectrons can be studied with respect to their distribution of the kinetic energies (photoemission, Section 3.4.5) or to their energy-dependent scattering at neighboring atoms (X-ray absorption fine structure, XAFS, Section 3.4.4). The analogous process with electron excitation is called collision (or impact) ionization. It is less important in catalysis research because of its risk to cause sample damage.

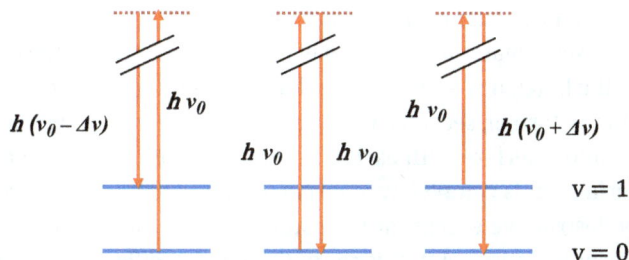

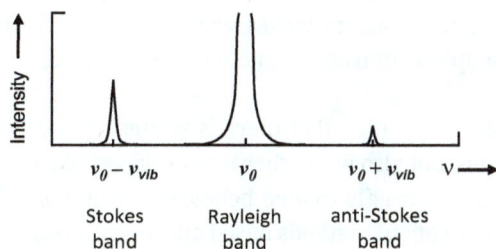

Figure 3.61: An elementary model for the inelastic scattering of electromagnetic waves. Adapted from ref. [90].

If the quantum $h\nu_0$ transferred to an electron from an incoming wave is not suited to promote it to an empty bound state or to the continuum, the attempted energy transfer is aborted (cf. Figure 3.61). The electron returns to its original level or, with very low probability, to levels differing from the latter by energies in the range of vibration modes. As a result, a secondary wave is emitted. If the electron returns to its initial level, exciting and secondary waves are identical in wavelength, but differ in phase. This is elastic scattering, which occurs likewise also during the interaction of electron waves: a secondary electron wave is emitted with identical energy, but in a different phase.

If the electron excited by a photon returns to an excited level, a photon with slightly lower energy is emitted (Figure 3.61). Only vibrational modes involving the corresponding atom allow this process with sizeable probability, i.e., they are excited by inelastic photon scattering. By superposition of many analogous events, a vibrational spectrum of the sample appears on a wavenumber scale, starting in the elastic peak and extending to lower energies (Figure 3.61). This is the principle of Raman excitation. In Raman terminology, the elastic peak is the Rayleigh band, and the lines of the spectrum at its low-energy side are the Stokes bands. The spectrum appearing at the high-energy side of the Rayleigh band with the wavenumber axis extending to higher energies (Figure 3.61) exhibits the anti-Stokes bands. They result from the transition of electrons from the first excited states of the vibrational modes to the ground states via inelastic photon scattering. As vibrational modes are mostly in the ground state at room temperature, the intensity of the anti-Stokes bands is significant only at higher temperatures.

When a primary wave passes a material, it excites elastic and inelastic scattering events along its path. Due to the latter, it is attenuated and disappears completely if the optical path is too long. From the former, secondary waves of the same wavelength are emitted, which superimpose and interfere with each other with a result that depends on the structure of the material. If it is completely *amorphous*, the interference of secondary waves is completely destructive except for the components in the direction of the primary wave; the material is transparent. If it contains *domains differentiated by boundaries* (grains, pores, micelles, colloids), or just density fluctuations, scattering intensity offside the primary wave direction is not cancelled completely. As a result, scattered light can be observed from the side, which allows studying the dispersion of entities delimited by the boundaries. Secondary waves scattered by atoms of *ordered arrays* cause regular patterns of amplification and extinction called diffraction (see also Box 3.4).

For the sake of completeness, we mention that there is scattering and diffraction not only with photons and electrons, but also with other matter waves like neutrons or He atoms. While neutron diffraction is briefly treated below, the reader is referred to reference [106] for more information about methods based on He scattering.

3.4.3.2 Elastic scattering at ordered arrays: diffraction

In this section, the relation between scattering and diffraction is discussed in some detail on the basis of an earlier presentation in reference [107]. While such detail may not be required for understanding the basics of catalysis, the concepts to be introduced are highly useful for a true comprehension of important characterization methods, in particular, X-ray absorption techniques.

We use the mathematical description of waves propagating in space by wave functions. Near the origin, the spherical waves are described by

$$\Psi = \Psi_0 \, \exp\left(i \, \vec{k} \, \vec{r}\right) \Big/ |\vec{r}| \tag{3.56}$$

In (3.56), $\vec{r}$ is the position vector of the observer (relative to the origin of the wave), $\vec{k}$ is the wave vector, with an absolute value of $2\pi/\lambda$ (cf. Section 2.1), Ψ_0 is the amplitude, and $1/|\vec{r}|$ accounts for the distribution of the intensity I on a spherical surface growing with

$\vec{r}$: $I \propto \Psi^2 \propto 1/r^2$. The expression $\exp i\,\vec{k}\vec{r}$ describes the phase of the wave at point $\vec{r}$, relative to that in the origin (cf. its relation to sine and cosine via the Euler formula in footnote 3, Section 2.1). The development of the phase in time ($\exp i\omega t$, with ω – angular frequency) has been dropped. The argument $\vec{k}\vec{r}$ operates similar to ωt: while the angular frequency $\omega = 2\pi\nu$ expresses the number of oscillations per second in the circular space, the absolute value of the wave vector $2\pi/\lambda$ does the same for the number of oscillations per unit length. As $\vec{k}$ and $\vec{r}$ may point to different directions, the scalar product makes sure that the operation is performed with the projection of one on the other vector.

Far from the origin, the bending of the wave front becomes negligible. Such a plane wave propagates along $\vec{k}$, the unit vector of which is $\vec{s}$. Wave fronts become planar only when $|\vec{r}|$ is very large and does not change significantly within the space of the experiment. Therefore, $|\vec{r}|$ in (3.56) is incorporated into the amplitude, and the wave function becomes

$$\Psi = \Psi_0 \exp\left(i\,\vec{k}\,\vec{r}\right) \tag{3.57}$$

a

b

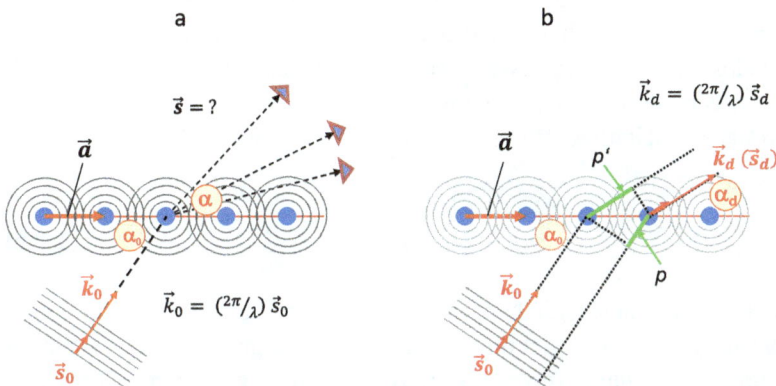

Figure 3.62: Scattering a plane wave on a one-dimensional lattice: how to derive one of the Laue equations. (a) Detector moved through space; (b) path difference resulting from scattering at adjacent atoms.

In Figure 3.62, the interaction between a one-dimensional lattice with a lattice vector $\vec{a}$ and a plane wave of a wavelength λ propagating along is $\vec{s}_0$ explained. The angle between $\vec{a}$ and $\vec{s}_0$ is α_0. Varying the angle α (i.e., the unit vector, $\vec{s}$), an observer tries to find out the intensity distribution resulting from the superposition of the secondary waves (Figure 3.62a, for the moment, the search is confined to the plane defined by $\vec{a}$ and $\vec{s}_0$). Figure 3.62b shows that the optical path length differs when the wave front hits neighboring atoms. An analogous path difference, which depends on α, emerges with the secondary waves. The differences are highlighted in green and denoted as p and p', respectively. The total path difference is $p' - p$ (actually $p' + (-p)$). When it is a multiple of the wavelength λ, the interference between the secondary waves originating from each pair of neighboring atoms is constructive.

$$p' - p = n\,\lambda \tag{3.58}$$

If the coincidence is not perfect, interference may be still nearly constructive for secondary waves from the next neighbors. However, disagreement grows with the distance between scatterers, and as the lattice has an infinite length, the result is total extinction for all directions violating condition (3.58). The intensity peaks resulting from eq. (3.58) are diffraction patterns. Therefore, vectors and angles providing a solution to this equation are indexed "d" in the following.

Writing p' ($= |\vec{a}|\,cos\,\alpha$) and p ($= |\vec{a}|\,cos\,\alpha_0$) as scalar product of vectors, the first Laue equation is obtained

$$\vec{a}\,(\vec{s}_d - \vec{s}_0) = h\,\lambda \tag{3.59a}$$

where h is an integer (the order of diffraction) and $\vec{s}_d - \vec{s}_0$ indicates a change in the propagation direction.

When the planar confinement of the search is dropped, each unit vector obtained by rotating $\vec{s}_d$ around $\vec{a}$ complies with the condition (3.59a). Hence, the diffraction pattern of the one-dimensional lattice is a cone with a cone angle of $2\alpha_d$. According to eq. (3.59a), α_d changes when α_0 is varied, but diffraction is always possible.

For a two-dimensional lattice, diffraction is additionally confined by a second Laue equation

$$\vec{b}\,(\vec{s}_d - \vec{s}_0) = k\,\lambda \tag{3.59b}$$

which defines a diffraction cone with the rotation axis $\vec{b}$. The intersections of the cones, defined by (3.59a and b), are particular directions that are thought to start at a common top of the two diffraction cones. This is the basis of Low-energy electron diffraction (LEED), a technique for the structural analysis of single-crystal facets (cf. Section 3.4.3.4). Although the directions of the diffracted beams vary with the angle between the incident electrons and the surface, diffraction is always possible.

Diffraction by a three-dimensional lattice is additionally confined by a third Laue equation

$$\vec{c}\,(\vec{s}_d - \vec{s}_0) = l\,\lambda \tag{3.59c}$$

With all three equations, the orientation between the lattice and the incoming radiation is fixed as well; diffraction does not occur any more at any orientation. Therefore, single crystal diffraction by the Laue technique involves the rotation of the crystal around all axes during irradiation. The diffractogram consists of single points in specified diffraction directions. In powder diffraction, which is more relevant for catalysis, such a rotation is unnecessary due to the random orientation of crystallites (see below).

Up to now, diffraction has been described using a geometrical model. In the follow-ing, a glimpse into the kinematic diffraction theory and operation with wave functions will confirm the Laue equations, and provide interesting additional insight.

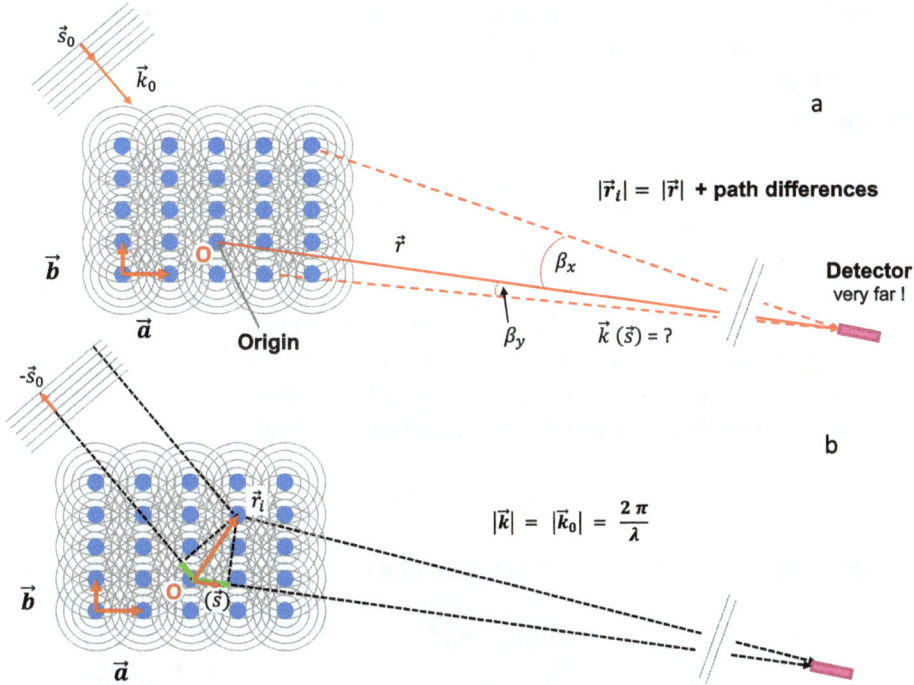

Figure 3.63: Kinematic theory of scattering and diffraction: the two-dimensional lattice. (a) Description of quantities involved; (b) path difference resulting from scattering at a pair of atoms.

From a two-dimensional lattice with translational vectors $\vec{a}$ and $\vec{b}$, secondary waves are induced by a primary wave $\vec{k}_0 = 2\pi\,\vec{s}_0/\lambda$ (Figure 3.63a). A detector at a position, $\vec{r}$, relative to an origin O in the lattice, records the result of the interaction. Notably, the detector is very far in terms of lattice dimensions (decimeters vs. nanometers). Hence, angles between the position vector $\vec{r}$ and connections between the lattice points and the detector (e. g. β_i in Figure 3.63a) are all nearly zero. Therefore, the distance between the lattice points and the detector can be approximated by adding the path differences resulting from the lattice geometry to $|\vec{r}|$; the error introduced by neglecting the angles β_i is small compared to the lattice dimensions.

A secondary wave induced in the origin O can be expressed as

$$\Psi(O) = \Psi_0\ f_N\!\left(\vec{k}_0\ \vec{k}\right)\frac{\exp\!\left(i\ \vec{k}\ \vec{r}\right)}{|\vec{r}|} \tag{3.60}$$

where Ψ_0 and $\vec{k}_0$ are the amplitude and the wave vector, respectively, of the incoming plane wave and f_N $(\vec{k}_0,\vec{k})$ is the scattering efficiency ("form factor") of the atom in O. For reasons outlined below, it depends on the change in directions between $\vec{k}$ and $\vec{k}_0$. In the waves emitted by the other atoms with their position vectors $\vec{r}_i = n_1 \vec{a} + n_2 \vec{b}$ (Figure 3.63b), $\vec{r}_i$ is expressed as the sum of $\vec{r}$ and the optical path differences $\Delta \vec{r}_i$. As λ does not change, it makes no difference if contributions to $\Delta \vec{r}_i$ are from the incoming or the outgoing wave. Likewise, the phase shift due to scattering is neglected because it is the same for all atoms. The secondary waves are therefore

$$\Psi(R_i) = \Psi_0 \, f_N \, (\vec{k}_0, \vec{k}) \, \exp(i \, \vec{k} \, \vec{r}) \, \exp(i \, \vec{k} \, \Delta \vec{r}_i)/|\vec{r}| \tag{3.61}$$

Ψ_0 and $\exp(i \, \vec{k} \, \vec{r})/|\vec{r}|$, which do not depend on the lattice, will be dropped in the following. f_N $(\vec{k}_0,\vec{k})$, which is not related to the lattice periodicity, but is a property of the lattice points (or the unit cells related to them, cf. Section 2.1), will be retained.

The optical path difference $\Delta \vec{r}_i$, which is highlighted in green in Figure 3.63b, consists of two pieces in the direction of $\vec{s}$ and $\vec{s}_0$, which can again be expressed by scalar products involving these vectors (cf. example in Figure 3.63b):

$$\Delta \vec{r}_i = \vec{s} \, \vec{r}_i + (-\vec{s}_0) \, \vec{r}_i$$

The contribution of point $\vec{r}_i$ to eq. (3.61) is therefore

$$\exp(i \, \vec{k} \, \Delta \vec{r}_i) = \exp(i \, \vec{r}_i \, (\vec{k} - \vec{k}_0))$$

The scattered wave function results from a summation over the whole lattice

$$\Psi \propto f_N\left(\vec{k}_0, \vec{k}\right) \sum_i \exp\left(i \, \vec{r}_i \left(\vec{k} - \vec{k}_0\right)\right) \tag{3.62}$$

or, representing the lattice vectors by their components

$$\Psi \propto f_N\left(\vec{k}_0, \vec{k}\right) \sum_{n_1, n_2} \exp\left(i \, n_1\vec{a}\left(\vec{k} - \vec{k}_0\right) + i \, n_2\vec{b}\left(\vec{k} - \vec{k}_0\right)\right)$$

With simple algebra $(\exp(x + y) = \exp(x) \cdot \exp(y))$, this becomes

$$\Psi \propto f_N\left(\vec{k}_0, \vec{k}\right) \sum_{n_1} \exp\left(i \, n_1\vec{a}\left(\vec{k} - \vec{k}_0\right)\right) \cdot \sum_{n_2} \exp\left(i \, n_2\vec{b}\left(\vec{k} - \vec{k}_0\right)\right) \tag{3.63}$$

In eq. (3.63), f_N describes the properties of the lattice points (or of the secondary lattice), while both the remaining factors depend only on the periodicity, i.e. on the primary lattice. Designating them as G and f_N as F, eq. (3.63) becomes $\Psi \propto F \cdot G$. The distribution of the intensities described by Ψ^2 is then

$$\Psi^2 \propto F^2 \cdot G^2 \tag{3.64}$$

Equation (3.64) highlights an important feature of X-ray diffraction: in the diffractograms, the influences of primary and secondary lattices are well separated; the translational lattice determines the direction of the diffracted intensity and the secondary lattice determines the intensity. There is no geometry of the secondary lattice that could change the angles at which the diffraction signals can be found. The secondary lattice does, however, modulate the intensity, including even complete the extinction of reflections (cf. Box 3.4).

The sums in eq. (3.63) may be replaced by

$$\sum_{n_1=0}^{M_1} exp\left(i\, n_1\vec{a}\,\left(\vec{k}-\vec{k}_0\right)\right) = \frac{sin\left(0.5\; M_1\;\vec{a}\;\left(\vec{k}-\vec{k}_0\right)\right)}{sin\left(0.5\;\vec{a}\;\left(\vec{k}-\vec{k}_0\right)\right)} \tag{3.65}$$

The G^2 term of eq. (3.64) contains the sines in the numerator and the denominator squared. This gives easy access to the peaks of the intensity Ψ^2, i.e. to the directions of the diffraction signals – they occur where $sin(0.5\,\vec{a}\,(\vec{k}-\vec{k}_0)) = 0$, i.e., $0.5\,\vec{a}\,(\vec{k}-\vec{k}_0)$ should be a multiple of π:

$$0.5\,\vec{a}\,(\vec{k}_d-\vec{k}_0) = h\,\pi.$$

With $\vec{k} = 2\,\pi\vec{s}/\lambda$, this becomes

$$\vec{a}\,(\vec{s}_d-\vec{s}_0) = h\,\lambda,$$

i.e., the first Laue equation (3.59a) also results from the kinematic wave theory.

While the argument $0.5\,\vec{a}\,(\vec{k}-\vec{k}_0)$ of the sine (cf. eq. (3.65)) determines the direction of the diffraction peaks, their widths depend on the number of scatterers available in the corresponding lattice extension (M_1 for $\vec{a}$, by analogy M_2 for $\vec{b}$, and M_3 for $\vec{c}$). Peaks become narrow when the M_i are large. When the Laue equation for a given lattice vector is slightly violated, interference of waves from the neighboring scatterers may be still largely constructive. However, phase misfit adds up with increasing distance between the scatterers: the resulting intensity penalty grows with the particle size. This is the basis for domain size determination via peak widths by the semi-empirical Scherrer equation (see Section 3.4.3.3).

Apart from the domain size (or coherence length), line broadening may also be caused by lattice imperfections. Although they do not totally break the structural coherence as do domain boundaries, they introduce sites scattering out of the phase, thus interfering with the diffraction conditions.

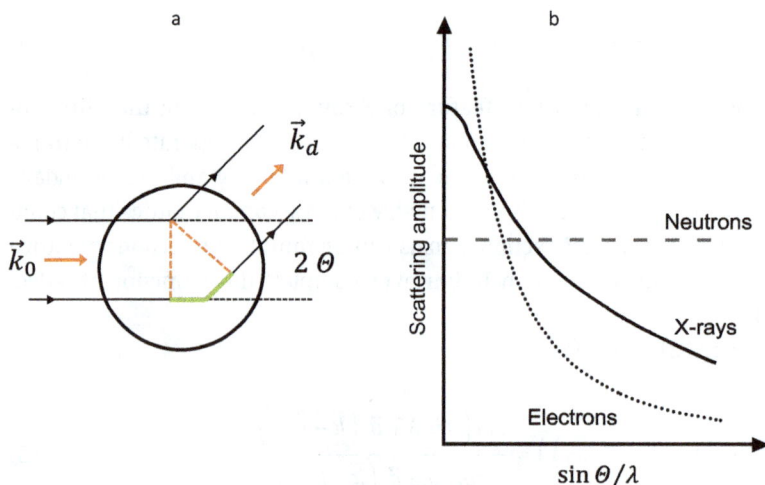

Figure 3.64: Angular dependence of the form factor for a primitive lattice (atomic form factor $f_N(\vec{k}_0,\vec{k})$): (a) path difference caused by scattering in an extended electron cloud; (b) angular dependence of f_N for the scattering of X-rays, compared with analogous relations for electrons and neutrons. (b) Adapted from ref. [108].

In a primitive unit cell, $f_N(\vec{k}_0,\vec{k})$ reflects the scattering power of a single atom, which increases with the number of electrons available for scattering: it is proportional to the atomic number. Light elements such as H and Li are therefore not easily detected by XRD, and the neighbors in the periodic table cannot be differentiated. The angular dependence expressed by the argument $(\vec{k}_0,\vec{k})$ results from the extension of the scattering atoms in space. Scattering at different positions within the electron cloud (cf. Figure 3.64a) causes a distribution of optical path differences; the larger the scattering angle Θ, the wider is the optical path difference. Atoms are small and so are path differences; therefore, f_N is always attenuated. The angular dependence of the scattering amplitude (analogous to f_N) is depicted in Figure 3.64b.

If the secondary lattice contains several atoms, each of them is part of a sub-lattice with the same translational vectors (cf. Figure 2.2). Each sub-lattice diffracts intensity into the same directions, defined by the Laue equations (3.59a–c). These parallel contributions, however, differ in phase due to the optical path differences within the secondary lattice, which depend on the scattering angle (for a visualization, see Figure 3.67). The resulting interferences within the secondary lattice are covered by its form factor (cf. eq. (3.71) below).

Finally, diffracted intensities from ordered lattices depend on temperature, because thermal motion of the scatterers results in a permanent violation of the diffraction conditions. This influence is described by the Debye-Waller term (eq. (3.66)), which compares the intensity at a temperature T, $I(T)$, with that of a rigid lattice, I_0, via the mean square displacement of the scatterers from their equilibrium positions $\vec{r}_{i,0}$, $\overline{u}^2$.

$$I(T) = I_0 \ exp\left(-\left|\vec{k}-\vec{k}_0\right|^2 \overline{u^2}\big/3\right)$$

$$\overline{u^2} = \int\limits_{x,\,y,\,z}^{0} (\vec{r}_i - \vec{r}_{i,0})^2 dx\ dy\ dz \tag{3.66}$$

Intensity losses due to thermal motion and displacement of scatterers are common to all methods based on the interference of scattered waves.

With a few exceptions, e.g., in Figure 3.64, the lattice has been treated as a point array in space so far. In reality, it is, however, a regular three-dimensional spatial pattern of electron density. Equation (3.62) can be easily modified to include this aspect and to derive a more general treatment. For this purpose, the discrete form factor f_N is dropped in favor of an electron density ρ, depending on the position vector $\vec{r}$, and the summation over all $\vec{r}_i$ is replaced by an integral over the whole space:

$$\Psi\left(\vec{k}\right) \propto \int\limits_{x,\,y,\,z} \rho(\vec{r}) \exp\left(i\ \vec{r}\left(\vec{k}-\vec{k}_0\right)\right) dx\ dy\ dz \tag{3.67}$$

In doing so, we drop even the assumption of an ordered structure: eq. (3.67) holds for any spatial distribution $\rho(\vec{r})$ of electron density. Equation (3.67) actually describes a Fourier transformation: the wave function results from a Fourier transformation of the electron density $\rho(\vec{r})$ in the r-space (with length dimensions) into a k-space with (length)$^{-1}$ dimensions, which is called reciprocal space accordingly. This conclusion is worth being repeated: when an incident (electromagnetic or electron) wave is scattered at a spatial electron distribution, $\rho(\vec{r})$, the resulting wave is the Fourier transformation of $\rho(\vec{r})$ into the reciprocal space. When the electron density is patterned in periodical order, the resulting wave function and the intensity distribution resulting from it exhibit well-defined maxima. These diffraction patterns can be registered by sampling the k-space (actually – sampling the change of the propagation directions $\vec{s}_d - \vec{s}_0$, because $\lambda = \text{const.}$).

Equation (3.67) suggests that the reverse relation between a given wave function $\Psi(\vec{k})$ and an unknown electron density map $\rho(\vec{r})$ may be obtained by just reverting the Fourier transformation, i.e.,

$$\rho(\vec{r}) \propto \int\limits_{x*,\,y*,\,z*} \Psi\left(\vec{k}\right) \exp\left(-i\ \vec{r}\left(\vec{k}-\vec{k}_0\right)\right) dx*\ dy*\ dz* \tag{3.68}$$

where the asterisk labels the dimensions of the reciprocal space. This is, indeed, the case, but unfortunately, there is no experiment that gives access to the complex wave function, Ψ: only intensities can be measured, which correspond to Ψ^2. Although the absolute value of Ψ can be determined, its phase remains unknown, which prevents a direct determination of the electron density distributions via eq. (3.68). This is known as the "phase dilemma of XRD." It is, however, no more a problem today, where extremely complicated structures can be resolved with powerful approximate approaches, e.g., the Patterson method or the direct methods. For more details, the reader is referred to pertinent textbooks, e.g., reference [109].

For each real lattice with translational vectors $\vec{a}$, $\vec{b}$, and $\vec{c}$, there is a lattice in the reciprocal space, the translational vectors $\vec{a}^*$, $\vec{b}^*$, and $\vec{c}^*$ of which are related to the real ones as follows:

-   the scalar product of each real with one reciprocal translational vector is 2π.[20] This may be expressed as $\vec{a}\vec{a}^* = \vec{b}\vec{b}^* = \vec{c}\vec{c}^* = 2\pi$.
-   the scalar product of a vector with the remaining reciprocal vectors is zero, i.e., they are perpendicular to each other.

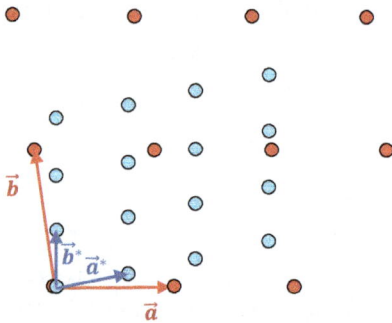

Figure 3.65: A two-dimensional lattice (red) and its reciprocal lattice (blue).

As an example, Figure 3.65 shows a two-dimensional (oblique) lattice and its reciprocal lattice. The reciprocal vectors $\vec{b}^*$ and $\vec{a}^*$ are normal to the real vectors $\vec{a}$ and $\vec{b}$, respectively. As the real vector $\vec{b}$ is longer than $\vec{a}$, its reciprocal vector $\vec{b}^*$ is shorter than $\vec{a}^*$. Both lattices extend to infinity. It can be shown that each reciprocal lattice vector ($\vec{r}_{hkl}^* = h\vec{a}^* + k\vec{b}^* + l\vec{c}^*$) is normal to a real lattice plane hkl, and its length $|\vec{r}_{hkl}^*|$ is the reciprocal of the lattice plane distance, i.e. d_{hkl}^{-1}.

The importance of the reciprocal lattice for diffraction results from the fact that the wave vectors of beams diffracted from a plane hkl are obtained by adding the reciprocal vector related to this plane to the vector of the incident wave, or

$$\vec{k}_d - \vec{k}_0 = \vec{r}_{hkl}^* \tag{3.69}$$

For checking this, (3.69) is divided by $2\pi/\lambda$, which results in an expression for $\vec{s}_d - \vec{s}_0$

$$\vec{s}_d - \vec{s}_0 = \frac{\lambda}{2\pi}\vec{r}_{hkl}^*$$

---

[20] There is an alternative definition setting the scalar product to 1; in this system, the absolute value of the wave vector is $1/\lambda$ instead of $2\pi/\lambda$.

This expression is inserted into a Laue equation (3.59) to make sure that eq. (3.69) deals with diffracted beams:

$$\vec{a}(\vec{s}_d - \vec{s}_0) = \vec{a}\ \frac{\lambda}{2\ \pi}\ \vec{r}_{hkl} = \vec{a}\ \frac{\lambda}{2\ \pi}\ \left(h\vec{a}^* + k\vec{b}^* + l\vec{c}^* \right)$$

Multiplying $\vec{a}$ into the parentheses results in

$$\vec{a}\ \frac{\lambda}{2\pi}\vec{r}_{hkl} = \frac{\lambda}{2\ \pi}\ \left(h\vec{a}\vec{a}^* + k\vec{a}\vec{b}^* + l\vec{a}\vec{c}^* \right) = h\,\lambda$$

In the parentheses, the first summand is $2\pi h$, the remaining ones are zero (cf. definition above). Therefore, the expression allows to reproduce the Laue equation, i.e. eq. (3.69) is correct.

As the reciprocal space extends to infinity while the diffraction vectors have a finite length, only some reciprocal lattice vectors fulfill eq. (3.69). In the reciprocal space, these solutions can be identified with a simple geometric approach, the Ewald sphere, which is, however, beyond the scope of this book. The reader is referred to pertinent textbooks [109].

Important take-home messages about scattering and diffraction are summarized in Box 3.4.

Box 3.4: Scattering and Diffraction: what to take home.
- Diffraction results from interference of waves emitted by elastic scattering of a primary wave at a periodic lattice.
- Elastic scattering at non-crystalline materials produces diffuse intensity offside the primary beam if there are density fluctuations or grain boundaries. The scattered radiation can be used to characterize the underlying inhomogeneities. If these inhomogeneities are periodic, the interference of scattered waves results in diffraction (e.g., in OMM).
- Electromagnetic radiation is scattered at the electron shell of atoms, electrons at both electron shell and nuclei, neutrons exclusively at the nuclei.
- When (electromagnetic or electron) waves are scattered at a spatial distribution of electron density, the resulting wave is the Fourier transform of the electron density from the real into the reciprocal space. For ordered patterns of electron density, this wave will have pronounced maxima in specific reciprocal-space directions on an almost zero background: the diffraction patterns.
- In a diffractogram, diffraction angles are determined by the primary (translational) lattice, diffracted intensities by the secondary lattice (the unit cell).

3.4.3.3 X-ray diffraction (XRD)

XRD is a standard tool for the structural analysis of crystalline solids all over solid-state chemistry and related areas. Although the focus is on surfaces in catalysis, XRD is applied in this field as well, because the solids exposing the surfaces may be crystalline. It is, however, not used at its full potential, because materials or phases applied in catalysis are rarely completely new. XRD is rather employed as a tool for quality control during catalyst preparation and as a probe for specific properties, like domain sizes, crystallinity or disorder, composition of mixed phases, texture, or for tracking solid-state reactions of crystalline phases. Applied in *in situ* environments, this can reveal important processes during preparation, pretreatment, activation, or deactivation.

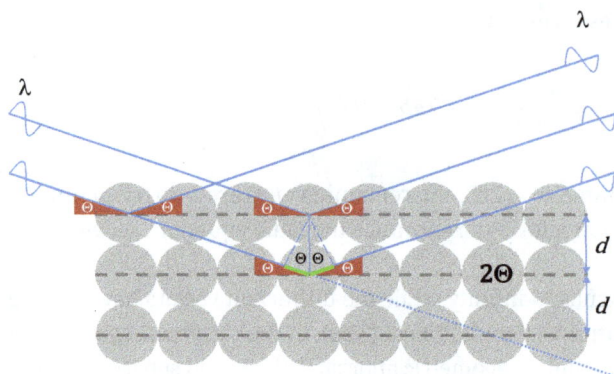

Figure 3.66: The Bragg equation: diffraction as reflection of waves at lattice planes.

In XRD, reference is frequently made to the Bragg equation, which is a diffraction condition completely equivalent to the Laue equations. The well-known geometric setting used to derive this equation is shown in Figure 3.66. Diffraction is considered as a reflection of the incident wave at lattice planes: the scattering (or Bragg) angle Θ between the incident radiation and the lattice plane is repeated between the lattice plane and the diffracted beam. The optical path difference caused by reflection at the adjacent planes with a spacing d is $2d \sin\Theta$, which must be equal to a multiple of the wavelength λ for achieving constructive interference:

$$2d \sin\Theta = n\,\lambda \qquad (3.70)$$

As discussed for the Laue equations before, this Bragg equation results in a narrow diffraction signal only if it is fulfilled for a large number of lattice planes at once. It should be noted that the deflection angle, i.e., the angle between the incident and the diffracted beam is 2Θ, which is per convention, the abscissa in powder diffractograms.

In Figure 3.66, a primitive lattice is tacitly assumed: there is one set of lattice planes in each direction. In Figure 3.67, the unit cell has three atoms, differentiated by shape and color. In each direction, there are now three sub-sets, each related to one constituent of the secondary lattice. When the Bragg equation is fulfilled, it holds for all three sub-sets, but there are now minor path differences between the beams diffracted by the sub-sets, which modify the diffracted intensity. As illustrated by the two examples in Figure 3.67, this effect strongly depends on the lattice plane causing the signal.

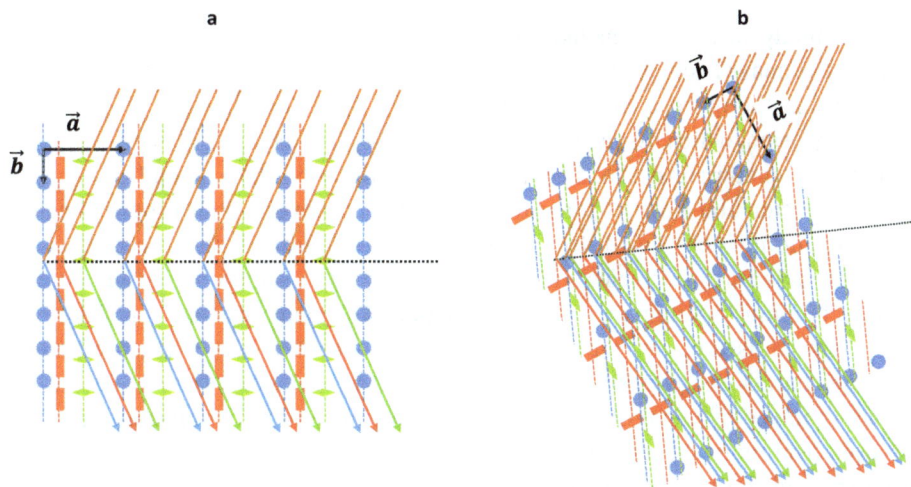

Figure 3.67: Diffraction at a non-primitive lattice: reflection at different sub-sets of lattice planes (cf. Figure 2.3) causing path differences due to the distribution of atoms in the unit cell; (a) and (b) exemplify the case for different lattice planes (diffraction angles).

Figure 3.67 illustrates what has been derived already in Section 3.4.3.2: the directions of the diffracted beams are determined by the translational lattice, and the diffracted intensities by the secondary lattice. The lattice-plane oriented treatment with the Bragg equation reveals why weights in the addition of the atomic contributions f_j depend on the diffracting lattice plane hkl:

$$F_{hkl} = \sum_j f_j exp\left(2\pi i\left(hx_j + ky_j + lz_j\right)\right) \tag{3.71}$$

j counts the atoms in the unit cell located at the atomic coordinates x_j, y_j, and z_j along the translational vectors $\vec{a}$, $\vec{b}$, and $\vec{c}$.

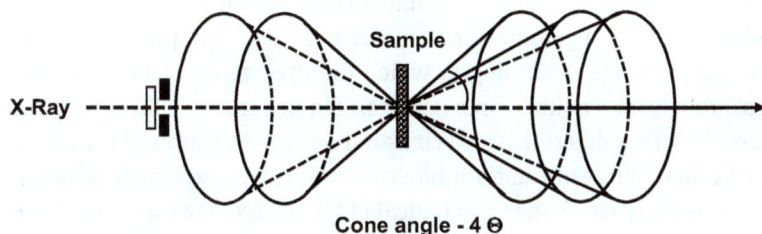

Figure 3.68: Diffraction patterns in powder diffraction.

In catalysis, XRD is always applied to powders, for which a random orientation of crystallites will be assumed for the moment. A lattice plane hkl diffracts when the angle between it and the incident beam is the Bragg angle Θ_{hkl}, according to eq. (3.70). This condition is actually fulfilled for the hkl plane in any orientation accessible by rotating Θ_{hkl} around the primary beam. In a powder, all these orientations are available. The diffraction pattern is, therefore, a cone formed by rotating the deflection angle $2\Theta_{hkl}$ around the primary beam, which results in a cone angle of $4\Theta_{hkl}$, as depicted in Figure 3.68.

A diffractometer consists of an X-ray source combined with beam optics, a sample holder, a detector, and a device for scanning the detector over a wide range of angles, relative to the incident beam. In powder diffraction, most studies are performed with Cu$K\alpha$ radiation, which combines good intensity and resolution with advantages in production and maintenance of the X-ray tube. For elements that may fluoresce under Cu$K\alpha$ radiation, e.g., Co or Fe, other sources (Co$K\alpha$, Mo$K\alpha$) are preferable. When the characteristic Cu$K\alpha$ radiation ($\lambda = 1.542$ Å) is employed, more distant Cu lines are removed by metal filters. The width of the incident radiation can be further decreased by a monochromator crystal set to Cu$K\alpha_1$ at 1.54056 Å. In a synchrotron, the source and the monochromator are part of the station (cf. Figure 3.84 below), while diffractometers may be provided by users or by the facility. Wavelengths can be chosen from a wide range. The extremely high intensity of synchrotron radiation allows wasting intensity for cutting line widths of incident radiation. Therefore, synchrotron XRD often reveals diffraction lines that would be superimposed in laboratory XRD.

According to Figure 3.68, the diffraction signals can be recorded by rotating the detector around the sample in a plane cutting the diffraction cones in half. In the Debye-Scherrer cameras used for taking diffractograms in the old days, the detector was just an X-ray film at the periphery of a circular camera (Figure 3.69a). The capillary containing the sample was rotated around its axis, and the cone angles were evaluated directly from the traces on the film. With position-sensitive detectors, this principle has meanwhile returned as transmission geometry. In synchrotron work, it allows recording diffractograms of appropriate samples in less than a second. It may not be suitable for strongly absorbing samples.

The Bragg-Brentano geometry (also called Θ–2Θ geometry, Figure 3.69b) operates using reflection. The sample is deposited on a flat surface, which is rotated, and the detector is moved around it by a goniometer, with twice the angular speed. The setup uses the parafocusing principle, according to which the divergence of the radiation arriving at a flat sample from a point source (exaggerated in Figure 3.69b) is reverted in the diffracted beam(s) if the detector is on a circle defined by the source and an osculation point on the sample. While the detector moves on the goniometer circle (black in Figure 3.69b), the focusing circle changes, as indicated for the two measurement positions in the figure, where the diffracted beams and the focusing circle are differentiated by color. Measurements in Θ-2Θ geometry are most familiar in catalysis research. They may take from minutes to days, depending on the sample properties. For measurements at elevated temperature or under reaction conditions, special high-temperature or environmental cells are available, both for Bragg-Brentano and transmission geometry. For other measurement principles, the reader is referred to ref. [109].

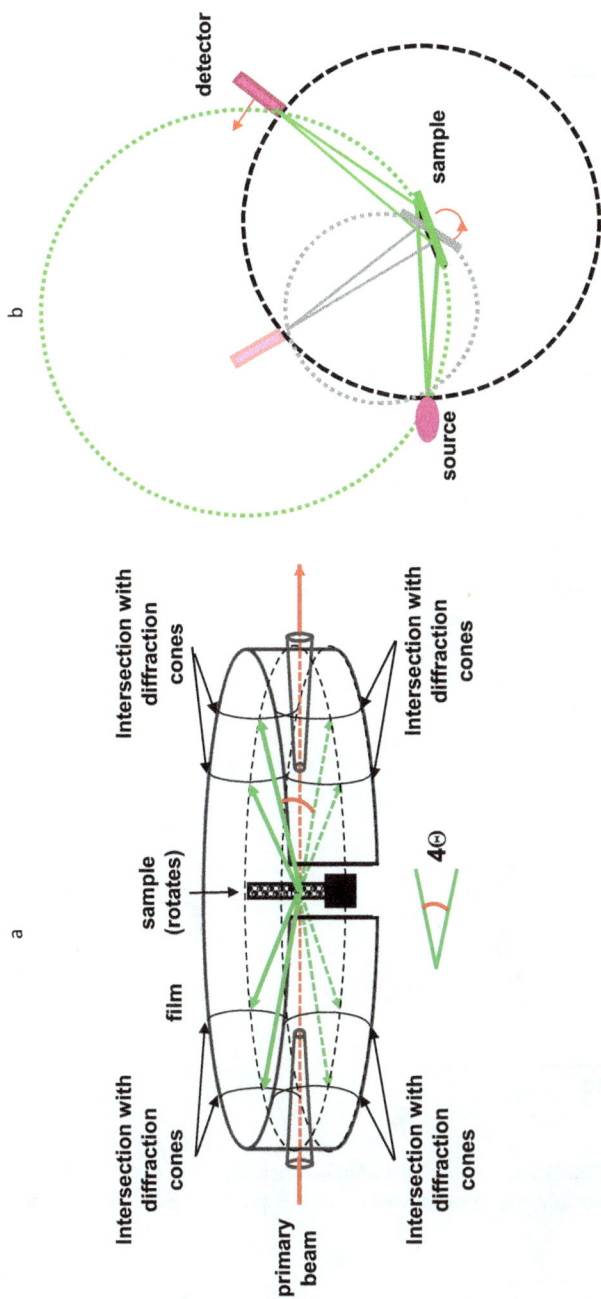

Figure 3.69: Geometric arrangements of the diffraction experiment: (a) Debye-Scherrer geometry (transmission); (b) Bragg-Brentano geometry (θ-2θ geometry, reflection).

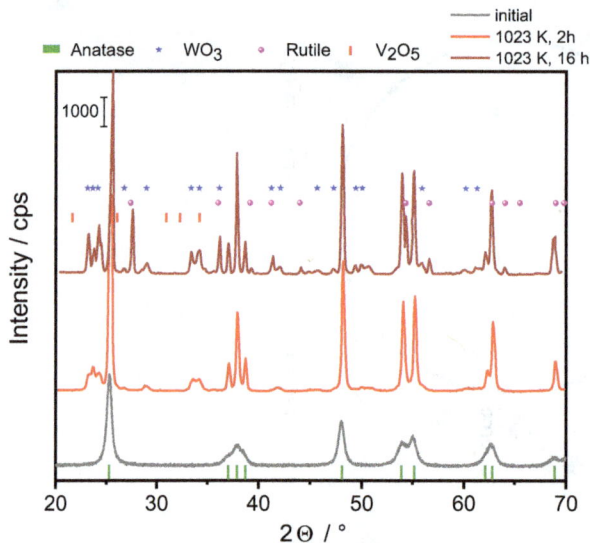

Figure 3.70: X-ray diffractograms of a V_2O_5-WO_3/TiO_2 catalyst after different thermal treatments. Reproduced from ref. [110] with permission from Elsevier.

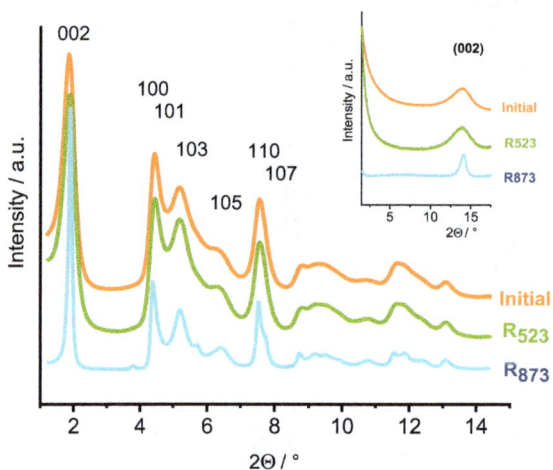

Figure 3.71: Synchrotron XRD ($\lambda = 0.20728$ Å) of a nanocrystalline MoS_2 sample after different reductive treatments (inset – low 2Θ-range of laboratory XRD measured with CuKα). Reproduced from ref. [111] with permission from Elsevier.

Figures 3.70 and 3.71 show examples of diffractograms taken with CuKα radiation (Figure 3.70 and inset of Figure 3.71) and with synchrotron radiation (Figure 3.71). Due to the lower wavelength used in the synchrotron, all peaks appear at lower 2Θ values (cf. (002) reflection in inset and main panel). The diffractograms of single phases (Figure 3.71 and "initial" in Figure 3.70) illustrate some typical features: owing to the narrower spacing of planes with higher Miller indices (hkl), there are more reflections at higher Bragg angles; and due to the decrease of the form factor f_N with growing Θ (Figure 3.64b), peaks tend to be less intense in this range.

The interpretation of diffractograms starts with the assignment ("indexing") of the reflections. This is conveniently performed by comparing experimental patterns with reference patterns from data bases (e.g., the powder diffraction file (PDF) of the International Center for Diffraction Data (ICDD)). A selection of these patterns is usually provided by software packages supplied with new diffractometers. In Figure 3.70, the initial pattern matches well with the reference pattern of anatase. Such comparison may, however, reveal characteristic differences: (i) although there are no additional reflections, peak intensities may differ from that in the reference, (ii) there may be shifts between the experimental and the reference peaks on the 2Θ scale, and (iii) there may be additional reflections not assignable to the reference selected. Such peaks indicate additional crystalline components, as for instance WO_3 and rutile in Figure 3.70 after calcination at 1,023 K for 2 h and 16 h, respectively.

As reflections of the catalytic materials are often broad due to the small particle sizes, strain, and/or structural disorder (cf. discussion of eq. (3.65) and below), intensities should always be evaluated by peak areas, instead of peak heights. Extreme line broadening, which is quite familiar with catalysts, may result in loss of intensity in the background. Platelet-like or needle-like particle morphologies may cause a preferential orientation of crystallites in the sample. This leads to an experimental artifact, most likely to occur in flat samples as employed in reflection geometry (Figure 3.69b). In such case, reflections of lattice planes parallel to the extended facets are drastically more intense than all other signals.

Systematic deviations of peak positions from the expected values may be due to the formation of mixed crystals. When lattice positions in a crystal of component A are occupied by B atoms with a different size, lattice dimensions change approximately in proportion to the amount of B (Vegard's rule). Lattice-plane distances of an alloy A_xB_{1-x}, for instance, can be estimated by

$$d_{AB} = x \, d_A + (1 - x) \, d_B \tag{3.72}$$

where d_A and d_B are the distances between the same lattice planes in pure A and B, respectively. Vegard's rule is often employed to derive the composition of alloy particles from peak positions. It can also be applied to other kinds of solid solutions, e.g., mixed oxides, as long as the structure is preserved upon the substitution of constituents.

Diffractograms of highly anisotropic and disordered structures may contain more specific information. In the patterns of nanocrystalline MoS_2 shown in Figure 3.71, the (002) reflection arises from the lattice plane across the MoS_2 slabs (cf. Figure 2.17), its width is related to the stacking height of the slabs. (100) and (110) are related to the diameter of the slabs. Other reflections ((h01), e.g., (101)) indicate stacking faults or turbostratic disorder (wrong orientation of successive S layers in the stacks). They are not well resolved in Figure 3.71 and rather cause a tailing of (100) toward higher scattering angles. Dramatic increase of intensity at very low angles ("Debye scattering," only in the inset) may result from few-layer stacks or uncorrelated single-layer material.

Once the nature of the phases causing the pattern is clear, XRD offers the opportunity to access more information about them, in particular about quantities present in mixtures, about particle sizes, and about strain due to structural defects. In the following, popular methods for extracting such information are discussed before full pattern analysis (or Rietveld refinement), the most universal method of quantitative analysis, is briefly introduced.

If only a single crystalline phase is present, its crystallinity may be evaluated by comparing its reflection intensity in the diffractogram with that obtained with a well-crystallized sample of the same phase. The analysis of mixture compositions takes advantage of the fact that intensities are, in principle, proportional to the amounts of the corresponding phase present. It is, however, complicated by the varying scattering efficiencies of the phases, which depend on the concentrations of the heavier atoms in them and on their atomic numbers. Attempts to derive mixture compositions by just comparing intensities of the major reflections neglect the influence of scattering efficiencies. In addition, samples may contain X-ray-amorphous material, which escapes detection in such procedure.

Analysis by comparison with reference materials is, therefore, the better choice. In this approach, the constituents identified in the sample are mixed with a reference material, e.g., α-Al_2O_3 or SiO_2, in a known relation (e.g., 1:1 w/w). The diffractograms of these mixtures calibrate the scattering efficiencies of the mixture components, which gives access to their amounts in unknown mixtures, based on diffractograms measured with the reference admixed. A deficit to the sample weight may be assigned to X-ray-amorphous material.

Beyond amorphous phases, there are some other types of materials that escape detection by XRD or prevent detection of components.

Apart from disorder, insufficient coherence length is the second major reason that allows sample components to escape detection by XRD. In Figure 3.70, 10 wt.-% of WO_3 and 1.5 wt.-% of V_2O_5 are missing in the initial sample because they form a monolayer of oxide species bound to the titania surface (cf. Sections 2.3.3 and 3.4.2.5). They can be observed only after being segregated from the TiO_2 surface, which decreases due to thermal stress. In ref. [110], Figure 3.70 was used to illustrate a conclusion drawn from Raman spectra, according to which the TiO_2 surface expels WO_3 rather than V_2O_5 upon thermal stress. If only XRD had been available to support this idea, it would have been mandatory to make sure that an amount of ca. 1 wt.-% V_2O_5 segregated from 10% WO_3/TiO_2 can be indeed detected by the method because sample constituents may also be overlooked due to small concentrations.

When reflections grow broader due to decreasing particle size (Section 3.4.3.2, see also below), they finally disappear. For metals, this happens typically below 2 nm, and for materials with large unit cells, e.g., zeolites, at much larger dimensions. If particle sizes are distributed, the lower end of the distribution may not contribute to signal intensity, and the reflection may rather represent the higher end. Signal broadening due to lattice defects gives rise to similar phenomena: highly defective phases may escape detection by XRD because the coherence in ordered domains is deteriorated although such phases may still cause signals in other techniques as in Raman spectroscopy or XAFS (see below). Phases of light elements are more likely to be overlooked by XRD than phases containing heavy elements.

Signal broadening caused by reduced coherence in scattering is the basis for the analysis of both particle size and crystal strain. While, however, decreased domain sizes always result in line broadening, strain affects peak shapes only if it is non-uniform: uniform strain (compression, elongation), which just modifies the translational vectors, results in shifted signal positions. Structural defects causing non-uniform strain may be dislocations, stacking faults, vacancies, etc.

The equations relating peak widths β to the domain size d or the microstrain ε are semiempirical. In the forms reported below, peak widths (here – the full width at half the peak maximum, FWHM, for alternatives cf. [112]) should be inserted in $°$ and after correction for instrumental broadening. The Scherrer equation

$$d_{hkl} = 57.3 \ K \ \frac{\lambda}{\beta \ cos\Theta} \tag{3.73}$$

is applied throughout the catalysis literature for evaluating particle sizes. d_{hkl} is the extension normal to the hkl plane, averaged over the whole particle. Differences in this averaging related to particle shape (e.g., for a cube or a sphere) are covered by a correction factor K, which is close to unity. The factor 57.3 is dropped when β is inserted in arc. The Scherrer equation is suited for particles between 2 and 50 nm. At the lower end, the reflections disappear in the background; at the upper end, sensitivity of β for further growth of d_{hkl} is too small to allow for acceptable resolution on the size scale.

Coherence lengths are always related to specific lattice planes. In particles with anisotropic shape, e.g., platelets, coherence lengths differ, depending on the lattice plane orientation. Therefore, anisotropic particle shapes reveal themselves by broadening of only some reflections, if the involved coherence lengths are short enough to induce line broadening.

Notably, the particle size determined by the Scherrer equation (3.73) always refers to monocrystalline domains. This might be grains in polycrystalline particles, as depicted in Figure 3.72. Conversely, particle size determination by chemisorption (eq. (3.37), Figure 3.45), which probes the external surface, would reflect the size of the whole aggregate. Sizes of anatase particles were determined by XRD using the Scherrer equation with the patterns shown in Figure 3.70, and via the BET surface areas S_{BET} ($d_{BET} = 6/(\rho_{TiO_2} \ S_{BET})$, spherical shape assumed). While the values were close in the initial state ($d_{BET} = 11$ nm, $d_{XRD} = 14$ nm), real particle sizes had grown much larger than domain sizes after 16 h at 1,023 K ($d_{BET} = 315$ nm, $d_{XRD} = 42$ nm), which suggests that neighboring particles had been merged to bigger aggregates.

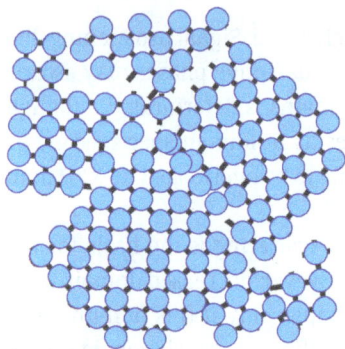

Figure 3.72: Polycrystalline particle (schematic).

In the diffractograms of MoS$_2$ (Figure 3.71), only the (002) reflection is sufficiently isolated to allow applying the Scherrer equation. From the lab data (inset), the average stack height was estimated to increase from 3.6 nm (initial state) to 13 nm (after reduction at 873 K). The highly defective nature of the material suggests, however, that strain cannot be neglected. Further studies on this topic are outlined below.

The influence of strain ε on the peak width is described by the Wilson equation

$$\beta = 4 \, \varepsilon \, tan\Theta \tag{3.74}$$

where $\varepsilon = 100 \, \Delta d_{hkl}/d_{hkl}$ is the percent variation in d spacing. In catalysis literature, strain is often neglected, although recent studies reveal its potential relevance [112]. The influences of size and strain can be separated making use of their different relations between β and the Bragg angle Θ (eqs. (3.73) and (3.74)). The Williamson-Hall plot, resulting from such an approach, requires, however, the availability of evaluable peaks in a wide range of 2Θ. Alternatively, use can be made of the different character of broadening, which is Gaussian for size-broadening and Lorentzian, for strain-broadening. With high-quality data, domain size and strain influences can be elucidated by fitting a single peak with a Voigt function (or a pseudo-Voigt function), which is a convolution (or a linear combination) of a Gaussian and a Lorentzian function.

All aspects of powder diffraction data treatment discussed so far are covered simultaneously by the Rietveld refinement, in which free parameters of a theoretical powder diffractogram are fitted versus all measured data points in a least-square procedure. An initial guess of a structural model is refined *versus* the experiment. The free parameters include lattice parameters in assumed crystal systems, atom positions in the unit cell, which allow modeling diffractograms of the candidate phases, displacement parameters, and parameters describing line shapes and the background. Instrumental parameters, zero-shift, and diffraction optics are included in the procedure. Depending on the system, between 10 and 100 parameters are varied in order to minimize the weighted mean square deviation S

$$S = \sum_i \frac{\left(y_{i,exp} - y_{i,mod}\right)^2}{\sigma_i^2}$$

where i counts the data points and σ_i is the standard deviation of the measurement in the corresponding range of the 2Θ scale. The method does not depend on experimental data from well-crystalline references.

The results of a successful refinement, e.g., structure and quantities of phases, their crystallinities, particle sizes and anisotropies, and crystal strain, are much more reliable than any result derived from individual peaks or a few of them, because they are extracted using the complete diffractograms of involved phases. Fit residuals plotted over the whole 2Θ range may indicate the presence of minor phases, the peaks of which were completely superimposed by those of major constituents.

More information about Rietveld refinement and related instructive examples may be found in ref. [112].

3.4.3.4 Other diffraction and scattering techniques

Scattering and diffraction are features familiar to all kinds of waves: electromagnetic, matter, and even acoustic waves. For matter waves, the relation between energy and wavelength is described by the de Broglie equation (see eq. (3.75) below). The energies required for neutron and electron wavelengths to match the dimensions in crystals are very different from those of X-rays. To obtain a wavelength of 1.54056 Å (CuKα1), energy of 8.04 keV is required for electromagnetic waves. Electron waves have this wavelength at 62.7 eV, neutron waves at 0.023 eV.

While the theory of scattering and diffraction is similar for all these waves, the intensity parameters (form factors) are not. X-rays are scattered by the electrons of an atom, electrons by both electrons and nucleus, and neutrons by the nucleus exclusively. As nuclei are extremely small, the dependence of scattering efficiencies on the diffraction angle, which is pronounced with X-rays and even more dramatic with electrons, is absent with neutrons (Figure 3.64b). Therefore, signals at high Bragg angles, which are relatively weak in XRD, can be well observed with neutrons.

Neutron diffraction offers attractive opportunities, which result from its particular relation between the scattering efficiencies and the atomic properties. In XRD, H atoms cannot be observed and neighbors in the periodic table or isotopes of an element cannot be differentiated due to the dependence of form factors on the atomic number. With neutrons, the variation of scattering properties is different, and not monotonic. Even isotopes can often be differentiated, because neutrons distinguish scattering nuclides. Neutron diffraction can often detect positions of light elements, such as H[21] in the pres-

---

21 Actually, due to problems with incoherent scattering by the proton, H is replaced by D in the measurements.

ence of heavy atoms, or differentiate isotopes and neighbors in the periodic table. The magnetic moment of the neutron opens attractive opportunities for studies on magnetic properties, which are beyond the scope of this book.

Experimental effort for realizing these opportunities is, however, high. Neutrons are generated either in particle accelerators (spallation sources) or in dedicated nuclear reactors. Access to the limited number of experimental stations available can be achieved in peer-reviewed selection processes as at other large research facilities. So far, neutron diffraction has remained a rarely used complementary technique in catalysis research. For more details about the method, the reader is referred to ref. [113].

Electron diffraction is applied in catalysis and surface science mainly in two versions, as LEED and as transmission electron diffraction. The latter occurs when crystalline matter is studied in the Transmission electron microscope (TEM, cf. Section 3.4.11.1).

In LEED, a planar surface is exposed to a monochromatic electron beam with energy of typically 100 eV, usually coming along the surface normal. Diffracted beams are registered above the sample by video equipment or a position-sensitive detector after passing an electrostatic grid optic. At this kinetic energy, the mean free path of electrons in the sample is on the order of 1 nm (cf. Figure 3.97 below), i.e., the method is highly surface-sensitive. To avoid surface charging, the sample must have an appropriate electric conductivity.

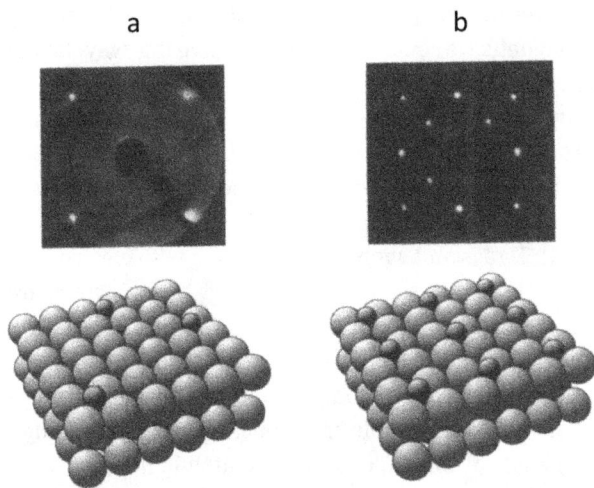

Figure 3.73: Low energy electron diffraction (LEED) patterns of a Ni(100) surface with O adsorbates: (a) disordered adsorption; (b) (2 × 2) superstructure of O atoms, causing a narrower spacing of the reciprocal lattice. Reproduced with permission from ref. [114], copyright IOP publishing, all rights reserved. The central reflection (00) is missing for technical reasons: its direction coincides with that of the incoming electron beam.

As shown in Section 3.4.3.2, diffracted beams are obtained for a two-dimensional lattice without constraints to the direction of the incident radiation. They directly visualize the structure of the reciprocal lattice. In Figure 3.73, this is exemplified by patterns of a Ni(100) surface with adsorbed O atoms. When the adsorbates are disordered, only the (100) pattern of Ni can be seen. The spots in Figure 3.73a are diffuse because of the presence of a diffuse H phase, which was in the focus of the paper [114]. The 2 × 2 superstructure shown in Figure 3.73b results in additional, now well defined spots, which cuts the translation vectors in half.

Due to the finite escape depth of the electrons, LEED can also probe subsurface structures, down to the fourth layer. It was, therefore, a key driver in the early years of surface science. Although completely inappropriate for catalytic materials, catalysis owes to this method a great wealth of fundamental information and understanding.

As mentioned in Section 3.4.3.1, scattering produces intensity offside the primary beam if there are grain boundaries or density fluctuations in the sample. The resolution on structural information grows with decreasing scattering angle. At angles between 5° and 0.1°, where scattering is sometimes referred to as "Debye scattering" (cf. Figure 3.71 inset), it is between 100 and 1 nm.

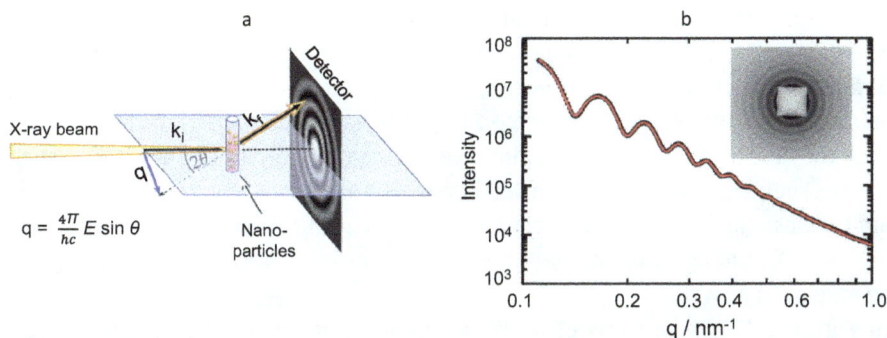

$$q = \frac{4\pi}{hc} E \sin\theta$$

Figure 3.74: Small angle X-ray scattering (SAXS): (a) schematic of experiment, with wave vectors of incident and scattered waves and scattering vector $\vec{q}$; to minimize angles with the (blocked) primary beam, the distance between the sample and the detector may be several meters; (b) scattering profile of a sample of poly(methyl methacrylate) particles of average 108 nm size. Adapted with permission from ref. [115].

In Small angle X-ray scattering (SAXS), information on particle or pore sizes and size distributions is extracted from the dependence of scattered intensity on the scattering angle. The sample is placed in a monochromatic collimated X-ray beam, and the scattered intensity is recorded by a flat-panel detector (Figure 3.74a). The experiment is preferably operated with synchrotron radiation. The symmetry of the scattering pattern allows reducing the problem to one coordinate of the reciprocal space, with the absolute value of the scattering vector $q = (4\pi/\lambda)\sin\Theta$. Figure 3.74b shows an intensity profile obtained with monodisperse spherical plastics particles of 108 nm size.

Although Fourier transformation of these profiles into real space shows the radial distribution of the general scattering potential, extracting more specific structural information usually requires fitting the scattering profiles of assumed structural models to the experimental data. SAXS is extensively used in various areas of materials characterization and in bioscience. For catalysts, which usually contain various scattering features (e.g., surfaces of size-distributed particles and pores), it is often difficult to derive reliable conclusions from scattering profiles. The use of SAXS has been therefore confined to specific problems for which some examples are given in ref. [112]. However, ongoing methodical development, in particular, the opportunities of elemental discrimination with synchrotron radiation (anomalous SAXS – ASAXS), suggest a rise of the method in catalysis research also.

Scattering data of whatever nature can also be evaluated by total scattering analysis. The method uses the intensity profile $I(\vec{k} - \vec{k}_0)$ over a wide range of scattering angles and models it as the superposition of all pairwise interferences of scattered waves possible in the material. Each structure can be considered as made up of pairs of atoms, with distances r ranging from a lower limit to infinity, which occur at specific frequencies depending on r: the atom pair distribution function (PDF). Diffraction lines result from the high frequencies of specific distances due to the presence of periodicity. However, all background features, which can be rather pronounced in a nanocrystalline material (cf. Figure 3.71) are covered by the summation as well.

The PDF $G(r)$ is obtained via a Fourier transformation of the experimental intensity data from the reciprocal into the real space. It is interpreted by constructing and varying structural models in order to try and fit the experimental PDF by those calculated for the model structures. As this involves a large number of free variables, such an analysis requires low-noise scattering data in a range extending to large scattering vectors (angles). Such data can be best generated by using synchrotron radiation (cf. Section 3.4.4.2), although data measured with high-energy lab sources (Mo$K\alpha$ or Ag $K\alpha$) can be employed as well.

In Figure 3.75, the analysis of a PDF is exemplified for a nanocrystalline MoS$_2$ (see diffractogram R873 in Figure 3.71). In the model employed, the MoS$_2$ unit cell was enlarged to 24 layers, which can accommodate all relevant kinds of structural defects and can, therefore, be considered to represent their averaged abundances. These types include 3R, instead of 2H stacking (stacking faults), random displacements between the layers in the a and b directions (turbostratic disorder), and a variation of the interlayer spacing (c-shift), which are all significant features of the MoS$_2$ microstructure.

In the model fits of Figure 3.75, the range below 6 Å agrees well with the experimental data for all models. Apparently, the MoS$_2$ slabs are largely intact, and disorder predominantly affects the relation between them. The introduction of stacking faults improves the agreement between the model and the experiment to some extent, but the effect of turbostratic disorder is much more significant. Finally, admitting variations in interlayer spacing brings the fit parameter in a range indicating very good agreement. The resulting structural parameters and their trends with the sample history may be looked up in ref. [111]. For more details about total scattering analysis, the reader is referred to ref. [116].

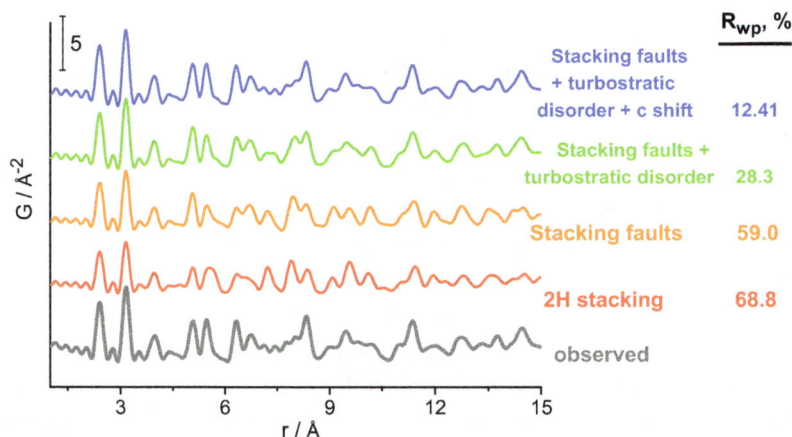

Figure 3.75: Pair distribution function (PDF) of a nanocrystalline MoS_2 (bottom profile, for diffractogram see sample R873 in Figure 3.71) and its analysis by models covering different types of disorder. Reproduced from ref. [111] with permission from Elsevier.

3.4.4 X-ray absorption fine structure (XAFS)

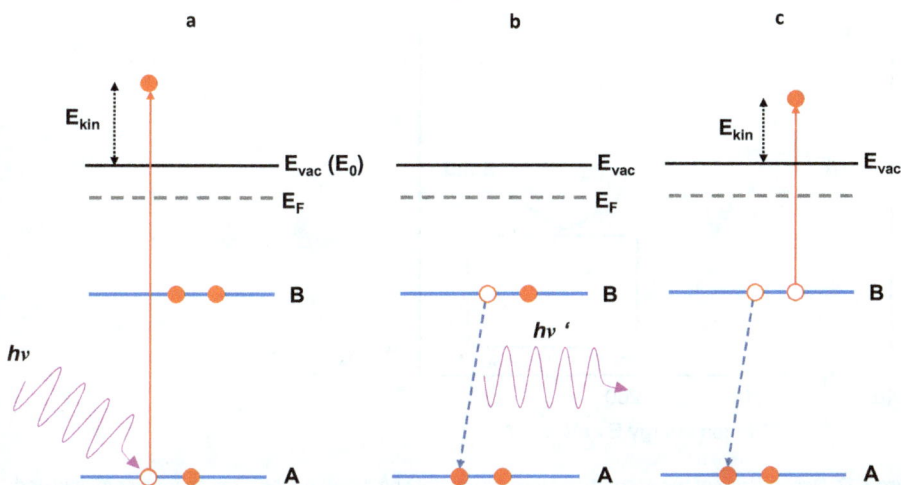

Figure 3.76: Basic processes of photoemission (a), X-ray fluorescence (b), and Auger electron emission (c).

The photoelectric effect is the basis of two important characterization techniques, XAFS and photoemission (cf. Section 3.4.3.1). In photoionization, a free electron with the kinetic energy E_{kin} is emitted, leaving behind a highly excited cation, with a hole in a core level (Figure 3.76a). To stabilize this core hole, negative charge is attracted during ionization: higher electron orbitals of the same atom are contracted; electron density of the neighboring atoms is attracted; and in metals, conduction electrons are displaced to shield the core hole charge. At the same time, secondary excitations may be induced: in metals, electrons at the Fermi level may populate empty levels in the conduction band; and in ionic compounds, electrons of the (anionic) valence band may cross over into empty cation levels. All these relaxation phenomena (or final-state effects), which must be differentiated from the subsequent de-excitation of the core hole, are reflected in the kinetic energy of the photoelectron. The secondary excitations are at the expense of this photoelectron energy, which results in tailings or satellite lines in photoelectron spectroscopy.

After a lifetime on the order of a femtosecond, the core hole is filled by an electron from a higher level. The energy gained by this can be used in two competing processes: for emitting a photon (X-ray fluorescence, Figure 3.76b) or for emitting an additional electron (Auger electron emission, Figure 3.76c). Both processes can be utilized for signal detection in XAFS. Auger electrons are used as complementary source of analytical information in photoelectron spectroscopy (cf. Section 3.4.5).

3.4.4.1 Fine structure in X-ray absorption

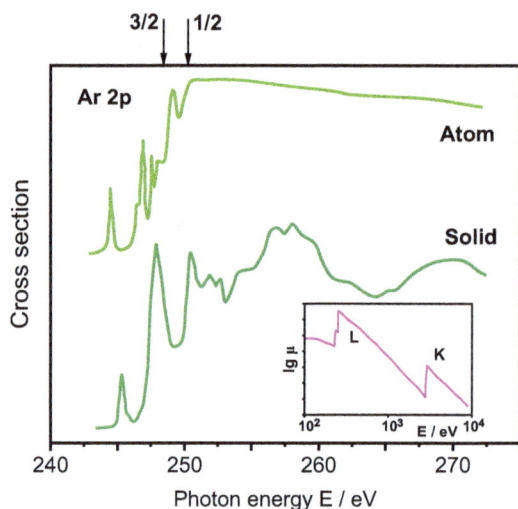

Figure 3.77: Dependence of the absorption cross section for soft X-rays on the X-ray energy: Ar in fluid and in solid form, $L_{2,3}$ absorption edges. The inset shows the development of the X-ray absorption coefficient over a wide range of X-ray energies in logarithmic coordinates, featuring dramatic increases at the K and L edge energies.

For each element i, the absorption coefficient μ_i for X-rays decreases with increasing X-ray energy E.[22] This decrease is superimposed by sharp steps, the absorption edges

---

22 because interactions become less likely with increasing energies: this is the reason why lead walls must be thicker to shield radiation of higher energy.

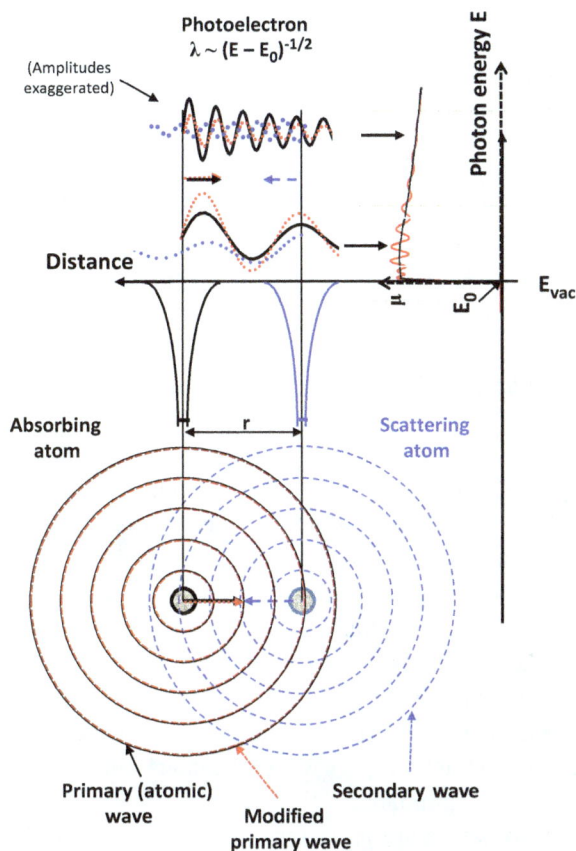

Figure 3.78: X-ray absorption fine structure (plotted along the "Photon energy E" axis) resulting from the backscattering of the emitted electron (along the "Distance" axis) and the interference of the outgoing and the backscattered electron wave in the emitting atom (absorber); for more detail see text. Reproduced from ref. [117] with permission from de Gruyter.

(Figure 3.77, inset). They reflect that radiation can be absorbed by a new channel, for which E is not sufficient below: the ejection of electrons from the K or the L shells (analogous edges of higher shells are less defined). When the energy is scanned across such an edge for a single atom, the edge is preceded by pre-edge peaks caused by transitions of the core electron into empty bound electron states (Figure 3.77, "Atom"). Above the edge, $\mu_i(E)$ goes on decaying smoothly. If the atom is part of a solid, the edge region changes in a characteristic way (Figure 3.77, "Solid"). The pre-edge features are modified,[23] and above the edge, the absorption coefficient exhibits fluctuations around the values of the atomic absorption coefficient, which decrease gradually at higher X-ray energies. X-ray absorption spectroscopy deals with these effects of neighboring atoms

---

23 e.g., because they arise now from a band structure instead of an atomic state.

on the course of $\mu_i(E)$ in the pre-edge region and above the edge, as far as the fluctuations can be detected (often more than 1,000 eV).

According to the de Broglie equation

$$\lambda = h/mv \tag{3.75}$$

(h – Planck's constant), an electron ejected from the absorber with a kinetic energy of $E_{kin} = E - E_0$ (where E_0 corresponds to the vacuum level, cf. Figure 3.78) has the wavelength

$$\lambda = h/(2m_e(E - E_0))^{1/2} \tag{3.76}$$

(m_e – electron mass). When the electron is scattered by a neighboring atom, a secondary wave is generated, part of which returns to the origin to interfere with the primary wave. The situation is illustrated in Figure 3.78 for an absorber with just one neighbor. The wave actually emitted from this pair of atoms results from this interference, and the resistance which the absorber experiences on this escape channel (and hence, the absorption coefficient μ_i) depends on its nature; higher μ_i resulting from constructive interference, and lower μ_i from destructive interference. The absorption coefficient is described by Fermi's golden rule, which relates the final to the initial wave function, i.e., to the 1s or 2p state of the initial electron. As they extend to just some pm around the nucleus (for more details see ref. [118]), it is this tiny space where the interference effect takes place, which is not easily illustrated in graphics.

The phase relation between the primary and the backscattered wave(s) is made up of contributions from the extra path length travelled by the secondary wave ($2r$ in Figure 3.78) and of phase shifts from scattering at the potentials of the neighboring and the central atoms. As the energy E increases, λ decreases (eq. (3.76)), which causes a continuous variation of the phase relation. For a single neighboring atom (or a shell of identical atoms at the same distance r), this results in the modulation of the atomic absorption coefficient μ_{i0} with a sine function, which is damped because the amplitude of the scattered waves, and thus their impact on μ_i has a generally decreasing tendency with increasing electron energy (see below, Figure 3.80a). This simple picture neglects the energy dependence of the phase shifts (see Figure 3.80b), but it illustrates how XAFS "measures" the distance between the absorber and the backscatterer. The neighbors may be identified via their scattering efficiencies and their contributions to phase shifts.

Notably, XAFS supplies structural information without relying on structural coherence, as required for diffraction. This potential arises from the double role of the absorber: it emits the electron, which probes the geometry and the population of the coordination sphere, and it communicates the result via its absorption coefficient. In diffraction, the source and the detector of the probing radiation are far from the sample; in XAFS, the absorber is both the emitter and the detector at the same time.

Although the structural information is less accurate than in XRD, and limited to a few shells around the absorber, XAFS is a powerful method for structural analysis just in regions where diffraction fails: for finely divided or disordered forms of matter, as frequently encountered in catalysis.

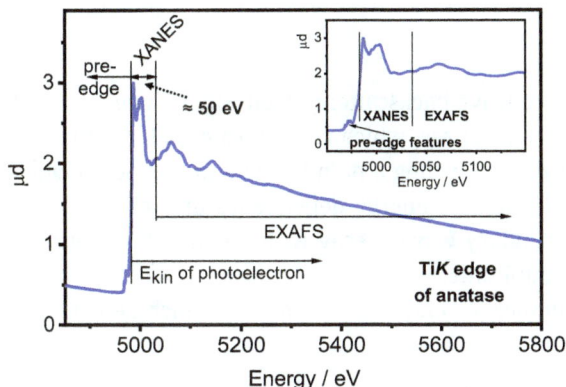

Figure 3.79: Energy regions of an X-ray absorption spectrum: pre-edge, XANES, and EXAFS. Reproduced from ref. [117] with permission from de Gruyter.

Structural information can also be obtained from the pre-edge features (cf. Figure 3.77), though on a different physical basis (see below). The region beyond the edge is divided into two sub-regions (Figure 3.79): the Extended X-ray absorption fine structure (EXAFS, >50 eV above the edge), which is caused by single and weak multiple scattering events, and the X-ray absorption near-edge structure (XANES). Due to the low electron energies, the XANES is dominated by intense multiple scattering. Therefore, pre-edge and XANES regions, which are often lumped under the term "XANES" in literature, are mostly used in the fingerprint mode in catalysis research, while coordination spheres around the absorbers can be modeled from the EXAFS on the basis of a relatively simple formalism (see below). XANES at soft edges like the K edges of O, N, or C, which is usually referred to as NEXAFS, is a method typically used in surface science. For a tutorial review, the reader may refer to ref. [119].

In the EXAFS region, the condensed matter effect is expressed as the EXAFS function $\chi(k)$ in which the difference between the solid-state and the atomic absorption coefficients, μ_i and μ_{i0}, respectively, is normalized by μ_{i0}:

$$\chi(k) = \frac{\mu_i(k) - \mu_{0i}(k)}{\mu_{0i}(k)} \tag{3.77}$$

where k stands for the absolute value of the electron wave vector

$$k = \left|\vec{k}\right| = \frac{2\pi}{\lambda} = \frac{2\pi}{h}\sqrt{(2m_e(E - E_0))} \tag{3.78}$$

In a simplified treatment, which assumes harmonic atomic oscillations and the arrangement of scatterers in shells at (approximately) equal distances r_j, $\chi(k)$ results from the summation of contributions from these shells:

$$\chi(k) = \sum_{j} \frac{A_j(k)}{r_j^2} \sin\left(2kr_j + \varphi_j(k)\right) \tag{3.79}$$

In eq. (3.79), A_j is the amplitude of the wave backscattered from shell j, which is evaluated by eq. (3.80). The sine describes its phase, which results from both the primary and the secondary (backscattered) wave traveling a path length r_j and experiencing a scattering phase shift of φ_j. Examples for the energy dependence of φ are shown in Figure 3.80b. Both primary and secondary waves decay with $1/r$, which results in a decay by $1/r_j^2$ for the backscattered amplitude.

The amplitude A_j contains a number of contributions, most of which depend on k as well:

$$A_j(k) = \frac{N_j}{k} \, F_j(k) \, exp\left(-2k^2\sigma_j^2\right) \, exp\left(-\frac{2r_j}{\lambda_{ie}(k)}\right) S_{0j}^2 \tag{3.80}$$

a   b

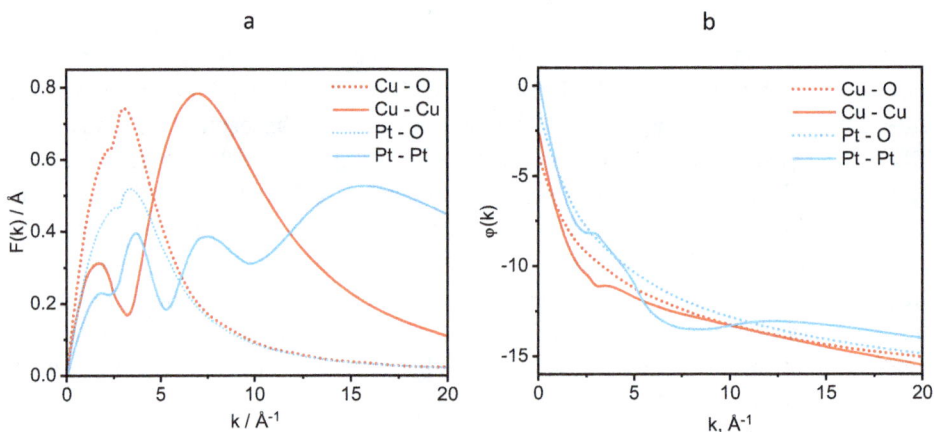

Figure 3.80: Dependence of the scattering efficiency F(k) and of the phase shift $\varphi(k)$ on k (i.e., via (3.78) on the X-ray energy). Effects of light and heavy scatterers are compared. Reproduced from ref. [117] with permission from de Gruyter.

In eq. (3.80), N_j is the coordination number (C.N.) of equal scatterers in shell j, $F_j(k)$ is the energy-dependent scattering efficiency of these neighbors (see examples in Figure 3.80a). The following Debye-Waller term describes that the intensity decreases with increasing temperature, which is usual in interference-based methods (cf. Section 3.4.3.2): σ_j is the mean square deviation from the equilibrium distance r_j, which grows with thermal motion. However, beyond thermal motion, such deviations may also arise from structural disorder, either due to structural defects or when in complicated structures, the atoms at slightly different distances from the absorber are lumped into a single shell. Thermal disorder can be largely suppressed by measuring at liquid nitrogen temperature (LNT) where excessive values of σ_j can be assigned to structural ("static") disorder.

The remaining two terms are minor corrections due to effects that are more relevant for photoemission (Section 3.4.5). They cover intensity losses by inelastic scattering and by secondary excitations (see above, final-state effects). The former are described by the decay function of electrons in matter, characterized by the energy-dependent inelastic mean free path λ_{ie}, and the latter are accounted for just by a correction term S_0^2.

For modeling experimental data with eqs. (3.79) and (3.80), the scattering behavior of all atoms that might be present in the coordination sphere must be known. This behavior is encoded in the energy dependence of scattering efficiencies F and of phase shifts ϕ. Such data are nowadays provided by software codes (e.g., FEFF), which are integrated into software packages for data reduction and modeling. These codes also supply the inelastic correction and the amplitude reduction term S_0^2. A successful fit defines the distance r between the absorber and the scatterer in a shell, the identity of the scatterer (via $F(k)$ and $\varphi(k)$), its coordination number, and the Debye-Waller term σ^2.

There is, however, a fourth fit parameter per shell, which is not related to the structure. The problem arises because of the unavailability of the exact edge energy E_0, which determines the electron wavelengths (eq. (3.76)), and therefore the zero of the k scale. It is usually estimated from the experimental data, e.g., as the inflection point of the rising μ_i (cf. Figure 3.79). In the case of metals, for instance, this indicates the Fermi level rather than the vacuum level. Therefore, the real position of the edge is established by fitting an edge-correction term ΔE_0, which is defined relative to the estimated edge.

Strategies to extract structural parameters from the EXAFS make use of the fact that the wave function is the Fourier transform (FT) of the scattering potential from the r-space into the k-space (cf. Section 3.4.3.2). Both are one-dimensional far from the edge. Reverse Fourier transformation indicates the radial distribution of the scattering potential (cf. Figure 3.86 below), although the nature of the scatterers remains undisclosed and the phase shifts induced by them, which cause shifts on the r axis, are not covered. Therefore, the r coordinate in presentations of the (reverse) FT is always classified as "uncorrected." Quantitative analysis requires fitting structure models to the experimental data, which is discussed in some detail in Section 3.4.4.2.

The diagnostic potential of the XANES is related to the pre-edge peaks, to the intensity just above the edge (the "white line"), to the edge energy, and to the shape of postedge features.

In Figure 3.81a, an intense pre-edge peak can be seen for Ti silicalite, where Ti is tetrahedrally coordinated (cf. Section 2.3.2.3) while anatase and rutile, with Ti in distorted octahedral environments, exhibit weaker and shifted pre-edge features. The signals arise from 1s-3d transitions, which are dipole-forbidden for orbitals of exclusively d character (O_h symmetry), but allowed for orbitals of partial p character (distorted O_h, in particular T_d symmetry). The white line arises from vacant levels close to the Fermi

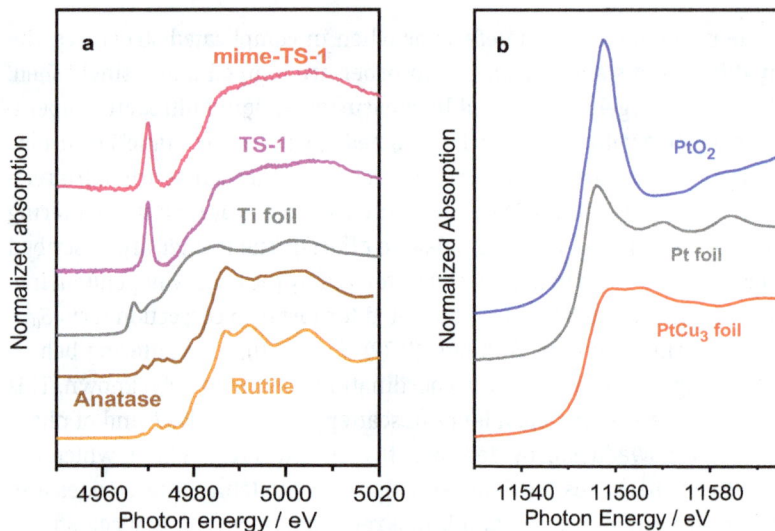

Figure 3.81: Major sources of analytical information in the XANES region: (a) pre-edge features; (b) the white line.

level, either in the conduction band or as free states of cations. Its intensity exhibits characteristic changes with the degree of d-band filling. In Figure 3.81b, the drastic difference between Pt metal and PtO_2 is due to the vacant d levels of the Pt^{4+} cation. When Pt is alloyed with Cu, the white line is attenuated, compared with Pt metal, apparently due to electron transfer from Cu to Pt.

The influence of the oxidation state on the (estimated) edge energy is illustrated in Figure 3.81a: the inflection points of the rising μ are at higher energies in TS-1 and in the TiO_2 polymorphs, than in Ti metal. In a very simple picture, such shifts, which are related to the binding-energy shifts in XPS (cf. Section 3.4.5), can be explained by the extra energy required for the leaving electron to overcome the attraction by the cation charge. In both cases, the relation between the oxidation state and the energy shift is useful, but not universal (cf. Section 3.4.5). The strong influence of the coordination geometry on the complex scattering phenomena causing the structure slightly above the edge is well illustrated by the differences between the spectra of anatase, rutile, and TS-1.

3.4.4.2 Acquisition and interpretation of X-ray absorption spectra

To measure the dependence of the absorption coefficient on the X-ray energy, X-rays of continuously varying energy are required. Electromagnetic radiation is emitted when charged particles change their velocities, for instance, in bremsstrahlung, which is the background under the characteristic X-ray lines in conventional X-ray generators (cf. Figure 3.83 below). Although laboratory instruments employing bremsstrahlung are still in use nowadays, most XAFS work is performed at synchrotrons.

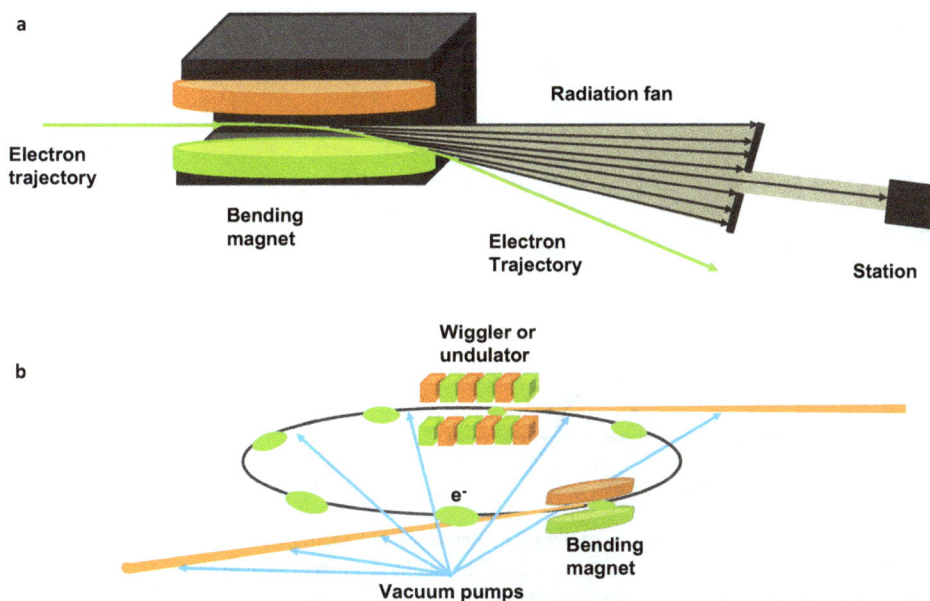

Figure 3.82: Electron storage ring and sources for synchrotron radiation: (a) bending magnet with beam to experimental station; (b) schematic of storage ring with wiggler and undulator. Unlike depicted in the figure, the electron trajectories are linear outside the magnetic fields.

Synchrotron radiation is emitted when such change of velocity is imposed on electrons travelling with relativistic velocities. It is caused by a magnetic field, e.g., in a bending magnet, and refers to the direction rather than to the absolute value of electron velocity. The radiation, which is emitted in the plane defined by the vectors of the electromagnetic force and the (initial) velocity, is strongly collimated along the initial direction. Radiation is thus emitted tangentially to the electron trajectory along the whole angle imposed by the bending magnet. From this fan (Figure 3.82a), slits cut out the beam admitted to the experimental station.

Wigglers and undulators are insertion devices that are much more efficient than bending magnets (Figures 3.82b, 3.83). They feature a sequence of many magnetic dipoles with alternating polarities, which force the electrons on a trajectory similar to a sine function. As a result, narrow radiation cones are emitted along the average flight direction. Wigglers and undulators differ in amplitude and width of the displacements imposed on the electron trajectory (Figure 3.83). Both are wider in wigglers, which results in a mere addition of average intensities. Undulators produce interference of radiation generated by the (weaker) magnets, which results in very strong intensity peaks. The energy of these peaks can be changed by varying the gap between the magnet poles, which allows measuring the absorption spectra at the top brilliance provided by the peaks.

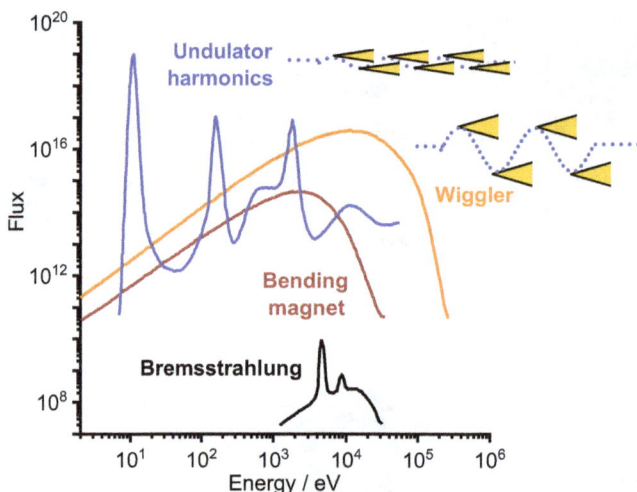

Figure 3.83: Photon flux generated by sources for synchrotron radiation, with difference between electron trajectories in the wiggler and the undulator highlighted. Bremsstrahlung given for comparison. Reproduced from ref. [117] with permission from de Gruyter.

Synchrotron radiation extends over a very wide range of energies. It is suitable for methods as different as IR or Mössbauer spectroscopy (Sections 3.4.8.3, 3.4.10), although the extremes may not be accessible at the same facility. The emitted radiation is polarized, and it is both highly intense and brilliant; the latter specifying the intensity per illuminated area. As electrons are travelling in small packets in the ring (bunches), radiation is pulsed with pulse lengths of 10–100 ps.

In the experimental stations, the radiation is conditioned for the use in specific measurements, e.g., of X-ray absorption. As shown in Figure 3.84, this includes shaping by slits, and monochromatization by diffraction on two large perfect crystal facets, e.g., Si(111), but also focusing and suppression of radiation that fulfills eq. (3.70) with n > 1 (for more details see ref. [117]).

The X-ray absorption coefficient can be recorded either directly by measuring X-ray intensities before and after the sample (transmission mode), or indirectly, utilizing processes induced by photoionization (X-ray fluorescence and Auger emission (Figure 3.76)), because they produce signals specific to the absorbing element. As Auger electrons lose their diagnostic information by interactions with gas-phase molecules, this mode is not used for work with catalysts where studies of samples exposed to the feed are of great importance (see below).

In transmission (Figure 3.84a), the beam intensities are measured by ionization chambers before and after the sample (I_0 and I, respectively), and the absorption coefficient is

$$\mu = \frac{1}{d}\ln\frac{I_0}{I}$$

(3.81)

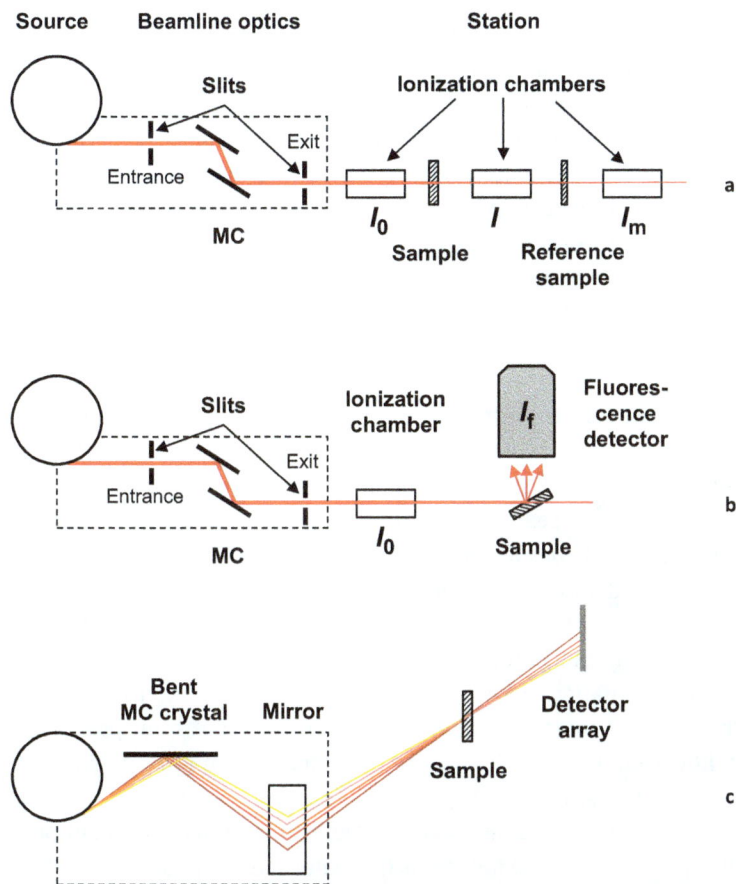

Figure 3.84: Experimental stations at a storage ring with equipment for measuring XAFS in different acquisition modes: (a) in transmission mode; (b) in fluorescence mode; (c) energy dispersive EXAFS (EDEXAFS). Reproduced from ref. [117] with permission from de Gruyter. MC - monochromator.

In eq. (3.81), d is the optical path length through the sample. As it is constant, μd is usually plotted instead of μ. It should be noted that μ is different from μ_i: it includes absorption by all components in the sample, and by all their channels accessible at the given X-ray energy. Between the second and the third ionization chamber, a reference sample is placed, which contains the absorbing element in a well-known state (e.g., a metal foil). It is used to calibrate the monochromator scale by relating the monochromator angle at the edge to the well-known edge energy.

Samples measured in transmission are powders pressed into pellets or beds of small catalyst particles. As the beam is often slightly displaced during the energy scan, uniform sample thickness and density (in beds) is of paramount importance. Sample thickness needs to be optimized: too thin samples produce too small intensity changes, and too thick samples absorb the beam completely.

Due to its simplicity, the transmission mode is applied whenever the height of the edge step is sufficient for reasonable statistics of the μ data. If the absorber i contributes only a small part to μ either due to its low concentration or due to the presence of heavier elements, the spectrum can be recorded using element-specific fluorescence lines of the i atoms. The intensity of the fluorescent radiation I_F is measured at an angle (often a right angle) to the beam (Figure 3.84b). The absorption coefficient is:

$$\mu \propto I_F/I_0 \tag{3.82}$$

Fluorescence signals are measured with multichannel semiconductor devices that reject elastically scattered incident radiation. Yields of fluorescence radiation may be quite small. Therefore, intensities of several fluorescence lines had to be lumped to obtain the absorption spectra in a reasonable time at older storage rings. At modern synchrotrons, fluorescence intensities are so high that lines can be differentiated by powerful crystal analyzers and individual, even weak lines can be picked for detecting the absorption. Actually, this allows measuring the full fluorescence (or X-ray emission) spectrum at each incident X-ray energy level during the scan across the absorption edge (X-ray absorption). This combination of X-ray absorption and X-ray emission has paved the way to new methods for the characterization of structural and electronic properties of solids, which are, however, beyond the scope of this book (for an introduction and reference to pertinent literature, see ref. [117]).

In modern synchrotrons, a scan across an edge takes 5–10 min in the transmission mode, but most of this time is consumed for monochromator manipulation rather than for signal accumulation. With continuously operating monochromators performing a periodically oscillating movement, this time can be decreased to 50 ms/scan. The full-time resolution of this quick EXAFS (QEXAFS) can, however, be realized only under favorable conditions and in the highly intense XANES region rather than for full EXAFS scans. Experiments with fluorescence detection are much more time-consuming than transmission experiments because of the limitations in counting rates that the present detectors can handle.

Energy-dispersive EXAFS (EDEXAFS) (Figure 3.84c) is a completely different approach to speed up XAFS experiments. In EDEXAFS, a bent grating is applied instead of a double-crystal monochromator to pick the radiation sent to the sample. As synchrotron radiation also has a slight divergence, (Bragg) angles between the incoming radiation and the bent crystal surface vary across the beam cross section, and so do wavelengths and angles of the diffracted beams (cf. the Bragg equation (3.70)): the bent grating is a polychromator sending light of different energies into different directions. The light is then focused in the sample, and finally analyzed by a position-sensitive detector (Figure 3.84c). In principle, this allows recording a spectrum in a single shot: for XANES measurements, time resolution in the microsecond range has been claimed. Due the polychromatization principle, the method does not allow measuring initial intensities I_0 for each energy, which complicates its use for EXAFS.

Conventional measurements, both in transmission and fluorescence mode, are frequently made in cryostats at LNT to minimize the influence of thermal disorder. Automated multi-position sample holders, combined with automated change between the edges accessible with the same monochromator crystal, may allow many hours of data acquisition without attendance.

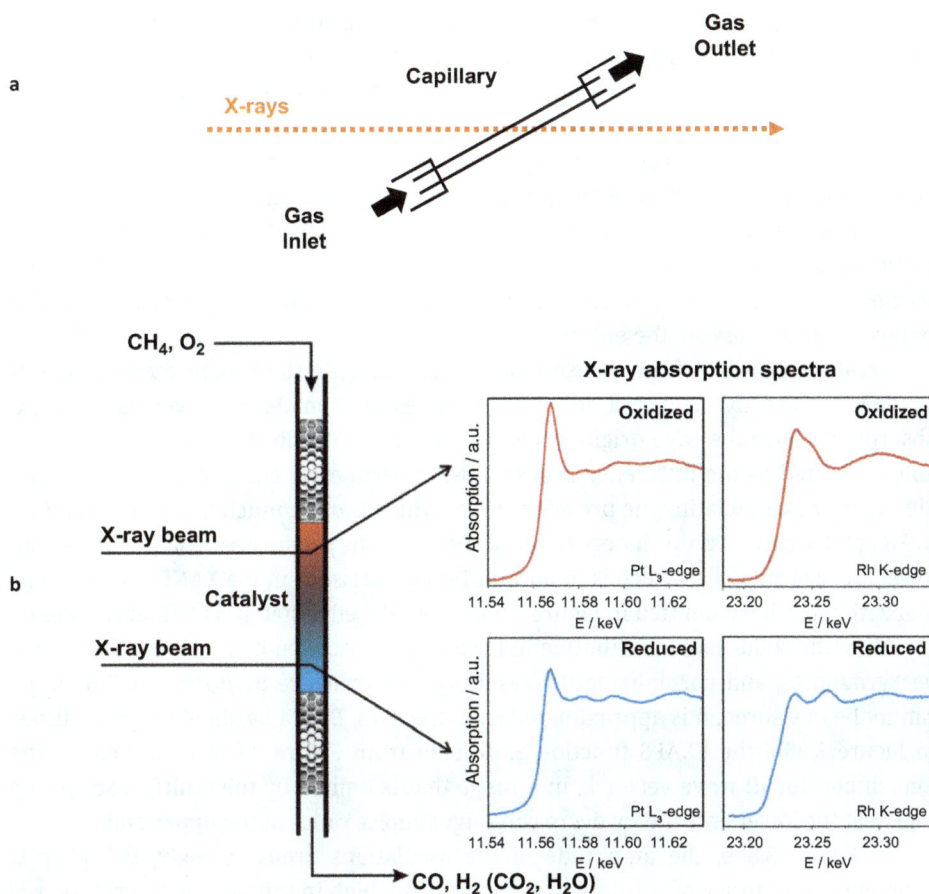

Figure 3.85: XAFS as a technique for *operando* characterization of catalysts: (a) schematic of a capillary cell; (b) length-resolved analysis of the oxidation state of Pt and Rh in a Pt-Rh/Al$_2$O$_3$ catalyst along a flow reactor during the partial oxidation of CH$_4$ by O$_2$. Reproduced from ref. [120] with permission from Springer Nature.

In catalysis research, there is, however, major interest in studying the catalyst state under typical reaction conditions. This is possible only with hard X-rays,[24] because soft X-rays require evacuated beam paths. Combining acquisition of the spectra under such *in situ* conditions with meaningful measurements of the catalytic reaction rates (*operando* principle) even offers the promise to correlate structural changes with trends of the reaction rate. Dedicated cell constructions are available for this purpose, which allow for parallel rate and XAFS measurements for catalysts or catalytic electrodes [117].

As an example of an environment suited for *operando* work, a capillary cell is depicted in Figure 3.85a. It is connected to a gas supply and to devices for concentration analysis. The cell is horizontally held by a frame, which can be moved in all directions and can be heated by a hot air blower from below. Absorption can be measured with an ionization chamber behind the cell or with a fluorescence detector normal to the beam (with the cell at 45°). The space above is available for detectors of other methods, e.g., XRD. Figure 3.85b shows XANES spectra, measured at different points of a capillary reactor filled with a Pt-Rh/Al_2O_3 catalyst and charged with a CH_4/O_2 mixture at 592 K. The reaction, which is relevant for the production of syngas (cf. Appendix A1), exhibits pronounced ignition/extinction phenomena (cf. Section 2.6.2). The spectra were taken in the temperature range of ignition: both noble metals are (partly) oxidized near the reactor entrance, but are reduced along the reactor. With increasing temperature, the reduction starts closer to the entrance [120].

Treatment of data from transmission measurements starts with the isolation of μ_i, which is directly measured only in the fluorescence mode. However, as the extra absorption above the edge originates exclusively from the absorber, μ_i can be reasonably estimated as the difference between the experimental curve and a background obtained by extrapolating the pre-edge curve (which is often much more inclined than in Figure 3.86a) across the energy range of the scan. The μ_i data are usually normalized to get an edge step of 1, which is important for comparisons in the XANES region. Normalization has been omitted in Figure 3.86, where the edge step is $\approx$1.7 (Figure 3.86a).

Next, the solid-state contribution to absorption is isolated by removing the atomic background μ_{0i} and normalizing the resulting fine structure by μ_{0i} (eq. (3.77)). As μ_{0i} cannot be measured, it is approximated from the μ_i vs. E data by standard procedures. In Figure 3.86b, the EXAFS function χ, derived from Figure 3.86a is plotted *vs.* the (one-dimensional) wave vector k, in a range that is limited by the multiple-scattering region at the lower end and by decreasing signal/noise ratios at the upper end.

In Figure 3.86b, the amplitude of the oscillations drops strongly at higher k, although the accuracy of the data may be still quite high in this region. Therefore, χ is

---

24 There is no standard differentiation between hard and soft X-rays. The border is usually set according to practical considerations, e.g. the need for high vacuum and the inadequacy of windows along soft X-ray beam paths. The dividing energy is somewhere between 2 and 5 keV.

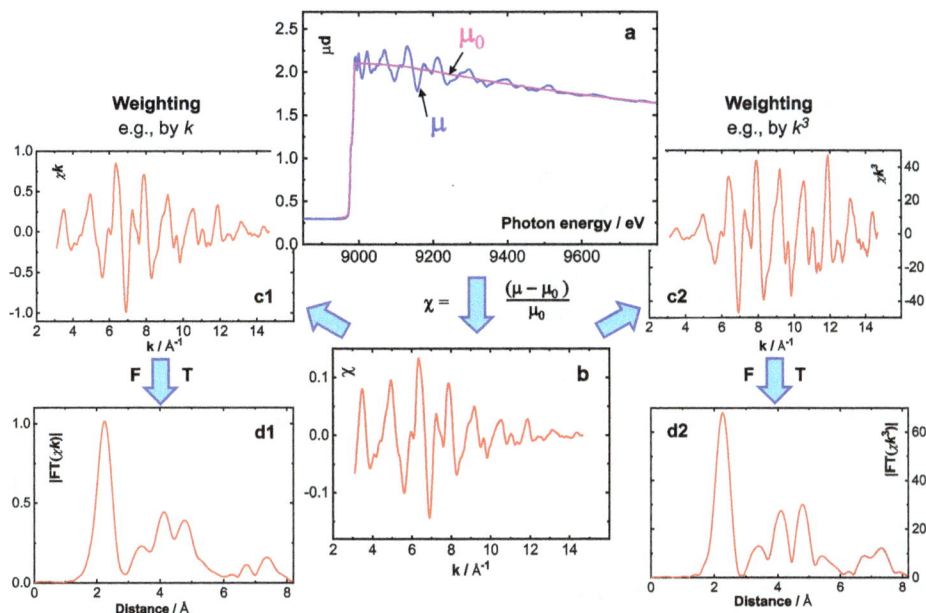

Figure 3.86: Data reduction in EXAFS. Reproduced from ref. [117] with permission from de Gruyter.

usually weighted by k to a power n, which is shown in Figures 3.86c1 and c2 for $n = 1$ and 3, respectively. For the strong scatterer Cu, weighting by k^3 emphasizes the fluctuations at high wavenumbers, which would be hardly detected in the original data by the naked eye, but are clearly significant. Figures 3.86d1 and d2 show the FT of these data, which indicate the distribution of the scattering potential along the (uncorrected) radial coordinate around the absorber. It arises from the atoms of the first coordination spheres, but the distances and the populations of these spheres can be found out only by modeling them against the experimental data.

Such modeling means building coordination spheres around the absorber, calculating their scattering properties using eqs. (3.79) and (3.80), comparing the results with the measured data (in k space, in r space, or both), and optimizing the geometry of the coordination sphere guided by the deviations obtained. Due to the large number of fit parameters (4 per shell), such modeling requires experience and critical attitude.[25] The analysis is facilitated when r is confined to suitable ranges by back FT with appropriate windows. Usually, this allows for reliable analysis of the first shell, even if it is composed of different elements. Although less confident, analysis of the second shell can still supply reliable results; shells further out can rarely be fitted with statistical significance. Fit results should be checked against plausibility criteria, as outlined in ref. [117].

---

25 There is a considerable risk that the fit ends up in a local minimum if too many shells are fitted simultaneously and/or if the fit parameters are not confined to physically reasonable ranges.

In literature, the nature of the adjacent atoms and their C.N.s are often employed to derive geometric models for the coordination sphere around the absorber. This is, however, justified only if the sample contains the absorber in just one coordination. In the absorption data, contributions from the absorber in all existing coordinations are superimposed. When structural homogeneity is sure for a sample, the identified coordination geometry can be a valuable quantitative result. For samples with structural heterogeneity, such generalization is just a flaw.

Checking the structural homogeneity is, therefore, an important aspect of work with XAFS. Often, relevant hints on this topic can be found already in the XANES or the EXAFS data. It is, however, advisable to combine XAFS work with structural studies by other methods to sort out this question. By a careful interpretation, XAFS can also supply highly useful evidence on samples with structural heterogeneities.

The interpretation of XANES spectra is closely related to the physical origin of the characteristic features (cf. Section 3.4.4.1). Intensity and position of the pre-edge peak at the K-edge of the first-row transition metals are used to identify the symmetry of the coordination sphere (see example in 3.4.4.3). The white line at the L_{III} edge of late transition metals can be used to study the electronic effects of various environments on metal atoms; in particular, their number of d-band vacancies. In physical mixtures, the contributions of constituents are superimposed, which can be used to determine the mixture compositions. Meanwhile, there are chemometric tools that allow deducing not only mixture compositions, but even (unknown) XANES of mixture components from suitable data sets [121].

The capability of XANES to reveal vacant states near the Fermi level even offers an opportunity to detect atomic hydrogen adsorbed on metal surfaces, which is not possible with most other spectroscopies. The approach uses the fact that the bond between the H atom and the metal results from antibonding states extending above the Fermi level: In the spectra of bulk Pd hydrides, they can be observed with the naked eye. When the interaction is confined to the surface atoms, the signals from the unmodified bulk metal atoms must be subtracted, which is difficult, but possible for very small metal particles. The $\Delta\mu$-XANES method operating on this basis is challenging, because it includes theoretical coverage of the particle structure that is influenced by the adsorbate, but it also offers a unique analytic potential also for other adsorbed molecules.

For some more detail on these methods and literature to deal with them, the reader may refer to ref. [117].

3.4.4.3 XAFS in examples

The EXAFS spectra in Figure 3.87a (r-space) and b (k-space) were measured during a study on the reduction of Pd^{2+} ion-exchanged onto an active carbon from a solution of $Pd(NH_3)_4(NO_3)_2$. They were recorded at LNT after reduction in dilute H_2 at different temperatures (given in °C in sample designations), flushing in He to destroy the hydride, where necessary, and cooling in He. Comparing the r-space spectra of the initial mate-

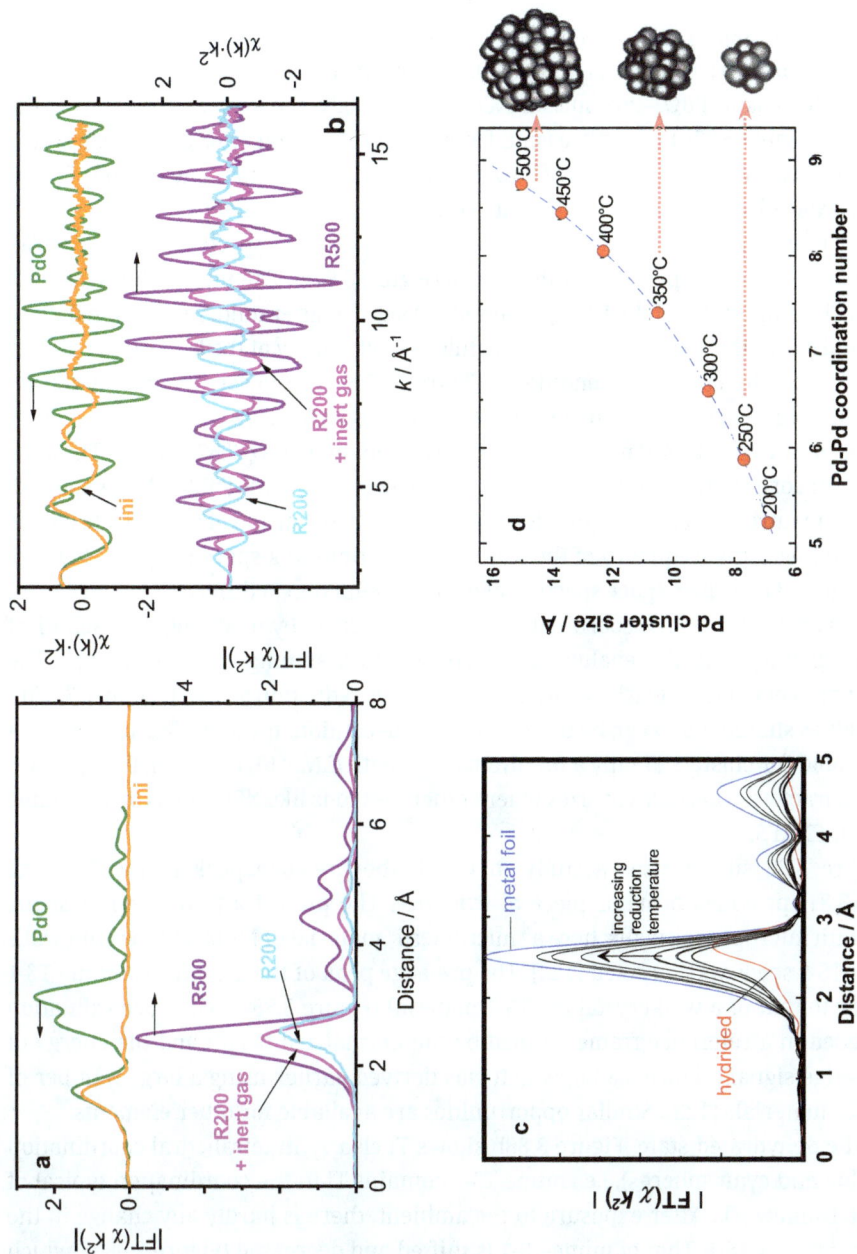

Figure 3.87: EXAFS analysis of carbon-supported Pd particles; (a, b) spectra of initial sample compared with PdO, and of reduced samples (incl. a hydride phase) in r-space (a) and in k-space (b); (c) development of r-space spectra with the reduction temperature; (d) Pd-Pd coordination numbers deduced from the spectra and Pd cluster sizes deduced from the coordination numbers. Reproduced from ref. [117] with permission from de Gruyter.

rial and PdO (upper spectra in Figure 3.87a) shows that Pd is indeed atomically dispersed in the precursor; there is a shell at ca. 1.5 Å (uncorrected), apparently Pd-O, but higher shells – as in PdO – are missing. Correspondingly, the k-space spectrum of the precursor (Figure 3.87b), mostly arising from the weak scatterer oxygen, is close to a sine and decays fast. The spectrum of PdO has much more structure, and extends far out on the k-scale, obviously due to contributions of Pd in the next shell(s).

After reduction, the first signal is shifted to 2.45 Å (lower spectra in Figure 3.87a), in agreement with the spectrum of the Pd foil (Figure 3.87c). The higher shells are well defined. The amplitudes of the FT signal are much smaller after milder reduction (Figure 3.87a), and so are the k-space amplitudes, while the periodicity of the k-space signal does not depend on the reduction conditions (Figure 3.87b). The lower first-shell intensity after mild reduction indicates smaller particles, in which surface atoms lacking a significant part of the coordination sphere are more abundant. Figure 3.87a and b also show the spectrum of a sample prepared without destroying the hydride ("R200"). As expected, all r-space features appear at larger distances due to the expansion of the hydride lattice. Correspondingly, the frequency of the sine-like oscillations in k-space is slightly higher.

In Figure 3.87c, the r-space spectra, taken after reduction at different temperatures, are summarized. First-shell coordination numbers derived by modeling the back-FT of the first signal were used to evaluate the particle sizes, assuming a spherical shape. The particles are very small, clearly below 2 nm even after reduction at 500 °C (Figure 3.87d). These values should, however, be treated with some caution: the spherical model is not likely to hold for clusters of only a few atoms (e.g., with $C.N. < 6$). However, the opportunity to characterize particles at sizes where other methods like XRD fail, is an attractive feature of EXAFS.

Figure 3.88 summarizes a study in which the pre-edge peak at the TiK-edge (Figure 3.81) provided a major piece of evidence. The project aimed at synthesizing OMMs with microporous walls: here a "micromeso" material of MCM-41 type with walls made of TS-1 seeds was targeted [122]. The pre-edge peak of the dehydrated mime-TS-1 is similar to that of a well-crystalline TS-1 material (Figure 3.88a). The Ti coordination is discussed in a reference frame defined by the normalized height and the energy of the pre-edge signal (Figure 3.88b), which was derived earlier using a large number of reference materials [123]. Similar opportunities are available for other elements.

In the dehydrated state, Figure 3.88b shows Ti clearly in tetrahedral coordination (dark blue and cyan spheres), i.e., mime-TS-1 contains Ti in the coordination typical of a zeolite framework. After exposure to the ambient, there is hardly any change in the pre-edge peak of TS-1. That of mime-TS-1 is shifted and decreased (Figure 3.88a), which indicates an almost sixfold coordination of Ti (cyan star in Figure 3.88b), apparently due to the adsorption of water. Crystalline TS-1 is very hydrophobic; water adsorbs only in structural defects or at silanol groups of the external crystallite surfaces. The adsorption of water on Ti sites of mime-TS-1 suggests that the regions with TS-1 structure are very thin (which was confirmed by TEM [122]) and that they are nearly completely accessible from the gas phase.

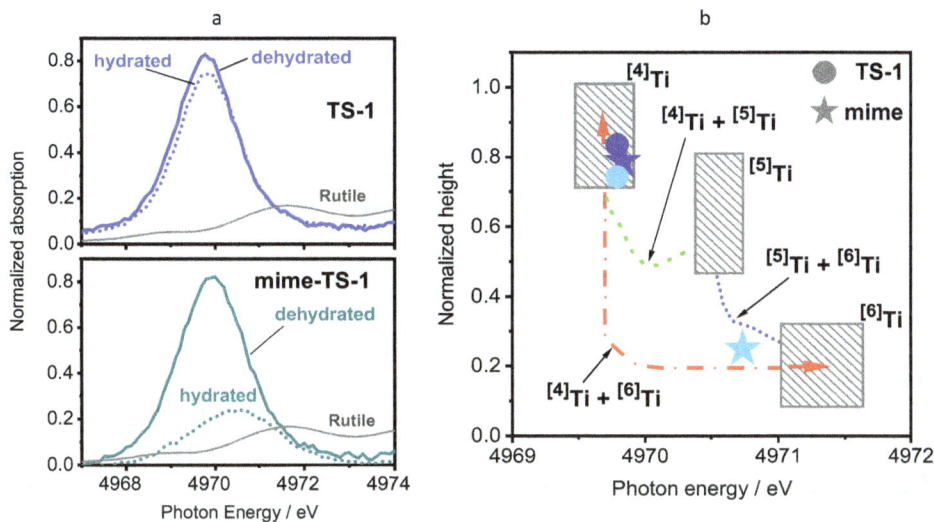

Figure 3.88: A XANES study on the nature and accessibility of Ti sites in MCM-41 pore walls, which were formed from TS-1 crystal nuclei: (a) pre-edge peak of well-crystalline, hydrophobic TS-1 (upper panel) and of TS-1-like MCM-41 pore walls (lower panel), signals of rutile are given for comparison, (b) reference frame for identifying the Ti^{4+} coordination via the height and the energy of the pre-edge peak; in mime-TS-1 (light blue symbols), Ti^{4+} is tetrahedrally coordinated (dehydrated state: circle) and accessible (hydrated state, star). Reproduced from ref. [122] with permission from Elsevier.

In summary, XAFS is a powerful tool for the structural analysis of materials, which probes the coordination sphere of elements of interest (identity of neighbors, coordination numbers, distances) without requiring structural coherence. The analysis of highly dispersed or of disordered states of matter is a stronghold of the method. Using hard X-rays, it is suitable for *in situ* and *operando* work in the gas phase, as also with liquid reactants, provided the optical path in the liquid is kept short. XAFS is, therefore, a major tool for fundamental studies in catalysis and electrocatalysis.

Structural information is extracted from EXAFS data by fitting models with a considerable number of free parameters. Therefore, reliability of results decreases strongly in the higher shells. For the first shell, distances can be established with errors of typically $\pm 0.005-0.01$ Å, *C.N.* with $\pm$ 10–20%$_{rel}$. In the coordination spheres, it is usually not possible to distinguish elements, which are neighbors in the periodic table of elements. In the second shell, the error of the *coordination number* may reach $\pm 50\%_{rel}$; the third shell often cannot be quantified with any statistical significance. When the initial structure of a sample is known, the use of a more extended model may be justified, because its parameters (identity of neighbors, r, *C.N.*) may be fitted, starting from the known values. Any change in the material may then be analyzed by releasing the constraints on some of the structural features while fitting the spectra obtained.

Dealing with catalyst characterization, one should keep in mind that XAFS in transmission or fluorescence mode is a bulk technique, which shows surface effects on the background of information from the bulk of the particles. Properties of surface atoms can be used or detected only if the particle dimensions are on the order of 1–2 nm (e.g., for particle size determination or $\Delta\mu$-XANES).

3.4.5 Surface analysis by photoemission techniques

3.4.5.1 Photoemission and surface sensitivity

The sketch of the photoemission process in Figure 3.76a shows that the kinetic energy of the photoelectron bears information about the emitting atom, because it is the difference between the excitation energy $h\nu$ and the ionization energy of its original atomic state. Photoelectrons ejected from samples originate from all levels of the elements present, which are accessible for the incident photon, including valence levels and bonding orbitals. Their study is the subject of photoelectron spectroscopy.

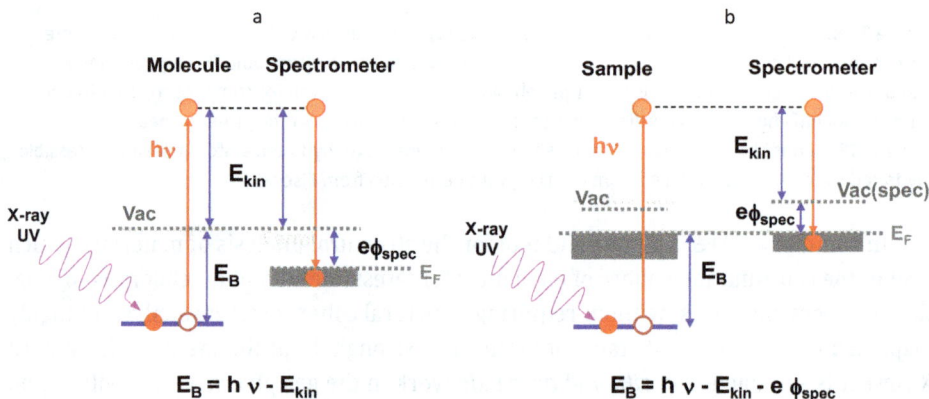

Figure 3.89: Energy balances during the analysis of photoelectrons from gas molecules (a) and from a metal (b). Reproduced from ref. [124] with permission from Wiley-VCH Verlag GmbH & Co. KGaA, Weinheim, Germany.

In Figure 3.89a, the sample is combined with a detector that retards the electrons to measure their kinetic energy E_{kin}. The difference between the absorbed photon energy and E_{kin} is the binding energy E_B:

$$E_B = h\nu - E_{kin} \tag{3.83}$$

In simplified accounts on photoemission, eq. (3.83) is sometimes treated as the fundamental energy balance of photoelectron spectroscopy. It holds, however, only for molecules in the gas phase, because the spectrometer and the gas molecules, which are not in electrical contact with each other, have the same vacuum level. This vacuum level is the zero of the binding-energy scale, i.e., E_B is indeed identical to the ionization energy of the corresponding core level.

Solid samples must be fixed to the spectrometer. For metals, this results in an alignment of the Fermi levels between the sample and the spectrometer (Figure 3.89b), which have, on the other hand, usually different work functions. As a result, there is no universal vacuum level in the system, which could serve as a starting point of the E_B scale. Instead, binding energies are referred to the Fermi level, which is universal for metals in contact with the spectrometer. Other kinds of solid samples are related to it via internal standards (see below). The resulting energy balance is (cf. Figure 3.89b)

$$E_B = h\nu - E_{kin} - e\,\phi_{spec} \tag{3.84}$$

where $e\phi_{spec}$ is the spectrometer work function. It is not necessary to determine its value, because the E_B scale can be calibrated with known binding energy values of reference metals. Obviously, binding energies on this scale differ from the ionization energies of the ionized shells by the (usually unknown) work function of the sample.

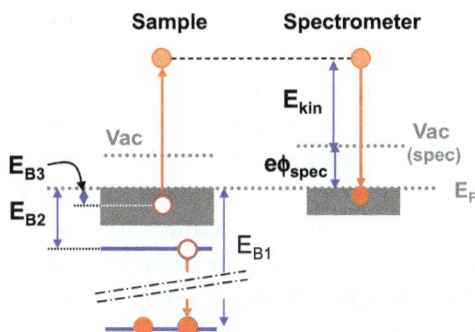

Figure 3.90: Energy balance during the analysis of Auger electrons. Reproduced from ref. [124] with permission from Wiley-VCH Verlag GmbH & Co. KGaA, Weinheim, Germany.

As mentioned before (cf. Figure 3.76), the Auger process is a mode of core hole de-excitation: Auger electrons are ejected from samples, together with photoelectrons. In Figure 3.90, the Auger process is placed in the environment of a spectrometer. The kinetic energy E_{kin} of the Auger electron measured against the spectrometer Fermi level is the excess of the energy gain by core hole stabilization, $E_{B1} - E_{B2}$, over the binding energy of the Auger electron E_{B3} and the spectrometer work function $e\phi_{spec}$:

$$E_{kin} = (E_{B1} - E_{B2}) - E_{B3} - e\,\phi_{spec} \tag{3.85}$$

A comparison of eq. (3.85) with the rearranged eq. (3.84) $E_{kin} = h\nu - E_B - e\,\phi_{spec}$ illustrates that photoemission is driven by an external source that can be varied (by changing the excitation energy), while the driving force of Auger emission is invariable: it is encoded in the electron structure of the atom. When the excitation energy is varied, only the photoelectron kinetic energies are changed, which can be used to establish the origin of unassigned peaks and to resolve the relatively rare superpositions between photoelectron and Auger lines.

Both photoelectron and Auger electron spectroscopy (AES) raise diagnostic information via the energy of electron levels, which are specific to each atom. For photoemission, opportunities strongly depend on the excitation energy; therefore, X-ray photoelectron spectroscopy (XPS) and Ultraviolet photoelectron spectroscopy (UPS) are differentiated. In the older literature, the former was often designated as ESCA (Electron spectroscopy for chemical analysis).

Auger lines can be excited by any process that creates core holes. Actually, AES is typically excited by electron beams, which are much more effective than X-rays and can be focused into nm-sized cross sections.[26] Well before the advent of the present high-performance electron analyzers, Auger spectra were acquired in minutes, while XPS was very time consuming. Due to the strong impact of electron beams on reactive materials, AES is rarely used for catalysts. However, the Auger lines resulting from the decay of photoinduced core holes provide a complementary source of analytical information in XPS (see Section 3.4.5.3). Their use is sometimes referred to as X-ray-induced AES (XAES).

Inelastic scattering annihilates the analytical information of photo or Auger electrons. While X-rays penetrate deeply into the solid even at the relatively low energies used in typical equipment (see below), the inelastic mean free path λ_{ie} of the electrons in solid samples is on the order of a few nm. Only electrons emitted by atoms at or near the external surface leave the solid without loss of kinetic energy. Therefore, XPS, UPS, and (X)AES probe the surface region of solids, with an average sampling depth that corresponds to λ_{ie}. The percentage of electrons escaping from the solid unaffected decreases with increasing distance from the external surface, according to an exponential decay function (see eqs. (3.90) and (3.91) below). The remaining photoelectrons end up in the background or are increasingly decelerated and trapped in the solid.

Photoelectrons lose their analytical potential also by inelastic events outside the solid: at a kinetic energy of 1,000 eV, λ_{ie} of electrons in 1 mbar of oxygen is on the order of just 10 mm. This is one of the reasons why photoelectron and Auger electron spectroscopy are traditionally performed in UHV environment. The other reason is that reactive surfaces of just a few cm^2 size can be kept stable only by a very tight control of the gas phase, ideally under vacuum. Although XPS can nowadays be measured under reactant pressures of several mbar, working under conditions near to real catalytic processes remains most challenging for all methods that rely on low-energy electrons.

---

26 The Auger process is also utilized for an element-sensitive version of electron microscopy (see Section 3.4.11.2).

3.4.5.2 Measuring and assigning photoelectron spectra

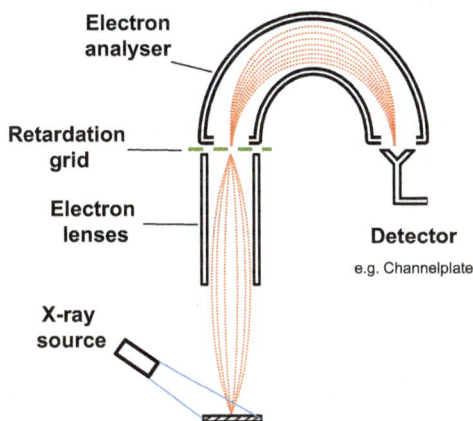

Figure 3.91: Components of a photoelectron spectrometer (without vacuum system and sample handling facilities).

A photoelectron spectrometer is a UHV system combined with an electron analyzer with detector, a source for X-rays or UV light, and sample handling facilities. In the scheme of Figure 3.91, the sample is already in the measurement position, and the vacuum system is omitted as well. While the sample is exposed to radiation, the photoelectrons are focused on the entrance of the analyzer, which differentiates them according to their kinetic energy. Those finding the analyzer exit are counted by a detector. Laboratory instruments may have a UV source, in addition to the X-ray source. With reverse polarity, electron analyzers can also handle positively charged particles. Therefore, photoelectron spectrometers may be additionally fitted with an ion source to measure LEIS spectra (cf. Section 3.4.6).

Traditional laboratory instruments work with dual X-ray sources, which allow for easy change between Al$K\alpha$ and Mg$K\alpha$ radiation (1,486.6 eV and 1,253.6 eV, respectively) for shifting Auger signals, relative to photoelectron lines on the E_B-scale (see Section 3.4.5.1). Due to the doublet character of the $K\alpha$ transitions, the excitation lines are rather broad, slightly below 1 eV, and come with a set of X-ray satellites and a bremsstrahlung background. The former must be removed from the spectra by standard algorithms; the latter is suppressed by an Al foil between the source and the sample to avoid radiation damage at the samples. This is not necessary with modern instruments, which work with monochromated Al$K\alpha$ sources and provide excitation radiation with line widths of 0.2–0.3 eV. In many synchrotron stations, photoelectron spectrometers are available for measurement with customized excitation energies. Due to the large intensity reserves in synchrotrons, spectra can be recorded with excitation line widths of 0.1 eV or even below.

UPS is measured using UV radiation from gas discharge lamps, typically operated with He. The widths of the most popular excitation lines (He I at 21.2 eV, He II at 40.8 eV) are in the meV range and do not affect the spectral resolution.

Samples are introduced into the UHV chamber via load locks in a staged pumping procedure, which may be time-consuming with porous materials inclined to adsorb moisture. Pressures during measurement are preferably in the 10^{-9} mbar range or below. Spectrometers are often equipped with reactor systems, in which samples can be pretreated or even operated as catalysts under realistic conditions before being pumped down and introduced into the UHV chamber for surface analysis. The virtue of such a quasi *in situ* approach is discussed below.

In the analyzer, the electrons interact with an electric field in a way that only those with a specific kinetic energy find the detector. In the hemispherical analyzer, which is an eye-catching attribute of photoelectron spectrometers, the field extends between two concentric hemispheres (Figure 3.91). In a typical operation mode, this field is kept constant, allowing electrons of only a specific kinetic energy to be registered, the so-called pass energy (PE). The scan along the kinetic-energy scale is realized by a retardation potential E_{ret}, which decelerates the photoelectrons at the analyzer entrance, and can be scanned to record the spectrum, because $E_{kin} = PE + E_{ret}$.

Modern analyzers use the fact that the trajectories of electrons entering with energies slightly different from the PE end up in a well-defined pattern around the spot related to the PE trajectory. By using position-sensitive detectors for simultaneous accumulation of intensity in finite energy ranges, data acquisition can be enormously sped up.

Working with low-energy electrons, XPS and AES are most handicapped in the attempt to keep catalysts under realistic conditions during the measurements. As mentioned above, "*quasi in situ*" technology was developed as a first step, which allows treating catalysts under a wide variety of conditions, and transferring them to the UHV without contact with the ambient. This is promising in cases where the gas-phase is clearly oxidizing or reducing, because both cooling and pumping reduce the driving force to achieve higher or lower oxidation states and decrease mobility in the solid phase. In mixed feeds, as applied, for instance, in selective hydrocarbon oxidation, there may, however, be a strong impact of the transfer process on the state of the surface. It has also been found that the quasi *in situ* approach fails to freeze metastable surface states, such as subsurface oxides or carbides, which exist only in oxidizing or in hydrocarbon-containing environments.

The effort to use XPS for studying solid surfaces while they interact with gases (e.g., during the oxidation of metals) or even for investigating liquid surfaces started in the early days of the method. The problem of minimizing inelastic collisions in the gas phase was addressed by arranging differential pumping stages along the electron trajectories toward the analyzer, and while the principle was proven, dramatic intensity losses by the small acceptance angles, defined by the apertures between the pumping chambers (Figure 3.92a), rendered work with these systems tedious. A new area of Ambient-pressure photoelectron spectroscopy (APPES) was born when differential pumping and electron optics were combined in a way that electron trajectories were focused in the apertures between the pumping stages (Figure 3.92b). Starting with instruments in synchrotrons around 2000, the technique has been meanwhile developed to a level

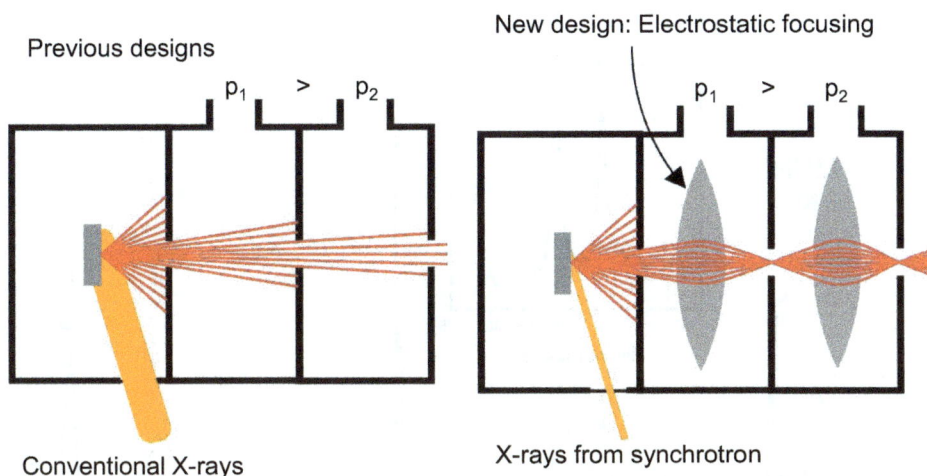

Figure 3.92: Combination of differential pumping and electrostatic focusing in Ambient pressure photoelectron spectroscopy (APPES). Reproduced from ref. [125] with permission from Elsevier.

that allows marketing even laboratory APPES instruments. Beyond the combination of differential pumping with electron optics, they differ from traditional instruments also in the availability of environmental cells that allow feeding and recovering gases or the continuous drain of liquid films.

As far as instrumentation is concerned, the following discussion is, however, confined to conventional instruments. For more information about APPES, the reader is referred to ref. [125].

When dealing with a new sample, a survey scan covering most of the kinetic energy range[27] is recorded first in a regime favoring intensity at the expense of energy resolution. This spectrum is used to identify lines and to decide which regions should be studied in more detail, i.e., with high energy resolution.

As an example, Figure 3.93 shows a survey spectrum of silver recorded with MgKα radiation. It is plotted vs. a decreasing binding energy scale.[28] In the inset, a detailed spectrum of the Ag 3d doublet is depicted.

The photoelectron signals in the survey originate from the 3s and higher levels of silver (Figure 3.93); lower shells cannot be ionized by MgKα radiation. Except for the lines from s levels, all signals are doublets. The valence-band region, comprising the 4d signal and weak 5s intensity extending to the Fermi level, can be better studied with UPS because of its narrow excitation line width and high photoemission cross sections of these levels for UV excitation. The broad signals labeled MNN are Auger lines. The

---

27 Very low kinetic energies are omitted due to huge intensities caused by unspecific secondary electrons.
28 E_B is a secondary quantity deduced from the photoelectron kinetic energy. Therefore, spectra are often plotted vs. E_{kin}, but with the axis numbered with E_B values, which therefore decrease from the left to the right (cf. Figure 3.93). This convention is, however, not followed by all groups.

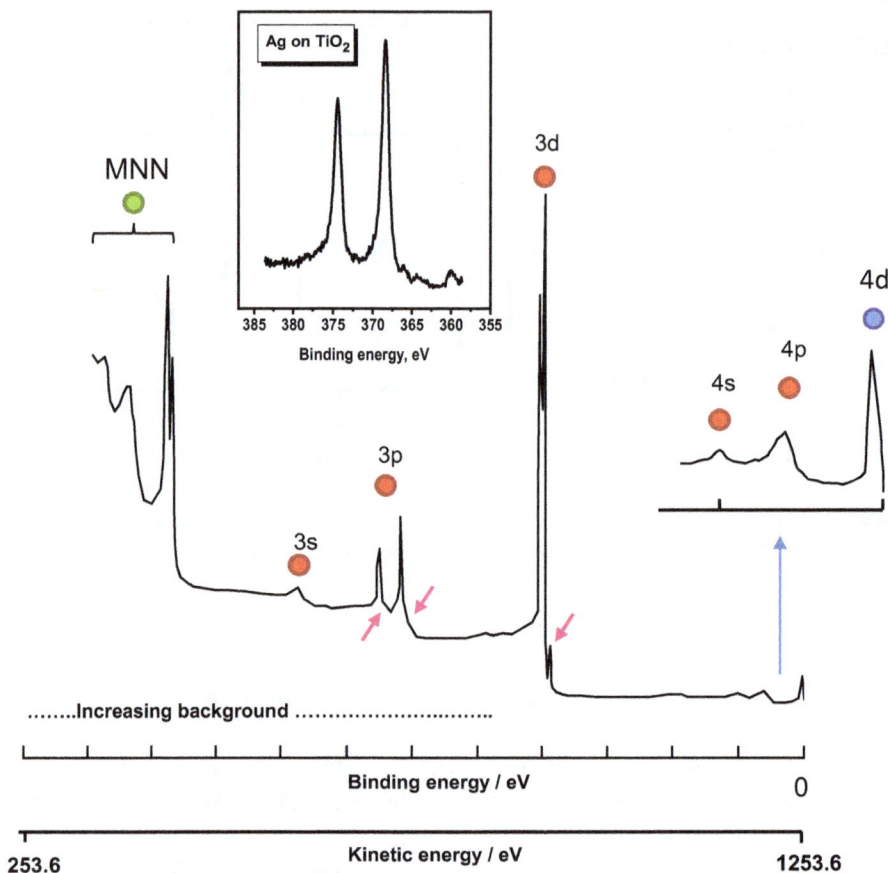

Figure 3.93: Features in a photoelectron spectrum, illustrated with MgKα spectra of silver metal. Main panel – survey spectrum of a silver foil; inset – detail spectrum: 3d doublet of supported Ag nanoparticles; red dots – photoemission lines; blue dot – valence region; green dot – various Auger signals; arrows indicate X-ray satellites.

electrons arise from N levels; their emission is driven by an N → M transition. The line width is due to a large number of sublevels in these high shells. The spectrum was recorded with characteristic radiation; therefore, X-ray satellites of the most intense lines can be observed at slightly higher binding energies (see arrows).

At the left of each signal, electrons that have lost kinetic energy in inelastic events increase the background level. In the survey scan of massive Ag (Figure 3.93), this stepwise increase is most pronounced to the left of the most intense signals, but in the inset originating from 2–3 nm Ag particles on a TiO_2 support [126], it is almost absent at the same signal. This observation bears already information on the type of sample: when it is massive, deeper layers of the same composition (here – pure Ag) contribute to the signal and, notably, to the background, which comprises electrons that have started with the correct E_{kin} deeper in the sample. In the supported sample, where we assume thick pore walls for the sake of simplicity, there is a different material below the small particles, which are very poor sources of decelerated electrons. Therefore, there is hardly an increase of the background level at lower kinetic energy levels.

During photoemission, electrons are continuously removed from the sample. When the sample is a metal in contact with the spectrometer, the missing electrons are replenished by a sample current from the latter. This is not possible for insulators or semiconductors with low conductivity; their surfaces tend to charge positively. The consequences of this charging effect depend on the instrumental configuration. In conventional instruments, where bremsstrahlung is attenuated by an Al foil and the irradiated area often exceeds the sample, a permanent flow of secondary and stray electrons is available to partly neutralize the surface charge. In the resulting stationary state, the E_B-scale is shifted. This shift is accounted for by an internal standard: the C 1s line of "adventitious carbon" (the carbonaceous impurities usually present in UHV systems and adsorbing on samples in minor amounts). Despite occasional flaws, this works well for almost all kinds of samples; of course, except for those containing carbon as an ingredient or adsorbate. By setting the C 1s line of adventitious carbon to the value where it is found on metals, the E_B–scale is pinned to the one that is valid for Fermi levels aligned between the sample and the spectrometer.[29]

In work with monochromatized sources, which focus the radiation into mm-sized or even smaller spots, and with synchrotron radiation, stray electrons that could limit surface charging are missing. In these configurations, studying samples with low conductivity requires an extra piece of equipment – an electron flood gun. The C 1s reference is still used to make sure that the correct dose of low-energy electrons has been applied.

UPS is also obstructed by surface charging, and here, the flood gun cannot help because of the very low kinetic energy of the photoelectrons. In some cases, the required electrical conductivity may be achieved by heating samples during the measurement as long as this does not interfere with the vacuum conditions or cause sample damage [124].

3.4.5.3 Sources of analytical information

Binding energies, line intensities, special line shape features, and the combination of XPS with XAES are the major sources of diagnostic information in photoelectron spectra. Arising from the ionization of core levels, the lines typically studied in XPS reflect atomic properties. XPS can tell the oxidation state and the abundance of an atom in the near-surface region, but is often unable to reveal to which neighbor it is coordinated. Information on binding structures may be found in the valence region, which is better examined by UPS. Structural information may also be drawn from the response

---

29 The C 1s binding energy is set in the range between 284.5 and 285 eV, slightly differing between the groups. This reference must be communicated when results are presented.

of the system on the ionization of the core level – the final-state effects (see introduction to 3.4.4 and below), which are most pronounced in Auger transitions.

3.4.5.3.1 Binding energies

In photoemission, the absorbed quantum $h\nu$ transfers an atom (or ion) with a full configuration of n electrons into a state with a hole in the core level i and an electron moving in space with the kinetic energy E_{kin}. The energy balance for this is

$$E_{ini}(n) + h\nu = E_{fin}(n-1,i) + E_{kin} \tag{3.86}$$

Combining this with eq. (3.84) yields

$$E_B(i) = E_{fin}(n-1,i) - E_{ini}(n) - e\phi_{spec} \tag{3.87}$$

i.e., apart from the work function $e\phi_{spec}$, the binding energy is the energy difference between the total energy of the final state with a core hole and the initial state. Binding energies are difficult to calculate theoretically, because the final state is a highly excited state (for more information, see ref. [124]).

In practical work, binding energy shifts are discussed rather than the absolute E_B values. Such "chemical" shifts compare the binding energies of an element in the sample and in a reference state (often the elemental state). As the work function cancels by subtraction, they result from the differences in the initial-state and final-state energies of the element in the sample and in the reference.

Initial-state energies may change, because the orbital i emitting the photoelectron may have different energies ε_i in the sample and in the reference. Final-state effects comprise the stabilization energies for the core hole, which the atom raises during excitation, and secondary excitations (cf. introduction to 3.4.4). Stabilization is achieved by attracting electron shells of the same atom and electron density from the vicinity. The former, intra-atomic relaxation cancels by subtraction. Extra-atomic relaxation effects usually differ between the states to be compared and, hence, affect binding energies. As they include the polarization of orbitals in the adjacent atoms and electron shifts in the conduction band of metals or in delocalized bonding orbitals, they belong to the few sources for the sensitivity of photoemission to structural properties. They are more pronounced in Auger excitation, which creates a double-charged ion from a single-charged one (cf. Figure 3.90). Therefore, a combination of Auger lines with photoemission lines can be used to identify compounds of the emitting atom (cf. Auger parameter, Section 3.4.5.3.4). Secondary excitations cause satellite lines or line asymmetries rather than binding energy shifts. Their use is briefly discussed in Section 3.4.5.3.2.

Beyond orbital energies and final-state effects, there is yet another quantity affecting the kinetic energy of the photoelectron. This energy is measured outside the solid, but most photoelectrons actually originate from below the surface where electric fields may be quite pronounced in ionic lattices. The correction covering the Coulomb interaction of the photoelectrons with these fields is referred to as Madelung term (V_M), alluding to a traditional method for the evaluation of lattice energies. The contributions to E_B shifts are summarized in eq. (3.88)

$$\Delta E_B = \Delta \varepsilon_i + \Delta E_{rel} - \Delta V_M \tag{3.88}$$

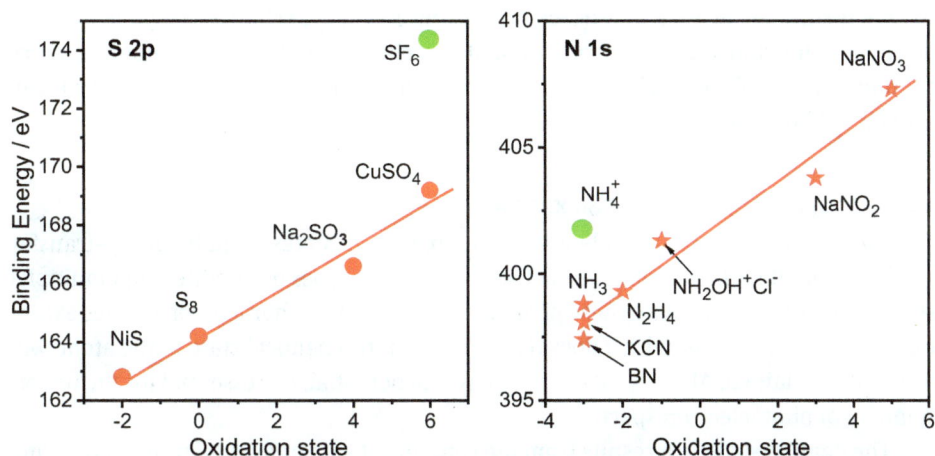

Figure 3.94: The relation between XPS binding energies and oxidation states: typical examples. Reproduced from ref. [124] with permission from Wiley-VCH Verlag GmbH & Co. KGaA, Weinheim, Germany.

Orbital energies change with the charge on the emitting atom. This change is the origin of the well-known rule, according to which the XPS binding energies increase by 0.8–1 eV per unit increase of the (formal) oxidation state. Illustrating this rule, Figure 3.94 also shows that there may be drastic exceptions. Deviations, as in the case of solid SF_6, are discussed in terms of a "ligand" effect, which operates most likely via all three summands of eq. (3.88). The high electronegativity of F atoms results in a higher charge on sulfur and a higher polarity in the crystal. The former influences the initial state (via the orbital energy), and the latter the final-state (via the polarizability) and the Madelung term (via the electrical fields in the crystal).

The discussion reveals a number of reasons for the frequent failures of the linearity rule. Only sometimes, such a failure can be assigned to just one of the influences summarized in eq. (3.88). The Ru^{IV} oxide RuO_2 may be such example: its $3d_{5/2}$ line peaks at 280.9 eV, while Al_2O_3-supported Ru^{IV} was found at 283.6 eV (see Figure 3.99 below). Most of the difference is probably due to the metallic conductivity of RuO_2, which allows for a very efficient core hole shielding in the final state. In zeolites, the binding energies of Si 2p, O 1s, and Na 1s exhibit characteristic trends with the Si/Al ratio, which can be well explained by changes of the Madelung potential due to the dependence of the framework polarity on the Si content [127].

In literature, binding energy shifts are often simply correlated with electron transfer from or toward the emitting atom. Due to the diversity of influences on E_B shifts (eq. (3.88), see also [124]), such conclusion, though often justified, is by no means safe and should be supported by parallel studies with other methods. Likewise, assignment of signals to particular oxidation states of the emitting atom is not always straightforward and should be backed by additional evidence wherever possible, e.g., satellite features (3.4.5.3.2) or Auger signals (3.4.5.3.4). For well-studied cases like O 1s of oxygen species on metal surfaces or C 1s of surface groups on polymers, reliable assignments were developed in extensive work that included correlations with data of complementary techniques. In applications in catalysis, such a basis is not always available, and conclusions should be drawn more cautiously.

3.4.5.3.2 Opportunities based on XPS line shapes

Line-shape effects that allow identifying the oxidation states include charge-transfer satellites and the related line asymmetries, some other types of satellites, and multiplet splitting. Before dealing with these phenomena, the doublet character of all lines except for signals from s levels, which does not depend on the chemical state of the atom, will be briefly explained. Although it has no analytical potential, it is essential for the understanding of photoelectron spectra.

The doublet structure results from the coupling of spin and angular moments. Photoemission leaves behind an electron, the spin s of which may be parallel or antiparallel to the angular momentum l. The resulting energies are identical for s-levels ($s + l = 1/2$ or -1/2), but different for p, d, f levels, etc. ($s + l = 1/2$ or 3/2; 3/2 or 5/2; 5/2 or 7/2, respectively) which gives rise to two lines. Both energy differences and intensity ratios are well-defined. The latter reflects the multiplicity of states, i.e. 2:4 for p, 4:6 for d levels, etc. In the analysis of superimposed signals by line fitting, both energy distances and intensity ratios must be kept constant unless these properties are modified by other effects, e.g. multiplet splitting (see below).

Charge transfer (CT) satellites arise from the excitations of valence-band electrons into empty cationic states, which are stabilized by the emerging core hole. By hybridization with the valence band, they form final states, which are populated with a probability, depending on their energy difference to the valence band and the degree of hybridization between them. Due to the energy required to populate the empty cation state, the satellites appear at higher binding energies.

Cation states involved may be empty d levels (first-row transition metals, e.g. Cu^{2+} with, Cu^+ without satellite, dramatically different satellite intensities for Co^{3+} and Co^{2+}) or f levels of lanthanides. The narrow spaced cationic levels in Ce^{4+} give rise to the only well-known example of two satellites per main line. As the satellites occur at both spin-orbit components and Ce^{3+} comes with its own satellite, Ce 3d spectra usually contain 10 components, which are not always correctly assigned and treated in literature (for more details – see ref. [128]). Arising from an interaction between the cation

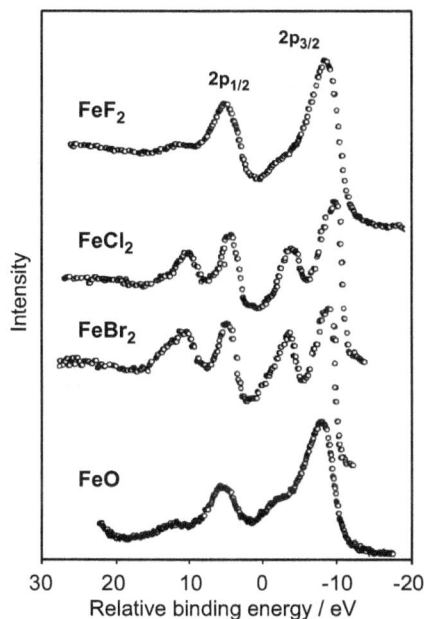

Figure 3.95: Influence of the counter ion on the cation satellite structure: Fe 2p of Fe(II) halides. Adapted with permission from Ref. [129]. Copyright 1992 The Physical Society of Japan.

states and the anion forming the valence band, CT satellites are among the few opportunities for structural sensitivity in XPS. This is nicely illustrated in Figure 3.95, where the Fe 2p CT satellites of Fe(II) halides are compared to that of FeO on a scale related to the center of gravity of the signal. The satellite intensity grows for larger anion sizes due to the increased spatial overlap between the cation and the anion states.

In metals, secondary excitations populate the free electron states right above the Fermi level, which results in a line asymmetry instead of a discrete satellite. The intensity of the effect depends on the density of states at the Fermi level. Therefore, d metals have much more asymmetric lines than metals with the Fermi level in an s or an sp band. Figure 3.99 below shows spectra with the 3d doublet of Ru supported on different oxides, which suggest that the asymmetry of lines from metal nanoparticles may also depend on the environment of the particles. The effect was ascribed to an electron transfer to or from the metal species [130], but a systematic investigation of the variability in line asymmetries is pending.

The unpaired electron remaining after photoemission also has a magnetic moment, which couples with that of unpaired valence electrons. In cations with unpaired electrons, this creates a complicated pattern of additional final states: the multiplet splitting. The effect is most intense between the adjacent shells, e.g. for 3s and 3p lines of first-row transition metal cations, which are rarely measured for intensity reasons. However, the influence of multiplet splitting on the most intense 2p lines is still strong enough to cause intensity ratios and E_B differences between the spin-orbit components ($2p_{3/2}$, $2p_{1/2}$) to differ significantly from the ideal values. For more detail on the analytical potential of multiplet splitting, the reader is referred to refs. [124, 131].

3.4.5.3.3 Surface sensitivity and quantitative analysis

For calculating the intensity of an XPS line, contributions originating from different sampling depths, z, and escaping from the solid with a depth-dependent probability, $\varphi(z)$ must be summarized. Disregarding a number of quantities such as incident flux, acceptance angle, etc., this can be expressed as:

$$I \propto \sigma\, S \int_0^{\infty} \rho(z)\ \varphi(z) dz \tag{3.89}$$

where $\rho(z)$ is the concentration of the emitting atom at the depth z, σ is its interaction cross section for the ionization of the level involved, and S is the instrumental sensitivity at the corresponding kinetic energy.

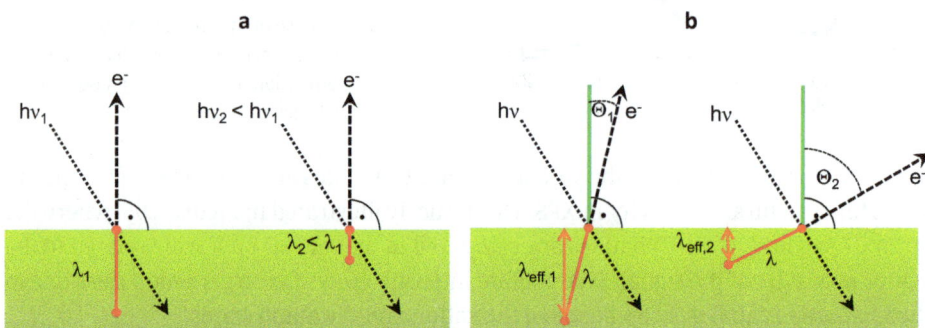

Figure 3.96: Opportunities for the variation (reduction) of the XPS sampling depth: (a) variation of the excitation energy; (b) variation of the angle between surface and analysis direction. Adapted from ref. [124].

The escape function is usually expressed in the exponential form of the Lambert-Beer law,

$$\varphi = \frac{I(z)}{I(z=0)} = exp\left(-{}^z\!/_{\lambda\, \cos\,\Theta}\right) \tag{3.90}$$

where Θ is the angle between surface normal and analyzer direction (see Figure 3.96b). $z/\cos\Theta$ takes into account that the way to escape from a depth, z is longer when the trajectory deviates from the surface normal. In usual spectrometer settings, the analyzer is in the direction of the sample normal (Figure 3.96a), and $\cos\Theta$ is 1.

In eq. (3.90), λ is the attenuation length. As electrons can change their direction (and thus miss the analyzer) also by elastic scattering, it is smaller than the inelastic mean free path, λ_{ie}, typically by 20–30%. However, owing to the better accessibility of λ_{ie} data, eq. (3.90) is usually applied with this quantity. In typical spectrometer geometries, the decay of intensity contributions with the depth, z is therefore,

$$I(z) = I(z = 0) \ exp\left(-\frac{z}{\lambda_{ie}}\right) \tag{3.91}$$

The depth $z = \lambda_{ie}$, from which the intensity decays by the factor e (to ca. 36.8%) is the average sampling depth, but as mentioned before, the signal also contains information from below: ca. 95% of the intensity originates from a depth of $3\ \lambda_{ie}$.

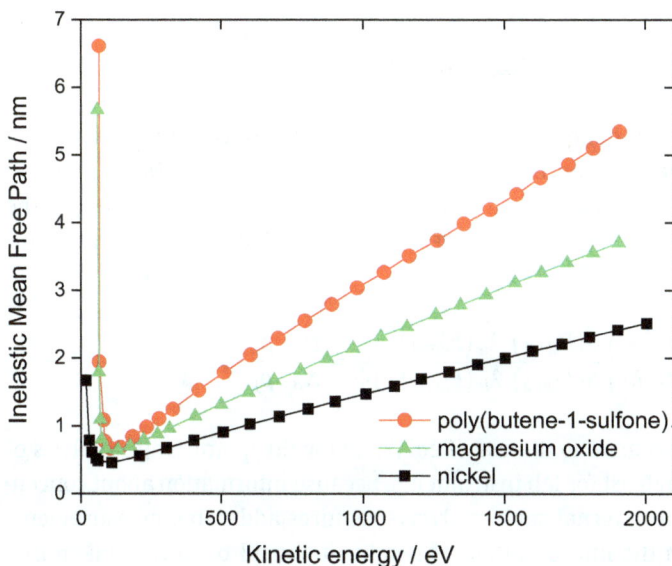

Figure 3.97: The relation between the inelastic mean free path and the kinetic energy of the photoelectrons, exemplified for three solids of different chemical nature. Adapted from ref. [132] with permission of IM publications open.

The dependence of λ_{ie} on the kinetic energy is well studied. Figure 3.97 exemplifies it for a metal, a polymer, and an oxide (see ref. [124] for more details). There is a minimum in the curves at $E_{kin} \approx 20$ eV, where $\lambda_{ie} \approx 0.5$ nm. In the kinetic energy range of typical laboratory spectrometers (200–1,500 eV), λ_{ie} scales with $(E_{kin})^n$, where $0.5 < n < 1$, which results in average sampling depths between 1 and 3 nm. In porous solids, sampling depths are higher according to their void fractions. The free choice of excitation energies at synchrotrons allows high surface sensitivity for all signals: even levels at $E_B \approx$ 100 eV, which emit electrons at 1,380 eV with an Al$K\alpha$ source, can be measured with escape depths of 1 nm or below with synchrotron radiation.

Unfortunately, knowing the decay function, $\varphi(z)$, is not sufficient for relating intensities to concentrations, because according to eq. (3.89), I depends also on how an element, i, is distributed along the depth coordinate ($\rho(z)$). To escape this dilemma, XPS intensity data are usually interpreted neglecting any concentration changes along z, i.e., assuming a homogeneous sampling region. With $\rho_i(z) =$ const. $= \rho_{av,i}$, the integral in eq. (3.89) becomes:

$$\int_0^\infty \rho(z)\ \phi(z)dz = \rho_{av,i} \int_0^\infty exp\left(-z/\lambda_{ie}\right) = \rho_{av,i}\ \lambda_{ie}$$

and

$$I_i \propto \sigma_i\ S(E_{kin,i})\ \rho_i\ \lambda_{ie}(E_{kin,i})$$

The parameters on the right-hand side (except for ρ_i), which are repeated in eq. (3.92) with their main dependences, can be lumped into sensitivity factors N_i. In practical XPS work, these are used to evaluate ratios of elemental concentrations, ρ_A/ρ_B in the near-surface region or, when all elements present are included, compositions of this space.

$$\frac{I_A}{I_B} = \frac{\sigma_A(hv)\ S(E_{kin,A})\ \lambda_{ie}(E_{kin,A})}{\sigma_B(hv)\ S(E_{kin,B})\ \lambda_{ie}(E_{kin,B})}\ \frac{\rho_A}{\rho_B} = \frac{N_A}{N_B}\ \frac{\rho_A}{\rho_B} \tag{3.92}$$

There is no practical alternative to the use of eq. (3.92) for the quantitative analysis of surface concentrations, at least for lab instruments, because information about concentration profiles below the external surface always requires additional measurements (see below). However, in the interpretation of results, it should be always taken into account that these concentrations are averaged over the sampling depth, λ_{ie}. The contribution of the external surface layer to the signal intensity is of the order of only 30%.

For intensity evaluation, peaks must be integrated over backgrounds, which are sometimes strongly inclined (cf. Figure 3.93). The most popular procedure for background determination takes into account the intensity at each kinetic energy in an iterative procedure (Shirley background), without, however, being universally applicable. Satellites belong to their main lines. While superposition of signals originating from different elements is rare in XPS, peak fitting to differentiate oxidation states of an element is standard. In such peak fitting, physical facts like E_B differences and intensity ratios between spin-orbit components must be observed and independent information (e.g., from spectra of the element in presumably uniform oxidation states) must be invested as much as possible to reduce ambiguities in the results. For more details about backgrounds and analysis of complex signals, the reader is referred to ref. [124].

The depth-averaged character of concentration data provided by XPS affects its potential for (electro)catalysis, which happens at the outermost atomic layer of solids. Surface sensitivity of XPS can be enhanced by reducing the average sampling depth. As mentioned above, this can be directly achieved by measuring with low excitation energies (and, hence, low E_{kin} and λ_{ie}) at a synchrotron (Figure 3.96a), or indirectly, by tilting the sample relative to the analyzer (Figure 3.96b). With growing angle Θ

between the surface normal and the analyzer direction, the photoelectron trajectory in the solid has an increasing component parallel to the surface. Therefore, the effective sampling depth decreases even in measurements with laboratory X-ray sources. In thin film technology, this Angle-resolved XPS (ARXPS) is a standard analysis technique. As its potential is strongly affected by surface roughness, it is, however, rarely used for catalysts. Instead, synchrotron XPS plays an ever-growing role in catalysis research.

3.4.5.3.4 Structural sensitivity by combining XPS and XAES

Probing the polarizability of the surrounding, extra-atomic relaxation effects, which are most pronounced in Auger transitions (see above), are of particular importance for the surface analysis of unknown samples. This analytical potential is used in the form of the Auger parameter, α, which is the sum of the binding energy and the kinetic energy of the most intense XPS line and Auger transition, respectively[30]:

$$\alpha = E_B(i) + E_{kin}(jkl) \tag{3.93}$$

Notably, due to the opposite orientation of the scales for E_B and E_{kin} (cf. Figure 3.93), the summation cancels any error introduced by the referencing of energy scales.

The particular influence of extra-atomic relaxation on the Auger parameter can be easily illustrated, starting from eq. (3.88). Specifying $\Delta E_{rel}(1)$ as the external relaxation energy for a singly charged final state, the E_B shift is:

$$\Delta E_B = \Delta \varepsilon_i + \Delta E_{rel}(1) - \Delta V_M(XPS)$$

The analogous equation for an Auger transition is:

$$\Delta E_{kin} = -\Delta \varepsilon_i - \Delta E_{rel}(2) + \Delta E_{rel}(1) - \Delta V_M(AES)$$

When a dielectric medium is polarized by a point charge q, the energy scales with q^2, i.e., $\Delta E_{rel}(2) \approx 4\,\Delta E_{rel}(1)$, and

$$\Delta E_{kin} = -\Delta \varepsilon_i - 3\,\Delta E_{rel}(1) - \Delta V_M(AES)$$

$$\Delta \alpha = \Delta E_B + \Delta E_{kin} \approx -2\,\Delta E_{rel}(1) \tag{3.94}$$

where differences in the Madelung contribution are neglected (cf. ref. [124]).

---

30 Strictly speaking, the Auger transition should result from the decay of just the core hole created by the excitation of the XPS line. These transitions are, however, often not sufficiently intense. Equation (3.93) actually describes the "modified" Auger parameter, which relies on the assumption that the relaxation effects are similar for all Auger transitions.

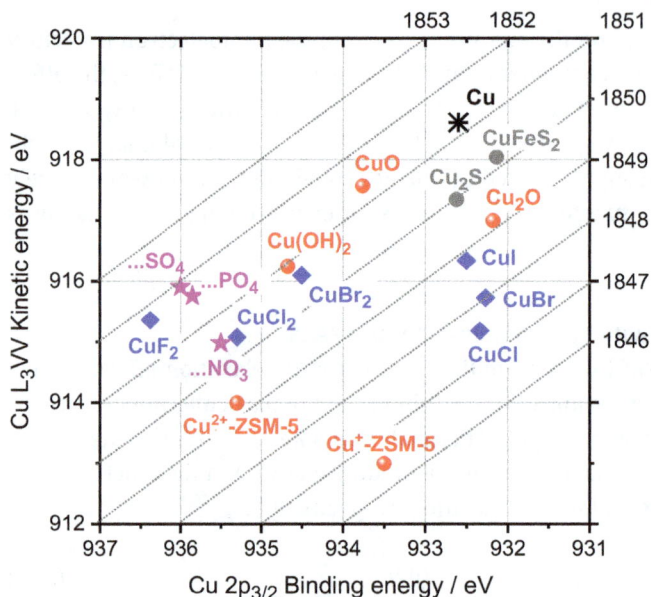

Figure 3.98: Chemical state plot for Cu compounds. Compiled with data from refs. [133, 32].

The diagnostic potential of the Auger parameter (3.93) for an element is sometimes summarized in chemical state plots [134], as exemplified in Figure 3.98 for copper. In the chemical state plot, the kinetic energy of the Auger transition is plotted vs. the binding energy of the related XPS signal. Lines indicating identical α cross the coordinates at an angle of 45°.

Figure 3.98 shows how the Auger signal extends the diagnostic potential of XPS. Cu metal, Cu_2O, CuI, CuBr, CuCl, and the ores Cu_2S and $CuFeS_2$, for instance, cannot be distinguished when only the Cu 2p signal is used. How well α differentiates these compounds depends on the resolution between the somewhat broader Auger signals. The wide spread of Auger parameters for Cu(II) compounds is, however, a remarkable opportunity for chemical differentiation.

Unfortunately, the application of the chemical state plot finds some limitation in the intensity of Auger lines. As mentioned above, Auger transitions compete with X-ray fluorescence, and Auger transitions tend to lose with increasing atomic numbers. While Figure 3.93 shows the Auger signals of silver to be well detectable, application of the Auger parameter for 3rd row transition metals has not been reported, as yet.

3.4.5.3.5 Ultraviolet photoelectron spectroscopy

In the early years of surface science, Ultraviolet Photoelectron Spectroscopy (UPS) was a powerful tool for studying highly ordered atomically flat surfaces of monocrystals, particularly when applied with angular resolution (ARUPS). Much of our knowledge on the spatial extension of electron bands in metals originates from this work. UPS with polarized light can also elucidate the orientation of adsorbates to the surface.

Application of UPS to catalysts is complicated by its failure with insulators. With conducting samples (e.g., semiconductors or some carbon materials), the method is attractive due to its high surface sensitivity, e.g., for the detection of defects near the Fermi level. The method can also be employed to determine the work function of samples (for details see ref. [124]). As UPS shows the structure of the valence band region, interpretation beyond the fingerprint level requires support by theoretical calculations. For the same reason, an excessive complexity of samples undermines the perspectives for a conclusive interpretation of the results.

3.4.5.4 XPS: examples for the interpretation of spectra

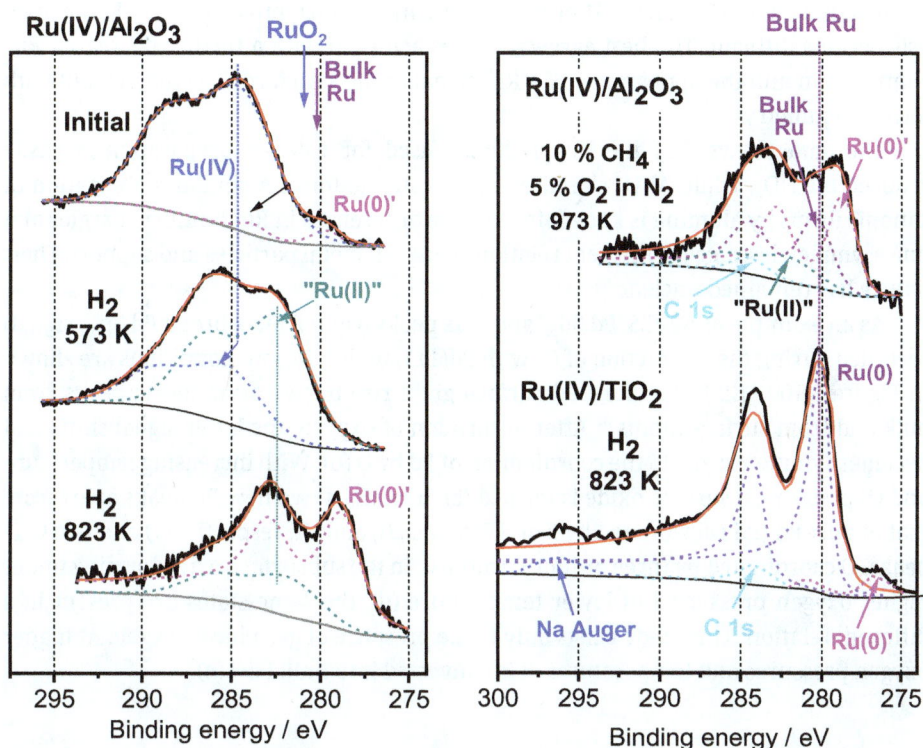

Figure 3.99: Ru 3d spectra of supported Ru catalysts after different treatments.

To exemplify a typical situation encountered in conventional XPS with real catalysts, Figure 3.99 shows Ru 3d spectra of two supported Ru samples after different treatments.

The work was part of a study on the partial oxidation (POX) of methane to synthesis gas (see Appendix A1).

The comparison of spectra after reduction in H_2 at 823 K reveals a strong influence of the supports. On TiO_2, a highly asymmetric doublet of Ru^0 appeared already after less severe reduction [130]; on Al_2O_3, the lines remained much broader, and a shoulder around 287 eV suggests the presence of residual Ru ions.

On TiO_2, the Ru $3d_{5/2}$ binding energy was ≈280 eV, similar to bulk Ru, but there is also a minor signal below 280 eV.[31] The huge asymmetry of the Ru 3d lines and the course of the baseline approaching the signal very gradually may appear arbitrary, but during the analysis, this turned out to be the only way to comply with the standard intensity ratio of 4 : 6 between $3d_{3/2}$ and $3d_{5/2}$ [130]. On Al_2O_3, the Ru^0 signal appeared clearly below 280 eV, and it was less asymmetric than the signals on the TiO_2 support. Although made with reference to the literature, the assignment of the cationic state as Ru^{II} should be considered tentative [130]. After treatment with a CH_4/O_2 mixture at 973 K, the analysis became difficult. The best agreement was achieved with a further decreased Ru^{II} contribution and the appearance of a Ru^0 signal at the standard E_B (>280 eV), but with lower asymmetry.

The data shows that the Ru precursor used for catalyst preparation is easily reduced on TiO_2, while Al_2O_3 stabilizes Ru in cationic form. A similar stabilization of cationic forms by alumina is known for Rh. The differences in Ru^0 binding energies and line asymmetries suggest different relations between metal particles and support. Their nature has remained unclear.

As an example of APPES, Pd $3d_{5/2}$ and O 1s peaks (right from a broad Pd $3p_{3/2}$ signal) recorded during the interaction of O_2 with Pd(111) under various conditions are shown in Figure 3.100 [125]. The pure Pd surface gives rise to two asymmetric peaks from bulk and from surface atoms.[32] After adsorption of oxygen, the latter signal shifts and becomes symmetric due to the coordination of Pd by O (b). With increasing temperature and O_2 pressure, a surface oxide is formed (c). It exhibits two new Pd peaks apart from that of bulk Pd (Pd coordinated by 2 or 4 O atoms, II, and III, respectively) and two O 1s peaks (O coordinated by three or four Pd atoms). In the subsurface oxide forming under higher oxygen pressure, but lower temperature (d), the same states are present in a different relation. This oxide exists only in the presence of gas-phase oxygen. At higher oxygen pressures and temperatures, it is converted into bulk PdO (e).

---

31 Ru 3d superimposes the C 1s signal usually employed as an energy reference. These spectra were calibrated with secondary references: the main lines of the supports employed – Ti 2p and Al 2p. The Na Auger signal is probably from a contamination.
32 Due to the upshift of the d-band center at the external surface (cf. Section 4.2.2) the surface atoms of late transition metals cause signals at lower E_B values (surface core level shift).

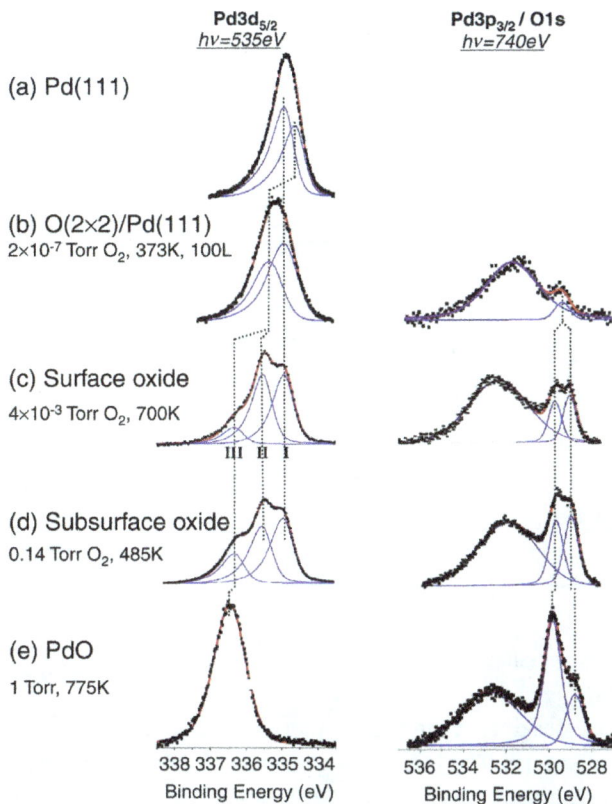

Figure 3.100: APPES: Adsorption and reaction of oxygen with a Pd(111) surface characterized by the Pd 3d$_{5/2}$ and the O 1s lines, the latter superimposed by the broad Pd 3p$_{3/2}$ signal. Reproduced from ref. [125] with permission from Elsevier.

3.4.6 Surface analysis with ions: Secondary ion mass spectrometry (SIMS) and Low-energy ion scattering (LEIS)

3.4.6.1 Interactions between low-energy ions and solids

The methods discussed in this section probe surfaces with ions, i.e., matter packages, the particulate aspect of which dominates their interaction with the environment. Their behavior is described by classical mechanics. The ion charge is also irrelevant for their trajectories; ions are used instead of atoms just because they are easier to accelerate and to detect. Collisions are, however, accompanied by electronic interactions, which are briefly summarized below.

Figure 3.101 illustrates the mechanical processes that happen when low-energy ions interact with solid surfaces. They may be scattered by surface atoms (as shown in the figure) or by atoms below. In these collisions, the projectiles transfer energy to the collision partner (s. eq. (3.96) below), and both partners move apart in different direc-

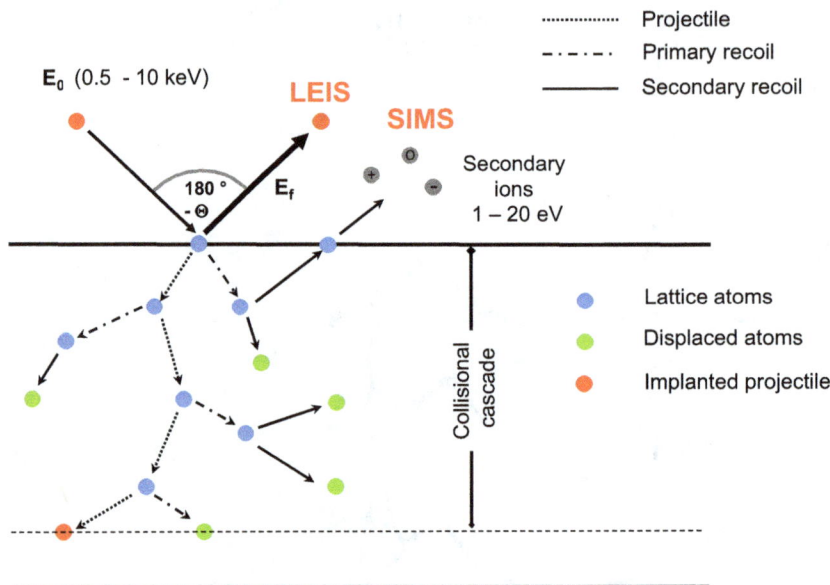

Figure 3.101: Interactions of low-energy ions with solids: the collisional cascade and the related techniques, Secondary ion mass spectroscopy (SIMS) and Low-energy ion scattering (LEIS).

tions. Due to neutralization processes during the collision, projectiles may return from the surface as ions or as neutral species. The ions among them are registered in LEIS, which is also referred to as Ion scattering spectroscopy (ISS). Other projectiles proceed into the solid where they go on colliding with atoms, displacing them and inducing secondary collisions with related displacements. Thus, the energy of an incident projectile is dissipated in a collision cascade resulting in a large number of displaced atoms, until they end up as an implant. Projectiles first scattered below the surface may also find a way out of the solid, but due to highly efficient neutralization processes, they come as neutrals. At lower kinetic energies, some re-ionization may occur.

Branches of the collision cascade return to the external surface (Figure 3.101), where they cause the desorption of monoatomic and clustered species, both charged and neutral. These surface fragments, which have energies of just 1–20 eV, can be studied by mass spectrometry. The corresponding methods are Secondary ion mass spectrometry (SIMS) or, with a supplementary ionization of the neutrals, Secondary neutrals mass spectrometry (SNMS).

While the energy transfer between projectiles and sample atoms is determined by collision mechanics, the approach of involved electron states down to extremely small distances induces a variety of electronic processes, which result in the large extent of projectile neutralization, additionally in the emission of electrons and X-rays, and even in the re-ionization of neutrals. Repulsion by the extended states of the solid, for instance, destabilizes the core levels of the projectile, e.g., 1s in He atoms approaching Al to <3 Å [135]. Below 1 Å, the upshifted 1s level becomes resonant with the Al conduction band, which allows an electron transfer resulting in a collision-induced neutralization of He ions: $He^+ + e^-_{sample} \rightarrow He^0$. Already at larger distances, excited projectile levels (e.g., He 2s) resonate with extended sample states. Such interaction results in excited states, which decay by (Auger) emission of an electron from projectile or sample, or by fluorescence. More detail about these processes and about the most complex situation during the detachment of secondary ions or neutrals from the surface can be found in ref. [135].

Noble gas ions are neutralized to >99.9% per collision. This is the reason for the extreme surface sensitivity of LEIS, which probes the outermost surface layer of samples exclusively. Contributions of projectiles first scattered below the surface are suppressed by energy losses in consecutive collisions or, if they manage to directly leave the solid, by neutralization at larger distances (see above) in the outermost layer. Regarding SIMS, its high surface sensitivity is obvious from the mechanism illustrated in Figure 3.101: the sputtered atoms or clusters result from the tongues of collision cascades. The majority of desorbed species originates from a depth of 3–6 Å [136].

Their exclusive surface sensitivity should open plenty of opportunities to both methods in catalysis research. Regarding SIMS, serious problems in the quantitative use of spectral intensities (see below) are probably a major reason behind its rather rare use in this field. In the absence of such annoyance, the still limited application of LEIS in catalyst characterization may arise from a lack of up-to-date information on the method. The incompatibility of low-energy ions with *in situ* conditions however, remains a disadvantage of both techniques. The following discussion includes a brief summary of SIMS as well, because the method is highly important in other fields like semiconductor and thin film technology, and its basics illustrate relevant features of ion-surface interactions.

3.4.6.2 Secondary ion mass spectrometry (SIMS)

A SIMS instrument consists of a UHV chamber fitted with an ion gun, with sample handling facilities, a secondary ion extraction lens, and a mass filter. For work with insulating samples, a flood gun (cf. Section 3.4.5) is required. Ion guns consist of a source and optical elements for accelerating, focusing, and rastering of the ions. While the most popular projectiles are cations (Ar^+, O_2^+, Cs^+), anions can be used as well. The mass filter may be a magnetic sector field analyzer, a quadrupole filter, or the most powerful time-of-flight (TOF) analyzer. The latter requires a pulsed primary ion flux, which is achieved by placing a chopper into the primary ion beam. TOF-SIMS allows recording all secondary cations or anions simultaneously over a wide range of masses and with a mass resolution of $m/\Delta m > 1000$ [135].

Due to the extreme sensitivity of mass analyzers, SIMS is very sensitive, as well. Although the ionization probability, α (the ratio between ions and neutrals of a species type) is only of the order of 10^{-3} on average, the detection limit can be as low as 10^{-5} of a monolayer [135]. The ionization probabilities can be influenced by the choice of projectile.

SIMS is, of course, destructive, but its extremely high detection sensitivity allows keeping the projectile dose very low. Under such "static SIMS" conditions, only 0.1–1% of the surface is affected during the measurement. Opposed to that, dynamic SIMS, with ion doses (per cm^2) in the mA instead of nA range, aims at studying differences between the surface and deeper layers of the material. Conclusions can be drawn from changing intensity distributions between signals of monoatomic species, but also of more complex secondary ions. Such cluster ions are the most attractive feature of SIMS, because they arise from structural elements of the surface, which are desorbed as such or result from recombination right at the surface. Recombination of arbitrary sputtered fragments is negligible.

Intensities of SIMS signals (here, for a monoatomic species M) depend on a number of parameters:

$$I_M^{\pm} \propto I_0 \ Y \ \rho_M \ \alpha_M^{\pm} \tag{3.95}$$

Equation (3.95) contains quantities that describe the yield of fragments in any sputter process: the flux of primary ions I_0, the sputter yield Y (the number of M atoms released per projectile impact), and the concentration of M, ρ_M. The ionization probability, $\alpha_M^{\pm}$, converts this into the yield of detectable ions. The proportionality covers some other quantities, e.g., instrumental factors. Sputter yields depend on the nature of the material. While the processes are well understood for metals where simple sputter yield calculators are available, other classes of materials like oxides are subjects of ongoing research [137]. Major problems arise from the ionization probability, $\alpha_M^{\pm}$. Its variation with the nature of the element, with its chemical state, with its environment ("matrix effect"), and with the nature of the projectile may cover 3–4 orders of magnitude [135] and is not well understood. Therefore, discussion of SIMS intensities usually remains on a qualitative level, although semiquantitative analysis of surface concentrations can be achieved with suitable standard materials of known surface composition. SIMS applications relevant for surface catalysis, some of which are described in ref. [135], mostly deal with idealized model systems, although work with real catalysts has been also reported.

3.4.6.3 Low-energy ion scattering (LEIS)

In LEIS (or ISS), the energies of ions deflected from their original direction by the scattering angle, Θ (cf. Figure 3.101) are recorded. For an element M, this energy depends on Θ and on the mass ratio A between target atom (M) and projectile p ($A = M_M/M_p$).[33] As shown in ref. [138], the energy ratio between outgoing and incoming projectiles, E_f/E_0, can be derived from combining the energy and impulse balances during the collision. For the most relevant case of A > 1, this ratio is:

---

33 As the interaction time during the collision is far below characteristic times of bond vibrations, the environment of the target atoms, M, is irrelevant.

$$\frac{E_f}{E_0} = \left(\frac{\cos\Theta + \sqrt{(A^2 - \sin^2\Theta)}}{A + 1} \right)^2 \tag{3.96}$$

In spectrometers, the energy E_f of the scattered ions depends only on the mass ratio A, because the angle $1 - \Theta$ between incoming and scattered projectiles is usually constant. Therefore, LEIS is analogous to a mass spectrometric analysis of the surface composition with complete surface sensitivity. The high analytical sensitivity of mass spectroscopy is, however, lost by the use of projectiles for probing the surface. Its blindness to oxidation states is a real disadvantage of LEIS.

LEIS requires a UHV chamber fitted with an ion gun, with sample handling facilities, an electron flood gun to neutralize surface charging on nonconducting samples, an energy analyzer, and a detector. As mentioned in Section 3.4.5.2, these components are compatible with the setup of photoelectron spectrometers, and LEIS is, indeed, marketed as an option in surface analysis equipment. Measurement geometry and analyzer are, of course, optimized for photoemission in these instruments. They definitely allow recording meaningful LEIS data, but cannot tap the full potential of the method. With optimized measurement geometry and design of the (electrostatic or TOF) analyzer, signal intensities can be increased (or ion doses for acceptable intensities can be decreased) by orders of magnitude [135]. In addition, resolution between elements of similar atomic number can be significantly improved.

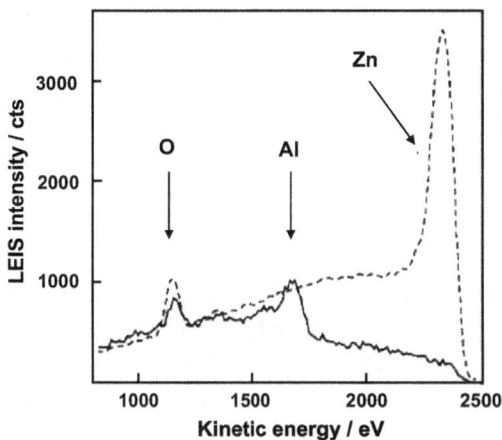

Figure 3.102: LEIS spectra of ZnO (dotted) and ZnAl$_2$O$_4$, measured with He ions. Reproduced from ref. [139] with permission of Elsevier.

Figure 3.102 shows LEIS spectra of ZnO and ZnAl$_2$O$_4$ measured with 3 keV He$^+$ ions. As the energy transfer to the target decreases with increasing target mass, signals of heavier elements appear at higher kinetic energies. The figure illustrates nicely the uneven distribution of elements on the kinetic energy scale (cf. eq. (3.96)): while more than 1,000 eV are available for elements below oxygen, the whole periodic system beyond Zn is crowded on ≈700 eV above 2,300 eV, because the incremental changes of the mass ratio, A, become minor at high target masses. The resolution between heavy elements can be dramatically improved, though at the expense of resolution between light elements, by using heavier projectiles, e.g., Ne^+ (preferably $^{20}Ne^+$) or Ar^+.

In Figure 3.102, the Zn signal is large in the spectrum of ZnO, but missing in $ZnAl_2O_4$: there is no Zn at the external surface of the aluminate. At the same time, there is intensity still beyond the Al signal. It decreases and disappears just where the Zn signal is expected. LEIS signals often exhibit a pronounced tailing to lower kinetic energies (cf. Zn peak of ZnO), mostly due to projectiles reionized after (multiple) scattering and neutralization below the surface. In the case of $ZnAl_2O_4$, the intensity below the energy of the missing Zn signal has the same origin and indicates that there is Zn in the bulk until just below the surface, though not exposed.

The spectra in Figure 3.102 are from a dedicated instrument, where low ion doses mitigate problems with surface charging and sputtering. These instruments allow measuring under static conditions (see Section 3.4.6.2), and surface charging is easily removed by a flood gun. In the add-on configuration, spectra of the same samples would differ in characteristic details due to the higher ion doses applied: they would come with broader lines, which might appear on a higher, gradually decreasing background, and the spectrum of $ZnAl_2O_4$ might show a minor signal also of Zn.

The increased line width mostly results from suboptimal neutralization of the surface charge. Due to the permanent impact of cations, the surfaces of insulating materials become positively charged similar as in XPS (cf. Section 3.4.5.2). At high ion doses, the flood gun may fail to achieve a uniform charging state of the surface. Therefore, add-on instruments, though well suited for a wide range of applications, cannot achieve a performance that would allow differentiating neighbors in the periodic system, e.g., Pd and Ag, as achieved with Ar^+ ions under optimized conditions, according to ref. [135].

Differential charging also broadens a feature common to any LEIS measurement: the huge intensities caused by sputtered secondary cations (cf. Section 3.4.6.2), which appear at low kinetic energies.[34] At inadequate charge neutralization, this signal may tail off over hundreds of eV, creating a decreasing background for the spectra (cf. Figures 3.103, 3.104). High ion doses emphasize the destructive character of LEIS, which is then performed in dynamic rather than in static mode. In an analysis of $ZnAl_2O_4$ (cf. Figure 3.102) under dynamic conditions, a sizeable damage may have been inflicted to the external Al/O layer before the scan arrives at the kinetic energy of the Zn signal – even though the ion beam is usually rastered across the surface during the measure-

---

34 An energy range usually omitted in the experimental settings.

ment. This sample damage problem may be handled by observing the spectra over a number of scans and extrapolating the resulting data (e.g., Zn/Al intensity ratios) to the start of the experiment (cf. Figure 3.103).

LEIS was long suspected to resist quantitative work in a similar way as SIMS, in particular via matrix influences on the probability of projectile neutralization. There is, however, no reason to expect such analogy due to the very different nature of the outgoing species (low-energy surface fragments at the tongue of collision cascades vs. ions after a single collision, with orders of magnitude higher kinetic energies).

By analogy to eq. (3.95), the intensity of an LEIS signal can be expressed as:

$$I_M \propto I_0\ \rho_M\ R\ P_n\ \frac{d\sigma_M}{d\Omega} \tag{3.97}$$

In eq. (3.97), $d\sigma_M/d\Omega$ is the differential cross section for scattering the projectile ion into a solid angle $d\Omega$ (for I_0 and ρ_M, see eq. (3.95)). R is a factor accounting for surface roughness and P_n is the neutralization probability. The proportionality covers instrumental factors.

While scattering cross sections are available from literature [135], the application of (3.97) is complicated by the influences of roughness and neutralization. The influence of roughness may be appreciated by considering an atom on the base of a narrow pit. It will be found only by projectiles impinging parallel to the pit walls, which will, on the other hand, defy the escape of most projectiles after the collision. The response of flat surfaces depends on their orientation toward projectile flux and detector. On rough surfaces with random (and unknown) orientations of surface facets, superposition of these influences results in signal reductions by 10–40% on average, except for surfaces with abundant pit structures [135].

Based on the knowledge on mechanisms of neutralization and reionization (see above) and on a rich experimental experience, it has been meanwhile concluded that matrix influences on the neutralization efficiency of target atoms should be rare and confined to typical exceptions, e.g., surfaces with very low work functions [135]. Some older claims for matrix effects were ascribed to experimental flaws, e.g., the presence of hydrogen. H atoms, which cannot be detected by LEIS, mask the atoms they are bound to. They can, however, be removed by very mild sputtering, after which the masked atoms become detectable. Still, generalized sensitivity factors for the most popular projectiles/target pairs, which might be used by a wide community of users, are not available as yet, probably because of large differences in instrumental factors. Instead, quantitative analysis is typically performed using standards, i.e., by comparing experimental intensities with intensities delivered by a standard sample of known surface concentration for each element present, under exactly the same measurement conditions.[35]

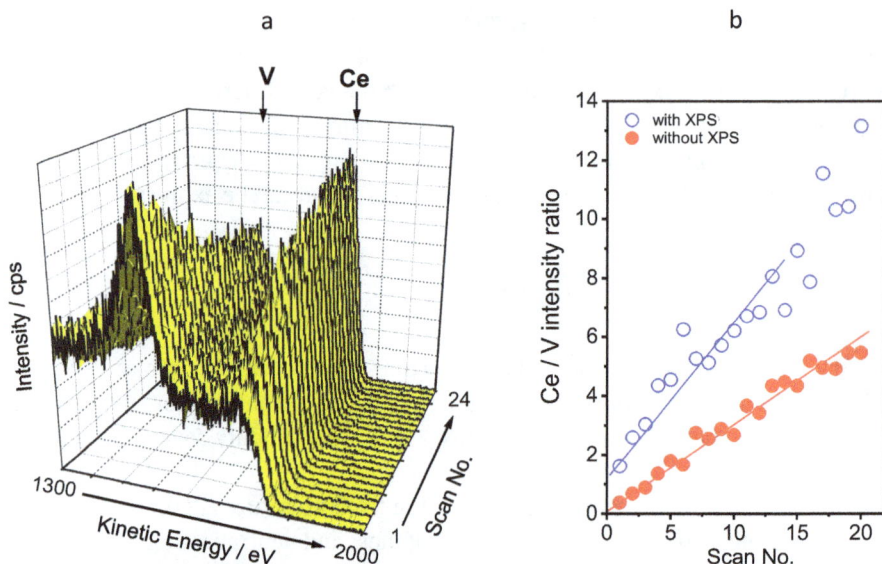

Figure 3.103: Dynamic LEIS with a V_2O_5/CeO_2 catalyst: (a) sputter series, uncovering the CeO_2 support; (b) development of the Ce/V intensity ratio along the sputter series of a sample previously studied by XPS (cf. panel (a)) and for a fresh sample. Reproduced with permission from ref. [140]. Copyright 2004 American Chemical Society.

Some examples for the application of LEIS to (electro)catalysts are shown in Figures 3.103 and 3.104. Figure 3.103 is from a study with supported V oxide catalysts, which aimed at finding out if an increase of the vanadium loading results in a completely covered support before three-dimensional structures are formed, or if parts of the support remain exposed. Only LEIS has the surface sensitivity to decide this question. As suggested by the decreasing baseline in Figure 3.103, the study was performed with an older add-on instrument in dynamic mode, following the approach to extrapolate the results of sputter series to the start of the experiments.

Figure 3.103a shows a series of spectra of a well calcined V_2O_5/CeO_2 catalyst, the V loading of which (≈ 22 V nm^{-2}) was far above the theoretical monolayer capacity of ≈ 8 V nm^{-2}. The spectra were measured with 2 keV He ions subsequent to XPS measurements of the sample. The V signal can be seen to decrease with ongoing sputtering, while the Ce peak grows dramatically. However, the Ce signal was present already in

---

35 Standards are typically metal surfaces, for O, Cl etc., also oxides and other compounds. When sample and standard are known or likely to differ significantly in surface roughness, results may be corrected for this influence.

the first scan, and it did not disappear even upon extrapolation of the trend to "Scan 0" (Figure 3.103b, open symbols). While this seems to confirm models according to which the supported oxide grows into the third dimension before the monolayer is closed, it was also observed that there is an influence of the XPS measurement on the subsequent LEIS analysis. In Figure 3.103b, the results of a run with a fresh catalyst are plotted as well (full symbols), and they confirm that the monolayer was, indeed, closed initially.

On the whole, the study showed that the surface V oxide monolayer is closed before the V oxide starts to extend into the third dimension.[36] As mentioned in Section 2.3.3, this refers to well-prepared catalysts in the calcined state.

In a later LEIS study along the same methodology, it was shown that the surfaces of many crystalline bulk vanadates and molybdates do not expose their cation, but only surface V (or Mo) oxide species [141]. This shows that the surface of these compounds cannot be derived by truncating the bulk lattice in directions of interest, because such surfaces may be reconstructed as known from metal surfaces (cf. section 2.2.1).

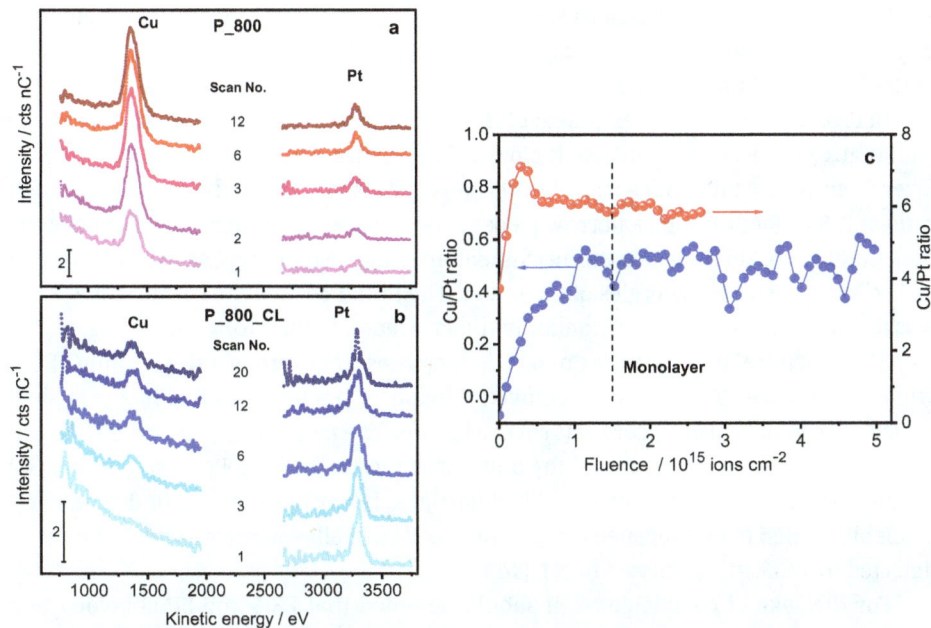

Figure 3.104: Dynamic LEIS: study of a carbon-supported Pt-Cu alloy catalyst in the initial state and after chemical leaching in H_2SO_4: (a, b) LEIS spectra (5 keV Ne); (c) development of the Cu/Pt atomic ratio deduced from the spectra using reference measurements with massive Cu and Pt samples. Reproduced from ref. [142] with permission from Wiley-VCH Verlag GmbH & Co. KGaA, Weinheim, Germany

---

36 The impact of XPS originated most likely from radiation damage causing some reduction of V^V to V^{IV} with concomitant breakup of V-O-support bridges.

The spectra displayed in Figure 3.104 were measured in a project dealing with the particle architecture in carbon-supported Pt-Cu alloy catalysts for the oxygen reduction reaction (ORR), which is the most critical step in the use of low-temperature fuel cells (cf. Section 4.9). The catalyst was prepared via a route combining co-impregnation of the support with the nitrates and their reduction with ethanol, followed by treatment in dilute H_2 at 1,073 K to obtain alloyed particles. Chemical and electrochemical leaching procedures described in literature to result in significant improvement of the ORR activity were performed with this precursor (P_800), which contained ≈44 wt-% of metal with a Cu/Pt atomic ratio of 2.6.

Figure 3.104 compares LEIS results obtained before and after chemical leaching of this precursor in H_2SO_4 at 353 K for 36 h. The leached material still contained 33 wt.-% of metal, with a Cu/Pt ratio of 0.33. Measurements with 5 keV Ne ions were performed using a dedicated instrument, but again in dynamic mode, as changes of the alloy composition below the surface were also of interest. Surface compositions were determined using Pt and Cu foils as standards. Figure 3.104a shows the spectra measured with P_800: both the Cu and Pt signals were present from the first scan on and grew with ongoing sputter time. After chemical leaching (P_800_CL, Figure 3.104b), Cu was absent in the first scan, but appeared while the surface was sputtered.

In Figure 3.104c, the development of the Cu/Pt atomic ratios with increasing fluence (cumulative ion dose per unit area) is plotted. The fluence required to remove one monolayer from the metal surface (ca. $1.5 \cdot 10^{15}$ ions/cm^2) is marked in the figure. The Cu/Pt ratio of P_800 goes through a narrow peak at low fluence and stabilizes at ≈6, above the bulk Cu/Pt ratio. Such surface Cu enrichment after thermal processing can be expected on the basis of surface energies of Cu and Pt. The initial peak of the surface Cu/Pt ratio was ascribed to the removal of contaminations, predominantly from Cu.

After chemical leaching, the Cu/Pt ratio increases from zero to a stable value of 0.5, while a monolayer of material is removed. Obviously, leaching resulted in a core-shell structure in which an alloy core was covered by just one monolayer of pure Pt. The deviation between the composition of the alloy core and the bulk Cu/Pt ratio (0.5 *vs.* 0.33) cannot be explained on the basis of LEIS data alone. Likely reasons for this discrepancy, probably related to heterogeneity in the composition of alloy particles, which cannot be detected by LEIS, are discussed in ref. [142].

For the sake of completeness, it should be noted that LEIS and SIMS belong to a large group of analytical methods working with ion bombardment, most of which are without direct benefit for catalysis. The sputtering effect caused by such bombardment is, for instance, used to analyze material compositions down to micrometers below the surface. The basics of these sputtering techniques are discussed in refs. [143, 144]. Increasing the energy of light projectiles, and thus, their operating distance in the solid can be used to analyze buried structures without extensive sputtering: in Rutherford backscattering spectrometry, the scattering of high-energy H^+ or He^+ (with energy transfer according to eq. (3.96)) can be tracked for targets that are micrometers below the surface. Their distance from the surface can be determined from the energy losses along the projectile trajectory to the target and back to the detector. For an introduction to the method, the reader may consult ref. [145].

3.4.7 Spectroscopy of electronic transitions: absorption of UV and visible light

This chapter deals with spectroscopy based the absorption of photons, not to eject electrons, but to raise them to an empty bound state. The wavelengths used range from ≈200 nm to ≈900 nm, covering UV (except for vacuum UV), visible light[37], and part of the near-IR range (NIR) . They can be extended far into the NIR by optional NIR analyzers. Referring to the electron levels involved, UV-Vis spectroscopy is sometimes designated as "electron spectroscopy." The NIR, where low-energy electronic transitions, as also overtones and combination modes of vibrational transitions appear, can be accessed with combined UV-Vis-NIR spectrometers and (in part) with IR spectrometers.

UV-Vis spectra are most frequently presented on a wavelength scale, but spectra plotted vs. the energy or on a wavenumber scale, which is more familiar in vibrational spectroscopy, can also be found. The wavenumber, $\bar{v}$, typically reported in cm^{-1}, is:

$$\bar{v} = 1/\lambda \tag{3.98}$$

It is closely related to the wave energy:

$$E = h\,v = h\,c/\lambda = h\,c\,\bar{v} \tag{3.99}$$

where c is the velocity of light (in vacuum – $3 \cdot 10^8$ m s^{-1}), which can be expressed by the wave properties as $c = v\lambda$.

A typical wavelength spectrum ranging from 200 to 800 nm extends between 50,000 and 12,500 cm^{-1} on the wavenumber scale. The latter is preferable for studies focused on the UV range, because the UV range from 200 to nearly 400 nm (25,000 cm^{-1}) covers two-thirds of the wavenumber abscissa.

3.4.7.1 Electron transitions and their diagnostic potential

Electron transitions in the UV and visible regions take place between orbitals in molecules or complexes. In solids, they can occur between bands, between bands and atomic (cation) states, and between cation states. In addition, photons in this range can excite surface plasmons (collective vibrations of the free electron plasma relative to the atomic core) in metals.

---

37 UV – 100–380 nm, visible – 380–780 nm, NIR – 780–3,000 nm (differentiation between ranges differs between sources).

Transitions in molecular species are often studied in the liquid phase to examine processes in precursor solutions. They can be also used for the investigation of adsorbates and of carbonaceous deposits on surfaces. Electrons excited in molecules may originate from occupied bonding (σ, π) and nonbonding (n) orbitals and end up in unoccupied orbitals, e.g., π^* or σ^*. The final states of such HOMO-LUMO transitions may be further differentiated by spin orientation (singlet, triplet), or symmetry, etc. π-π^* transitions in molecules or structures with conjugated double bonds are typical examples for this type of signal. Both wavelength and absorption coefficients of these transitions increase with the extension of the π-system. Signals in the UV-Vis range can be also expected from molecules with functional groups containing double bonds, e.g., –C=O, –N=O, or with incomplete valence shells. Band positions of organic molecules and related absorption coefficients can be found in standard works [146]. In the liquid phase, the lines are broadened due to poorly resolved rotational and vibrational fine structure, which can be observed in the gas phase.

Absorption coefficients of electron transitions can vary over ten orders of magnitude depending on selection rules (change of parity according to the Laporte rule, conservation of total spin). The impact of chemisorption on molecular symmetry may induce major intensity changes: transitions may become less prohibited, thanks to lower symmetry. As a result, intensity distributions between signals characteristic of a molecule may change dramatically, while band positions are less affected as long as the molecule is adsorbed unchanged. As in the liquid phase, lines are broad due to rotational and vibrational fine structure, which is resolved only in favorable cases.

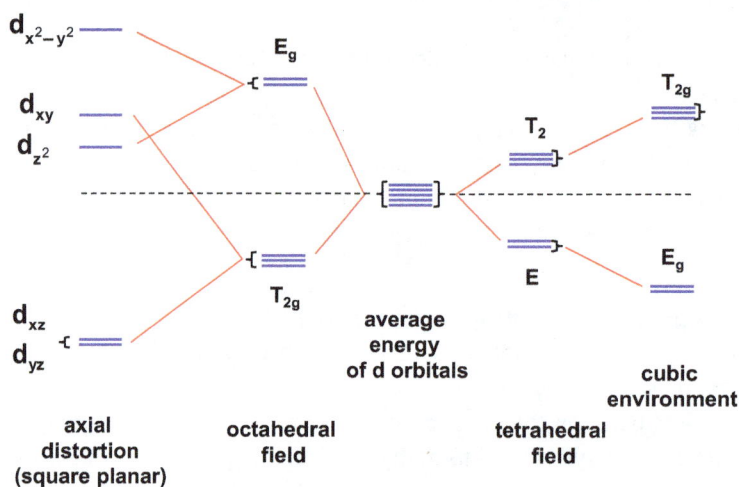

Figure 3.105: Splitting of degenerate d orbitals in ligand fields (d^1 case).

In ions coordinated by ligands, the energies of levels involved in HOMO-LUMO transitions are modified by ligand influences. Such influences can be described by ligand field theory, which is discussed in detail in advanced textbooks [147] or specialist texts [148].

An octahedral arrangement of negative ligands around a d^1 cation, for instance, results in a stabilization of the d_{xy}, d_{xz}, and d_{yz} levels, with concomitant destabilization d_{z^2} of and $d_{x^2-y^2}$ (Figure 3.105). An axial elongation toward square planar geometry differentiates both groups. The stabilization of d_{xz} and d_{yz} relative to d_{xy}, and of d_{z^2} relative to $d_{x^2-y^2}$ may give rise to three different electron transitions. On the other hand, tetrahedral and cubic ligand fields stabilize d_{z^2} and $d_{x^2-y^2}$ and destabilize d_{xy}, d_{xz}, and d_{yz}. f levels degenerated in free f^1 ions are split in ligand fields in a similar way. The extent of splitting depends on the charge on the cation and on the nature of the ligands, which are ranked according to their ligand strength in the spectrochemical series.

The situation becomes much more complex in multielectron systems, which develop nondegenerate states already in the free ion due to Russell-Saunders or j-j coupling. The interaction of these terms with the ligand field and the development of splitting dependent on field strength are summarized in correlation diagrams, which allow determining the nature of the ground state and the empty states accessible by electron transfer. When the nature of the transition(s) can be clarified, their number and their energies are of high diagnostic potential with respect to the oxidation state of the cation and the coordination geometry.

In complexes, transitions within the ligands (ligand-centered, LC transitions) may contribute to the spectrum as well as charge-transfer transitions from the ligand to the metal (LMCT) or vice versa (MLCT). In multinuclear complexes, electrons can change over to the next nearest cation, which results in (homo- or heteronuclear) intervalence charge transfer (IVCT) bands.

Many of these transitions can be also observed in solids. As the choice of ligands (anions) is limited in catalytic materials, transitions within or toward the ligand (LC, MLCT transitions) are rare. LMCTs occur between anions and the central cation of a coordination sphere in highly dispersed (isolated) structures, or between the valence band formed by the anions and the conduction band formed by the cations in crystals. As energy bands are broader than the related discrete levels in molecular structures, a crystalline material absorbs at a significantly lower energy (higher wavelength) than the analogous mononuclear complex or an isolated coordination sphere. This relation between wavelengths of LMCT bands and cluster sizes (e.g., mono-, oligo-, and polynuclear oxide aggregates) can be employed in the characterization of dispersed oxide systems (see below).

LMCT bands, which are broad already in the spectra of molecular species, are wide and more intense than d-d transitions in solid materials. However, many d-d transitions are broad as well, which is caused by the coupling between electron excitation and lattice vibrations (electron-phonon coupling). Band widths grow with increasing temperature.

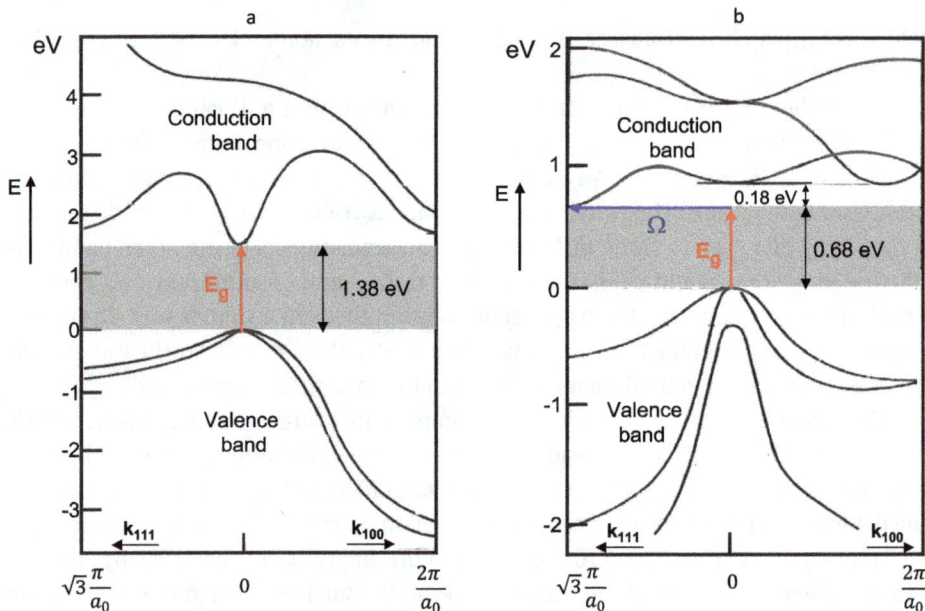

Figure 3.106: Band structure diagram of GaAs and Ge, featuring a direct bandgap (a, GaAs) or an indirect bandgap (b, Ge).

LMCT transitions attract much attention, because their absorption onset marks the width of the bandgap in the sample. Knowledge of bandgaps is of paramount importance for semiconductor technology as also for photocatalysis (cf. Figure 2.58, Section 4.8). Utilizing UV-Vis spectroscopy for bandgap determination, one needs, however, to have in mind that there are two different types of bandgaps, which require different data handling.

Bandgaps may be direct or indirect (Figure 3.106). In a direct bandgap, the highest valence band level and the lowest conduction band level are at the same k in the Brillouin zone, i.e., the coefficients, with which the basis functions at the lattice points contribute to the total wave function (cf. eqs. (2.2), (2.3), Figure 2.4) have the same spatial patterning in both levels. In GaAs (Figure 3.106a), the coefficients are all +1 in the total wave function of both emitting and receiving level. In materials with direct band gaps, absorption increases steeply, once the energy exceeds the bandgap energy E_g.

In indirect bandgaps, the lowest energy difference between valence and conduction band is at different k (e.g. in Ge, Figure 3.106b), i.e., the periodicities in the spatial pattern of basis wave function coefficients are not identical. A direct transition is possible also in this case, but it requires higher energy. Indirect transitions involve a coupling of the electron excitation with lattice vibrations (phonon-assisted electron transition, phonon indicated by Ω in Figure 3.106b). As they are not very efficient, the increase of absorption should be slow right above E_g, which can, however, be followed by a steep increase when the direct transition becomes possible – in Ge, for instance, only 0.18 eV above E_g. These differences in the dependence of the absorption coefficient on the radiation energy are not always to be seen clearly (cf. Section 3.4.7.3).

The abovementioned increase of LMCT band wavelengths with growing cluster size can be also considered from the solid-state viewpoint (cf. Figure 2.4): in the range of quantum-size effects, spectra are blue-shifted and bandgaps are widened with increasing particle dispersion. A method to derive particle sizes and even particle size distributions from bandgap energies is outlined below (Section 3.4.7.3, eq. (3.105)).

The interaction of electromagnetic waves with metal nanoparticles of sizes close to the wavelength results in an absorption mode quite different from those discussed so far. Under the influence of the field, the conduction electrons in the particle are displaced relative to the nuclei, which causes restoring forces. Absorption of a photon thus initiates an oscillation of the electron cloud relative to the nuclear framework at an eigenfrequency that depends on the density of electrons (i.e., the nature of the material), the effective electron mass, the size, and the shape of the particles. The phenomenon is called localized surface plasmon resonance. Resonance absorptions of some metals like Cu, Ag, and Au, appear in the visible range and cause the intense colors of metal colloid solutions. In bimetallic particles, the resonance position can be related to alloy composition, and core-shell structures can be differentiated from well-mixed particles. The resonance is sensitive to the dielectric properties of the environment, e.g., to the solvent in which colloids are dispersed. For more details, the reader is referred to ref. [149], for a survey on theory to ref. [150].

3.4.7.2 Measuring UV-Vis spectra

Scattering and absorption are the major interaction modes with matter also for UV radiation and visible light. In addition, specular reflection occurs at cuvette windows or at phase boundaries. The dependence of absorption on wavelength or energy can be established by measuring transmitted intensities, but in the case of solid samples also via the intensity reflected after scattering events. The absorbed energy is usually dissipated as thermal energy. In some cases, it also induces fluorescent radiation, which interferes with any mode of spectroscopy.

Absorption measurements in transmission are treated under the assumption that contributions of scattering and specular reflection are low. This holds certainly for liquid samples, where reflection by the cuvette may be corrected for. In the thin wafers used for transmission measurements with powders, scattering and reflection may become significant. Strategies suitable for handling these problems are described in ref. [149], the most relevant being, probably, the dilution of samples by a material with similar refractive index to reduce the impact of scattering.

In optical spectroscopy, the ratio between transmitted and incident intensities is called transmittance (τ). Its negative decadic logarithm is traditionally designated as absorbance (A):

$$A = -\log \tau = -\log {}^{I_t}\!/\!_{I_0} \tag{3.100}$$

$$A = -a\,d = -\varepsilon\,c\,d \tag{3.101}$$

There is an alternative definition with the natural logarithm. Instead of "absorbance," the term "attenuance" is now recommended by IUPAC, because attenuation results not only from absorption as discussed above. In the limits of weak absorption, the absorbance is proportional to the optical path length, d (eq. (3.101), compared with eq. (3.81) for X-ray absorption). The proportionality factor, a, which is analogous to μ in (3.81), is the linear decadic absorption (or attenuation) coefficient. For signals resulting from only one absorber, it is proportional to the absorber concentration, c and can be expressed as εc, where ε is the molar decadic extinction (attenuation) coefficient (Lambert-Beer law, cf. (3.101)).

Reflected intensity arises from complicated processes that include multiple scattering events at atoms, scattering and specular reflection at grain boundaries, and diffraction at crystalline arrays within the penetration depth. Some radiation leaves the sample as diffusely reflected intensity, the angular distribution of which is no more correlated with the direction of the incident beam. Due to the enhanced optical path length in the sample, diffusely reflected radiation is an attractive source of spectroscopic information. For its appropriate use in diffuse reflection spectroscopy (DRS), some problems must be dealt with. On the one hand, diffuse reflection should be differentiated from intensity resulting from specular reflection of the incident beam at the sample surface, which results from just one interaction with the solid. On the other hand, the optical path length in samples or sample beds depends on their scattering properties, which often varies with the wavelength. Therefore, the absorption coefficient (usually denoted as K in DRS) is always related to a scattering coefficient, S.

The relation of this normalized absorption coefficient K/S to the reflectance measured at infinite sample thickness ($R_\infty = I_{refl}/I_0$) was derived from a phenomenological model employing forward and reverse photon fluxes between differential slices of the solid. Within this model, which uses a number of assumptions like illumination by diffuse light, weak absorption, and scattering as the only mode of photon re-direction (for more details – see ref. [151]), the relation between K/S and R_∞ over the whole spectral range is:

$$\frac{K}{S} = \frac{(1 - R_\infty)^2}{2\,R_\infty} = F(R_\infty) \tag{3.102}$$

$F(R_\infty)$ is called Kubelka-Munk function as a tribute to the authors of the model. K and S have the dimensions of an inverse length. They can be separately determined by additional measurements at each wavelength of interest [151], which is, however, rarely attempted in catalysis. Instead, spectra are reported as $F(R_\infty)$ vs. λ or $\bar{\nu}$.

In practice, UV-Vis spectrometers do not determine absolute reflectivities. Instead, R_∞ is compared with the reflectivity, R_{st} of a white standard like MgO, BaSO$_4$, or Spectralon[R]: $R_{\infty,exp} = R_\infty/R_{St}$. The spectra of the most convenient standards are known and usually stored in instrument software, which allows evaluating R_∞ and F(R_∞). For systems, in which the validity of assumptions for the Kubelka-Munk model is doubtful, spectra are sometimes plotted in units of apparent absorption instead, which is the negative decadic logarithm of the (measured) reflectance:

$$-\log R_{\infty,exp} = \log {R_{St}}/{R_\infty} \tag{3.103}$$

Equation (3.102) may suggest that the Kubelka-Munk function measures the concentration, c, of the absorber in a similar way as the absorbance, A, characteristic of the transmission mode (cf. eq. (3.101)). However, due to the paramount role of scattering in DRS, the relation between F(R_∞) and the absorber concentration, c, is much more complicated. Already within the Kubelka-Munk model, a number of conditions (e.g., small K, no influence of c on the scattering coefficient, S) must be fulfilled to keep F(R_∞) proportional to c. Sample series, in which species of different scattering efficiency (e.g., quasi-molecular monomeric and particulate structures of the absorber) coexist in varying concentrations, will hardly comply with the latter assumption. In addition, the relation between F(R_∞) and c is modified when S changes across the spectral range, which is usually the case. Interactions neglected in deriving eq. (3.102) may further complicate the problem.

Thus, quantitative DRS work is possible only in narrow limits, highly reproducible packing of powdered samples being a crucial prerequisite for any success. Although such studies have been reported (cf. ref. [149]), the method is mostly used in a semiquantitative way. It can be employed to reveal trends in series of samples, e.g., with varying concentration of the absorbing atom, or to follow the evolution of species over time.

Figure 3.107 shows the main elements of a UV-Vis spectrometer: a radiation source, a monochromator, a sample compartment (here, an integrating sphere for DR measurements), and a detector. The gas discharge lamps supplying the required radiation do not cover the whole spectral range, which has interfered with the application of Fourier Transform technology familiar in IR spectroscopy (Section 3.4.8) to UV-Vis spectroscopy. The switch from the deuterium lamp to the tungsten halogen lamp at ≈320 nm can be usually discerned in DR spectra.

In scanning spectrometers, the beam is directed to a grating monochromator, the rotation of which scans the energy over the desired range. The beam proceeds to the sample compartment, and the reflected intensity is recorded by the detector. Figure 3.107 shows a double-beam spectrometer, in which the signals from sample and reference[38] are measured simultaneously. For this purpose, a chopper sends the beam to

---

38 In transmission, the reference might be a cuvette with the pure solvent.

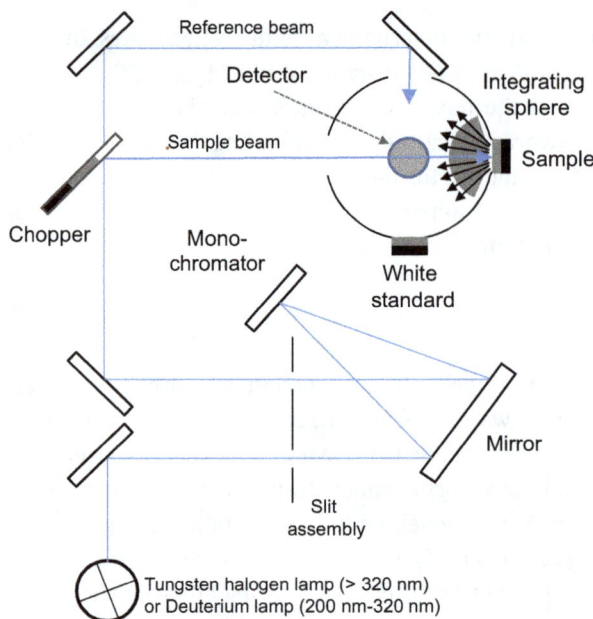

Figure 3.107: UV-Vis spectrometer for measurements with an integrating sphere. The detector is behind the plane of projection (cf. Figure 3.108).

sample and reference, alternately. In single-beam spectrometers, the spectra of sample and reference are measured sequentially. The detector is usually a photomultiplier in the UV-Vis range. NIR radiation is rather detected by its effect on photoconducting materials such as PbS.

Alternatively, wavelengths are differentiated after the sample by a bent grating, which sends different wavelengths into different directions (cf. Figure 3.84, Section 3.4.4.2). In such diode array spectrometers, position-sensitive detectors allow simultaneous analysis of whole wavelength ranges and, therefore, extremely fast acquisition of the spectra.

The transmission mode is employed for liquid samples in simple cuvettes and for powders pressed into self-supporting wafers. For measurements with scattering media, e.g., turbid liquids or powders, the detector needs to be placed very close to the sample to allow collecting intensity from a large solid angle. Intense scattering resulting in low yields of transmitted light impedes work with samples dominated by heavy atoms. Although transmission UV-Vis studies with oxides of light elements, e.g., zeolites, can be found in literature [149], solid catalysts are almost exclusively investigated in the DR mode.

Arrangements for measuring the diffuse reflectance from solid samples are also designed to collect intensity from a large solid angle. In addition, they aim to block contributions of specular reflection from the sample surface. The options available are the integrating sphere, mirror-optical arrangements, and fiber optical probes.

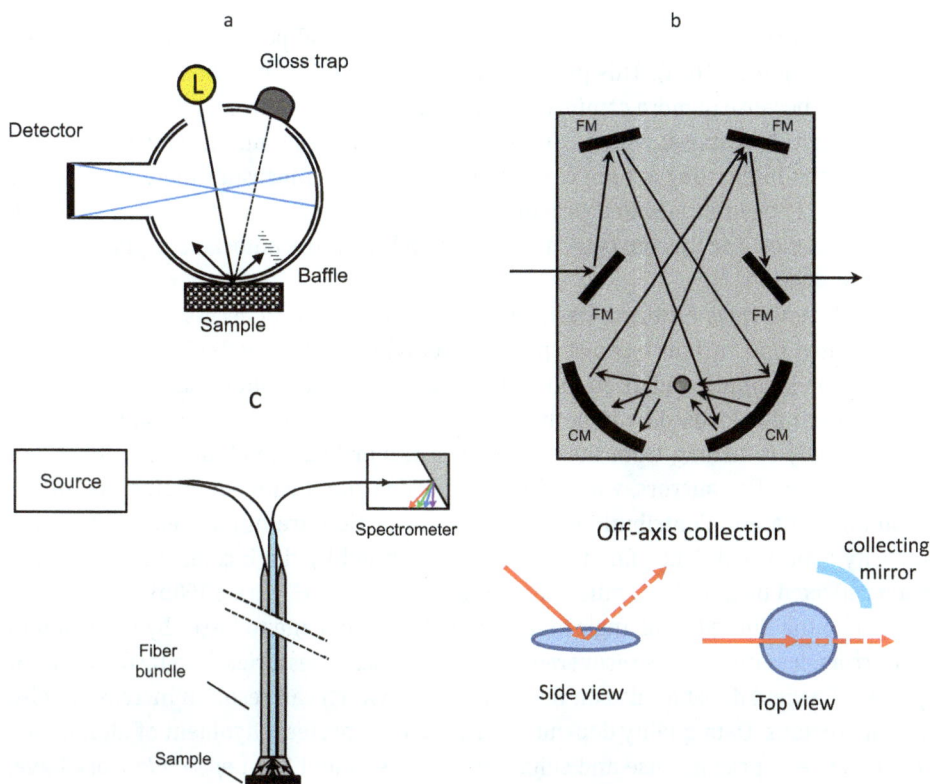

Figure 3.108: Sample environments for UV-Vis diffuse reflectance measurements: (a) integrating sphere (L – lamp); (b) mirror optics accessory. FM – flat mirror; CM – collecting mirror; and c – fiber optic probe.

The walls of integrating spheres (or Ulbricht spheres, Figures 3.107, 3.108a) are coated with a highly reflecting material, e.g., $BaSO_4$ or Spectralon[R]. Apertures for sample, white standard, incident light, and detector occupy not more than 5% of the spherical surface. Radiation returning from sample and white standard is scattered many times by the walls, before being registered by the detector. In the version shown in Figure 3.107, specular reflection from the sample surface sends most of the light back toward the source. At inclined sample illumination (Figure 3.108a), a gloss trap even allows assessing the specularly reflected intensity. Its influence is further suppressed by baffles or by placing the detector in a recess behind the sphere surface (Figure 3.108a). Although the detector images only a small part of the sphere surface, intensity losses are very low. Therefore, integrating spheres are suited for measurements with high sensitivity and precision.

On the other hand, studies of catalysts under reaction conditions or even combined with meaningful rate measurements (*in situ* or *operando* work, cf. Section 3.4.4.2) are difficult to realize with integrating spheres, which are very sensitive to heating or temperature gradients. Although spheres may be filled with gaseous reactants, heating samples requires retracting them from the ideal tangential position (as shown for the detector in Figure 3.108a). This principle was proven [149], but it has not found wide application because of more comfortable alternatives (see below) and the lack of forced flow through the catalyst which is required for reliable rate measurements. Suitably modified, the integrating sphere can, however, be used to measure the reflectance of turbid liquids even at elevated pressures, e.g., in studies of nucleation and crystallization processes. For this purpose, a suitable window in the autoclave is placed at the sampling position.

DRS is nowadays routinely performed with mirror optical accessories, which can be integrated into various types of spectrometers (Figure 3.108b). Deflected by two flat mirrors, the incoming radiation falls into a curved (usually ellipsoidal) mirror above the sample and is focused by it on the sample surface. A second ellipsoidal mirror collects intensity returning from the sample in a rather large solid angle and directs it to the next two flat mirrors, which direct the beam into a path continuing that of the incoming radiation. Contributions of specular reflection are minimized by collecting intensity "out of axis," i.e., far from the plane defined by the incoming light and the beam reflected by a (virtual) mirror in the sample surface (Figure 3.108b).

Due to the limited solid angle employed for detection and to losses by reflection in six mirrors, less intensity is recovered by mirror optics accessories than by the Ulbricht sphere. The yield depends also on the wavelength, which can result in increased noise in some regions. Data quality depends, crucially, on a perfect alignment of all mirrors. Convenience in practical use and suitability for *in situ* and even *operando* work have, however, strongly favored the application of such accessories in catalysis research.

In commercially available environmental cells, the sample cup is included in a heatable chamber with ports for gas flow and cooling water, the top face of which emplaces a dome with windows right above the catalyst (for incoming and outgoing radiation, cf. Figure 3.108b, and for operator view). Such cells can be used in wide ranges of pressure and temperature, although problems with the exact determination of reaction temperatures have been reported. The spectra may contain signals from species in the gas phase above the catalyst. An optional gas outlet below the sample cup allows rate measurements with forced flow through the catalyst. In *operando* studies with this arrangement, it should be, however, kept in mind that the spectra represent only a thin layer at the top of a catalyst bed, which, as a whole, achieves the conversions measured at its end.

While mirror optics accessories require the catalyst to be placed into a sample cup, flexibility of *in situ* work has been greatly enhanced by the advent of fiber optic probes. Such probes consist, at the minimum, of an illumination fiber, which guides the radiation to the analysis spot, and a read fiber collecting the response of the catalyst surface and guiding it to monochromatization and/or analysis (Figure 3.108c). Often, several fibers are integrated into a bundle, in which illumination and read fibers may be assigned. Bundle diameters may be a few mm, but fiber diameters may be as small as 50–100 μm allowing even spatially resolved measurements of single catalyst pellets.

The influence of specular reflection can be reduced by strutting the illumination fibers apart from the read fiber(s), right at the probe head.

Fiber optic probes withstanding high temperatures and pressures can be used to characterize operating catalysts in laboratory reactors, either by introducing them into the interior or by measuring through quartz windows. Likewise, studies in turbid media, e.g., during the crystallization of zeolites from precursor gels, may be performed with the probe immersed or by measuring through a window. Instructive examples may be found in refs. [152, 153]. Fiber optic probes are also used in installations that combine several spectroscopic methods for the simultaneous characterization of catalysts during catalytic operation; see, for instance refs. [152–155].

3.4.7.3 Typical applications of UV-Vis spectroscopy in catalyst research

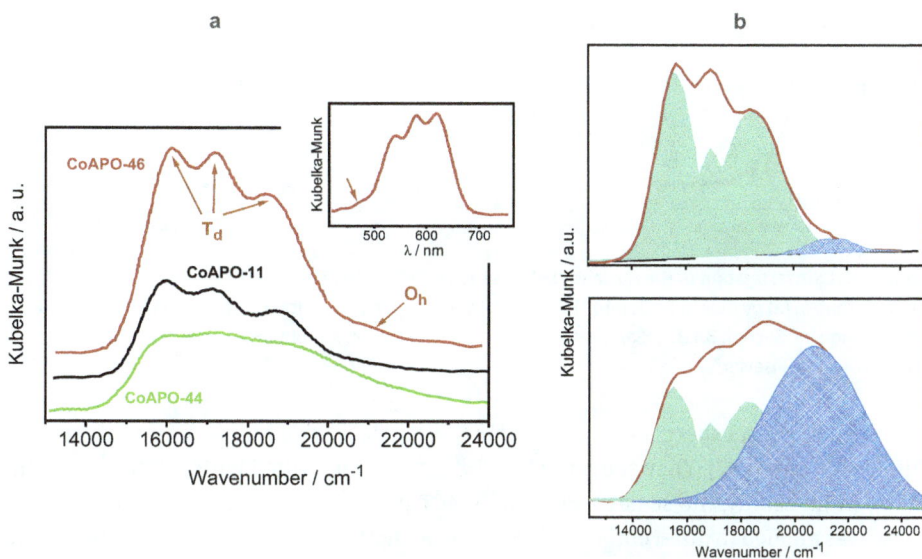

Figure 3.109: Analysis of Co-substituted AlPO zeolites: differentiation of extra- and intra-framework Co^{2+}: (a) spectra of different CoAPO materials; the triplet labeled T_d (see arrows) arises from Co^{2+} in the framework, a broad singlet around 21,000 cm^{-1} (O_h, see arrow) from extra-framework oxide species; (b) spectra of well-prepared and of ill-prepared CoAPO-46. Adapted from ref. [156] with permission from Elsevier.

Monitoring d-d transitions allows determining the oxidation state and the coordination of many transition metal cations. Figure 3.109 shows how this was employed to assess the quality of Co-substituted AlPO molecular sieves (cf. Section 2.3.2.3). As in alumosilicate zeolites, cations on framework positions are tetrahedrally coordinated, while their coordination in amorphous oxides or competing dense phases is octahedral. The spectra of three samples from preparations targeting different framework types

(CoAPO-11, -44, -46) are shown in Figure 3.109a. They are plotted on the wavenumber scale. For comparison, one of them is repeated on the wavelength scale in the inset. The three correlated signals at ca. 15,000, 17,000, and 18,500 cm^{-1} arise from transitions in the high-spin d^7 Co^{2+} cation in tetrahedral environment; the shoulder at ca. 21,000 cm^{-1} can be assigned to Co^{2+} in octahedral environment. The signals were fitted as illustrated in Figure 3.109b for CoAPO-46 and for a failed preparation of this phase. Low intensity ratios between signals of Co^{2+} in octahedral and tetrahedral coordinations indicate a high quality of the material. As the scattering properties of Co^{2+} in the framework and in oxide aggregates are likely to differ significantly, there should be no simple way to derive atomic ratios from intensity ratios.

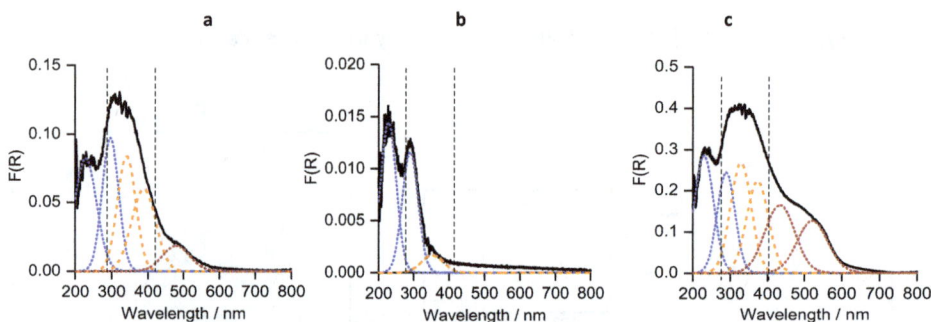

Figure 3.110: Investigation of the nuclearity of Fe sites in ZSM-5: UV-Vis spectra of samples prepared via different routes: (a) by CVD of FeCl$_3$ into H-ZSM-5; (b) via a mechanochemical route; and (c) by solid-state ion exchange (cf. Section 3.1.3.7). Samples studied in hydrated state. Adapted with from ref. [157] with permission from Elsevier.

Figure 3.110 shows spectra of ZSM-5 zeolite, into which iron had been introduced via different routes [157]. They show the LMCT band, which is fitted with sub-bands, exclusively. In this study, the sensitivity of the LMCT band to the extension of the detected phase was applied to differentiate Fe^{III} oxo species in the catalysts obtained. Considering earlier literature and own work with other techniques, bands below 300 nm ($\approx$230 nm, $\approx$290 nm) were assigned to isolated Fe-oxo sites in tetrahedral and higher coordination, respectively. Bands between 300 nm and 400 nm were ascribed to oligomeric structures containing 2, 3, or more Fe^{III} ions (cf. Section 3.1.3.1), signals beyond to Fe oxide aggregates, which may be highly disordered in this kind of sample. While the spectra suggest major differences in the speciation of Fe sites, any effort to convert the intensity distributions into wt.-% of sites remains doubtful. Indeed, a recent comparison of Fe site distributions in similar catalysts by "quantitative" diffuse reflectance UV-Vis and by high-end Mössbauer spectroscopic analysis revealed major discrepancies [34], confirming the failure of UV-Vis DRS in concentration analyses, except for very specific situations (see above).

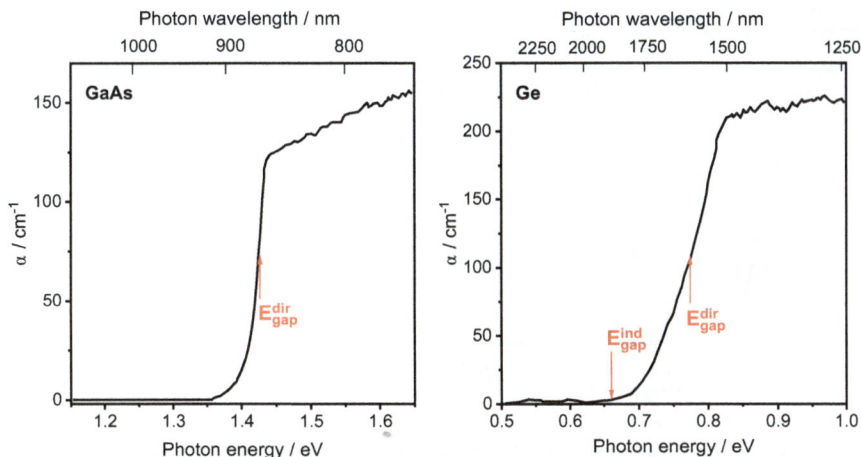

Figure 3.111: Dependence of the absorption coefficient, α, on the energy in the vicinity of the bandgap: GaAs – direct bandgap, Ge – indirect bandgap. Adapted from ref. 158.

As indicated above, UV-Vis spectroscopy is used to determine semiconductor bandgaps, which may be direct or indirect. In Figure 3.111, the energy dependence of the absorption coefficient α is reported for the examples, GaAs and Ge, used to illustrate these bandgap types in Figure 3.106. α, determined from diffuse reflectance measurements, is analogous to K in eq. (3.102). Its increase above the bandgap energy is less pronounced in the indirect semiconductor, Ge, than in GaAs (Figure 3.111a), but there is no obvious rise of the gradient above the direct bandgap energy of Ge (Figure 3.111b).

A popular method for bandgap determination is based on the observation that the course of the absorption coefficient above the edge depends on the difference between wave and edge energies:

$$(\alpha\, h\, v)^{1/n} = A\,(h\, v - E_g) \tag{3.104}$$

In eq. (3.104), A is a constant, and n is 0.5 for direct and 2 for indirect bandgaps. Instead of α, the Kubelka-Munk function $F(R_\infty)$ is often employed, because S (cf. eq. (3.102)) is unlikely to vary strongly in the narrow energy range involved. In the so-called Tauc plots of $(F(R_\infty)hv)^2$ or $\sqrt{F(R_\infty)}\,hv$ vs. hv for direct or indirect bandgaps, respectively, the linear range around the inflection point is extrapolated toward the abscissa, and the bandgap energy is identified by the value at the intersection point. The method is illustrated in the inset of Figure 3.112 with the example of a (0001)-oriented ZnO single crystal. With polycrystalline samples, the definition of the range used to define the extrapolation line, and thus the exact E_g value, may, to some extent, depend on the operator's choice [159]. An instructive overview over UV-Vis-based methods for the determination of band gaps can be found in ref. [158].

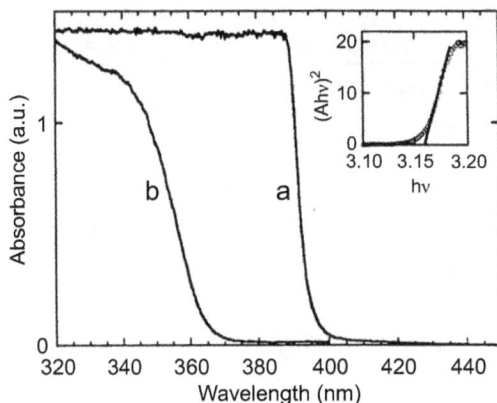

Figure 3.112: Particle-size dependence of LMCT bands for particle-size determination exemplified with ZnO; absorption spectrum of (a) a (0001) ZnO single crystal; (b) a suspension of ZnO nanoparticles; the inset shows the single-crystal data arranged in a Tauc plot (3.105) for bandgap determination. For the determination of particle sizes and size distributions, see text and ref. [160]. Reproduced with permission from ref. 160. Copyright 2003 American Chemical Society.

The blue shift resulting from bands narrowing in quantum-size particles can be used to assess their radius, R. The bandgap energy, $E_g{}^*$ of the quantum-size particle is approximately [158]:

$$E_g^* \approx E_g + \frac{h^2}{8\,R^2}\left[\frac{1}{m_e}+\frac{1}{m_h}\right]-\frac{1.8\,e^2}{4\,\pi\,\epsilon\,\epsilon_0\,R} \tag{3.105}$$

where m_e (m_h) are the effective electron (hole) masses in the crystal of the material, ϵ is its relative permittivity, ϵ_0 is the permittivity of the free space, and e is the elementary charge. In Figure 3.112, curve b was obtained from a suspension of ZnO nanoparticles. Its smaller gradient suggests that it arises from the superposition of different band edges originating from particles of different sizes. In ref. [160], it is shown how the particle size distribution can be extracted from such data.

More examples for the application of UV-Vis spectroscopy in catalysis, e.g., for monitoring redox processes of transition-metal ions or for studying adsorbates (e.g., hydrocarbon pools characteristic for the MTH reaction (cf. Appendix A1)), can be found in ref. [149].

3.4.8 Vibrational spectroscopy

Energy consumption for the excitation of vibrations in molecules or solid structures is the source of information in vibrational spectroscopy. Transitions between vibrational

Table 3.2: Ranges of IR radiation.

	near–IR	mid–IR	far–IR
Wavelength λ/µm	1–2.5	2.5–50	50–1,000
Wavenumber λ^{-1}/cm^{-1}	10,000–4,000	4,000–200	200–10
Energy / eV	1.240–0.496	0.496–0.025	0.025–0.0012

$$1\ eV\ \approx\ 8{,}040\ cm^{-1}$$

$$1\ eV \cdot N_A \approx 96.5\ kJ/mol \qquad 1\ cm^{-1} \cdot N_A \approx 12\ kJ/mol$$

energy levels require energies in the infrared range of the electromagnetic spectrum. Table 3.2 shows the customary subdivision of this range into near-IR, mid-IR, and far-IR, together with some conversions between familiar energy units. Vibrational spectroscopy with catalysts uses almost exclusively mid-IR. Surface science methods extending into far-IR will also be briefly discussed below.

Vibrations may be excited by electromagnetic radiation or by electrons. In the former case, excitation may be the result of photon absorption or of inelastic scattering (cf. Section 3.4.3.1, Figure 3.61), which is the basis of infrared spectroscopy and Raman spectroscopy, respectively. Excitation by collision with electrons is utilized by measuring the energy losses of incident (monochromatic) electrons (HREELS – High-resolution electron energy loss spectroscopy, see below). HREELS is an important method in surface science, but not applicable to real catalysts.

3.4.8.1 Vibrations and vibrational spectra

Bond vibrations in chemical systems can be described by the harmonic oscillator model (Figure 3.113). It is based on Hooke's law, where restoring force, F and potential energy, V of the oscillator are proportional to the displacement, q and to its square, q^2, respectively (cf. eqs. (3.106) and (3.107), where k is the force constant). The model is unrealistic insofar as it does not allow the oscillator to split, which happens with chemical bonds, but it is useful for spectroscopy, because the energy levels accessible by the excitation are far below the bond dissociation energies.

The solution to the Schrödinger equation for the potential energy described by (3.107) can be found in textbooks of physical chemistry [162]. The energy levels of oscillators are the eigenvalues in a set of eigenfunctions (of the form of Hermite polynomials), which are itemized by their vibrational quantum numbers, n. The eigenvalue of the nth function is given by eq. (3.108), where v is the eigenfrequency of the oscillator. For most oscillators, the ground state ($n = 0$) is predominantly populated at ambient conditions. According to eq. (3.54), radiation of the frequency, v_{rad}, can be absorbed when the difference between initial and final energy levels is equal to hv_{rad}. In the harmonic

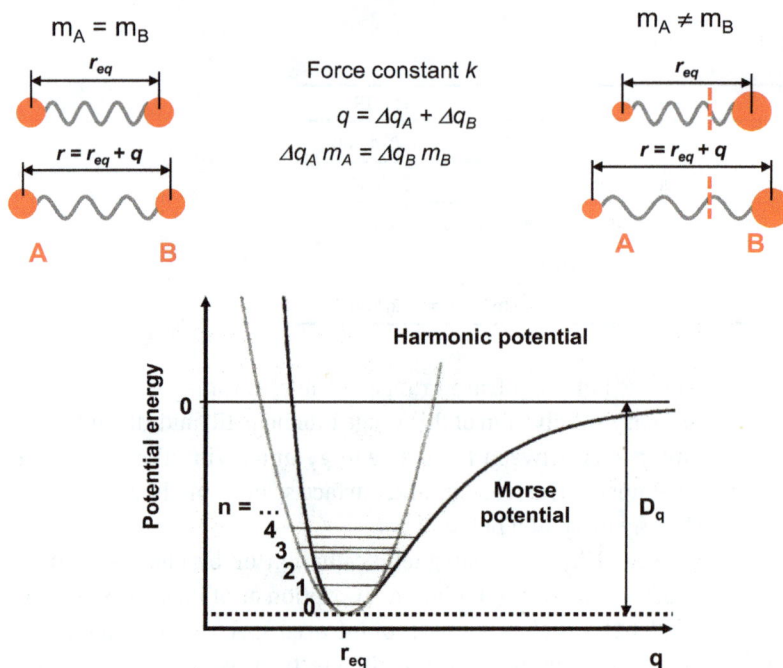

Figure 3.113: Harmonic and anharmonic oscillator. In potential diagrams, the energy zero is usually assumed at infinite distance, which has not been complied with in eqs. (3.107) and (3.112). Potential diagram adapted from ref. [90].

oscillator model, this difference ΔE_{harm} is $h\nu$ (cf. (3.109)), i.e., the eigenfrequency ν of the oscillator is equal to the radiation frequency.

$$F = -k\,q \tag{3.106}$$

$$V = k\,q^2/2 \tag{3.107}$$

$$E_n = (n + 0.5)\,h\nu \tag{3.108}$$

$$\Delta E_{harm} = h\nu \tag{3.109}$$

In the simplest chemical application, the harmonic oscillator consists of two equal A atoms, the bond between which acts like a spring with a force constant k (Figure 3.113). The eigenfrequency of this oscillator is:

$$\nu = \frac{1}{2\,\pi}\,\sqrt{k/\mu} \tag{3.110}$$

Equation (3.110) already covers the case that the atoms connected by the bond have different masses, m_A and m_B. The oscillator mass, μ is then the reduced mass:

$$\frac{1}{\mu} = \frac{1}{m_A} + \frac{1}{m_B} \tag{3.111}$$

In the A_2 molecule, $\mu = 2\, m_A$, and the movement of the two atoms is mirrored by the plane between them. In the AB molecule, the movement is relative to the center of gravity, and the contribution of the two atoms to the total displacement, q, is reciprocal to their masses (cf. Figure 3.113).

The harmonic oscillator model fails already for transitions from the ground state to the second excited state. While such overtones, which should appear at twice the wavenumber of the fundamental vibration $\bar{v}_F$, are strictly forbidden in it, they appear at wavenumbers slightly smaller than that ($\bar{v}_O < 2\, \bar{v}_F$) in reality, though at low intensities. Indeed, the symmetry of the harmonic potential is at variance with the different resistances of bonds to compression and elongation and with the fact that they can dissociate. The potential of a real oscillator might rather be described by the Lennard-Jones potential (cf. eq. (3.149) below), but in IR spectroscopy, the Morse potential (3.112) is preferred, maybe because an analytical solution of the Schrödinger equation with it has been known for long.

$$V = D_e\,(1 - exp(-\beta q))^2 \tag{3.112}$$

In (3.112), D_e is the sum of dissociation and zero-point energy (cf. Figure 3.113) and β is a constant related to the stiffness of the potential. The energy levels, E_n, resulting as eigenvalues from the solution of the Schrödinger equation with (3.112) are:

$$E_n = (n + 0.5)\,h\,v_e - (n + 0.5)^2\,h\,v_e x_e \tag{3.113}$$

v_e and x_e are related to the potential parameters as:

$$v_e = \frac{\beta}{\pi}\,\sqrt{D_e/2\mu} \tag{3.114}$$

$$x_e = h\,v_e/4\,D_e \tag{3.115}$$

and the energies required for transitions to the next level (i.e., from (n-1) to n) are:

$$\Delta E_{n-1 \rightarrow n} = h\,v_e - (2n-1)\,h\,v_e x_e \tag{3.116}$$

From (3.116), it is obvious that v_e is not the frequency of the fundamental mode and that the differences between subsequent levels decrease with growing n. The Morse parameters, D_e and β, of an oscillator can be determined when both fundamental mode and first overtone are accessible.

According to eq. (3.110), the frequency of a vibration can be correlated with the strength of the oscillating bond. While this is correct, it is, at the same time, important to realize that force constants probe bond strengths under conditions that are not very relevant for chemists: they report the resistance of the bond to a differential deviation from its equilibrium length. Chemists, instead, gauge the strength of bonds by their dissociation energy, which is accessible only via the anharmonic oscillator, e.g., (3.112). Correlations between band frequencies and strengths of the oscillating bonds that can be sometimes found in literature (often implicitly!) must be treated with caution. There is some probability that "stiff" bonds may be more difficult to be split up, but many exceptions are known from this rule. A better way to characterize bond strengths, which utilizes the anharmonicity of the vibration will be discussed in Section 3.4.8.3.

Equation (3.110) implies also that eigenfrequencies, ν (and wavenumbers, $\tilde{\nu}$) decrease as the vibrating masses increase. This is the reason why modes involving more than two atoms tend to vibrate at lower frequencies. It is also the basis of a well-known approach to shift signals by exchanging an atom of the oscillating bond with one of its isotopes, e.g., H with D, ^{14}N with ^{15}N. The resulting frequency shifts can be well forecast by eq. (3.110). Such an experiment can contribute to the assignment of unidentified bands. In addition, shifting intense signals by modifying the vibrating group(s) sometimes allows detecting weaker signals normally superimposed by the shifted intensity.

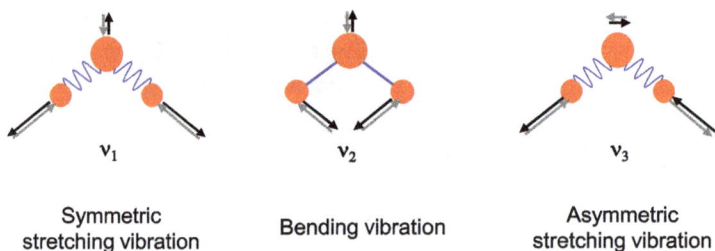

ν_1	ν_2	ν_3
Symmetric stretching vibration	Bending vibration	Asymmetric stretching vibration

Figure 3.114: Normal modes of a bent triatomic oscillator.

The scope of eqs. (3.106) through (3.116) extends far beyond the diatomic oscillator. In a molecule at rest, all movements of atoms can be described as a superposition of vibrations, according to its normal modes. A normal mode is a vibration, which can be described by a single displacement quantity, q. All atoms move at the same frequency and in the same phase; amplitudes are weighted with masses similar as in the AB oscillator (Figure 3.113), and the center of gravity remains at rest.

In Figure 3.114, normal modes of a bent triatomic molecule B_2A (e.g., H_2O) are illustrated. In mode ν_1, the AB bonds are simultaneously elongated and compressed. To keep the center of gravity, this oscillation is combined with a movement of the A atom along the angle bisector, which slightly decreases and increases the bond angle in the stretching and compression phases, respectively. In v_2, the widening and narrowing of the bond angle is again accompanied by a movement of A for preventing shifts of the center of gravity. In v_3, the AB bonds elongate and contract in antiphase, while the A atom moves normal to the angular bisector to balance the center of gravity. In phenomenological nomenclature (see below), v_1 and v_3 are symmetric and asymmetric stretching modes, respectively; v_2 is a bending mode. A molecule has as many normal modes as vibrational degrees of freedom, therefore, Figure 3.115 shows the complete set of B$_2$A. For identifying the normal modes of a molecule, its symmetry properties and bonding structure are of major importance.

For each normal mode, the probability distribution of the displacement quantity, q, and the energies of ground and excited states can be obtained by inserting the description of its (harmonic or anharmonic) potential into the Schrödinger equation. q is related to the movement of the individual atoms via vector operations and weighting; the force constant is obtained by vector averaging of contributions from all bonds.

Orientation in the wealth of spectroscopic information is greatly facilitated by the fact that normal modes are often dominated by the vibration of a particular bond or angle: its frequency is only moderately affected when atoms beyond are changed. The C–C stretching vibration of the double bond in propene, for instance (actually: the mode dominated by this vibration), is shifted by just 19 cm^{-1} when an allylic H atom is substituted by Br (from 1,651 cm^{-1} to 1,632 cm^{-1}). Such weak influence of the environment allows considering functional groups as individual oscillators causing signals in specific wavenumber ranges. Box 3.5 exemplifies some of these group frequency ranges. Comprehensive information can be found in specialized textbooks [163, 164].

Box 3.5: Typical wavenumber ranges for some stretching vibrations.

> 4,000 cm^{-1}	various overtones, combination modes[a]
4,000–2,500 cm^{-1}	X-H stretching vibrations (X = C, N, O, S)
2,500–1,800 cm^{-1}	triple bond stretching (C≡C, C≡N, C≡O)
2,000–1,500 cm^{-1}	double bond stretching (C=C, C=O, aromatic ring)
1,500–500 cm^{-1}	C-X single bond stretching (X = C, N, O, Cl, S)
< 1,200 cm^{-1}	lattice vibrations in solids
400–250 cm^{-1}	metal-adsorbate vibrations (Me-C, Me-O, Me-N)

[a]simultaneous excitation of two modes

Vibrations may be differentiated by the relative motion of atoms. As mentioned above, stretching varies bond lengths; bending or deformation modes vary bond angles. They are designated by v and δ, respectively. When more than two atoms are involved, stretching vibrations may be symmetric or asymmetric, as in the bent triatomic molecule shown in Figure 3.114 or in the linear triatomic molecule in Figure 3.115. The latter has four normal modes; the two deformation modes, one of which is depicted in Figure 3.115, are degenerate. In planar molecules, deformation modes may be in-plane or out-of-plane. Figure 3.115 shows four deformation modes of CH$_2$ groups that coexist

Linear molecules

stretching (ν)

symmetric

asymmetric

stretching (ν)

bending (δ)

Nonlinear molecules

Bent triatomic molecule – s. Figure 3.115

Motions relative to a plane

bending   rocking

in-plane

twisting   wagging

out-of-plane

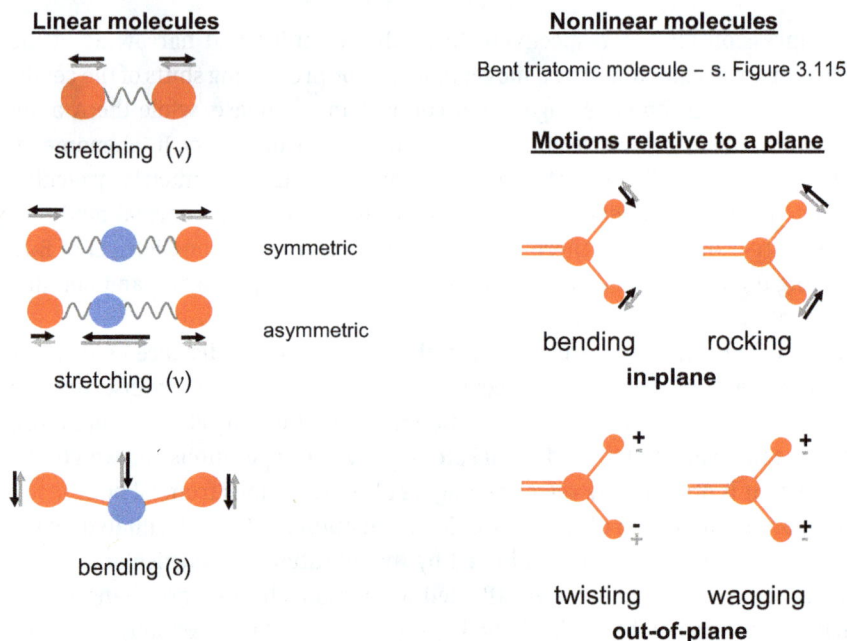

Figure 3.115: Phenomenological nomenclature of molecular vibrations.

with the (symmetric and asymmetric) stretching modes illustrated in Figure 3.114. Two of these deformation modes are in-plane — the bending and the rocking modes; the remaining ones (twisting and wagging modes) are out-of-plane.

3.4.8.2 Excitation processes and spectroscopies

IR spectroscopy tracks the consumption of photons by vibrational modes, in which the dipole moment, μ of the molecule changes along the displacement coordinate, q:

$$\left(\frac{d\,\mu}{d\,q}\right)_{q=0} \neq 0 \tag{3.117}$$

Absorption probabilities grow with increasing variation of q during the oscillation. In molecules with point symmetry, modes symmetrical toward the center of symmetry are IR-silent (e.g., the symmetric stretching modes of CO_2, of all A_2-type molecules, and of CH_4), but not, where existent, the asymmetric counterparts. Chemisorption may break such symmetry. O_2, for instance, remains silent when adsorbed via a π interaction of its double bond, but becomes detectable when coordinated via one of the O atoms.

Raman spectroscopy utilizes the inelastic scattering of electromagnetic radiation. According to its classical model, the incident radiation with a frequency, ν, polarizes the oscillator and thus becomes modulated with the oscillator eigenfrequency, ν_1, acquiring some intensity at the frequencies, $v - v_1$ and $v + v_1$. Such interaction happens only if the polarizability, a, of the mode changes along the displacement coordinate, q:

$$\left(\frac{d\,a}{d\,q}\right)_{q=0} \neq 0 \tag{3.118}$$

The intensity of Raman signals increases with growing change of a during the oscillation.

As there is no influence of molecular symmetry on polarizability, modes that are IR-silent for reasons of symmetry are well detectable by Raman spectroscopy. On the other hand, as the polarizability and, consequently, its possible changes are very small in various important oxide supports as Al_2O_3, SiO_2, or MgO, Raman spectroscopy allows detecting the signals of oxide species supported on their surfaces, which are superimposed by their intense lattice vibrations in IR spectroscopy.

HREELS vibrations are excited by inelastic scattering of electrons, the energy losses of which are recorded by energy analyzers. The method gives access to wavenumbers deep in the far-IR range (down to ca. 50 cm^{-1}), where the bonds between adsorbates and surfaces can be studied. Electrons are applied as a monochromatic beam. Differentiating weak loss signals at a few meV off the elastic peak (typically at 100 eV) requires powerful analyzers, good electrical conductivity of the sample, and UHV conditions.

Inelastic scattering of electrons may proceed by dipole scattering, which requires a change of the dipole moment along the oscillation (cf. eq. (3.117)), and by impact scattering. In the direction of specular reflection, the former dominates the latter by orders of magnitude, while impact scattering has no preferential orientation similar to diffuse reflection. As any dipole moment parallel to a metal surface is cancelled by the image dipole, dipole scattering is selective for modes (and components) perpendicular to the surface. It can, therefore, be used to establish the spatial orientation of adsorbates relative to the metal surface. Selection rules of impact scattering are complex, but modes forbidden in dipole scattering are detectable. The full potential of HREELS is accessible in studies with flat ordered surfaces by comparing spectra dominated by the different selection rules. The method is extremely sensitive, with a detection limit of 10^{-4} monolayers for strong absorbers like CO, and still 0.01 monolayer for H atoms. For more information, the reader is referred to ref. [165].

Vibrational spectroscopy can also be performed utilizing second-order nonlinear optical phenomena. When media scatter very intense electromagnetic radiation of the frequency v, its strong oscillating field, $E = E_0 \cos 2\pi v t$, polarizes bonds to an extent that the response of the electrons is influenced by the anharmonicity of the molecular potentials. The polarization P is no longer proportional to E. In a Taylor expansion, $P = a_1 E + \frac{1}{2} a_2 E^2 + \ldots$, the second term becomes

$$P_2 = \frac{1}{2}\, a_2\, E_0^2 \cos^2(2\pi v t) = \frac{1}{4}\, a_2\, E_0^2\, (1 + \cos(4\pi v t))$$

i.e., the scattered radiation contains contributions of doubled frequency. In environments with inversion center (isotropic phases, crystals with point symmetry), a_2 is, however, zero: the second-order nonlinear optical effect operates only in crystals of a particular symmetry and at surfaces.

While frequency doubling (or Second harmonic generation, SHG) is important in laser technology, spectroscopy benefits from a related effect in Sum frequency generation (SFG): when the medium is exposed to radiation of different frequencies, v_1 and v_2, the scattered radiation contains also the combination mode, $v_1 + v_2$. To develop spectroscopy on this basis, tunable lasers are required, which are presently available only for wavenumbers >2,000 cm^{-1}. In SFG, the variable IR frequency (v_2) is combined with a carrier frequency, v_1, which is in the visible range, because in this range, the output intensity at $v_1 + v_2$ can be easily differentiated from contributions of the two excitation radiations and analyzed by high-gain photodetectors.

Allowing vibrational spectroscopy of adsorbates on small samples at realistic gas-phase pressures and even in liquid phases, SFG has become a major tool of surface science [166]. Its application to the study of the Horiuti-Polanyi mechanism of alkene hydrogenation on monocrystalline metal facets has been exemplified in Section 2.5.1, but it can also be applied to polycrystalline samples [166]. It is also applied to the study of electrode processes. After three decades, it has, however, remained a demanding and expensive technique. The reader can find more details on it in ref. [166].

3.4.8.3 Measuring IR spectra of heterogeneous catalysts

Just like any other spectrometer, IR spectrometers contain a radiation source, a facility for differentiating the radiation according to its energy, a sample stage, and a detector. Unlike in UV-Vis and X-ray absorption spectroscopy, wavelength differentiation is typically accomplished by the Fourier-Transform (FT) technology in instruments used for catalysis research, because in the IR range, grating monochromators are used only for high-end precision measurements.

In the Michelson interferometer at the heart of FT spectrometers, radiation is split by a semitransparent mirror (a beam splitter, Figure 3.116). Beams are reunited after visiting mirrors, one of which seesaws relatively to the incoming partial beam in order to vary its optical path, δ. For a monochromatic radiation of a wavenumber $\bar{v}_1$, the phase is thus modified according to $cos\,(2\pi\bar{v}_1\delta)$, and the reunited beam arriving at the detector exhibits a cosine dependence of its intensity, I, on the beam path, δ – the interferogram of the monochromatic radiation:

$$I'(\delta) = 0.5\ I(\bar{v}_1)\ (1 + cos\ (2\pi\bar{v}_1\delta))$$

In reality, the ideal cosine response is modified by a number of experimental deficiencies, most of which are frequency-dependent, such as nonideal beam splitting or detector sensitivities. Summarizing these corrections, which are well known for each instrument as $C(\bar{v})$, the interferogram becomes:

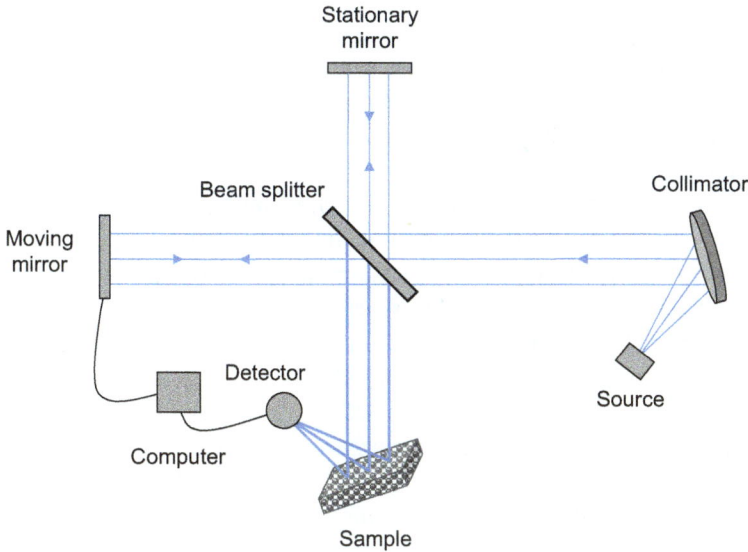

Figure 3.116: FTIR spectrometer with Michelson Interferometer.

$$I(\delta) = 0.5 \; C(\bar{\nu}_1)I(\bar{\nu}_1) \; cos \; (2\pi\bar{\nu}_1\delta) = I_{exp}(\bar{\nu}_1) \; cos \; (2\pi\bar{\nu}_1\delta) \qquad (3.119)$$

where $I_{exp}(\bar{\nu}_1)$ is the actual response of the spectrometer to monochromatic radiation of intensity $I(\bar{\nu}_1)$. In eq. (3.119), the invariable part of the interferogram, which bears no spectral information, has been dropped.

When the interferometer is exposed to the intensity distribution $I(\bar{\nu})$ of a radiation source (or its response in the spectrometer $I_{exp}(\bar{\nu})$), (3.119) needs to be integrated over all wavenumbers, which results in a Fourier transformation of the intensity distribution from the reciprocal (wavenumber) space into the real (optical path length) space:

$$I(\delta) = \int_{0}^{\infty} I_{exp}(\bar{\nu}) \; cos \; (2\pi\bar{\nu}\delta)d\bar{\nu} \qquad (3.120)$$

Reverse FT of $I(\delta)$ restores the initial intensity distribution:

$$I_{exp}(\bar{\nu}) = \int_{-\infty}^{\infty} I(\delta) \; cos \; (2\pi\bar{\nu}\delta)d\delta \qquad (3.121)$$

The experimental spectrum of a sample, $I_{s,exp}(\bar{\nu})$, is obtained by measuring and comparing $I_{exp}(\bar{\nu})$ with and without the sample in the beam. The latter is referred to as background spectrum. Correction of $I_{s,exp}(\bar{\nu})$ by $C(\bar{\nu})$ to access the real sample spectrum $I_s(\bar{\nu})$ is performed automatically.

FT spectrometers operate without slits, which improves optical throughput and sensitivity. Measurements are fast, because a wide range of wavenumbers is covered simultaneously, and their wavenumber axis is very accurate, because it can be calibrated with a reference laser (Jaquinot, Fellgett, and Connes advantages, respectively). The evaluation of spectra is, however, affected by the finite limits of the integration in (3.121), which tends to introduce line broadening and side bands. Approaches to mitigate such effects remove side bands, but retain some line broadening. Resolutions of 2 cm^{-1}, as typically achieved with FTIR instruments are, however, sufficient for work with solid samples including studies on adsorbed species.

For most applications, the required radiation can be provided by a thermal light source (globar), an SiC rod electrically heated to >800 K, which behaves approximately like a Planck black body. Covering the mid-IR, it extends well into near- and far-IR at higher temperatures. Tungsten halogen lamps can be used for the full NIR and high-pressure Hg discharge lamps for the far-IR.

Detectors utilize the thermal effects of IR radiation (thermal detectors) or count photons that induce excitations in materials with very small bandgaps (quantum detectors). The DTGS detector uses the pyroelectric effect of deuterated triglycine sulfate which produces capacity changes; the more expensive MCT detectors (mercury cadmium telluride), which are much more sensitive and faster, generate a current proportional to the IR intensity. They need to be cooled by liquid nitrogen.

Sample compartments in IR spectroscopy are more diverse than in UV-Vis spectroscopy (cf. Section 3.4.7.2). In catalysis research, diffuse reflection is preferred also in the IR range ("DRIFTS"), but the transmission mode has its particular merits and applications. For studies of surface catalytic processes in the liquid phase, attenuated total reflection has become most important. In surface science, IR spectra are also measured in reflection mode, which will be briefly outlined, as well.

Basics of the transmission experiment have been described in Section 3.4.7.2. As scattering and reflection are less influential in the IR range, transmission was the standard mode for solid-state IR spectroscopy, for a long time. Nowadays, it is used, in particular, for quantitative work, because the Lambert-Beer law (3.101) allows determining site concentrations, once the extinction coefficients of the vibrations involved are known. They are certainly unknown for the majority of the myriad signals in the IR spectra of catalysts, but they are available for important standard applications, and more are going to be explored for relevant signals.

To avoid interaction with the ambient, e.g., with moisture, the self-supporting wafers for transmission measurements are housed in spaces with suitable windows that can be evacuated or flushed with gas. Due to limited thermal stability of the window materials, the setup must contain a second, heatable sample position, where the catalyst may be pretreated, e.g., degassed. Quantitative adsorption studies at different coverages or identification of adsorbates during a reaction are typical experiments that can be performed after moving the catalyst to the measurement position. *Operando* work, even at the low temperatures tolerated by the window materials, is possible only for very slow reactions, due to unfavorable mass transfer conditions in such cells (missing forced flow).

DRIFTS (for basics of diffuse reflection, see Section 3.4.7.2) is usually measured with mirror optics accessories, which stand out for straightforward sample preparation – just by filling the sample cup. Unlike in the UV-Vis region, there are no significant changes in optical yield, and hence in data quality, across the IR range. Fiber optical probes are now available also for the mid-IR and will play a greater role in the future. As scattering is less intense in the IR range than in UV-Vis, DRIFTS requires very sensitive detectors. On the other hand, due to multiple interactions between radiation and sample, absorption effects and, hence, signal intensities are larger. Unlike transmission, DRIFTS also allows detecting weak overtones and combination modes in the NIR. Transmittance of the sample is indispensable for transmission, but not for DRIFTS, which can provide acceptable spectra, even from colored or black materials when they are mixed with diluents (KBr or diamond powder). The eligibility of DRIFTS for *in situ* and *operando* work (cf. Section 3.4.7.2) is an additional advantage of this approach.

Due to the paramount influence of particle size and packing on scattering (scattering coefficient, S, in eq. (3.102)), DRIFTS is, however, inadequate for the determination of species concentrations. Trends in intensity ratios between bands of different species or of a species and a reference material, e.g., the support can, however, be used to establish tendencies of concentrations and concentration ratios.

Attenuated total reflection (ATR) uses the evanescent wave phenomenon. Evanescent waves extend into a space where the wave cannot propagate, e.g., behind an interface that causes its total reflection due to large refractivity differences between the adjacent media (Figure 3.117a). Electron wave functions that extend outside metal surfaces, which cause electron tunneling and the surface dipole contributing to the work function (Section 2.1, Figure 2.6b), are evanescent waves as well. They exist, because the continuity of waves would otherwise be violated at the turning points (or at the metal surface, Figure 2.6b). Although they do not transport energy, it is possible to tap them as illustrated by total light reflection within a prism in Figure 3.117a: from the right-hand side, the light in the prism is not visible, unless a second prism is placed near enough to touch the space with nonzero evanescent wave intensity. In this arrangement, some light can "tunnel" through the interface, causing total reflection. The respective intensity is missing in the reflected light.

In ATR-IR spectroscopy (Figure 3.117b), the evanescent IR waves extend from a special internal reflection element (IRE) into a fluid medium, where they are tapped by the absorption processes to be studied. The penetration depth, d_p, of the evanescent wave depends on the wavelength, λ, on the angle between incident beam and surface, and on the refractive indices of IRE and sample, n_{IRE} and n_s. In the mid-IR, d_p is typically between 0.5 and 2 µm. ATR-IR is comparable with transmission at very small optical path lengths, which are of the same order as d_p, mostly somewhat larger. It is, therefore, most suitable for studies of strongly absorbing liquids (Figure 3.117b), of powdered samples, which are deposited on the IRE and can be contacted with

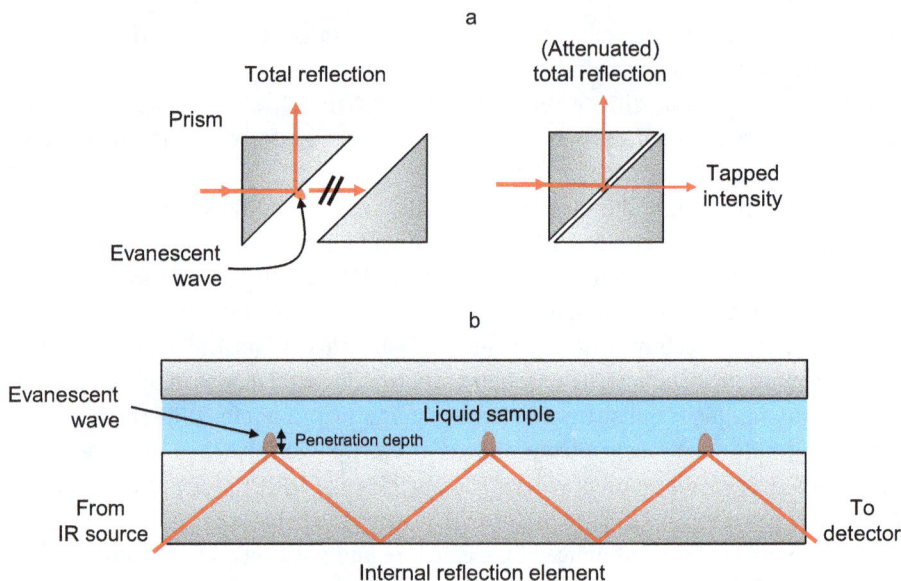

Figure 3.117: Principle of attenuated total reflection (ATR): (a) tapping intensity from an evanescent wave by moving a prism into its space; Adapted from a graphics published under https://de.wikipedia.org/wiki/Evaneszenz (accessed Jan 20, 2023), (b) ATR setup: tapping intensity from evanescent waves by oscillators within their space.

liquid reactants, and for interactions between liquids and electrodes, which closely approach the IRE.

Materials for the IRE should be transparent over a wide wavenumber range and stable in chemical media. Typical examples are ZnSe (n_{IRE} = 2.42), Si (n_{IRE} = 3.45), and Ge (n_{IRE} = 4.0). A choice of materials with different refractive indices is required, because unfortunate combinations of n_{IRE} and n_S may result in excessive decrease or growth of d_p, causing insufficient or complete absorption or a sensitivity of d_p to particle packing in the deposited sample layer. For more detail, the reader is referred to ref. [167]. IREs are available in different versions, from single-bounce elements (one active total reflection) to poly-bounce elements, with up to 25 reflections. The former are preferred for studies with a theoretical background, because the optical properties of the reflected light are better defined in them. As absorption by molecules adsorbed at catalyst surfaces is usually rather weak, poly-bounce elements are better suited for catalysis research.

Samples are prepared by dropping slurries of catalyst powders onto the IRE and drying them. Catalyst layers may be stabilized by a binder as familiar in electrochemistry. The porous layer should be thicker than d_p to avoid excessive intensity from species in the supernatant liquid. For studying electrode processes, metal films of some nm thickness can be deposited on the IRE by physical vapor deposition (cf. Section 3.4.12). The evanescent wave permits observing adsorbates even through the metal layer that quenches electromagnetic radiation. Adsorbates on single-crystal facets have been probed by pressing them onto a semispherical single-bounce IRE after previous contact with the gaseous or liquid adsorptive [168].

ATR is well suited for *in situ* and even *operando* work as temperatures are low in liquid-phase processes (cf. Section 3.4.12). Commercial cells are on the market; flow cells with minimized liquid volume above the solid sample have been described in literature (for more detail see ref. [167]). For studies of photocatalytic processes, the catalyst can be illuminated by LEDs shining through a window in the lid of the liquid sample space [169] (cf. Figure 3.117b).

In surface science, IR spectra are measured by IR reflection-absorption spectroscopy (abbreviated RAIRS, IRRAS, sometimes IRAS). The method requires a mirror function of the sample, which is best provided by atomically flat facets of metal single crystals, but it can also be used with polycrystalline metal surfaces. Surfaces are exposed to the radiation under incident angles Θ (from the surface normal) between 10 and >80°, and intensity is probed in the direction of specular reflection. Absorption by adsorbates or deposited layers results from two interactions between radiation and surface species – before and after the reflection. Notably, RAIRS is possible also with very thin nonmetal layers grown on metal surfaces, which provide the mirror function for radiation transmitted by the supported layer. This allows IR spectroscopy on well-defined model systems for real catalysts, e.g., Al_2O_3 grown on an NiAl alloy, which models a typical support for catalytically active metal particles [170] (cf. Figure 1.7a). For more applications of RAIRS, the reader may consult ref. [168].

Unlike SFG, RAIRS covers the whole IR range, and it is less costly. It allows detecting very low coverages of adsorbates when their absorption cross sections are high, but its surface sensitivity is limited: the superposition of signals from components in the gas phase with those of surface species puts strict limitations on adsorptive partial pressures. The problem can be mitigated to some extent by working with polarized radiation.[39] For work under realistic adsorptive pressures or even in the presence of liquid adsorptives, SFG is, however, the only option.

3.4.8.4 IR spectroscopy with catalysts: typical applications

IR spectroscopy allows studying various aspects of catalysts and catalytic processes. Lattice vibrations, which are below 1,200 cm^{-1} for most solids, can be used to identify structures. The study of surface chemistry is another important application, which deals mostly with surface OH groups of oxides or relevant groups like -SH or disulfide in sulfides. Likewise, adsorbates and their transformations can be investigated under various conditions. Performed with reactants, such studies may target the identification

---

39 Due to the quenching of electromagnetic radiation parallel to metal surfaces, spectra of adsorbates are sensitive to the polarization of light, unlike spectra of gas-phase molecules. The former can be, therefore, isolated by comparing data obtained with different polarizations. At high gas-phase pressures, such treatment may produce inacceptable noise.

of surface intermediates and reaction mechanisms. Conditions are chosen to allow the reaction of interest to proceed. Adsorbates can also be used to probe properties of the catalyst surface, e.g., acidity and basicity or oxidation states. In such studies, conditions are chosen to facilitate the experiment, without compromising on the complete and unmodified adsorption of the probe.

Due to their high intensity, lattice vibrations can be recorded only when samples are strongly diluted, e.g., by KBr. While this restrains the informational value with respect to other features, e.g., surface functional groups, the structural information often justifies the extra experiment, because it allows identifying phases at particle sizes too small to cause diffraction, or even structural motifs in amorphous materials. The frequency of lattice modes can depend on the abundance of particular atoms in the structure, e.g., of Al^{3+} in zeolites. On this basis, linear relations between band frequencies and Al content of the zeolite framework were established for different modes and zeolite types [171], which also allow the study of isomorphous substitutions (e.g., B for framework Al).

Present-day knowledge on surface functional groups and adsorbates has very much resulted from decades of IR work accompanied by perpetual effort for reliable band assignment, more and more backed by theoretical studies. Only for carbon materials, XPS has contributed the bulk of evidence.

In the following, IR characterization of surface functional groups is discussed with a focus on OH groups, which may be present at the surfaces of oxide catalysts or supports in various structural environments. In Section 2.2.2, type I through IV OH groups have been mentioned, as well as geminal and vicinal silanol groups. They all can be differentiated by their vibrational frequencies. Additional differentiation arises when the cations bridged by OH groups of type II and higher are in different coordination geometries. The ranking of five OH vibrations observed on γ-Al_2O_3 between 3,790 and 3,680 cm^{-1} was, for instance, explained by referring to the coordination numbers of the adjacent Al ions (IV or VI, cf. Section 2.2.2) and to the structure of preferentially exposed planes [172]. In a later DFT-based examination [173], the basic ideas were confirmed, though with some significant modifications.

In zeolites, Brønsted sites (OH groups bridging between framework Si and Al atoms, cf. Figure 1.7b) coexist with other OH groups: silanol groups at the external surface or in structural defects and Al-OH groups from extra-framework alumina species not integrated into the lattice during preparation or extracted from it by dealumination. These species can be observed in Figure 3.119b below, which shows bands of external silanol groups at 3,745 cm^{-1}, of framework silanol defects at 3,725 cm^{-1}, of Brønsted sites at 3,612 cm^{-1} and of bridging extra-framework OH groups at 3,665 cm^{-1}.[40] Undisturbed Brønsted OH groups appear between 3,600 and 3,650 cm^{-1}, but OH groups pointing into narrow cavities often vibrate at lower frequencies because the OH bond is weakened by H-bond interactions offered by nearby framework O atoms. The low-frequency OH

---

40 The signals shift by ca. 5 cm^{-1} upon cooling to 77 K.

band in Y zeolite at 3,550 cm^{-1}, which arises from OH groups in the sodalite cages, illustrates this case. Bands of Brønsted sites can be also red-shifted by interactions with O atoms of extra-framework species.

The interaction of molecules with oxide surfaces includes both chemisorption and surface reactions, and it is sometimes difficult to draw a line between adsorbates and new surface functional groups. Some rather unreactive reaction products (e.g., carbonates, nitrates, sulfates) are candidates for spectator species or even poisons in catalytic reactions. They may also block the surface during contact with the ambient (e.g., carbonates on basic catalysts) and need to be removed by an activation procedure. Other surface species as NH_4^+, nitrites, methoxy, formates, bicarbonates, and amides are candidates for reaction intermediates.

The diversity of surface species on oxide surfaces is exemplified in Scheme 3.4 by reaction products formed in the interaction of CO_2 or CO with basic sites, including some species that result from a consecutive interaction with hydrogen. Some of them occur during the hydrogenation of CO or CO_2 to valuable products, e.g., methanol (cf. Figure 4.49), the carbonates rather as spectators or poisons. Strategies for their differentiation sometimes allow assignment in surprising detail. In the free carbonate ion, for instance, there is only one IR-active stretching mode – the asymmetric one at 1,415 cm^{-1}. The less symmetric surface-bound carbonates absorb via both symmetric and asymmetric modes, which appear above and below 1,415 cm^{-1}. It was observed that the wavenumber difference between these modes ("Δv_3 shift") is related to the basicity of the surface O atom involved and the resulting carbonate structure: $\Delta v_3 \approx 100$ cm^{-1} indicates monodentate species formed with the most basic oxygen, $\Delta v_3 \approx 300$ cm^{-1} suggests bidentate, and $\Delta v_3 \approx 400$ cm^{-1} indicates bridged structures [174]. The Δv_3 shift needs, however, to be complemented by other pieces of evidence. There are, for instance, fairly symmetric polydentate structures (cf. Scheme 3.4), which exhibit small Δv_3 shifts, but high thermal stabilities. Therefore, thermal stability needs to be included in assignment effort [174].

Studying adsorbates during catalytic reactions in order to identify steps of the reaction mechanism or to detect reasons of poisoning phenomena is a major application of IR spectroscopy in catalysis research. It should be noted, however, that claims derived from IR work are abundant in literature, but not always justified. A credible basis for mechanistic conclusions includes differentiation between reaction intermediates and spectators and a correlation between surface concentrations of intermediates and measured reaction rates, which must be close to (or allow extrapolation to) rates measured in conventional reactors. Such effort requires a complex program including various spectroscopic and kinetic experiments (nowadays, usually combined in *operando* spectroscopy), the discussion of which is beyond the scope of this text. For instructive examples, the reader may consult refs. [175–178]. Instead, an experiment that revealed an unexpected poisoning mechanism is briefly outlined in the following.

Carbonates

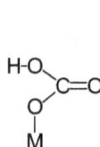

| Free ion | Monodentate | Bidentate | Bidentate bridging | Polydentate |

Bicarbonates   Formates

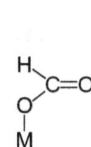

| Monodentate | Bidentate | Monodentate | Bidentate | Bidentate bridging |

Dioxomethylene   Methoxy

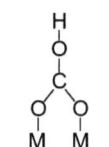

Bidentate

Scheme 3.4: Surface species detected on oxide surfaces interacting with CO or CO_2.

The example is based on a project dealing with the potential of gold for three-way catalysis (TWC), i.e., the simultaneous conversion of CO, unburnt hydrocarbons, and NO to CO_2, H_2O, and N_2 (cf. Appendix A1). Au nanoparticles supported on various oxides are extremely active for the oxidation of CO [179]. Hydrocarbons are oxidized at smaller, but still appreciable rates. While the rather low melting point of gold (Figure 3.19) is opposed to its use under TWC conditions, the project was meant to find out if Au might contribute its excellent oxidation activity to combinations with other (high-melting) metals.

In an initial study with monometallic Au catalysts, the fabulous CO oxidation activity of gold was found to be severely poisoned by propene, which represented the unburnt hydrocarbons in the TWC model feed. Over Au/La_2O_3-Al_2O_3 ("LaAl"), for instance, CO conversion was 90% already at ≈380 K both in a binary CO/O_2 and in the TWC model feed. In the latter, however, CO conversion started to decrease above 400 K, falling to 40% at ≈500 K, where it turned again to surpass 95% around 600 K. The effect was mitigated when Au was supported on more reducible oxides such as TiO_2 or $CeZrO_2$, where 95% conversion were achieved at 510 K and 470 K, respectively [180].

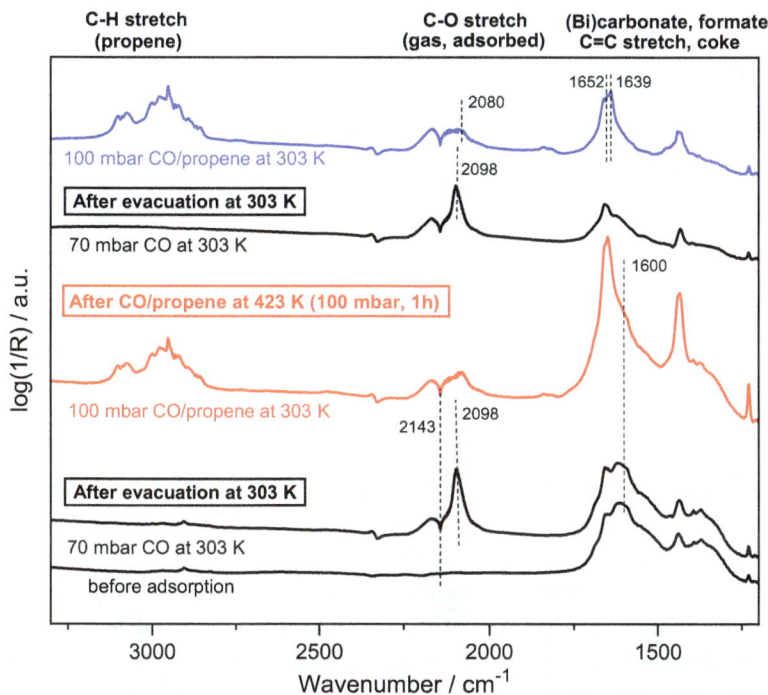

Figure 3.118: Three-way catalysis over supported gold nanoparticles: poisoning of Au/La$_2$O$_3$-Al$_2$O$_3$ in propene-containing feed studied by DRIFTS. Sequence of experiments is top-down (see text). Reproduced from ref. [181] with permission from Elsevier.

Figure 3.118 shows IR spectra, in which the interaction of CO with Au/LaAl at 303 K is compared in the absence and in the presence of propene (CO/propene[41] = 9). Basic bands assignments are indicated above the diagram. The upper spectrum was recorded after exposing the catalyst to these weakly poisoning conditions for 1 h. The signals of gas-phase species mask some adsorbate signals. The presence of adsorbed propene can be deduced only from the band at 1,639 cm^{-1} in the strongly superimposed 1,600–1,800 cm^{-1} region, because this wavenumber deviates from that of gas-phase propene (1,652 cm^{-1}). Adsorbed CO causes a tiny peak at 2,080 cm^{-1} sticking out of the rotational structure of gas-phase CO. Based on literature, the adsorption sites are thought to be at the perimeter between Au particles and the support.

Subsequent evacuation removed all adsorbate bands, but signals remained that arise from various (bi)carbonate species difficult to be differentiated (not shown). Upon dosing only CO on this surface, the adsorbate band was much more intense and blue-shifted, which reveals adsorption competition between CO and propene in the mixture.

---

41 Under dry conditions, as poisoning was more severe in the absence of H$_2$O.

The wavenumber shift of the CO signal suggests an electronic influence of propene on the co-adsorbed CO, most likely via the Au nanoparticles (see below and [181]).

After exposing the sample to the CO/propene mixture at 423 K (strongly poisoning conditions), CO adsorption was studied again. In the presence of propene, significant differences were noted only below 1,800 cm^{-1}. While not all details of this region were completely understood, it is obvious that a broad signal at $\approx$1,600 cm^{-1} remained from the previous step. IR signals around 1,600 cm^{-1} arise from carbonaceous deposits, which are typical reasons for catalyst deactivation. Notably, however, in the absence of propene, CO adsorbs on this surface at the same wavenumber and intensity[42], as before. Apparently, poisoning does not result from blockage of CO adsorption, but rather of the oxygen supply. As discussed in Section 2.3.3, surface oxygen of supports can be activated along the perimeter with the supported metal particles. This region around the Au particles becomes blocked by coking. At higher temperatures, the active oxygen cleans the surface by combusting the deposits, which explains that poisoning severity decreases with increasing support reducibility.

In the example, wavenumber shifts of the CO stretching mode were used to conclude on electron density changes at the adsorption site. The capability of adsorbates to reflect properties of the adsorption site in their IR spectra is the basis for surface characterization methods with probe molecules. Using an adsorption process, they are ideally surface-sensitive. Probe molecules can reveal properties of acid and basic sites. With respect to redox catalysis, the nature of the adsorbing metal site and its oxidation state can be gauged. In this respect, IR competes with XPS, outperforming it in surface sensitivity, but falling short in universality and, due to fragmentary data on extinction coefficients, with respect to quantitative analysis.

Probe molecules for acid sites are always bases. While they adsorb at Lewis sites (L) via a donor-acceptor interaction without dramatic changes in the molecule, the interaction with Brønsted sites, (B), depends on the strengths of the partners. Probes may be protonated, but the interaction with less acidic OH bonds may result in just an elongation and weakening of the O-H bond, which shifts of the O-H band to lower wavenumbers.

Probes that become protonated allow distinction between B and L sites, because the cationic and neutral adsorbates are easily differentiated by their spectra. The standard technique for this purpose is pyridine adsorption, where the adsorbates are distinguished by ring vibrations. The most relevant signals appear at $\approx$1,540 cm^{-1}, $\approx$1,640 cm^{-1}, and 1,610 cm^{-1} for the pyridinium ion, but at $\approx$1,450 cm^{-1} and in the 1,620–1,610 cm^{-1} range for coordinated pyridine (the third mode at $\approx$1,580 cm^{-1} is very weak). A fourth

---

42 Assessed by comparing peak heights with the intensity of the gas-phase signal.

mode at ≈1,490 cm^{-1} is insensitive to the type of acidity. The signal around 1,620 cm^{-1} was reported to respond to the strength of the L sites, with lower wavenumber indicating lower acidity. There is no such opportunity for B sites. Their strength can be explored by recording spectra after desorbing pyridine at different temperatures. Adsorption is usually performed at 323–473 K to avoid signals of H-bonded pyridine. Pyridine adsorption is suited for quantitative analysis, because extinction coefficients are available in literature for the nonsuperimposed bands at 1,540 cm^{-1} and 1,450 cm^{-1}.

Like pyridine, ammonia can also differentiate Brønsted and Lewis sites: NH_4^+ ions absorb at 1,450 cm^{-1} and at 3,130 cm^{-1}; coordinated NH_3 at 1,250 cm^{-1}, 1,610 cm^{-1}, and 3,330 cm^{-1}. NH_3 is easier to handle, and is a frequent reactant in catalysis, unlike pyridine. There are also extinction coefficients in literature for relevant bands. All the same, the use of NH_3 is sometimes discouraged because of lower specificity, higher complexity of its adsorption process, and its reducing properties.

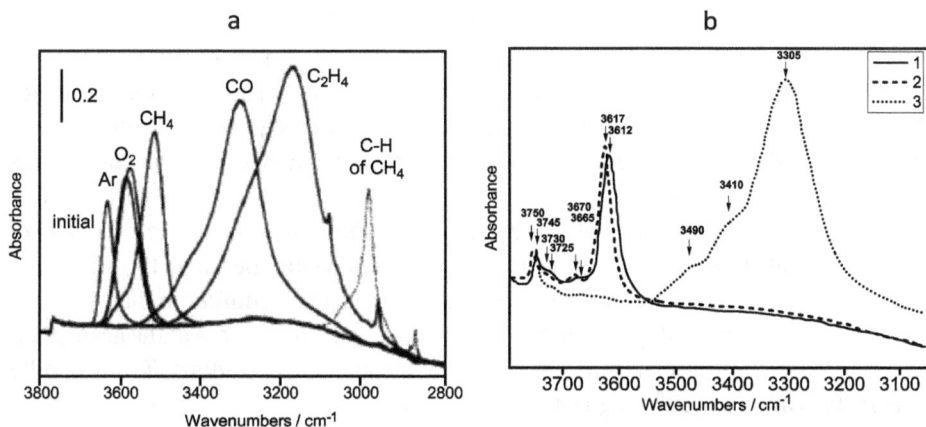

Figure 3.119: Displacement of O-H signals by the interaction with weak bases: (a) interaction of Brønsted sites in H-ZSM-5 with various bases (Ar, O_2, etc., see labels); (b) correlation of wavenumber shifts and intensity effects with the acidity of interacting OH groups. 1 – bare H–ZSM–5 at 300 K; 2 – bare H–ZSM–5 at 77 K; 3 – adsorption of CO at 77 K. Adapted with permission from refs. [182] (a) and [183] (b). Copyright 1994 (a) and 1987 (b) American Chemical Society.

The strength of Brønsted sites can be studied by their interaction with weak bases at low temperatures (e.g., 77 K), which is exemplified in Figure 3.119a. In these experiments, the OH band is not only shifted, but becomes more intense and broader and is sometimes distorted. Its intensity grows due to the larger variation of the dipole moment in the elongated oscillator. The change in band width is related to the anharmonicity of the vibration. In materials other than zeolites, distribution of acid site strengths may contribute to broadening and distortion.

For a particular probe, e.g., CO, shifts and effects on intensity and line width are linearly correlated with the acidity of the OH group. As mentioned above, four OH signals are identified in the spectrum of H-ZSM-5 in Figure 3.119b. Upon cooling to 77 K, they shift by 5 cm^{-1}; upon interaction with CO, three of them are distorted to different extents. The shift is largest for the Brønsted site at 3,612 cm^{-1} ($\approx$310 cm^{-1}), and smaller for the Al-OH vibration at 3,670 cm^{-1} (260 cm^{-1}) and the internal Si-OH group at 3,725 cm^{-1} (240 cm^{-1}). For terminal OH groups in the weakly acidic SiO$_2$, a shift of 90 K was found in a parallel experiment. For more details on the analysis of acidity with probe molecules, the reader is referred to refs. [184, 185].

Basic sites are characterized by acidic probe molecules, the most popular being CO$_2$. However, due to the various reaction options of basic surfaces with CO$_2$ (cf. Scheme 3.4), complicated spectra are obtained, which do not allow quantitative analysis. In an approach analogous to the characterization of acidic OH groups by their interactions with weak bases, the basicity of sites can also be studied by observing their effects on the C-H vibration of probe molecules, e.g., of pyrrole, chloroform, or propyne. With the latter, a correlation between the size of alkali cations exchanged into Y zeolite and the basicity of the framework oxygen was demonstrated [185].

For less fundamental work, methanol is an easy-to-handle probe for basic surface sites. In typical experiments, methoxy groups from dissociative adsorption, sometimes on-top and bridged forms, coexist with undissociated methanol, which can all be differentiated by their C-O vibrations. For some important surfaces, extinction coefficients are available in literature.

CO is a valuable probe not only for Brønsted sites, but also for the electronic state of transition metal atoms. When it adsorbs on them, the wavenumber of its C-O stretching vibration, $\bar{v}_{CO}$, exhibits characteristic shifts relative to its wavenumber in the gas-phase (2,143 cm^{-1}). These shifts depend on the nature and the oxidation state of the adsorption site: they reflect varying electron densities at the metal atom, excluding, of course, fully coordinated cations, i.e., the highest oxidation state.

CO adsorbs on metals on top and in twofold and threefold bridged sites. Signals of linear and bridged forms are easy to distinguish (see Figure 3.122 below), and conclusions on particle size or on surface composition in alloy surfaces can be derived from their intensity ratios. In the linear form, the sensitivity of the C-O vibration for the electronic state of the metal results from the bonding between the molecule and the metal (Figure 3.120b). Two interactions contribute to the bond: (1) a σ interaction between the 5σ orbital of CO (dominated by the lone pair at C, cf. Figure 3.120a) and a state of the metal, e.g., d_{z^2}, which results in an electron transfer toward the metal, and (2) a π interaction of the 2π^* orbitals of CO (Figure 3.120a) with metal orbitals of suitable symmetry, e.g., d_{xz} and d_{yz}, through which electron density is backdonated to the adsorbed CO molecule. As this electron density populates orbitals, which are antibonding with respect to the C-O bond, the capability of the metal to backdonate electrons influences the strength of this bond and can be therefore gauged by $\bar{v}_{CO}$.[43]

---

43 This explanation is sometimes referred to as Blyholder model. Another model for CO adsorption on metals is discussed in Section 4.2.2.

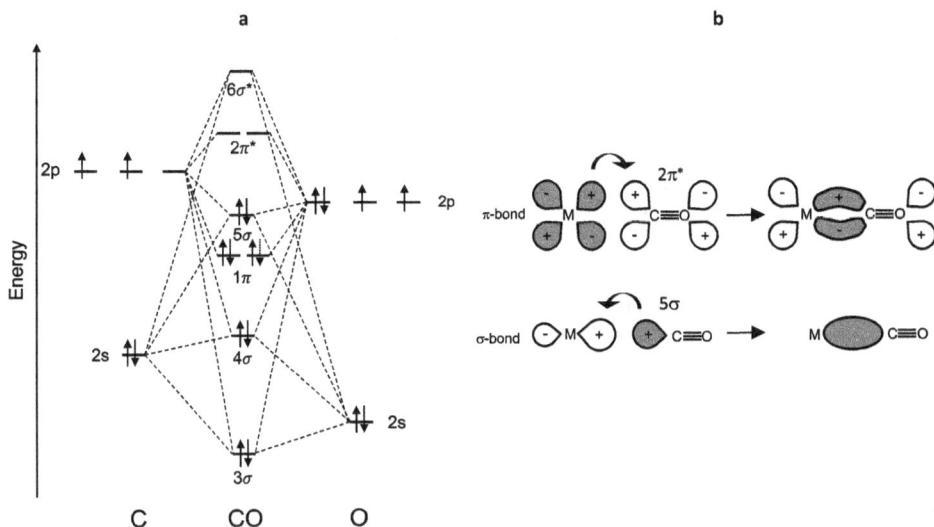

Figure 3.120: Molecular orbitals in the CO molecule (a) and in the bond between CO atop adsorbed on a metal surface atom (b).

The tendency of $\bar{\nu}_{CO}$ upon adsorption on cations suggests that another influence becomes relevant: $\bar{\nu}_{CO}$ increases with growing cation charge — even beyond the value of the free molecule. It was, therefore, proposed that shifts measured with CO on cations result from the Stark effect [186], which describes the influence of strong electric fields on molecular oscillations.

The differentiation of exposed elements by probe molecules is illustrated in Figure 3.121 by a classical example that paved the way to modern concepts of catalysts for hydrotreating reactions. The relation between Co and Mo in sulfided Co-Mo/Al$_2$O$_3$ was probed with NO, which gives rise to two signals.[44] Spectra a and b demonstrate how NO differentiates Mo and Co sites. In the spectrum of the sulfided Co-Mo catalyst, both doublets are superimposed, but with a clearly dominating Co signal, as can be deduced from the comparison with spectrum 4, for which the monometallic sulfide catalysts were mixed in the appropriate ratio. Applied to the layer structure of MoS$_2$ (Figure 2.17a), the observation that Co covers Mo in the bimetallic catalysts resulted in a model where Co replaces exposed Mo atoms in the edges of the MoS$_2$ slabs (CoMoS phase, see Figure 4.15).

---

44 Two NO molecules per site (dinitrosyl species) are adsorbed. They vibrate either in phase or in antiphase. NO is now rarely used because of its non-negligible reactivity towards the samples.

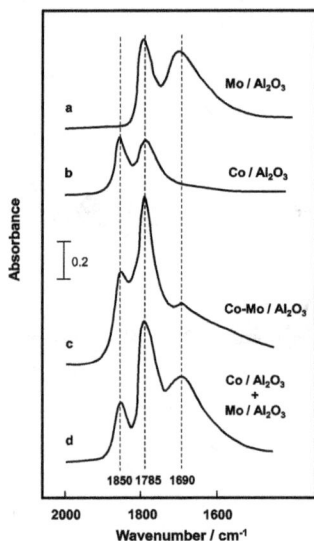

Figure 3.121: The relation between Co and Mo in sulfided Co-Mo/Al$_2$O$_3$ catalysts revealed by an IR study of adsorbed NO. Adapted from ref. [187] with permission from Elsevier.

Figure 3.122 reports the influence of liquid water on the IR spectra of CO adsorbed on a Pt/Al$_2$O$_3$ catalyst, a topic highly relevant in a time of growing attention to liquid-phase processes and electrochemistry. The spectra were measured in ATR mode with the same catalyst, either in flowing dilute CO or in CO-saturated water, and were corrected for the influence of the fluid phase.

In the gas phase (Figure 3.122a), the band of linear adsorbed CO appeared at 2,065 cm^{-1} and shifted to 2,071 cm^{-1} due to dipolar coupling effects at higher coverages. The broad band between 1,900 and 1,750 cm^{-1} arises from CO on twofold and threefold bridging sites. In water (Figure 3.122b), these bands were much more intense and red-shifted by almost 50 cm^{-1}. The authors noted also an influence of the pH value on band positions (but not on intensities) and a higher rate of CO oxidation in water than in the gas phase.

The changes in the spectra were attributed to a combination of two effects (Figure 3.122b, inset): a H-bond interaction between water and the CO molecule, which weakens the C-O bond and causes a shift and a growing intensity of its vibration similar to interactions of weak bases with O-H bonds (Figure 3.119), and an electronic effect of the water on the metal particle. The latter enhances electron backdonation to CO, which contributes to the red shift, but not to signal intensity. The electronic effect can be further varied by changing the pH value.

3.4.8.5 Measuring Raman spectra of heterogeneous catalysts

Raman scattering is rather inefficient: only 1 of 10^8 photons is scattered into the Stokes or anti-Stokes series, which appear as weak signals close to the intense elastic (Rayleigh) peak. Thanks to the use of laser sources, holographic notch filters, grating monochromators, and position-sensitive detectors that allow simultaneous registration of

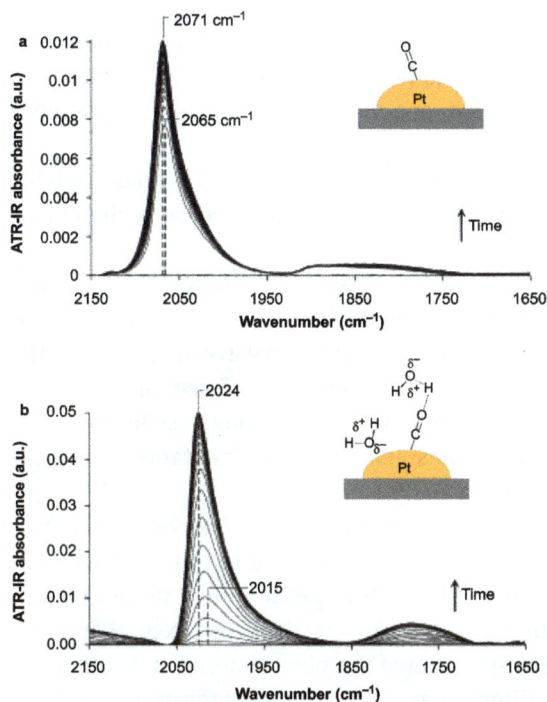

Figure 3.122: The influence of water on the adsorption of CO on Pt: ATR-IR spectra measured in the gas phase (a) and in water (b), with models for explaining the effects observed. Adapted from ref. [188] with permission from Elsevier.

extended wavelength regions, it is utilized, however, in a highly effective and stable spectroscopy nowadays, which offers attractive opportunities for the study of catalysts. Its specificity for transition metal oxide sites attached to supports of low polarizability like Al_2O_3, SiO_2, or zeolite has been already mentioned. Due to negligible Raman scattering in the gas phase, by water, by glass and quartz, it allows *in situ* work in both liquid and gas phases.

To make use of these promises, some problems introduced by solid samples must be solved. Sample fluorescence, which exceeds Raman intensities by orders of magnitude, must be avoided. At high temperatures, the influence of thermal radiation must be suppressed. Both problems can be mitigated by shifting the excitation wavelength.

Fluorescent sites in solids typically emit in the 300–800 nm region. At an excitation wavelength of 532 nm, which is familiar in work with liquid samples, this radiation may drown the Raman spectra. With excitation radiation in the near-IR or in the UV, the spectral lines appear in these regions as well. The UV option has the additional advantage of a higher Raman efficiency: the inelastic yield increases with v_{excit}^4. UV excitation is also favorable in work at elevated temperatures, because this spectral region is less affected by blackbody radiation of the sample. Due to the high energy deposited by the UV laser, care must be taken to avoid sample damage in these experiments. NIR excitation, which is applied in spectrometers operating with FT technology, is more commonly used with organic materials.

Most recently, pulsed time-gated measurement protocols have become available that reject fluorescence radiation due to its delayed temporal response to the pulsed laser excitation [189].

Figure 3.123a shows the typical configuration of a UV Raman spectrometer. The laser is directed on the sample via a beam splitter, and the scattered intensity is collected by focusing optics. The notch filter reflects the excitation wavelength back to the sample and allows the remaining intensity passing to spectral differentiation. Similar to EDEXAFS (Figure 3.84c), this is achieved by a slightly bent grating: it sends radiation of different wavelengths into different directions, enabling their simultaneous registration on the surface of charge-coupled devices (CCD).

The application of Raman spectroscopy to operating catalysts has very much benefited from the invention of fiber-optical light guides. Due to the huge differences between excitation and Raman signal intensities, fiber-optical Raman probes (Figure 3.123b) are designed to meet specific challenges. In the fiber, the excitation beam becomes contaminated with the Raman spectrum of the silica material and with other contributions. Therefore, a short-pass filter is placed at the outlet of the excitation fiber. Then, the beam passes a ball lens, which guides it to the sample and diverts the backscattered light into a concentric space probed by the collection fibers. Before entering these fibers, the scattered light passes the notch filter.

Although Raman spectroscopy is an important tool for the study of catalysts and catalytic materials (cf. Section 3.4.8.6), it is not suited well for determining concentrations of the observed species. Reasons for this failure, which include matrix effects and the complex changes of scattering properties with sample packing, coloring, etc. (cf. Sections 3.4.7.2 and 3.4.8.3), are discussed in ref. [190].

The basic version of Raman spectroscopy discussed so far, which is sometimes referred to as spontaneous Raman scattering, is predominant in catalysis research. Modified Raman processes like resonance Raman scattering and Surface-enhanced Raman scattering (SERS) are likely to gain importance in the future. In resonance Raman scattering, effects induced by exciting the Raman transitions at an energy coinciding with that of an electronic transition within the sample are used. SERS utilizes metal particles (typically Ag) as secondary probe heads, because the electromagnetic field of the excitation radiation can be extremely enhanced by plasmon resonances in them (cf. Section 3.4.7.1). The reader may find more information on these techniques in refs. [191, 192].

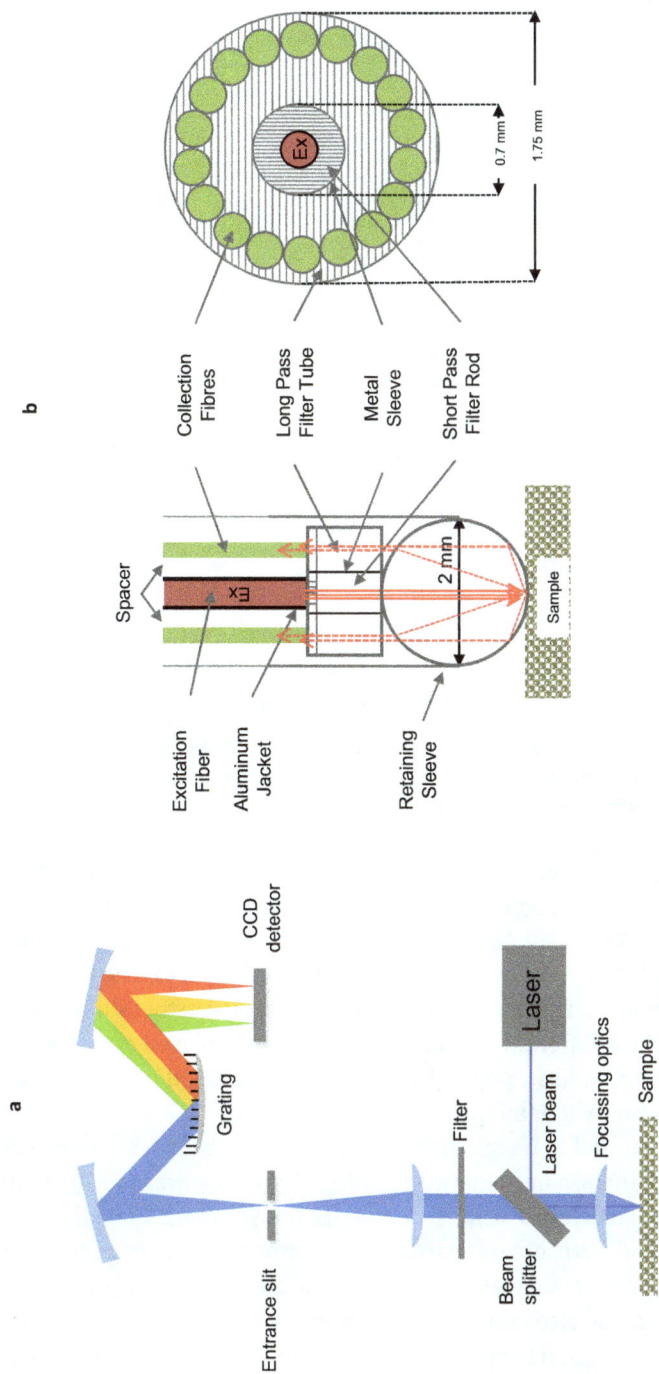

Figure 3.123: Instrumentation for Raman spectroscopy: (a) general scheme of an instrument using a grating spectrometer, (b) sampling tip of an optical fiber probe. Adapted from ref. [193] with permission of the Optica Publishing Group.

3.4.8.6 Raman spectroscopy in catalysis: typical applications

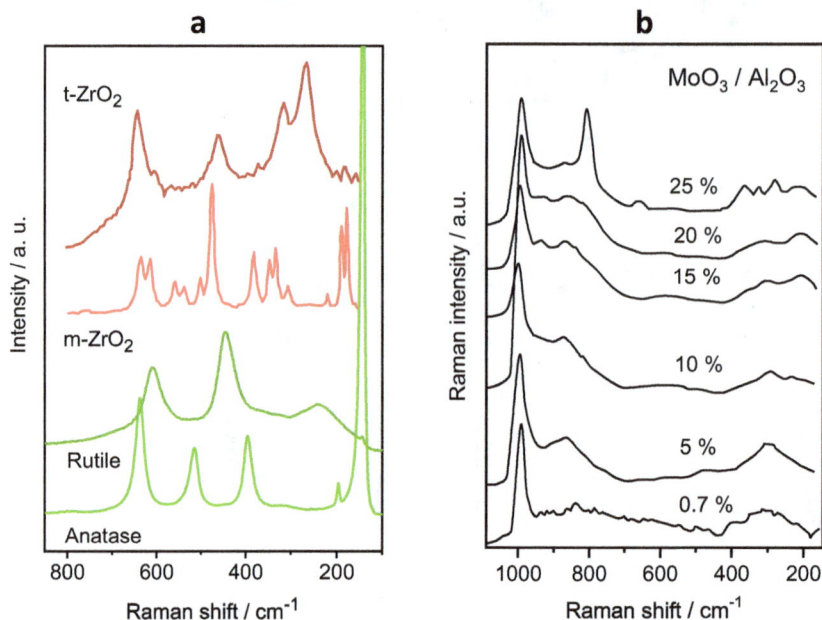

Figure 3.124: Raman spectroscopy for the characterization of solid phases: (a) differentiation of monoclinic and tetragonal ZrO_2, of anatase and rutile; (b) characterization of supported oxide species: Mo sites on Al_2O_3, (b) Adapted from ref. [194] with permission from Elsevier.

Like IR spectroscopy, Raman spectroscopy allows identifying phases or structural motifs at particle sizes (or degrees of disorder) interfering with diffraction. Figure 3.124a shows that tetragonal and monoclinic ZrO_2 are easily differentiated, which causes problems in XRD, if reflections are broadened due to small particle sizes. The differentiation of anatase from rutile (cf. Figure 3.124a) was used in an interesting study about the phase transition, in which different excitation lasers were employed to probe different depths: due to the strong absorption of TiO_2 (only) in the UV, the near-surface region has more influence on UV Raman spectra than on spectra excited at 532 nm. The phase transformation was found to be initiated at contact points between nanoparticles in the typical porous anatase agglomerates (with BET surface areas >150 m^2/g), where the critical anatase particle size could be exceeded (cf. Section 2.3.2.5). Increasing temperature accelerated both phase transition and sintering, which proceeded inward out until rutilization was completed in the external surface region [191].

Raman spectroscopy can be also used to observe nucleation processes in aqueous systems, which employs its potential to identify structural motifs in disordered phases. Despite low scattering efficiencies, this works even for siliceous and alumosilicate systems, as demonstrated in studies of mechanisms operating in the synthesis of zeolites (cf. ref. [191]).

In Figure 3.124b, the characterization of supported transition metal oxide species is exemplified by a study of dehydrated MoO_3/Al_2O_3 catalysts. As indicated in Sections 2.3.3 and 3.4.2.5 (cf. Figure 3.59), Mo(VI) forms two-dimensional surface oxide species under these conditions. At low surface coverages, isolated species dominate and coexist with oxide islands. While the sharp band just above 1,000 cm^{-1} was assigned to the Mo=O bond in isolated molybdate species, the broad signal extending to ca. 800 cm^{-1} originates from Mo=O bonds in various oligomeric surface oxide structures, the bridging Mo-O-Mo structures of which absorb below 400 nm. Only at high Mo content, MoO_3 was detected by the signal slightly above 800 cm^{-1}.

Scattering efficiencies of three-dimensional oxide particles, though not known exactly, are considered to exceed those of surface-bound species by an order of magnitude. Therefore, the corresponding bands mark an endpoint in what might be perceived as a titration of the capacity of supports to accommodate monolayer surface oxide species [194].

Supported transition metal oxide catalysts were also characterized in different reaction atmospheres. While the surface structure is significantly perturbed in the ambient due to the formation of a surface water film resulting in hydration and even dissolution, changes are minor upon heating. Combining Raman spectroscopy with rate measurements, it was concluded that these catalysts engage in various redox reactions, not via the Me=O double bond, but via their Me-O-support bonds, which explains pronounced dependences of rates on the nature of the support [195]. In a similar way, Raman spectroscopy was used to study transition metal oxide species introduced into zeolites in framework or extra-framework positions (ref. [191]).

Figure 3.125 illustrates how Raman spectra can be employed to assess defect structures. Such opportunities are related to the fact that highly symmetrical structures give rise to a low number of lines. Ideal graphite, for instance, has two major bands referred to as G and 2D (or G') at 1,580 cm^{-1} and 2,690 cm^{-1}. Structural defects cause an additional signal at $\approx$1,350 cm^{-1}, the D band. While the G band results from in-plane vibrations of sp^2-hybridized C atoms, the D band is a ring-breathing mode that can be activated only adjacent to a structural defect. The G/D intensity ratio is employed to assess the perfection of graphite-related structures (incl. graphenes or CNTs, cf. Figure 3.125a).[45]

CNTs can exhibit G/D values >150, but these types are not of interest for catalysis due to their low surface reactivity. G/D values of 1 and 0.8 evaluated from the spectra in Figure 3.125a for the undoped (commercial) CNTs and NCNT-L, respectively, reflect a high defect density, particularly in the N-doped CNTs, which were made by CVS with

---

45 Notably, although the 2D band is the overtone of the D mode, it can be activated *without* nearby defects: it is, therefore, intense also in perfect graphitic structures and is usually omitted in studies dealing with defects. Shape and position of G and 2D (G') signals, however, bear much more information on the properties of graphene materials.

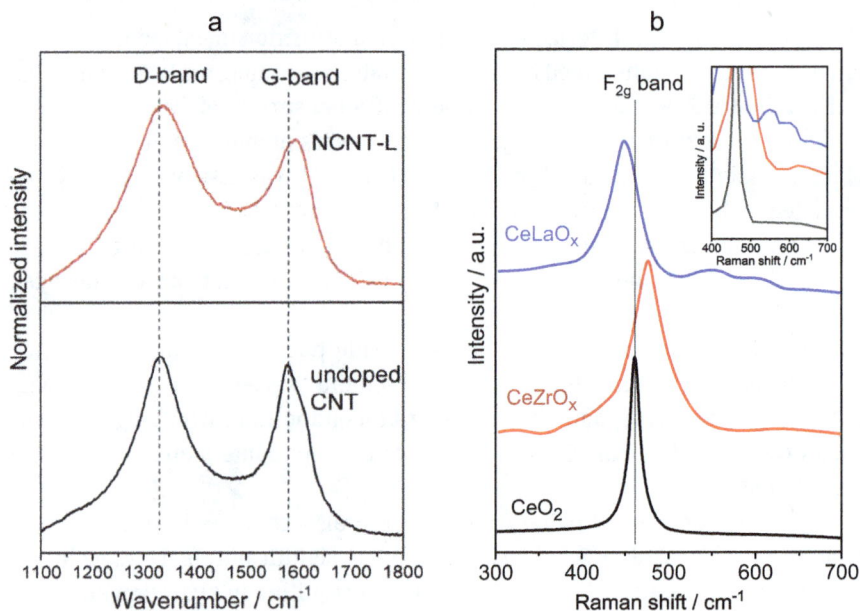

Figure 3.125: Raman spectroscopy of defect sites: (a) structural defects in graphene and related systems (here, CNTs). Reproduced with permission from ref. [196], copyright American Chemical Society 2009. (b) Lattice expansion/contraction and oxygen vacancies in CeO_2 and Ce-based mixed oxides. Adapted from ref. [197] with permission from Elsevier.

an N-containing source (acetonitrile) using a Co catalyst (cf. Section 3.1.1.5). Nitrogen-doped CNTs turned out to be promising noble-metal-free electrocatalysts for the oxygen reduction reaction (cf. Section 4.9) in both acidic and alkaline media, although problems remained.

The highly symmetric fluorite structure of CeO_2 gives rise to just one Raman signal, a symmetric breathing mode of the O^{2-} ions around Ce^{4+} (F_{2g} mode) at ≈460 cm^{-1}. In Figure 3.125b, this signal is shown as a reference to be compared with spectra of Ce-Zr and Ce-La mixed oxides. The simplicity of the mixed-oxide spectra shows that the fluorite structure is preserved, with the foreign cations occupying Ce positions (compare with ZrO_2, Figure 3.124a). Shifts of the F_{2g} signal reflect lattice expansion or contraction due to differences in ion sizes between Ce^{4+}, Zr^{4+} and La^{3+}. The weak signals around 600 cm^{-1} (see inset) arise from lattice defects, mostly oxygen vacancies. They are strongly enhanced by the doping, and two different bands can be distinguished – at ≈600 cm^{-1} and at ≈550 cm^{-1}. The former arises from oxygen vacancies compensating for Ce^{3+} in the structure; the latter from vacancies related to trivalent cations (i.e., La^{3+}) substituted for Ce^{4+}. Obviously, the latter are absent in $CeZrO_x$. However, the lattice contraction due to substitution of Ce^{4+} by the smaller Zr^{4+} apparently creates oxygen vacancies, which are compensated by reduction of Ce^{4+}.

3.4.9 Magnetic resonance spectroscopy

3.4.9.1 Magnetism and magnetic resonance

Before dealing with magnetic resonance, it is useful to recall some basic information about magnetism and magnetic materials. Magnetism describes forces mediated by magnetic fields, which originate from (and act on) electrical currents and atoms with inherent magnetic properties. In magnetic materials, the action of such magnetic atoms may be combined in patterns discussed below, which results in specific types of response to the magnetic field. Magnetism is a dipolar (i.e., directional) property: fields originating from a source are related to the orientation of the source in space. Correspondingly, the force (due to the dipolar nature of the magnets, a torque, rather than a force) imposed by an electrical current or a strong magnet on a weaker one depends on the mutual orientation of the partners.

The situation may be described considering a magnetic field of the strength $\vec{H}$[46] originating from a current (e.g., in a coil) or from a permanent magnet. It forces weaker magnets into an orientation that enhances the total magnetic field, although this alignment is counteracted by thermal motion. The joint effect of the original field, $\vec{H}$, and the aligned smaller magnets is described by the magnetic induction (or flux density), $\vec{B}$. It is related to $\vec{H}$ and the properties of the aligned magnets by:

$$\vec{B} = \mu_0 (1 + \chi_V) \vec{H} \tag{3.122}$$

In (3.122), χ_V is the magnetic susceptibility of the material that contains the smaller magnets, i.e., its capability to enhance the magnetic field. μ_o is the permeability of the vacuum, which is a natural constant with the dimension $N\ A^{-2}$. B has the dimension of T (Tesla): $1\ T = 1\ Vs\ m^{-2}\ (= 1\ kg\ A^{-1}\ s^{-2})$.

The following treatment is confined to the case of small magnets interacting with a strong one. A small magnet is described by its magnetic moment, $\vec{\mu}$. The magnetic moment determines the torque, $\vec{M}$, created in the small magnet by the flux density, $\vec{B}$ raised by the strong magnet:

$$\vec{M} = \vec{\mu} \times \vec{B} \tag{3.123}$$

The absolute value of the torque is $|\vec{M}| = |\vec{\mu}||\vec{B}|\ sin\alpha$, where α is the angle between the magnetic field and the magnetic dipole in it. The torque acts on an axis perpendicular to both $\vec{\mu}$ and $\vec{B}$. The potential energy of the (not-yet-aligned) dipole in the field is:

$$E = -\vec{\mu}\ \vec{B} \tag{3.124}$$

---

46 With the absolute value H, measured in $A\ m^{-1}$.

Dipoles change orientation to minimize this, which is fulfilled when $\vec{\mu}$ and $\vec{B}$ are parallel.

Magnetic properties of atoms or ions are caused by spinning elementary particles (electrons or nucleons). They can be cancelled under certain conditions, e.g., in electron pairs. The magnetism of electrons and nucleons is sometimes illustrated by the comparison of the spinning charge with a circular current. When an electron is unpaired, not only its spin, $\vec{s}$, is unbalanced by a second one, but also its orbital momentum, $\vec{l}$. Both add vectorially to a total angular momentum $\vec{j}$:

$$\vec{j} = \vec{s} + \vec{l}$$

This gives rise to $(2j + 1)$ degenerate energy levels, which can be differentiated by their response to the magnetic field (Zeeman effect, see below). When an atom (ion) contains more than one unpaired electron, various coupling principles become effective, which the reader may find in textbooks of physical chemistry [162].

The coupling between spin and angular momentum, which is well defined in free atoms (ions), is strongly modified when a cation is surrounded by a coordination sphere of anions. The theoretical description of this situation is focused on the interaction of the field with the spin, while the modification of the spin by its coupling with the orbital momentum is encoded in a spectral parameter (g, see below).

Table 3.3: Important magnetic nuclei and their nuclear spins I.

	1/2	1	3/2	5/2	7/2
Isotope	^1H, ^{13}C, ^{29}Si, ^{31}P, ^{15}N	^2H	^{23}Na	^{27}Al	^{51}V

Like electrons, nucleons have both a spin and an angular momentum. They add to a total nuclear magnetic momentum (the "nuclear spin") in a way that does not depend on the chemical environment. Therefore, only the existence and the size of nonzero nuclear spins are relevant from the chemical point of view. Table 3.3 reports the most important magnetic nuclei with their spin I (see below).

The magnetic moment, $\vec{\mu}_G$, of electrons or nucleons is proportional to their spin (generalized as $\vec{G}^{47}$), and the ratio between them is the gyromagnetic ratio, γ:

$$\vec{\mu}_G = \gamma \, \vec{G} \tag{3.125}$$

---

47 The generalized vector $\vec{G}$ has been introduced in this treatment to emphasize the analogies between nuclear and electron magnetism, but also, because it is not always clear in literature if $\vec{S}$ and $\vec{I}$ are used with units (Js) or without, as in this text.

For electrons, $\vec{G}$ is the spin vector, $\vec{S}$ ħ resulting from the coupling of all unpaired spins available, for nuclei, it is $\vec{I}$ ħ. The gyromagnetic ratio γ of charged particles is related to their charge, q, and their mass, m by

$$\gamma = g\,\frac{q}{2m}\left(= g\,\frac{\mu_P}{\hbar}\right) \tag{3.126}$$

g is the gyromagnetic factor (g-factor), μ_P is the magneton of a particle P ($\mu_P = \hbar\, q/2\, m_P$). For electrons, it is called the Bohr magneton μ_B. Equation (3.126) shows how quantities may be lumped in different ways. From eqs. (3.125) and (3.126), the magnetic moments of atoms and nuclei are

$$\vec{\mu}_e = g\,\frac{\mu_B}{\hbar}\,\vec{G} \qquad \vec{\mu}_N = \gamma\,\vec{G} \tag{3.127}$$

Both equations labeled (3.127) are completely equivalent, but preferred in different fields: the one with μ_B for electron and the one with γ for nuclear magnetism. Due to the inverse relationship of γ (or μ_P) to the particle mass m_P (cf. (3.126)), which is not balanced by the nuclear g values,[48] the magnetic moments of nuclei are smaller than that of the electron by more than three orders of magnitude. As a result, the magnetic properties of materials are dominated by the magnetism of the electron shell.

The mutual interaction of magnetic atoms results in characteristic coupling phenomena. If these are negligible, the atoms behave as independent elementary magnets: they have no preferential orientation in the absence of a magnetic field (Figure 3.126), but align with a field vector, $\vec{H}$. This enhances the magnetic flux density, $\vec{B}$ (cf. eq. (3.122)), i.e., the susceptibility, χ_V, of such paramagnetic materials is >0. Due to the competition between the effects of the magnetic field and of thermal energy, χ_V changes reciprocally with T (Curie law).

Materials without magnetic atoms are not void of magnetism. An external field $\vec{H}$ causing polarization of the electron shells induces in them a very weak flux $\vec{B}$ opposite to $\vec{H}$, i.e., a negative susceptibility χ_V. Such materials are diamagnetic. As the induction of the reverse flux consumes energy, diamagnetic materials avoid spaces with higher field intensities, H, i.e., diamagnetic materials are repelled by the field.

Cooperative magnetic phenomena cause much more drastic effects than paramagnetism or diamagnetism. They arise from exchange interactions between the magnetic atoms, which align their magnetic moments either in a parallel or in an antiparallel way (ferro- or antiferromagnetism, respectively, Figure 3.126). When the antiparallel atomic magnets have different absolute values, the material is ferrimagnetic (Figure 3.126).

In ferromagnetic materials, the magnetic order extends over domains sized in the µm range. Energetically unfavorable strong fields outside the material, which would

---

48 g value of free proton – 5.586, of free neutron – -3.826, of free electron – 2.002.

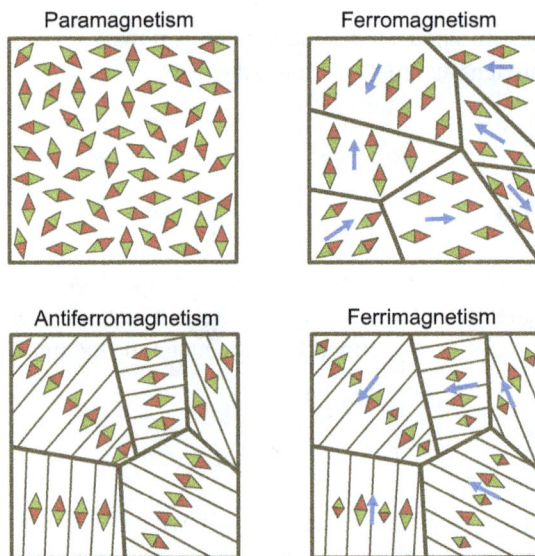

Figure 3.126: Differentiation of magnetic materials according to the orientation of elementary magnets, in the absence of a magnetic field.

result from larger domain sizes, cause domain splitting, while the energy required for creating domain walls puts lower limits on the size. When a ferromagnetic material heated to a temperature at which magnetic ordering is disrupted by thermal energy (the Curie temperature, T_c), is cooled in the absence of a field, the magnetic domains form without preferential orientation (Figure 3.126): the material is nonmagnetic. Upon exposure to a growing field, $\vec{H}$, the domains order under annihilation of the domain walls, which finally results in one-domain grains of the material. Due to the quantum mechanical exchange interaction, the resulting susceptibility χ_V is orders of magnitude higher than the value obtained by simple addition of the same quantity of independent (paramagnetic) moments. When the field is released, the magnetization does not break down completely. Depending on the material, the remaining "remanent" magnetization may be small (nonpermanent, soft magnetism) or near the value under the field (permanent magnets).

Ferromagnetism is linked to specific ranges of temperature and particle size. Above T_C, where thermal energy annihilates the exchange interaction, the material becomes paramagnetic. χ_V, which exhibits a complex behavior below T_C, decreases with T^{-1} above (Curie-Weiss law). When particles of a ferromagnet are below the critical domain size, the magnetic moments in these one-domain particles are still coupled by the exchange interaction, but the orientation of magnetization is influenced by thermal energy. The whole of such "superparamagnetic" particle behaves like an elementary magnet of a paramagnetic material, but with a huge magnetic moment and, correspondingly, high magnetization in the field.

Antiferromagnetic materials are also subdivided into magnetic domains, though for different reasons than ferromagnetic materials. Their magnetic moments cancel in all volume elements ($\chi_V = 0$, cf. Figure 3.126), they become magnetic only when the exchange interaction becomes disrupted by thermal energy. Once this interaction is annihilated (above the Néel temperature), antiferromagnets behave like paramagnetic materials, similar to ferromagnets above the Curie temperature. When the Néel temperature is approached by heating, χ_V increases gradually, until it starts to decrease with T^{-1}, above the Néel temperature.

Ferrimagnets have an antiferromagnetic orientation of the coupled magnetic moments, which are, however, of different magnitude. Therefore, ferrimagnets behave like ferromagnets, though with lower magnetization within the domains. Above their Néel temperature, they behave like paramagnetic materials.

The principles of magnetic resonance are based on the behavior of individual magnetic atoms or nuclei. It is described by the Schrödinger equation with the appropriate energy operator substituted, the solutions of which are wave functions with related energies as eigenvalues. For the electron spin $\vec{S}$ and the nuclear spin $\vec{I}$ (both dimensionless), the energy operators (Hamiltonians) $\hat{H}$ are (cf. eq. (3.124))

$$\hat{H}_e = -g\,\mu_B\,\widehat{\vec{S}}\vec{B} \qquad \hat{H}_N = -\gamma\,\hbar\,\widehat{\vec{I}}\vec{B} \tag{3.128}$$

where $\widehat{\vec{S}}$ and $\widehat{\vec{I}}$ are spin operators. The resulting eigenvalues (energies) of the mth eigenfunction are:

$$E_{e,m} = g\,\mu_B\,B\,m_S \qquad E_{N,m} = \gamma\,\hbar\,B\,m_I \tag{3.129}$$

m_I (m_S) are the quantum numbers labeling the solutions. They are, at the same time, the component of the nuclear spin $\vec{I}$ (or of $\vec{S}$) in the direction of the magnetic flux $\vec{B}$, e.g., $I_z = m_I = -I, -I+1, \ldots, I-1, I$, likewise for S_z (m_S). I and S can assume positive integer or half-integer values, which results in the well-known ($2I + 1$) or ($2S + 1$) multiplicities of the states. According to eq. (3.129), all these states have the same energy in the absence of a field; they are degenerate.

A magnetic field splits these degenerate states energetically, as illustrated in Figure 3.127 for two examples (Zeeman effect). Transitions between the states can be achieved by irradiation with electromagnetic waves of suitable frequency ν:

$$\Delta E = h\,\nu = g\,\mu_B\,B \qquad \Delta E = h\,\nu = \gamma\,\hbar\,B \tag{3.130}$$

Frequencies are in the microwave range in Electron paramagnetic resonance (EPR), and in the radio wave range in Nuclear magnetic resonance (NMR). It should be noted that due to the tiny energy differences between the magnetic states, differences in their population are very small. In a field of 0.3 T, around which B is typically scanned in an

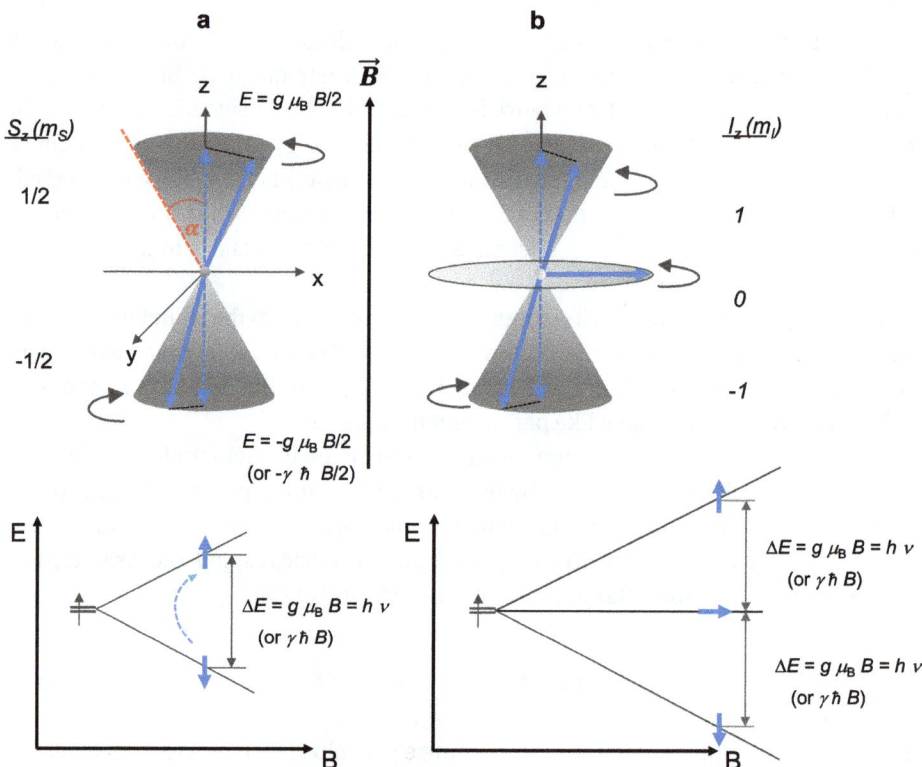

Figure 3.127: Elementary magnets in the magnetic field: (a) S = 1/2 state (or I = 1/2); (b) S = 1 state (or I = 1).

X-band EPR spectrometer, the excess population of the ground state is just 0.1% at room temperature [198]. Due to the much weaker nuclear magnetism, energy differences achieved by the much stronger magnets in NMR spectrometers are even smaller, and so are the excess populations: at 10 T, for instance, the excess of parallel spins is just 1 in 10^5 [199].

The quantization of the angular momentum defines only its z component in the field direction, while x and y normal to the field remain free. As a result, the elementary magnets perform a precession movement around the field direction, in which their axis moves on a cone, the angle of which is defined by the ratio of S_z to $|\vec{S}|$ (I_z to $|\vec{I}|$, cf. Figures 3.127, 3.129, 3.135). The angular frequency ω of this precession, which is called Larmor frequency, is proportional to the field B via a factor of proportionality, which can be shown to be the gyromagnetic ratio γ:

$$\omega = 2\pi\nu = \gamma B \tag{3.131}$$

According to eq. (3.130), this is also the angular frequency $\omega = \Delta E/\hbar$ of the radiation, a quantum of which allows the transition between the states. As a result, the transition causes the dipole to change its orientation to the field, but its precession frequency remains the Larmor frequency (3.131).

By now, the magnetic moments of electrons or nuclei have been considered in isolation. However, various interactions complicate the picture, providing, at the same time, diagnostic opportunities. The magnetic moment of an unpaired electron, the size and direction of which is influenced by the coordination sphere around the ion in a way to be discussed below, may interact with the moments of a magnetic nucleus or of other unpaired electrons. The nuclear magnetic moment, the external effect of which is modified by diamagnetic electron shells in a way to be discussed below, may interact with nearby paramagnetic sites or with nearby magnetic nuclei of the same or of different nature.

When electron and nuclear magnetic moments coexist, the field raises additional energy contributions by the nuclear Zeeman effect and by the coupling between electron and nuclear moments. The latter results in the hyperfine splitting of the signals. Together with the (electron) Zeeman effect, the Hamiltonian for this situation is (cf. (3.128)):

$$\widehat{H} = g\,\mu_B\,\widehat{\vec{S}}\vec{B} - \gamma\,\hbar\,\widehat{\vec{I}}\vec{B} + a\,\widehat{\vec{S}}\,\widehat{\vec{I}} \tag{3.132}$$

where a is the hyperfine splitting constant. Due to a selection rule allowing only transitions without change of m_I, the nuclear Zeeman effect does not influence the pattern. Figure 3.128a shows, how an electron spin and a nuclear spin of I = 1/2 are coupled into two states with parallel and antiparallel dipole orientation, which are slightly different already without the field. In the field, they provide two signals, the field difference between which can be shown to be equal to a. Coupling of electrons with nuclear spins, I, results in ($2I + 1$) lines, all spaced by the same distance, a (see Figure 3.132a below).

Likewise, when an atom contains more than one unpaired electron, their coupling may also cause significant energy differences. Figure 3.128b depicts the case of a triplet state in ligand fields of isotropic and of axial symmetry. The splitting is described by a splitting parameter, D and the Hamiltonian is:

$$\widehat{H} = \widehat{\vec{S}}\,D\,\widehat{\vec{S}} \tag{3.133}$$

This interaction, which is sometimes referred to as fine structure, depends on the coordination geometry (Figure 3.128b). In anisotropic geometries, it results in significant splitting already without magnetic field and is, therefore, often called zero–field splitting (ZFS).

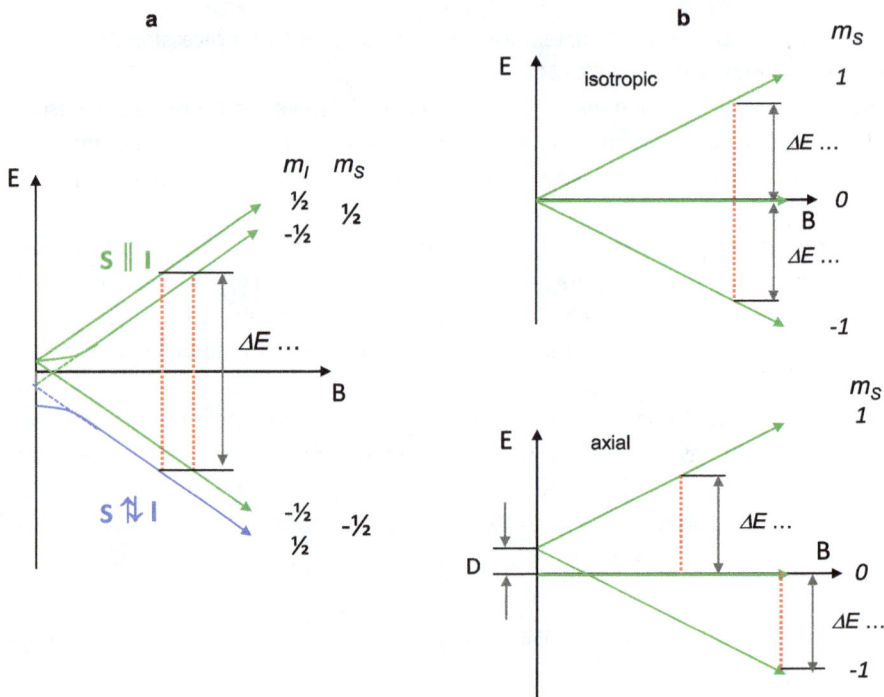

Figure 3.128: Examples for the interaction between electron and nuclear spins (a) and between two electron spins at one atom (b).

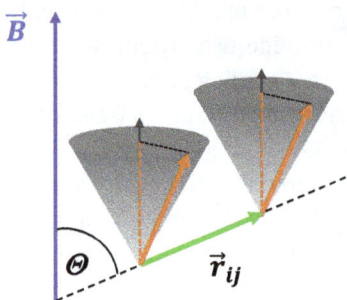

Figure 3.129: Magnetic dipole-dipole interactions: definition of the connecting vector $\vec{r}_{ij}$.

When unpaired electrons are on different ions, the magnetic interaction between these dipoles modifies the external magnetic field, B_0. The difference ΔB experienced by one of them as a result of the other one being located at a distance of r_{ij} decays with r_{ij}^{-3}:[●]

$$\Delta B \sim \frac{\mu_B^2}{|\vec{r}_{ij}|^3} \left(3\,cos^2\Theta - 1\right) \tag{3.134}$$

In (3.134), Θ is the angle between the vector $\vec{r}_{ij}$ connecting the dipoles and the field (cf. Figure 3.129). In catalytic materials, the arrangement of paramagnetic atoms is usually random. At low concentrations, the sites are magnetically isolated. With increasing concentration, the distribution of distances, r_{ij}, creates a range of field contributions ΔB at the sites, which broadens the line, up to a complete loss of the signal. On the other hand, when paramagnetic sites are very close to each other in ordered structures (<1.5 nm), exchange narrowing may dominate dipolar broadening, and very sharp signals may appear (cf. Figure 3.132 below).

For NMR, the influence of a paramagnetic site on a nuclear magnet is deleterious: while the nucleus adds a slight contribution to the field at the electron, the drastic influence in the reverse direction causes prohibitive NMR line widths. ZFS has no direct analogy in NMR, because there is only one nuclear spin per atom. Dipole-dipole interactions in disordered arrays, which are treated similar to eq. (3.134), also interfere with the acquisition of NMR spectra. In solid-state NMR, they are averaged out by magic angle spinning (MAS, see below). NMR spectra often exhibit fine structure due to interactions between identical magnetic atoms distributed in regular patterns (e.g., ^{1}H in organic molecules). They are mediated by the diamagnetic electron shell and result in signal splitting under the influence of neighboring (coupled) nuclear spins, according to their multiplicity.

Magnetic resonance spectra can be measured in two different modes: (i) with a constant frequency, scanning the magnetic field to record the transitions (continuous wave (cw) technique) or with constant magnetic field, employing radiation pulses to vary the frequency. The cw technique is standard in EPR, but does not perform well with the much weaker effects of nuclear magnetism. NMR is recorded with pulse techniques. Some of them are, meanwhile, also used for EPR, though confined to conditions less useful for catalysis research.

3.4.9.2 Electron paramagnetic resonance spectroscopy (EPR)

Figure 3.130a depicts major elements of a cw spectrometer. The sample is in a microwave resonator between the poles of an electromagnet, which provides a highly homogeneous field, B_0. The resonator is part of a microwave bridge with a reference and a signal arm, which is fed with microwave radiation by a source (a Klystron or a Gunn diode). Frequencies for cw EPR measurements are available in different ranges (bands): between 9.2 and 10 GHz for the most familiar X-band spectrometers, or around 35 GHz in Q-band spectrometers, which allow better resolution of signals on the B scale (cf. eq. (3.130)). This variability, together with the opportunity to vary the intensity by an attenuator, is required for tuning the resonator before the measurement. The circulator directs the microwaves to the resonator and sends waves reflected by it to the detector (a microwave diode). Like the waveguides, the resonator is a cavity with highly reflecting walls. In the initial tuning of the resonator to the microwave length, a standing wave is formed in the cavity: all wave energy is converted to heat in the resonator; no waves reach the detector.

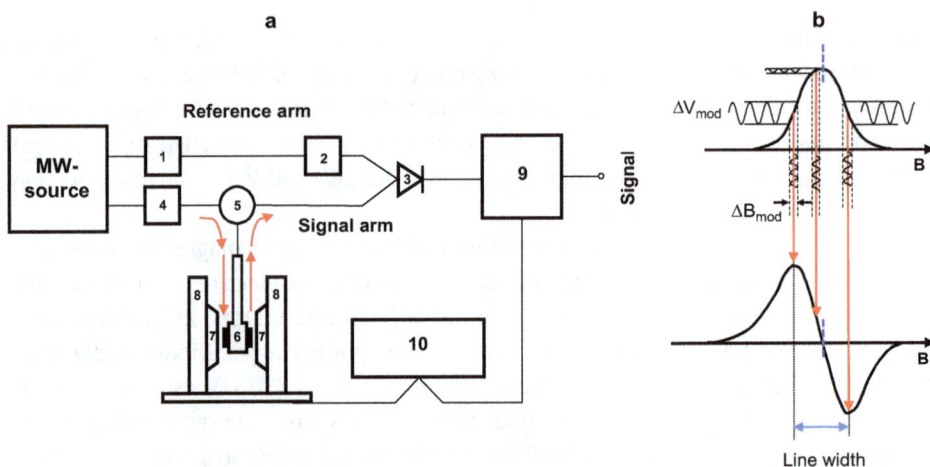

Figure 3.130: Cw magnetic resonance spectrometer: (a) general scheme: microwave bridge with sample cell as resonator; 1 – bias control, 2 – phase control, 3 – microwave diode, 4 – attenuator, 5 – circulator, 6 – resonator, 7 – modulation coil, 8 – magnet, 9 – phase-sensitive detection (PSD), 10 – modulator. (b) Principle of PSD using a modulated auxiliary field.

When B_0 is scanned, absorption of microwave energy due to transitions in the sample causes an energy mismatch between incoming and outgoing microwaves. As a result, microwaves are sent toward the detector diode via the circulator. The diode is kept under optimum working conditions by input, which is tuned in the reference arm by a bias and a frequency modulator. The signal is isolated from much of the experimental noise by phase-sensitive detection (PSD). For this purpose, the magnetic field, B_0, is modulated with an oscillating component, B_1 (typically of 100 kHz), which imprints the frequency also on the resonance signal ΔV and allows its separation from noise of different frequencies. As shown in Figure 3.130b, the signal obtained by this procedure is the first derivative of the original signal. Indeed, EPR spectra are always reported in the first-derivative mode, which emphasizes spectroscopic features in the often broad and overlapping signal shapes (see Figure 3.133).

Due to the low excess population of the ground state under normal conditions, the rate at which spins relax from the excited to the ground state is important for the measurement. When relaxation rates are low or microwave power is high, absorption may stop because of saturation: populations of excited and ground states have become identical. Relaxation to the ground state is a first-order process characterized by the relaxation time, T_1, its inverse rate constant. It is nonradiative and occurs by interactions with vibrational modes of the lattice (spin-lattice relaxation).[49] Absorption is also

---

49 When there are excited states already close to the ground state, relaxation becomes very fast. As a result, signals become extremely broad; they are often observable only at very low temperatures.

accompanied by an increase in entropy, which decays by interactions between the spins (spin-spin relaxation, characterized by T_2). This feature will be discussed in more detail in Section 3.4.9.3.

EPR spectra are specified by g-values, by hyperfine splitting constants a, and by zero-field splitting parameters D (cf. eqs. (3.130), (3.132),and (3.133), Section 3.4.9.1). The g-value of a free electron is 2.0023. The g-values of paramagnetic free cations can be predicted by theory using, for instance, the Lenné formula, in cases where Russell-Saunders coupling prevails.

g	0	0		$g_{||}$	0	0		g_{zz}	0	0
0	g	0		0	$g_\perp$	0		0	g_{xx}	0
0	0	g		0	0	$g_\perp$		0	0	g_{yy}

Cubic symmetry   Axial symmetry   Arbitrary structures

Scheme 3.5: g Matrices of structures with different symmetry.

In the presence of ligand fields, spin-orbit coupling is strongly modified. Their influence translates the spin vector $\vec{S}$ into a magnetic moment vector $\vec{\mu}_e$ differing not only in the absolute value, but also in the direction. In eq. (3.127), g, now, has to operate the transformation of $\vec{S}$ into a vector of any direction and size determined by the properties of the coordination sphere. Such transformation of a vector is produced by its multiplication with a matrix, i.e., g becomes a (3 × 3) matrix $\underline{g}$ in the solid, in a more general notion – a tensor. By appropriate choice of the coordinate system (direction $\vec{z}$ along the external field parallel to the axis of the highest symmetry), $\underline{g}$ becomes a diagonal matrix: depending on the type of symmetry, the spectrum contains signals, which are related to up to three principal g values (Scheme 3.5).

In the following, relations between g and the ligand field are briefly illustrated. For an electron in a nondegenerate ground state, the deviations of g from the free electron value g_e are:

$$g_i = g_e \pm \frac{n\,\lambda}{\Delta E} \tag{3.135}$$

In eq. (3.135), λ is the spin-orbit coupling constant, ΔE is the energy difference between the ground state and the state involved in the coupling (see below), and n an integer between 2 and 8 that depends on the pair of states involved. Equation (3.135) with the negative sign refers to coupling with empty states (e.g., in d^1 ions: $g < g_e$), with the positive sign to coupling with filled states (e.g., d^9 ions: $g > g_e$). As λ increases with the atomic number, the g values of heavier elements tend be more distant from 2.0023. However, the deviation is also influenced by the energy splitting of states in the ligand field, ΔE.

The use of eq. (3.135) can be illustrated for a d^1 ion in an axial ligand field with compression along z. Its term scheme can be derived from Figure 3.105, which shows the opposite distortion – axial elongation: upon compression, d_{xy} becomes the nondegenerate ground state and d_{z^2} the highest level. For evaluating $g_\parallel$, coupling of d_{xy} with $d_{x^2-y^2}$ is included, because these orbitals commute about the z axis: $g_\parallel = g_e - 8\lambda/(E_{x^2-y^2} - E_{xy})$. For evaluating $g_\perp$, d_{xz} and d_{yz} are relevant, because they commute with d_{xy} by sequential rotation around x and y: $g_\perp = g_e - 2\lambda/(E_{xz/yz} - E_{xy})$. Equation (3.135) can be employed also for other coordination geometries provided that they result in a nondegenerate ground state. Magnitudes of $g_\parallel$ and $g_\perp$ and the relation between them can be used to narrow down the range of coordination geometries for unknown structures. If optical spectra of the material are available, their combination with g values from EPR improves the chances for unambiguous assignment. For more details on the interpretation of anisotropic g matrices, the reader may refer to ref. [198].

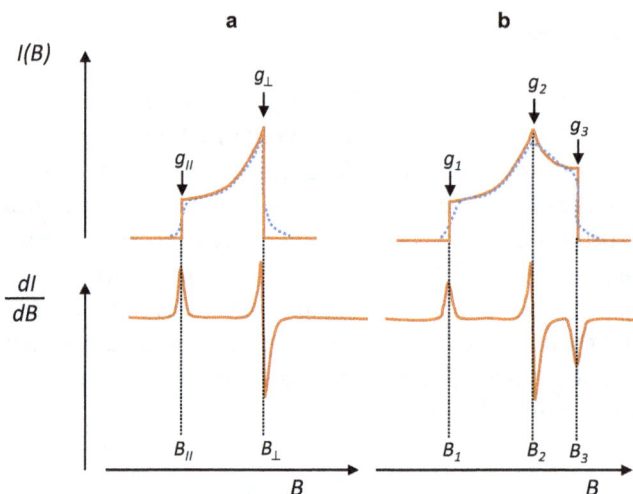

Figure 3.131: EPR spectra of powdered samples: (a) – coordination spheres with axial symmetry, (b) – anisotropic coordination spheres.

Due to the anisotropy of the g values, absorption intensities of EPR transitions depend on the orientation of the coordination sphere to B_0: if B_0 is parallel to the z direction of an axial coordination sphere, radiation is absorbed at the energy corresponding to $g_\parallel$; if it is perpendicular to z at the energy related to $g_\perp$. At intermediate orientations, the g values result from a weighted superposition of $g_\parallel$ and $g_\perp$. In powdered samples, absorption starts at the higher of the two values (cf. eq. (3.130) with constant ΔE; in the example of Figure 3.131a at $g_\perp$) and breaks down below the lower one ($g_\parallel$), but it extends over the whole range in between. Absorption increases toward $g_\perp$, because in a random powder the perpendicular orientation is more abundant than the parallel one. In the first derivative, also shown in Figure 3.131a, infinite values are replaced by maxima/minima. Figure 3.131b depicts the situation for an example of anisotropic symmetry, i.e., three g values. The intermediate one is indicated by the zero crossing and the highest one again by the negative peak.

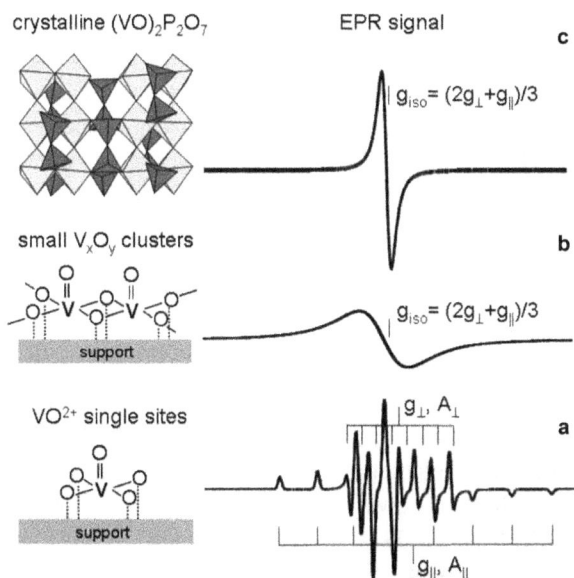

Figure 3.132: Influence of magnetic interactions on EPR spectra, exemplified with signals of V^{IV} oxo species: (a) isolated VO^{2+} site, signal exposes hyperfine structure caused by the 7/2 nuclear spin of ^{51}V; (b) dipole-dipole interactions in small V_xO_y clusters; and (c) exchange narrowing at small V-V distances. Reproduced from ref. [200] with permission from the Royal Society of Chemistry.

Additional structure in EPR spectra arises from influences of magnetic nuclei and of additional unpaired electrons at the same atom (Section 3.4.9.1). Both features are modified by the ligand field as well, and the corresponding parameters are, therefore, matrices like $\boldsymbol{g}$. Regarding hyperfine splitting, the $\boldsymbol{A}$ matrix becomes diagonalized when z is defined parallel to the axis of the highest symmetry (Scheme 3.5 with $g_{\alpha\beta}$ replaced by $A_{\alpha\beta}$). As $\boldsymbol{g}$ refers to the same axis, relations between features caused by g and by hyperfine splitting (A) are simple: the $A_{\alpha\beta}$ series extends to both sides of the $g_{\alpha\beta}$ features. In Figure 3.132a, this is illustrated by a spectrum of magnetically isolated VO^{2+} sites ($I(^{51}V) = 7/2$). There are two sets of eight hyperfine structure lines. Each is centered at one of the components of the axial $\boldsymbol{g}$ matrix ($g_\parallel$ or $g_\perp$), and the spreading of the series is determined by the corresponding component of the axial $\boldsymbol{A}$ matrix ($A_\parallel$ or $A_\perp$). While it is sometimes difficult to identify the atom hosting the unpaired electron on the basis of g values, hyperfine structure immediately clarifies this point for magnetic nuclei.

Figure 3.132b exemplifies the effect of dipole-dipole interactions on the spectrum (cf. Section 3.4.9.1). The signal is strongly broadened by the interactions within a poorly ordered, mostly two-dimensional supported surface oxide phase: the hyperfine structure is lost, but the isotropic g value is still related to the components of the underlying $\underline{g}$ matrix. In the ordered material exhibiting the spectrum of Figure 3.132c, the sites are even closer to each other. As a result, the signal is dominated by exchange narrowing, which averages out the hyperfine structure as well as the g anisotropy.

The fine-structure (or ZFS) likewise depends on the coordination geometry, and $\underline{D}$ is a matrix like $\underline{g}$ and $\underline{A}$. The components of $\underline{D}$ are sometimes lumped into two quantities, D and E ($D = 1.5\,D_{zz}$, $E = 0.5\,(D_{xx} - D_{yy})$), where $E = 0$ for axial coordination geometry. Due to the selection rule $\Delta m_S = 1$, the $2S + 1$ states of a system with S > 1/2 give rise to $2S$ transitions, which are at identical fields in isotropic geometry, but at different fields for anisotropic systems (Figure 3.128b). At high values of the splitting parameters, D and E, signals may be very far from g_e. Such signals are usually labeled with effective g values (g'). More information on the treatment and the diagnostic potential of fine structure interactions can be found in ref. [198].

EPR spectra of catalytic materials can be very complex, and reliable assignments usually require longstanding experience. Sites may be magnetically isolated, subject to dipolar broadening, or even to exchange narrowing. Isolated sites can by identified by their hyperfine structure if associated with a magnetic atom; if not, by the temperature dependence of their intensity obeying the Curie law. It is also important to differentiate surface sites from sites in the bulk. This can be achieved by comparing coordination symmetries suggested by the spectral parameters (usually lower at the surface) or by the response of signals to probe molecules offered. When spectral resolution is insufficient to distinguish features appearing in a narrow field range or superimposed signals from different sites, problems may be solved by measuring at higher frequencies (Q-band spectroscopy, see earlier in this section).

In Figure 3.133a, a temperature series performed to differentiate Fe^{III} species in Fe-ZSM-5 is depicted. The catalyst was prepared by CVD of FeCl$_3$ into H-ZSM-5, with subsequent hydrolysis of the resulting Z-FeCl$_2$ species, drying, and calcination. At 90 K, three signals can be discerned, located at effective g-values of $g' \approx 6$, 4.3, and 2, the latter of low intensity. They arise from magnetically isolated Fe^{III} sites, because their intensity decreases when spectra are measured at higher temperatures, in agreement with the Curie law. There is, however, also intensity from broad features around $g' = 2$. Upon heating to 673 K, they increase gradually, until a huge signal appears at $g' = 2$, which is shown in the figure on a compressed intensity scale.

The signals at $g' \approx 6$ and 4.3 arise from ($m_S =$) $-1/2 \rightarrow +1/2$ transitions in ligand fields of strong axial or rhombic distortion, respectively, and large ZFS.[50] They were assigned to Fe^{III} in axially distorted and tetrahedral environments, respectively [201]. The effective g value of the third isolated site (≈ 2) suggests a highly symmetric environment with

---

50 At X-band microwave frequencies, the allowed transitions between other levels of the S=5/2 state (between ±5/2 and ±3/2, and between ±3/2 and ±1/2) cannot be excited in the available B_0 range.

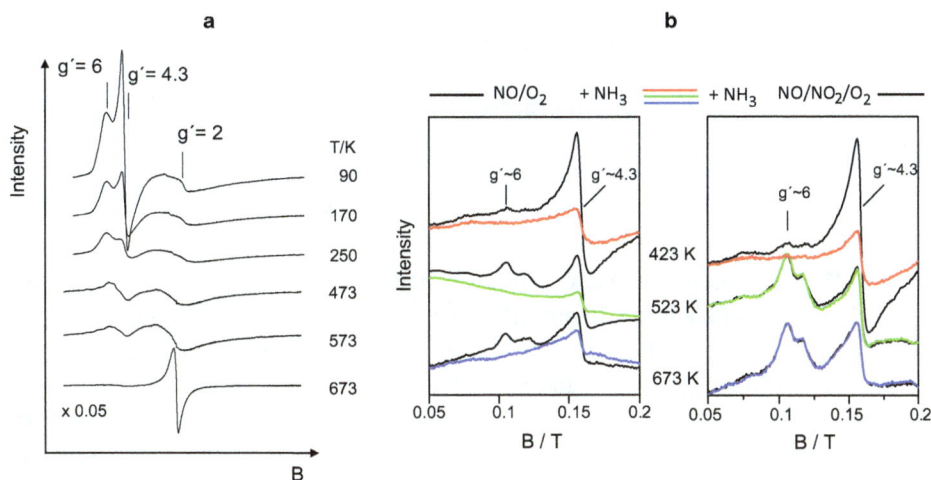

Figure 3.133: EPR spectroscopy with Fe species in ZSM-5 catalysts: (a) differentiation of Fe sites present by measuring EPR spectra at different temperatures; (b) *operando* EPR spectroscopy during selective catalytic reduction (SCR): standard SCR (left) and fast SCR (right); black spectra – feeds without NH_3, colored spectra – feeds with NH_3. Adapted from refs. [157] (a) and [201] (b) with permission from Elsevier.

negligible ZFS (cf. Figure 3.128b). In Fe-ZSM-5 materials of various sources, the site characterized by $g' \approx 4.3$ was always populated, the signal at $g' \approx 6$ was found frequently, but not always, while the isotropic signal was rarely observed.

The huge signal at $g' \approx 2$ appearing at 673 K (Figure 3.133a) originates from a large amount of Fe^{III} oxide aggregates, which had apparently formed during drying and calcination of the hydrolyzed samples. These antiferromagnetic aggregates are EPR-silent at low temperatures. Signals of the spins in them appear only above their Néel temperature (cf. Section 3.4.9.1). Smaller, more disordered oxide clusters, in which the antiferromagnetic coupling is stronger affected by thermal energy, also cause broad signals at $g' \approx 2$, which may vary considerably in the details and start growing at room temperature.

The spectra in Figure 3.133a suggest that such oxide clusters coexist with isolated sites and oxide aggregates in the sample. The huge differences in magnitudes and temperature dependences of paramagnetic sites and exchange-coupled arrays (oxide clusters) prevent any quantitative analysis that covers all Fe^{III} sites. It is, however, justified to determine ratios between the paramagnetic sites, which all obey the Curie law (for more detail about concentration analysis, see below).

EPR is suited for *in situ* and *operando* measurements, provided that the reactor is accommodated in the resonator cavity and any interference with the magnetic field that could arise from heating and improper choice of materials is avoided. Recent *operando*-EPR equipment allows even simultaneous Raman and UV-Vis diffuse reflection measurements [200].

The study summarized in Figure 3.133b deals with the SCR of NO in flue gases of car effluents by added NH_3:

$$4\,NO + 4\,NH_3 + O_2 \rightarrow 4\,N_2 + 6\,H_2O \qquad \text{Standard SCR} \qquad (3.136)$$

Over some catalysts, the reaction proceeds more than an order of magnitude faster, when NO_2 is in the mixture:

$$2\,NO + 2\,NO_2 + 4\,NH_3 \rightarrow 4\,N_2 + 6\,H_2O \qquad \text{Fast SCR} \qquad (3.137)$$

Figure 3.133b is focused on the low-field range; no relevant changes were noted beyond. In the black spectra (SCR feed without NH_3), the g' $\approx$ 4.3 signal decreases at higher temperatures in agreement with the Curie law, but the signal at g' $\approx$ 6 increases, particularly in the presence of NO_2 (right panel). In the full standard SCR feed (left panel), both signals decrease significantly, because Fe^{III} is reduced to EPR-silent Fe^{II}, the one at g' $\approx$ 6 disappears even completely. The presence of NO_2 changes this drastically, except for T = 423 K, where fast SCR rates are still low: at 523 K, all isolated sites remain Fe^{III} in the full fast SCR feed.

The anti-Curie temperature dependence of the g' $\approx$ 6 signal suggests that the initial sample contained Fe^{II} even after calcination at 873 K, which was confirmed by Mössbauer spectroscopy [201]. These sites, which are probably stabilized by two framework Al ions (Box 3.2), can be oxidized only by NO_2. When different catalysts were compared, the excess intensity at g'$\approx$6 in the presence of NO_2 changed the same way as the activity in fast SCR. The active sites for fast SCR may be, therefore, Fe^{III} ions created by NO_2 in an environment better suited for Fe^{II}.

The example shows how EPR contributes valuable insight into surface catalytic phenomena. More examples can be found in refs. [198, 200]. EPR stands out as one of the most sensitive analysis techniques available: with narrow signals, down to 10^{11}–10^{12} spins per sample volume (usually $\approx$0.1 cm^3) can be safely detected. This is both an advantage and a risk: for reactions where the active site is suspected to be a tiny, highly active minority among spectator species, EPR may offer the selectivity required to single out the minority. On the other hand, the risk of being fooled by contaminants or by dead or sluggish minority sites calls for rate measurements under the conditions used for spectroscopy and comparison with rates obtained under real conditions.

Obviously, EPR can be applied only to magnetic ions. Although the number of such species is large due to Hund's rule, the scope of EPR is narrowed down by the fact that ions with an even number of spins can be detected only at low temperatures ($\leq$ 77 K), due to very short relaxation times. Such lacking universality, which may be considered a disadvantage, holds also with respect to quantitative EPR analysis. As mentioned earlier, signals of exchange-coupled spins are not eligible for direct concentration analysis, and strong dipolar broadening may result in loss of intensity. Concentrations of magnetically isolated cations can be determined by comparison with spin standards, e.g., frozen dilute solutions of the same cation. As baselines are not always clear to establish and the double integration required for intensity evaluation tends to enhance inaccuracies, such analyses are, however, rarely performed in catalysis research.

3.4.9.3 Nuclear magnetic resonance spectroscopy (NMR)

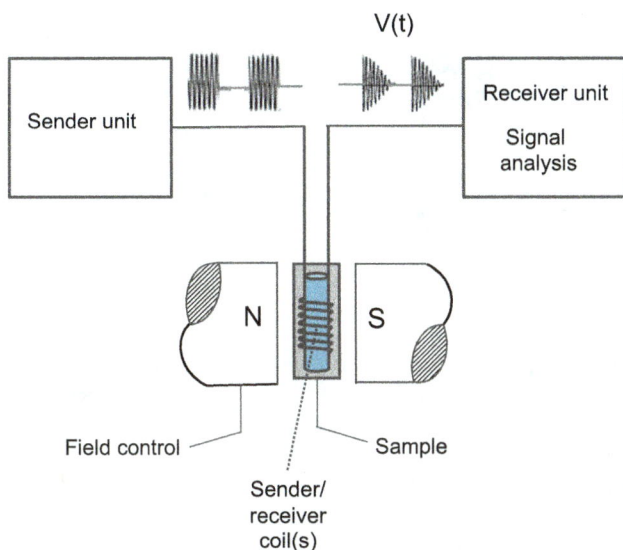

Figure 3.134: General scheme of a pulse NMR spectrometer.

Figure 3.134 shows basic components of a pulse (or FT-) NMR spectrometer, which allows fast recording of the spectra and improvement of their signal-to-noise ratio by accumulating multiple repeats. The sample is placed in a constant field of typically 5–20 T. A coil surrounding the sample perpendicular to the field serves to expose the sample to radiofrequency pulses (300–900 MHz) and for receiving its response, which is then analyzed and processed in the detector system. Unlike in cw spectrometers, the field is constant in pulse spectrometers. As will be shown below, pulsing is equivalent to frequency variation. Transitions are specified by their (Larmor) frequencies (3.131) instead of the field required to excite them.

Frequency variation for permanent radiation is technically challenging. However, it can be shown by Fourier analysis that rectangular pulses of a nominal angular frequency, ω_N, consist of a continuous range of lower and higher frequencies around ω_N, which is broader than required for recording transitions of a nucleus expected at ω_N. A strong pulse excites all spins available in the sample and turns their magnetic moments around an axis perpendicular to the field (Figure 3.135b). The extent of spin rotation depends on the pulse length. Typical extents are $\pi/2$ and π. Notably, a $\pi/2$ pulse creates magnetization perpendicular to the field, because it also collects the I_x/I_y components of the spins, which are at random in the unperturbed system, into a particular direction (Figure 3.136 c). It is only this "transverse" magnetization, which allows tracking processes in the sample, because the spins continue their precession in the new orientation, which results in an induction signal, V(t), in the receiver coil.

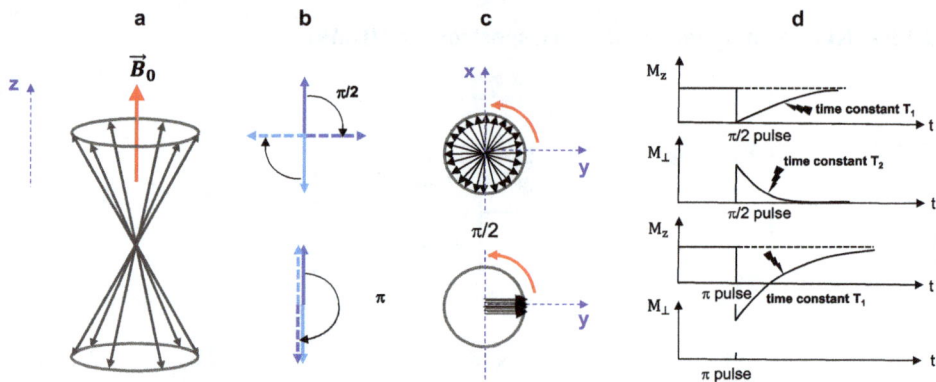

Figure 3.135: Effects of $\pi/2$- and π-pulses on the magnetization and their relaxation: (a) ensemble of spins in precession movement around $\vec{B}$; (b) effect of $\pi/2$- and π-pulses on the z components of the elementary magnets; (c) effect of a $\pi/2$-pulse on the x/y components of the elementary magnets; and (d) effects of pulses on the magnetization parallel and perpendicular to $\vec{B}$ and their decay by relaxation; only $M\perp$, the transversal magnetization can be detected. Adapted from ref. [199].

For the sake of simplicity, a system allowing just a single transition with the Larmor frequency ω_0 will be considered first. If only one spin were rotated into the xy-plane, its precession would produce a sinusoidal, nondecaying free induction. This does not hold, however, for an ensemble of many spins, which starts precessing with identical orientation after the $\pi/2$-pulse. Their phase coherence right after the pulse is a state of high order; therefore, coherence is lost by spin-spin interactions, which add new frequencies near ω_0 to the system. As a result, transverse magnetization and, hence, the free induction $V(t)$ decay gradually for entropy reasons. This transverse relaxation is characterized by the relaxation time, T_2 (Figure 3.135d, cf. Section 3.4.9.2).

In real systems, resonance frequencies are distributed, because the local magnetic fields are not homogeneous, for instance, due to spatial fluctuations of the external field or to dipole-dipole interactions between identical or different nuclei at random distances. Such widening of the frequency distribution accelerates the loss of coherence between precessing spins, causing a faster free induction decay (FID). The system may, however, also contain nuclei resonating at different frequencies for reasons discussed below. These patterned inhomogeneities of the local magnetic field result in complex FID patterns. What is suggested by these examples can be confirmed by theory: the NMR spectrum, $P(\omega)$, extending in the frequency domain and the free induction decay, $V(t)$, in the time domain are related as forward and reverse Fourier transformation:

$$P(\omega) = \int_0^\infty V(t) \cos \omega t \, dt \qquad\qquad V(t) = \int_0^\infty P(\omega) \cos \omega t \, d\omega \qquad (3.138)$$

Relaxation phenomena place limits also on pulse NMR: pulsing on incompletely relaxed systems would result in saturation. A $\pi/2$-pulse balances the initial excess magnetization along the field; a π-pulse turns it around. Relaxation of these energetic transitions requires spin-lattice interactions (relaxation time T_1, Figure 3.135d, cf. Section 3.4.9.2). It is favored by molecular mobility with fluctuations in the range of the Larmor frequencies, i.e., T_1 is short in fluids, but long in solids. On the other hand, inhomogeneities of internal magnetic fields favor dephasing of spin precession in solids, which results in small values of T_2.

Pulse NMR is slow on the time scale of elementary processes. After a $\pi/2$-pulse, spins need time to develop the characteristic pattern that reflects the underlying frequency distribution. The result depends on whether the local magnetic fields are stable during this period or not. To define a line of a width $\Delta\omega$, a "correlation time" τ_c is required that allows the fastest of the involved spins to "lap" the slowest during precession. This results in the condition:

$$\tau_c \Delta\omega \geq 1 \tag{3.139}$$

τ_c is in the μs - ms range, as compared to fs in radiative processes. When the local field variations fluctuate faster than that, the field experienced by the spin ensemble is averaged out. Lines become narrow, and the transverse relaxation time, T_2 increases. Due to this dependence on correlation times, pulse NMR is an important tool for studying mobility on the molecular scale, in particular, diffusion.

Factors influencing the position of resonances on the frequency scale will be only briefly mentioned here, because they should be well known from the standard use of NMR. Frequency shifts due to diamagnetic shielding by the electron shell, which modifies the interaction between nuclear spin and external field, depend on the chemical environment of the nucleus. Actually, NMR spectra are usually reported on chemical shift scales rather than on the frequency scale. Homonuclear interactions at well-defined distances give rise to fine structure with high diagnostic potential (multiplet splitting). In solids of other than isotropic structure symmetries, however, all these features are subject to magnetic flux anisotropies that depend on the orientation of the field relative to the axis of the highest symmetry (z). The chemical shift $(1-\sigma)$,[51] for instance, becomes a matrix like $\underline{g}$ in EPR, and NMR lines of powdered samples have similar shapes as the EPR lines shown in Figure 3.131.

Quadrupole splitting (QS) is another phenomenon relevant for solid-state NMR. Nuclei with $I > 1/2$ (e.g. ^2H, ^{27}Al) have a nuclear quadrupole moment, which interacts with gradients in anisotropic local electrical fields. For $I = 1$ in an axial field, for instance, this causes energy shifts of the $m_I = 0$ level relative to the $m_I = \pm 1$ levels, similar to the situation in ZFS (Figure 3.128b). As the spacing between the resulting two transitions

---

51 σ represents the diamagnetic shielding effect on a scale from 0 to 1.

depends on the orientation of the magnetic field relative to the electric field gradient, a broad signal with a shallow intensity minimum is obtained. When the field gradients are washed out by mobility on time scales below t_C, the lines collapse into one.

In NMR, a number of approaches are available to remove the influence of field inhomogeneities created by the interactions between sites. In EPR, where Larmor frequencies are three orders of magnitude higher, time-resolved detection, on which these methods are based, has been developed with some delay, due to considerable experimental challenges. Owing to short T_2 relaxation times of electron spins, such methods can be applied only at very low temperatures. While classical EPR thus provides insight in coordination geometries, NMR is rather focused on the chemical differentiation of atoms in (quasi)molecular structures, although methods for gaining structural information (atom-atom distances) are also available (see below).

The slow measurement process, in which field conditions need to be stable over the correlation time, τ_c, to be "imprinted" in the spectrum allows effective strategies to suppress the influence of local field inhomogeneities. To simulate conditions as in fluids, where the spectra are determined by an averaged field, solid samples are rotated at spinning frequencies $\gg \tau_c^{-1}$.

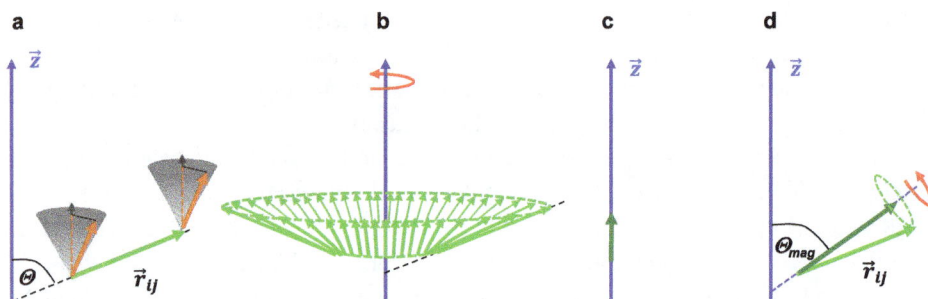

Figure 3.136: Cancellation of dipolar broadening by magic angle spinning (MAS): (a) interacting dipoles with a connecting vector $\vec{r}_{ij}$, at any angle Θ from $\vec{z}$; (b) rotation of this vector about $\vec{z}$; (c) non-zero averaged contribution of the dipole pair to the external field; and (d) rotation of $\vec{r}_{ij}$ about an axis inclined to $\vec{z}$ by the magic angle Θ_{mag} (s. text). Any average from such rotation is parallel to the rotation axis, which cancels its contribution to the external field due to eq. (3.134).

Rotation averages all vectors in space into their projections on its axis. Figure 3.136a–c illustrates this for the case that the vector $\vec{r}_{ij}$ between two nuclei (cf. Figure 3.129) is rotated about the z axis. The length of the projection on z depends on the angle Θ (Figure 3.136c), but it remains a nonzero contribution. In reality, the sample is rotated not about z, but about an axis inclined to z by the "magic" angle Θ_{mag}, which fulfils the condition $3 \cos^2 \Theta_{mag} = 1$, i.e., $\Theta_{mag} \approx 54.74°$ (Figure 3.136d). Projections of $\vec{r}_{ij}$ on this axis may have any length: due to eq. (3.134), their influence on the spectrum is zero.

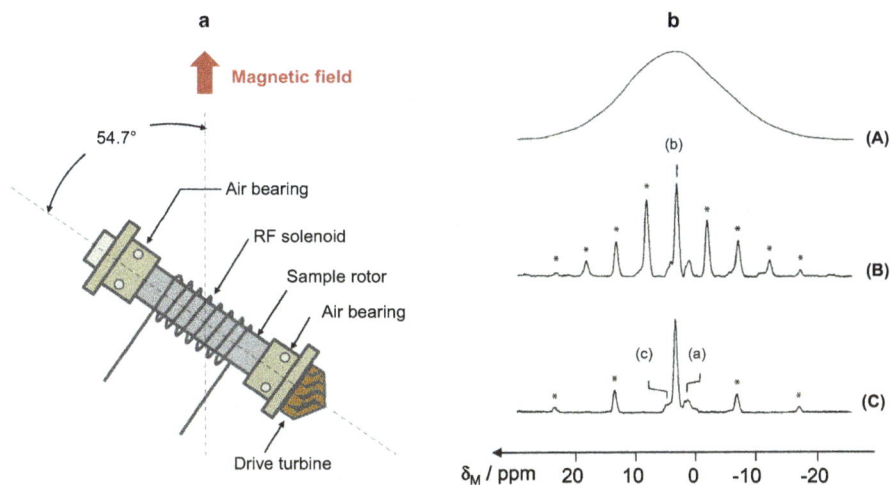

Figure 3.137: MAS-NMR: (a) sample rotor, suspended and driven by pressurized air; (b) transition from static NMR (A) to MAS-NMR (C); (B) shows the effect of an insufficient rotation frequency. See text for the nature of the sample and the interpretation of signals. Reproduced from ref. [202] with permission from Springer Nature.

Figure 3.137a illustrates the arrangement of the sample between the poles of the magnet. The effect of the rotation is shown in Figure 3.137b, where ^{1}H spectra of a Y zeolite containing both H and Na ions are compared without spinning and at different spinning frequencies, ν_{rot}. At low ν_{rot}, the broad static signal is resolved into a line pattern in which the isotropic line is surrounded by pairs of spinning side bands (labeled *). The spacing between them is proportional to ν_{rot}, which can be varied to avoid superpositions between side bands and relevant features. The three signals detected arise from bridging protons (b and c) and from H in silanol groups (a). The large b signal is from Brønsted sites in the supercages, from where the Na ions are easier exchanged than from the narrow sodalite cages, where their electrostatic stabilization is more efficient.

MAS-NMR completely averages heteronuclear dipolar interactions and chemical shift anisotropies. Homonuclear interactions may include flip-flop spin exchange processes at very low distances, which require very high spinning frequencies. They may be reduced to some extent by diluting interacting nuclei with noninteracting ones (e.g., H by D). QS is reduced by MAS, but not removed, which may result in considerable line widths in ^{27}Al-MAS NMR. In principle, QS may be completely suppressed by double-rotation (DOR) techniques, which are, however, limited with respect to their spinning frequencies. Therefore, multi-quantum NMR, which will be briefly summarized below, is nowadays preferred for this purpose. Heteronuclear interactions may be decoupled by exposing the sample to the resonance frequency of one of the nuclei (usually ^{1}H): when the states with $m_I = +1/2$ and $-1/2$ are equally populated, the interaction is quenched.

Cross-polarization (CP), which allows detecting minority nuclei with low gyromagnetic ratios γ, e.g., ^{13}C, in the presence of a majority of nuclei with high γ (usually ^{1}H), utilizes the otherwise undesired heteronuclear interactions between them. The method takes advantage of the phenomenon that polarization can be transferred between different nuclei, 1 and 2, when their resonance frequencies are identical ($\omega_1 = \omega_2 = \omega$). To achieve this, the nuclei are exposed simultaneously to radio waves of different field strengths (double excitation), the magnetic components B_1 and B_2 of which obey the Hartmann-Hahn condition:

$$\left(\frac{\Delta E}{\hbar} = \right)\omega = \gamma_1 B_1 = \gamma_2 B_2$$

(3.140)

A typical experiment starts by exciting the ^{1}H nuclei with a $\pi/2$-pulse, which is followed by "spin locking," i.e., exposure of the sample with radiofrequency according to (3.140) to transfer polarization from the majority to the minority nuclei. The ^{1}H nuclei are then decoupled, and the FID originating exclusively from the minority nuclei is recorded. The method can be combined with MAS-NMR, which modifies, however, the Hartmann-Hahn condition by a term that includes the rotation frequency.

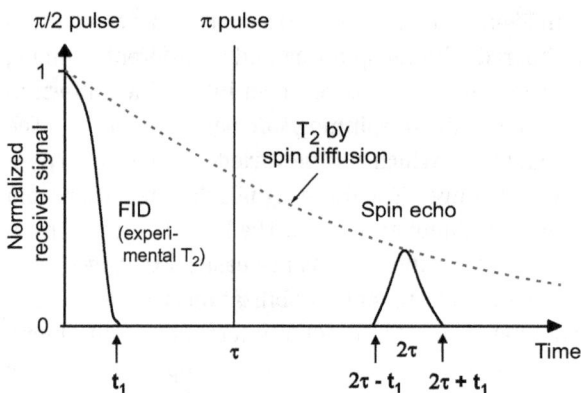

Figure 3.138: The Hahn echo, revealing the entropic contribution (entropic dephasing) to an experimental free-induction decay (FID). Adapted from ref. [199].

Many advanced NMR techniques employ sequences of pulses. Although their details are beyond the scope of this book, the echo approach, which is inherent in most of them, will be briefly discussed using the Hahn echo, which was the first procedure of this kind. The Hahn echo sequence starts with a $\pi/2$-pulse, after which the spins are given a time τ to dephase, before they are turned around by a π-pulse (Figure 3.138). The spins may be compared with runners in a race, all running with different speeds. When they are turned around at the time τ to return toward the starting point with their specific speeds, they all arrive there at $2\,\tau$. The spins do not. Although the spin inhomogeneities that are cancelled by echoing (the different "speeds") include even the effect of chemical shifts, the echo is smaller than the initial spin ensemble, because the irreversible entropic dephasing (e.g., flip-flop exchanges), remains active all the time. Thus, the Hahn echo shows the contribution of spin-spin interactions to experimental T_2 values deduced from the FID (Figure 3.138). The spin diffusion contribution to T_2 can be gauged by varying the dephasing interval τ.

The Hahn echo is the basis of the Pulsed field gradient (PFG) technique, a powerful method for studying self-diffusion processes. Up to now, fields applied were implicitly assumed to be homogeneous, i.e., it did not matter at which point in space a spin experiences a field pulse. In PFG, the sequence is started with a field gradient pulse, in which the field strength varies along the main field direction z. This field gradient pulse labels the spins, imprinting a dependence of their precession frequency on z. After allowing for a time interval Δ, the spins are reverted by a π-pulse like in the Hahn experiment, and the field gradient pulse is repeated. If all spins are still at their original positions, they will form just a Hahn-type echo, based on diffusional dephasing. If, however, a spin has moved to another z position, its Larmor frequency is changed by the second field gradient pulse, and it will miss the echo. Relating the missing intensity to the sampling interval, Δ, gives access to diffusion rates. PFG-NMR was developed for gas-phase diffusion, but has been meanwhile extended to diffusion in liquids [203]. As mentioned above, it tracks self-diffusion, which needs to be distinguished from mutual diffusion in concentration gradients. More information about diffusion can be found in ref. [199].

Distances between nuclei of different elements can be determined by advanced spin-echo techniques. In SEDOR (Spin-echo double resonance), echoing sequences with one of the nuclei (I) are compared with the second one either at random or with oriented spin (by administering a π pulse to nucleus, S, together with the reversal π-pulse for I). The impact of the S orientation on the echo of I gives access to the intensity of the I-S coupling and, using the I-S coupling constant, to the distance between the nuclei. For solids, SEDOR has been, meanwhile, largely replaced by its extension to MAS conditions – REDOR (Rotational-echo double resonance).

Multi-quantum (MQ)NMR, which is another powerful tool of modern NMR, tracks the response of magnetic systems to the simultaneous excitation by more than one quantum. Such multi-quantum excitation of nuclei with $I > 1/2$ is achieved by a very short and strong radiofrequency pulse. After a development time, t_1, the system is returned to its initial single-quantum level by a reconversion pulse and fed with a detection (i.e., $\pi/2$-)pulse, which allows monitoring the echo as a function of t_2. After performing this procedure with different development times, t_1, the responses in the time domains, t_1 and t_2, are Fourier transformed into the frequency domains, ω_1 and ω_2. In the two-dimensional MQ NMR spectrum obtained, the projection of the ω_2 dimension (x-axis of the plot) corresponds to the usual one-dimensional spectrum, while the projection of the ω_1 dimension (y-axis of the plot) shows the spectrum without quadrupolar broadening.

For solid samples, MQ NMR can be combined with magic angle spinning (MQ MAS-NMR, sometimes also referred to as CRAMPS - Combined rotation and multi-pulse spectroscopy). Such methods allow, for instance, eliminating residual ^{1}H coupling that remains at the rotation frequencies usually available for MAS.

Spin-echo and double-excitation techniques also allow combining paramagnetic with nuclear magnetic resonances. In ENDOR (Electron nuclear double resonance), a paramagnetic transition is nearly saturated by intense microwave irradiation. Under such conditions, resonance of an adjacent magnetic nucleus by simultaneous irradiation with radio waves influences the absorption coefficient of the EPR transition: This allows recording the NMR transition via the EPR intensity, by sweeping the radiofrequency. ESEEM (Electron spin echo envelope modulation) utilizes the influence of magnetic nuclei in the vicinity of unpaired electrons on paramagnetic spin echoes of the latter. It allows measuring NMR spectra without any radio waves. Distances between paramagnetic sites (e.g., transition metal ions) and magnetic nuclei, e.g., ^{1}H, can be determined by both ENDOR and ESEEM.

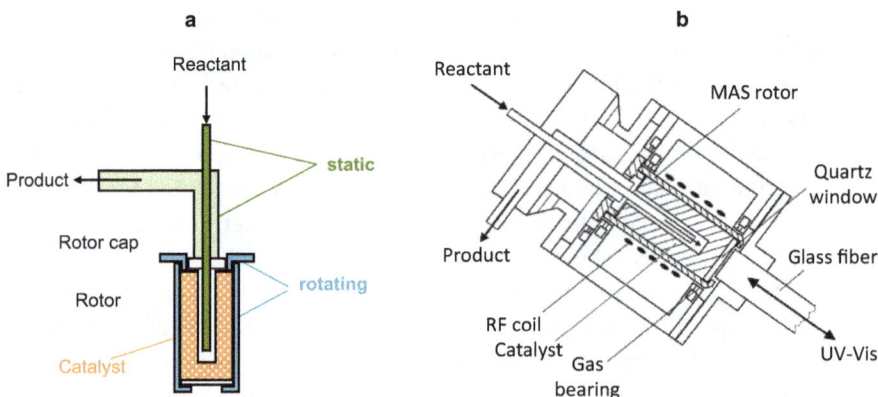

Figure 3.139: *in-situ* and *operando* studies with MAS-NMR: (a) standard rotor for continuous flow experiments; (b) rotor allowing combined MAS-NMR and UV-Vis measurements, integrated into air turbine. (b) Reproduced from ref. [204] with permission from the Royal Society of Chemistry.

Unfortunately, sample spinning, which is crucial for solid-state NMR, extremely handicaps the study of catalysts under reaction conditions. While early approaches like batch reaction with reactants enclosed in the rotor or freezing reactive states for post-catalytic analysis are subject to severe limitations, the delicate technique illustrated in Figure 3.139 makes the best of the situation. The catalyst is placed at the walls of the rotor. The gaseous reactant is fed through a static glass tube projecting into the rotor along its axis, while the products leave the rotor through the concentric space around the tube (Figure 3.139a). The arrangement is heated by the bearing gas and can be used at temperatures up to 423 K. Figure 3.139b shows an advanced version with a quartz window at the bottom allowing additional UV-Vis measurements (e.g., of hydrocarbon adsorbates) with a fiber optical probe (cf. Section 3.4.7.2).

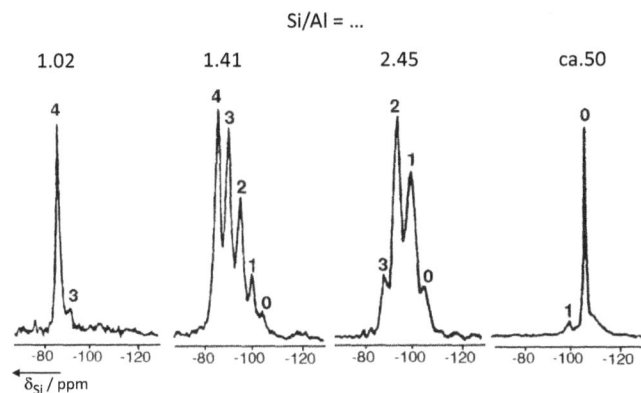

Figure 3.140: ^{29}Si MAS-NMR spectra of Na-FAU zeolites with different Si/Al ratios; labels at signals report the average number of framework Al sites adjacent to the probed Si site. Adapted from ref. [205] with permission from Elsevier.

Figure 3.140 illustrates a typical example for the characterization of catalytic materials by MAS-NMR. In zeolites, ^{29}Si-NMR is able to resolve framework Si atoms with different numbers of Al neighbors. Signal intensities are proportional to site concentrations via the same factor. Taking into account the Loewenstein rule (no Al-O-Al elements, cf. Section 2.3.2.3), the framework Al content can be evaluated from such a spectrum and can be compared with chemical analysis data to judge on the presence of extra-framework Al. The spectra in Figure 3.140 are from a series of FAU zeolites (Na-X, Na-Y) with various Si/Al ratios. From them, the authors concluded on crystallographic locations preferred by Al in FAU frameworks at low Al contents.

Due to the nuclear magnetism of ^{1}H, ^{27}Al, and ^{29}Si, zeolites have always been a preferred field for NMR studies. Apart from detailed structural characterization, this has resulted in methods for the investigation of acid and basic sites. ^{27}Al-NMR allows differentiating framework from extra-framework Al, although some forms[52] can be detected only on instruments working with very high magnetic fields. Many of the methods applied to zeolites can be also used with materials of different structures, e.g., silicas, aluminas, amorphous alumosilicates, etc.

With respect to other magnetic nuclei, ^{51}V has received much attention, because vanadium is the most prominent redox component in catalysis. Due to the quadrupole moment of ^{51}V, signals are broad even at high spinning rates, and the presence of par-

---

52 Al partly detached from the framework, which exhibits very high quadrupolar broadening due to a strongly asymmetric coordination geometry.

amagnetic V^{IV} can further complicate measurements. Examples include the characterization of surface V oxide species on various supports and the study of n-butane interaction with vanadyl phosphates, which are precursors of important catalysts for n-butane oxidation to maleic anhydride (cf. Appendix A1). In the latter case, V in different oxidation states was indirectly detected by ^{31}P spin echo mapping (cf. ref. [203]).

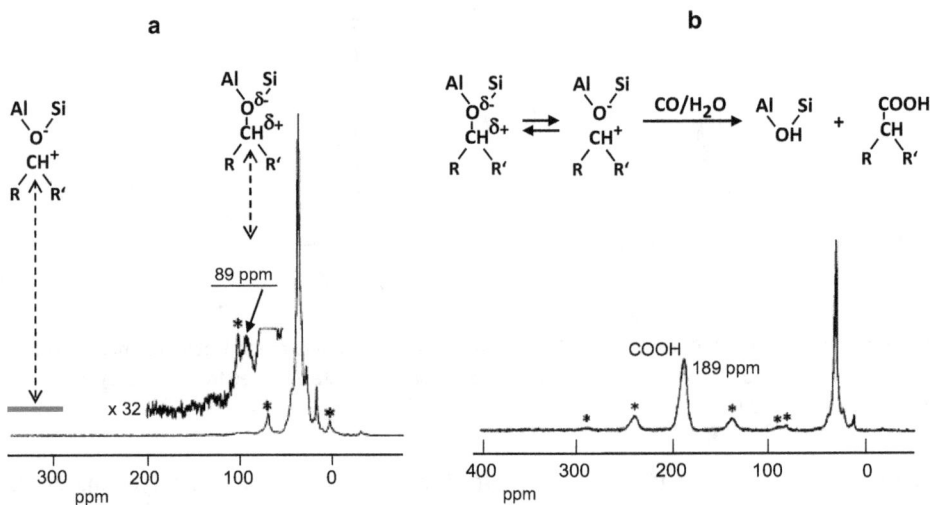

Figure 3.141: MAS-NMR as a tool for *operando* studies with catalysts: intermediates of acid-catalyzed hydrocarbon reactions studied by ^{13}C-CP-MAS-NMR: (a) Detection of alkoxide groups in the interaction of $^{13}CH_2=CH_2$ with H-ZSM-5; (b) trapping of intermediates produced with $^{12}CH_2=CH_2$ by ^{13}CO and H_2O, using a protocol known from carbenium ion chemistry. Adapted from ref. [206] with permission from Elsevier.

The study of adsorption processes and reaction mechanisms using ^{13}C-NMR is another important use of solid-state NMR in catalysis. Figure 3.141 shows an example dealing with the nature of intermediates in reactions catalyzed by solid Brønsted acids. As discussed in Section 4.5.2, there is disagreement on whether carbenium ions can be stable on surfaces in the gas phase (i.e., without stabilization by solvation) or if they are only transition states. The ^{13}C CP-MAS-NMR spectrum in Figure 3.141a was measured with H-ZSM-5 exposed to ethene-1-^{13}C, under static conditions. No signal appears in the 300–350 ppm range, where carbenium ions were observed in liquid-phase studies. Instead, there is a signal at ca. 90 ppm, which can be assigned to C in the α position of an ether. The sample was treated with CO/H_2O, which is known to convert carbenium ions into carboxyl groups. The carboxyl signal in Figure 3.141b is convincing. As carbenium ions could not be detected before (Figure 3.141a), this suggests that the alkoxy groups may be in equilibrium with a small amount of carbenium ions, as indicated in Figure 3.141b.

More NMR studies of reaction intermediates, e.g., in acid-catalyzed reactions like the conversion of methanol to hydrocarbons (MTH), or of methanol with isobutene to methyl-tert-butyl ether (MTBE) can be found in refs. [202, 207, 208].

Finally, ^{129}Xe atoms can be employed as a probe for properties of microporous materials, which is read out by the NMR spectra of this nucleus. Due to its large electron cloud, which is easily polarized, ^{129}Xe has a wide range of chemical shifts and reflects interactions sensitively. The chemical shift depends on the type of cavity (cage, pore, dead-end pore), on the cavity dimensions, it is influenced by adsorption on neutral sites (metal surfaces), by electric and magnetic fields, and by Xe-Xe interactions (i.e., by p_{Xe}) [209]. On the basis of appropriate experimental strategies, this may result in deep insight into structural features of microporous materials.

3.4.10 Mössbauer spectroscopy

While magnetic resonance utilizes the softest radiation among all spectroscopies, Mössbauer spectroscopy works with the most energetic one – with γ-rays. It studies transitions between nuclear levels, which gauge influences of the electron shell and from outside the atom on the nucleus. While it is suited for only a few elements, its diagnostic potential is extraordinary where it is applicable. Therefore, the basic features of Mössbauer spectroscopy are briefly discussed in the following.

3.4.10.1 The Mössbauer effect

In the late 1950s, Rudolf Mössbauer discovered the recoil-free emission and absorption of γ-photons in solids. In this energy range, the energy of an emitted photon, $h \nu_{em}$, is not identical with the energy released by the driving transition, $h \nu_{trans}$, because the emitting atom suffers a recoil to which an energy, $E_{R,em} = E_{em}/m_s c^2$ (c – velocity of light, m_s – mass of source), is allotted. This energy misses in the photon energy, $h \nu_{em}$, because it remains at the source. Likewise, the photon needs (almost) the same energy of $E_{R,abs} = E_{abs}/m_s c^2$ in excess of the transition energy, $h \nu_{trans}$, to allow for both the transition and the recoil of the target. Therefore,

$$h \nu_{trans} = h \nu_{em} + E_{R,em} = h \nu_{abs} - E_{R,abs} \tag{3.141}$$

In optical spectroscopy, the recoil energy is small compared to the natural line widths on the energy scale, and therefore it can be neglected. This fails, however, for nuclear levels emitting γ-radiation: the recoil caused by the 14.4 keV photons emitted by free ^{57}Fe nuclei, for instance, is six orders of magnitude larger than the emission/absorption line width.

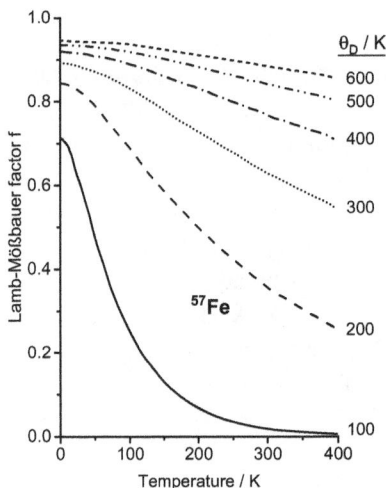

Figure 3.142: The 14.4 keV transition of ^{57}Fe: temperature dependence of the Lamb-Mössbauer factor for solids characterized by different Debye temperatures. Reproduced from ref. [210] with permission from Wiley-VCH Verlag GmbH & Co. KGaA, Weinheim, Germany.

While recoil prevents spectroscopy with nuclear transitions in fluid phases, Mössbauer found that integration of the atom in a solid lattice makes a difference: as its movement is limited by the lattice forces; it cannot recoil individually. In simple terms, it is now the whole lattice which recoils, and E_R is dramatically reduced by replacing m_S by the mass of the lattice.

In solids, the recoil energy is used to excite lattice vibrations. Therefore, the probability of recoil-free emission/absorption (the Lamb-Mössbauer factor, f) results from a comparison between the recoil energy and the quantum of the lattice vibrations (the phonon) and grows with increasing lattice forces. Figure 3.142 shows how the Lamb-Mössbauer factor for the 14.4 keV transition of Fe depends on the temperature and on the Debye temperature, θ_D. For Fe metal, for instance (θ_D = 473 K at 0 K), f is >0.9 at the temperature of liquid He and still ≈0.8 at room temperature (θ_D = 373 K). For Sb metal (θ_D ≈ 200 K between 0 and 300 K), the Lamb-Mössbauer factor for the 37.1 keV transition used for Mössbauer spectroscopy is ≈0.6 at cryogenic temperatures, but decays to 0.25 at room temperature [210]. Due to their dependence on lattice forces, Lamb-Mössbauer factors vary with the binding situation of the element studied (e.g., cations on zeolite exchange positions vs. cations in oxides, halides, etc.).

3.4.10.2 Spectroscopy with the Mössbauer effect

Radiation sources for Mössbauer spectroscopy contain radioactive nuclei that emit suitable γ-radiation upon decay. This excludes energies above 100 keV, which would result in extremely low Lamb-Mössbauer factors. The natural width of the excitation line needs to be small enough to allow for sufficient spectral resolution, and the source needs to have an adequate half-life to avoid its frequent replacement.

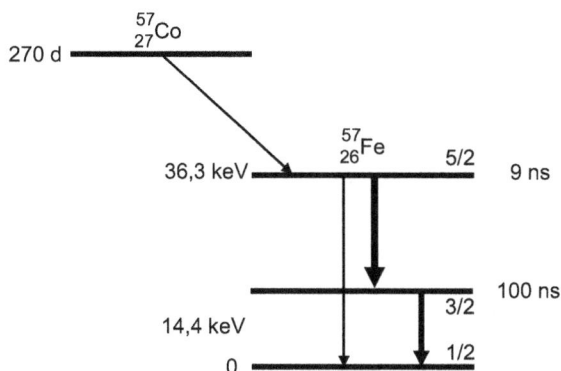

Figure 3.143: Nuclear transitions leading to the emission of the 14.4 keV radiation used in ^{57}Fe Mössbauer spectroscopy.

Figure 3.143 illustrates the nuclear reaction producing the γ-radiation used for spectroscopy with ^{57}Fe. The parent isotope $^{57}_{27}Co$ is converted to $^{57}_{26}Fe$ via electron capture with a half-life of 270 d. The $^{57}_{26}Fe$ nuclei are formed in an excited state ($I = 5/2$), which decays via two channels. The majority channel via the $I = 3/2$ state gives rise to radiation of 14.41 keV energy with an extremely low line width of $\approx 5 \times 10^{-9}$ eV.

Table 3.4: Important Mössbauer isotopes and their properties.

Isotope	Energy / keV	Natural abundance / %	Parent isotope	Half-life
^{57}Fe	14.4	2.1	^{57}Co	270 d
^{99}Ru	89.4	12.8	^{99}Rh	16 d
^{119}Sn	23.9	8.6	^{119m}Sn[a]	250 d
^{121}Sb	37.1	57.3	^{121m}Sn[a]	76 y
^{193}Ir	73.0	62.7	^{193}Os	31 h
^{197}Au	77.3	100	^{197}Pt	18 h

[a]Metastable isotope

While almost 90 Mössbauer transitions are known in isotopes of more than 40 elements, sources of reasonable flux and lifetime that can be used without intricate preparations are available for only a few of them (Table 3.4). In the sources, the parent isotope is present in a known form, often as a metal, which may be diluted in a matrix element to suppress magnetic interactions (e.g., ferromagnetism of Fe). Almost 90% of all Mössbauer work relates to Fe, often on samples made with ^{57}Fe-enriched iron. Tin is also frequently studied, while work with Sb can be affected by the substantial width of the excitation line. Due to short lifetimes, sources for work with Ru, Ir, and Au need to be prepared specifically for each project, and due to the high source energies, spectra must be recorded at cryogenic temperatures.

Mössbauer spectroscopy actually compares the environments in which nuclei of the same element are in the known source and the unknown sample. These roles can be also swapped: the unknown sample can be used as the source, which is compared with a sample of known structure. In the case of iron, this gives access to the study of Co samples: prepared from the radioactive isotope, $^{57}_{27}Co$, they are used as the source, the radiation of which is probed by a known Fe "sample" (e.g., a decayed ^{57}Fe source). Although studies with this so-called Mössbauer emission spectroscopy are relatively rare, we owe them key insight into the structure of important Co-containing catalysts, e.g., Co-Mo/Al$_2$O$_3$ for HDS [211] or into deactivation mechanisms of Co catalysts for Fischer–Tropsch synthesis [212].

Synchrotrons also provide highly brilliant radiation at energies typical for nuclear transitions. Monochromatization of this radiation to linewidths not drowning the minute energetic effects monitored in Mössbauer spectroscopy is very challenging, but possible [213]. Nuclear levels can, however, also be probed by a dynamic technique, which comprises their excitation by a very strong pulse of suitable energy and the registration of de-excitation profiles on the time scale (nuclear (resonant) forward scattering). These profiles exhibit structure, which allows concluding on the same features that shape Mössbauer spectra – ionic charges, charge gradients, and magnetic fields (see below). Though being just only explored, nuclear forward scattering is likely to become an important technique, due to the high source brilliance and because it can be used with nuclei, for which manageable Mössbauer sources are not available.

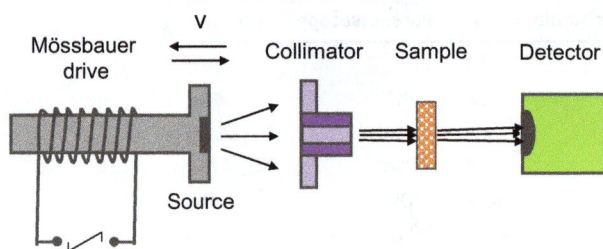

Figure 3.144: Basic elements of a Mössbauer spectrometer.

Unlike a synchrotron, monochromatic Mössbauer sources lack energy dispersion, which is required for any spectroscopy. In Mössbauer spectroscopy, dispersion is achieved by the Doppler effect, according to which the frequency arriving at a moving target differs from the frequency emitted by a source at rest:

$$E(v) = E_0 (1 + v/c) \qquad \text{or} \qquad \nu(v) = \nu_0 (1 + v/c) \tag{3.142}$$

(E_0, $E(v)$ – energies emitted by a source at rest and received by a moving target; v, c – velocity of target, of light). In spectrometers (Figure 3.144), it is usually the source that is moved, while the sample is at rest. The typical relative velocities, v, of some cm s^{-1}

correspond to a ≈1 µeV spread to both sides of the excitation line of keV energies. Due to two fortunate circumstances, this is sufficient to measure spectra: the spectroscopic effects (see below) are of the same order of magnitude, and the excitation lines are extremely narrow, thanks to relatively long lifetimes of the decaying states (cf. Table 3.4). The sample head can be cooled. In advanced spectrometers, it can be subjected to strong magnetic fields. The spectrometer bench is completed by a detection device, a scintillation counter or a Geiger-Müller tube. Pulses counted are amplified, discriminated from noise by suitable equipment, and related to the source velocity in a multi-channel analyzer fed with velocity data from the source drive electronics.

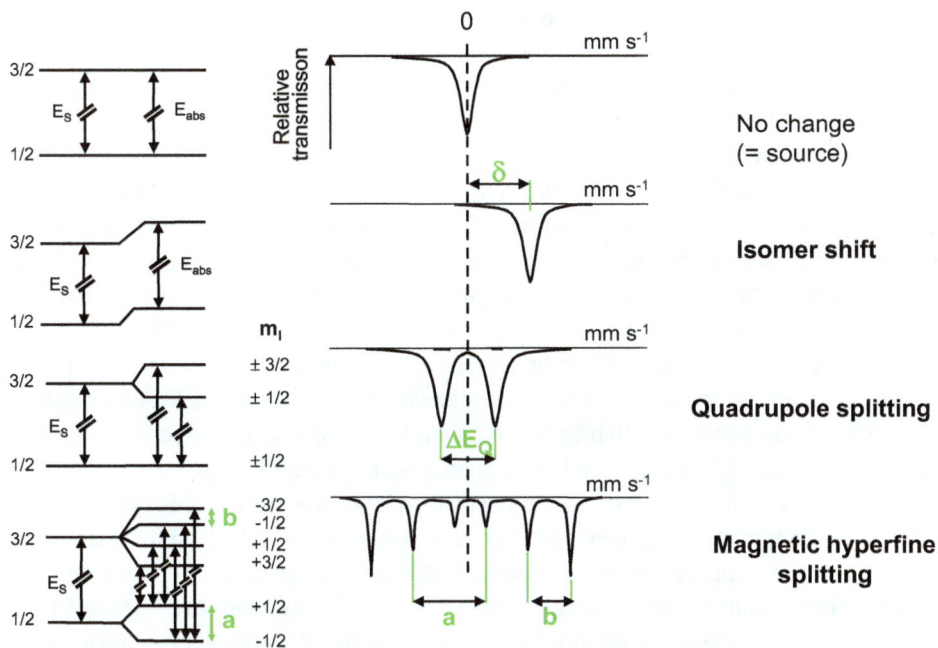

Figure 3.145: Sources of analytical information in Mössbauer spectra: left – transitions in the source (E_S) compared with transitions in the sample (E_{abs}); right – resulting spectra; the magnetic hyperfine splitting (hfs) allows determining the energy separation between ground states (a) and excited states (b) caused by the local magnetic field.

Mössbauer spectra are analyzed in terms of three quantities: the isomer shift δ, the quadrupole splitting QS, ΔE_Q, and the local magnetic field, B. The isomer shift is caused by the Coulomb interaction between electronic and nuclear charge in the nucleus. Only wave functions of s levels have a nonzero amplitude at the position of the nucleus (in nonrelativistic approximation). Perturbations in p-, d-, or f-levels are indirectly reflected in the nucleus via changes in the screening capacity of the respective levels toward the outer s orbitals. When a nucleus changes from the ground state to an excited state, its charge distribution changes as well, and so does the Coulomb energy from its interaction with the electron charge. When the electron system around the nucleus is the same in source and absorber, the changes cancel. When the electron systems differ, an isomer shift δ arises (Figure 3.145).

δ depends on the changes of both the nuclear radius during the excitation (it may increase or decrease depending on the element) and of the electron density in the nucleus, which may be caused, for instance, by the oxidation state differing between source and sample. Relations between isomer shift and oxidation states depend on the elements. In Fe compounds, δ decreases from 1.6-1.9 mm s^{-1} for Fe^{I} to -0.9 mm s^{-1} for Fe^{VI}, while Fe^0 in alloys, steels, etc. appears in a range of some tenths mm s^{-1} around v = 0 mm s^{-1}. With Ru, Au, and Ir, the isomer shift grows with increasing oxidation state.

Any nuclear level with $I > 1/2$ has a quadrupole moment. When it interacts with an electrical field gradient, its degenerate states split up into $(2I + 1)$ or $(I + 1/2)$ states for integer or half-integer I, respectively, which are separated by a splitting energy, ΔE_Q. This also splits the Mössbauer signal, e.g., into a doublet for ^{57}Fe or ^{119}Sn (cf. Figure 3.145). In powdered samples without preferential orientation, the intensities of the doublet components are identical.[53] Field gradients may result from the charge distribution within the absorbing ion or from the coordination sphere around it. The latter dominates in cations with spherical charge distribution, e.g., high-spin Fe^{III} (d^5) or in Sn^{IV} (d^{10}). When the charge distribution in the cation is asymmetric (e.g., in high-spin Fe^{II}, d^6), it dominates the effect of the coordination sphere and causes larger splittings, ΔE_Q.

The degeneracy of the nuclear levels can also be removed by interactions of the nuclear magnetic moment, μ_N, with a local magnetic field, B. Levels with spin I produce $(2I + 1)$ sublevels, between which excitations can be excited with the selection rules, $\Delta I = \pm 1$ and $\Delta m_I = 0$ or ± 1. The situation is shown for ^{57}Fe in Figure 3.145. Due to the selection rules, six transitions are excited, the intensity distribution between which is $3 : 2 : 1 : 1 : 2 : 3$ in random particle samples.[53] If there is no isomer shift, δ, the first three lines are mirrored at the signal center at v = 0 mm s^{-1}, else at v = δ. Distances between lines on each side correspond to the splitting between the excited-state levels (b in Figure 3.145); only, the distance between the central lines 3 and 4, which depends also on the ground-state splitting a, is different. The local magnetic field B can be derived from the width of the sextet. Additional quadrupole splitting results in characteristic shifts between some of the lines (not shown), which give access to ΔE_Q.

The local magnetic field B may be an external field (as employed in NMR; up to 10 T in usual Mössbauer spectrometers). Internal fields in magnetic materials may be, however, much stronger than this, e.g., 50 T in some iron oxides. They arise from three contributions: from the interaction of the nuclear moment with the spin of unpaired electrons (spin-dipolar field), likewise with the orbital momentum of these electrons,

---

53 For monocrystalline samples, the intensity distribution in QS doublets (or in magnetic sextets, see below), depends on the orientation of the electrical field gradient (or the magnetic field) relative to the direction of the γ-photons.

and from the Fermi-contact interaction, if there is a net spin in the s-electron density at the nucleus. Such spin arises from spin polarization of inner filled s-shells by lone electrons in outer shells.

Like in NMR, the detection of magnetic interactions is subject to limitations by a correlation time, which is in the nanosecond range in Mössbauer spectroscopy. This is well above the timescale of spin relaxation processes in paramagnetic materials. Paramagnetic sites appear, therefore, as singlets or, if surrounded by distorted coordination spheres, as doublets. At low temperatures, however, where spin relaxation may be frozen, line shapes of paramagnetic sites may become rather complicated (see Figure 3.146). In materials exhibiting cooperative magnetic phenomena (ferro-, ferri-, and antiferromagnetism), spin relaxation is blocked, and the magnetic dipole interaction gives rise to the characteristic sextet structure of signals.

Superparamagnetic ferro- or ferrimagnetic particles, the size of which is below that of the magnetic domains (cf. Section 3.4.9.1) behave like paramagnetic materials: their spectra are singlets or, in the presence of QS, doublets. At lower temperatures, however, the coupling interaction overcomes the impact of thermal energy, and the characteristic sextets appear in the spectra. An external magnetic field supports the ordering of the atomic magnets, and the sextet appears at higher temperatures. From the extent of splitting at different external fields, the magnetization of the particles can be determined, which gives access to their average volume and size.

Mössbauer spectroscopy is not a fast method: depending on the activity of the source and the concentration of the absorber in the sample, minutes, hours, or days may be required for a spectrum. When temperatures and magnetic fields are to be varied, a week per sample may be required. On the other hand, it is not expensive: radiation protection is straightforward and not costly, at least, in work with ^{57}Fe, ^{119}Sn, and ^{121}Sb sources. Cooling of the sample stage is standard in Mössbauer spectrometers. There are now even cells that allow measurements under reaction conditions (see literature cited in ref. [210]). In view of the temperature dependence of the Lamb-Mössbauer factor (cf. Figure 3.142), such work is promising only for Fe-containing catalysts.

For assigning nature and abundance of species in the sample, the experimental spectra are fitted with Lorentzian signals of the appropriate structure (singlets, doublets, and sextets, see above), the parameters of which (δ, ΔE_Q, B, line widths, signal intensities) are varied to reproduce the experimental data. The parameters δ, ΔE_Q, and B provide a parameter space that allows differentiation of a large number of phases by comparison with reference data. To evaluate species concentrations from signal intensities, identical Lamb-Mössbauer factors are often assumed. As shown in Figure 3.142, this assumption holds best at cryogenic temperatures. At high temperatures, differences in Lamb-Mößbauer factors between coexisting species may cause significant errors.

For paramagnetic sites with spin states of S > 1/2 (i.e., all Fe ions relevant for catalysis), the complex patterns observed when spin relaxation is frozen at low temperatures cannot be captured by such models. To avoid serious errors in the analysis, such signals should be treated with the spin Hamiltoninan formalism [214], which is also employed for modeling EPR signals (cf. eq. (3.133)). The interaction between the electrons modified by ligand field and local magnetic field and gauged by their influence on the nuclear states is described by the matrix $\underline{D}$ of ZFS parameters. As mentioned in Section 3.4.9.2, the elements of $\underline{D}$ may be combined to characteristic quantities - the axial ZFS, D ($= 1.5\, D_{zz}$), and the rhombicity parameter, E ($= 0.5(D_{xx} - D_{yy})$). In the spin Hamiltonian formalism, experimental data are fitted to find suitable values for D and E and draw conclusions about the symmetry of the ligand field. It is, however, even more attractive to fix these quantities to values found in independent EPR studies of the same samples, which warrants compatibility between the results of both EPR and Mössbauer spectroscopy. The powerful constraints imposed by this on the signals of paramagnetic species strongly increases the credibility of the Mössbauer fits, which may be doubtful if too many components are included.

In Figure 3.146, this approach is exemplified by the analysis of a ^{57}Fe-ZSM-5 catalyst studied before its use in the SCR of NO by NH$_3$. A parallel EPR measurement revealed that two paramagnetic Fe^{III} oxo species must be considered; their ZFS parameters were known from earlier work. Beyond them, four more structures were differentiated: Fe^{III} oxide nanoparticles, Fe^{II} (despite previous calcination at 873 K), and Fe^{III} oxo oligomers, which could be split into magnetic and diamagnetic oligomers. At high field, the former join the nanoparticle sextet, while the latter do not exhibit a sextet even under these conditions. They can be identified with the binary Fe-O-Fe oxo structures lively discussed but never proven in the past, while the magnetic oligomers contain more Fe atoms. From the site distribution given in Figure 3.146, a magnetic moment of 3.29 $N_A\, \mu_B$ (N_A – Avogadro number, μ_B – Bohr magneton) was predicted, while a moment of 3.35 $N_A\, \mu_B$ was found in a measurement with a SQUID magnetometer. No other technique is able to resolve a comparable complexity in material structure into a realistic quantitative analysis of coexisting sites. Unfortunately, Mössbauer spectroscopy provides this opportunity only for a few elements and in a narrow range of measurement conditions.

Mössbauer spectroscopy is often used to study the development of phases during treatments or under catalytic conditions. Where Fe or Sn are employed as promoters, their interaction with the active component can be tracked. Recent studies deal with catalysts for CNT synthesis (cf. Section 3.1.1.5) or with Pt-free fuel cell cathodes, where Fe is substituted into graphene structures. More applications can be found in ref. [216].

3.4.11 Imaging

While the techniques discussed so far in this chapter give a rather indirect access to properties of catalytic materials, morphological aspects can be directly addressed by imaging. Imaging methods available to catalysis research are now capable of resolving structure, morphology, and composition on the atomic scale. They may be considered

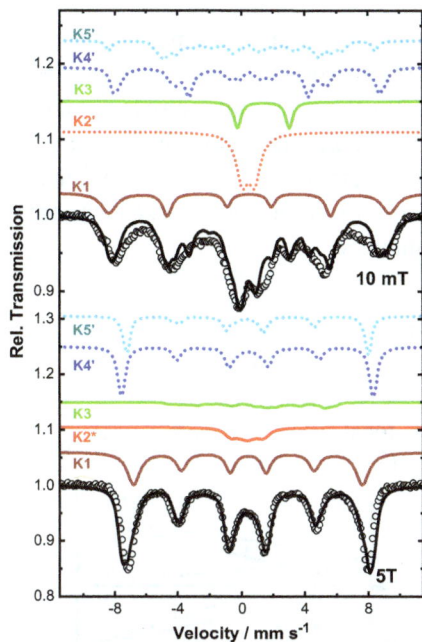

Figure 3.146: Mössbauer analysis of a ^{57}Fe-ZSM-5 catalyst; spectra were measured at 5 K in low and high magnetic fields (10 mT and 5 T, respectively); signals and assignments: K1 – Fe^{III} oxide aggregates (23%), at 5 T additionally magnetic Fe^{III} oxo oligomers (12%); K2' – non-blocked superparamagnetic structures (total of Fe^{III} oxo oligomers, 20%); K2* - diamagnetic Fe^{III} oxo oligomers (8%); K3 – Fe^{II} (7%); K4' – paramagnetic Fe^{III}-oxo species, D = 0.3 cm^{-1}; E = 0.1 cm^{-1} (30%); K5' – paramagnetic Fe^{III}-oxo species, D = 0.5 cm^{-1}, E = 0.025 cm^{-1} (20%). Reproduced with permission from ref. [215]. Copyright 2020 American Chemical Society.

as advanced versions of vision and of touch. In TEM, a probe sent to the object from far (electrons instead of photons in optical microscopy) provides full images of it at any point of time. In the more recent scanning probe microscopy, a probe is scanned across the object surface at a few nm distance and records the interactions at each point. The image is constructed sequentially by relating probe positions to observed interaction intensities. Already earlier, the scanning principle had been combined with electron microscopy: in the Scanning electron microscope (SEM), a focused electron beam is sent to the surface from far to probe its responses.

Although micrographs from both electron and scanning probe microscopy may be extremely revealing, it should be always kept in mind that the samples studied are extremely small. In scanning probe microscopy, which is used for fundamental studies in surface science rather than in catalysis research, this creates no problems, because the surface can be well characterized by the method itself. Evidence from electron micrographs is, however, sometimes questioned because of doubts that the few grains imaged can represent kilograms of a real catalyst. The problem is real: to be representative, electron micrographs should result from appropriate sampling on two levels: the abundance of structures in the few micrograms of sample taken from a macroscopic catalyst batch should be correctly represented in the micrographs, and the properties of this catalyst batch should be correctly represented by the microgram sample. While the former can be achieved by imaging a large number of points (or grains) in the sample[54], the latter remains a point of concern. Quantitative data obtained by electron microscopy (particle sizes, alloying degrees, etc.) are, therefore, often cross-checked with techniques that allow averaging over large sample amounts.

3.4.11.1 Transmission electron microscopy (TEM)

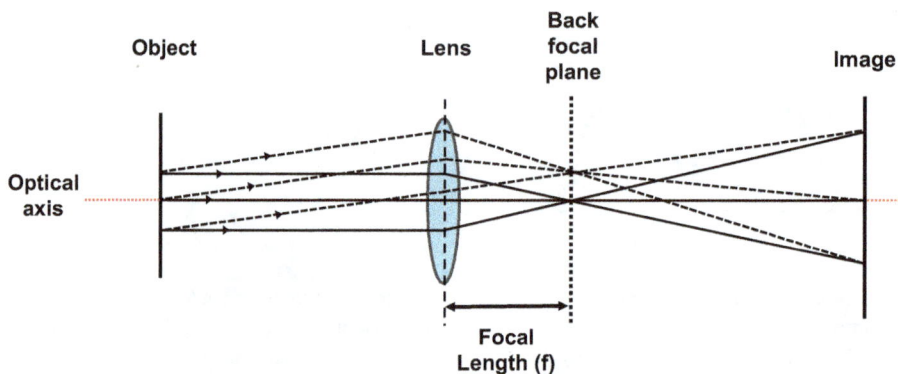

Figure 3.147: Examples of beam paths in a simple optical imaging system.

In an optical image, light originating from points in the object is focused onto related image points by an optical system. This is achieved by a lens, as illustrated in Figure 3.147 by the pairs of solid and dashed lines representing light beams. The figure shows that lenses offer another important opportunity: light arriving at parallel incidence is focused by them onto points within the back focal plane. The location of these points is related to the angle between the incident light bundle and the optical axis of the arrangement. We shall come back to this feature below (Figure 3.152). In general, the back focal plane exhibits a continuous intensity distribution related to the object by a Fourier transformation (cf. eq. (3.146) below).

Beyond the unambiguous correlation between object and image points depicted in Figure 3.147, imaging requires a contrast that allows distinguishing the latter: image points must not be all of the same color or intensity. In light microscopy, where intensity losses between object and image plane are negligible due to the large aperture angles

---

54 As an example of this strategy, a catalogue of nanostructures observed in $(Mo,V)O_x$ catalysts for the oxidative dehydrogenation of ethane to ethene has been recently established. It relates structure and composition of 19 nanostructures to information on location and abundance [217].

of optical lenses, the contrast originates exclusively from the interaction of the incident light with the sample, by absorption and scattering. In electron microscopy, where aperture angles of lenses are very small, the situation is different (see discussion of Figures 3.152 and 3.153).

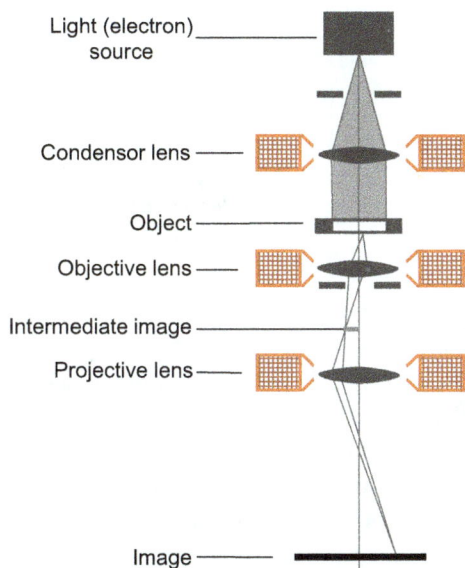

Light (electron) source

Condensor lens

Object

Objective lens

Intermediate image

Projective lens

Image

Figure 3.148: Beam paths in transmission microscopes – comparison of optical microscopy (glass lenses) and electron microscopy (electromagnetic lenses).

To achieve magnifications targeted in microscopes, multistage optical systems are required. Figure 3.148 compares the two-stage system of transmission microscopes operating with light and with electrons. Light originating from the source is collimated by a condenser lens, which turns the directions of photon propagation parallel over the whole object cross section. The image is produced from the transmitted light, which is exemplified in the figure by light starting from only one object point. As in Figure 3.147, the light from this point is focused by a lens (the objective lens) onto a point of an image, which is the intermediate image in the two-stage arrangement. Beyond, the light is dispersed in a similar way as behind the object and is focused in the image plane by the projective lens.

Images are never ideal. As a result of diffraction at the rims of lenses and of various kinds of lens imperfections (aberrations), points in the object appear as diffuse disks surrounded by weak diffraction rings: the Airy disks (Figure 3.149a). The Airy disk radius, r_D, is specified as the distance between center and first minimum. Overlap of adjacent airy discs limits the point-to-point resolution. According to the Rayleigh criterion, disks can be differentiated when their intensity peaks are not closer than the Airy disk radius. (Figure 3.149b, c). It can be shown that this results in a resolution limit of:

$$r_{D,min} = 0.61\, \lambda/NA \tag{3.143}$$

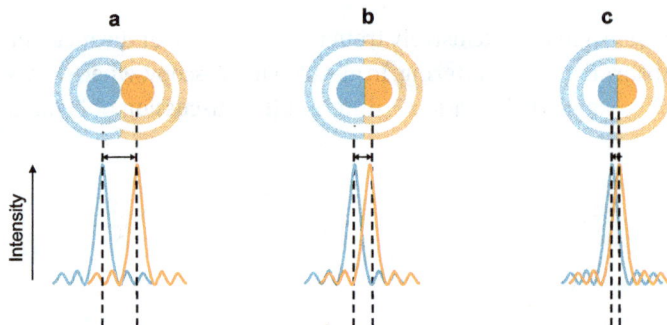

Figure 3.149: Influence of lens aberrations and diffraction on the resolution: intensity in Airy discs representing two object points in a real image: (a) points are well resolved; (b) Rayleigh resolution limit; (c) points are not resolved.

where the numerical aperture NA is an optical parameter related to aperture and refractive index of the lens. With typical values for NA between 1 and 1.5, the resolution limit is around 200 nm for visible light ($\lambda > 400$ nm). It can be further decreased by imaging with radiation of shorter wavelengths (UV, X-ray), however, at high technological expense. The visible light resolution limit can also be broken by techniques using fluorescence dyes[55] or by scanning nearfield optical microscopy (SNOM), a near relative of scanning probe microscopy (Section 3.4.11.3). These methods, however, have not yet been adapted to catalytic materials.

Imaging with electrons instead of photons results in drastic improvement of resolution. In a TEM (Figure 3.148), electrons are accelerated by a potential V to a linear velocity v determined by the energy balance:

$$e\,V = m_e v^2 / 2$$

Their wavelength, λ, results from the de Broglie equation (3.75):

$$\lambda = h / \sqrt{2\,m_e e\,V} \tag{3.144}$$

At the velocities achieved, relativistic corrections are, however, required, which changes (3.144) into (3.145):

$$\lambda = h / \sqrt{2m_{0,e}eV\left(1 + eV/2m_{0,e}c^2\right)} \tag{3.145}$$

---

55 Techniques applying **RESOLFT** (reversible saturable optical fluorescence transitions). They are based on methods to prevent adjacent molecules from fluorescing simultaneously, which allows their sequential detection.

($m_{0,e}$ - electron mass at rest). At the typical acceleration voltage of 200 kV, the wavelength is below 3 pm, which no longer limits resolution. Instead, resolution is mostly limited by imperfections of electron sources and lens aberrations. In the late 1990s, when the resolution limit was at ca. 0.2 nm, aberration correctors combining electrostatic and magnetic fields became available, which have, meanwhile, allowed shifting resolution limits of the TEM down to 0.05 nm.

Figure 3.150: Pitfalls in the interpretation of two-dimensional projections of three-dimensional objects; courtesy Dr. Thomas Lunkenbein, Fritz-Haber-Institut Berlin.

Despite the enormous gain in resolution, imaging by electrons cannot remove a fundamental limitation of all transmission microscopy: images are two-dimensional; they show a projection of the three-dimensional sample structure on the image plane. Figure 3.150 illustrates which error can result from hasty interpretation of such projections. The limitation can be removed by the tomographic approach, which is briefly outlined in Section 3.4.11.2.

Electron microscopes are analogous to optical microscopes with respect to the beam path (Figure 3.148), with glass lenses replaced by electromagnetic lenses. Objects should be penetrated by the electron beam, without sizeable attenuation by inelastic interactions. This limits sample thickness to <100 nm, except for special instruments working with very high acceleration voltages (>1,000 keV), which requires intricate sample preparation techniques in many fields of materials research. Most catalysts are, however, available as powders and can be exposed just spread on a suitable sample grid. To minimize interactions of imaging electrons with species other than sample material, electron microscopes work in high vacuum, typically at $\approx 10^{-7}$ mbar.

In the electron source, electrons are extracted from a cathode using thermal emission or the tunnel effect. In thermoionic emission, electrons emitted by heated V-shaped tungsten wires or by a LaB_6 single crystal are attracted by a potential of some kV. It is applied to a Wehnelt cylinder, which releases the electrons as a preconditioned beam for further acceleration. Modern microscopes employ field emission sources based on Schottky or cold field emission. Both types differ in extraction voltages and operation temperatures. Cold field emission taps the electron density extending beyond the metal surface in evanescent waves (cf. Figure 2.6b and Section 3.4.8.3) by applying high extraction voltages to extremely fine tungsten tips. Such sources provide the highest brilliance and the narrowest energy distribution of the electrons (<0.5 eV), but they need to be periodically refreshed by flushing adsorbed deposits.

Electrons are further manipulated by various components: electromagnetic lenses, electrostatic deflectors, which move the imaging beam across the sample holder, and stigmators, which correct lens astigmatisms. The condenser lens (system) transforms the incoming electrons into a parallel beam; the objective lens focuses the electrons leaving the sample onto the intermediate image, which is then imaged by the projective lens (Figure 3.148). There are rigid or variable apertures at various positions, e.g., behind lenses for cutting off intensity at high aperture angles where lens aberrations are strong, for assigning electrons their role in image formation (contrast aperture, see below), or for confining the sample region to be imaged. The image is displayed on a ZnS-based phosphor screen for direct observation. In addition, it can be recorded by various techniques — photographic film in the past, nowadays, rather CCD or semiconductor (CMOS) detectors.

While beam paths in TEMs and optical microscopes are similar, the nature of contrasts is completely different. In electron microscopes, imaging is based exclusively on elastic scattering, at the electron energies applied predominantly at the potentials of the nuclei. Elastic scattering deflects electrons by a wide distribution of scattering angles, shifting from forward toward backscattering with increasing target mass and thickness.

In the TEM, inelastically scattered electrons contribute to a diffuse background and need to be kept away from the image. On the other hand, processes induced by inelastic scattering can be utilized for various purposes in the TEM, and even more in the SEM (cf. Section 3.4.11.2). Ionization of electron levels in the sample raises element-specific spectra of fluorescence photons and of Auger electrons (cf. Figure 3.76). Fluorescence may even extend to the visible region. Secondary electrons are another characteristic feature resulting from the interaction of high-energy electrons with solids.[56] Their abundance is very high at low energies and tends to zero beyond 50 eV. Secondary electrons escape into the vacuum only from a layer extending to a few nm below the external surface (cf. Figure 3.97). As their production requires high-energy electrons, their inten-

---

56 Secondary electrons may originate from scattering of valence or conduction electrons with the primary beam or with outgoing core electrons, which in turn suffer energy losses, or from the decay of plasmons excited by the primary beam.

sity is highest where the primary beam hits and leaves the object. Scattered high-energy electrons cause diffuse secondary electron emission all over the sample surface.

In a simplified model, imaging is ideal when scattered and non-scattered intensity coming from an object point is completely refocused onto an image point, without detuning of the phase. In reality, elastic scattering results in phase shifts, and intensity is lost because electrons scattered to larger angles are not captured by the electron lenses. In a crystalline object, such deflection can be also caused by diffraction. As a result, image regions related to strongly scattering (or diffracting) parts of the sample collect less intensity than other parts imaged by non-scattered or weakly deflected electrons. Such contrast is referred to as **amplitude contrast**. It is further differentiated into mass-thickness contrast caused by scattering and diffraction contrast.

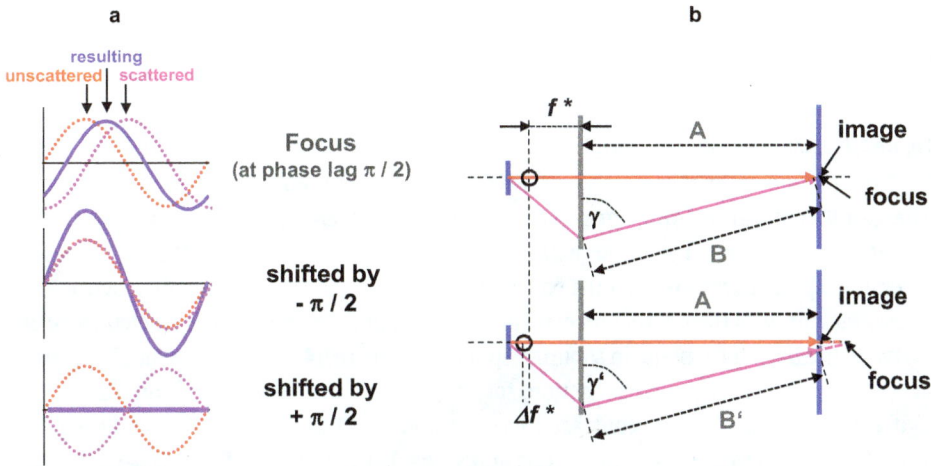

Figure 3.151: Phase contrast – nature and realization by intentional defocusing: (a) interference of scattered and unscattered waves in an image point, spontaneous case and after artificial phase shifts by − π/2 or + π/2; (b) shifting the phase relation between unscattered and scattered waves: changing the path length for the latter by intentional defocusing ($\gamma' \neq \gamma$, B' ≠ B).

Scattered light usually arrives at the image point phase-shifted relative to non-scattered light. Apart from scattering events, lens aberrations contribute to such difference in phases, $\Delta\varphi$. Intensity is enhanced by constructive interference between scattered and non-scattered waves, ($\Delta\varphi = 0$ or 2π), but decreased by destructive interference ($\Delta\varphi = \pi$). In the more likely case of intermediate $\Delta\varphi$, intensity remains all the same, irrespective if the imaged point caused strong or weak scattering (Figure 3.151a). When the phase difference is manipulated to achieve $\Delta\varphi = 0$, 2π, or π (for procedure see below), scattering regions change intensity accordingly (Figure 3.151a), while non-scattering regions remain unchanged (**phase contrast**). Notably, phase contrast does not show object features directly, but only interference patterns of intensity scattered with different phase shifts.

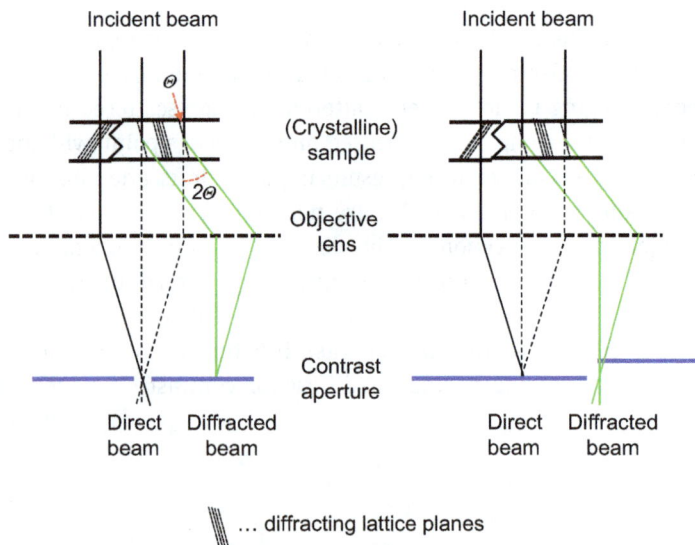

Figure 3.152: Diffraction contrast, imaging in bright-field mode (left) and in dark-field mode (right).

The spontaneous **amplitude contrast** described above includes electrons captured by the lenses only at the expense of large aberrations. Image quality can be strongly improved by operating the instrument with a contrast aperture in the back focal plane. Figure 3.152 illustrates this with the diffraction contrast, where non-diffracted and diffracted electrons arrive in the back focal plane in separate spots (cf. Figure 3.147). This holds for all electrons across the primary beam cross section, for which only some rays are exemplified in the figure. In Figure 3.152a, the contrast aperture blocks all but the non-diffracted electrons. Therefore, diffracting parts of the sample appear darker than non-diffracting parts. Usually, the orientation of most crystallites does not comply with the Bragg condition, which results in a bright image background, on which diffracting parts appear in dark tones. This kind of image is, therefore, referred to as bright-field (BF) image. In Figure 3.152b, the contrast aperture blocks the non-diffracted electrons and gives way to electrons diffracted into a particular direction. This reverses the contrast: diffracting crystals appear bright on a dark background of non-diffracting sample regions (dark-field (DF) images).

The benefit of the contrast aperture for mass-thickness contrast is very similar. Electrons scattered by an amorphous sample into various directions off the primary beam may be conceptually grouped into parallel rays as well, but their directions are not discrete as in the case of diffraction, but distributed. In the resulting intensity distribution in the back focal plane, intensity on a concentric circle around the optical axis originates from electrons scattered at the same angle to the primary beam (cf. Figure 3.147). Placed in the optical axis of the microscope, the contrast aperture lets pass non-scattered or, depending on its diameter, slightly scattered electrons to image their region of origin in bright tones. Stronger scattering regions appear dark (BF image). With the contrast aperture off-axis, the contrast is reversed.

BF and DF imaging are general techniques applied to deflection-based contrasts. They are also used in Scanning transmission electron microscopy (STEM, cf. Section 3.4.11.2), which combines features of SEM and TEM. In Figure 3.160 below, BF and DF imaging are exemplified by micrographs showing the same catalyst region in BF and DF STEM. While TEM micrographs published in literature are mostly BF images, DF imaging is preferred in STEM.

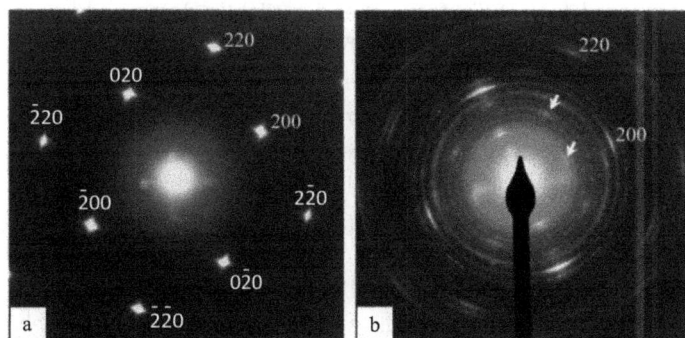

Figure 3.153: Diffraction images of a monocrystal and of dispersed nanocrystals: (a) Pt crystallites of identical orientation producing a diffractogram identical with that of a monocrystal; (b) Pt nanoparticles of various orientations, mixed with Pt silicide particles (for explanation of arrows consult original paper). Reproduced from ref. [218] with permission from Elsevier.

As mentioned above (Figures 3.147, 3.152), diffracted electron beams create points in the back focal plane of the objective lens. In microscopes, these diffraction patterns can be projected on the observation screen to allow structure identification of imaged crystals. Figure 3.153a shows a pattern as expected from a monocrystal, although it originates from a set of oriented nanocrystals.[57] Other lattice planes may be brought to diffraction by tilting the sample. In polycrystalline samples, the cone-shaped diffraction patterns (cf. Figure 3.68) create concentric circles in the back focal plane when the orientation of crystallites is random. This can be nicely observed in Figure 3.153b taken after reduction of the sample in H_2 at 873 K. The pattern results from a polycrystalline mixture of phases, although remnants of the initial Pt crystallites are still there: the intense reflections from Figure 3.153a are now pulled apart into circular arcs, indicating a spread in orientation, probably due to thermal and chemical perturbation of the SiO_2 support layer. The new phases were identified as Pt silicides [218] – two Pt_3Si phases and $Pt_{12}Si_5$.

In polycrystalline samples, crystallites can be selected for structural analysis by an aperture inserted at a suitable location in the optical system (Selected area diffraction,

---

57 Diffraction pattern of a Pt/SiO_2 model catalyst. Pt crystallites were eptaxially grown on an NaCl(001) facet and covered by a 25 nm SiO_2 film, before the NaCl was dissolved.

SAED). Alternatively, the electron beam can be narrowed down for probing any part of the sample. Beam sizes down to 20 nm can, nowadays, be achieved without critical loss of parallelism.

As shown in Figure 3.151a, the phase difference, $\Delta\varphi$, between scattered and non-scattered electrons superimposing in image points needs to be manipulated to utilize the **phase contrast.** This may be achieved by inserting phase-shifting plates or by a slight intentional defocusing of the image. In Figure 3.151b, the optical paths between object and image points are compared for a fully focused and a slightly defocused image. There are no differences in front of the lens and in the further path of the non-scattered electrons (A). Defocusing, however, changes the angle γ between the lens plane and the leaving wave into γ' and, therefore, also the distance to be traveled between lens and screen. In the dimensions of electron microscopes, the difference between B and B' obtained by this slight defocusing of the image is of the order required to tune the phase difference, $\Delta\varphi$, in the image point (Figure 3.151a). In a defocus series, the focused image point is displaced from above the screen to behind, due to a variation of the lens focus, which allows covering the full range of phase differences, $\Delta\varphi$, from 0 to π. This should result in a reversal of contrast (cf. Figure 3.151a), which can, however, rarely be observed, because phase contrast is additionally modified by a number of influences, e.g., sample thickness and defocus dependence of aberrations.

Figure 3.154 shows a polycrystalline MgO sample imaged under different focus conditions. A reversal of intensity can be detected only in the Fresnel fringes enframing some crystal edges (see arrows), which are a by-product of phase contrast. The figure illustrates, however, that many details of the object can be discerned much better under defocus conditions.

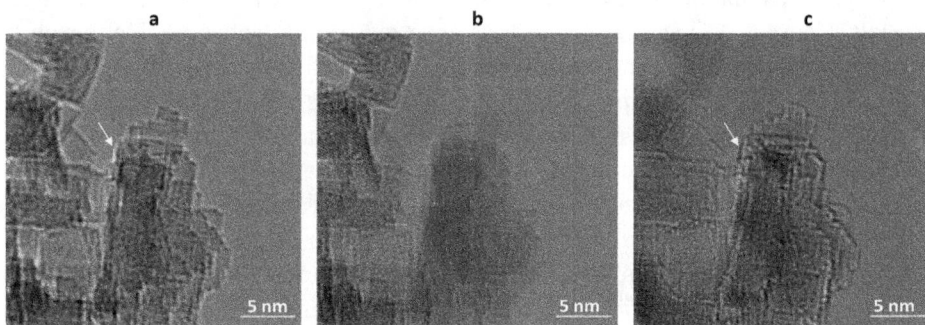

Figure 3.154: Defocus series on polycrystalline MgO: (a) underfocus, (b) focus, and (c) overfocus. Courtesy Dr. Gerardo Algara-Siller, Fritz-Haber-Institut Berlin.

Phase contrast is the basis for the highest resolutions accessible in the TEM: for High-resolution transmission electron microscopy (HRTEM), which allows resolving the dimension of atoms. To achieve this, samples must be thin enough to minimize multiple scattering events. Due to the complex nature of phase contrast, a reliable identification of atom positions requires support by mathematical modeling of images obtained under various defocus conditions.

The methods used for this are beyond the scope of this book, only the strategy will be briefly outlined. It is based on Abbe's theory of imaging, according to which an optical image results from a sequence of Fourier and reverse Fourier transformations of the primary planar (electromagnetic or electron) wave. As discussed earlier (eq. (3.67)), scattering of an incident wave at an electron density $\rho_0(x,y)$ results in an FT of ρ_0 into the reciprocal space ($\rightarrow \Psi(u, v)$). The intensity distribution caused by $\Psi(u, v)$ can be viewed in the back focal plane. An ideal lens collects the scattered waves completely and produces a reverse FT in the image plane, which establishes an imaged electron density map, $\rho_{im,id}(X, Y)$:

$$\rho_0(x, y) \xrightarrow{FT} \Psi_{id}(u, v) \xrightarrow{r-FT} \rho_{im,id}(X,Y) \tag{3.146}$$

$$\rho_0(x, y) \xrightarrow{FT} \Psi_{id}(u, v) \xrightarrow{T(u,v)} \Psi_{id}(u, v) \otimes T(u, v) \xrightarrow{r-FT} \rho_{im}(X,Y) \tag{3.147}$$

Non-idealities of real lenses (finite extension, aberrations) are represented by transfer functions, $T(u,v)$, which are convoluted with the ideal image function to obtain the real one.

Being the target of the effort, $\rho_0(x,y)$ is unknown. Images are therefore evaluated assuming structural models and improving them by comparing modeled and real images. The evaluation of the wave $\Psi_{id}(u,v)$ released by the object is complicated by the three-dimensional nature of real samples, in which scattering potentials vary along the primary beam direction, z. Wave propagation in them can be modeled with multi-slice algorithms, where potentials are kept constant in z direction within a slice and outgoing waves are used as input for the next slice.

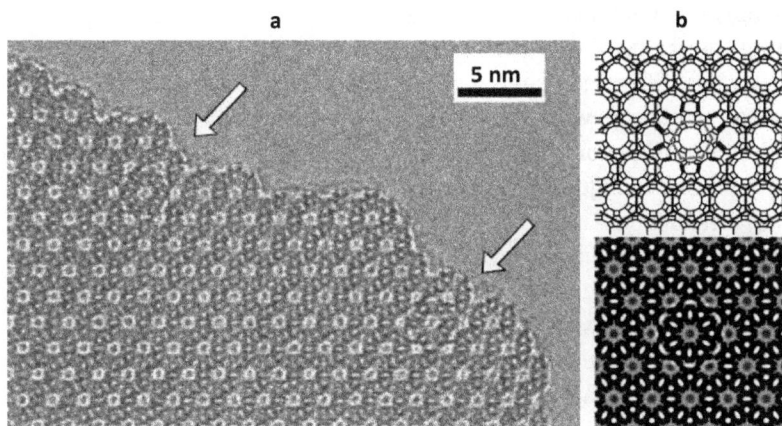

Figure 3.155: HRTEM imaging of structural defects in the framework of zeolite L: (a) micrograph; (b) schematic model and simulated image. Reproduced from ref. [219] with permission from Wiley-VCH Verlag GmbH & Co. KGaA, Weinheim, Germany.

Micrographs featuring predominant phase contrast should be always supported by results of model calculations as shown, for instance, in Figure 2.15b. A more complex example, in which structural defects in zeolite L are imaged, is given in Figure 3.155. It is remarkable not only for its quality and agreement between model and image, but also for achieving such quality for a zeolite known to be prone to structural damage by the imaging electrons.

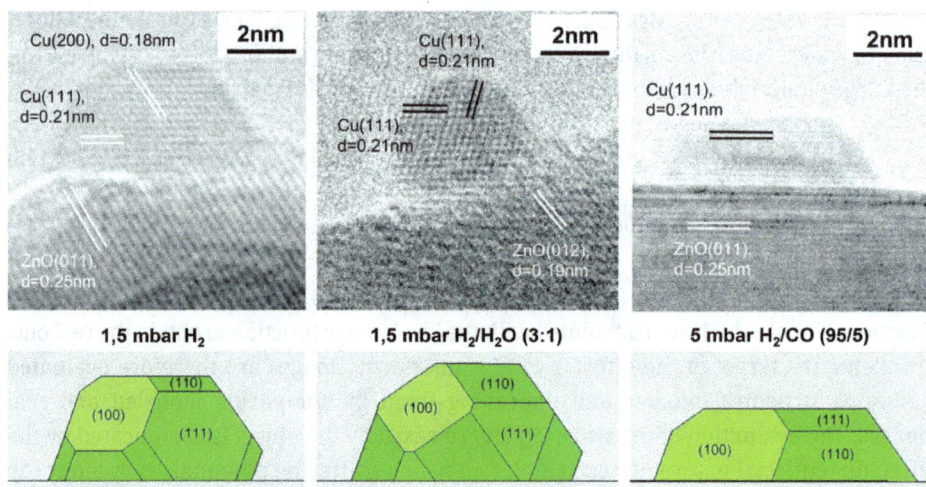

Figure 3.156: Environmental TEM micrographs: a Cu/ZnO catalyst in different reactive atmospheres: dry H_2, moist H_2, and H_2 with CO added, at 490 K. From ref. [220]. Reprinted with permission from AAAS.

HRTEM is often operated with a large or even without any contrast aperture, in order to utilize much of the scattered/diffracted intensity. When only two diffraction spots (a diffracted and a non-diffracted one) are selected instead, parallel equidistant dark lines can appear on the image, which are referred to as lattice fringes (Figure 3.156). Selecting an additional diffraction beam can produce simultaneous lattice fringes from two sets of lattice planes. In Figure 3.156, the fringes are even from different components and distinguish Cu from ZnO. As indicated in the figure, distances between lines are identical with the lattice plane distances. Lattice fringes are no direct image of the lattice planes though: their position changes with the defocus conditions. The phenomenon was first reported in 1956, when point-to-point resolution of microscopes was still far from the actual spacing in the set of lattice planes that caused the fringes. The lattice fringe phenomenon is not amenable to simple explanations, but it may be derived from wave theoretical treatment of image formation.

Transmission electron microscopes usually contain an Electron energy loss spectrometer (EELS) for studying compositions of imaged spots. In Section 3.4.8.2, electron energy loss spectroscopy of vibrational modes (HREELS) has been mentioned. In EELS, the loss signals employed arise from the ionization of core levels similar to the edges used in XAS. Similar to the edge energies, they indicate the nature of an atom and allow conclusions on its oxidation state. Its bulk concentration may be deduced from signal intensities. Unlike Energy dispersive X-ray spectroscopy (EDX), which is typically used in the scanning mode (see Section 3.4.11.2), EELS has a good sensitivity also for light elements.

The high-vacuum requirement penalizes electron microscopy in experiments with working catalysts in a similar way as XPS. There are two approaches to deal with this problem: differentially pumped environmental cells and, more recently, cells separated from the vacuum system by windows transparent for electrons.

The differential pumping stages of aperture-based cells cannot be accommodated in conventional microscopes: the Environmental TEM (ETEM, also referred to as Controlled-atmosphere electron microscope, CAEM) is a dedicated instrument. The high energy of the imaging electrons allows reactant pressures in the environmental cell of 10–20 mbar along optical path lengths of ca. 10 mm. The micrographs in Figure 3.156 are actually from a CAEM study: the ZnO-supported Cu nanoparticles are imaged in three different atmospheres at 490 K and pressures up to 5 mbar. They illustrate the influence of the redox potential on the shape of the Cu crystallites, which had been indirectly observed by XAFS before. Under the most reducing conditions (i.e., the lowest H_2O partial pressure), Cu crystallites are flattened, most likely by an interaction with oxygen vacancies created by a slight reduction of the ZnO surface (cf. Figure 4.18a).

Closed cells allow electron microscopy with the catalyst under pressures up to a few bars, and even in a liquid phase. The required dramatic reduction of the optical path lengths of the electrons in these media is achieved by manufacturing the cells by microelectromechanical systems (MEMS). A typical gas cell is shown in Figure 3.157. The height of the gas channel is of the order of 1 μm. The windows typically made of Si_3N_4 are only 10-50 nm thick, but withstand pressures of several bars due to their small area of 0.1–1 mm^2.

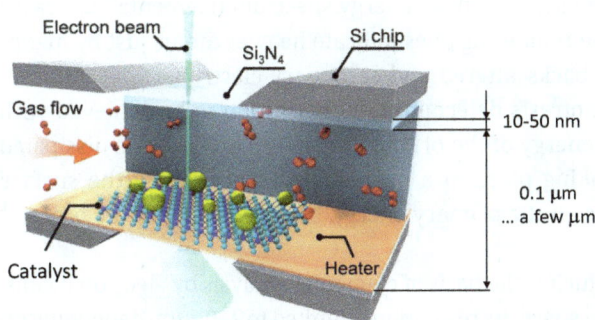

Figure 3.157: A MEMS-based gas cell for *in-situ* and *operando* (S)TEM studies on catalysts. Reproduced with permission from ref. [221]. Copyright 2020 American Chemical Society

Sample holders fitted with closed gas cells can be used with any TEM or SEM. Closed cells allow feeding reactants at well-defined weight-feed ratios (W/F, cf. Section 3.2.1), although these will be much higher than in any fixed-bed reactor, due to the tiny catalyst samples. Reaction rates can be measured during imaging (*operando* principle), although they are unlikely to agree with rate data from fixed-bed reactors for a number of reasons [222]. The contact between feed and catalyst is, however, much better defined than in the CAEM.

These strong points of closed gas cells come together with a few limitations, e.g., some influence of the windows deteriorating resolution particularly for light elements and some restrictions in the choice of samples due to the cell dimensions. Atomic resolution at atmospheric pressure has been, however, demonstrated for closed cells with catalysts containing Pd, Pt, and Co. Instructive examples can be found in ref. [222], where opportunities and limitations of closed liquid cells for electrocatalytic studies are also discussed.

3.4.11.2 Scanning electron microscopy (SEM)

Figure 3.158a shows the beam path in a scanning electron microscope. As mentioned above, electrons are focused in the sample surface in SEM – not in the image as in TEM. This is accomplished by a combination of condenser and objective lenses. Together with the operation mode of the condenser lens, the objective aperture determines size and current of the probing beam. A scanning coil moves this beam across the sample surface. The sample response at each point is recorded by detectors. Objects may be thick, because detectors are above the sample in most SEM versions. With thin samples, transmitted electrons can be collected below the sample. This STEM has developed into an extremely powerful technique in recent decades.

There are various sample responses that create contrast in the SEM: backscattered electrons, secondary and Auger electrons, fluorescent X-rays, and transmitted electrons. Signals detected above the sample can originate from its surface (secondary and Auger electrons) or from the bulk (backscattered electrons and X-rays). Some of them reveal the nature of the emitting atom by their energy spectrum (elemental contrasts: X-rays and Auger electrons). The remaining ones indicate heavier atoms just by higher intensities (material contrasts: backscattered and secondary electrons). The differentiation into surface and bulk contrasts determines the resolution to be achieved with the corresponding signals. The energy of the primary beam (up to 30 keV) is dissipated in a cascade of interactions taking place in a pear-shaped space below the surface (Figure 3.158b). Its size depends on the primary energy, but is typically of the order of micrometers.

X-ray fluorescence (XRF), which is the basis of elemental analysis by electron beams, arises from the whole interaction space. Its resolution is limited to 2–0.1 μm depending on the primary energy. Due to the competition of XRF with Auger emission (cf. Section 3.4.5), it is best suited for elements beyond Na. Under the name EDX (Energy dispersive X-ray

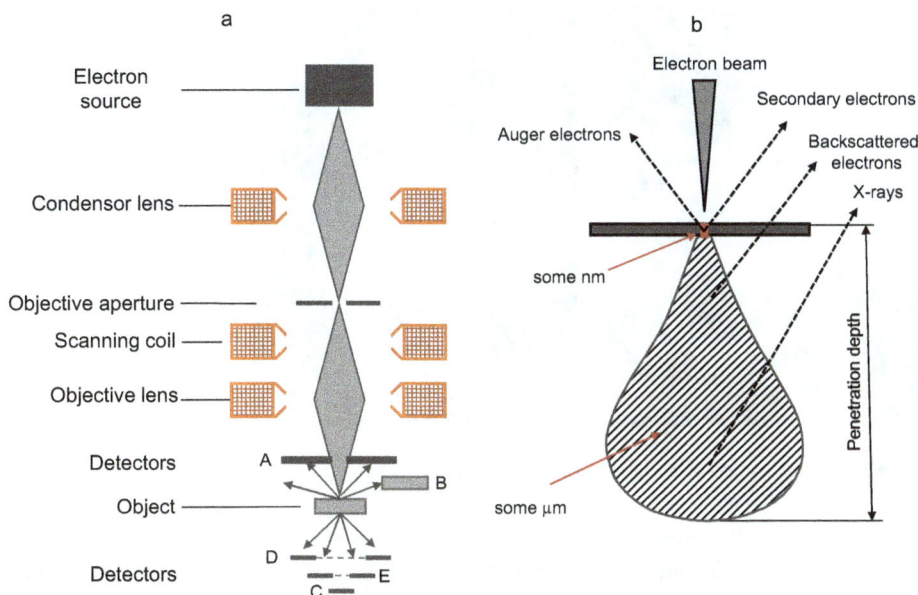

Figure 3.158: Scanning electron microscopy. (a) Beam path and detectors in a SEM: A – detector for backscattered electrons; B – Everhard-Thornly detector for secondary electrons; C, D, E – detectors for STEM: bright-field detector (C), HAADF detector (D), intermediate annular detector (E). (b) Surface and bulk contrasts. The bulk contrasts originate from a space, which is pear-shaped, because the interaction cross sections of the atoms grow with decreasing energy of the interacting electrons.

analysis), it is a supplementary operation mode, also in TEMs. Specialized instruments for elemental analysis with electron beams are optimized for analysis rather than for imaging (electron microprobe). Backscattered electrons may be redirected and further slowed down on their way out of the solid; therefore, their sampling depth is somewhat smaller than that of X-rays. Spatial resolution remains, however, in the μm range.

Arising from the touchpoint of the primary beam with the surface, surface contrasts allow resolutions in the nm range. As mentioned in Section 3.4.11.1, flux of low-energy (secondary and Auger) electrons is orders of magnitude higher when excited by the primary beam than by diffuse backscattered electrons. On surfaces inclined relative to the primary beam, more low-energy electrons are emitted per beam cross section unit than on surfaces perpendicular to the beam (edge effect).

Secondary electron contrast is the standard operation mode of SEM. It provides point-to-point resolution of 5 nm in routine work, but down to 1 nm under favorable conditions. The Everhard-Thornly detector (B in Figure 3.158a) attracts secondary electrons by a potential of the order of 1 keV, after which they can be registered by a scintillation counter. Due to the edge effect, edges of objects appear brighter than the top facets, which generates the impression of a three-dimensional image. In Figure 3.159 showing V_2O_5 in different crystal morphologies, objects appear to be mildly illuminated from one side, because electrons originating from facets at the detector side have the

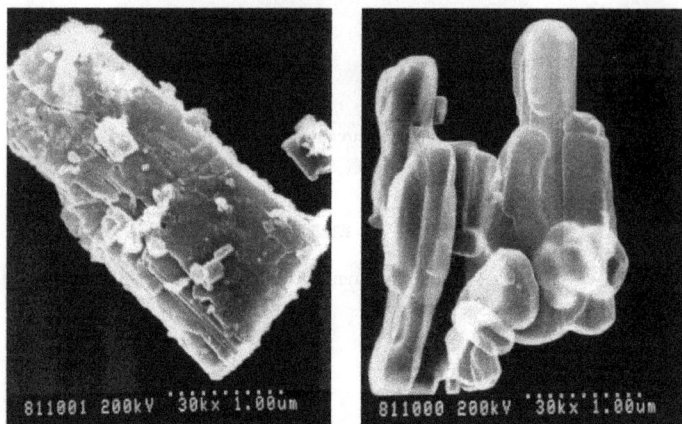

Figure 3.159: SEM images of V_2O_5 prepared with different crystal morphologies.

largest chance to be counted. Secondary electron contrast of nonconducting samples is seriously blurred by surface charging[58] (cf. Section 3.4.5). To avoid these problems, sample surfaces are often coated with a thin gold film.

Contrast by Auger electrons allows elemental surface mapping, but requires an additional electron energy analyzer. It is, therefore, rarely implemented in standard instruments. There are, however, specialized scanning Auger electron microscopes. In these instruments, secondary electron contrast is used for general imaging of the surfaces, which are then imaged with respect to elemental distribution with suitable Auger electrons.

Figure 3.160: STEM images showing the structure of $MnWO_4$ in atomic resolution: (a) dark-field image (HAADF-STEM); (b) bright-field image. Courtesy Dr. Thomas Lunkenbein, Fritz-Haber-Institut Berlin.

---

58 The electron balance usually results in a negative surface charge, but at high attraction potential, the charge can even become positive. The charging effect may be removed by varying the attraction voltage.

All previous discussions related to contrast do not apply to STEM, where limitations on sample thickness are similar to those in TEM (≤ 100 nm, for atomic resolution < 50 nm). This confines the interaction space (Figure 3.158b) to a tiny region around the beam, suppressing thus scattering cascades which might deteriorate resolution. As there are no lenses behind the sample, STEM utilizes the full potential of amplitude contrasts. Imaging intensity is simultaneously collected by a central detector (BF), by a high-angle annular detector (for High-angle annular dark field HAADF), sometimes additionally by an intermediate annular detector (Annular dark field, ADF), which are indicated by C, D, and E, respectively, in Figure 3.158a.

Figure 3.160 provides a comparison between HAADF and BF images showing the structure of $MnWO_4$ in atomic resolution. In the HAADF mode, the intensity arising from Rutherford scattering is proportional to $\approx Z^2$ (Z - elemental number), i.e., heavy atoms appear brighter than lighter ones. The white dots representing the W atoms can be clearly discerned, while the Mn atoms appear darker (Figure 3.160a). In the BF image (Figure 3.160b), the contrast is reversed. Signals arising from O atoms are often over-whelmed by the intensity of heavier atoms, such as W.

As resolution in the STEM is based on deflection by scattering, support by image sim-ulation is not required to interpret STEM images in atomic resolution. While the $MnWO_4$ in Figure 3.160 is crystalline, which allows optimizing contrast by finding crystal ori-entations in which atom columns are parallel to the beam (as in Figures 2.14a, 2.15b), HAADF-STEM can also discern individual atoms of heavy elements. Examples of this, showing Au clusters of different sizes below 2 nm coexisting with isolated gold atoms on TiO_2 can be found in ref. [223].

The resolution achieved by STEM is determined by the size of the probe at the sample. For high resolving power, electrons with the lowest possible energy spread such as obtained by cold field emission, and accelerated by potentials higher than used with the conventional SEM (up to 300 kV) are sent through lenses operating with most efficient correctors. Under these conditions, STEM competes with TEM in point-to-point resolution.

STEM instruments are usually equipped with an EDX facility, sometimes also with an electron spectrometer for EELS. As in all transmission techniques, analyses average over the material penetrated by the beam. Spatial resolution of EDX is somewhat less than that of imaging, because scattered electrons excite fluorescence also around the beam path. In the STEM, electrons can also be diffracted by crystals, but due to the var-iation of incident angles in the focused beam, diffracted beams also exhibit an angular spread. The patterns of this Convergent beam electron diffraction (CBED) are diffrac-tion discs, which provide a great wealth of structural information in one shot and with spatial resolution in the nm range.

The use of STEM for studying catalysts at atmospheric pressure or even in the liquid phase has been outlined already in Section 3.4.11.1. Electron tomography is another intrigu-ing application of STEM. In tomography, the object is imaged under a wide range of angles to the imaging beam. The projections obtained allow the construction of a three-dimen-sional model of the object, which can be explored from all directions and in various cross sections. As TEM-based electron tomography is complicated by the intricacies of phase contrast, the more straightforward image interpretation favors STEM, for this purpose.

3.4.11.3 Scanning probe microscopy

Scanning probe microscopy (SPM) images samples in the contrast of short-range interactions with a tip, which is moved across their surfaces. To track interactions perceptible only at nanometer distances to the surface and to achieve atomic resolution, positioning of the tip must be controlled on sub-nm scales. This is achieved by combining conventional systems for coarse positioning with three-dimensional piezoelectric actuators. Some actuator types (tripod, tube scanner) are illustrated in Figures 3.161a and b.

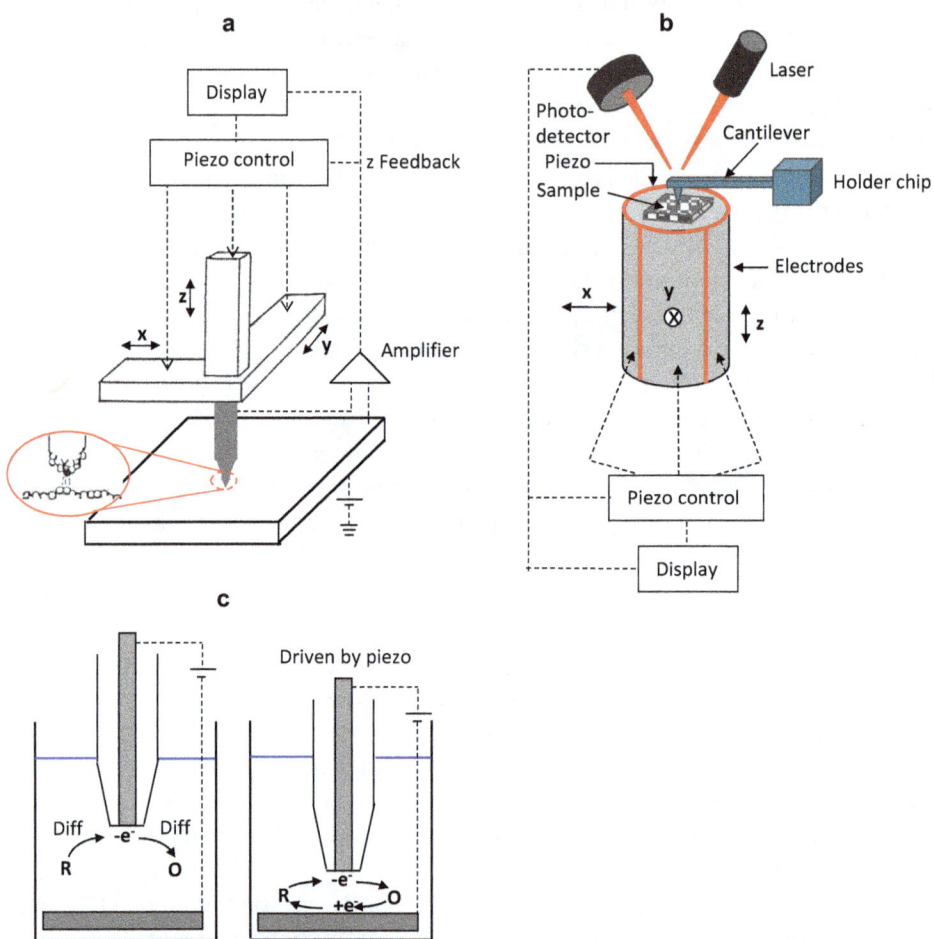

Figure 3.161: Scanning probes across planar surfaces: (a) Scanning tunneling microscope (STM); (b) Atomic force microscope (AFM); and (c) Scanning electrochemical microscope (SECM).

With respect to the sample, SPM requires macroscopically flat surfaces, which can be oriented at right angles to the tip. The techniques are, therefore, first and foremost, tools of surface science, where surface processes like reconstruction, faceting, adsorption, and reaction can be studied with well-defined single-crystal facets in a wide range of conditions, even in electrolytes. For this reason, some SPM techniques are also highly relevant for electrocatalysis. Given the irregular shapes and high roughness of catalyst grains, SPM has been rarely applied to real catalysts, although interesting studies with crystallites of zeolites have been reported (see below).

Among the versions of SPM, Scanning tunneling microscopy (STM), Atomic force microscopy (AFM), and Scanning electrochemical microscopy (SECM) are most important for surface catalysis. In STM, AFM, and SECM, contrast is provided by tunnel currents, by atomic forces, and by Faradaic currents, respectively. The dramatic dependence of these effects on the distance, d, between sample and tip (see below) is the basis of the imaging capabilities down to atomic resolution. Indeed, one may wonder how tips featuring curvatures of tens of nm radius can differentiate atoms in the sample surface below. It becomes possible, because tiny differences in tip-sample distance result in significant effects in the contrast quantity (tunnel current, atom force, cf. detail in Figure 3.161a). Even if the tip atom nearest to the sample surface, maybe at a kink in the tip curvature, hardly protrudes from the tip surface, it may dominate the imaging process unless other atoms compete, e.g., at very rough spots of the surface. In the SECM, both electrode diameter and distance between tip and surface are of the order of µm rather than of nm, and spatial resolution is correspondingly lower.

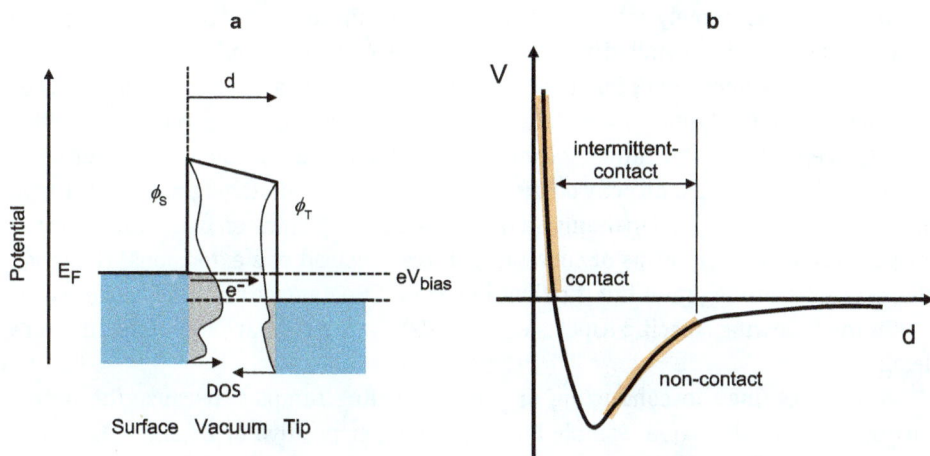

Figure 3.162: Tip-surface interactions used for STM and AFM: (a) tunneling of electrons (here - from surface to tip); (b) tip-surface interaction potential in AFM and ranges of operation modes.

When electrons tunnel between two metals under a bias potential, V_{bias} (see Figure 3.162a), the current decays exponentially with d, the separation between tip and sample surface:

$$I_T(d) = C\,exp\left(-\,\frac{2d\,\sqrt{2\,m_e\,\phi_{av}}}{\hbar}\right)$$ (3.148)

In eq.(3.148), ϕ_{av} is the average work function, i.e., $(\phi_S + \phi_T)/2$, and C is a constant containing the influence of the bias voltage. In AFM, the interaction potentials, $V(d)$, between tip and surface may be approximated by the Lennard-Jones potential (σ, ε – potential parameters, cf. Figure 3.162b):

$$V(d) = 4\,\varepsilon\left\{\left(\frac{\sigma}{d}\right)^{12} - \left(\frac{\sigma}{d}\right)^{6}\right\}$$ (3.149)

SECM uses a redox mediator (R in Figure 3.161c), which is reduced/oxidized at an ultra microelectrode (the tip) and records the influence of a nearby counter electrode (the sample) on the diffusion-limited Faradaic current. Driving the reverse reaction, the counter electrode increases the mediator concentration in the boundary layer and creates a feedback loop that enhances the reaction rate. The effect depends on the rate provided by the counter electrode: with decreasing activity, the latter starts disturbing the diffusion processes around the ultra microelectrode instead of driving the feedback loop, which decreases the reaction rate.

Due to the extremely small distances between tip and surface, scanning probe microscopes must be installed free of mechanical vibrations and well-isolated thermally. All microscopes can be run in two different modes – in a constant height mode or in a constant contrast mode. In the latter, the z coordinate of the tip is regulated at each point to keep the contrast function (tunnel current, force, electrode current) constant, i.e., the surface is imaged directly on the z coordinate. In constant height mode, the contrast varies at each point, depending on the actual distance between tip and surface: the image is established in terms of currents or forces, instead of z extensions. The mode allows faster scanning rates but is riskier in terms of tip damage on rough surfaces.

In the following, specific aspects of STM, AFM, and SECM are illustrated in more detail.

STM is confined to conducting or semiconducting samples, because the tunnel current must be dissipated. Sample bias can be either positive or negative. Electrons tunnel from the filled density of states (DOS) of the negative part (in Figure 3.162a - the sample) to the Fermi level of the positive part (- the tip). On flat facets, images are, therefore, largely based on the spatial distribution of the DOS in the sample surface. In mixed surfaces, it may be even possible to distinguish components, if their electron structure is very different, although there is no well-defined elemental sensitivity. z excursions indicating surface steps or pits arise from a convolution of electronic and morphological influences and do not always reflect real heights. STM is also possible with metals covered by thin isolating oxide layers, as long as the layer does not prevent tunneling. Notably, STMs can also be operated in gases and in electrolytes.

STM has been extended into scanning tunneling spectroscopy, where information on the surface electron density can be derived from the dependence of the tunnel current, I_T, on the bias voltage, V_{bias}. The tip can also be employed to manipulate features of the surface, e.g., by displacing adsorbate species.

As an example, Figure 3.163 shows STM images of an MoS$_2$ model catalyst after different treatments [224]. MoS$_2$ was prepared in the form of monolayer slabs supported on an Au(111) facet. Different from morphologies under other conditions (cf. Figure 2.17), the slabs are triangular, but their internal hexagonal structure can be clearly seen in the figure. The edges are terminated by Mo, which is saturated by S$_2$ dimers oriented perpendicular to the image plane (Figure 3.163a). A bright rim proceeding on the basal plane along the slab edge is most conspicuous. The bright contrast suggests metallic conductivity in this one-dimensional "brim" state, which has been confirmed by model calculations. The brim has been proposed to catalyze hydrogenation reactions.

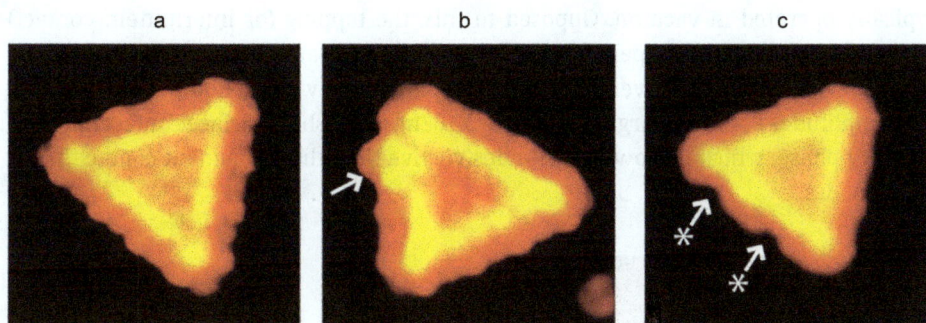

Figure 3.163: STM images of an MoS$_2$ monoslab deposited on an Au(111) facet after different treatments, with tip positions encoded in colors: (a) initial state; (b) after treatment with atomic hydrogen, exposure to NO, and annealing at 473 K; and (c) after annealing at 573 K. Reproduced from ref. [224] with permission from Elsevier.

In Figure 3.163b, two bright blobs protrude from the left-hand slab edge. This sample was treated with atomic hydrogen, which is known to produce sulfur vacancies at the slab edges, where NO can adsorb. The image suggests the presence of two adjacent NO molecules, which was supported by model calculations. From TPD experiments with real catalysts, it is known that NO desorbs between 473 K and 573 K. Indeed, the adsorbate could not be observed any more after heating to 573 K (Figure 3.163c). Notably, the remaining vacancies are separated, which suggests significant mobility of the terminating sulfur species at 573 K.

AFM is not confined to conducting samples. The AFM probe consists of a flexible cantilever (typical dimension 500×100 μm), which carries the tip on one end and is operated by an mm-sized holder chip on the other (cf. Figure 3.161b). Atomic forces on the tip, which change when the tip-sample distance is varied, are recorded via changes of cantilever deflection or eigenfrequency. The deflection is typically measured by a laser beam reflected by the cantilever surface onto a position-sensitive photodiode. Tip curvatures are similar to those in STM.

AFM can be operated in static and dynamic modes. The static (or contact) modes are similar to constant-height or constant-force modes in other SPM versions. They are operated in the repulsive branch of the interaction potential (Figure 3.162b), extremely close to the surface, actually in contact with it.

Despite lower imaging rates, dynamic modes are more popular because they are less stressful for the surfaces studied. The non-contact (NC) mode operates 10–100 nm from the surface, where attractive forces too low to be directly measured still influence cantilever vibrations (Figure 3.162b). The cantilever is excited by an external source near (but not at) its eigenfrequency. The varying force from the surface modifies the vibrational pattern sustained by the external excitation, and a feedback loop restores it by varying the z coordinate. The NC mode achieves the highest resolutions and is typically operated in vacuum. Opposed to this, the tapping (or intermittent contact) mode, which also uses the perturbation of oscillator vibrations by the surface, can be applied even to surfaces covered by a hydrate film or fully in the liquid phase. Oscillation amplitudes are much larger (cf. Figure 3.162b), and a short surface contact happens in every phase, which is, however, less destructive than the permanent contact in the static modes.

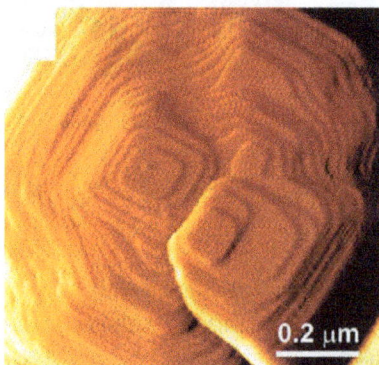

Figure 3.164: AFM image of a zeolite A crystal. Reproduced with permission from ref. [225]. Copyright 2011 American Chemical Society.

Figure 3.164 shows AFM images of zeolite A crystallites taken in contact mode during a study of the crystallization mechanism. The height of the steps is of the order of 1 nm.

The AFM has been developed into modifications utilizing different interactions, e.g., magnetic and chemical force microscopy, or electrochemical force microscopy, which is AFM under electric potentials between tip and surface. In spectroscopic modes, the dependence of forces on the distance d is measured, or bonding interactions within macromolecules, e.g., proteins, are studied (single molecule force spectroscopy).

Due to the role of liquid-phase diffusion in the operation mode of SECM, this technique works at larger distances from the surface (up to 200 µm) and with lower spatial resolution (down to 200 nm). On the other hand, the variability of the redox mediator provides rich opportunities beyond the imaging of electrode surfaces or the rapid screening of electrocatalysts. When the intermediate (O in Figure 3.161c) is instable, its lifetime or its reaction with a reactant offered in the electrolyte can be studied by varying the tip-surface distance, d. In this spectroscopic mode, SECM allows determining diffusion coefficients and can contribute to the elucidation of electrocatalytic reaction mechanisms [79].

3.4.12 Electrochemical characterization techniques

Electrocatalysts can be characterized by most of the methods discussed so far. In this section, techniques are discussed that allow examining the behavior of electrodes under electric potentials in electrolytes. They comprise methods specific for electrochemistry, the basics of which can be found in ref. [226], and *in situ* or *operando* versions of spectroscopy and microscopy.

Potential sweep methods are a standard starting point, when potential-dependent interactions between electrodes and electrolytes (ad/desorption processes, reactions) are to be explored. In Linear sweep voltammetry (LSV), the potential of the working electrode is ramped from a starting point to an end point, with the electrode at rest. Typical scan rates are between 10 and 1,000 mV s^{-1}, in special cases beyond. When the electrolyte or species dissolved in it are electroactive in the potential window, currents caused by the induced redox processes can be detected. Faraday currents resulting from them increase as the potential approaches the standard potential of the redox couple, until mass transfer limitations come into play, which deplete the reactant and enrich the product close to the electrode. With further increasing potential, the depletion sphere extends, and the Faraday current eventually decreases, i.e., the redox reaction is indicated by a peak. Its location is determined by the sum of the electrode potential of the redox couple and a diffusion excess potential. LSV may be considered an electrochemical analogue to TPR or TPO (cf. Section 3.4.2.5).

In Cyclic voltammetry (CV), the linear potential ramp is followed by a reverse one at identical scan rate. As a result, (reversible) redox reactions and adsorption/desorption processes are reverted; Faraday currents flow into the opposite direction. As the diffusional delay of redox peaks complies with sweep orientation, peak potentials in the two scans differ, but the underlying electrode potential can be deduced from their average. Deviations from ideal behavior, peak potentials and currents, and the overall shape of the CV allow conclusions on the electrochemical system studied. CVs can be used to identify potentials required to initiate a reaction or to achieve a specific current density under given conditions (T, reactant concentration). Information on diffusion processes, on deactivation or corrosion phenomena can be obtained, the latter with a series of repetitive CVs. For more detail, see ref. [79].

While CV deals with changes of the current caused by a gradually increasing potential, step techniques investigate the response of the electrochemical system to a sudden step of either the potential or the current. Recording the response of the current to a potential jump is referred to as chronoamperometry (CA), the reverse configuration as chronopotentiometry (CP). The transient current response in CA usually comprises a fast decaying non-Faradaic component and a more extended contribution of the Faradaic current due to different rates of redox reactions before and after the step. As in CV, these rates are influenced by reactant diffusion: the Faradaic current tends to decrease, because the concentration gradient near the electrode fades out as the reactant is depleted. According to the Cottrell equation, it changes with $t^{-1/2}$ (t – time after step). CA is the standard technique for durability studies with catalytic electrodes. Stability at relevant potentials documented by CA is the prerequisite for proceeding from the three-electrode setup to technical environments (electrolyzer, fuel cell, etc.) with new catalysts. Integration of the current yields the charge, which can be used to determine the quantity of reaction product formed, if the reaction stoichiometry is known.

CP records the response of the potential after applying a current to the working electrode: the galvanostat finds a potential at which the system can consume or provide the specified current by driving a redox reaction. Again, the response is influenced by diffusion, because the required potential increases as the reactant is depleted near the electrode. When there are no limitations on the potential, it may grow into a range where electrons can be consumed or provided by another redox reaction. CP provides insight into reaction mechanisms and kinetics. It is most familiar in battery research, where it is used to study charge/discharge behavior as well as electrodeposition.

The electrochemically active surface area (ECSA) is a key quantity in electrocatalysis, because it is required to determine the specific activity of catalytic materials (cf. eq. (1.10)). It is analog to the number of active sites used to specify the turnover frequency in thermal catalysis (eq. (1.4)). For supported metal particles, it is analog to the metal surface area accessible by chemisorption techniques (cf. Section 3.4.1.6), but not necessarily identical with it.

The ECSA is determined by measuring charges consumed during reactions of surface atoms or of adsorbates, or in adsorption processes. Reduction of an oxide (mono)layer on metal particles is an example for the first approach. The catalyst is oxidized in a forward scan, and the charge determined during the reduction wave is compared with the monolayer reduction charge (e.g., 405 μC cm^{-2} for PdO [227]). In CO stripping, the electrolyte is purged with CO under potential control to adsorb a monolayer of CO on the metal. After removing dissolved CO from the electrolyte by flushing, the potential is ramped to oxidize the adsorbate, and the charge obtained is compared with the specific charge of a CO monolayer, e.g., 420 μC cm^{-2} on Pt. Alternatively, the ECSA can be measured via the adsorption of H atoms at a potential more positive than the equilibrium potential of H_2 evolution (underpotential deposition, H_{upd}). For polycrystalline Pt, the specific charge for this one-electron transfer is 210 μC cm^{-2}. For more details on CO stripping and H_{upd}, see refs. [228, 229].

Electrochemical impedance spectroscopy (EIS) is an important tool for the study of electrical systems, which is also applied to electrocatalytic processes. The impedance, Z, is a generalized concept of resistance (i.e., the ratio between potential, E and current, I) adapted to alternating current (AC). It is a complex quantity depending on the AC frequency, $\nu = \omega/2\pi$, and can be written as $Z(\omega) = Z_0 \, exp(i\phi)$, where Z_0 is the resistance and ϕ the phase shift between potential and current. Impedance spectra are measured by recording the current response of a system while the frequency of a sinusoidal potential excitation of small amplitude is varied. Information is extracted from them by finding the equivalent circuit model best representing the data. The models give access to relevant quantities like rates of charge transfer and mass transport, properties of catalyst layers (e.g., resistance, homogeneity), to corrosion phenomena. For details of the method, the reader may consult ref. [230].

Operando spectroscopy with electrocatalysts has been mentioned, so far, only for methods using hard X-rays (XAFS, Section 3.4.4.2). Absorption by the aqueous electrolyte becomes the more challenging the lower the energy of the applied radiation. IR spectra of working electrocatalysts can be measured in the ATR mode and with thin-film techniques. In ATR-IR, the working electrode is deposited on the IRE as a very thin layer (Figure 3.165a). The method takes advantage of the μm extension of the evanescent waves (Section 3.4.8.3), which allows tapping them even at the interface between electrolyte and thin metal layers, but it is limited to materials allowing the preparation of such thin layers. With some elements, e.g., Pt, surface enhancement effects can be utilized (Surface-enhanced IR absorption spectroscopy, SEIRAS, cf. ref. [231] for details). Spectra obtained in the ATR mode are always dominated by the adsorbates. Information on species just desorbed, which may be important to elucidate reaction mechanisms, is hardly accessible.

In the IRAS mode (Figure 3.165b, cf. Section 3.4.8.3), working electrode and prism form a narrow slit to minimize the electrolyte layer thickness. The method is feasible only with smooth electrodes of high reflectivity, which decreases surface enhancement effects. Due to the still appreciable path length through the electrolyte, spectra are dominated by the signals of water and of dissolved species. The low amount of electrolyte fixed between prism and electrode results in species concentrations changing during *operando* runs, which interferes with intensity accumulation for high-quality spectra. This disadvantage is even more pronounced in an approach working with the electrode pressed toward the prism until contact is established, which has also been adapted to ATR-IR.

Figure 3.165c shows an approach in which this dilemma is removed by introducing forced flow between IRE and electrode, through a central bore in the latter. To obtain a homogenous radial flow pattern, electrode and IRE surfaces must be exactly parallel, which is achieved by a microelectrode positioning system [232]. In the configuration

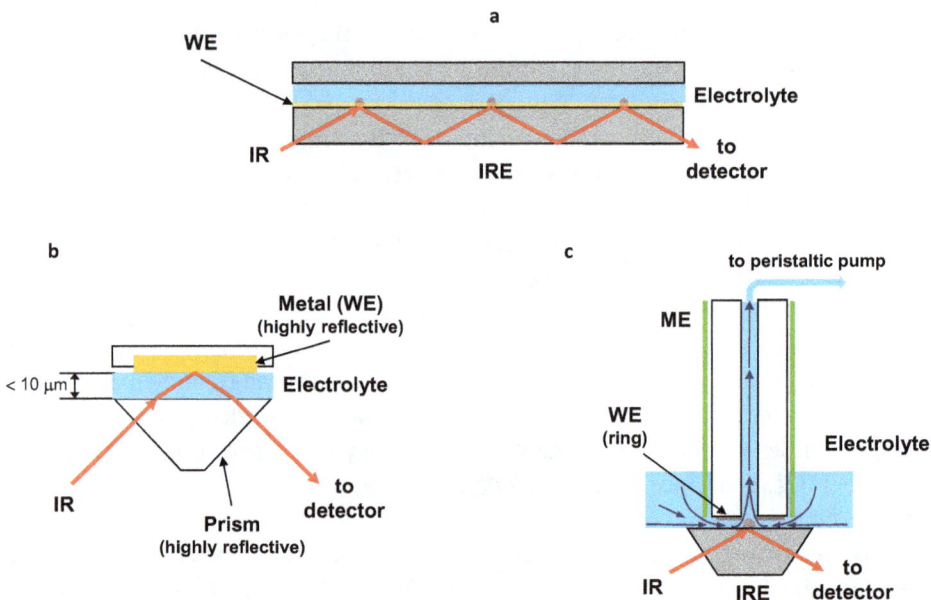

Figure 3.165: Approaches for *operando* IR spectroscopy with electrodes: (a) ATR-IR with electrode deposited on IRE; (b) IR reflection absorption spectroscopy (IRAS) at an electrode surface; and (c) ATR-IR with a ring electrode opposite to IRE. WE – working electrode, IRE – internal reflection element, ME – microelectrode (employed in c for controlling the orientation of the electrode relative to the IRE).

depicted in Figure 3.165c, the evanescent wave tracks the electrolyte stream just after interacting with the working electrode.

Due to its selection rules, Raman spectroscopy is less constrained by the water present in *in situ* and *operando* work with electrocatalysts. *In situ* cells have been described in literature and have been applied, for instance, to monitor potential-dependent phase changes in Ni oxide catalysts for oxygen evolution [233].

Although electrocatalysts operate with ensembles of nanoparticles distributed in size and shape, great effort has been made recently to elucidate and understand the behavior of individual particles, which may help optimizing real catalysts. Working in these dimensions requires a number of methodical advances, which are only briefly mentioned here. They include, for instance, the growth of individual nanoparticles ready for electrochemical characterization and (*ex situ*) microscopy right on the tip of nano-electrodes [234], the use of the Scanning electrochemical cell microscope (SECCM)[59] for electrochemistry with individual nanoparticles on flat electrodes [235] (cf. Figure 3.166) or for the deposition of individual particles, e.g., on fragile glassy carbon platelets [236] eligible for

---

59 A scanning probe microscope (cf. Section 3.4.11.3) developed from the electrochemical droplet cell (EDC), in which an electrolyte droplet extending from the tip of a nanocapillary makes contact with the surface of the working electrode (cf. Figure 3.166). The tip can be positioned at different spots of the working electrode with piezo-actuators.

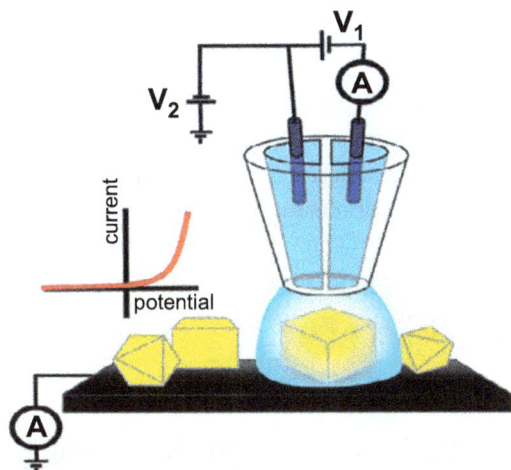

Figure 3.166: Study of facet-dependent electrochemical behavior with the Scanning electrochemical cell microscope (SECCM). Reproduced with permission from ref. [235]. Copyright 2020 American Chemical Society.

operando imaging in the liquid-cell TEM,[60] and single entity electrochemistry with catalyst powders dispersed in the electrolyte, without binders or additives used to stabilize electrodes. The latter method works with the short current pulses caused by the particles when hitting an inert microelectrode, in the course of their Brownian trajectories [237].

3.5 Computational chemistry in catalysis research

Contributions of theory to catalysis research have been outlined already in Section 1.5. Theory claims to reveal catalyst surfaces and processes proceeding on them in a detail inaccessible to experiment for various reasons. The development of quantum chemistry has resulted in a performance that more and more justifies this claim, although theoretical models usually need comparison with measured "supporting points" so far, in order to judge the suitability and accuracy of approaches or functionals applied. For simple metal-catalyzed reactions, quantum chemical activation energies and turnover frequencies can be accurate enough to make safe comparisons between elements or mixed phases, which is the basis for computational (or *in silico*) screening methods. For the acid-catalyzed reaction of methanol with small alkenes in a zeolite (H-ZSM-5), where dispersive interactions play an important role, activation energies and rate constants have been recently obtained with "chemical accuracy" (± 4 kJ/mol for E_A, ± one order of magnitude for k) [238].

---

60 A version of the closed-cell technology highlighted in Section 3.4.11.1.

Computational chemistry covers various methods of theoretical chemistry. Beyond ab initio quantum chemistry based on wave functions or density functionals, it includes molecular mechanics, where spatial structures of systems comprising very large number of atoms are investigated by energy minimization based on empirical force fields around the atoms, and molecular dynamics, which studies the temporal evolution of such systems. Although the latter has also been applied to catalytic processes, this presentation is focused on quantum chemical methods, the basics of which are summarized only briefly, because a more profound treatment can be found in textbooks.

3.5.1 Wave function-based quantum chemistry and density functional theory

All information about the properties of chemical systems can be obtained by solving the Schrödinger equation:

$$\hat{H}\,\Psi = E\,\Psi \tag{3.150}$$

The Hamilton operator $\hat{H}$ and the wave function Ψ depend on the space and time coordinates of all nuclei and electrons.

The first step in treating complex systems as familiar in catalysis is to solve the electronic Schrödinger equation, which is eq. (3.150) with the coordinates of the nuclei frozen (Born-Oppenheimer approximation) and the time dependence of all quantities excluded.

$$\hat{H}_{el}\,\Psi_{el} = E_{el}\,\Psi_{el} \tag{3.150a}$$

The resulting electronic wave function Ψ_{el} describes the electronic structure of the system. The electronic Hamilton operator $\hat{H}_{el}$ contains the kinetic energy of all electrons ($\hat{T}_e$), the electrostatic interactions between nuclei and electrons ($\hat{V}_{ne}$), the electrostatic repulsion between electrons ($\hat{V}_{ee}$), and between nuclei ($\hat{V}_{nn}$). Other interactions can be included, e.g., the energy of a magnetic field (cf. eqs. (3.128), (3.132), and (3.133)).

While there are analytical solutions of (3.150a) for small systems (single-electron systems, very small atoms and molecules like He, H_2), wave functions Ψ_{el} for large systems are extremely complex, already because of the number of variables: in a finite system with n electrons (e.g., a molecule), their number is $4n$ — three space coordinates and a spin coordinate per electron.

The electronic Schrödinger equation (3.150a) can be solved by expressing the n-electron wave function, Ψ_{el}, of the system (hereafter designated as Ψ) by means of single-electron wave functions, $\varphi_i(\vec{r}_i)$, where $\vec{r}_i$ is the position vector of electron i. The $\varphi_i(\vec{r}_i)$ are referred to as molecular orbitals. In the Hartree approximation, Ψ is the product of the one-electron wave functions, $\Psi(\vec{r}_i, \ldots, \vec{r}_n) = \Pi_{i=1}^{n}\,\varphi_i(\vec{r}_i)$. This neglects, however, the electronic spin and the Pauli principle. Therefore, spin orbitals, $\chi_i(x_i)$ are formed instead by multiplying $\varphi_i(\vec{r}_i)$ with the spin factor $\sigma_i(s_i)$ (= α or β), and the n-electron wave function is constructed in the form of a determinant (Slater determinant):

$$\Psi(x_1; \ldots, x_n) = \frac{1}{\sqrt{n!}} \begin{bmatrix} \chi_1(x_1) & \cdots & \chi_n(x_1) \\ \vdots & \ddots & \vdots \\ \chi_1(x_n) & \cdots & \chi_n(x_n) \end{bmatrix} \tag{3.151}$$

In (3.151), Ψ changes its sign when any two electrons are exchanged and becomes zero when two electrons occupy the same spin orbital (Hartree-Fock (HF) approximation).

The one-electron molecular orbitals $\varphi_i(\vec{r}_i)$ can be obtained as solutions of a set of coupled one-electron Schrödinger equations, the Hartree-Fock equations:

$$F\varphi_i = (T_e + V_{ne} + V_{ee} - K)\varphi_i = \varepsilon_i \varphi_i \tag{3.152}$$

where F is referred to as Fock operator. In (3.152), ε_i is the one-electron energy related to the orbital, $\varphi_i(\vec{r}_i)$,[61]; T_e and V_{ne} are the one-electron operators for the kinetic energy and the nuclear-electron interaction, respectively. The electron-electron interaction is described by two operators: V_{ee} represents the classical electrostatic interaction, but a nonclassical "exchange operator" K remains, as a consequence of using Ψ in the form of a Slater determinant.

Equation (3.152) is coupled via the electron-electron interaction, $V_{ee} - K$: to solve one of them for an orbital φ_i, one has to know all orbitals φ_j occupied by the other electrons in the system. Equation (3.152) is, therefore, solved by iteration, starting with an initial guess $\varphi_j(0)$ for all electrons and constructing the electron-electron interaction $(V_{ee} - K)$ (0) from this. Solving eq. (3.152) with this operator yields a new set of orbitals, $\varphi_j(1)$, from which the next repulsion operator $(V_{ee} - K)(1)$ is obtained. The iteration is carried on, until changes in the orbitals and the total energy are below defined thresholds. Therefore, the Hartree-Fock method is also called "self-consistent field" (SCF) approach. The final HF wave function is a Slater determinant (3.151) containing the spatial orbitals (φ_i) with the lowest orbital energies (ε_i), each multiplied with a spin factor. The total HF energy is obtained by evaluating the expectation value of the electronic Hamilton operator $(\hat{H}_{el})$ with this HF determinant. More detail can be found in textbooks.

In the HF approach, the application of the classical electron-electron operator, $\hat{V}_{ee,HF}$, to the wave function in the form of a single Slater determinant describes this interaction only averaged, which is a major cause of its approximate nature. Total energies obtained from this are always higher than the real energies, which is ascribed to elec-

---

61 The spin orbitals $\chi_i(x_i)$ resulting from them by multiplication with the spin factors have the same energies. To cover interactions specific to unpaired electrons, additional operators have to be included, cf. eqs. (3.128), (3.132), (3.133).

tron correlation effects not covered by the HF approach. Deviations in total energies are normally just a few percent. However, as bond energies typically amount to a few percent of total energies as well, such error is inacceptable for applications in chemistry. Errors for other properties may be rather large.

Notably, the HF method is quite time consuming, because the construction of the K operator scales with n^4 in the simplest traditional implementation. More recent procedures are, however, able to reach a better scaling.

Wave function-based post-HF methods add flexibility to the form of the wave function, Ψ. In the configuration interaction (CI) approach, the wave function is a linear combination of Slater determinants representing not only the ground state, but also excited states.[62] Inserting wave functions of this type into the multielectron Schrödinger equation (3.150a) results in a linear eigenvalue problem, the solution to which are the coefficients of the included determinants in the final wave function and an improved total energy. CI with the full set of excited configurations (full CI) allows solving the Schrödinger equation at any accuracy level, however at enormous computational cost. The linear combinations are, therefore, often truncated after the single or double excitations.

In CI calculations, the orbitals making up the configurations are kept constant; in the alternative multiconfiguration SCF approach (MC-SCF), they are varied as well. Both the coefficients of the determinants and the orbitals in them are optimized in successive or simultaneous strategies. In the family of coupled cluster (CC) methods, linear combinations of determinants are also employed to construct wave functions. They are, however, obtained here by applying an exponential operator e^T to the HF determinants, where the operator, T, describes single, double, and higher excitations. The "gold standard" of present day quantum chemistry is CCSD(T), i.e., CC with single and double and perturbationally treated triple excitations. In many cases, it yields nearly chemical accuracy at reasonable computational cost. It scales with n^7 in the simplest implementation, but much more efficient codes are becoming available, meanwhile.

When correlation effects are small, perturbational approaches can be used instead of the variational methods just mentioned. In perturbational methods, the real Hamiltonian $\widehat{H}^\lambda$ is composed of a known Hamiltonian $\widehat{H}^{(0)}$ and a perturbation operator $\widehat{V}$:

$$\widehat{H}^\lambda = \widehat{H}^{(0)} + \lambda \, \widehat{V}$$

so that $\widehat{H}^\lambda$ is obtained at $\lambda = 1$. In the Møller-Plesset method, for instance, $\widehat{H}^{(0)}$ is the sum of all one-electron Hamiltonians (F in (3.152)). Instead of $\widehat{V}_{ee}$, however, the averaged HF repulsion operator $\widehat{V}_{ee,HF}$ is used, and the perturbation operator $\widehat{V}$ is the difference of $\widehat{V}_{ee}$

---

62 In such Slater determinants, occupied spin orbitals are replaced by virtual, non-occupied solutions of the HF equation (3.152). The ground state is described including formally excited states: single-excited, double-excited, in principle, up to n-fold excited. Similar variations are employed when excited states are to be described.

and $\hat{V}_{ee,HF}$. Wave functions and energies are expanded in power series of λ, which are truncated suitably. The energy obtained at the 0th order is just the sum of all one-electron energies ε_i, and the sum of 0th and 1st order reproduces the HF total energy. The second-order Møller-Plesset approximation (MP2), however, provides good accuracy, requiring the smallest computer capacity among the post-HF methods.

Density functional theory was developed to find energies and electron distributions for the ground states of complex systems, without solving the Schrödinger equation. It is based on the statements that (1) the external potentials[63] of a system are accessible to within a constant once the spatial distribution of its electron density is known, and that (2) the electron density distribution of the ground state results in the lowest possible energy, which sets a target function for iterative methods. These statements are referred to as Hohenberg-Kohn theorem.

Regarding (1), it should be noted that the external potential V_{ext} is the quantity that introduces the materials properties into the Hamiltonian of the Schrödinger equation (cf. $\hat{V}_{ne}(\vec{r}_i)$ in (3.152)): when V_{ext} is known, all relevant properties (nature of atoms, number of electrons, energy, spatial electron density distribution ρ, etc.) are, in principle, accessible. Moreover, the total energy E can be considered a functional (a function of a function) of the electron density, ρ: E is a function of ρ, which is a function of the position vector $\vec{r}$ on its part. Following a presentation in ref. [239], E is again split into contributions of the kinetic energy, T_e, and of the interactions of the electrons with the external potential, E_{ne}, and with other electrons E_{ee}:

$$E_0[\rho_0] = T_e[\rho_0] + E_{ne}[\rho_0] + E_{ee}[\rho_0]$$

where the index, 0, indicates the ground state. When $\rho_0(\vec{r})$ is known, E_{ne} can be obtained by integrating $\rho_0(\vec{r}) \, V_{ext}(\vec{r})$ over all position vectors. The classical contribution to E_{ee} can be described (in atomic units) as:

$$E_{ee-cl} = \frac{1}{2} \iint \frac{\rho(\vec{r}_1) \, \rho(\vec{r}_2)}{\vec{r}_2 - \vec{r}_1} \, d\vec{r}_1 d\vec{r}_2$$

The nonclassical part of E_{ee}, i.e., the consequences of the Pauli principle, remains as an additional term referred to as exchange correlation energy (or functional) E_{XC}:

$$E_0[\rho_0] = T_e[\rho_0] + E_{ne}[\rho_0] + E_{ee-cl}[\rho_0] + E_{XC}[\rho_0] \tag{3.153}$$

Unfortunately, the precise form of the energy functional, $E_{XC}[\rho_0]$ is not known. Already, the kinetic energy T cannot be expressed as a functional of the density. The problem is usually handled by approximating T with an expression accessible via classical

---

[63] Nuclear potentials + long-range influences, e.g., an electric field.

approaches and shifting the difference to the real kinetic energy into $E_{XC}[\rho]$. This leaves the problem of identifying an acceptable functional form for $E_{XC}[\rho]$, which would allow optimizing the density $\rho(\vec{r})$, targeting the minimization of $E[\rho]$ according to eq. (3.153).

In nearly all numerical applications of DFT, the optimization of the nonclassical part of E_{ee} is achieved by using one-electron wave functions, φ_i, as introduced by Kohn and Sham. They result from the solution of a one-electron equation of the form:

$$\left(-\frac{1}{2} \, \nabla^2 + V_{KS}(\vec{r}) \right) \varphi_i = \varepsilon_i \, \varphi_i(\vec{r})$$

(3.154)

where $-\nabla^2/2 \dots$ is the kinetic energy operator ($\nabla^2 \dots = (\partial^2 \dots /\partial x^2 + \partial^2 \dots /\partial y^2 + \partial^2 \dots /\partial z^2)$). In the Kohn-Sham equations (3.154), V_{KS} is an effective potential containing the external potential V_{ext}, the classical and the nonclassical contributions to electron-electron interaction:

$$V_{KS}[\vec{r}] = V_{ext}[\vec{r}] + \int \frac{\rho(\vec{r}')}{\vec{r} - \vec{r}'} \, dr' + \frac{\partial E_{XC}}{\partial \rho}$$

As in the HF approach, (3.154) is a set of coupled equations, because ρ in V_{KS} is the total electron density. Therefore, the Kohn-Sham equations are solved self-consistently, starting with a guess of $\rho(\vec{r})$, which determines all components of V_{KS}, solving eq. (3.154) to obtain a set of wave functions φ_i, from which a new electron density map can be evaluated, etc. The wave functions φ_i and their energies ε_i have little physical meaning, but the total electron density resulting from the optimized φ_i should be equal to the real electron density $\rho_0(\vec{r})$, which gives access to the real external potential, V_{ext}, and the total energy.

It should be noted, that the Kohn-Sham equations closely resemble the HF equations (3.152). The difference is that the exchange-correlation part $\partial E_{XC}/\partial \rho$ covers the HF exchange operator K, all correlation effects, and also the difference between the correct kinetic energy, $T_e[\rho]$ and the kinetic energies of the one-electron Kohn-Sham orbitals, φ_i. The great advantage of DFT over HF is that the time-consuming construction of the exchange operator K is avoided and correlation effects need not be calculated by very slow procedures like CI or CC. For q atoms, the computational effort scales to $\approx q^3$.

At this point, the exchange correlation functional $E_{XC}(\rho)$ is still unknown, which prevents applying DFT (i.e., solving the Kohn-Sham equations). Defining $E_{XC}(\rho)$ is actually the crucial step in the application of DFT, because all unknown effects have been shifted into this functional as pointed out above. So far, no functional has been found that gives satisfactory results in all situations. Therefore, the choice of the functional determines achievable accuracies. In the treatment of catalytic systems, this choice may result from a tradeoff between accuracy and system size, due to limitations of computer capacities.

Being the simplest exchange correlation functional, the local density approximation (LDA) makes use of the fact that $E_{XC}(\rho)$ is known for the uniform electron gas. LDA is based on the assumption that in each volume element indicated by $\vec{r}$, the relation between electron density $\rho(\vec{r})$ and exchange-correlation $\varepsilon_{XC}[\rho(\vec{r})]$ is as it would be in the electron gas. Therefore,

$$E_{XC}^{LDA}[\rho] = \int \rho(\vec{r})\, \varepsilon_{XC}[\rho(\vec{r})]\, d\vec{r}$$

LDA is suited for systems with minor or gradual variations of the electron density, which fails, for instance, at phase boundaries. Although it is quite accurate for bond geometries, it overestimates bond energies and fails with weak interactions.

The general gradient approximation (GGA) improves the LDA by including the gradient $\nabla\rho(\vec{r})$ in order to account for more pronounced spatial variations of the electron density. For various reasons, the gradient cannot be used directly, but is included into a more complicated function F_{XC}:

$$E_{XC}^{GGA}[\rho] = \int \rho(\vec{r})\, \varepsilon_{XC}[\rho(\vec{r})]\, dr + \int F_{XC}[\rho(\vec{r}), \nabla\rho(\vec{r})]\, dr$$

Different choices for F_{XC} resulted in a family of GGA functionals, e.g., PW91, PBE, and BLYP. They improve markedly on binding energies in molecules and reproduce adsorption geometries of molecules on surfaces. On the other hand, they often fail with adsorption energies and, in particular, with van der Waals interactions.

GGA functionals have been further improved by including expressions for the exchange energy (the dominant part of E_{XC}) from wave function-based calculations on Hartree-Fock level. B3LYP is an example of such hybrid functional, which can be used to model weak interactions.

3.5.2 Cluster and periodic models

All theoretical modeling requires initial guesses, which are refined by the computational methods just highlighted. Such optimization should target the minimum of the Gibbs free energy. However, in solid materials, effects of entropy changes along such structure optimization are often negligible compared to variations of energy. Therefore, models are usually optimized with respect to the total energy of the structure.

As any model is restricted in size and detail, it is a permanent point of concern if the limited model structure captures the key aspects of the real system. Results from models of different sizes are often compared to judge the effects of model truncation. Checking predicted against measured values of observables is the ultimate test for model validity. While disagreement in such comparisons was a powerful stimulus for the improvement of theory in the past, it may also call for reviewing methods of experimentalists nowadays, particularly when wave function-based calculations are involved.

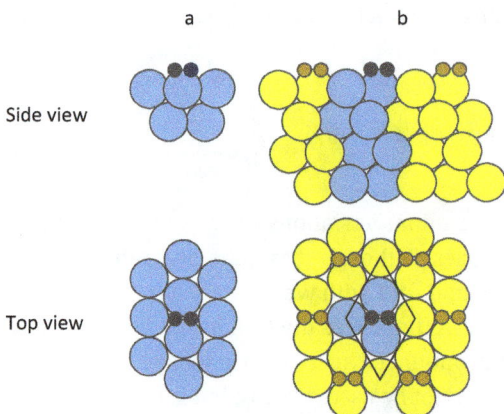

Figure 3.167: Cluster and periodic models: H atoms (in black) on a metal surface. (a) Cluster model: 15 metal atoms; (b) periodic model: basic unit with 12 metal atoms (blue) and two H atoms (black), transitional replicas – metal atoms in yellow, H atoms in brown. Adapted from ref. [240].

Models for solids are categorized as cluster and periodic models. Cluster models represent solid structures as pseudomolecules. Examples can be found in Figure 1.7b, where clusters employed to model active sites in zeolites are depicted, and in Figure 3.167a, which shows a H_2 molecule interacting with a cluster of metal atoms. Bonds extending beyond the cluster boundaries need to be saturated to avoid perturbations of the wave functions. In zeolite models, clusters are usually terminated with OH groups (instead of O-Si) because the electronegativities of H and Si are similar. There is no such opportunity for metals and this contributes to a slow convergence of cluster energies with size. During geometry optimization of zeolite models, the external atoms may be fixed at known positions to ensure compatibility of the models with the long-range structure.

Cluster models are convenient because they are flexible and can be evaluated with codes available for molecular quantum chemistry. Their limitations are related to influences from beyond their boundaries, e.g., a slow convergence of energies when metal clusters grow across the nonmetal/metal transition, influences of Coulomb interactions, or of mechanical stress. Coulomb interactions can be accounted for by embedding the cluster in appropriate arrays of point charges. For more advanced treatments, the cluster model can be combined with a force-field-based molecular mechanics model. Figure 3.168 shows such a QM/MM model representing an acid site of a high-silica ferrierite zeolite interacting with an isobutene molecule. In ref. [241], the total zeolite model contained 71 Si, 1 Al, and 144 O atoms, while the quantum-mechanical treatment dealt

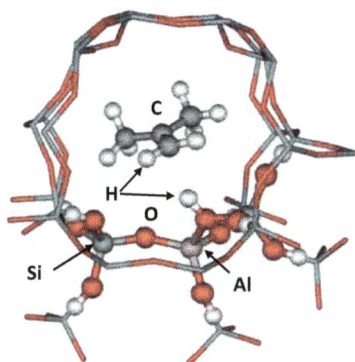

Figure 3.168: QM/MM cluster model of isobutene adsorbed on an acid site in H-FER. The QM cluster is depicted as a ball-and-stick model; atom positions in the MM cluster are indicated by sticks. Reproduced from ref. [241] with permission from Elsevier.

only with three T atoms of the zeolite (together with related O atoms, acidic and truncating protons) and the isobutene.

Periodic models make use of the Bloch theorem, according to which the wave function $\Psi_{Per}(\vec{r})$ of particles in a periodic system, e.g., a crystal, is obtained by modulating the wave function, $\Psi_{EC}(\vec{r})$, of a small number of particles in an "elementary cell" with a plane wave factor, $e^{i\vec{k}\vec{r}}$ (cf. eq. (3.57); $\vec{k}$ is the vector describing the translational lattice in the reciprocal space):

$$\Psi_{Per}(\vec{r}) = e^{i\vec{k}\vec{r}}\ \Psi_{EC}(\vec{r}) \tag{3.155}$$

These periodic wave functions can be treated with methods, which are, in most cases, based on DFT. Figure 3.167b illustrates how periodic models allow describing extended materials with very small repetitive units. The infinite metal surface is treated with a repetitive unit of just 12 atoms covering three layers below, which is enough to capture spatial structures and energies of extended electron states and to remove truncation errors at cluster boundaries, except for the lowest layer. The relaxation of the subsurface metal structure in response to the adsorption can be better described than in the two-layer cluster in Figure 3.167a. Periodic models become problematic when disorder phenomena in real phases are to be studied, e.g., surface or bulk defects, because the small repetitive units result in unrealistic defect densities.

In the treatment of adsorption phenomena, clusters are preferred for the low coverage limit, which would require large repetitive units per adsorbate molecule in periodic models. For studies of adsorbate-adsorbate interactions, periodic models are easily arranged, while clusters need to be fairly large to keep both adsorbates away from the cluster edges.

3.5.3 Quantum chemistry in catalysis research: typical applications

DFT currently dominates theoretical modeling of catalytic materials and surface processes due to its capability to treat large systems with acceptable accuracy. Although its accuracy is permanently improved by progress in the development of DFT functionals, wave function-based calculations are still often used as benchmarks, just in methodical work. DFT, which was originally confined to the ground state (Section 3.5.1), has been, meanwhile, extended to excited electronic states (Time-dependent DFT, T-DFT). Still, wave function-based methods are superior particularly for highly excited states like the final-state energies in photoemission, which are required to predict XPS binding energies (cf. Section 3.4.5.3.1). For calculating the rate constant of the reaction between alkenes and methanol over H-ZSM-5, Piccini et al. [238] combined DFT- and wave function-based spheres in a QM/QM hybrid approach (see below).

Total energy calculations are the basis for geometry optimization, which results in detailed structural models of the surface or the near-surface region of catalytic materials. For catalysts capable of reacting with gas-phase components, e.g., metals with O_2, sulfides with H_2S, the state of the surface can be, meanwhile, predicted depending on the chemical potential of the reactive components (e.g., O_2, H_2S; see below). With suitable restrictions during optimization, such surface phase diagrams may also include features of faceted surfaces containing steps, kinks, structure defects, etc., which are metastable relative to the low-index planes.

Total energies can also be calculated for systems containing molecules at any distance and orientation to a surface. This allows studying adsorption processes or activation barriers of hypothetical reaction steps, which grants theoretical chemistry an enormous influence in the discussion of catalytic reaction mechanisms. By calculating second derivatives of the total energy vs. the spatial coordinates in the equilibrium state (cf. eqs. (3.106) (3.107)), force constants of inter- and intramolecular vibrations are accessible, and frequencies of vibrational modes can be predicted. While even UV-Vis spectra can, nowadays, be treated by evaluating electronically excited states with T-DFT, highly excited states (core hole spectra) are still better described by wave function-based methods.

It should be noted, however, that reliable results for vibrational modes and, in particular, rate constants may require considerable complexity of the model. Theory-derived vibrational data often help assignment of complex spectra without, however, being able to reproduce experimental wavenumbers exactly. An empirical scaling factor has been introduced often to improve the agreement. To exemplify such complexity, a study combining DFT with wave function-based treatment of a model comprising 300 atoms in order to predict activation energies and rate constants of the acid-catalyzed conversion of small alkenes with methanol in a zeolite will be briefly highlighted.

In this study, Piccini et al. [238] obtained the potential energy surfaces describing the interaction between alkenes and methanol adsorbed on zeolite sites by periodic DFT calculations of the full model, using a functional capable of covering dispersive interactions. Regarding the reaction coordinate, the energies of initial and transition states were improved by modeling the reactants in a smaller section of the zeolite (a ten-ring containing the Brønsted site) with a post-HF approach (MP-2). The residual error was corrected by comparing with calculations of an even smaller cluster (3 T-atoms) with a high-end coupled cluster method. Entropic contributions to adsorption and reaction were evaluated from partition functions in harmonic approximation, but with anharmonic corrections. They were obtained by solving one (one-dimensional) Schrödinger equation per normal mode (i.e., 3n for n atoms in the cluster).

For zeolites with benign structures, reaction rates per g zeolite can be directly deduced from the rates per site (i. e., TOFs) obtained in such calculations.

Adsorption on metal surfaces and metal-catalyzed reactions proceeding via a rate-determining bond scission are less demanding with respect to model detail and computational effort. In their highly influential work on ammonia synthesis (summarized, e.g., in ref. [242], cf. Figure(s) 4.1b, 4.2, 4.16) and on reactivity trends across the transition metals [240] (cf. Section 4.2), Nørskov et al. applied periodic DFT calculations with slab models comprising up to 4 layers, which resulted in repetitive units of 15–30 atoms, depending on the roughness of surfaces. Productivities of NH_3 synthesis could be well captured by inserting the rate constants obtained into a microkinetic model of the reaction, although the concentration of active sites was derived from experimental data.

In this work and in parallel studies of other groups, it was found that activation energies of dissociative M_2 adsorptions are linearly correlated with the adsorption enthalpies, as long as only metal facets of the same roughness (atomically flat or with equal density of steps) are included. These reactions follow the Brønsted-Evans-Polanyi (BEP) equation (3.156), which relates activation energies of reactions to their reaction enthalpies:

$$E_A = \alpha \, \Delta H_R + b \qquad\qquad (3.156)$$

Depicting the reaction coordinate for dissociative adsorption of M_2 molecules, Figure 3.169a illustrates why eq. (3.156) may hold as long as the bonding pattern in the transition state is not severely influenced by the change of the metal. There is even a universal BEP relation followed in a 0th order approximation by a great number of reactions proceeding via a rate-limiting splitting of a M_1M_2 bond (M_1 or M_2 = C, N, or O, Figure 3.169b) [243]. BEP relations for reactions on corrugated surfaces differ from those on flat surfaces just by the b values of eq. (3.156) [244]. Calculation of transition states is tedious even with relatively small DFT-based models. In recent work on computational screening of catalytic activities in large parameter fields (*in silico* screening, see below), activation energies were, therefore, interpolated by BEP relations. Ref. [245] provides more details on BEP relations and their use.

Surface phase diagrams are obtained by minimizing the free surface energy of surface slabs under given chemical potentials μ_i of all elements in the solid (e.g., metal, O, S) and in reactive gas-phase components (e.g., H). The expression for the Gibbs free energy of a solid, $G = \Sigma n_i \mu_i$, neglects that any solid is confined by surfaces. The free surface energy γ at pressure p and temperature T is the difference between the Gibbs free energies of such confined amount of material (e.g., a surface slab with

a

b

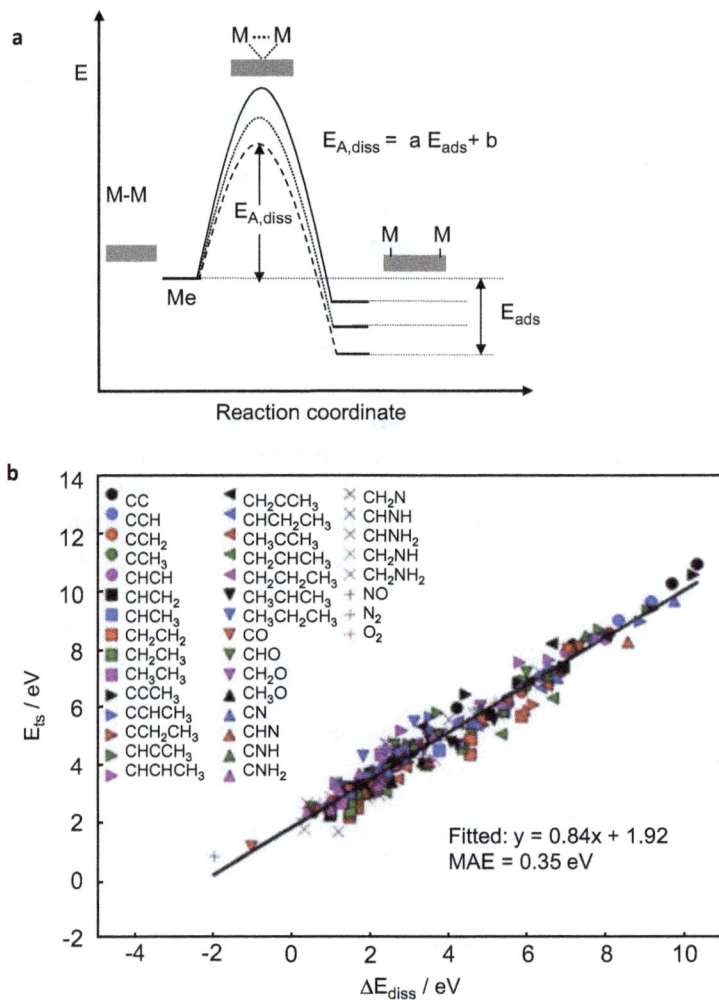

Figure 3.169: The Brønsted-Evans-Polanyi (BEP) relation applied to the catalytic dissociation of molecules: (a) reaction coordinate for the dissociation of M_2 molecules suggesting an approximately linear relation between $E_{A,diss}$ and E_{ads}; (b) universal BEP relation for reactions with a rate-determining dissociation of a bond between C, N, or O atoms; reproduced from ref. [243] with permission from Springer Nature. MAE – mean absolute error.

the area A_{tot}) and the same quantity thought be part of a larger block in the bulk, related to the area A_{tot}:

$$\gamma(p, T) = \frac{1}{A_{tot}} \left(G(p, T, n_1 \ldots n_i) - \sum n_i \mu_i(p, T) \right)$$

(3.157)

In binary compounds, e.g., MeX_n, one of the chemical potentials in (3.157) can be replaced, because their relation is fixed by the bulk potential, e.g.,

$$\mu_{MeX_n} = \mu_{Me} + n\,\mu_X \tag{3.158}$$

As mentioned in Section 3.5.2, entropic contributions can be often neglected in work with solid structures (see, for instance, refs. [239, 246]). Therefore, total energies from model calculations are used instead of Gibbs free energies. In the (periodic) models, the repeat unit is a thin slab of a geometrical surface A, and $G(p,\,T,\,n_1 \ldots n_i)$ is replaced by E^{slab}, e.g. by $E^{slab}(N_{Me},\,N_X)$ in the MeX_n example, where N_{Me} and N_X are the numbers of Me and X atoms, respectively, in the model slab. Likewise, the chemical potential of the bulk is replaced by the total energy of the material related to the formula unit, e.g., $E^{f.u.}_{MeX_n}$. For the MeX_n example, this converts eqs.(3.157) and (3.158) into:

$$\gamma(p,\,T) = \frac{1}{2A}\left(E^{slab}(N_{Me},\,N_X) - N_{Me}\left(E^{f.u.}_{MeX_n} - n\mu_X(p,\,T)\right) - N_X\mu_X(p,\,T)\right)$$

and after some rearrangement into:

$$\gamma(p,\,T) = \frac{1}{2A}\left(E^{slab}(N_{Me},\,N_X) - N_{Me}E^{f.u.}_{MeX_n} + (nN_{Me} - N_X)\mu_X(p,\,T)\right) \tag{3.159}$$

The temperature- and pressure-dependent chemical potential of X (e.g., O or S) can be evaluated from tabulated thermodynamic data. It can be also expressed in terms of gas-phase pressures or compositions. A component not present in the bulk solid (e.g., H) contributes to the slab energy, $E^{slab}(N_{Me},\,N_X,\,N_H)$, while its contribution to the bulk system ($MeX_n + H_2$) is subtracted as a part of $\Sigma n_i\,\mu_i\,(p,T)$, i.e.,

$$\gamma(p,\,T) = \frac{1}{2A}\left(E^{slab}(N_{Me},\,N_X,\,N_H) - N_{Me}E^{f.u.}_{MeX_n + H_2} + (nN_{Me} - N_X)\mu_X(p,\,T) - N_H\mu_H\right) \tag{3.160}$$

The surface structure stable in a given environment can be found by minimizing γ according to (3.159) and (3.160) for a suitable set of chemical potentials μ_X/μ_H, which is, however, extremely laborious, because it involves the variation of slab structures. In a less stringent approach, the influence of chemical potential variation on γ is evaluated for a realistic choice of slab models. This approach has so far been preferred, although it depends on chemical intuition in the selection of models and penalizes structures of low or without any surface symmetry.

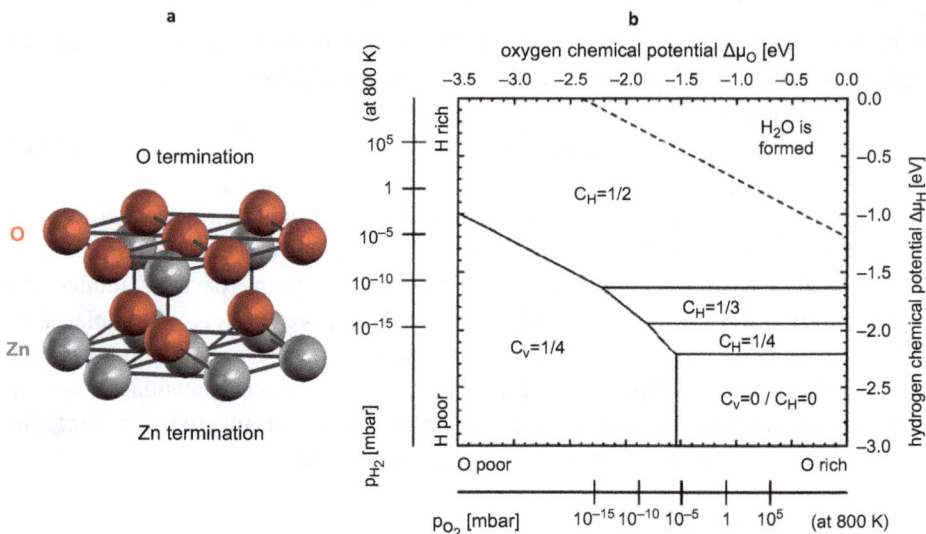

Figure 3.170: The polar O-ZnO(0001) surface at 800 K: (a) model of the unreconstructed surface structure; (b) surface phase diagram, comprising stabilization of the surface by acquisition of H atoms or by formation of O vacancies. Reproduced with permission from ref. [247]. Copyright 2004 American Physical Society.

The approach is nicely illustrated in a study on the stabilization of the O-terminated ZnO(0001) surface [247] (Figure 3.170a), which has been mentioned in Section 2.2.1 as an example of a Tasker type III facet. The required reduction of the negative surface charge can be achieved by charge transfer from O toward Zn atoms within the structure, optionally combined with surface distortion or reconstruction (i), by loss of oxygen, i.e., formation of O vacancies (ii), or by interaction with other components (iii), e.g., formation of OH groups via reaction with H_2 or H_2O.[64]

After discarding option (i), the author evaluated models with a wide range of O vacancy concentrations using an equation similar to (3.159) and found that the loss of surface O atoms is negligible, even at very low chemical potentials μ_O: down to μ_O = -1.54 eV ($\approx 10^{-5}$ mbar O_2 at 800 K), the surface remains saturated. Below, the surface with the lowest assumed vacancy concentration (1/4 monolayer) becomes stable. In an analogous study of saturated surfaces with varied concentrations of OH groups, it was found that an H-free surface exists only below μ_H = -2.2 eV, which corresponds to $<10^{-15}$ mbar H_2 at 800 K. Between μ_H = -2.2 eV and -1.63 eV (10^{-10} mbar H_2 at 800 K), the stable surface takes up hydrogen up to half a monolayer and does not change any more, at all accessible values of μ_H.

---

64 Special aspects arising from the problem that Tasker III slabs cannot have identical terminations at their opposite sides are discussed in ref. 247 as well as the conceptual problem of imagining a simultaneous thermodynamic reservoir of H_2 and O_2 at p and T, as references for μ_H and μ_O.

Together with data from studies of mixed structures containing both O vacancies and OH groups, these results are summarized in Figure 3.170b. The data shows, for instance, that the surface takes up hydrogen even under UHV conditions, where the residual atmosphere is rich in H_2: at 800 K and 10^{-10} mbar O_2, a H_2 pressure below 10^{-15} mbar would be required to exclude OH groups. In the absence of H_2 (and H_2O), the bare stoichiometric O-ZnO(0001) surface would stabilize according to option (i) [247]. However, as the concentrations of H_2 (or H_2O) allowed for this are below accessible limits, stabilization by formation of OH groups is preferred even at very high O_2 pressures.

In the last two decades, computational tools have been developed that allow finding surface compositions of promising catalysts (*in silico* screening). So far, this has succeeded only for simple reactions, for which profound knowledge of the reaction mechanism including microkinetic models are available [248]. As mentioned above, the use of the BEP relation for interpolating activation energies on the basis of adsorption energies helps limiting the computational effort for screening a great number of candidates. It has been complemented by the identification of linear relations also between adsorption energies, e.g., between the adsorption energies of CH_x species (x = 1, 2, or 3) and those of C, more generally, between those of XH_y (X = C, N, O, S) and those of the X adsorbate on many metal and alloy surfaces [249].

The identification of quantities that provide easy access to key parameters of the catalytic process is at the heart of computational screening. Such "descriptors" can be derived from an analysis of the reaction mechanism. They target steps that determine rates and selectivities, for instance, the adsorption energies of C and O (or of CO and OH) for the synthesis of methanol and higher alcohols, where methanation is an undesired side reaction, or the free Gibbs energy of H adsorption for the electrochemical hydrogen evolution. Kinetic parameters calculated by DFT (or obtained via the scaling relations) are inserted in microkinetic models to obtain turnover frequencies.

Descriptor-based computational screening achieved its initial breakthroughs with reactions, the key steps of which proceed on one site type. Bifunctional catalysis is much more difficult to handle with this concept. On the other hand, the use of descriptors is not confined to activity and selectivity: descriptors for stability of a mixed-metal surface in an aqueous electrolyte may be, for instance, the free energy of water splitting followed by adsorption of oxygen on the surface, or the free energy of dissolution. Even material prices have been employed as descriptors in a setting that included economic aspects of the process under investigation.

The opportunity to identify the most active catalyst(s) for a simple reaction using a single descriptor is based on a principle proposed by Paul Sabatier in 1911. According to this Sabatier principle, which will be discussed in more detail in Section 4.2.1, a catalyst provides the highest reaction rates when its interactions with the reactant and the resulting product are neither too strong nor too weak. Insufficient interaction prevents the activation of the reactant; interaction that is too strong prevents the release of the product. Taking ammonia synthesis as an example (cf. Figures 1.6b, 2.57 for the reaction mechanism), the catalyst should be able to split N_2, but it should also allow NH_3 to desorb from the surface. A plot of the reaction rate vs. the descriptor $E_{ads,N}$ exhibits a distinct maximum (volcano plot) not far from the adsorption energies on Fe and Ru, which are used in technical catalysts (see Figure 4.1).

Computational screening will be exemplified with an electrocatalytic application: the hydrogen evolution reaction (HER). Although oxygen evolution with its high overpotentials (cf. Section 4.9) is certainly the more attractive research field, there is a demand on research also for HER, because its state-of-the-art Pt catalyst contributes much to the as yet intolerable cost of PEM[65] water electrolysis (cf. Section 2.8.4). As mentioned above, the standard free Gibbs energy of H adsorption from molecular hydrogen ΔG_H^0 was selected as a descriptor because of the obvious role of adsorbed H atoms in the reaction: their strong adsorption might retard the release of H_2; insufficient interaction with the surface might retard the surface redox step. The highest rate may be expected at $\Delta G_H^0 \approx 0$.

Figure 3.171a confirms this expectation showing a volcano relationship between exchange currents (cf. Section 2.7) measured with a number of largely polycrystalline materials and ΔG_H^0 on them, as obtained from DFT calculations. One may wonder why reaction rates in an aqueous phase can be correlated with interaction energies calculated without the presence of an electrolyte. In a study on H adsorption on Pd (a Pd monolayer on Au(111)), the influence of water on the adsorption energy was found to be small [250]. Therefore, the electrolyte influence can be neglected as long as the adsorbed hydrogen is not charged.

Figure 3.171b summarizes a computational screening by Greeley et al. [251], which comprised 736 samples consisting of a monometallic or binary alloy monolayer that covered a host metal. 16 metals were included in roles of both host metal and "solute metal," which formed the overlayer mixed with the host in different concentrations (33, 66, or 100 atom-% of the solute). In Figure 3.171b, points representing these samples are plotted in a field defined by descriptors for activity and stability, the former being the absolute value of ΔG_H^0. With respect to stability, free energies of four deactivating processes were examined, but only the most unfavorable among these results was used to place a sample on the stability scale.

In Figure 3.171b, no samples appear in the lower left corner representing the best parameters in terms of both activity and stability. The frequency of samples in this quadrant is rather low, quite opposite to the right upper quadrant exposing the materials failing with respect to both properties. The figure shows that the customary Pt catalyst is almost ideal in terms of stability, but, by far, not in terms of activity: a large number of samples are predicted to be more active, though most of them are likely to fail for stability reasons. In the figure, some points are specified by disclosing the solute metals involved. The line indicates materials the activities of which were not exceeded at their level of stability.

---

65 PEM – proton exchange membrane.

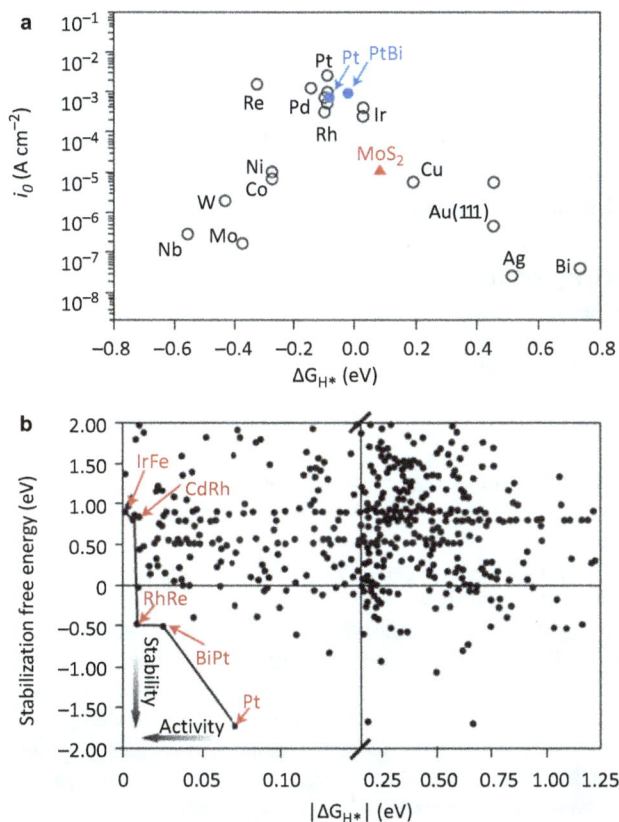

Figure 3.171: Computational screening for the hydrogen evolution reaction (HER): (a) Volcano relation between experimental exchange currents and the Gibbs free energy of H formation for (mostly) polycrystalline metals; for Pt, data are shown from polycrystalline samples and from single facets (the non-labeled crowd of points between those labeled "Pt" and "Rh"), the HER activity of MoS_2 was found following analogies of its structure to those of active sites in hydrogen-releasing enzymes (biomimetic approach); (b) properties of metals and surface alloys examined, plotted *vs.* activity and stability descriptors. The line connects solutions of the Pareto-optimal set. In the labels, the solute metal is given first: BiPt designates a PtBi surface alloy on a Pt sample. Reproduced from ref. [248] with permission from Springer Nature.

The high activity of the BiPt system is in remarkable disagreement with earlier literature that reported irreversibly adsorbed Bi to poison the HER activity of Pt, which was even reproduced by the authors [251]. The positive effect was obtained only after annealing the sample at 770 K, which was apparently required to integrate the adsorbed Bi atoms into the surface alloy layer.

In silico screening has been so far largely applied to catalysis by metals or metallic phases, because their behavior can be well reproduced by DFT, which is not always the case for other types of materials. It has, however, already become an important component of catalyst development. In the future, models may be extended to cover other important aspects of catalysis such as the effect of promoters and poisons (see Section 4). Still, the preparation of catalysts has remained a largely experimental challenge, their examination in real catalytic environments may reveal unexpected problems, and as in any other field, breakthrough of the innovation may be impeded by technological or economical handicaps. The application of theory to catalysis on industrial scales has been, however, strongly encouraged by a new industrial catalyst originating right from the first extended computational screening (a NiFe alloy catalyst for the methanation step in ammonia syngas production, cf. Appendix A1) and by a number of earlier innovations on technical catalysts, which were based on ideas generated in theoretical studies [248].

References

[1]   Schüth, F.; Unger, K., Precipitation and Coprecipitation. In *Handbook of Heterogeneous Catalysis*, Ertl, G.; Knözinger, H.; J., W., Eds. Wiley-VCH: Weinheim, 1997; Vol. 1, pp 72–86. in doi. org/10.1002/9783527619474.ch2a

[2]   Baerns, M.; Behr, A.; Brehm, A.; Gmeling, J.; Hofmann, H.; Onken, U.; Renken, A., *Technische Chemie*. Wiley-VCH: Weinheim, 2006. ISBN: 978-3527310005

[3]   Levenspiel, O., *Chemical Reactor Engineering*. 3rd ed.; John Wiley & Sons: New York, 1998.

[4]   Behrens, M.; Lolli, G.; Muratova, N.; Kasatkin, I.; Haevecker, M.; Naumann d'Alnoncourt, R.; Storcheva, O.; Köhler, K.; Muhler, M.; Schlögl, R., The effect of Al-doping on ZnO nanoparticles applied as catalyst support. *Physical Chemistry Chemical Physics* **2013**, *15* (5), 1374–1381. doi 10.1039/c2cp41680h

[5]   Beck, J. S.; Vartuli, J. C.; Roth, W. J.; Leonowicz, M. E.; Kresge, C. T.; Schmitt, K. D.; Chu, C. T. W.; Olson, D. H.; Sheppard, E. W.; McCullen, S. B.; Higgins., J. B.; Schlenker, J. L., A new family of mesoporous molecular sieves prepared with liquid crystal templates. *Journal of the American Chemical Society* **1992**, *114* (27), 10834–10843. doi: 10.1021/ja00053a020

[6]   https://en.wikipedia.org/wiki/Mesoporous_silica. (accessed Mar 3, 2023).

[7]   Van Der Voort, P.; Vercaemst, C.; Schaubroeck, D.; Verpoort, F., Ordered mesoporous materials at the beginning of the third millennium: New strategies to create hybrid and non-siliceous variants. *Physical Chemistry Chemical Physics* **2008**, *10* (3), 347–360. doi 10.1039/B707388G

[8]   Mintova, S.; Barrier, N., *Verified Syntheses of Zeolitic Materials*. International Zeolite Association; available on www.iza-online.org: 2016.

[9]   Jansen, J. C., The preparation of oxide molecular sieves. A. Synthesis of zeolites (Chapter 5). Studies in Surface Science and Catalysis 2001, 137, 175–227. doi.org/10.1016/S0167-2991(01)80246-9.

[10]   Oleksiak, M. D.; Rimer, J. D., Synthesis of zeolites in the absence of organic structure-directing agents: Factors governing crystal selection and polymorphism. *Reviews in Chemical Engineering* **2014**, *30* (1), 1–49. doi:10.1515/revce-2013-0020

[11]   van Bekkum, H.; Flanigan, E.; Jacobs, P. A.; Jansen, J. C., *Introduction to Zeolite Science and Practice*. 2nd ed.; Elsevier: Amsterdam, 2001; Vol. 137. ISBN: 9780080534794

[12]   Mercury, 1.0; https://www.ccdc.cam.ac.uk/solutions/csd-core/components/mercury/ (accessed Dec 4, 2022): Cambridge, 2022.

[13]   Kalmutzki, M. J.; Hanikel, N.; Yaghi, O. M., Secondary building units as the turning point in the development of the reticular chemistry of MOFs. Science **2018**, *4* (10), eaat9180. doi:10.1126/sciadv. aat9180

[14]   Stock, N.; Biswas, S., Synthesis of Metal-Organic Frameworks (MOFs): Routes to Various MOF Topologies, Morphologies, and Composites. *Chemical Reviews* **2012**, *112* (2), 933–969. doi 10.1021/ cr200304e

[15] Bitzer, J.; Kleist, W., Synthetic Strategies and Structural Arrangements of Isoreticular Mixed-Component Metal–Organic Frameworks. *Chemistry – A European Journal* **2019**, *25* (8), 1866–1882. doi.org/10.1002/chem.201803887

[16] Tanabe, K. K.; Cohen, S. M., Postsynthetic modification of metal–organic frameworks – A progress report. *Chemical Society Reviews* **2011**, *40* (2), 498–519. doi 10.1039/C0CS00031K

[17] Mackenzie, J. D.; Bescher, E. P., Chemical routes in the synthesis of nanomaterials using the sol–gel process. *Accounts of Chemical Research* **2007**, *40* (9), 810–818. doi 10.1021/ar7000149

[18] Landau, M. V., Sol–gel process. In *Handbook of Heterogeneous Catalysis*, Ertl, G.; Knözinger, H.; Schüth, F.; Weitkamp, J., Eds. Wiley-VCH: Weinheim, 2008; Vol. 1, pp 119–160. doi. org/10.1002/9783527610044.hetcat0009

[19] Innocenzi, P., Understanding sol–gel transition through a picture. A short tutorial. *Journal of Sol-Gel Science and Technology* **2020**, *94* (3), 544–550. doi 10.1007/s10971-020-05243-w

[20] Okuyama, K.; Kousaka, Y.; Tohge, N.; Yamamoto, S.; Wu, J. J.; Flagan, R. C.; Seinfeld, J. H., Production of ultrafine metal oxide aerosol particles by thermal decomposition of metal alkoxide vapors. *AIChE Journal* **1986**, *32* (12), 2010–2019. doi.org/10.1002/aic.690321211

[21] Becker, M. J.; Xia, W.; Xie, K. P.; Dittmer, A.; Voelskow, K.; Turek, T.; Muhler, M., Separating the initial growth rate from the rate of deactivation in the growth kinetics of multi-walled carbon nanotubes from ethene over a cobalt-based bulk catalyst in a fixed-bed reactor. *Carbon* **2013**, *58*, 107–115. doi. org/10.1016/j.carbon.2013.02.038

[22] Mleczko, L.; Lolli, G., Carbon nanotubes: An example of multiscale development – A mechanistic view from the subnanometer to the meter scale. *Angewandte Chemie International Edition* **2013**, *52* (36), 9372–9387. doi.org/10.1002/anie.201302791

[23] Zhang, Q.; Huang, J.-Q.; Zhao, M.-Q.; Qian, W.-Z.; Wei, F., Carbon nanotube mass production: Principles and processes. ChemSusChem **2011**, *4* (7), 864–889. doi.org/10.1002/cssc.201100177

[24] Tricoli, A.; Nasiri, N.; Chen, H.; Wallerand, A.; Righettoni, M., Ultra-rapid synthesis of highly porous and robust hierarchical ZnO films for dye sensitized solar cells. *Solar Energy* **2016**, *136*, 553–559. doi 10.1016/j.solener.2016.07.024

[25] Strobel, R.; Baiker, A.; Pratsinis, S. E., Aerosol flame synthesis of catalysts. *Advanced Powder Technology* **2006**, *17* (5), 457–480. doi 10.1163/156855206778440525

[26] Che, M., Interfacial coordination chemistry - Concepts and relevance to catalysis phenomena. *Studies in Surface Science and Catalysis* **1993**, *75*, 31–68. doi.org/10.1016/S0167-2991(08)64004-5

[27] Munnik, P.; de Jongh, P. E.; de Jong, K. P., Recent developments in the synthesis of supported catalysts. *Chemical Reviews* **2015**, *115* (14), 6687–6718. doi 10.1021/cr500486u

[28] Wang, L.; Hall, W. K., The preparation and genesis of molybdena-alumina and related catalyst systems. *Journal of Catalysis* **1982**, *77* (1), 232–241. doi.org/10.1016/0021-9517(82)90163-4

[29] Carrier, X.; Lambert, J. F.; Kuba, S.; Knözinger, H.; Che, M., Influence of ageing on MoO_3 formation in the preparation of alumina-supported Mo catalysts. *Journal of Molecular Structure* **2003**, *656* (1–3), 231–238. doi 10.1016/s0022-2860(03)00328-4

[30] Lee, J. M.; Seo, S. M.; Suh, J. M.; Lim, W. T., Synthesis and single-crystal structures of fully dehydrated fully Sr^{2+}-exchanged zeolite Y (FAU) and its benzene sorption complex, $|Sr_{37.5}|[Si_{117}Al_{75}O_{384}]$-FAU and $|Sr_{37.5}(C_6H_6)_{33}(H_2O)_{15}|[Si_{117}Al_{75}O_{384}]$-FAU. *Journal of Porous Materials* **2011**, *18* (5), 523–534. doi 10.1007/s10934-010-9415-z

[31] Marturano, P.; Drozdova, L.; Kogelbauer, A.; Prins, R., Fe/ZSM-5 prepared by sublimation of $FeCl_3$: The structure of the Fe species as determined by IR, ^{27}Al-NMR and EXAFS Spectroscopy. *Journal of Catalysis* **2000**, *192*, 236–247. doi: 10.1006/jcat.2000.2837

[32] Grünert, W.; Hayes, N. W.; Joyner, R. W.; Shpiro, E. S.; Siddiqui, M. R. H.; Baeva, G. N., Structure, chemistry, and activity of Cu-ZSM-5 catalysts for the selective reduction of NOx in the presence of oxygen. *The Journal of Physical Chemistry A* **1994**, *98* (42), 10832–10846. doi: 10.1021/j100093a026

[33] Grundner, S.; Markovits, M. A. C.; Li, G.; Tromp, M.; Pidko, E. A.; Hensen, E. J. M.; Jentys, A.; Sanchez-Sanchez, M.; Lercher, J. A., Single-site trinuclear copper oxygen clusters in mordenite for selective conversion of methane to methanol. *Nature Communications* **2015**, *6*, Art. No. 7546. doi 10.1038/ncomms8546

[34] Grünert, W.; Padmalekha, K. G.; Ellmers, I.; Pérez Vélez, R.; Huang, H.; Bentrup, U.; Schünemann, V.; Brückner, A., Active sites of the selective cataytic reduction of NO by NH_3 over Fe-ZSM-5; Combining reaction kinetics with postcatalytic Mössbauer spectroscopy at cryogenic temperatures. *ACS Catalysis* **2020**, 10 (5), 3119–3030. doi 10.1021/acscatal.9b04627.

[35] Grünert, W., Active sites for selective catalytic reduction. In *Urea-SCR Technology for deNOx After Treatment of Diesel Exhausts*, Nova, I.; Tronconi, E., Eds. Springer: Berlin-Heidelberg-New York, 2014; pp 181–220. doi 10.1007/978-1-4899-8071-7_7

[36] Děděcek, J.; Sobalik, Z.; Wichterlová, B., Siting and distribution of framework aluminium atoms in silicon-rich zeolites and impact on catalysis. *Catalysis Reviews – Science and Engineering* **2012**, *54* (2), 135–223. doi 10.1080/01614940.2012.632662

[37] Espinosa-Alonso, L.; de Jong, K. P.; Weckhuysen, B. M., A UV-Vis micro-spectroscopic study to rationalize the influence of Cl^-(aq) on the formation of different Pd macro-distributions on gamma-Al_2O_3 catalyst bodies. *Physical Chemistry Chemical Physics* **2010**, *12* (1), 97–107. doi 10.1039/b915753k

[38] Pan, X. L.; Bao, X. H., The effects of confinement inside carbon nanotubes on catalysis. *Accounts of Chemical Research* **2011**, *44* (8), 553–562. doi 10.1021/ar100160t

[39] Kortewille, B.; Wachs, I. E.; Cibura, N.; Pfingsten, O.; Bacher, G.; Muhler, M.; Strunk, J., Photocatalytic methanol oxidation by supported vanadium oxide species: Influence of support and degree of oligomerization. *European Journal of Inorganic Chemistry* **2018**, *2018* (33), 3725–3735. doi 10.1002/ejic.201800490

[40] Ermakov, Y. I.; Kuznetsov, B. N.; Karakchiev, L. G.; Derbenev, S. S., Supported molybdenum catalysts based on tetrakis (π-allyl/-molybdenum). *Reaction Kinetics and Catalysis Letters* **1974**, *1* (3), 307–313. doi.org/10.1007/BF02070899

[41] Iwasawa, Y., Chemical design surfaces for active solid catalysts. *Advances in Catalysis* **1987**, *35*, 187–264. doi 10.1016/s0360-0564(08)60094-2

[42] Grünert, W.; Feldhaus, R.; Anders, K.; Shpiro, E. S.; Minachev, K. M., Reduction behavior and metathesis activity of WO_3/Al_2O_3 catalysts III. The activation of WO_3/Al_2O_3 catalysts. *Journal of Catalysis* **1989**, *120*, 444–456. doi 10.1016/0021-9517(89)90284-4

[43] Grünert, W.; Stakheev, A. Y.; Mörke, W.; Feldhaus, R.; Anders, K.; Shpiro, E. S.; Minachev, K. M., Reduction and metathesis activity of MoO_3/Al_2O_3 catalysts II. The activation of MoO_3/Al_2O_3 catalysts. *Journal of Catalysis* **1992**, *135*, 287–299. doi 10.1016/0021-9517(92)90286-Q.

[44] Coperet, C.; Basset, J. M., Strategies to immobilize well-defined olefin metathesis catalysts: Supported homogeneous catalysis vs. surface organometallic chemistry. *Advanced Synthesis & Catalysis* **2007**, *349* (1–2), 78–92. doi 10.1002/adsc.200600443

[45] Samantaray, M. K.; Pump, E.; Bendjeriou-Sedjerari, A.; D'Elia, V.; Pelletier, J. D. A.; Guidotti, M.; Psaro, R.; Basset, J.-M., Surface organometallic chemistry in heterogeneous catalysis. *Chemical Society Reviews* **2018**, *47* (22), 8403–8437. doi 10.1039/c8cs00356d

[46] Vidal, V.; Théolier, A.; Thivolle-Cazat, J.; Basset, J.-M., Metathesis of alkanes catalyzed by silica-supported transition metal hydrides. *Science* **1997**, *276* (5309), 99–102. doi 10.1126/science.276.5309.99

[47] Vidal, V.; Théolier, A.; Thivolle-Cazat, J.; Basset, J.-M.; Corker, J., Synthesis, characterization, and reactivity, in the C-H bond activation of cycloalkanes, of a silica-supported tantalum(III) monohydride complex: ($\equiv$SiO)$_2$Ta^{III}–H. *Journal of the American Chemical Society* **1996**, *118* (19), 4595–4602. doi 10.1021/ja953524l

[48] Basset, J.-M.; Coperet, C.; Soulivong, D.; Taoufik, M.; Cazat, J. T., Metathesis of Alkanes and Related Reactions. *Accounts of Chemical Research* **2010**, *43* (2), 323–334. doi 10.1021/ar900203a.

[49]   Margitfalvi, J. L.; Jalett, H. P.; Tàlas, E.; Baiker, A.; Blaser, H. U., Enantioselective hydrogenation of ethyl pyruvate on tin promoted Pt/Al$_2$O$_3$ catalysts. *Catalysis Letters* **1991**, *10* (5), 325–333. doi 10.1007/bf00769167

[50]   Chupin, C.; Candy, J. P.; Basset, J. M., Surface organometallic chemistry on metals: Influence of organometallic fragments grafted to the surface of Rh particles on the competitive hydrogenation of terminal and internal double bonds of unsaturated primary alcohols. *Catalysis Today* **2003**, *79–80*, 15–19. doi.org/10.1016/S0920-5861(03)00037-3

[51]   Jia, C.-J.; Schüth, F., Colloidal metal nanoparticles as a component of designed catalyst. *Physical Chemistry Chemical Physics* **2011**, *13* (7), 2457–2487. doi 10.1039/c0cp02680h

[52]   Liang, C. H.; Xia, W.; Soltani-Ahmadi, H.; Schlüter , O.; Fischer, R. A.; Muhler, M., The two-step chemical vapor deposition of Pd(allyl)Cp as an atom-efficient route to synthesize highly dispersed palladium nanoparticles on carbon nanofibers. *Chemical Communications* **2005**, (2), 282–284. doi 10.1039/b412150c

[53]   Kurtz, M.; Bauer, N.; Büscher, C.; Wilmer, H.; Hinrichsen, O.; Becker, R.; Rabe, S.; Merz, K.; Driess, M.; Fischer, R. A.; Muhler, M., New synthetic routes to more active Cu/ZnO catalysts used for methanol synthesis. *Catalysis Letters* **2004**, *92* (1–2), 49–52. doi: 10.1023/B:CATL.0000011085.88267.a6

[54]   Binder, A.; Seipenbusch, M.; Muhler, M.; Kasper, G., Kinetics and particle size effects in ethene hydrogenation over supported palladium catalysts at atmospheric pressure. *Journal of Catalysis* **2009**, *268* (1), 150–155. doi 10.1016/j.jcat.2009.09.013

[55]   Puurunen, R. L., Surface chemistry of atomic layer deposition: A case study for the trimethylaluminum/water process. *Journal of Applied Physics* **2005**, *97* (12). , Art. No. 121301. doi 10.1063/1.1940727

[56]   Xie, Y.-C.; Tang, Y.-Q.; Eley, D. D.; Pines, H.; Weisz, P. B., Spontaneous Monolayer Dispersion of Oxides and Salts onto Surfaces of Supports: Applications to Heterogeneous Catalysis. In *Advances in Catalysis*. 1990, 37, 1–32. doi doi.org/10.1016/S0360-0564(08)60362-4

[57]   Margraf, R.; Leyrer, J.; Knözinger, H.; Taglauer, E., Study of molybdate dispersion on supported catalysts using ion scattering and Raman spectroscopy. *Surface Science* **1987**, *189–190*, 842–850. doi.org/10.1016/S0039-6028(87)80520-4

[58]   Beyer, H. K.; Karge, H. G.; Borbely, G., Solid-state ion-exchange in zeolites. 1. Alkaline Chlorides/ZSM-5. *Zeolites* **1988**, *8* (1), 79–82. doi 10.1016/s0144-2449(88)80035-6

[59]   Karge, H. G.; Wichterlová, B.; Beyer, H. K., High-temperature interaction of solid Cu chlorides and Cu oxides in mixtures with H-forms of ZSM-5 and Y zeolites. *Journal of the Chemical Society, Faraday Transactions* **1992**, *88* (9), 1345–1351. doi 10.1039/ft9928801345

[60]   Buffat, P.; Borel, J. P., Size effect on the melting temperature of gold particles. *Physical Review A* **1976**, *13* (6), 2287–2298. doi 10.1103/PhysRevA.13.2287

[61]   Baltes, C.; Vukojevic, S.; Schüth, F., Correlations between synthesis, precursor, and catalyst structure and activity of a large set of CuO/ZnO/Al$_2$O$_3$ catalysts for methanol synthesis. *Journal of Catalysis* **2008**, *258* (2), 334–344. doi.org/10.1016/j.jcat.2008.07.004

[62]   Albonetti, S.; Cavani, F.; Trifirò, F.; Venturoli, P.; Calestani, G.; López Granados, M.; Fierro, J. L. G., A Comparison of the Reactivity of "nonequilibrated" and "equilibrated" V-P-O catalysts: Structural evolution, surface characterization, and reactivity in the selective oxidation of n-butane and n-pentane. *Journal of Catalysis* **1996**, *160* (1), 52–64. doi.org/10.1006/jcat.1996.0123

[63]   Dong, Y.; Geske, M.; Korup, O.; Ellenfeld, N.; Rosowski, F.; Dobner, C.; Horn, R., What happens in a catalytic fixed-bed reactor for n-butane oxidation to maleic anhydride? Insights from spatial profile measurements and particle resolved CFD simulations. *Chemical Engineering Journal* **2018**, *350*, 799–811. doi.org/10.1016/j.cej.2018.05.192

[64]   Laudenschleger, D.; Ruland, H.; Muhler, M., Identifying the nature of the active sites in methanol synthesis over Cu/ZnO/Al$_2$O$_3$ catalysts. *Nature Communications* **2020**, *11* (1), 3898. doi 10.1038/s41467-020-17631-5

[65]   Kartheuser, B.; Hodnett, B. K.; Zanthoff, H.; Baerns, M., Transient experiments on the selective oxidation of methane to formaldehyde over V_2O_5/SiO_2 studied in the temporal-analysis-of-products reactor. *Catalysis Letters* **1993**, *21* (3), 209–214. doi 10.1007/bf00769472

[66]   Buyevskaya, O. V.; Rothaemel, M.; Zanthoff, H. W.; Baerns, M., Transient studies on reaction steps in the oxidative coupling of methane over catalytic surfaces of MgO and Sm_2O_3. *Journal of Catalysis* **1994**, *146* (2), 346–357. doi 10.1006/jcat.1994.1073

[67]   Pérez-Ramírez, J.; Kondratenko, E. V.; Novell-Leruth, G.; Ricart, J. M., Mechanism of ammonia oxidation over PGM (Pt, Pd, Rh) wires by temporal analysis of products and density functional theory. *Journal of Catalysis* **2009**, *261* (2), 217–223. doi.org/10.1016/j.jcat.2008.11.018

[68]   Morgan, K.; Maguire, N.; Fushimi, R.; Gleaves, J. T.; Goguet, A.; Harold, M. P.; Kondratenko, E. V.; Menon, U.; Schuurman, Y.; Yablonsky, G. S., Forty years of temporal analysis of products. *Catalysis Science & Technology* **2017**, *7* (12), 2416–2439. doi 10.1039/c7cy00678k

[69]   Burch, R.; Shestov, A. A.; Sullivan, J. A., A transient kinetic study of the mechanism of the $NO+H_2$ reaction over Pt/SiO_2 catalysts: 1. Isotopic transient kinetics and temperature programmed analysis. *Journal of Catalysis* **1999**, *186* (2), 353–361. doi.org/10.1006/jcat.1999.2566

[70]   Ledesma, C.; Yang, J.; Chen, D.; Holmen, A., Recent Approaches in mechanistic and kinetic studies of catalytic reactions using SSITKA Technique. ACS *Catalysis* **2014**, *4* (12), 4527–4547. doi 10.1021/cs501264f

[71]   Kisch, H.; Bahnemann, D., Best practice in photocatalysis: Comparing rates or apparent quantum yields? *The Journal of Physical Chemistry Letters* **2015**, *6* (10), 1907–1910. doi 10.1021/acs.jpclett.5b00521

[72]   Qureshi, M.; Takanabe, K., Insights on measuring and reporting heterogeneous photocatalysis: Efficiency definitions and setup examples. *Chemistry of Materials* **2017**, *29* (1), 158–167 doi 10.1021/acs.chemmater.6b02907.

[73]   Minero, C.; Bedini, A.; Minella, M., On the standardization of the photocatalytic gas/solid tests. *International Journal of Chemical Reactor Engineering* **2013**, *11* (2), 717–732. doi:10.1515/ijcre-2012-0045

[74]   Rath, T.; Bloh, J. Z.; Lüken, A.; Ollegott, K.; Muhler, M., Model-based analysis of the photocatalytic HCl oxidation kinetics over TiO_2. *Industrial & Engineering Chemistry Research* **2020**, *59* (10), 4265–4272. doi 10.1021/acs.iecr.9b05820

[75]   Bloh, J. Z., A holistic approach to model the kinetics of photocatalytic reactions. *Frontiers in Chemistry* **2019**, *7* Art. No. 128. doi 10.3389/fchem.2019.00128.

[76]   Denuault, G.; Sosna, M.; Williams, K.-J., 11 - Classical Experiments. In *Handbook of Electrochemistry*, Zoski, C. G., Ed. Elsevier: Amsterdam, 2006; pp 431–469. ISBN: 9780444519580

[77]   Masa, J.; Zhao, A.; Xia, W.; Sun, Z.; Mei, B.; Muhler, M.; Schuhmann, W., Trace metal residues promote the activity of supposedly metal-free nitrogen-modified carbon catalysts for the oxygen reduction reaction. *Electrochemistry Communications* **2013**, *34* (0), 113–116. dx.doi.org/10.1016/j.elecom.2013.05.032

[78]   Stulik, K.; Amatore, C.; Holub, K.; Marecek, V.; Kutner, W., Microelectrodes. Definitions, characterization, and applications (Technical report). *Pure and Applied Chemistry* **2000**, *72* (8), 1483–1492. doi:10.1351/pac200072081483

[79]   Sandford, C.; Edwards, M. A.; Klunder, K. J.; Hickey, D. P.; Li, M.; Barman, K.; Sigman, M. S.; White, H. S.; Minteer, S. D., A synthetic chemist's guide to electroanalytical tools for studying reaction mechanisms. *Chemical Science* **2019**, *10* (26), 6404–6422. doi 10.1039/C9SC01545K

[80]   Argyle, M. D.; Bartholomew, C. H., Heterogeneous catalyst deactivation and regeneration: A review. *Catalysts* **2015**, *5* (1), 145–269. doi 10.3390/catal5010145

[81]   Thommes, M.; Kaneko, K.; Neimark, A. V.; Olivier James, P.; Rodriguez-Reinoso, F.; Rouquerol, J.; Sing Kenneth, S. W., Physisorption of gases, with special reference to the evaluation of surface area and pore size distribution (IUPAC Technical Report). *Pure and Applied Chemistry*, **2015**; 87, 1051. doi 10.1515/pac-2014-1117

[82]   Cychosz, K. A.; Guillet-Nicolas, R.; García-Martínez, J.; Thommes, M., Recent advances in the textural characterization of hierarchically structured nanoporous materials. *Chemical Society Reviews* **2017**, *46* (2), 389–414. doi 10.1039/c6cs00391e

[83]   Anderson, J. R.; Pratt, K. C., *Introduction to Characterization and Testing of Catalysts*. Academic Press: Syndey, 1985. ISBN: 9780120583201

[84]   Kleitz, F.; Schmidt, W.; Schüth, F., Calcination behavior of different surfactant-templated mesostructured silica materials. *Microporous and Mesoporous Materials* **2003**, *65* (1), 1–29. doi. org/10.1016/S1387-1811(03)00506-7

[85]   Wagner, M., *Thermal Analysis in Practice: Fundamental Aspects*. Hanser Publications: Cincinnati, 2017. ISBN: 978-1-56990-643-9

[86]   Hemminger, W. F.; Cammenga, H. K., *Methoden der Thermischen Analyse*. Springer: Berlin, Heidelberg, 2011. ISBN: 9783540150497

[87]   Xia, X.; Naumann d'Alnoncourt, R.; Strunk, J.; Litvinov, S.; Muhler, M., Coverage-dependent kinetics and thermodynamics of carbon monoxide adsorption on a ternary copper catalyst derived from static adsorption microcalorimetry. *The Journal of Physical Chemistry B* **2006**, *110* (16), 8409–8415. doi 10.1021/jp0609481

[88]   Auroux, A., Acidity characterization by microcalorimetry and relationship with reactivity. *Topics in Catalysis* **1997**, *4* (1), 71–89. doi 10.1023/A:1019127919907

[89]   Mekki-Berrada, A.; Auroux, A., Thermal methods. In *Characterisation of Solid Materials: From Structure to Surface Reactivity*, Che, M.; Vedrine, J. C., Eds. Wiley-VCH: Weinheim, 2012; Vol. 2, pp 747–852. doi. org/10.1002/9783527645329.ch18

[90]   Niemantsverdriet, J. W., *Spectroscopy in Catalysis*. 1st ed.; VCH: Weinheim, 1995. ISBN: 9783527287260

[91]   Friedel Ortega, K.; Arrigo, R.; Frank, B.; Schlögl, R.; Trunschke, A., Acid–base properties of N-doped carbon nanotubes: A combined temperature-programmed desorption, X-ray photoelectron spectroscopy, and 2-propanol reaction investigation. *Chemistry of Materials* **2016**, *28* (19), 6826–6839. doi 10.1021/acs.chemmater.6b01594

[92]   Wilmer, H.; Genger, T.; Hinrichsen, O., The interaction of hydrogen with alumina-supported copper catalysts: A temperature-programmed adsorption/temperature-programmed desorption/isotopic exchange reaction study. *Journal of Catalysis* **2003**, *215*, 188–198. doi: 10.1016/S0021-9517(03)00003-4

[93]   Schwidder, M.; Santhosh Kumar, M.; Bentrup, U.; Perez-Ramirez, J.; Brückner, A.; Grünert, W., The role of Brønsted acidity in the SCR of NO over Fe-MFI catalysts. *Microporous and Mesoporous Materials* **2008**, *111* (1), 124–133. doi 10.1016/j.micromeso.2007.07.019

[94]   Jeroro, E.; Lebarbier, V.; Datye, A.; Wang, Y.; Vohs, J. M., Interaction of CO with surface PdZn alloys. *Surface Science* **2007**, *601* (23), 5546–5554. doi.org/10.1016/j.susc.2007.09.031

[95]   de Jong, A. M.; Niemantsverdriet, J. W., Thermal desorption analysis: Comparative test of ten commonly applied procedures. *Surface Science* **1990**, *233* (3), 355–365. doi.org/10.1016/0039-6028(90)90649-S

[96]   Redhead, P. A., Thermal desorption of gases. *Vacuum* **1962**, *12* (4), 203–211. doi.org/10.1016/00 42-207X(62)90978-8

[97]   Genger, T.; Hinrichsen, O.; Muhler, M., The temperature-programmed desorption of hydrogen from copper surfaces. *Catalysis Letters* **1999**, *59* (2–4), 137–141. doi 10.1023/a:1019076722708

[98]   Peter, M.; Fendt, J.; Wilmer, H.; Hinrichsen, O., Modeling of temperature-programmed desorption (TPD) flow experiments from Cu/ZnO/Al$_2$O$_3$ catalysts. *Catalysis Letters* **2012**, *142* (5), 547–556. doi 10.1007/s10562-012-0807-3

[99]   Falconer, J. L.; Schwarz, J. A., Temperature-programmed desorption and reaction: Applications to supported catalysts. *Catalysis Reviews* **1983**, *25* (2), 141–227. doi 10.1080/01614948308079666

[100] Monti, D. A. M.; Baiker, A., Temperature-programmed reduction. Parametric sensitivity and estimation of kinetic parameters. *Journal of Catalysis* **1983**, *83*, 323–335. doi: 10.1016/0021-9517(83)90058-1

[101] Thomas, R.; van Oers, E. M.; de Beer, V. H. J.; Medema, J.; Moulijn, J. A., Characterization of γ-alumina-supported molybdenum oxide and tungsten oxide; reducibility of the oxidic state versus hydrodesulfurization activity of the sulfided state. *Journal of Catalysis* **1982**, *76*, 241–253. doi 10.1016/0021-9517(82)90255-X

[102] Tkachenko, O. P.; Klementiev, K. V.; van den Berg, M. W. E.; Gies, H.; Grünert, W., The reduction of copper in porous matrices. The role of electrostatic stabilisation. *Phys. Chem. Chem. Phys.* **2005**, *8*, 1539–1549. DOI: 10.1039/B514744A

[103] Tkachenko, O. P.; Klementiev, K. V.; van den Berg, M. W. E.; Koc, N.; Bandyopadhyay, M.; Birkner, A.; Wöll, C.; Gies, H.; Grünert, W., Reduction of copper in porous matrices. Stepwise and autocatalytic reduction routes. *The Journal of Physical Chemistry B* **2005**, *109* (44), 20979–20988. doi: 10.1021/jp054033i

[104] Besselmann, S.; Freitag, C.; Hinrichsen, O.; Muhler, M., Temperature-programmed reduction and oxidation experiments with V_2O_5/TiO_2 catalysts. *Physical Chemistry Chemical Physics* **2001**, *3* (21), 4633–4638. doi 10.1039/B105466J

[105] Arnoldy, P.; van den Heijkant, J. A. M.; de Bok, G. D.; Moulijn, J. A., Temperature-programmed sulfiding of MoO_3/Al_2O_3 catalysts. *Journal of Catalysis* **1985**, *92* (1), 35–55. doi.org/10.1016/0021-9517(85)90235-0

[106] Graham, A. P., The low energy dynamics of adsorbates on metal surfaces investigated with helium atom scattering. *Surface Science Reports* **2003**, *49* (4), 115–168. doi.org/10.1016/S0167-5729(03)00012-8

[107] Ertl, G.; Küppers, J., *Low Energy Electrons and Surface Chemistry.* 2nd ed.; VCH: Weinheim, 1985. ISBN: 978-0895730657

[108] Shull, C. G.; Wollan, E. O., X-ray, electron, and neutron diffraction. *Science* **1948**, *108* (2795), 69. doi 10.1126/science.108.2795.69

[109] Klug, H. P.; Alexander, L. E., *X-ray Diffraction Procedures.* John Wiley & Sons: New York-Sydney-Toronto, 1985.

[110] Kompio, P. G. W. A.; Brückner, A.; Hipler, F.; Manoylova, O.; Auer, G.; Mestl, G.; Grünert, W., V_2O_5-WO_3/TiO_2 catalysts under thermal stress: Responses of structure and catalytic behavior in the selective catalytic reduction of NO by NH_3. *Applied Catalysis B* **2017**, *217*, 365–377. doi: 10.1016/j.apcatb.2017.06.006

[111] Bekx-Schürmann, S.; Mangelsen, S.; Breuninger, P.; Antoni, H.; Schürmann, U.; Kienle, L.; Muhler, M.; Bensch, W.; Grünert, W., Morphology, microstructure, coordinative unsaturation, and hydrogenation activity of unsupported MoS_2: How idealized models fail to describe a real sulfide material. *Applied Catalysis B: Environmental* **2020**, *266*, 118623. doi.org/10.1016/j.apcatb.2020.118623

[112] Behrens, M.; Schlögl, R., X-ray diffraction and small angle X-ray scattering. In *Characterisation of Solid Materials: From Structure to Surface Reactivity*, Che, M.; Vedrine, J. C., Eds. Wiley-VCH: Weinheim, 2012; Vol. 1, pp 537–583. doi.org/10.1002/9783527645329.ch15

[113] Jobic, H., Neutron scattering. In *Characterisation of Solid Materials: From Structure to Surface Reactivity*, Che, M.; Vedrine, J. C., Eds. Wiley-VCH: Weinheim, 2012; Vol. 1, pp 105–209. doi.org/10.1002/9783527645329.ch5.

[114] Heinz, K., LEED and DLEED as modern tools for quantitative surface structure determination. *Reports on Progress in Physics* **1995**, *58* (6), 609–653. doi 10.1088/0034-4885/58/6/003

[115] Krumrey, M.; Garcia-Diez, R.; Gollwitzer, C.; Langner, S., Größenbestimmung von Nanopartikeln mit Röntgenkleinwinkelstreuung. *PTB-Mitteilungen* **2014**, *124* (4), 13–16.

[116] Petkov, V., Nanostructure by high-energy X-ray diffraction. *Materials Today* **2008**, *11* (11), 28–38. doi.org/10.1016/S1369-7021(08)70236-0

[117] Grünert, W.; Klementiev, K., X-ray Absorption spectroscopy - Principles and practical use in materials analysis. *Physical Sciences Reviews* **2020**, *5*, Art. No 20170181. doi 10.1515/psr-2017-0181.

[118]   Rehr, J. J.; Albers, R. C., Theoretical approaches to X-ray absorption fine structure. *Reviews of Modern Physics* **2000**, *72* (3), 621–654. doi 10.1103/RevModPhys.72.621

[119]   Hähner, G., Near edge X-ray absorption fine structure spectroscopy as a tool to probe electronic and structural properties of thin organic films and liquids. *Chemical Society Reviews* **2006**, *35* (12), 1244–1255. doi 10.1039/B509853J

[120]   Grunwaldt, J. D.; Baiker, A., Axial variation of the oxidation state of Pt-Rh/Al$_2$O$_3$ during partial methane oxidation in a fixed-bed reactor: An in situ X-ray absorption spectroscopy study. *Catalysis Letters* **2005**, *99* (1), 5–12. doi 10.1007/s10562-005-0770-3

[121]   Martini, A.; Alladio, E.; Borfecchia, E., Determining Cu-Speciation in the Cu-CHA Zeolite Catalyst: The Potential of Multivariate Curve Resolution Analysis of In Situ XAS Data. *Topics in Catalysis* **2018**, *61* (14), 1396–1407. doi 10.1007/s11244-018-1036-9

[122]   Reichinger, M.; Schmidt, W.; van den Berg, M. W. E.; Aerts, A.; Martens, J. A.; Kirschhock, C. E. A.; Gies, H.; Grünert, W., Alkene epoxidation with mesoporous materials assembled from TS-1 seeds – Is there a hierarchical pore system ? *Journal of Catalysis* **2010**, *269*, 367–375. doi 10.1016/j. jcat.2009.11.023

[123]   Farges, F.; Brown, G. E.; Rehr, J. J., Ti K-edge XANES studies of Ti coordination and disorder in oxide compounds: Comparison between theory and experiment. *Physical Review B* **1997**, *56* (4), 1809–1819. doi: 10.1103/PhysRevB.56.1809

[124]   Grünert, W., Auger Electron, X-ray and UV photoelectron spectroscopy. In *Characterisation of Solid Materials: From Structure to Surface Reactivity*, Che, M.; Vedrine, J. C., Eds. Wiley-VCH: Weinheim, 2012; Vol. 1, pp 537–583. doi.org/10.1002/9783527645329.ch13

[125]   Salmeron, M.; Schlögl, R., Ambient pressure photoelectron spectroscopy: A new tool for surface science and nanotechnology. *Surface Science Reports* **2008**, *63*, 169–199. doi 10.1016/j. surfrep.2008.01.001

[126]   Grünert, W.; Brückner, A.; Hofmeister, H.; Claus, P., Structural properties of Ag/TiO$_2$ catalysts for acrolein hydrogenation. *The Journal of Physical Chemistry B* **2004**, *108* (18), 5709–5717. doi 10.1021/ jp049855e

[127]   Grünert, W.; Muhler, M.; Schroeder, K.-P.; Sauer, J.; Schlögl, R., Investigations of zeolites by photoelectron and ion scattering spectroscopy. 2. A new interpretation of XPS binding energy shifts in zeolites. *The Journal of Physical Chemistry* **1994**, *98* (42), 10920–10929. doi: 10.1021/ j100093a039

[128]   Mullins, D. R.; Overbury, S. H.; Huntley, D. R., Electron spectroscopy of single crystal and polycrystalline cerium oxide surfaces. *Surface Science* **1998**, *409* (2), 307–319. doi.org/10.1016/ S0039-6028(98)00257-X

[129]   Okada, K.; Kotani, A., Interatomic and intra-atomic configuration interactions in core-level X-ray photoemission spectra of late transition-metal compounds. *Journal of the Physical Society of Japan* **1992**, *61* (12), 4619–4637. 10.1143/jpsj.61.4619

[130]   Elmasides, C.; Kontarides, D. I.; Grünert, W.; Verykios, X. E., XPS and FTIR study of Ru/Al$_2$O$_3$ and Ru/ TiO$_2$ catalysts: Reduction characteristics and interaction with a methane-oxygen mixture. *The Journal of Physical Chemistry B* **1999**, *103* (2), 5227–5239. doi 10.1021/jp9842291

[131]   Biesinger, M. C.; Payne, B. P.; Grosvenor, A. P.; Lau, L. W. M.; Gerson, A. R.; Smart, R. S., Resolving surface chemical states in XPS analysis of first row transition metals, oxides and hydroxides: Cr, Mn, Fe, Co and Ni. *Applied Surface Science* **2011**, *257* (7), 2717–2730. doi 10.1016/j.apsusc.2010.10.051

[132]   Tanuma, S., Electron attentuation lengths. In *Surface Analysis by Auger and X-Ray Photoelectron Spectroscopy*, Briggs, D.; Grant, J. T., Eds. IM Publications and Surface Spectra Ltd.: Chichester, 2003; pp 259–294. ISBN: 978-1-901019-04-9

[133]   Biesinger, M. C., Advanced analysis of copper X-ray photoelectron spectra. Surface and Interface Analysis 2017, *49* (13), 1325–1334. doi 10.1002/sia.6239

[134] Wagner, C. D.; Gale, L. H.; Raymond, R. H., Two-dimensional chemical state plots: A standardized data set for use in identifying chemical states by x-ray photoelectron spectroscopy. *Analytical Chemistry* **1979**, *51* (4), 466–482. doi 10.1021/ac50040a005

[135] Brongersma, H. H.; Draxler, M.; de Ridder, M.; Bauer, P., Surface composition analysis by low-energy ion scattering. *Surface Science Reports* **2007**, *62* (3), 63–109. doi.org/10.1016/j.surfrep.2006.12.002

[136] Kruse, N.; Chenakin, S., Low energy ion scattering and secondary ion mass spectrometry. In *Characterisation of Solid Materials: From Structure to Surface Reactivity*, Che, M.; Vedrine, J. C., Eds. Wiley-VCH: Weinheim, 2012; Vol. 1, pp 453–510. doi.org/10.1002/9783527645329.ch11

[137] Baer, D. R.; Engelhard, M. H.; Lea, A. S.; Nachimuthu, P.; Droubay, T. C.; Kim, J.; Lee, B.; Mathews, C.; Opila, R. L.; Saraf, L. V.; Stickle, W. F.; Wallace, R. M.; Wright, B. S., Comparison of the sputter rates of oxide films relative to the sputter rate of SiO_2. *Journal of Vacuum Science & Technology A* **2010**, *28* (5), 1060–1072. doi 10.1116/1.3456123

[138] Niehus, H.; Heiland, W.; Taglauer, E., Low-energy ion scattering at surfaces. *Surface Science Reports* **1993**, *17* (4), 213–303. doi.org/10.1016/0167-5729(93)90024-J

[139] Brongersma, H. H.; Jacobs, J.-P., Application of low-energy ion scattering to studies of growth. *Applied Surface Science* **1994**, *75* (1), 133–138. doi.org/10.1016/0169-4332(94)90149-X

[140] Briand, L. E.; Tkachenko, O. P.; Guraya, M.; Gao, X.; Wachs, I. E.; Grünert, W., Surface-analytical studies of supported vanadium oxide monolayer catalysts. *The Journal of Physical Chemistry B* **2004**, *108*, 4823–4830. doi 10.1021/jp037675j

[141] Merzlikin, S. V.; Tolkachev, N. N.; Briand, L. E.; Strunskus, T.; Wöll, C.; Wachs, I. E.; Grünert, W., Anomalous surface compositions of stoichiometric mixed oxide compounds. *Angewandte Chemie International Edition* **2010**, *49* (43), 8037–8041. doi 10.1002/anie.201001804

[142] Petrova, O.; Kulp, C.; Pohl, M.-M.; ter Veen, R.; Veith, L.; Grehl, T.; van den Berg, M. W. E.; Brongersma, H.; Bron, M.; Grünert, W., Chemical leaching of Pt–Cu/C catalysts for electrochemical oxygen reduction: Activity, particle structure, and relation to electrochemical leaching. *ChemElectroChem* **2016**, *3* (11), 1768–1780. doi 10.1002/celc.201600468

[143] Hofmann, S., Quantitative depth profiling in surface analysis: A review. *Surface and Interface Analysis* **1980**, *2* (4), 148–160. doi 10.1002/sia.740020406

[144] Hofmann, S., Sputter depth profiling: Past, present, and future. *Surface and Interface Analysis* **2014**, *46* (10–11), 654–662. doi 10.1002/sia.5489

[145] Wang, Y. Q.; Nastasi, M., *Handbook of Modern Ion Beam Materials Analysis*. Materials Research Society: Warrendale, Pennsylvania, 2009; Vol. 1. ISBN: 978-1605112169

[146] Perkampus, H. H., *UV-vis Atlas of Organic Compounds*. 2nd ed.; VCH: Weinheim, 1992. ISBN: 9783527285105

[147] Cotton, F. A.; Wilkinson, G.; Murillo, C. A.; Bochmann, M., *Advanced Inorganic Chemistry*. 6th ed.; Wiley-Interscience: New York, 1999. ISBN: 978-0-471-19957-1

[148] Figgis, B. N.; Hitchman, M. A., *Ligand Field Theory and Its Applications*. Wiley-VCH: New York, 1999. ISBN: 978-0-471-31776-0

[149] Jentoft, F., Electronic spectroscopy: Ultra violet-visible and near-IR spectroscopies. In *Characterisation of Solid Materials: From Structure to Surface Reactivity*, Che, M.; Vedrine, J. C., Eds. Wiley-VCH: Weinheim, 2012; Vol. 1, pp 89–148. https://doi.org/10.1002/9783527645329.ch3

[150] Kelly, K. L.; Coronado, E.; Zhao, L. L.; Schatz, G. C., The optical properties of metal nanoparticles: The influence of size, shape, and dielectric environment. *The Journal of Physical Chemistry B* **2003**, *107* (3), 668–677. doin 10.1021/jp026731y

[151] Jentoft, F. C., Ultraviolet-visible-near infrared spectroscopy in catalysis: Theory, experiment, analysis, and application under reaction conditions. *Advances in Catalysis* **2009**, 52, 129–211. doi 10.1016/s0360-0564(08)00003-5

[152] Nijhuis, T. A.; Tinnemans, S. J.; Visser, T.; Weckhuysen, B. M., Towards real-time spectroscopic process control for the dehydrogenation of propane over supported chromium oxide catalysts. *Chemical Engineering Science* **2004**, *59* (22), 5487–5492. doi.org/10.1016/j.ces.2004.07.103

[153] Grandjean, D.; Beale, A. M.; Petukhov, A. V.; Weckhuysen, B. M., Unraveling the crystallization mechanism of CoAPO-5 molecular sieves under hydrothermal conditions. *Journal of the American Chemical Society* **2005**, *127* (41), 14454–14465. doi 10.1021/ja054014m

[154] Brückner, A.; Kondratenko, E., Simultaneous operando EPR/UV-vis/laser-Raman spectroscopy: A powerful tool for monitoring transition metal oxide catalysts during reaction. *Catalysis Today* **2006**, *113* (1), 16–24. doi.org/10.1016/j.cattod.2005.11.006

[155] Bentrup, U., Combining in situ characterization methods in one set-up: Looking with more eyes into the intricate chemistry of the synthesis and working of heterogeneous catalysts. *Chemical Society Reviews* **2010**, *39* (12), 4718–4730. doi 10.1039/b919711g

[156] Gao, Q.; Weckhuysen, B. M.; Schoonheydt, R. A., On the synthesis of CoAPO–46, –11 and –44 molecular sieves from a $Co(Ac)_2 \cdot 4H_2O \cdot Al(iPrO)_3 \cdot H_3PO_4 \cdot Pr_2NH \cdot H_2O$ gel via experimental design. *Microporous and Mesoporous Materials* **1999**, *27* (1), 75–86. doi.org/10.1016/S1387-1811(98)00274-1

[157] Santhosh Kumar, M.; Schwidder, M.; Grünert, W.; Brückner, A., On the nature of different iron sites and their catalytic role in Fe-ZSM-5 DeNOx catalysts: New insights by a combined EPR and UV/VIS spectroscopic approach. *Journal of Catalysis* **2004**, *227* (2), 384–397. doi 10.1016/j.jcat.2004.08.003

[158] Zanatta, A. R., Revisiting the optical bandgap of semiconductors and the proposal of a unified methodology to its determination. *Scientific Reports* **2019**, *9* (1), 11225. doi 10.1038/s41598-019-47670-y

[159] Viezbicke, B. D.; Patel, S.; Davis, B. E.; Birnie III, D. P., Evaluation of the Tauc method for optical absorption edge determination: ZnO thin films as a model system. *Physica Status Solidi (B)* **2015**, *252* (8), 1700–1710. doi 10.1002/pssb.201552007

[160] Pesika, N. S.; Stebe, K. J.; Searson, P. C., Relationship between absorbance spectra and particle size distributions for quantum-sized nanocrystals. *The Journal of Physical Chemistry B* **2003**, *107* (38), 10412–10415. doi 10.1021/jp0303218

[161] Brus, L., Electronic wave functions in semiconductor clusters: Experiment and theory. *The Journal of Physical Chemistry* **1986**, *90* (12), 2555–2560. doi 10.1021/j100403a003

[162] Atkins, P.; de Paula, J., *Atkins' Physical Chemistry*. 8th ed.; Oxford University Press: New York, 2006. ISBN: 978-0198700722

[163] Günzler, H.; Gremlich, H., *IR-Spektroskopie: Eine Einführung*. 4th ed.; Wiley-VCH: Weinheim, 2003. ISBN: 9783527308019

[164] Stuart, B. H., Infrared spectroscopy: Fundamentals and applications. John Wiley and Sons: Chichester-Hoboken, 2004. ISBN: 9780470854273

[165] Soriaga, M. P.; Chen, X. L.; Li, D.; Stickney, J. L., High resolution electron energy-loss spectroscopy. In *Application of Physical Methods to Inorganic and Bioinorganic Chemistry*, Scott, R. A.; Lukehart, C. M., Eds. Wiley Interscience: 2005. ISBN: 978-0-470-03217-6

[166] Vidal, F.; Tadjeddine, A., Sum-frequency generation spectroscopy of interfaces. *Reports on Progress in Physics* **2005**, *68* (5), 1095–1127. doi 10.1088/0034-4885/68/5/r03

[167] Andanson, J. M.; Baiker, A., Exploring catalytic solid/liquid interfaces by in situ attenuated total reflection infrared spectroscopy. *Chemical Society Reviews* **2010**, *39* (12), 4571–4584. doi 10.1039/b919544k

[168] Zaera, F., New advances in the use of infrared absorption spectroscopy for the characterization of heterogeneous catalytic reactions. *Chemical Society Reviews* **2014**, *43* (22), 7624–7663. doi 10.1039/c3cs60374a

[169] Almeida, A. R.; Moulijn, J. A.; Mul, G., In situ ATR-FTIR study on the selective photo-oxidation of cyclohexane over anatase TiO_2. *The Journal of Physical Chemistry C* **2008**, *112* (5), 1552–1561. doi 10.1021/jp077143t

[170] Wolter, K.; Seiferth, O.; Libuda, J.; Kuhlenbeck, H.; Bäumer, M.; Freund, H. J., Infrared study of CO adsorption on alumina supported palladium particles. *Surface Science* **1998**, *402–404*, 428–432. doi. org/10.1016/S0039-6028(97)01053-4

[171] Thibault-Starzyk, F.; Maugé, F., Infrared spectroscopy. In *Characterisation of Solid Materials: From Structure to Surface Reactivity*, Che, M.; Vedrine, J. C., Eds. Wiley-VCH: Weinheim, 2012; Vol. 1, pp 3–48. https://doi.org/10.1002/9783527645329.ch1

[172] Knözinger, H.; Ratnasamy, P., Catalytic aluminas - Surface models and characterization of surface sites. *Catalysis Reviews-Science and Engineering* **1978**, *17* (1), 31–70. doi 10.1080/03602457808080878

[173] Digne, M.; Sautet, P.; Raybaud, P.; Euzen, P.; Toulhoat, H., Hydroxyl groups on γ-alumina surfaces: A DFT study. *Journal of Catalysis* **2002**, *211* (1), 1–5. doi.org/10.1006/jcat.2002.3741

[174] Lavalley, J. C., Infrared spectrometric studies of the surface basicity of metal oxides and zeolites using adsorbed probe molecules. *Catalysis Today* **1996**, *27* (3), 377–401. doi.org/10.1016/0920-5861(95)00161-1

[175] Topsøe, N. Y.; Topsøe, H.; Dumesic, J. A., Vanadia/titania catalysts for selective catalytic reduction (SCR) of nitric-oxide by ammonia: I. Combined temperature-programmed in-situ FTIR and on-line mass-spectroscopy studies. *Journal of Catalysis* **1995**, *151* (1), 226–240. doi.org/10.1006/jcat.1995.1024

[176] Topsøe, N.-Y.; Dumesic, J. A.; Topsøe, H., Vanadia-titania catalysts for selective catalytic reduction of nitric-oxide by ammonia. 2. Studies of active-sites and formulation of catalytic cycles. *Journal of Catalysis* **1995**, *151*, 241–252. doi: 10.1006/jcat.1995.1025

[177] Zheng, S. R.; Jentys, A.; Lercher, J. A., Xylene isomerization with surface-modified HZSM-5 zeolite catalysts: An in situ IR study. *Journal of Catalysis* **2006**, *241* (2), 304–311. doi 10.1016/j. jcat.2006.04.026

[178] Kähler, K.; Holz, M. C.; Rohe, M.; van Veen, A. C.; Muhler, M., Methanol oxidation as probe reaction for active sites in Au/ZnO and Au/TiO₂ catalysts. *Journal of Catalysis* **2013**, *299*, 162–170. doi 10.1016/j. jcat.2012.12.001

[179] Haruta, M., Catalysis of gold nanoparticles deposited on metal oxides. *CATTECH* **2004**, *6* (3), 102–115. doi 10.1023/A:1020181423055

[180] Ulrich, V.; Moroz, B.; Sinev, I.; Pyriaev, P.; Bukhtiyarov, V.; Grünert, W., Studies on three-way catalysis with supported gold catalysts. Influence of support and water content in feed. *Applied Catalysis B: Environmental* **2017**, *203*, 572–581. doi 10.1016/j.apcatb.2016.10.017

[181] Ulrich, V.; Froese, C.; Moroz, B.; Pyrjaev, P.; Gerasimov, E.; Sinev, I.; Roldan Cuenya, B.; Muhler, M.; Bukhtiyarov, V.; Grünert, W., Three-way catalysis with supported gold catalysts: Poisoning effects of hydrocarbons. *Applied Catalysis B: Environmental* **2018**, *237*, 1021–1032. doi.org/10.1016/j. apcatb.2018.06.063

[182] Makarova, M. A.; Ojo, A. F.; Karim, K.; Hunger, M.; Dwyer, J., FTIR study of weak hydrogen bonding of Brønsted hydroxyls in zeolites and aluminophosphates. *The Journal of Physical Chemistry* **1994**, *98* (14), 3619–3623. doi 10.1021/j100065a013

[183] Kustov, L. M.; Kazanskii, V. B.; Beran, S.; Kubelkova, L.; Jiru, P., Adsorption of carbon monoxide on ZSM-5 zeolites: Infrared spectroscopic study and quantum-chemical calculations. *The Journal of Physical Chemistry* **1987**, *91* (20), 5247–5251. doi 10.1021/j100304a023

[184] Lercher, J. A.; Gründling, C.; Eder-Mirth, G., Infrared studies of the surface acidity of oxides and zeolites using adsorbed probe molecules. *Catalysis Today* **1996**, *27* (3), 353–376. doi. org/10.1016/0920-5861(95)00248-0

[185] Knözinger, H.; Huber, S., IR spectroscopy of small and weakly interacting molecular probes for acidic and basic zeolites. *Journal of the Chemical Society, Faraday Transactions* **1998**, *94* (15), 2047–2059. doi 10.1039/a802189i

[186] Zaki, M. I.; Vielhaber, B.; Knözinger, H., Low-temperature CO adsorption and state of molybdena supported on alumina, titania and zirconia. An infrared spectroscopic investigation. *The Journal of Physical Chemistry* **1986**, *90*, 3176–3183. doi: 10.1021/j100405a026

[187] Topsøe, N.-Y.; Topsøe, H., Characterization of the structures and active sites in sulfided Co-Mo/ Al$_2$O$_3$ and Ni-Mo/Al$_2$O$_3$ catalysts by NO chemisorption. *Journal of Catalysis* **1983**, *84* (2), 386–401. doi: 10.1016/0021-9517(83)90010-6

[188] Ebbesen, S. D.; Mojet, B. L.; Lefferts, L., In situ ATR-IR study of CO adsorption and oxidation over Pt/ Al$_2$O$_3$ in gas and aqueous phase: Promotion effects by water and pH. *Journal of Catalysis* **2007**, *246* (1), 66–73. doi.org/10.1016/j.jcat.2006.11.019

[189] Kogler, M.; Heilala, B., Time-gated Raman spectroscopy - A review. *Measurement Science and Technology* **2021**, *32* (1), Art. No. 012002, doi 10.1088/1361-6501/abb044.

[190] Mestl, G., In situ Raman spectroscopy - A valuable tool to understand operating catalysts. *Journal of Molecular Catalysis A: Chemical* **2000**, *158* (1), 45–65. doi.org/10.1016/S1381-1169(00)00042-X

[191] Fan, F. T.; Feng, Z. C.; Li, C., Raman and UV Raman spectroscopies. In *Characterisation of Solid Materials: From Structure to Surface Reactivity*, Che, M.; Vedrine, J. C., Eds. Wiley-VCH: Weinheim, 2012; Vol. 1, pp 49–87. doi.org/10.1002/9783527645329.ch2

[192] Jones, R. R.; Hooper, D. C.; Zhang, L. W.; Wolverson, D.; Valev, V. K., Raman techniques: Fundamentals and frontiers. *Nanoscale Research Letters* **2019**, *14*, Art No. 231. doi 10.1186/s11671-019-3039-2

[193] Motz, J. T.; Hunter, M.; Galindo, L. H.; Gardecki, J. A.; Kramer, J. R.; Dasari, R. R.; Feld, M. S., Optical fiber probe for biomedical Raman spectroscopy. *Applied Optics* **2004**, *43* (3), 542–554. doi 10.1364/ ao.43.000542

[194] Wachs, I. E., Raman and IR studies of surface metal oxide species on oxide supports: Supported metal oxide catalysts. *Catalysis Today* **1996**, *27* (3–4), 437–455. doi 10.1016/0920-5861(95)00203-0

[195] Banares, M. A.; Wachs, I. E., Molecular structures of supported metal oxide catalysts under different environments. *Journal of Raman Spectroscopy* **2002**, *33*, 359–380. doi 10.1002/jrs.866

[196] Kundu, S.; Nagaiah, T. C.; Xia, W.; Wang, Y.; Dommele, S. V.; Bitter, J. H.; Santa, M.; Grundmeier, G.; Bron, M.; Schuhmann, W.; Muhler, M., Electrocatalytic activity and stability of nitrogen-containing carbon nanotubes in the oxygen reduction reaction. *The Journal of Physical Chemistry C* **2009**, *113* (32), 14302–14310. doi 10.1021/jp811320d

[197] Sudarsanam, P.; Mallesham, B.; Reddy, P. S.; Großmann, D.; Grünert, W.; Reddy, B. M., Nano-Au/CeO$_2$ catalysts for CO oxidation: Influence of dopants (Fe, La and Zr) on the physicochemical properties and catalytic activity. *Applied Catalysis B: Environmental* **2014**, *144*, 900–908. doi.org/10.1016/j. apcatb.2013.08.035

[198] Pietrzyk, P.; Sojka, Z.; Giamello, E., Electron paramagnetic resonance. In *Characterisation of Solid Materials: From Structure to Surface Reactivity*, Che, M.; Vedrine, J. C., Eds. Wiley-VCH: Weinheim, 2012; Vol. 1, pp 343–406. doi.org/10.1002/9783527645329.ch9

[199] Kärger, J.; Ruthven, D. M.; Theodorou, D. N., *Diffusion in Nanoporous Materials*. Wiley-VCH: Weinheim, 2012. ISBN: 978-3-527-31024-1

[200] Brückner, A., In situ electron paramagnetic resonance: A unique tool for analyzing structure-reactivity relationships in heterogeneous catalysis. *Chemical Society Reviews* **2010**, *39* (12), 4673–4684. doi 10.1039/b919541f

[201] Pérez Vélez, R.; Ellmers, I.; Huang, H.; Bentrup, U.; Schünemann, V.; Grünert, W.; Brückner, A., Identifying active sites for fast NH$_3$-SCR of NO/NO$_2$ mixtures over Fe-ZSM-5 by operando EPR and UV-vis spectroscopy. *Journal of Catalysis* **2014**, *316*, 103–111. doi 10.1016/j.jcat.2014.05.001

[202] Hunger, M.; Brunner, E., NMR spectroscopy. In *Molecular Sieves - Science and Technology*, Karge, H. H.; Weitkamp, J., Eds. Springer: Berlin, 2004; Vol. 4, pp 201–294. doi 10.1007/b94236

[203] Gladden, L. F.; Lutecki, M.; McGregor, J., Nuclear Magnetic resonance spectroscopy. In *Characterisation of Solid Materials: From Structure to Surface Reactivity*, Che, M.; Vedrine, J. C., Eds. Wiley-VCH: Weinheim, 2012; Vol. 1, pp 289–342. doi.org/10.1002/9783527645329.ch8

[204] Hunger, M.; Wang, W., Formation of cyclic compounds and carbenium ions by conversion of methanol on weakly dealuminated zeolite H-ZSM-5 investigated via a novel in situ CF MAS NMR/ UV-Vis technique. *Chemical Communications* **2004**, (5), 584–585. doi 10.1039/b315779b

[205] Engelhardt, G., Chapter 9: Solid state NMR spectroscopy applied to zeolites. In *Studies in Surface Science and Catalysis*, van Bekkum, H.; Flanigen, E. M.; Jacobs, P. A.; Jansen, J. C., Eds. Elsevier: 2001; Vol. 137, pp 387–418. doi.org/10.1016/S0167-2991(01)80251-2

[206] Stepanov, A. G.; Luzgin, M. V.; Romannikov, V. N.; Sidelnikov, V. N.; Paukshtis, E. A., The nature, structure, and composition of adsorbed hydrocarbon products of ambient temperature oligomerization of ethylene on acidic zeolite H-ZSM-5. *Journal of Catalysis* **1998**, *178* (2), 466–477. doi. org/10.1006/jcat.1998.2172

[207] Hunger, M., Brønsted acid sites in zeolites characterized by multinuclear solid-state NMR spectroscopy. *Catalysis Reviews – Science and Engineering* **1997**, *39* (4), 345–393. doi. org/10.1080/01614949708007100

[208] Hunger, M., Solid-state NMR spectroscopy. In *Zeolite Characterization and Catalysis*, Chester, A. W.; Derouane, E. G., Eds. Springer: Dordrecht: 2009; pp 65–106. doi 10.1007/978-1-4020-9678-5_2

[209] Springuel-Huet, M. A.; Bonardet, J. L.; Gédéon, A.; Fraissard, J., ^{129}Xe NMR for studying surface heterogeneity: Well-known facts and new findings. *Langmuir* **1997**, *13* (5), 1229–1236. doi 10.1021/la951566h

[210] Stievano, L.; Wagner, F. E., Mössbauer spectroscopy. In *Characterisation of Solid Materials: From Structure to Surface Reactivity*, Che, M.; Vedrine, J. C., Eds. Wiley-VCH: Weinheim, 2012; Vol. 1, pp 407–452. doi.org/10.1002/9783527645329.ch10

[211] Bouwens, S.; van Zon, F. B. M.; van Dijk, M. P.; van der Kraan, A. M.; De Beer, V. H. J.; van Veen, J. A. R.; Koningsberger, D. C., On the structural differences between alumina-supported CoMoS type-I and alumina-supported, silica-supported, and carbon-supported CoMoS type-II phases studied by XAFS, MES, and XPS. *Journal of Catalysis* **1994**, *146* (2), 375–393. doi 10.1006/jcat.1994.1076

[212] Bezemer, G. L.; Remans, T. J.; van Bavel, A. P.; Dugulan, A. I., Direct evidence of water-assisted sintering of cobalt on carbon nanofiber catalysts during simulated Fischer–Tropsch conditions revealed with in situ Mössbauer spectroscopy. *Journal of the American Chemical Society* **2010**, *132* (25), 8540–8541. Doi 10.1021/ja103002k

[213] Potapkin, V.; Chumakov, A. I.; Smirnov, G. V.; Celse, J.-P.; Ruffer, R.; McCammon, C.; Dubrovinsky, L., The ^{57}Fe synchrotron Mossbauer source at the ESRF. *Journal of Synchrotron Radiation* **2012**, *19* (4), 559–569. doi:10.1107/S0909049512015579

[214] Schünemann, V.; Winkler, H., Structure and dynamics of biomolecules studied by Mössbauer spectroscopy. *Reports on Progress in Physics* **2000**, *63* (3), 263–353. doi 10.1088/0034-4885/63/3/202

[215] Grünert, W.; Ganesha, P. K.; Ellmers, I.; Velez, R. P.; Huang, H. M.; Bentrup, U.; Schünemann, V.; Brückner, A., Active sites of the selective catalytic reduction of NO by NH$_3$ over Fe-ZSM-5: Combining reaction kinetics with postcatalytic Mössbauer spectroscopy at cryogenic temperatures. *ACS Catalysis* **2020**, *10* (5), 3119–3130. doi 10.1021/acscatal.9b04627

[216] Li, X.; Zhu, K.; Pang, J.; Tian, M.; Liu, J.; Rykov, A. I.; Zheng, M.; Wang, X.; Zhu, X.; Huang, Y.; Liu, B.; Wang, J.; Yang, W.; Zhang, T., Unique role of Mössbauer spectroscopy in assessing structural features of heterogeneous catalysts. *Applied Catalysis B: Environmental* **2018**, *224*, 518–532. doi.org/10.1016/j.apcatb.2017.11.004

[217] Masliuk, L.; Heggen, M.; Noack, J.; Girgsdies, F.; Trunschke, A.; Hermann, K. E.; Willinger, M. G.; Schlögl, R.; Lunkenbein, T., Structural complexity in heterogeneous catalysis: Cataloging local nanostructures. *The Journal of Physical Chemistry C* **2017**, *121* (43), 24093–24103. doi 10.1021/acs. jpcc.7b08333

[218] Wang, D.; Penner, S.; Su, D. S.; Rupprechter, G.; Hayek, K.; Schlögl, R., Silicide formation on a Pt/SiO$_2$ model catalyst studied by TEM, EELS, and EDXS. *Journal of Catalysis* **2003**, *219* (2), 434–441. doi.org/10.1016/S0021-9517(03)00219-7

[219] Ohsuna, T.; Slater, B.; Gao, F.; Yu, J.; Sakamoto, Y.; Zhu, G.; Terasaki, O.; Vaughan, D. E. W.; Qiu, S.; Catlow, C. R. A., Fine structures of Zeolite-Linde-L (LTL): Surface structures, growth unit and defects. *Chemistry - A European Journal* **2004**, *10* (20), 5031–5040. doi.org/10.1002/chem.200306064

[220] Hansen, P. L.; Wagner, J. B.; Helveg, S.; Rostrup-Nielsen, J. R.; Clausen, B. S.; Topsøe, H., Atom-resolved imaging of dynamic shape changes in supported copper nanocrystals. *Science* **2002**, *295*, 2053–2055. doi 10.1126/science.106932

[221] Song, B.; Yang, T. T.; Yuan, Y.; Sharifi-Asl, S.; Cheng, M.; Saidi, W. A.; Liu, Y.; Shahbazian-Yassar, R., Revealing sintering kinetics of MoS$_2$-supported metal nanocatalysts in atmospheric gas environments via operando transmission electron microscopy. *ACS Nano* **2020**, *14* (4), 4074–4086. doi 10.1021/acsnano.9b08757

[222] Boniface, M.; Plodinec, M.; Schlögl, R.; Lunkenbein, T., Quo Vadis micro-electro-mechanical systems for the study of heterogeneous catalysts inside the electron microscope? *Topics in Catalysis* **2020**, *63* (15), 1623–1643. doi 10.1007/s11244-020-01398-6

[223] Grünert, W.; Großmann, D.; Noei, H.; Pohl, M. M.; Sinev, I.; De Toni, A.; Wang, Y.; Muhler, M., Low-temperature CO oxidation with TiO$_2$-supported Au^{3+} ions. *Angewandte Chemie International Edition* **2014**, *53*, 3245–3249. doi 10.1002/anie.201308206

[224] Topsøe, N.-Y.; Tuxen, A.; Hinnemann, B.; Lauritsen, J. V.; Knudsen, K. G.; Besenbacher, F.; Topsøe, H., Spectroscopy, microscopy and theoretical study of NO adsorption on MoS$_2$ and Co-Mo-S hydrotreating catalysts. *Journal of Catalysis* **2011**, *279* (2), 337–351. doi 10.1016/j.jcat.2011.02.002

[225] Cubillas, P.; Stevens, S. M.; Blake, N.; Umemura, A.; Chong, C. B.; Terasaki, O.; Anderson, M. W., AFM and HRSEM invesitigation of zeolite A crystal growth. Part 1: In the absence of organic additives. *The Journal of Physical Chemistry C* **2011**, *115* (25), 12567–12574. doi 10.1021/jp2032862

[226] Bard, A. J.; Faulkner, L. R.; White, H. S., *Electrochemical Methods: Fundamentals and Applications*. 3rd ed.; Wiley-Blackwell: Hoboken, New Jersey, 2022. ISBN: 978-1-119-33406-4

[227] Hiltrop, D.; Masa, J.; Maljusch, A.; Xia, W.; Schuhmann, W.; Muhler, M., Pd deposited on functionalized carbon nanotubes for the electrooxidation of ethanol in alkaline media. *Electrochemistry Communications* **2016**, *63*, 30–33. doi.org/10.1016/j.elecom.2015.11.010

[228] Binninger, T.; Fabbri, E.; Kötz, R.; Schmidt, T. J., Determination of the electrochemically active surface area of metal-oxide supported platinum catalyst. *Journal of the Electrochemical Society* **2013**, *161* (3), H121-H128. doi.org/10.1016/j.elecom.2015.11.010

[229] Rudi, S.; Cui, C.; Gan, L.; Strasser, P., Comparative study of the electrocatalytically active surface areas (ECSAs) of Pt alloy nanoparticles evaluated by H$_{upd}$ and CO-stripping voltammetry. *Electrocatalysis* **2014**, *5* (4), 408–418. doi 10.1007/s12678-014-0205-2

[230] Choi, W.; Shin, H.-C.; Kim, J. M.; Choi, J.-Y.; Yoon, W.-S., Modeling and applications of electrochemical impedance spectroscopy (EIS) for lithium-ion Batteries. *Journal of Electrochemical Science and Technology* **2020**, *11* (1), 1–13. 10.33961/jecst.2019.00528

[231] Wang, H. L.; You, E. M.; Panneerselvam, R.; Ding, S. Y.; Tian, Z. Q., Advances of surface-enhanced Raman and IR spectroscopies: From nano/microstructures to macro-optical design. *Light-Science & Applications* **2021**, *10* (1) Art. No. 161. doi 10.1038/s41377-021-00599-2.

[232] Cychy, S.; Hiltrop, D.; Andronescu, C.; Muhler, M.; Schuhmann, W., Operando thin-layer ATR-FTIR spectroelectrochemical radial flow cell with tilt correction and borehole electrode. *Analytical Chemistry* **2019**, *91* (22), 14323–14331. doi 10.1021/acs.analchem.9b02734

[233] Yeo, B. S.; Bell, A. T., In situ Raman study of nickel oxide and gold-supported nickel oxide catalysts for the electrochemical evolution of oxygen. *The Journal of Physical Chemistry C* **2012**, *116* (15), 8394–8400. doi 10.1021/jp3007415

[234] Aiyappa, H. B.; Wilde, P.; Quast, T.; Masa, J.; Andronescu, C.; Chen, Y.-T.; Muhler, M.; Fischer, R. A.; Schuhmann, W., Oxygen evolution electrocatalysis of a single MOF-derived composite nanoparticle on the tip of a nanoelectrode. *Angewandte Chemie International Edition* **2019**, *58* (26), 8927–8931. doi.org/10.1002/anie.201903283

[235] Choi, M.; Siepser, N. P.; Jeong, S.; Wang, Y.; Jagdale, G.; Ye, X.; Baker, L. A., Probing single-particle electrocatalytic activity at facet-controlled gold nanocrystals. *Nano Letters* **2020**, *20* (2), 1233–1239. doi 10.1021/acs.nanolett.9b04640

[236] Tarnev, T.; Cychy, S.; Andronescu, C.; Muhler, M.; Schuhmann, W.; Chen, Y.-T., A Universal nano-capillary based method of catalyst immobilization for liquid-cell transmission electron microscopy. *Angewandte Chemie International Edition* **2020**, *59* (14), 5586–5590. doi.org/10.1002/anie.201916419

[237] El Arrassi, A.; Liu, Z.; Evers, M. V.; Blanc, N.; Bendt, G.; Saddeler, S.; Tetzlaff, D.; Pohl, D.; Damm, C.; Schulz, S.; Tschulik, K., Intrinsic activity of oxygen evolution catalysts probed at single $CoFe_2O_4$ nanoparticles. *Journal of the American Chemical Society* **2019**, *141* (23), 9197–9201. doi 10.1021/jacs.9b04516

[238] Piccini, G.; Alessio, M.; Sauer, J., Ab initio calculation of rate constants for molecule–surface reactions with chemical accuracy. *Angewandte Chemie International Edition* **2016**, *55* (17), 5235–5237. doi.org/10.1002/anie.201601534

[239] Sautet, P., Quantum chemical methods. In *Characterisation of Solid Materials: From Structure to Surface Reactivity*, Che, M.; Vedrine, J. C., Eds. Wiley-VCH: Weinheim, 2012; Vol. 2, pp 1119–1145. doi.org/10.1002/9783527645329.ch24

[240] Hammer, B.; Nørskov, J. K., Theoretical surface science and catalysis – Calculations and concepts. *Advances in Catalysis* **2000**, *45*, 71–129. doi 10.1016/S0360-0564(02)45013-4

[241] Nieminen, V.; Sierka, M.; Murzin, D. Y.; Sauer, J., Stabilities of C_3–C_5 alkoxide species inside H-FER zeolite: A hybrid QM/MM study. *Journal of Catalysis* **2005**, *231* (2), 393–404. doi.org/10.1016/j.jcat.2005.01.035

[242] Vojvodic, A.; Medford, A. J.; Studt, F.; Abild-Pedersen, F.; Khan, T. S.; Bligaard, T.; Nørskov, J. K., Exploring the limits: A low-pressure, low-temperature Haber–Bosch process. *Chemical Physics Letters* **2014**, *598*, 108–112. doi.org/10.1016/j.cplett.2014.03.003

[243] Wang, S. G.; Temel, B.; Shen, J. A.; Jones, G.; Grabow, L. C.; Studt, F.; Bligaard, T.; Abild-Pedersen, F.; Christensen, C. H.; Nørskov, J. K., Universal Bronsted-Evans-Polanyi relations for C-C, C-O, C-N, N-O, N-N, and O-O dissociation reactions. *Catalysis Letters* **2011**, *141* (3), 370–373. doi 10.1007/s10562-010-0477-y

[244] Logadottir, A.; Rod, T. H.; Nørskov, J. K.; Hammer, B.; Dahl, S.; Jacobsen, C. J. H., The Brønsted–Evans–Polanyi relation and the volcano plot for ammonia synthesis over transition metal catalysts. *Journal of Catalysis* **2001**, *197* (2), 229–231. doi.org/10.1006/jcat.2000.3087

[245] van Santen, R. A.; Neurock, M.; Shetty, S. G., Reactivity Theory of transition-metal surfaces: A Brønsted–Evans–Polanyi linear activation energy–free-energy analysis. *Chemical Reviews* **2010**, *110* (4), 2005–2048. doi 10.1021/cr9001808

[246] Reuter, K.; Scheffler, M., Composition, structure, and stability of $RuO_2(001)$ as a function of oxygen pressure. *Physical Review B* **2001**, *65* (3), 035406. doi 10.1103/PhysRevB.65.035406

[247] Meyer, B., First-principles study of the polar O-terminated ZnO surface in thermodynamic equilibrium with oxygen and hydrogen. *Physical Review B* **2004**, *69* (4), 045416. doi 10.1103/PhysRevB.69.045416

[248] Nørskov, J. K.; Bligaard, T.; Rossmeisl, J.; Christensen, C. H., Towards the computational design of solid catalysts. *Nature Chemistry* **2009**, *1* (1), 37–46. doi 10.1038/nchem.121

[249] Abild-Pedersen, F.; Greeley, J.; Studt, F.; Rossmeisl, J.; Munter, T. R.; Moses, P. G.; Skúlason, E.; Bligaard, T.; Nørskov, J. K., Scaling properties of adsorption energies for hydrogen-containing molecules on transition-metal surfaces. *Physical Review Letters* **2007**, *99* (1), 016105. doi 10.1103/PhysRevLett.99.016105

[250] Roudgar, A.; Groß, A., Hydrogen adsorption energies on bimetallic overlayer systems at the solid–vacuum and the solid–liquid interface. *Surface Science* **2005**, *597* (1), 42–50. doi.org/10.1016/j.susc.2004.02.040

[251] Greeley, J.; Jaramillo, T. F.; Bonde, J.; Chorkendorff, I.; Nørskov, J. K., Computational high-throughput screening of electrocatalytic materials for hydrogen evolution. *Nature Materials* **2006**, *5* (11), 909–913. doi 10.1038/nmat1752

4 A close-up to some important aspects of surface catalysis

4.1 A preface

Catalyst development aims at materials that combine high activity and selectivity for the desired product(s) with excellent stability. This statement already points out the complexity of the task: in noncatalytic reactions, highly reactive species are known to be not very selective when different reaction channels are available. As mentioned in Section 3.5.3, activity correlates with the capability of surfaces to bind molecules with forces strong enough to activate them for subsequent reaction steps, but weak enough to allow the desorption of the resulting products (Sabatier principle). Selectivity requires that the energy barriers for all the undesired reaction and desorption channels along this path[1] are substantially higher than those of the desired steps.

In reality, this has to be achieved with catalytic materials that usually offer more than one type of reactivity; metals employed for the (de)hydrogenation of hydrocarbons may also isomerize or crack them; acid sites can simultaneously catalyze a whole family of related reactions (cf. Section 4.5.2); and redox oxides can contribute different forms of lattice oxygen or adsorbed gas-phase oxygen for the activation of a hydrocarbon reactant (see Section 4.3.3). Many catalysts consist of different components, e.g., support and supported phase(s): both can contribute their own reactivity and influence each other. The picture is further complicated when catalysts contain promoters employed to improve various aspects of their performance, or poisons that may have been adsorbed during operation in real feeds.

As mentioned already in Section 1.5, this complexity has interfered with the identification of unifying concepts that are valid across the whole field and of algorithms that would allow predicting the most successful catalysts for divergent reaction types. While predictive models are becoming available for a number of reactions catalyzed by metals (cf. Section 3.5) and seem to be near for acid catalysis, such a level of understanding will require still considerable effort for the complex processes during selective hydrocarbon oxidation over mixed-oxide catalysts (cf. Section 4.3.3), for catalysis at the solid–liquid interface, and for many processes at catalytic electrodes. In the sections to come, relations between the structure, the catalytic activity and selectivity are, therefore discussed on rather different levels of scientific maturity, ranging from atomistic models that are supported by theory to heuristic models deduced from the generalization of experimental observations.

---

[1] Including undesired modes of reactant adsorption, e.g., activation of a C=C double bond although the C=O bond in the same molecule is targeted.

https://doi.org/10.1515/9783110632484-004

For exhibiting attractive catalytic properties, a material must offer high reactivity and specific surface area, which are both typical indicators for metastable states of matter. Thus, catalysts will inevitably deactivate even if there are no poisons or reaction channels that result in deposits, although the deactivation rate may be rather slow. It depends on the level of atom mobility in the near-surface region, which determines how and at which rate the solid material responds to the driving force set up by the difference of the chemical potentials in its bulk and at its surface.

The response may be confined just to the formation of adsorbed species, but may also include a variety of other processes, e.g., adsorbate-induced surface reconstructions, penetration of atoms from gas-phase species (H, O, C) below the surface, reactions of the surface with gas-phase components (cf. Section 2.5.1), reactions between catalyst components, structural perturbation of the near-surface layer, surface corrosion and leaching of components into the liquid phase, etc. Some of these processes may be required to create particular types of active sites (e.g., highly active sites for methanol formation at the Cu/ZnO perimeter by creation of surface defects in ZnO, cf. Section 4.2.4, or highly active oxidation sites on Pt surface oxides, Section 2.5.1), or they are even part of catalytic reaction mechanisms (e.g., oxygen anion diffusion between redox and reoxidation sites in Mars–van Krevelen-type oxidation reactions (Section 2.5.1)). Some other processes cause the loss of catalytic activity, e.g., formation of inactive material by the reaction between the catalyst and fluid-phase components or between catalyst components. As mentioned in Section 1.4.2, deactivation can occur on various time scales, and can be almost completely reverted in many cases by applying regeneration and reactivation technologies. These processes are discussed in some detail in Section 4.4.

The performance of catalysts is closely related to the structure of surface sites that interact with the reactants. The term "structure" is often implied to specify only the arrangement of atoms in space (the spatial structure), but it actually also covers information about which kind of atoms are arranged and what their oxidation states are, i.e., the electronic structure. Although the spatial electron density distribution completely determines all the other structural aspects (Hohenberg–Kohn theorem, Section 3.5.1), electronic and spatial aspects of structure are often discussed separately for practical reasons,[2] but their close relation should always be kept in mind.

Changes in electronic and spatial surface structure usually cause variations of both activity and selectivity. However, as selectivity typically results from activity rankings with respect to competing reactions, it makes sense to discuss relations between the structure and the activity separately.

---

2 Not the least being that the real electron density distribution in a material is usually not known a priori and more difficult to determine than individual aspects of spatial and electronic structure.

4.2 Structure and activity

4.2.1 The Sabatier principle

According to the Arrhenius equation (1.1), the highest activities (i.e., rates under reference conditions) can be achieved when the activation energy is low and the pre-exponential factor A is high. Small activation energies are favorable, but activity can also be limited by a low number of active sites. Ammonia synthesis is an example of such a case. Over Fe catalysts, the highly stable N_2 molecule is split with an apparent activation energy of just 4 kJ/mol (Figure 2.57). The sites required to achieve such a low activation energy are, however, sterically rather demanding (cf. Section 4.2.3) and, therefore, rare in Fe and also in Ru catalysts.

In simple reactions, requirements can be discussed in terms of a generalized Sabatier principle: the interaction with the site should be strong and perturbing the reactant sufficiently to initiate the desired reaction, but not as vigorous that it would interfere with the desorption of the products or induce undesired parallel or consecutive side reactions.

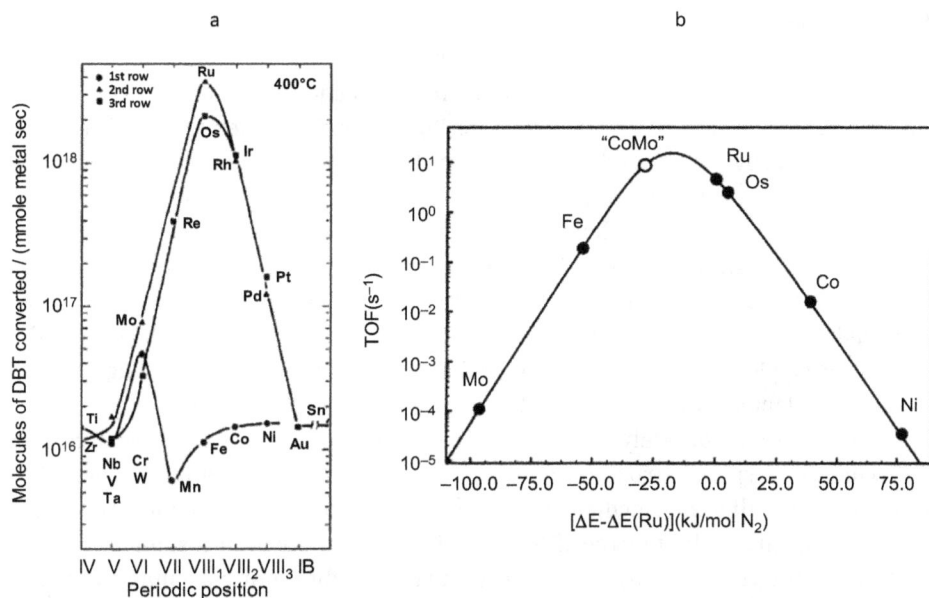

Figure 4.1: Volcano plots illustrating the activity trends according to the Sabatier principle: (a) activities (rates of sulfur removal from dibenzothiophene (DBT) under given conditions) over sulfide catalysts across the d elements (reproduced from ref. [1] with permission from Elsevier) and (b) ammonia synthesis activity versus binding energy of N (relative to E_{Ru-N}), both from theoretical calculations; conditions: 673 K, 50 bar, $H_2/N_2 = 3$, at 5% NH_3 (reproduced with permission from ref. [2]. Copyright 2001 American Chemical Society).

The Sabatier principle primarily describes the influence of the electronic structure on activity. In classical textbooks, it was often demonstrated by volcano curves that are obtained when activities were plotted versus formation energies of compounds across a number of catalytic elements, e.g., activities of formic acid decomposition versus formation enthalpies of formates. As more data became available, activities were plotted versus experimental adsorption enthalpies instead of formation enthalpies, e.g., activities of CO oxidation versus adsorption enthalpies of O_2. The key influence of the electronic structure was emphasized by plots of activities versus the position of elements in the Periodic Table. As an example, Figure 4.1a shows how HDS activities of sulfides vary across the d-elements and form pronounced maxima in the second and third period, while the differing behavior in the first period implies the importance of additional influences.

Meanwhile, activities can be evaluated for simple reactions by inserting rate constants from theoretical calculations into microkinetic models (cf. Section 3.5.3). In Figure 4.1b, turnover frequencies of NH_3 synthesis are plotted versus the binding energy between N and the catalytic elements (from DFT calculations). In the resulting volcano plot, Ru and Os are quite close to the top, while the TOF of Fe, which has been dominating industrial ammonia synthesis for more than a century, is lower by an order of magnitude. This surprising relation is due to the experimental conditions modeled, where 5% NH_3 in the reaction atmosphere were specified. As mentioned in Section 2.5.1, the strong adsorption of NH_x species by Fe shifts the rate-determining step (rds) from N_2 dissociation to the desorption of NH_x species under these conditions. As a result, reaction orders of NH_3 in the kinetic rate laws are stronger negative for Fe than for Ru. At low NH_3 contents, rates over Fe are even higher than over Ru. Therefore, Fe is still used in the first stages of technical ammonia synthesis, while the expensive Ru, if used at all, is employed in the final section only.

In a similar effort, it was shown that the volcano behavior of HDS activities exemplified in Figure 4.1a is also due to an optimum of interaction strengths between HS– (or RS–) groups and the surface in the best catalysts, although comparison with reality was somewhat affected by strong scatter in the experimental data [3]. The success of the Sabatier principle raises the question, how the variation of adsorbate binding energies across the d elements can be explained. We come back to this topic later.

In the early days of catalysis, a modified Sabatier principle was proposed to control the influence of spatial structure on reaction rates (Balandin multiplet theory). The formation of the di-σ intermediate in the Horiuti-Polanyi mechanism, for instance (cf. Figure 2.56a), should be favored if the Me–Me distance is within a suitable relation to the C–C bond length; an aromatic hydrogenation might require a suitable arrangement of surface atoms (a multiplet, now rather designated as ensemble). Although it is difficult to completely separate the electronic and steric influences,[3] it is obvious that specific atomic arrangements can offer specific electronic interactions, accelerating or even enabling particular reaction steps. At the same time, there are reactions known to

---

3 Bandwidths change with atom distances, cf. Section 2.1; further effects are discussed later.

require more than one or two bonds between the adsorbate and the surface for the rds to proceed (e.g., metal-catalyzed alkane hydrogenolysis). Such structure-sensitive reactions are more likely to respond to differences in the structure of the catalytic surface than reactions proceeding on single surface atoms (structure-insensitive reactions).

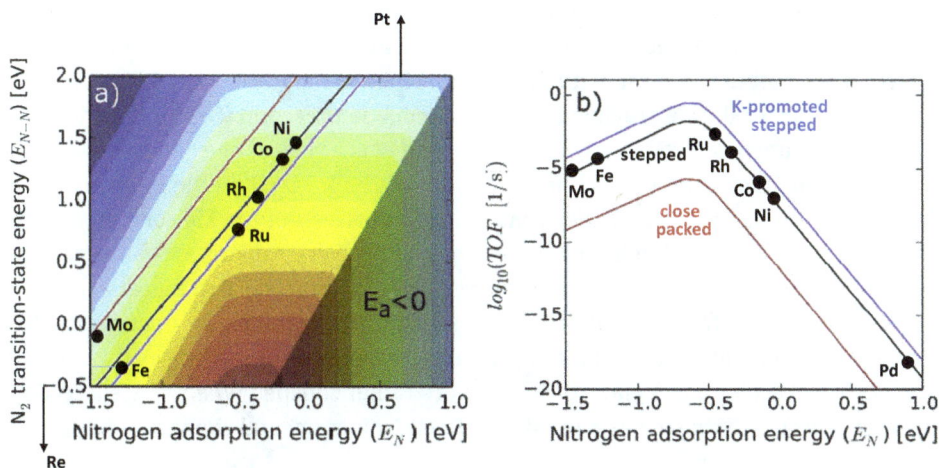

Figure 4.2: Influence of surface roughness and alkali promoter on the Brønsted-Evans-Polanyi relation for N_2 dissociation (a) and on catalytic activities predicted for a kinetic model, including N_2 dissociation and NH_3 desorption (b); calculated points included only for stepped surface. Adapted from ref. [4] with permission from Elsevier.

For simple reactions on metal surfaces, the influence of the spatial via the electronic structure can be nicely illustrated by comparing the Brønsted-Evans-Polanyi (BEP) relations between the binding energies of N and the activation energies of N_2 dissociation over flat and stepped metal surfaces (Figure 4.2a). The correlation line for the stepped surfaces is parallel to that for the close-packed surfaces (in red), and predicts activation energies of N_2 dissociation decreased by ca. 0.5 eV at sites comprising highly exposed metal atoms that are available at the steps (the so-called B5 sites, see later). Applied in the suitable microkinetic model, this results in almost parallel volcano curves for close-packed and stepped surfaces (Figure 4.2b), with turnover frequencies on the latter exceeding those on the former by 3–4 orders of magnitude.

While influences of the spatial structure on catalytic activities are obvious, further improvement of the given sites, e.g., by optimizing distances between atoms in facets, is difficult. Dimensions in a metal facet can be varied by depositing the metal as a monolayer on the same facet of a different metal, e.g., Ni on W(100). Catalytic and electronic properties of such strained facets can differ strongly from those of the analogous bulk metal facet, but the modified electron structure of the supported metal actually results from two sources: from variation of the dimensions in its (two-dimensional) lattice, and from electronic effects of the supporting metal [5].

While these effects can be well described by theory (see Section 4.2.2), influences of lattice strain on the activity of real catalysts have so far been suggested and detected only in a few cases. In Fe catalysts for NH_3 synthesis, the surface becomes nitrided under reaction conditions. Likewise, Fe carbide, rather than Fe metal surfaces, is considered to catalyze the Fischer–Tropsch (FT) reaction. There is, however, not much information on how nitridation or carbidization change activity and if this is related to the lattice expansion or to electronic effects by the intercalated atoms. Theory predicts strain effects to be most pronounced with the late transition metals (see later). Thus, they are quite plausible for the electrochemical oxygen reduction reaction (ORR) (cf. Section 4.9) over carbon-supported PtMe alloy cathodes where Pt in a shell of one or a few monolayers around an alloy core, is much more active than bulk Pt. When the Pt forms just a monolayer, it is subject to both electronic and strain effects of the Me component. For thicker layers, the latter dominates because mechanical strain operates over larger distances than electronic effects in metals (cf. ref. [6] and literature cited there).

Utilizing strain for reactivity control on purpose is certainly an opportunity for further development of the field, but it involves considerable challenges for catalyst synthesis and activation. So far, researchers are rather satisfied when they succeed in optimizing or modifying the existing strain effects by incremental changes or by special treatments.

In acid–base catalysis, a perfect matching of acidity and basicity between the reactant and the catalyst surface provides the highest rates per site. Reminiscent of the Sabatier principle, a weakly acidic site may be unable to bind the basic reactant at all, while excessive acidity may prevent the site to be vacated after the surface reaction. However, alternative reaction channels may be available, in particular, in catalysis with (solid) Brønsted acids. Sites of excess acidity may, for instance, be vacated by cracking the product, instead of keeping it adsorbed. This violates the Sabatier principle: activity goes on increasing beyond the optimum acidity, but at the expense of selectivity and yield.

Oxidation catalysis is more complex because of the diversity of reactants offered by catalysts. Requirements are very different for total and for partial (or selective) oxidation: the former involves highly reactive oxygen species, which are absolutely deleterious for the latter. O atoms on metal surfaces, for instance, are highly electron-deficient. If the metal withstands bulk oxidation, this electrophilicity dominates the interactions with reactants, which may, however, be modulated by the formation of subsurface O atoms or surface-oxide phases (cf. Section 2.5.1). Highly reactive oxygen species can be formed by the activation of O_2 on the defect sites of oxide surfaces (cf. Scheme 2.3) or by the interaction of O atoms with n-type semiconductor oxides (cf. Section 2.1[4]). Reactions

---

4 Figure 2.9a shows the case where the adsorbed oxygen withdraws electrons from the acceptor level of a p-type semiconductor. As these can be replenished by thermal excitation from the valence band, the O atom becomes fully charged and can be integrated into the lattice (nucleophilic oxygen). In n-type semiconductors (cf. Figure 2.8), oxygen can withdraw electrons only from donor levels,

involving the resulting electrophilic oxygen tend to be violent, breaking, for instance, C–C bonds in hydrocarbons. While such a reaction will be certainly accelerated by an enhanced reactivity of the surface oxygen, hydrocarbon conversion is not likely to decrease beyond the optimum oxygen electrophilicity, as required by the Sabatier principle. Rather, other oxidation reactions with more damage to the carbon skeleton will take over, finally resulting in combustion.

Catalysts for selective oxidation mostly operate with nucleophilic oxygen, which is O^{2-} stabilized by cations of the lattice. Although selectivity is the key property in this field, the aspect of activity cannot be neglected. Highest rates per site are achieved when the nucleophilicity of the lattice oxygen fits to the electrophilic properties of the reactant. Increased reactivity of the catalyst will again result in unselective side reactions rather than in a blocked catalyst surface. Selective oxidation is discussed in more detail in Section 4.5.2.

4.2.2 Electronic structure and catalytic activity on metal surfaces

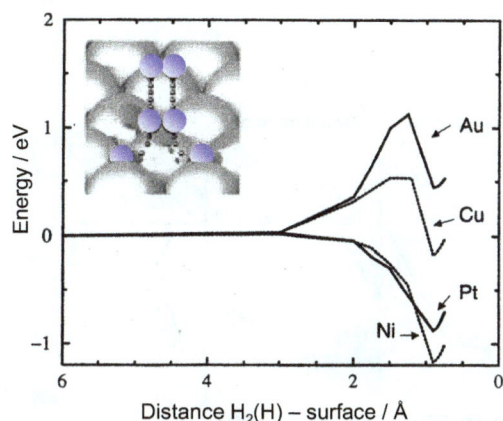

Figure 4.3: Reaction coordinates for the dissociative adsorption of H_2 on close-packed facets of different metals (from DFT calculations). Adapted from ref. [7] with permission from Elsevier.

In Figure 4.3, a H_2 molecule approaching a close-packed metal facet and dissociating on it (inset) is depicted, and the dependence of the energy on the distance from the surface is shown for the metals Ni, Pt, Cu, or Au. Dissociative adsorption of H_2 is well known to be fast and nonactivated on Ni and Pt, activated on Cu, and impossible on Au. The curves are from DFT studies, but it is of great interest to understand, which properties of the

which limits the amount of charge available, and is further confined by the emerging band bending. The O adsorbates remain, therefore, electrophilic.

metals cause the differences. This kind of question is omnipresent in surface catalysis. For transition metal surfaces, theory has provided a rather simple framework that
allows understanding and predicting trends in adsorption and catalysis. The concept is
introduced here largely following its presentation in ref. [7]. The reader will find more
information about its potential and its limitations there, alternatively in ref. [8].

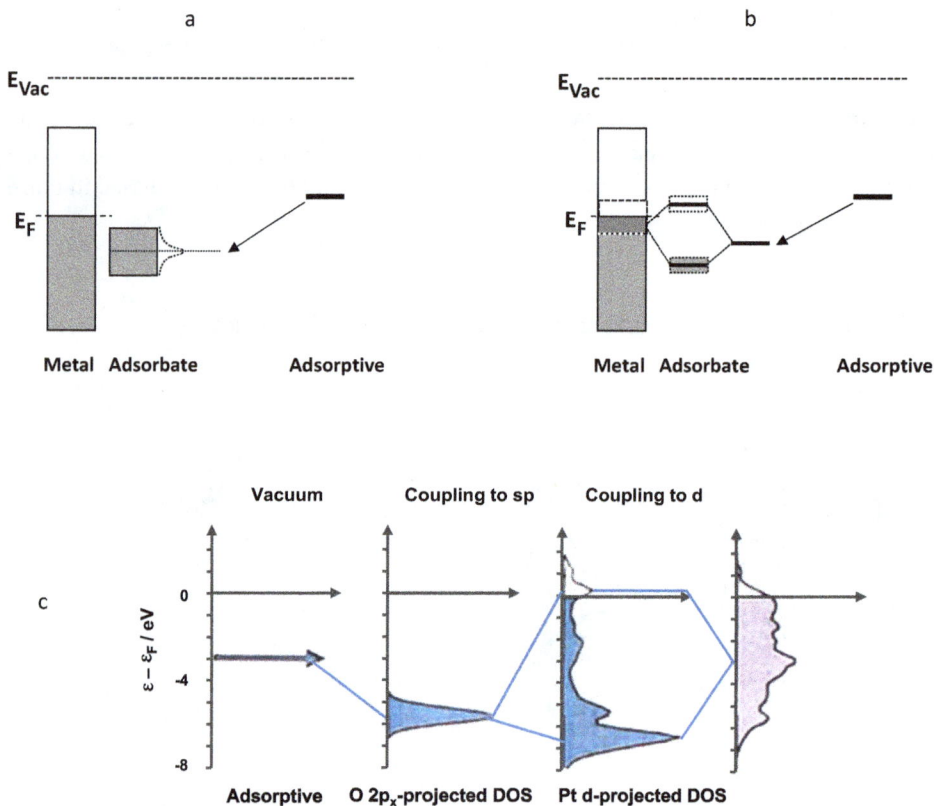

Figure 4.4: Formation of adsorbate states by the interaction of adsorptive levels with an s-band (a) or with
a d-band (b); in (c), the combination of both processes is shown from a density of states calculations for the
adsorption of an O atom on Pt. Reproduced from ref. [7] with permission from Elsevier.

To begin with, the adsorption of a monoatomic species is considered. When it approaches
the surface, its highest, most extended energy levels interact with those of the metal
surface. Interaction occurs with s- or sp-bands, both broad and diffuse, and with the
much narrower d-band, where orbitals of different symmetries feature different energies within the band width. As there are no symmetry requirements by the diffuse
bands, a broad range of levels interacts with the narrow electron states in the approaching species. As a result, the sharp levels in the adsorptive are transformed into a density
distribution around the average, and they are stabilized due to gains in exchange energy

(Figure 4.4a). The latter contribution can be considered constant across each period of the transition metals.

As the d-bands are narrow, their interaction with the levels of the approaching species results in a pair of binding and antibinding levels, which are also broadened by the interaction with nearby parts of the s or sp bands (Figure 4.4b). In the adsorbed state, the filling of these orbitals is determined by the Fermi level. The latter aspect causes a strong difference between the binding in molecules and on metal surfaces. When, for instance, a filled antibonding state is shifted above the Fermi level (cf. Figure 4.4b), it donates its electrons to the metal, which results in a bond. In a molecule, the same interaction scheme would describe a nonbinding situation. Likewise, when the binding orbital formed from states which are empty in both metal and adsorptive, is shifted below the Fermi level, it becomes populated by electrons from the metal and contributes to the binding of the adsorbate. The contributions arising from the interactions with the d levels play a crucial role in the explanations of reactivity trends across the periods.

Figure 4.4c illustrates these changes of the electron structure for the adsorption of an O atom at Pt(111) in a more realistic way by showing density-of-states data from DFT calculations [7]. In the first stage, the p_x level of oxygen is broadened and is shifted to a lower energy by coupling to the broad sp band. By the interaction of this state with a symmetry-allowed range in the d-band, covalent bonding and antibonding states emerge. The latter extends beyond the Fermi level and, therefore, creates a substantial contribution to the adsorption energy.

Figure 4.5a shows what happens to this d-band contribution when the d-band energy changes. When the d-band extends widely beyond the Fermi level, i.e., to the left in a period (cf. Figure 2.7), the antibonding orbital is fully above E_F and, therefore, empty, which gives rise to a strong adsorbate bond. Further right in the period, the d-band moves lower, until it is completely below the Fermi level, which is then located in the s-band. As a result, the antibonding orbital becomes successively populated, and the adsorbate bond becomes weaker, until the d-contribution in it disappears completely.

This complies with the well-known fact that elements become more noble to the right of the periods. At the same time, the underlying model is based on quantitative theoretical work, and has been employed for a great number of correct predictions and plausible explanations of phenomena. Among several choices, Hammer and Nørskov [7] selected the energy ε_d of the d-band center (related to the Fermi energy) as a descriptor for adsorption energies. Figure 4.5b shows that the adsorption energies of O atoms on the close-packed facets of 4d transition metals are indeed linearly correlated with ε_d. Similar correlations exist for other adsorbates, e.g., H, N, S, and Cl. Metals with the highest d-band center provide the strongest interactions with these species.

While ε_d is a powerful descriptor, correlating with properties relevant for catalysis and adsorption within a period (see later), relations between the periods are additionally influenced by the size of the metal atoms. Their electrons generate a repulsive interaction with filled adsorbate levels, in particular to the right in periods. As a result, 4d and 5d elements are more noble in many situations than 3d elements of the same group.

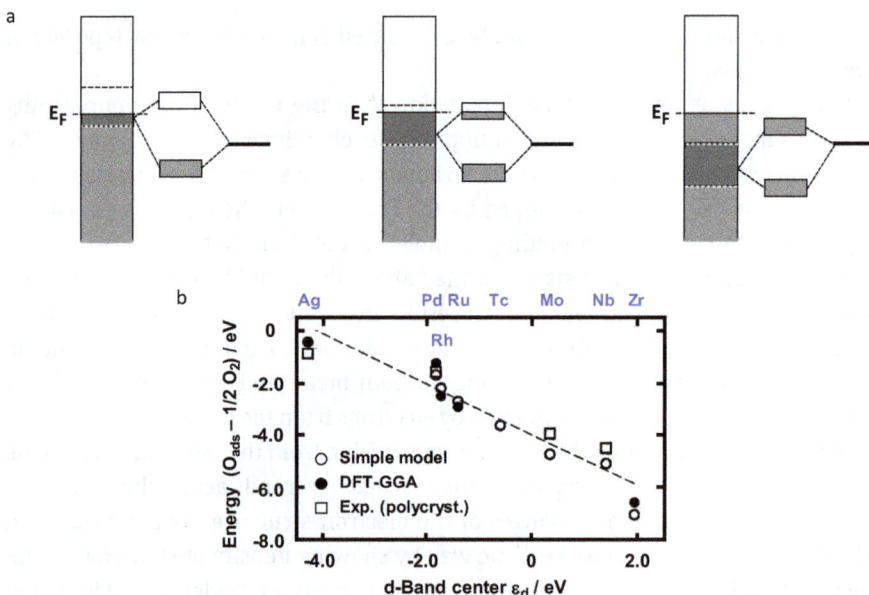

Figure 4.5: Monoatomic adsorbates on transition-metal surfaces: (a) influence of the d-band energy on the contribution of the d-band interaction to the adsorption energy and (b) correlation between the oxygen adsorption energy and the energy ε_d of the d-band center for close-packed facets of 4d elements. Adapted from ref. [7] with permission from Elsevier.

A descriptor for this influence[5] is discussed in ref. [7], where its numerical values and d-band center energies for close-packed surfaces are reported for all transition metals as well.

The specification "for close-packed surfaces" related to the ε_d data (e.g., in Figure 4.5b) is important because these energies critically depend on surface structures. The width of a band depends on the coordination number around the involved atoms and the overlap between their binding orbitals (cf. Section 2.1). Bands are wider in the bulk than at the surfaces, and band widths further decrease with increasing exposure of atoms, e.g., at the edges of the terraces or at kinks. Likewise, bands become narrower when the lattice is extended, e.g., by tensile strain. The reverse happens upon lattice compression, which is less relevant for catalysis.

The consequences of such narrowing of bands, which depend on their degree of filling, are illustrated in Figure 4.6. When ε_d remains constant, the compressed band in Figure 4.6b should be filled almost completely, that in Figure 4.6a should fall nearly empty, which would result in negatively or positively charged sites, respectively. Hammer and Nørskov rejected such charging in their theoretical studies. Instead, ε_d is changed until the original band filling is restored [7] (Figure 4.6). Hence, band compression leads to an

---

5 The squared adsorbate-metal d coupling matrix element, V_{ads}^2.

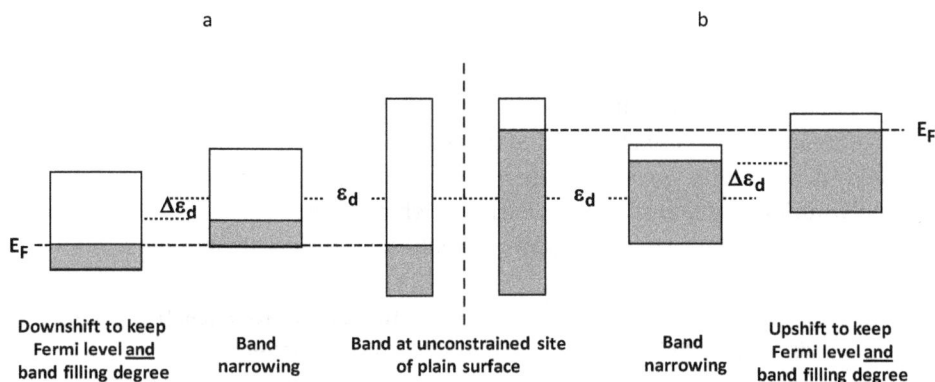

Figure 4.6: Shifts of the d-band center, as a consequence of band narrowing.

upshift of the d-band center in the late transition metals: the interactions offered by the corresponding sites are stronger than on close-packed facets. For surfaces under strain, this has already been mentioned in Section 4.2.1. For sites at the terrace edges and the kinks, there are numerous examples throughout this book. Effects become weaker with decreasing d-band filling and disappear in the middle of the period. They are reverted in the early transition metals, which are, however, rarely used in their metallic form in catalysis.

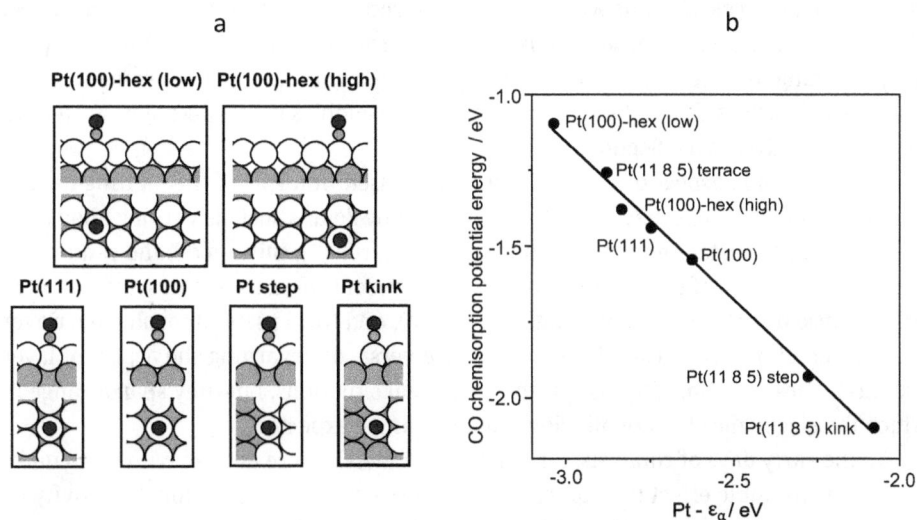

Figure 4.7: Surface structure, d-band center, and interaction with adsorbates: CO adsorbed on top of Pt, (a) geometrical models for various sites, (b) dependence of the d-band center on the site geometry and resulting CO adsorption energies. Reproduced from ref. [7] with permission from Elsevier.

The dependence of the d-band properties on the surface structure causes significant differences even between the flat facets of the same metal. This is illustrated in Figure 4.7, where the adsorption energies of CO on-top of various Pt sites are plotted versus the energy of the d-band center at these sites (most of them shown in Figure 4.7a). The study also includes two sites of the rare reconstructed Pt(100)-hex(1x5) surface [9], a slightly buckled structure, the Pt density of which exceeds even that of the "close-packed" Pt(111) surface by 4%. At steps and kinks, CO was evaluated for a Pt(11 8 5) surface.

Figure 4.7b confirms the difference between the flat Pt facets (ca. 0.1 eV between CO on Pt(100) and Pt(111)), which is, of course, much smaller than the differences between Pt on the flat facets and at steps or kinks (CO adsorption on step Pt being by ≈0.5 eV and on kink Pt by ≈0.7 eV stronger than on Pt(111)). Notably, on the (111)-oriented terraces of Pt(11 8 5), CO adsorbs even weaker than on the Pt(111) facet, probably because the atoms at the bottom of the steps have additional neighbors. The weakest CO adsorption was found on the lower site of the buckled Pt(100)-hex(1x5) surface. At the upper site, the adsorption is somewhat stronger, but not yet comparable with that of the unreconstructed Pt(100). Figure 4.7b shows the adsorption energy of the diatomic CO molecule to correlate with ε_d similar to that of monoatomic adsorbates (Figure 4.5b). The reasons are discussed later.

Alloying is an important tool for tailoring the catalytic behavior of metal catalysts. Its effects originate from both the electronic and structural changes: alloy surfaces offer mixed patterns of atoms with different electronic properties. The electronic effect of one component on the other one is sometimes referred to as the ligand effect, and the patterning as ensemble effect. Alloying an active with an inactive element to discourage undesired side reactions requiring large ensembles is a well-known strategy to improve selectivities (cf. Section 4.3.5).

The electronic aspect of alloying can be discussed in terms of the shifting d-band centers, although things may become complicated by details such as varying alloy compositions and/or surface enrichment phenomena. In ref. [7], shifts of d-band centers are given for a selection of transition metals in two different roles: for the first atom of A, in a close-packed surface layer of B (and vice versa), and for a pseudomorphic overlayer of A, on top of a close-packed B layer (and vice versa, disregarding the aspect of layer stability). These are limiting cases rather than real situations, but they span a range in which the electronic effect of alloying A with B can be expected.

In the early days of catalysis, for instance, 3d metals were alloyed with Cu to make use of the ensemble effect (to suppress cracking reactions) and to tailor the activity by varying the filling degree of their d-bands. The data in ref. [7] show why this approach, which was effective with respect to selectivity, had limited success in terms of activity: the d-band shifts of Fe, Co, or Ni dissolved in a Cu layer (*) or deposited as a monolayer on it (⁺) are small and even opposite (Fe: −0.13 eV* → −0.05 eV⁺; Co: −0.16 eV* → −0.06 eV⁺; Ni: +0.19 eV* → +0.05 eV⁺). The data imply that Au would have been a better choice instead, giving drastic effects with all three elements (Fe: −1.42 eV* → −2.19 eV⁺;

Co: -1.56^* eV → -2.39 eV$^+$; Ni: -1.13 eV* → -2.10 eV$^+$). Meanwhile, Au has indeed found industrial application in a surface alloy with Ni (Au and Ni are immiscible in the bulk): it is used to delay catalyst deactivation in the steam reforming of methane:

$$CH_4 + H_2O \leftrightarrows CO + 3\,H_2 \tag{4.1}$$

In this reaction, the dissociation of a H atom from CH_4 (i.e., a dehydrogenation) is rate-determining, but the Ni catalysts tend to deactivate mainly by the deposition of carbon. Due to the downshift of the d-band in Ni, Au destabilizes the Ni-C bond and renders the carbon more reactive to combine with surface oxygen. The gain in stability outbalances a moderate loss of dehydrogenation activity, which is also caused by the lower d-band at the Ni atoms.

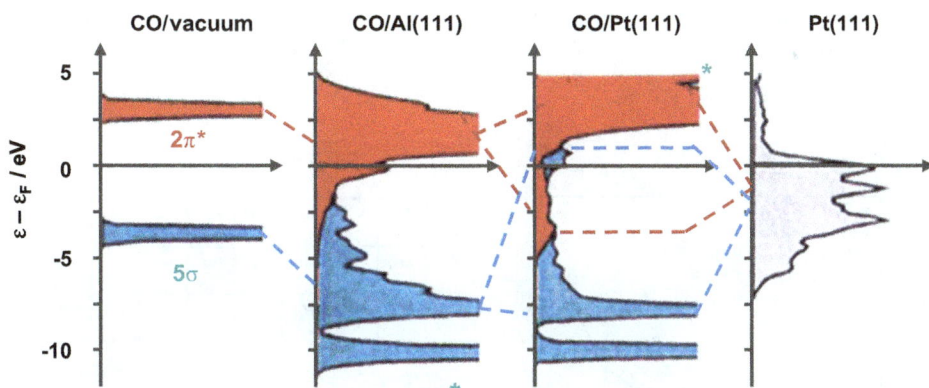

Figure 4.8: Molecular adsorption on metals: change of the DOS during the adsorption of CO on Pt(111); the level labeled * arises from mixing with the 4s state of CO. Reproduced from ref. [7] with permission from Elsevier.

For the adsorption of molecules on metals, the interactions of both the highest occupied and the lowest unoccupied molecular orbitals (HOMO/LUMO) with the surface need to be considered. As an example, the adsorption of CO on Pt(111) is illustrated in Figure 4.8 in a two-step scheme, similar to that used in Figure 4.4c to discuss the adsorption of monoatomic species. The 5σ and $2\pi^*$ orbitals of CO are broadened and stabilized by their interaction with the diffuse s band. Owing to their different symmetry properties, they interact with different ranges in the d-band of Pt, forming bonding and antibonding orbitals and exchanging electrons with the metal. The bonding state formed from the d band and the empty $2\pi^*$ orbital extends below the Fermi level, it becomes populated by electron density from the metal. The antibonding state derived from the d band and 5σ goes beyond the Fermi level, which causes electron transfer from the adsorbate to the metal. This is equivalent to what has been discussed as σ donation by CO and backdonation by the metal in the earlier Blyholder model of CO adsorption (Figure 3.120b). The electron density donated by the metal destabilizes the C-O bond, because $2\pi^*$ is antibonding within the molecule. Therefore, enhanced backdonation facilitates the splitting of this bond.

Figure 4.8 shows that there is a significant backdonation from the metal into a $2\pi^*$-derived orbital with Pt [7]. On the other hand, σ donation (the 5σ-derived density of states above the Fermi level) is relatively weak, and its bonding contribution is counteracted by the repulsion between the Pt atoms and the filled bond orbitals. To the right of Pt, both donation by 5σ and backdonation into $2\pi^*$ are diminished, because the antibonding components of both interactions are shifted below the Fermi level, similar to how it is illustrated for monoatomic adsorbates in Figure 4.5. To the left of Pt, the higher d-band results in an increased population of the bonding $2\pi^*$-derived orbital, while the bonding contribution of 5σ remains small [7]. This results in a stronger bond between the surface and the adsorbates, and an increased tendency to bond splitting toward the early metals in a period, due to the decreased repulsive interactions, also toward the earlier periods.

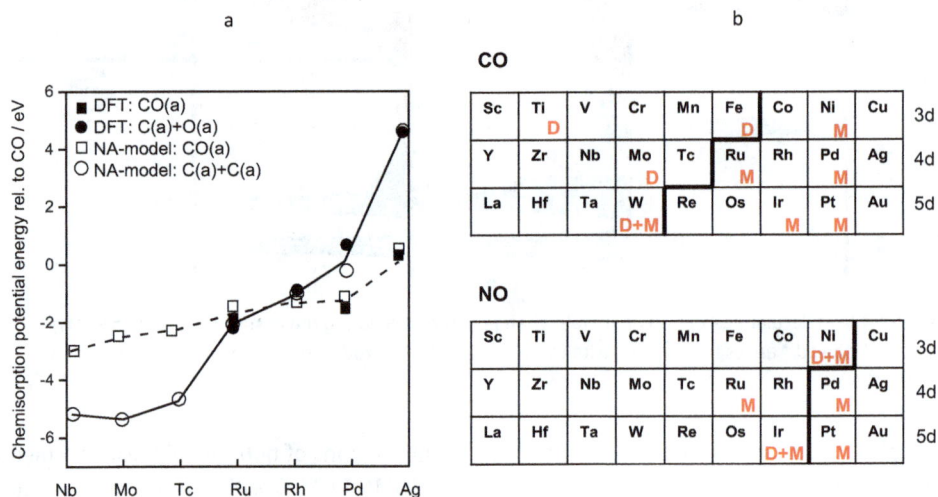

Figure 4.9: Molecular (M) or dissociative (D) adsorption: (a) comparison of adsorption energies of CO and of its fragments on close-packed facets of transition metals (NA – data from tight-binding model, originating from ref. [10]) and (b) molecular or dissociative adsorption of CO and NO detected at room temperature on different metals (compilation of experimental data). Adapted from ref. [7] with permission from Elsevier.

It has been found, however, that the influence of the d-band energy is less pronounced in diatomic than in monoatomic adsorbates. Figure 4.9a, in which the adsorption energies of CO or of C + O on close-packed facets of 4d metals are compared, shows the energies of the fragments responding more strongly to the position in the period than those of CO. To the left of Ru, the fragments are more stable than the molecule, i.e., CO adsorption is dissociative, and to the right of Rh, adsorption is molecular.

This is well in agreement with earlier knowledge on the adsorption modes of CO and of other molecules on transition metals (compiled for CO and NO in Figure 4.9b). The transition from dissociative to molecular adsorption is further to the right in the earlier periods because the repulsion between the adsorbate and the metal atoms is stronger in the 4d or 5d element of each group (see earlier). The transition also depends, of course, on the nature of the adsorptive. NO can be dissociated by relatively noble metals. The borderline for the adsorption of N_2 (not shown in Figure 4.9b) was similar to that for CO.

Beyond the prediction of stabilities, the d-band model allows also assessing tendencies among the kinetic barriers in reactions of bond dissociation (and, vice versa, of bond formation), and thus, among the reaction rates. The details differ depending on the transition state – if it is early or late on the reaction coordinate. In the first case, the molecular orbitals of the reactant are still intact, and they interact with the surface, much like those of a diatomic adsorbate. Late (or extended) transition states can be discussed by analogy to monomolecular adsorbates.

According to Hammer and Nørskov [7], early transition states occur in the dissociation of very short bonds, typically bonds involving hydrogen (e.g., H_2, C–H in methane). Therefore, the transition state of H_2 dissociation (cf. Figure 4.3) might be illustrated by schemes like that shown in Figure 4.8, where the bonding and antibonding states are the σ and σ^* orbitals of H_2 approaching the surface. The lower d-band in Cu is responsible for both the lower stability of the bond between H and the metal, and the higher activation energy of adsorption on Cu, compared with Ni (cf. Figure 4.3). The effect seems dramatic for neighbors in a period, but the differences in ε_d are pronounced just between these elements ($\approx$1.3 eV; $\Delta\varepsilon_d$ between Fe and Ni $\approx$0.4 eV [7]). Extended transition states are characteristic of the dissociation of molecules like CO, N_2, or NO.

A recent modification of the d-band model, which allows improved predictions in the range of the late transition metals, has been proposed in ref. [11].

4.2.3 Spatial structure and catalytic activity

As mentioned in Section 4.1 and exemplified in Section 4.2.2, spatial structure acts on surface catalysis via the underlying electronic structure. The present section deals with situations where this relation is less obvious. As catalysts expose the active element often in various structures that may catalyze different reactions, correlations between rates and active site structures may also be implicitly reflected in the selectivity data. While the present section is focused on explicit relations between the spatial structure and the activity, relations between selectivity and structure are discussed in Section 4.3.

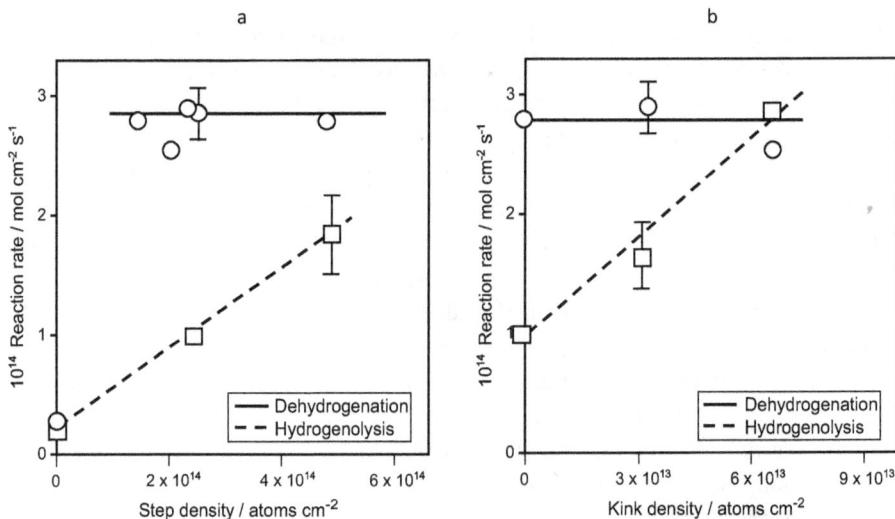

Figure 4.10: Dehydrogenation and ring opening (hydrogenolysis) of cyclohexane over high-indexed facets of Pt: reaction rates depending on step density (a) and on kink density (b). Conditions: H_2/cyclohexane = 20, $p(H_2) = 10^{-4}$ Pa, 423 K; step atom density in b: 2.4×10^{14} cm^{-2}. Adapted from ref. [12] with permission from Springer Nature.

Generalizing a wealth of observations on catalysts with varying metal particle sizes, Somorjai and Blakely [12] demonstrated in classical surface science experiments that different reactions may have different demands for the participation of the catalyst. Figure 4.10 shows how the rates of cyclohexane dehydrogenation and hydrogenolysis (to benzene and *n*-hexane, respectively) depend on the density of steps and kinks on high-index Pt facets, which were obtained by cutting Pt single crystals at appropriate angles (cf. Figure 2.13). While the dehydrogenation rate does not depend on the concentration of steps or kinks, there is a clear influence of both site types on hydrogenolysis.

The further discussion shows how pitfalls can be avoided by critical assessment of the data. According to Figure 4.10, the hydrogenolysis rate per kink site exceeds that per step site by an order of magnitude. An assignment of hydrogenolysis activity to step sites is, therefore, all but conclusive (and was not proposed by the authors of ref. [12]), because the influence of the steps in Figure 4.10a may also be due to a slight inaccuracy in facet orientation causing a nonzero kink density per step. Likewise, the insensitivity of dehydrogenation to step and kink concentrations seems to suggest that the reaction proceeds with the same rate on each exposed Pt atom. Notably, however, this rate is almost an order of magnitude higher than that on a flat Pt($\bar{1}$11) facet, which exposes the same pattern as the terraces of the high-index facets. Obviously, steps are required for the dissociation of C–H bonds, which does not, however, seem to be rate-determining. Instead, the reaction product benzene diffuses across the whole surface before being desorbed in the rds.

As mentioned earlier, reactions requiring only a single surface atom for their rds are referred to as structure-insensitive. Hydrogenation of small alkenes, as also of benzene, and CO oxidation over noble metals are examples of this reaction type. Among surface reaction steps, the hydrogenation of alkyl adsorbates has been found to exhibit only minor responses to structural changes. This explains the structural insensitivity of alkene hydrogenation, where the addition of hydrogen to the half-hydrogenated intermediate (cf. Horiuti-Polanyi mechanism, Section 2.5.1) is rate-determining [13].

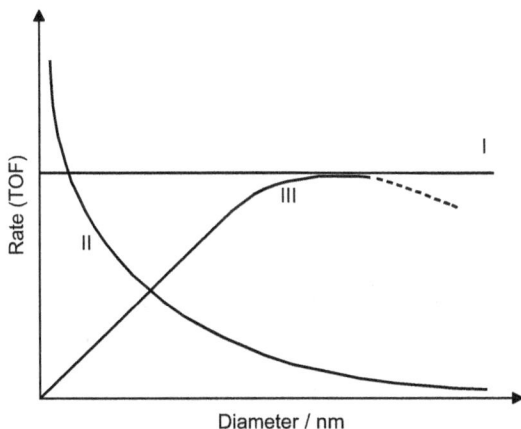

Figure 4.11: Influence of particle size on activities of metal-catalyzed reactions: (I) structure-insensitive reactions or reactions rate-limited by desorption from the whole surface; (II) reactions on exposed sites at step edges or kinks; and (III) reactions requiring large or specific ensembles.

Most catalytic reactions are structure-sensitive and require specific site structures to proceed at high rates. The trend to higher reactivity of metal atoms exposed at step edges or kinks, may suggest that TOFs of structure-sensitive reactions generally increase with decreasing particle size, because such sites are more abundant in small particles. This holds indeed for reactions requiring a low number of surface atoms. On the other hand, the probability to find ensembles with specific atom arrangements rather decreases at smaller particle sizes. Reactions requiring such ensembles follow the opposite trend with particle size, as indicated in Figure 4.11.

A particle-size dependence of type II is typical of reactions with rate-determining C–C or C–H splitting steps, for instance, the activation for methane for steam reforming, or alkane hydrogenolysis, which is a side-reaction in dehydrogenation processes like naphtha reforming (cf. Appendix A1). Notably, the most active sites of this kind are undesired in both processes – in naphtha reforming because of their competition with the target reaction (hyperactive sites, cf. Section 3.1.5), and in steam reforming, because of enhanced carbon formation (cf. discussion of eq. (4.1)). In both cases, the problem can be handled by poisoning these sites with sulfur. In steam reforming, it can be achieved by adding some ppm H_2S to the feed [14], which is an alternative approach to the application of Ni-Au surface alloy catalysts mentioned in Section 4.2.2. Stabilization of naphtha-reforming catalysts has been described in Section 3.1.5.

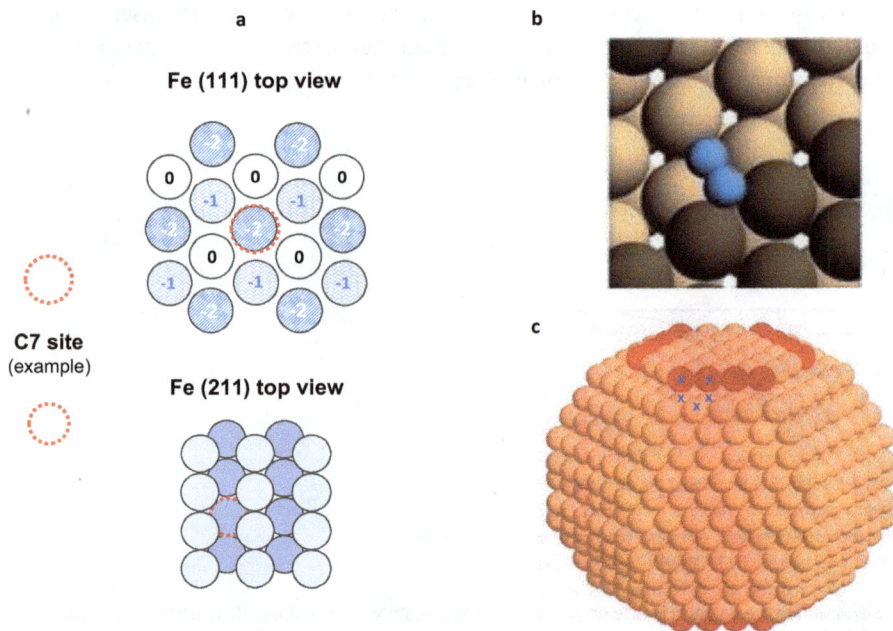

Figure 4.12: Multiatomic sites for the dissociation of triple bonds (e.g., in N_2 and CO): (a) C7 sites on Fe facets (adapted from ref. [15] with permission from Springer Nature); (b) transition state of N_2 dissociation on a B5 site at a step in a Ru(0001) site; and (c) B5 site on a metal nanocrystal with hcp lattice, from ref. [16], adapted with permission from the AAAS.

The splitting of C≡O and N≡N triple bonds as also the buildup of aromatic carbonaceous deposits from hydrocarbon precursors are reactions requiring larger ensembles. Therefore, these reactions are faster on larger particles (type III in Figure 4.11). N_2 and CO dissociation is by orders of magnitude faster on corrugated than on flat surfaces, but the activity does not correlate with roughness: over the rough Fe(211) facet, the rate of NH_3 synthesis (at 673 K, 20 bar and $N_2/H_2 = 3$) was 100-fold that over the dense Fe(110) facet, but over the less corrugated Fe(111) facet, it was 130fold [15]. N_2 splitting is, therefore, considered to proceed most swiftly on a type of site on which the molecule can interact with seven Fe atoms at the same time, the so-called C7 sites (Figure 4.12a).

For the hcp metals Ru and Co, a B5 site was identified, which is able to split N_2 or CO at low activation energy by allowing simultaneous interaction of the molecule with five surface atoms. B5 sites are not available on any of the ideal low-index facets, but appear at steps in them (Figure 4.12b), i.e., at characteristic artifacts of real facets. Such steps are not available on the surfaces of very small particles, but appear at particle sizes >5 nm, as illustrated in Figure 4.12c. Due to the unique activity of the B5 sites, NH$_3$ synthesis over supported Ru and FT synthesis over supported Co catalysts are dominated by them above the critical particle size (curve III in Figure 4.11).

Fe catalysts for NH$_3$ synthesis and for FT synthesis are very different with respect to microstructure and surface chemistry. FT synthesis is catalyzed by Fe carbide rather than by Fe metal. These catalysts are not covered by the tendencies shown in Figure 4.11.

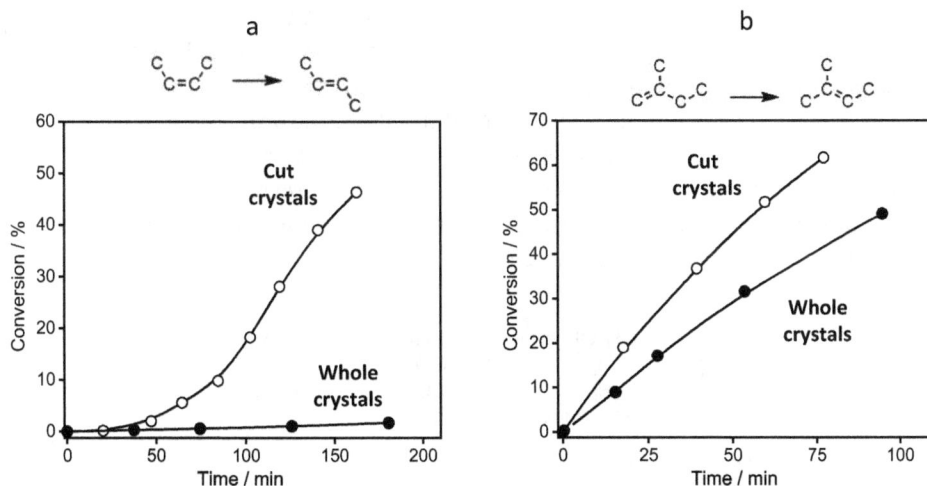

Figure 4.13: Catalytic anisotropy of MoS$_2$ identified by running reaction on macroscopic (cut and whole) MoS$_2$ crystals: (a) reaction proceeding on edge planes and (b) reaction proceeding on basal planes or on all exposed sites. Adapted from ref. [17] with permission from Taylor & Francis.

In catalysis with oxides and sulfides, early demonstrations of structure sensitivity took advantage of the structural anisotropies exhibited by many of these materials. In a classical experiment, Tanaka et al. [17] cut macroscopic MoS$_2$ crystals perpendicular to the basal planes (cf. Figure 2.17) to enhance the availability of edge planes, and compared the rates of several reactions over these samples. A substantially higher rate of *cis*-2-butene conversion into the *trans* isomer over the chopped MoS$_2$ (Figure 4.13a) indicates that this reaction proceeds on the edge planes. Similar results were obtained for ethene hydrogenation and for isotope scrambling between C$_2$H$_4$ and C$_2$D$_4$. Conversely, the double-bond shift in 2-methyl-1-butene was only slightly faster over the chopped than over the intact crystals (Figure 4.13b).

The importance of edge planes for catalysis over MoS$_2$ (and WS$_2$), including, in particular, the splitting of C–S bonds in HDS, is a matter of course nowadays. Ideas about active sites on these planes, however, have changed dramatically since that time, when most reaction mechanisms were thought to involve direct bonds between Mo and the reactants. This implied a key role of coordinative unsaturation, which is not compatible with the conditions of HDS, under which such catalysts operate. C–S bond breaking has been shown to indeed require a bond between the metal and sulfur [18], it proceeds in S vacancies [19, 20]. Alkene hydrogenation was found to proceed on saturated MoS$_2$ (WS$_2$) surfaces, first by indirect evidence (e.g., ref. [20]), and later by direct observation [21]. Sometime before 2000, Daage and Chianelli [22] localized the hydrogenation activity of MoS$_2$ at the crystal rim (Figure 2.17c) instead of the whole edge surface. After the observation of a one-dimensional metallic state on model MoS$_2$ monoslabs by STM (Figure 3.163), the hydrogenation activity was ascribed to this brim state, supported by theoretical studies [23]. This assignment is, however, under debate in literature. The activity of MoS$_2$ nanoslabs for electrochemical H$_2$ evolution could also be related to their rim, notably by experiments with model MoS$_2$ monoslabs, previously characterized by STM [24]. The activity was ascribed to the rim region without specifying the brim as the active site.

Catalytic anisotropies may also be expected for anisotropic oxide crystals (cf. Figures 2.18 and 2.19), but their mechanical properties prevent size reduction techniques like cutting. Instead, synthesis procedures and treatments had to be found that result in crystals of different habitus. Figure 3.159 shows results of such effort for V$_2$O$_5$.

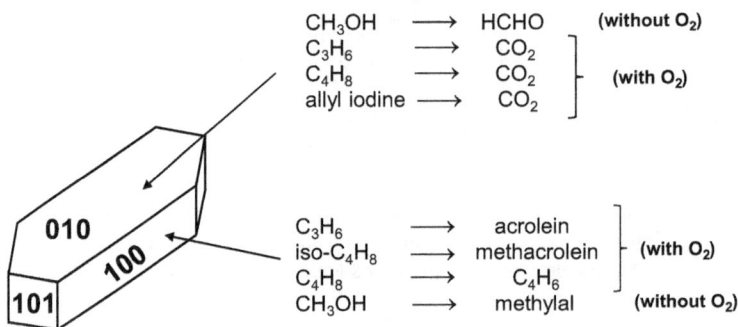

Figure 4.14: Catalytic anisotropy of MoO$_3$. Adapted from ref. [25] (for the reaction conditions, see ref. [26]).

For MoO$_3$, the catalytic anisotropy, probed by various test reactions, is summarized in Figure 4.14. The different reactivities of the (010) and the (100) facets are obvious. There does not seem to be a quasi-inert facet of MoO$_3$. The (010) facet appears to provide electrophilic oxygen (cf. Sections 4.2.1 and 4.3.3) by activating O$_2$ from the gas phase, whereas it dehydrogenates in the absence of O$_2$. Oxidation processes on the (100) facet are more selective. In the absence of O$_2$, this facet catalyzes both dehydrogenation and dehydration, i.e., it is more acidic. There are more examples for catalytic anisotropy of oxide phases in ref. [26].

Due to the presence of facets with total oxidation activity, satisfactory selectivity for target products of selective oxidation are achieved with simple inorganic phases only in rare cases (e.g., vanadyl phosphates for n-butane oxidation to maleic anhydride, cf. Appendix A1). Modern catalysts for selective oxidation are typically characterized by a pronounced complexity in composition and structure, and are designed to optimize the steps along the Mars–van Krevelen mechanism (Figure 2.55), rather than to utilize morphology effects (cf. Section 4.3.3).

The relation between structure and activity is the background for the everlasting effort to identify the active sites of catalysts for the given reactions, and thus a major aspect of catalysis research. Typical points of interest are, how many atoms of a catalytic element participate in the active site, which are their oxidation states, and what are their roles. The answers to these questions are also relevant for the discussion of reaction mechanisms: a mechanism requiring two surface atoms will differ from a mechanism doing with one only. Due to the importance of these questions, hypotheses on the structure of active sites have been part of scientific papers for decades, on a level ranging from pure speculation to serious interpretation of state-of-the-art data. As a result, there have been strong changes in active site models for some reactions. As there is more attention of the audience when a research topic is hot, the more mature models are not always better known among scientists.

In research on the selective oxidation of propene to acrolein for instance,

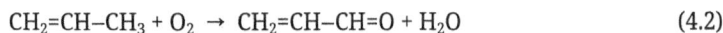

$$CH_2=CH-CH_3 + O_2 \rightarrow CH_2=CH-CH=O + H_2O \qquad (4.2)$$

a mechanism requiring the cooperation between Bi and Mo in the active sites of mixed molybdate catalysts was generally accepted for more than three decades [27]. It was based on the idea that the composition of mixed-oxide surfaces can be derived from their bulk structure. Meanwhile, it is known that mixed-oxide surfaces can be reconstructed (see Section 3.4.6.3) and that Bi can be strongly depleted just on the surfaces of Bi molybdates [28]. The observation that Bi does not change its oxidation state during reaction (4.2) finally resulted in a new model, in which Bi exerts only an indirect influence on the active sites that exclusively operate via Mo atoms [29] (for the mechanism, see Section 4.5.2.2).

When the promising activity of Fe zeolites for the SCR of NO by hydrocarbons (or by NH_3, eq. (3.136)) was discovered, it was soon ascribed to binary Fe-O-Fe structures, which were considered the predominant, if not exclusive site in the catalysts [30]. This was questioned by the observation of a highly complex site structure in Fe-ZSM-5 catalysts of comparable activity. In the following, models according to which the SCR could be catalyzed by all exposed Fe sites, among them most efficiently by binary (or oligomeric) structures [31, 32], competed with the view that the reaction proceeds on isolated Fe oxo sites exclusively. Recently, it was established by post-catalytic Mössbauer studies that the rate of NH_3-SCR over Fe-ZSM-5 in the temperature range around 600 K originates almost completely from oligomeric clusters [33]. Contrary to the early (largely correct!) guesses, they represent, however, not more than 10–15% of the Fe available.

For V_2O_5/TiO_2 catalysts, the binary structure of sites for NH_3-SCR had been concluded quite early from the approximately squared dependence of SCR rates on the V content [34]. The exclusive presence of isolated Cu sites in the most active SCR catalysts known to date (Cu chabazites) has long challenged views on the generally binary nature of the most efficient SCR sites. In the NH_3-containing reaction environment, these sites are, however, coordinated with NH_3, which attenuates their interaction with the charged framework. It has been found that these species form temporary dimers, which allow fast reoxidation of the Cu sites, thus removing a bottleneck that limits the total reaction rate with isolated Cu sites [35]: apparently, NH_3-SCR (eq. (3.136)) also requires paired sites in Cu zeolites, where they can be temporary phenomenon. This agreement on the nature of the active sites in the most important SCR catalyst types, should, however, be considered no more than a stage of a scientific process, which does not conclude the discussion. The similarity of site structures does not necessarily support claims with respect to an identity of the reaction mechanisms (cf. Section 4.5.6).

A change of paradigms may be underway just now with respect to the active sites in the Ti silicalite TS-1 (cf. Section 2.3.2.3). It is applied to activate H_2O_2 in several commercial processes (see Appendix A1), and its active site has been believed to be an isolated framework Ti atom for decades. In an ^{17}O-NMR study, supported by DFT-based interpretation of spectra, a close analogy was found between the activation of $H_2^{17}O_2$ by TS-1 samples of different Ti contents, and by the only Ti complex capable of efficiently using H_2O_2 as a primary oxidant in homogeneous systems, a binuclear Ti complex [36] (cf. Scheme 4.8). At the moment, evidence seems to be conflicting: in well-prepared TS-1, pairing of Ti has never been observed by XAFS, and likewise, the techniques employed in ref. [36] to characterize samples without H_2O_2 adsorbed detected clustering only in some of them. Among the potential solutions for this dilemma, a transient nature of the dimeric Ti sites has been brought into play [37].

The development of active site models can also change views on the role of catalyst ingredients as active components or promoters. In early concepts on Cu/ZnO-based methanol synthesis catalysts, ZnO favored the reaction mainly by scavenging feed impurities or by stabilizing Cu^+ ions considered as the active sites by some groups. After the spectacular influence of defective ZnO on the morphology of Cu crystallites had been disclosed (cf. Figure 3.156), its effect on reaction rates was explained by a varying exposure of the most active Cu facets. While ZnO is a promoter in the models mentioned so far, the highest reaction rates have, meanwhile, been shown to arise from contact points at the perimeter between the Cu particles and the defective ZnO, which enable a route via intermediates bound to both Cu and Zn (cf. Section 4.2.4). Zn has become part of the active site in such a model. Its oxidation state in the binary sites (Zn^0, i.e., a Cu–Zn surface alloy, or $Zn^{(2-\delta)+}$, i.e., cations with an oxygen-deficient coordination sphere) is still a matter of debate [38–41].

It has been known for long that the extraction of sulfur from aromatic molecules like thiophene or dibenzothiophenes is accelerated by an order of magnitude when Al_2O_3-supported MoS_2 is promoted by Co or Ni. After long scientific controversies, it is meanwhile accepted that the improved activity arises from Co (or Ni) atoms decorating the S-edge of MoS_2 slabs – the CoMoS (NiMoS) phase [42] (Figure 4.15, for parasitic competitive structures in real catalysts; see Figure 2.43b). The Mo edges remain the same as in unpromoted MoS_2 (Figure 2.17). At the S-edges, Co or Ni takes the positions

a   b

MoS₂(10ī0)
(Mo-edge)

(100 % S)

MoS₂(10ī0)
(Mo-edge)

(100 % S)

CoMoS(ī010)
(S-edge)

(50 % S)

NiMoS(ī010)
(S-edge)

(50 % S)

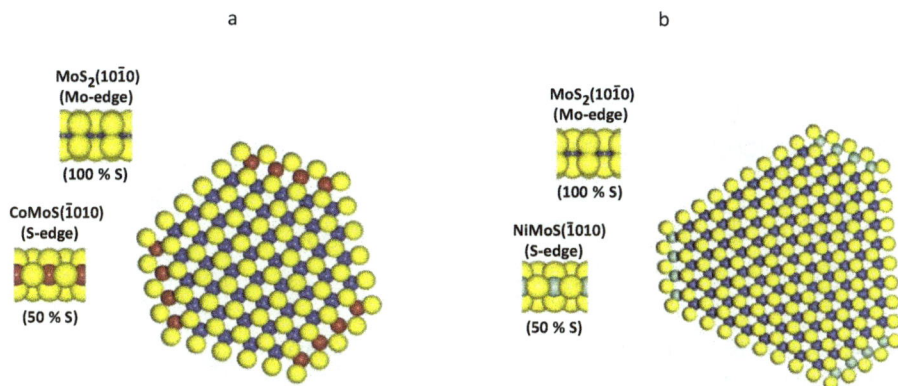

Figure 4.15: MoS₂ slabs with edges decorated by Co atoms (CoMoS phase, a) or by Ni atoms (NiMoS phase, b). Reproduced from ref. [43] with permission from Elsevier.

of Mo atoms, which also affects the termination by S species. The differences in slab shapes are due to the differences in surface energies of unpromoted and promoted edges.

While the actual termination of the promoted edges depends on the chemical potential of S, the Co (or Ni) sites tend to be more exposed than any unpromoted edge sites, which favors direct removal of sulfur from reactant molecules (DDS route, see Section 4.3.1). More reasons for the dramatic gain in activity can be found in ref. [43], but they all mean that Co (or Ni) interacts with reactants, at least in some steps of the reaction scheme. For these steps, Co(Ni) is part of the active site. For other steps, in particular, related to hydrogenation of reactants or products, it is just a promoter.

As mentioned in Section 3.1.5, stationary catalytic activity emerges from the initial interaction between the catalyst precursor and the feed. Among the various processes that can cause these transients, the formation of active sites from feed components (or of secondary species driving the reaction) represents a special aspect of the relation between the site structure and the activity. The oxidative dehydrogenation (ODH) of ethylbenzene to styrene, for instance,

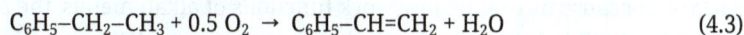

$$C_6H_5-CH_2-CH_3 + 0.5\ O_2 \rightarrow C_6H_5-CH=CH_2 + H_2O \qquad (4.3)$$

can be catalyzed by oxygen-containing carbonaceous deposits on surfaces, which do not catalyze this reaction themselves at all, e.g., Al₂O₃ or carbon materials like diamond or graphite, on which all active sites were previously removed by pretreatments [44]. Although these materials are only supports for the active coke phase, their properties like porosity, surface structure, are reflected in the performance of the resulting catalysts.

$$\text{CH}_3\text{OH} \longrightarrow (\text{CH}_2)_n \rightleftharpoons \text{C}_3\text{H}_6$$

with C_2H_4 above and C_4H_8 below $(\text{CH}_2)_n$, and arrows pointing to Saturated hydrocarbons and Coke.

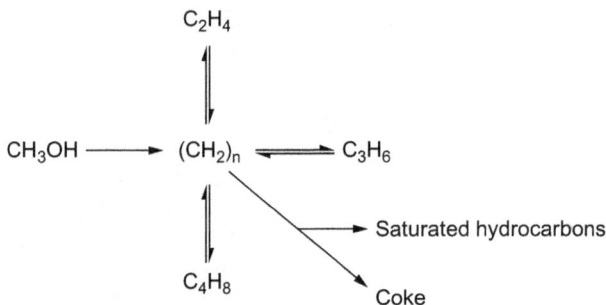

Scheme 4.1: Hydrocarbon pool mechanism of the methanol-to-hydrocarbon reaction. Reproduced from ref. [45] with permission from Elsevier.

The methanol-to-hydrocarbons (MTH) reaction is the most prominent example of a process requiring secondary species formed by conversions of the reactant on the active sites: a pool of olefinic or aromatic hydrocarbons (represented by $(\text{CH}_2)_n$ in Scheme 4.1). The catalytic cycle, which is discussed in more detail in Section 4.3.2, consists of repeated methylations of double bonds or aromatic rings, with the subsequent elimination of hydrocarbon products [46]. All catalytic steps proceed on acid sites of the zeolite, but the reaction does not proceed at any comparable rate without the intraporous hydrocarbon pool.

4.2.4 Promoting activity

Owing to the diversity of catalytic materials and reactions, there are very different options to promote catalysts for improving their activity. This section starts with the best understood phenomena of chemical (mostly electronic) and structural promotion, presents a number of examples illustrating the diversity of useful effects, and concludes with a case of promotion by a gas-phase component.

The influence of alkaline or alkaline earth ions on the splitting of triple bonds as in N_2 has been studied in some detail. Alkali ions are well stabilized on transition metal surfaces, because due to the low work functions of alkali metals, the state formed from their s level while interacting with the surface (Figure 4.4a) remains above the Fermi level. Alkali promoters are introduced into catalysts by impregnation with a suitable salt. After calcination and reduction, an O-deficient alkali oxide structure remains on the metal surface, which has an effect similar to that of a mere cation.

The electric field of the cations adds an electrostatic component to the adsorbate-metal bond, which has not been considered so far.[6] The sign and strength of this

---

6 Attenuating the surface dipole at the metal surface (Figure 2.6b), the cations create a local decrease of the work function. The resulting shift of the electron levels relative to the vacuum level concerns,

contribution depend on the charge of the adsorbates [7]. In a study of NH_3 synthesis on Ru surfaces [47], these interactions were examined for a field strength of the order expected for an adjacent alkali cation. The resulting energies were negligible for N (similar to the initial N_2 molecule [7]), but increasingly positive (i.e., destabilizing) for NH, NH_2, and NH_3. Moreover, the transition state of N_2 splitting was found to be stabilized by the positive charge, i.e., dissociation was accelerated.

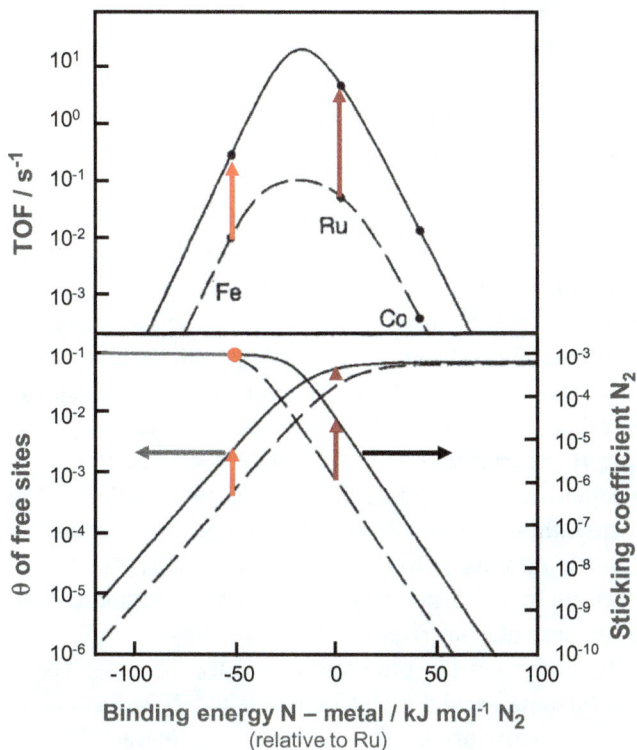

Figure 4.16: Influence of positive surface charges on ammonia synthesis: dependence of promoting effect on the nature of the catalytic element, evaluated for 673 K, 1 bar, H_2/N_2 = 3, and 0.2% NH_3. Adapted from ref. [47] with permission from Elsevier.

From microkinetic models with rate constants from DFT calculations and BEP correlations for promoted and unpromoted sites (Section 3.5.3, cf. Figure 4.2), interesting insight in how the promoter cooperates with different metals was obtained [47]. The upper panel of Figure 4.16 shows that the promoter increases the TOF by more than an order of magnitude on both Fe and Ru for the conditions specified in the legend.

---

however, both the d-band and the Fermi level, i.e., if there is any change of ε_d at all, it is inferior to the electrostatic effects discussed next.

The lower panel reports the changes of surface properties that determine the overall reaction rates: the abundance of free sites and the sticking coefficient of N_2 dissociation on them (the ratio between impinging and reacting molecules). On the Fe surface, which binds N stronger than Ru, the effect of promotion on N_2 splitting is small under the given conditions, but the quantity of free sites increases by almost an order of magnitude. This is absolutely crucial for iron, where less than one out of thousand surface atoms are available for interacting with N_2, even at only 1 bar and 0.2% NH_3 (cf. Figure 4.16), while real NH_3 synthesis reactors operate at $\geq$100 bar and up to 20% NH_3. Ru, on the other hand, benefits mostly via acceleration of the surface reaction, which is orders of magnitude slower without promoter than on (unpromoted) Fe (Figure 4.16). There is also some gain in the exposure of free Ru sites, but due to the weaker Ru-N bond, Ru surfaces are anyway less crowded with NH_x species than Fe surfaces under the same conditions.

The analogy in the electron structure of CO and N_2 may raise expectations of similar effects in the FTS, which is considered to require splitting of the C$\equiv$O bond in the most influential mechanistic models (Section 4.5.3). The case is, however, all but clear. Co-based FTS catalysts do not benefit from alkali ions at all. For Fe catalysts, where K is applied to promote selectivity and stability (see later), both positive and negative effects on activity have been reported, depending on the composition and conditions [48]. The problem is complex, because several quantities can be influenced by the promoter: the reduction degree of Fe, the rate of Fe carbidization, the rate of CO dissociation, the stability of Fe-C, Fe-O, and Fe-H bonds. As mentioned earlier, alkali promoters destabilize monoatomic adsorbates in NH_3 synthesis over Ru and Fe. However, in an experimental study of interactions between CO and pure polycrystalline Fe, the influence of K on the adsorption enthalpies of CO and of its fragments were not up to expectations from simple analogies: they increased, instead of decreasing [49].

The activity of Co-based FTS catalysts can be promoted by adding noble metal compounds [50] and/or precursors for Lewis-acidic oxides such as MnO, ZrO_2, La_2O_3, and TiO_2 [51]. The former are typical structure promoters. Co ions tend to interact strongly with catalyst supports. The noble metals easily formed during catalyst reduction activate hydrogen, which migrates to the Co ions (hydrogen spillover, cf. Section 4.5.4) and increases their reduction degree under the given conditions. On the other hand, the promoting effect of oxide species that are in contact with the surface of Co particles was found to scale with their Lewis acidity [51]. Therefore, they were suggested to favor the splitting of the CO molecule by interacting with its O atom: strictly speaking, they are part of the active site.

The classical unsupported iron catalyst for ammonia synthesis contains structure promoters as well. Its precursor almost completely consists of magnetite (Fe_3O_4, $\approx$95%). The remaining components like CaO, SiO_2, Al_2O_3, and K_2O play different roles in the generation and utilization of the catalyst.

The microstructure of such catalysts is very complex. Most of their catalytic effect originates from metastable Fe (nitride) platelets preferentially exposing the (111) facet,

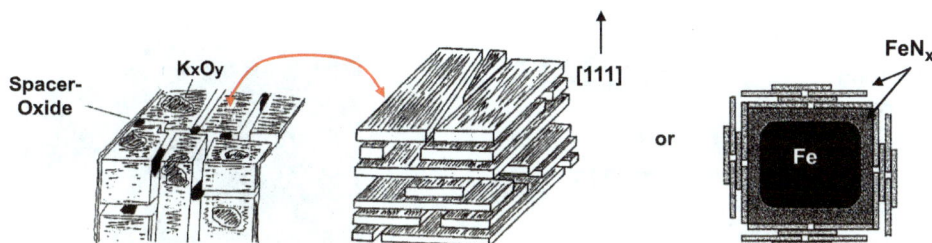

Figure 4.17: The microstructure of the iron catalyst for NH_3 synthesis. Reproduced from ref. [52] with permission from Springer Nature.

but it is not possible to convert the precursor into them completely. Figure 4.17 illustrates how they might be mixed with less active massive Fe crystallites. The tendency of the (111)-oriented platelets to stick together may be countered by clusters of spacer oxide squeezed in between them. Larger spacer oxide particles keep open pores between blocks of more or less parallel (111)-oriented platelets. Enabling access of the reactants to the active sites without having any relation to the latter, the spacers are typical structure promoters.

When the activation in the reaction mixture (not in H_2!) was preceded by a hydrothermal treatment at 723 K, inactive Fe facets were found to reconstruct into active ones, which relax, however, into the inactive, stable form during catalysis. Notably, if Fe was in contact with Al_2O_3 during the hydrothermal treatment, the (111)-oriented Fe was formed by reduction of the intermediate $FeAl_2O_4$ and remained stable under reaction conditions [53]. Thus, the promoting effect of Al_2O_3 exceeds its role as a spacer, and it concerns both activity and stability.

As mentioned in Section 4.2.3, the role assigned to ZnO in $Cu/ZnO–Al_2O_3$ catalysts for methanol synthesis has evolved from a scavenger for poisons via that of a structure promoter toward the view that Zn is part of the most active sites. The latter was strongly supported by an experiment, in which the activity of a Cu/Al_2O_3 catalyst was increased threefold despite some loss in accessible Cu surface area by just adding some Zn via the gas phase by MOCVD [54].

IR studies of adsorbed CO and measurements of the Cu surface area suggested that Cu becomes increasingly covered by (oxygen-deficient) ZnO at growing severity of the reduction conditions, as depicted by the model shown in Figure 4.18a. At the same time, severe reduction of $Cu/ZnO–Al_2O_3$ catalysts was found to boost their methanol synthesis activity: when a catalyst pretreated in a CO/H_2 feed was exposed to the standard $CO/CO_2/H_2$ mixture,[7] which is slightly less reducing, it produced a strong transient activity peak, the decay of which was related to some reoxidation of ZnO_{1-x} in the presence of CO_2 [57]. Figure 4.18a illustrates how the perimeter between Cu and the ZnO_{1-x} moieties

---

7 In Cu/ZnO catalysts, reaction rates are generally higher in CO_2-containing mixtures because methanol is formed from CO_2 (cf. Section 4.5.6).

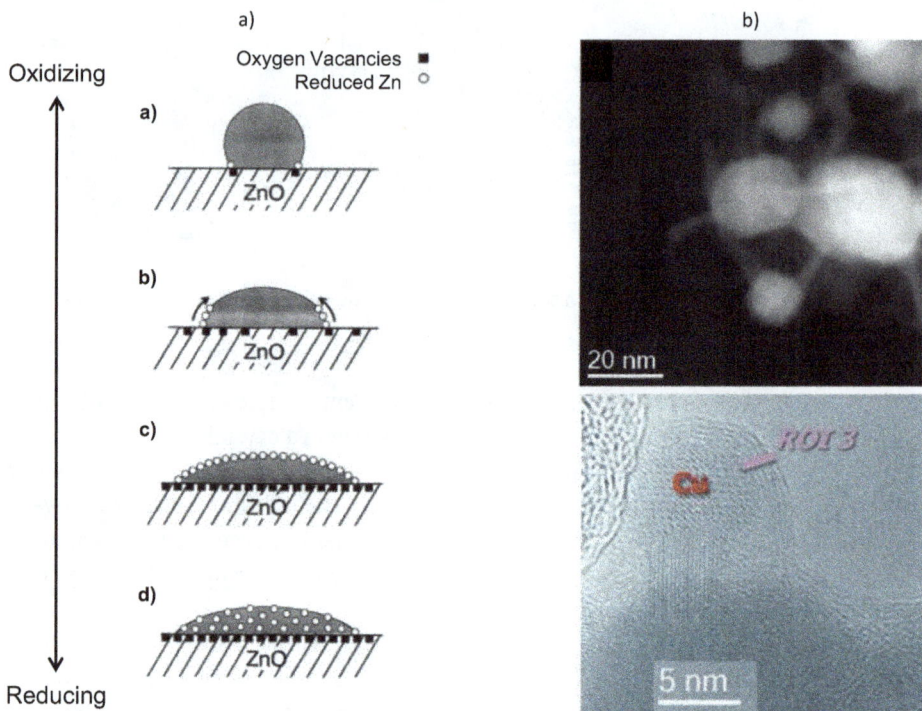

Figure 4.18: Decoration of Cu with ZnO under reducing conditions: (a) dependence of coverage on redox potential (schematic, lowest image designates bulk alloy), adapted from ref. [55] with permission from Elsevier, (b) micrographs of a catalyst prepared along industrial recipes and reduced in 20% H_2/Ar at 523 K; upper panel – HAADF-STEM; lower panel – HRTEM (for intensity scan along ROI3 – see original paper), reproduced from ref. [56] with permission from Wiley-VCH Verlag GmbH & Co. KGaA, Weinheim, Germany.

increases with growing oxygen deficiency of ZnO. The final form of a bulk alloy results, however, in a deactivation of the catalyst.

In Figure 4.18b, the overgrowth is visualized by an electron micrograph of a sample prepared along routes used also for industrial methanol synthesis catalysts, and activated by a typical reduction procedure. The micrographs are remarkable because they differentiate Cu and ZnO, despite their similar scattering properties. The upper panel shows ZnO in a rod-like morphology and the Cu particles surrounded by a diffuse sphere, in which Zn was detected. The HRTEM image shows ZnO in a graphite-like morphology covering the Cu – probably not densely.

Reducibility and mobility required for such dynamics under relatively mild conditions are not available to well-crystallized ZnO. Al^{3+} ions introduced into the catalyst in the initial co-precipitation step (cf. Section 3.1.1.1) prevent ZnO from crystallizing and stabilizing during subsequent thermal treatments, and keep it in a reactive form required for the interaction with Cu. While this is a typical effect of a structure promoter, Al^{3+} ions substituted on Zn sites should also be present on the reactive perimeter of ZnO_{1-x} with the Cu particles. Being unlikely to interact with the reactants like Zn, they exert an electronic influence on neighboring Zn and Cu atoms. Therefore, the Al component of the technical catalyst for methanol synthesis is now considered both a structural and an electronic promoter [58].

In the following, some more specific modes of activity promotion are briefly outlined. In supported noble metal catalysts, small particle sizes are important to achieve high exposure of the expensive active component. Naphtha reforming (cf. Appendix A1) is catalyzed by Al_2O_3-supported Pt nanoparticles, which are typically promoted by a second metal. At least two of these promoters, the main purpose of which is discussed in Section 4.3.5, are also structure promoters. Re and Sn cations are stabilized by the large alumina surface in nearly atomic dispersion. During reduction of the impregnated and calcined catalyst precursor, these cations, which are less reducible than the Pt precursor, provide nuclei for the formation of Pt particles. Therefore, the resulting Pt dispersion is better than that obtained along the same route without Re or Sn promoters.

As mentioned in Section 2.3.3, the K-promoted V_2O_5/SiO_2 catalyst for the oxidation of SO_2 to SO_3 in sulfuric acid plants is an SLP catalyst. The reaction is catalyzed also by V_2O_5, but its activity is by orders of magnitude lower. Only in the presence of alkali ions (in technical catalysts – K^+), a melt can be formed by the interaction of the precursors V_2O_5 and K_2SO_4 with the feed [59]. The high activity exhibited by the melt also gets lost when the melt solidifies below 700 K.

In acidic zeolites, e.g., H-ZSM-5, mild steaming is known to enhance the activity for protolytic hydrocarbon cracking (cf. Figure 4.31) by 1–2 orders of magnitude. The AlO_x species extracted from the framework by such treatment (cf. Section 3.1.4) are known to interact with adjacent Brønsted sites. The promoting effect was ascribed to an increased acidity of this paired site, but a recent study concluded that it is due to a modified transition state rather than to enhanced adsorption strengths: on the paired site, the transition state is closer to the product, which results in increased activation entropy $\Delta S^{\#}$, and hence, decreased $\Delta G^{\#}$ [60].

The final example deals with a promoting effect of gas-phase water on the oxidation of CO, which proceeds over supported gold nanoparticles with high rates, even far below 273 K. For temperatures around 300 K, it has been reported that addition of water in low amounts mitigates or even removes the pronounced differences in activities exhibited by gold on reducible and nonreducible supports under dry conditions [61]. In the presence of H_2O, the activation of O_2, which limits rates over Au on irreducible supports like Al_2O_3 in dry feed, can be bypassed by reaction (4.4)

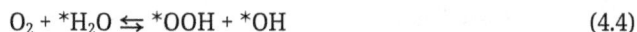

$$O_2 + {}^*H_2O \leftrightarrows {}^*OOH + {}^*OH \tag{4.4}$$

where * designates an Au surface site. The Au hydroperoxy species formed is known from other reactions catalyzed by Au nanoparticles. In eq. (4.5), it oxidizes an adsorbed CO molecule. The water invested in (4.4) is returned in eq. (4.6), while the remaining oxygen atom oxidizes a second CO molecule in eq. (4.7):

$$*OOH + *CO \rightarrow *OH + * + CO_2 \tag{4.5}$$

$$2\,*OH \rightarrow *H_2O + *O \tag{4.6}$$

$$*CO + *O \rightarrow CO_2 + * \tag{4.7}$$

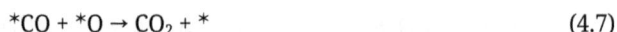

Comparing CO oxidation over Al_2O_3- and TiO_2-supported gold, the reaction was shown to obey the same rate law, which was derived from mechanism (4.4) through (4.7).

CO oxidation over supported Au nanoparticles is discussed in some more detail in Section 4.5.6.

4.3 Structure and selectivity

4.3.1 Side products and strategies

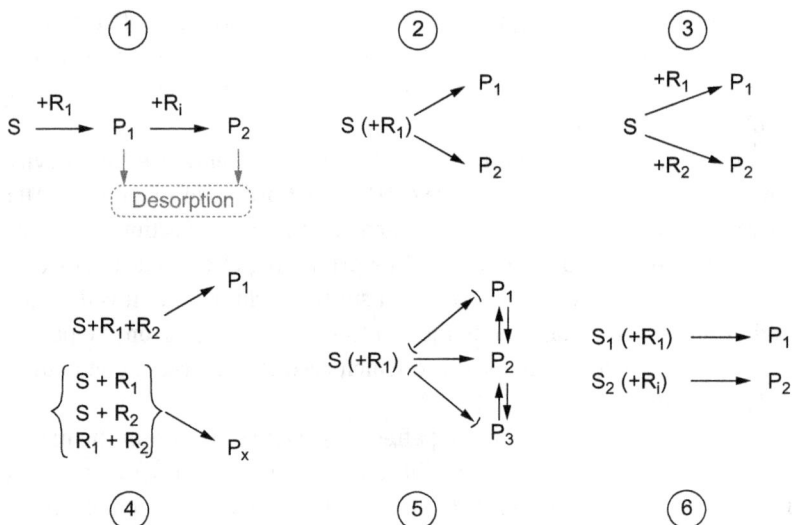

(1)

$$S \xrightarrow{+R_1} P_1 \xrightarrow{+R_i} P_2$$

Desorption

(2)

$$S\,(+R_1) \nearrow^{P_1}_{\searrow P_2}$$

(3)

$$S \begin{array}{c} \xrightarrow{+R_1} P_1 \\ \xrightarrow{+R_2} P_2 \end{array}$$

(4)

$$S+R_1+R_2 \nearrow P_1$$

$$\left\{ \begin{array}{c} S+R_1 \\ S+R_2 \\ R_1+R_2 \end{array} \right\} \searrow P_x$$

(5)

$$S\,(+R_1) \begin{array}{c} \nearrow P_1 \updownarrow \\ \rightarrow P_2 \updownarrow \\ \searrow P_3 \updownarrow \end{array}$$

(6)

$$S_1\,(+R_1) \longrightarrow P_1$$
$$S_2\,(+R_i) \longrightarrow P_2$$

Scheme 4.2: Mechanistic sources of by-products. Substrate S may be an intermediate in a reaction sequence; products P_i may be substrates in consecutive steps; contributions of reverse reactions neglected for the sake of simplicity.

Selectivity is a key issue in technical catalysis because it decides on raw material consumption and on the extent of waste production. Side products result from branching and consecutive steps in the reaction network. As any discussion of selectivity problems and opportunities for their solution remains abstract without examples, this section presents ample information about catalytic reaction mechanisms and networks, without, however, being comprehensive. Additional information on the topic can be found in

Section 4.5, which deals with types of catalytic reaction mechanisms, and in the appendix A1, which summarizes relevant information on important catalytic processes.

Typical situations resulting in side-product formation are shown in Scheme 4.2. Only the selectivity-determining step has been displayed: in the branching versions 2 through 5 and the parallel reaction 6, the starting material (substrate) **S** may result from a series of steps, and the products P_i may initiate further steps. Reverse reaction steps have been neglected for the sake of simplicity. In catalysis, situations 1 through 6 describe surface reactions: the products are surface species, and so are substrate and reactant, or at least one of them. Adsorption and desorption have been omitted, except for the consecutive case 1. In surface catalysis, each consecutive step competes with desorption.

In case 1, the reactants R_i may be identical or different. Sometimes, they are the same species, but with different reactivity, e.g., nucleophilic and electrophilic oxygen (cf. Section 4.3.3), and surface or subsurface hydrogen (cf. Section 4.3.4). Selective alkyne hydrogenation and selective hydrocarbon oxidation are typical examples for case 1 – P_1 being the targeted alkene or oxygenate and P_2, the alkane or an intermediate along the route to CO/CO_2. There are, however, also situations where the final product is the desired one, e.g., CO_2 in hydrocarbon combustion. CO, on the other hand, is almost always undesired, be it in catalytic combustion for energy production or for the abatement of air pollutants, be it as a side product in selective oxidation: it is targeted only in syngas production.

Cases 2 through 5 describe different versions of competitive reactions. Case 2 includes reactions with molecules offering different types of reactivity, e.g., a C=C and a C=O bond. High chemoselectivity in such reactions is a challenge in both catalytic and noncatalytic chemistry. Other realizations of case 2 are specific for catalysis. In selective oxidation of hydrocarbons, not only is the product at risk of combustion (case 1), the substrate is also at risk (case 2). Likewise, in the selective hydrogenation of alkynes, the selective step may compete with a direct route to the alkane. In the presence of H_2, hydrocarbons can be isomerized on surfaces of Pt and of other metals, but this hydro isomerization is often accompanied by the cleavage of the molecule, i.e., hydrocracking. In the HDS of oil fractions, sulfur can be either directly split from the contaminant, or the sulfur-containing molecules must be hydrogenated prior to sulfur removal. In literature, these routes are designated as direct desulfurization (DDS) and hydrogenation (HYD) pathway, respectively (cf. Appendix A1).

In case 3, different reactants compete for the same substrate. This is typical of the FTS, where hydrocarbon molecules are built up from C_1 units originating from CO. In each reaction step, the growing chain is either extended by a C_1 unit (R_1) or desorbed by reacting with surface hydrogen (R_2). On some catalysts, an additional competition between CH_x and C_1 units that still carry oxygen may result in the simultaneous formation of hydrocarbons and oxygenates. The conversion of syngas to (mixed) alcohols with more than one C atom by tuning the selectivity of this step (Higher alcohol synthesis, HAS) has drawn much attention in recent syngas chemistry.

Case 4 is typical for reactions in which a molecule is converted by both an oxidizing and a reducing component, e.g., the SCR of NO. Equation (3.136) shows the version with the reductant NH_3, but over suitable catalysts, the SCR proceeds also with other reductants such as hydrocarbons or even H_2. Hydrocarbon ammoxidation to nitriles, e.g., of propene to acrylonitrile (4.8), is another important reaction complying with case 4:

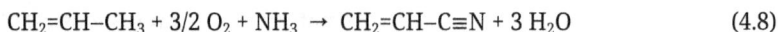

$$CH_2=CH–CH_3 + 3/2\ O_2 + NH_3\ \rightarrow\ CH_2=CH–C{\equiv}N + 3\ H_2O \qquad (4.8)$$

In (3.136) and (4.8), the catalytic oxidation of ammonia is an undesired side reaction interfering with high conversions of the substrates (NO, C_3H_6).

Case 5 is very common in acid catalysis. The products P_i may be carbenium ions on the surface[8] or desorbed products, which find new sites for consecutive reactions. The interconversion of products P_i may be fast, as for instance, the skeletal isomerization of carbenium ions. In other cases, the products may react in consecutive steps without an established equilibrium between them. In acid-catalyzed hydrocarbon cracking, for instance, the carbenium ion formed from the substrate is immediately isomerized to equilibrium, in which tertiary structures are most abundant. The resulting mechanisms of cracking are discussed in some detail in Section 4.5.2.

The intra-zeolite hydrocarbon pool, enabling the conversion of methanol to gasoline or to various olefins (MTH, cf. Sections 4.2.3 and 4.3.2) may be considered an example of case 5 as well. The products P_i are desorbed into the zeolite cavities where they find sites for further methylation, isomerization, or cracking steps that release hydrocarbons. Equilibrium between molecules in the pores is usually not established.

Box 4.1 – Autothermal reforming of methane.

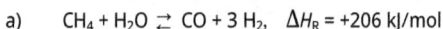

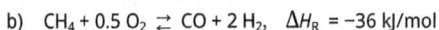

    a) $CH_4 + H_2O \rightleftarrows CO + 3\ H_2, \quad \Delta H_R = +206\ kJ/mol$

    b) $CH_4 + 0.5\ O_2 \rightleftarrows CO + 2\ H_2, \quad \Delta H_R = -36\ kJ/mol$

The reactions are coupled via their enthalpies: reaction b) feeds reaction a).

Processes of H_2 (or syngas) production from natural gas, e.g., the autothermal reforming of methane (Box 4.1) or the partial oxidation of hydrocarbons (see case b in Box 4.1 for methane) are quite different implementations of case 5. Box 4.1 shows only the desired routes, but hydrocarbon combustion to CO_2 and water, and the equilibration between products of partial and total oxidation proceed over the catalysts as well. With increasing temperature, combustion becomes disfavored for equilibrium reasons. Under conditions resulting in predominant total oxidation, equilibrium within the reaction mixture is often not established, but it is at the higher temperatures used in commercial processes to obtain large yields of CO and H_2.

Case 6 in Scheme 4.2 refers to competitions between the substrate molecules. The reactions may be completely independent ($i = 2$) or coupled via a reactant ($i = 1$ – substrate selectivity, with the special case of enantioselectivity). Case 6 is of paramount

---

8 For more details on carbenium ions and the problems of their stability, see Section 4.5.2.

importance because commercial feeds are often product mixtures (e.g., oil fractions). With catalysts capable of converting the desired feed component, exclusively or predominantly (i.e., exhibiting substrate selectivity), there may be no need for an expensive separation step upstream of the reactor. Examples of case 6 are the preferential cracking of n-alkanes (coming together with branched alkanes) for improving the octane number in the downstream gasoline production, the hydrogenation of alkyne contaminants to alkenes in C_2 or C_3 streams from steam crackers, and the preferential oxidation (PROX) of residual CO in hydrogen feeds for fuel cells. Creating enantiomeric excess by enantioselective conversion of racemic mixtures is another attractive field, in which, however, achievements of molecular catalysis have so far overshadowed those of surface catalysis. The reader may find information on enantioselective surface reactions in refs. [62, 63].

Reactions paired in schemes of type 6 can be rather diverse. In the abatement of NO in flue gases from coal-fired plants by SCR (eq. (3.136)), the catalyst should not oxidize SO_2 to SO_3, because $(NH_4)_2SO_4$ formed from it and the unconverted NH_3 may clog the catalyst. Situations of type 6 also occur in complex reaction networks: in naphtha reforming (cf. Figure 1.6 and Section 4.5.4), the target aromatics are formed by dehydrogenation of (substituted) cyclohexanes, but cyclopentanes are also present in the reaction mixtures. Their dehydrogenation to cyclopentadienes is the first step to coke formation.

There are a number of strategies to handle the various risks of losing the target product to side reactions. Before dealing with strategies specific for catalysis, those based on thermodynamics and on kinetic rate laws are briefly mentioned. In systems in which catalysts allow establishing or approaching equilibrium, such strategies are related to the Le Chatelier principle, which suggests, for instance, high temperatures to favor CO and H_2 over CO_2 and H_2O in syngas production. Far from equilibrium, selectivity in competitive reactions, differing with respect to activation energies and/or kinetic reaction orders, can be improved by appropriate choice of reaction temperature and feed concentrations or pressures. With consecutive reactions, a reaction regime with minimum backmixing and a well-defined residence time that allows trapping the target product at its concentration peak are both important.

Opportunities for selectivity control depend on the type of reaction network (cf. Scheme 4.2) and on the type of sites involved. Strategies are different for reactions using different sites or sites of the same type. The latter situation, which is most challenging, in particular for consecutive schemes (case 1), will be dealt with first. Consecutive schemes at the heart of selective hydrocarbon oxidation have raised a great amount of insight and expertise, which are reviewed in a separate Section 4.3.3. They are also important in selective hydrogenation, which is discussed in Section 4.3.4.

As in noncatalytic chemistry, recovering high yields of the intermediate is most difficult if it is more reactive than the primary substrate. If the consecutive product can be reinserted into the process, it may make no sense to completely suppress its formation. In the Friedel–Crafts alkylation of benzene with ethene to ethylbenzene, for instance,

$$C_6H_6 + CH_2=CH_2 \rightarrow C_6H_5-CH_2-CH_3 \tag{4.9}$$

diethylbenzene formation is hard to avoid. In the analogous alkylation of benzene with propene to cumene, consecutive alkylation can be suppressed only by keeping benzene conversion below 2%, which afflicts the process with the evil choice between large benzene recycle streams or a demanding separation of (poly)alkylbenzenes. Suitable catalysts allow solving this problem by offering both alkylation and transalkylation activity, which allows producing more monoalkylbenzene by reacting the higher alkylated products with benzene in an extra stage. Ethylbenzene and cumene are produced over zeolite catalysts nowadays (H-ZSM-5 and wide-pore zeolites like H-BEA, MCM-22, respectively).[9]

If competing reactions (case 2) use the same sites, it may be possible to influence their relation via the polarity or the spatial extension of their transition states. In the liquid phase, this can be handled via solvent properties. In the gas phase, such opportunities require confined spaces with specific spatial restrictions, electrical fields or wall interactions. Zeolites may confine the space for bulky transition states (cf. Section 4.3.2). A polar transition state will experience different interactions depending on the polarity of the framework polarity (i.e., its Al content). Significant field effects were also reported for the interior space of nm-sized CNTs. More information on the influence of spatial constraints on thermodynamics and kinetics can be found in refs. [64] and [65] and in Section 4.7.

In Fischer-Tropsch synthesis (FTS, (case 3)), the different identities of reactants ($R_1 = CH_x$, $R_2 = H$) offer additional opportunities to adjust selectivity. Although the growing chain can be desorbed as an α-alkene, high concentrations of surface hydrogen favor its termination by consecutive hydrogenation to the alkane. Product distributions in FTS depend crucially on the stability and surface coverages of CH_x and H adsorbates, which can be influenced by the feed composition, the reaction temperature, and the reactivity of the applied metal. The consequences are discussed in some detail in section 4.5.3.

Interconversions between intermediates (case 5) can be used to influence selectivity by steric constraints on the scale of molecular dimensions. Such restrictions, as encountered in zeolites, affect the nature and concentrations of intermediates P_i, and thus of the final products released. Bulky products kept from leaving the pore system via narrow exits may have no chance than to undergo conversion to molecules of slimmer shape. Aspects of shape selectivity are discussed in section 4.3.2.

When the target and side reactions proceed on sites of different types, information on their nature raises additional opportunities for addressing the selectivity problem by enhancing the target reaction and/or by discouraging the side reactions. Sites for side

---

9 It should be noted that there is no firm link between alkylation and transalkylation activity of a catalyst. In a former process for cumene production working with solid phosphoric acid (SPA, cf. Section 2.3.3), benzene conversion was indeed confined to 2 % due to insufficient transalkylation activity of SPA.

reactions may be rejected by optimizing catalyst preparation, or they may be selectively blocked by poisons. Examples of the latter strategy (sulfur poisoning of hyperactive sites in naphtha reforming or steam reforming) have been mentioned already in Section 4.2.3. More examples follow in Section 4.3.5, because poisons for the side reactions are promoters for selectivity at the same time. Parasitic oxidation of reactants, e.g., of NH_3 in NH_3-SCR (eq. (3.136), case 4), is a side reaction that can be suppressed by proper catalyst preparation. It is known to proceed on small disordered oxide aggregates where the adsorbing reactant finds an excess of labile lattice oxygen. Over well-prepared Fe-ZSM-5, NH_3 oxidation remains negligible at temperatures up to 770 K. Likewise, the ODH of propane to propene over V oxide species dispersed on silica surfaces is most selective when the formation of oxide clusters can be avoided in the preparation [66].

Rates of target reactions may be enhanced by a proper choice and treatment of the catalytic component and/or by suitable promoters. In Section 4.3.4, how metal surfaces can be modified to hydrogenate a C=O group rather than a nearby C=C bond is illustrated. In metal-catalyzed hydrocarbon reactions, product selectivity depends on the particle size: over very small particles, the abundance of highly exposed sites favors cracking, over reactions preserving the number of C atoms, which become predominant on larger particles.

Coke deposition is a side reaction of desired hydrocarbon conversions, which may be related to case 6 in Scheme 4.2. It links the selectivity problem with the stability issue in important industrial processes. Typical strategies to suppress coke formation on metal surfaces involve the destruction of extended sites (ensembles) by promoters (cf. Section 4.3.5) and/or increased H_2 pressures to hydrogenate coke precursors. Naphtha reforming, the conversion of C_7–C_9 open-chain hydrocarbons and of substituted cyclopentanes into aromatics (cf. Figure 1.6, Section 4.5.4) is a typical example of such a situation. Due to its endothermal and volume-expanding nature, thermodynamics suggest best performance at high temperatures and low pressures. To cope with coke formation, the reaction may be performed at relatively high H_2 pressures (20–40 bar at $\approx$770 K), which allows stable aromatization of pentacyclic structures, while open-chain molecules are only isomerized.

Responding to market demands, the process was completely rearranged in the 1990s. The temperature was raised by $\approx$50 K and the H_2 pressure was decreased to <10 bar in order to achieve aromatization of the open-chain components also. Catalysts were modified to handle the increased coke selectivity (cf. Section 4.3.5). The remaining deactivation was made up for by a new process structure. Instead of a stationary catalyst in a series of fixed-bed reactors, one of which would be regenerated at intervals of some months, the new process featured moving beds and a catalyst recycle. After passing the series of reactors in ca. 1 week, the deactivated catalyst is regenerated in this process and reactivated before being fed back to the top of the first reactor. Both process versions are still operative nowadays.

Catalytic cracking of hydrocarbon fractions is another important process involving strong coking tendencies. In FCC, cracking proceeds in a riser (cf. Section 2.8.2), at residence times on the order of a second, and in the absence of H_2. As there is no significant heat transfer in the riser, the catalyst is cooled and coked simultaneously. However, coking is part of the technology: the coke is combusted in a regenerator fluid bed where the catalyst is loaded with the heat required for the next pass through the riser. This kind of heat management is most effective, but places restrictions on the properties of feeds to be processed, in particular with respect to coke formation tendencies.

Catalysts that suppress coking contain a hydrogenating function that is stable toward H_2S because of the sulfur impurities in heavy feeds. NiMo sulfides are the typical choice. The resulting hydrocracking processes differ dramatically from FCC. Due to H_2 pressures up to 200 bar, reactants are liquid and are typically converted in trickle-bed reactors. While residence times are long under such conditions, temperatures required are much lower than in FCC. Though being more expensive than FCC in both investment and operating cost, hydrocracking plays a significant role in refineries because of its flexibility with respect to both feeds and products, in particular because of its capability to process even very heavy feeds.

Admission control is an elegant approach to achieve selectivity in the competition between substrates (case 6). Hydrophobic pores may discourage polar reactants, which are preferred by hydrophilic pores. Shape selectivity may exclude bulky molecules from the active sites. The latter is discussed in more detail in the next section. More information about polarity effects in the environment of intra-zeolite active sites can be found in ref. [64].

4.3.2 Shape selectivity

Shape selectivity requires a pore system with dimensions comparable to those of the reactants. It is therefore almost exclusively discussed for zeolites and related materials (zeotypes), but it was also reported for other microporous materials, e.g., heteropoly compounds [67] and MOFs [68]. There has also been some effort to extend the concept by synthesizing shaped cavities around active sites (see later).

The different modes of spatial influence on catalytic processes have been generalized into three types of shape selectivity (Figure 4.19). Reactant shape selectivity, based on molecular sieving effects (Figure 4.19a), is a tool for admission control to the active sites. Pore sizes <0.6 nm required for most applications are provided by 10-ring apertures in zeolites. Diffusivities of reactants in the narrow pores differ dramatically, depending on the molecular dimensions (kinetic diameters), which penalizes access of more spacious molecules by orders of magnitude, though not excluding them totally. FCC catalysts, the cracking activity of which is mainly contributed by ultrastabilized Y zeolite, often contain the shape-selective H-ZSM-5 as a minority component. Such catalysts are able to selectively crack *n*-alkanes, which improves the octane number of gasoline fractions produced by the cracker.

Product shape selectivity is based on intrapore diffusivities as well, but it operates by sieving the reaction products, which can leave the pore system only when their kinetic

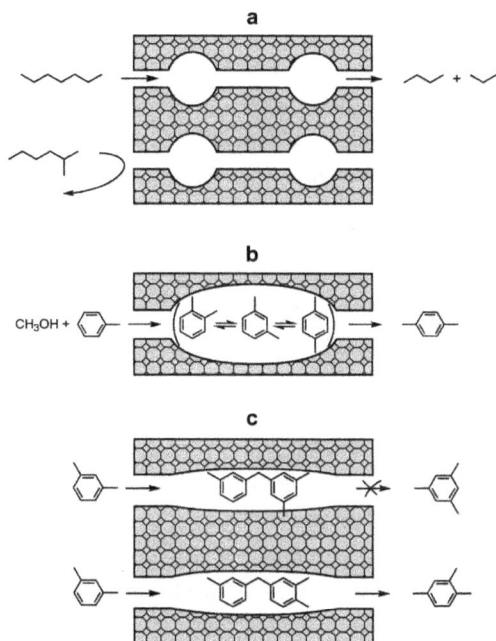

Figure 4.19: Types of shape selectivity: (a) reactant shape selectivity; (b) product shape selectivity; and (c) restricted transition state selectivity.

diameter complies with the pore size (Figure 4.19b). As the influx of reactants is not confined, a stationary reaction regime requires fast interconversion between the products (case 5 in Section 4.3.1), which allows the bulkier molecules to adapt a more suitable shape. Methylation of toluene with methanol over medium-pore zeolites, e.g., H-ZSM-5, is a well-studied example of this mode of shape selectivity. All xylene isomers were observed in the zeolite cavity by *in situ* IR spectroscopy [69]. *p*-Xylene selectivity can be further enhanced by working with large, benign zeolite crystals or by poisoning Brønsted sites on the external zeolite surface using a silylation treatment. Such treatment can also constrict pore mouths at the external surface, which enhances the sieving capability of this bottleneck.

Opposed to the modes of shape selectivity discussed so far, restricted transition state shape selectivity is not related to diffusion rates. It rather penalizes reactions, the transition states of which do not comply with the spatial conditions in the cavities. It is usually exemplified by the disproportionation of xylenes to toluene and trimethylbenzenes (Figure 4.19c), where the yield of pseudocumene is enhanced at the expense of mesitylene, which is formed via a transition state of prohibitive size. In a computational study of this reaction in different pore systems [70], influences of spatial restrictions on local isomerization rates were indeed confirmed. However, spatial restrictions may also operate via entropic effects and via changing contributions of competing reaction mechanisms. This may be superimposed by diffusional limitations with respect to reaction products escaping from the pores, i.e., product shape selectivity. To summarize, restricted transition state shape selectivity exists, but is more complex than suggested by traditional explanations. It covers all influences by spatial constrictions in pore systems, except for those operating via the rates of reactant access to it and of product release from it.

The MTH reaction (cf. Section 4.2.3, Appendix A1) is a typical example of transition state shape selectivity in this wider understanding. The nature of the hydrocarbon pool ($(CH_2)_n$ in Scheme 4.1), and hence the reaction products depend on the properties of the pore system [46]. In the most confined spaces (one-dimensional 10-ring pores, H-ZSM-22), the pool consists of linear and branched alkenes and releases alkenes, but almost no ethene. In H-ZSM-5, the wider channel intersections allow the additional formation of aromatics with low methylation degree, which are typical sources of ethene. Products range from alkylbenzenes to short-chain alkenes, featuring significant ethene selectivity. Zeolites offering larger voids, e.g., H-BEA or H-SAPO-34 (CHA topology), accommodate higher methylated benzenes also, which release propene preferentially. Due to the eight-ring apertures between its cages, H-SAPO-34 additionally offers a pronounced product shape selectivity, preventing the release of aromatic products. Thus, H-SAPO-34 has become the typical catalyst for MTO or MTP applications.

The lock-and-key principle of enzyme catalysis is shape selectivity in its ultimate perfection: reactants are guided to the site, oriented for easy access of the transition state. Efforts have been made to mimic such effects in catalysis using a combination of surface organometallic chemistry (cf. Section 3.1.3.4) with the molecular imprinting approach. In a typical synthesis, a placeholder ligand having a shape similar to the supposed transition state is coordinated to a surface organometallic catalyst. A SiO_2 matrix is piled up around these sites using, for instance, tetramethoxy silicon ($Si(OCH_3)_4$) as a polymerizing reagent. After removing the placeholder ligands, vacancies of suitable shape remain at the organometallic sites. Though being far from industrial practice, successful implementations of such a biomimetic approach as described in ref. [71] may open new perspectives for surface catalysis.

4.3.3 Selective oxidation in the gas phase

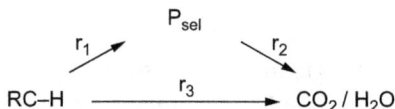

Scheme 4.3: The basic consecutive scheme of selective hydrocarbon oxidations.

Selective hydrocarbon oxidations are prime examples of processes targeting the intermediates of consecutive reactions. This holds, in particular, for gas-phase oxidations with O_2, which are discussed in the present section (for selective oxidation in the liquid phase, see Section 4.7). Scheme 4.3 depicts the typical structure of such reaction systems in a simplified way. To obtain the selective product P_{sel} in high yields, the rate r_1 of its formation needs to significantly exceed the rate of r_2 of its combustion, as also the rate r_3 of reactant combustion. A promising catalyst achieves high selectivity for P_{sel} up to high hydrocarbon conversions: the dependence of selectivity on conversion is an important criterion for ranking catalysts.

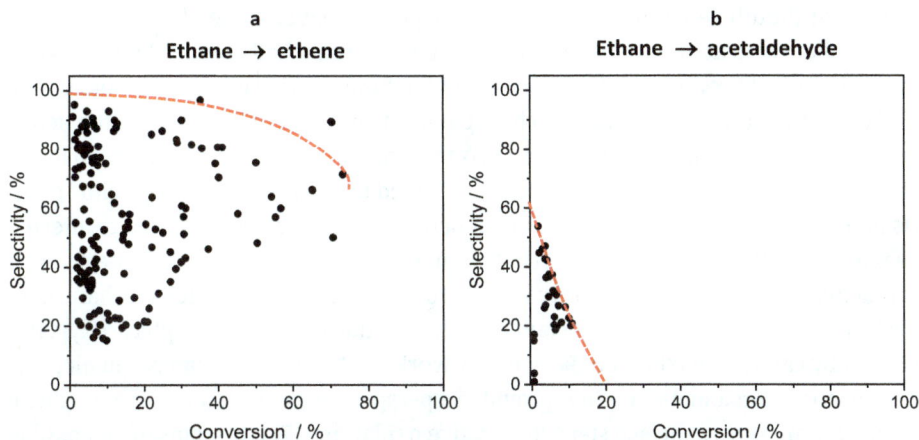

Figure 4.20: Selectivity/conversion data of two oxidative conversions of ethane as of 1996. Adapted from ref. [72] with permission from Elsevier.

Analyzing literature on relevant selective gas-phase oxidations ca. 25 years ago, Batiot and Hodnett [72] observed characteristic patterns in the relation between selectivity and conversion. Examples of these compilations are shown in Figure 4.20. There is strong scatter in the data because all pertinent papers were included, notwithstanding the quality of catalysts described. The diagrams for oxidative conversion of ethane to ethene (ODH, Figure 4.20a) and to acetaldehyde (Figure 4.20b) differ, however, strongly in the distribution of points. They are all in the left lower corner in the latter case, but extend to pretty high selectivities even at conversions >60% in the ODH of ethane. Obviously, it is extremely difficult to obtain acetaldehyde in such yields from ethane.

In ref. [72], these differences were ascribed to differences in the stabilities of the reactant and the product (RC-H and P_{sel} in Scheme 4.3). The selective product has to withstand the conditions required for the activation of the reactant, until it has escaped from the catalyst. The weakest bonds in acetaldehyde and in ethene have dissociation energies of 349 kJ/mol (the C–C bond) and 444 kJ/mol (the C–H bond), respectively. Obviously, ethene has the better chance to be recovered from the catalyst. For 14 relevant reactions, the authors related the differences between the dissociation energy of the weakest C–H bond in the reactant and the weakest (C–H or C–C) bond in the product to the selectivity/conversion pattern extracted from literature and the supposed limiting performance (cf. dashed lines in Figure 4.20). They concluded that catalysts with promising selectivity/conversion characteristics should exist for reactions where the dissociation energy of the weakest C–C or C–H bond in the product is not more than 30 kJ/mol smaller than that of the weakest C–H bond in the reactant. When this difference is >70 kJ/mol, the existence of catalysts allowing attractive yields of P_{sel} is unlikely.

It should be noted that these conclusions are based on thermal stabilities of molecules in the gas phase, while catalysis occurs with species activated by surfaces. They are rules of thumb generalizing a limited (though extensive) scope of literature along plausible heuristic assumptions, but they highlight a real challenge in the process: to minimize contact between labile products and the catalyst. This is why optimum porosity of partial oxidation catalysts, which originates from a trade-off between the active surface area and the chance for the products to escape from there rather tends to lower values, and why basic sites are deliberately poisoned in catalysts producing acidic products and vice versa. For the same reason, much attention has been devoted to alternative reaction regimes like ultra-short contact time catalysis, staged oxygen dosage along the reactor, e.g., via membranes, or realization of the Mars–van Krevelen mechanism in a cyclic reaction regime where reactant oxidation in the absence of gas-phase oxygen is followed by catalyst reoxidation. Recent DFT work on the ODH of alkanes and alcohols over various polyoxometallates has provided a perspective on descriptors that might be suitable for analyzing surface steps in selective oxidations [73]. It seems to be possible that systems breaking the rules of Batiot and Hodnett may be found by such effort in the future, but probably not soon due to the complexity of surface processes even in catalysts of well-defined structure.

As *in silico* screening is not around the corner in this field, it is worth considering the heuristic concepts that have allowed establishing a number of successful selective hydrocarbon oxidations in industry. One of these concepts is the differentiation of oxygen species encountered in oxidation catalysts and of mechanisms preferred by them.

As long as recombination to O_2 is not possible, O atoms tend to claim electrons from reaction partners in order to stabilize in the −2 oxidation state in products: they are strong electrophilic reagents. While the energy of O atoms or ions increases with their charge in the gas phase, this order is reversed in the ionic lattice of oxides due to the Coulomb interaction with cations, which turns lattice oxygen anions into nucleophilic agents. Their strength depends on the nature and distribution of cations and on the degree of coordinative unsaturation (cf. Figure 2.20), which provides opportunities to tune their reactivity. The actual role of the different types of lattice oxygen in oxidation reactions, e.g., of atoms bridging between identical or different cations or of terminal oxygen at short distance to a cation (e.g., V=O, Mo=O, cf. Figures 2.18 and 2.19), has been the subject of lively discussions [74]. Peroxide species, which can be observed by Raman spectroscopy, have been considered relevant for selective gas-phase oxidations only in rare cases.

O atoms or anions insufficiently stabilized by Coulomb interactions are strong electrophilic agents causing vigorous reactions, which often break the carbon skeleton of hydrocarbons. Oxygen adsorbed on metal surfaces does not respect C–C bonds at all: metal sites exposed under reaction conditions always catalyze total oxidation of hydrocarbons, with the notable exception of silver (see later).

Electrophilic attack in oxide-catalyzed reactions is usually ascribed to one of the radical anions, O_2^- or O^-, although other forms that are more difficult to detect cannot be ruled out. Adsorbed O_2^- and O^- can be prepared via recipes reported in the older literature. Both can be detected by EPR (O_2^- also by IR spectroscopy due to its asymmetric charge distribution) up to 570 K. Due to their extreme reactivity (O^- is known to oxidize H_2 already at 77 K!), they certainly influence catalytic reactions even at much higher temperatures, despite low concentrations and short lifetimes. Both radical anions are considered intermediates in the reduction of O_2 to lattice oxygen:

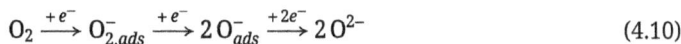

$$O_2 \xrightarrow{+e^-} O_{2,ads}^- \xrightarrow{+e^-} 2\,O_{ads}^- \xrightarrow{+2e^-} 2\,O^{2-} \tag{4.10}$$

When O_2^- or O^- find a molecule to react with, the resulting electrophilic oxidation complies with Scheme 2.3. Such a reaction may be selective if the product is sufficiently stable. The oxidation of benzene to maleic anhydride or of naphthalene to phthalic anhydride are examples of electrophilic oxidations applied in commercial processes. Their use is, however, limited because the products can meanwhile be obtained without losing part of the feedstock molecules to CO_2.

On the other hand, the oxidation of ethylene to ethylene oxide shows that electrophilic oxidations can be even gentle under suitable circumstances: selectivities are up to 90% now in commercial plants. The mechanism and the surface processes have been subject to long discussions in literature [75]. Under reaction conditions, Ag is strongly modified by subsurface oxygen, and the catalytic surface is that of an oxide rather than of a metal. According to ref. [75], the electrophilic attack of surface oxygen of $Ag_2O(001)$ on the C=C bond is a Eley-Rideal step, and completely selective. However, the bare Ag site left over offers routes that may end up in ethylene oxide, acetaldehyde, or even in unselective oxidation, when another ethene molecule gets access. Therefore, impeding ethene adsorption in this O vacancy, e.g., by working at high O_2 partial pressures, helps suppressing the side reactions. High O_2 pressures, together with a joint influence of promoters (Cs, Cl, Re), are also required to keep the oxide surface stable at the typical reaction temperatures [75].

Ethylene oxidation on silver surfaces is also special with respect to the mechanism of the competing combustion, which has been found to start with an attack of nucleophilic oxygen at the vinyl C–H bond [75]. Nucleophilic oxidation can be highly selective, provided the reactivities of the catalytic surface and the reactant are well matched. In most cases, the initial H abstraction from the reactant, which results in a surface OH group and an alkyl species (if starting with an alkane) is rate-determining. If another H is abstracted at a β-carbon atom, the reactant becomes dehydrogenated, and the water formed from the two OH groups is desorbed. To close the cycle, the removed oxygen atom is replaced and the cations, which are reduced during the previous steps, are reoxidized (cf. Figure 2.55). An example of such ODH has been given earlier in eq. (4.3).

Alternatively, a lattice O atom may be inserted into the adsorbate activated by the first hydrogen abstraction. After losing another H atom to the catalyst, the substrate leaves the surface as a carbonyl compound. As the H atoms obtained from the reactant are released from the surface as water, two oxygen vacancies need to be re-filled in the final reoxidation step. The allylic oxidation of propene to acrolein (eq. (4.2)) is a typical example of such a process. Over suitable catalysts, NH groups formed from gas-phase NH_3 may be inserted into activated feedstock molecules, instead of O atoms. Equation (4.8) shows how acrylonitrile can be obtained from propene by this ammoxidation (for more details see Section 4.5.2.2). Likewise, methyl groups on aromatic rings can be converted to nitrile groups.

Nucleophilic oxidation is the archetype of the Mars–van Krevelen mechanism (Figure 2.55). Oxidation of the starting material, maybe in multiple steps, and reoxidation of the surface by reduction of gas-phase oxygen are sequential. As shown in eq. (4.10), the latter proceeds via highly reactive electrophilic species, concentrations and lifetimes of which must be kept low. For achieving high selectivity, fast oxygen reduction is therefore as important as a good match between the reactivities of feedstock and lattice oxygen. Known catalyst components often fail complying with both requirements at the same time. Thus, feedstock oxidation and oxygen reduction proceed on different components in most selective oxidation catalysts. This adds mass and charge transfer processes to the mechanism, and the criteria of high oxygen anion mobility and sufficient electric conductivity to the requirements for a successful catalyst.

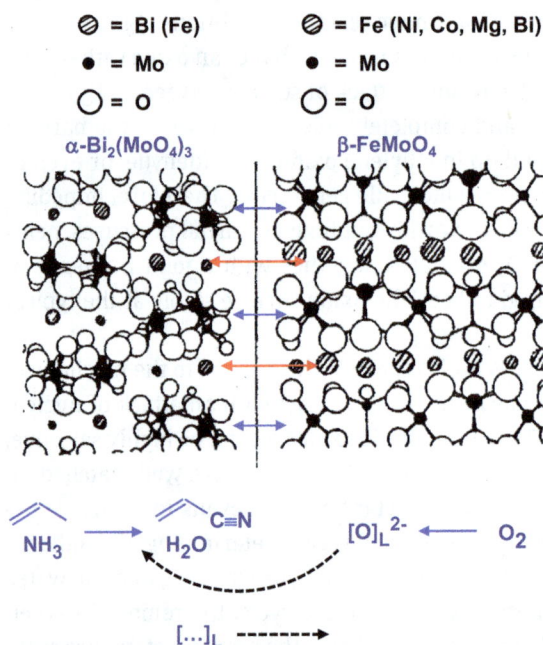

Figure 4.21: Cooperation between molybdate phases in catalysts for propene ammoxidation. Adapted from ref. [76] with permission from Springer Nature.

The complexity of reaction mechanisms in selective oxidation is reflected in complicated multi-component catalyst formulations. The Mars–van Krevelen mechanism requires redox systems tolerating permanent oscillation of the oxidation state without structural degradation. V, Mo, and Sb are the elements dealing with this part in most catalysts nowadays, but other elements such as Cu, Mn, Fe, and even U are suited as well. If oxygen reduction proceeds on a different catalyst component, grain boundaries may become obstacles for the oxygen transport to the oxidation sites. The problem can be solved with structures allowing epitaxial relations or intergrowth with close registry, between structural elements. Figure 4.21 illustrates this phase cooperation principle with molybdate phases from a catalyst for propene ammoxidation. Acrylonitrile is formed on α-$Bi_2(MoO_4)_3$ of Scheelite structure, the so-called reaction phase, while oxygen reduction proceeds on β-$FeMoO_4$, in which part of the iron is exchanged by other ions (Ni, Co, the so-called reoxidation phase). When these phases stick together with their (010) facets, molybdate and cation positions in their structures match with only a few percent deviation, which diminishes transport resistance at the grain boundary.

In technical catalysts, the two phases are often arranged in core–shell particles, in which the molybdate reoxidation phase supports a thin, porous layer of the reaction phase. Correspondingly, their Bi content may be quite low, in remarkable disagreement with their popular designation as "bismuth molybdate catalysts." They could as well be referred to as "Bi-promoted Fe or Co molybdate catalysts."

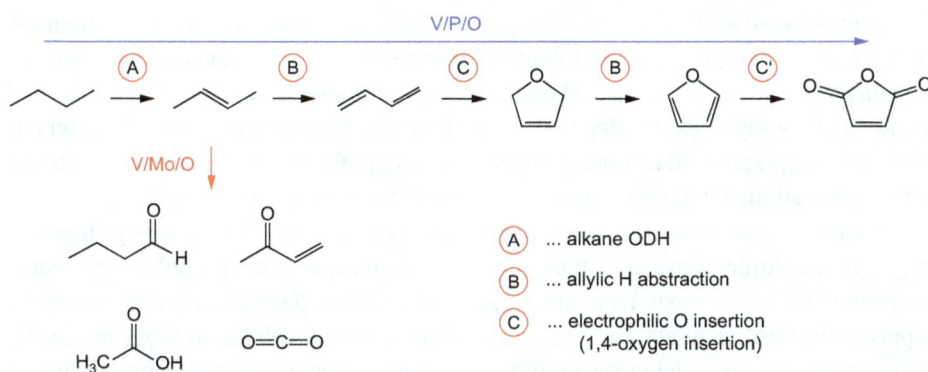

Figure 4.22: Selective oxidation of butane to maleic anhydride and catalytic functionalities required along the reaction sequence.

C–H bond activation is often only the initial step of a longer series of fast surface reactions. The intermediates may not be detectable in the gas phase, but their role in the mechanism can be corroborated by finding out that they react to the final product at higher rates than the starting material. Figure 4.22 shows the reaction network of butane oxidation to maleic anhydride on vanadyl phosphate catalysts, which was identified by this method. As any desorption and re-adsorption of the more labile intermediates bears the risk of total oxidation, all these steps should proceed at or near the site operating in the initial step.

The network requires different types and intensities of reactivity; abstraction of allylic hydrogen (e.g., from butene) is less demanding than activation of an alkane C–H bond in butane; oxygen insertion into electron-rich molecules like butadiene or furane requires electrophilic rather than nucleophilic oxygen. The catalyst needs to offer the required sites in the reach of the adsorbate, it must be multifunctional [66]. In addition, an appropriate relation between nucleophilic and electrophilic reactivities is required. If the electrophilic oxygen is too reactive, intermediates may be lost to combustion, even without desorption and re-adsorption. In Figure 4.22, this is exemplified by observations with more electrophilic V–Mo mixed oxides. On these catalysts, electrophilic oxygen is inserted into butene instead of butadiene, which results in a variety of products and a high extent of total oxidation [66].

As mentioned in Section 4.3.1, selectivity in partial oxidation may be wrecked when an excess of highly reactive oxygen is available near the adsorbate, e.g., in the form of oxide clusters. Selective sites, e.g., V^V or V^{IV} species, are therefore often isolated by matrices of lower reactivity like phosphate ions or MgO, or they are deposited on supports of high surface area, e.g., SiO_2 [66]. More instructive examples of this site-isolation principle, which is not at variance with the fact that suitable catalyst structures allow inserting several O atoms sequentially with high selectivity (cf. Figure 4.22), are shown in ref. [76].

As mentioned earlier, the selectivity of oxidation processes can also be influenced by the reactor design. Ultra-short contact time reactors are a proven technology for this purpose. Beyond the industrial applications mentioned in Section 2.8.2, it was also found to allow high selectivities ($\approx$85%) in the ODH of ethane to ethene [77], which might challenge standard technology of ethane or naphtha pyrolysis. The approach has not yet been studied in larger scales. For mechanistic aspects see Section 4.5.4.

Membrane reactors are a reaction engineering tool applicable for many purposes in gas-phase, liquid-phase, and biocatalysis. In membranes for gas-phase processes, the permeable layer is usually formed by zeolite crystals. Beyond staged O_2 dosage to suppress unselective oxidation, zeolite membranes can also be used to withdraw H_2 during (nonoxidative) dehydrogenation of hydrocarbons, for water removal during dehydrations, or for shape-selective removal of products in equilibrium-limited reactions [78].

The electrochemical membrane reactor is a very specific approach to the handling of selective oxidation reactions, in which oxygen is supplied in the form of oxygen anions ions pumped through an ion-conducting membrane by an electrostatic potential. On its two sides, the membrane is coated with catalysts for O_2 reduction and for delivering oxygen anions to the reactants. In ref. [79], the method is reviewed, together with other alternative reaction-engineering concepts for partial oxidation reactions.

4.3.4 Selective hydrogenation

a

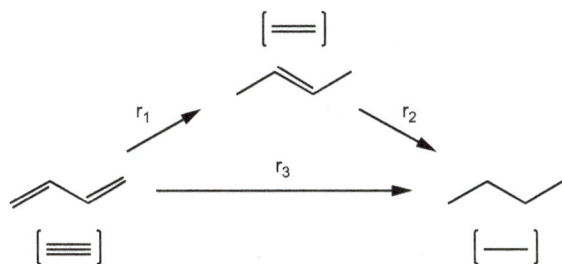

b

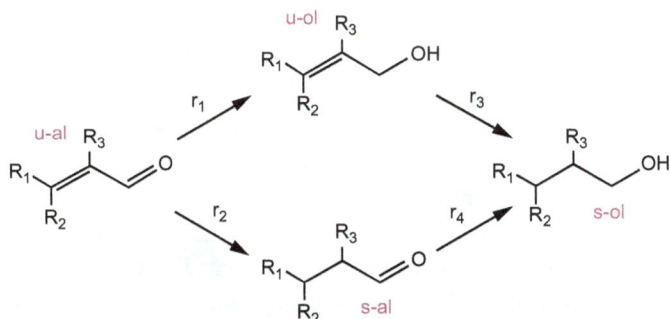

Scheme 4.4: Basic consecutive schemes of selective hydrogenation: (a) hydrogenation of alkynes or dienes and (b) hydrogenation of unsaturated carbonyl compounds.

Typical reaction systems of selective hydrogenation are shown in Scheme 4.4. While they resemble selective oxidation (Scheme 4.3) in their consecutive nature, selective hydrogenation always includes a recognition problem. In the hydrogenation of unsaturated carbonyl compounds (mostly aldehydes), the groups to be differentiated belong to the same molecule. In alkyne or diene hydrogenation, the requirement of substrate selectivity results from the typical technical application: the reactions are performed in mixed streams, e.g., traces of acetylene or propyne are to be converted in unseparated C_2 or C_3 steam cracker fractions, respectively, or butadiene in the C_4 fraction. Requirements for differentiating electron-rich triple (or conjugated double) from C=C double bonds or polar C=O bonds from nonpolar C=C bonds are different, and so are strategies for catalyst design in these fields.

In the hydrogenation of alkynes or dienes, successful catalysts differentiate reactants and products by their adsorption energies: electron-rich alkynes/dienes are adsorbed more strongly and displace the alkene products as long as they are present in the gas phase.[10] At the same time, reactivity and abundance of surface hydrogen need to be lowered to penalize consecutive hydrogenation of the alkene, relative to its displacement. In alkyne hydrogenation, oligomerization is another undesired side reaction (not included in scheme 4.4a).

Palladium is the only metal that achieves sizable alkene selectivities in alkyne and alkadiene hydrogenations. Its performance can be further improved by adding promoters, e.g., an additional metal (Ag, Au, Cu, Pb, or Bi), and by working with the so-called selectivity modifiers (CO in the gas phase, quinoline in liquid-phase reactions). Although such modifications have been reported to confer some selectivity also to Pt, alkynes and dienes are hydrogenated exclusively over modified Pd catalysts.

Pd dissolves hydrogen in the bulk. Subsurface hydrogen accessing the adsorbates from below the outermost Pd layer is highly reactive and unselective. With short-chain alkynes, pure Pd was shown to exhibit selectivity only at very low H_2 pressures, where carbonaceous deposits are formed on the surface. By ambient-pressure XPS, a metastable near-surface PdC_x phase was found to be formed from these deposits [81]. The carbide species intersect access of subsurface H to the adsorbates, which is selectively hydrogenated by hydrogen adsorbed at the surface instead. When the PdC_x phase disappears at increased H_2 pressures, alkene selectivity gets lost.

Promoters and modifiers allow keeping high alkene selectivities also at the elevated H_2 pressures required for stable catalyst operation. CO for instance, which adsorbs stronger than hydrogen, favors selectivity via various effects, according to a theoretical study [82]. It increases the preference of alkyne over alkene adsorption, expels subsurface H from the near-surface layer, and diminishes the availability of surface hydrogen and of free adsorption sites for the alkyne. The latter also impedes the oligomerization reaction.

Most of the promoters and modifiers mentioned earlier operate by separating the active sites. This was suggested in a theoretical study on the Lindlar catalyst, in which Pd is doped with Pb [83]. Applied with the modifier chinoline, it catalyzes stereospecific hydrogenations of long-chain alkynes to alkenes. In ref. [83], it was found that both Pb and quinoline increase the preference of alkyne over alkene adsorption. Pb, which is nearly atomically dispersed in the alloy with Pd, destabilizes hydride and expels surface hydrogen from its vicinity. Due to the limited solubility of H_2 in solvents, hydrogen supply is short anyway. Together with the surface Pb atoms, the spacious quinoline modifier forms a surface overlayer, limiting the space for the adsorption of alkyne and hydrogen, which results in the high selectivities known in this field since the 1950s.

Challenges and strategies in the hydrogenation of unsaturated aldehydes (u-al, Scheme 4.4b) differ strongly from those in alkyne or diene hydrogenation. While the

---

10 For this reason, pore-diffusion limitations must be avoided. They cause the concentration of the reactant (e.g., butadiene) to drop to zero in the pores, which allows the product (e.g., butene) to be adsorbed and hydrogenated. Therefore, the active component is deposited only near the pellet surface in this kind of catalysts (egg-shell profile, Figure 3.17).

unsaturated alcohol (u-ol) is the target product in most cases,[11] thermodynamics rather favor hydrogenation of the C=C bond, which is stronger adsorbed on most metal surfaces. Hence, selectivity for the u-ol product is low for most metals capable of activating hydrogen. However, their modification along suitable strategies has allowed developing efficient solutions for many applications. The field is rather diverse and alive, and in some details not yet well understood. Its status has been summarized in recent reviews [84, 85], from which much of the information presented further has been taken.

The effort required for achieving selectivity depends on the structure of the reactants. It decreases with increasing distance between C=C and C=O bonds. Hydrogenating α, β-unsaturated aldehydes like acrolein ($R_1 = R_2 = R_3 = H$ in Scheme 4.4b), methacrolein ($R_1 = R_2 = H$, $R_3 = CH_3$) or crotonaldehyde ($R_1 = CH_3$, $R_2 = R_3 = H$) are the touchstones of selectivity performance, but hydrogenation of cinnamaldehyde ($R_1 = $ phenyl, $R_2 = R_3 = H$) and furfural (R_1 and R_3 connected via $-O-CH=CH-$, $R_3 = H$) are popular test reactions as well.

Although selectivity for unsaturated alcohols could be improved by favoring their desorption over their consecutive hydrogenation (r_3), maximum u-ol yields have so far almost exclusively targeted by suppressing r_2 in favor of r_1 (Scheme 4.4b). Starting with low u-ol selectivities for pure metal surfaces due to their preferred interaction with the C=C bond,[12] the adsorption of u-al reactants at their polar end can be promoted by strategies based on electronic or steric effects.

The electrophilicity of the catalytic metal can be attenuated by adding an electropositive metal (Figure 4.23a), by adsorbing electron-rich modifiers, or by using an electron-donating support (i). The main purpose of this approach is to discourage C=C hydrogenation. Lewis-acid sites or O vacancies of an oxide additive, which offer an interaction with the lone electron pair of the carbonyl oxygen (Figure 4.23b), attract the C=O bond near to the surface (ii). The steric approaches operate with the influence of crowding on reactant adsorption on the catalyst surface (iii, Figure 4.23c) or with space confinements (iv, Figure 4.23d). In real catalysts, these influences may be superimposed and difficult to distinguish.

Electron transfer, as intended in strategy (i), is a local rather than a global effect. Sulfur imparts to copper an appreciable selectivity for the hydrogenation of C=O in crotonaldehyde. The effect was found to result from a strong bond of the carbonyl oxygen to the surface, becoming possible in the vicinity of S atoms rather than on a general repulsion of the C=C group [85, 87, 88]. Enhanced u-ol selectivities typically obtained

---

11 β-Unsaturated alcohols (u-ol) are intermediates for pharmaceuticals and fragrances. Many saturated aldehydes (s-al) are better accessible via the oxo synthesis. The exception is hydrocinnamaldehyde, which is obtained from cinnamaldehyde ($R_1 = $ phenyl, $R_2 = R_3 = H$ in Scheme 4.4 b) by hydrogenation of the C=C bond.

12 Only with group Ib elements, appreciable u-ol selectivity can be achieved with monometallic supported catalysts, though at low reaction rates due to the limited capability of these metals for hydrogen activation.

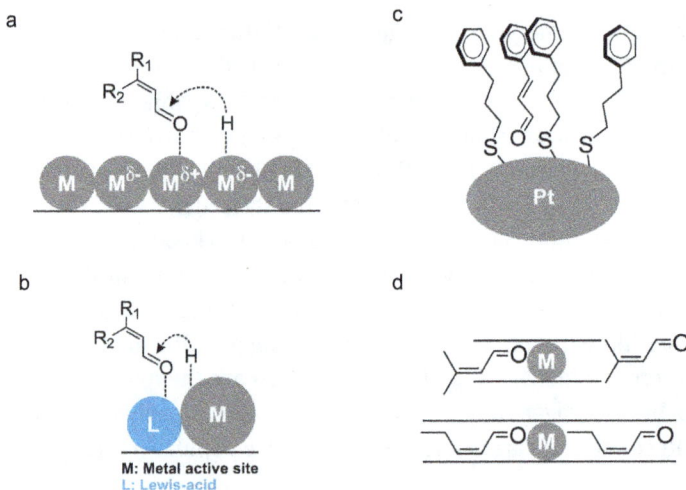

Figure 4.23: Strategies for the selective hydrogenation of α, β-unsaturated aldehydes to alcohols: (a) electropositive additives (adapted from ref. [85]); (b) Lewis-acidic modifiers (c) orienting substrates by co-adsorbates (adapted from ref. [86]); and (d) orienting substrates by confinement in pores.

with metals supported on carbon materials instead of inert supports such as SiO_2 have been ascribed to electron donation from aromatic π systems to the metal. The effect is more intense on the surface of carbon nanofibers and CNTs than on the flat graphite surfaces, and even a dependence of selectivity on CNT diameters (i.e., the bending of the π-system) was reported. On the other hand, surface oxygen groups, which may originate from the preparation process (cf. Section 2.3.2.6) interfere with this electron donation. They may be removed by thermal treatments, but the catalytic effects were found to differ, depending on the type of carbon material [85].

As mentioned in Section 3.4.8.4, electron transfer between surface metals atoms can be monitored via the IR spectra of adsorbed CO. In a study with unsupported Pt nanoparticles decorated with the more electropositive Co using methods of colloidal chemistry [89], such electron transfer was indeed observed. It was, however, significant only for small Pt particles, because the coverage by Co, which anchored only at highly exposed Pt sites, remained low on large particles. As expected, selectivity shifts in cinnamaldehyde hydrogenation were dramatic with the small particles, but had faded away at particle sizes of ≈25 nm. There was, however, a second origin for the strong effects with small particles, because the decorating Co poisoned their highly exposed Pt atoms, which had a particular preference for the C=C bond. In ref. [89], this contribution even exceeded that of electron transfer from Co to Pt. Such superposition of effects may explain why positive effects of Co, Fe, Ni, and Sn on hydrogen-activating components like Pt, Pd, and Ru have also been found in cases where electron transfer could not be identified by spectroscopy [84].

In surfaces of alloyed particles, incomplete reduction of the electropositive component may leave cations, which results in a combination of strategies (i) and (ii)

(Figure 4.23a and b). The case is typical for noble metals modified with Sn. There has been a discussion, which of the effects (frustrated C=C hydrogenation or favored adsorption via carbonyl oxygen) contributes most to the selectivity changes observed [85].

Strategies (iii) and (iv) take advantage of the different locations of the double bonds in the chain – C=O being always at the end and C=C usually not – with the exception of acrolein. In DFT studies, adsorption of acrolein and crotonaldehyde on Pt(111) was found to be planar. Adsorption of prenal (R_1 = R_2 = CH_3, R_3 = H in Scheme 4.4b) was planar only at low coverages, at θ = 0.25 – the molecule was adsorbed upright via the O atom. This may explain why achieving good u-ol selectivity is easier with cinnamaldehyde and furfural than with smaller molecules like crotonaldehyde or acrolein [84]. Selectivities tending to decrease with the degree of corrugation of facets (S_{u-ol} over Pt(111) > Pt(110)) and with decreasing metal particle size [85] can be ascribed to lower steric interference on curved surfaces, which neglects, however, the possible influences of exposed metal sites on them.

The steric constraint can be enhanced by populating surfaces with adsorbates, which enforce a favorable orientation of the substrate toward the surface (strategy iii). The approach is illustrated in Figure 4.23c where the phenyl rings of a thiol modifier offer dispersion interactions to the ring of cinnamaldehyde. Selectivities are highest when the structure of the modifier is adapted to that of the substrate, as in Figure 4.23c. The figure also suggests that the steric effect of the thiol modifiers may be superimposed by an electronic effect of type (i) as described earlier for the interaction of sulfur with Cu surfaces [85].

Confinement by molecular sieving pores is the most rigid form of steric constraint (strategy iv, Figure 4.23d). It was first achieved with Pt particles in zeolites, while most promising results have been recently published with metal or alloy nanoparticles encapsulated by thin layers of metal-organic frameworks (cf. Section 2.3.2.4) [85]. Encouraging improvements in u-ol selectivities have also been reported for metal particles in the interior space of CNTs, where spatial constraints are, however, too weak to suggest substrate orientation by pore walls. Instead, explanations have been sought in electronic effects [85].

4.3.5 Promoting selectivity

In the following, the numerous examples of selectivity promoters mentioned so far are briefly summarized under a different point of view and complemented with other relevant cases.

Promoting selectivity means enhancing the difference between the rates to the target and to the side (or consecutive) products. Strategies for achieving this goal depend on if and how the sites for the target and the side (or consecutive) reaction can be differentiated. They may aim at predominantly suppressing the side or consecutive reactions or at predominantly accelerating the target reaction. Promoters may operate via the modes introduced in Section 1.4.3: they may directly influence the active site – its electronic state or its spatial structure, or they may act upon the structure or morphology of the catalyst.

The selectivity issue in alkyne (diene) hydrogenation is most challenging because the target and the consecutive reaction proceed on the same sites. The Pd component is modified in order to poison complete hydrogenation and oligomerization stronger than the formation of the alkene. Expelling highly active subsurface H atoms from the near-surface layers and decreasing the availability of surface hydrogen, the modifier CO (Section 4.3.4) retards direct alkane formation and/or consecutive alkene hydrogenation. Metal additives incapable (or less capable) of activating H_2 limit the supply of surface hydrogen as well by destabilizing it in their vicinity. On the Pd surface crowded by adsorbed modifier molecules and additive atoms, alkyne adsorption sites are diluted, which suppresses oligomerization. Such destruction of sites for a side reaction is the typical effect of a geometrical promoter.

Selectivity changes achieved by patterning the active surface with inactive atoms, e.g., by alloying the active with an inactive metal, are usually categorized as ensemble effect (cf. Section 4.2.2). The approach is applied in naphtha reforming catalysts. To cope with process conditions favoring coke deposition (cf. Section 4.3.1), different manufacturers developed Pt–Sn or Pt–Ir alloy catalysts and reoptimized the existing Pt–Re–type catalysts. The second component can exert various promoting influences (cf. Section 4.2.4), but Sn and Re are known to operate mainly via the ensemble effect. Pt–Sn/γ-Al$_2$O$_3$ catalysts, for example, are somewhat less active than Pt/γ-Al$_2$O$_3$ catalysts due to the inertness of Sn. They are, however, much less prone to deactivation by coking, because Sn destroys the larger sites required for the growth of coke precursors.

Ir provides an example how a similar effect can be achieved along a different route, because Ir surfaces are extremely active for hydrocracking. In reforming catalysts, this activity is attenuated by the atomic dilution of Ir in the Pt surface in a sort of reverse ensemble effect, but the residual activity is sufficient to destroy coke precursors while they build up on the alloy surface. Selectivity is achieved here by controlled acceleration of a reaction that competes with a side reaction (coking). For the sake of completeness, it should be kept in mind that coke formation can be delayed even by electronic promoters, e.g., by alloying the Ni surfaces of steam-reforming catalysts with Au (Section 4.2.2). Mechanisms of coke deposition and coke properties are, however, very different in steam reforming and naphtha reforming.

When target and side reactions proceed on sites of different nature, poisoning the side reactions may be rather straightforward. In redox catalysis with hydrocarbons, acid sites are usually undesired. They can cause cracking reactions (e.g., in dehydrogenations, which proceed at relatively high temperatures) or attract unsaturated molecules to surfaces, where they may suffer total oxidation. Catalysts are therefore often promoted by alkali ions to suppress acidity (cf. Section 4.3.3).

Accelerating the target reaction may be achieved by activating existing sites for it, by creating new sites, which may include an additional element, or by changing surface coverages of relevant adsorbates. The latter may be exemplified by the effect of alkali cations in Fe-based FT catalysts. Destabilizing surface H atoms, the (electronic) alkali promoter favors higher coverages by CH_x units and larger chain growth probabilities. As a result, the molecular-weight distribution of hydrocarbon products is shifted to higher weights, and they are desorbed as alkenes rather than as alkanes.

Ethylene oxide selectivity in the electrophilic oxidation of ethene over Ag catalysts is strongly enhanced by chlorine, which is provided with the feed (cf. Sections 4.3.3 and Appendix A1). According to ref. [75], it blocks the access of ethene to vacancies in the dense O adsorbate layer (i.e., to Ag sites), where it would have a choice between selective and nonselective routes. At the same time, adsorbed Cl enhances the electrophilicity of the adjacent O atoms, which accelerates the Eley–Rideal-type oxidation of the ethene molecule, considered responsible for the highest selectivity. It was also argued that the high O coverages required to prevent contact between ethene and Ag cannot be stabilized under reaction conditions by unpromoted silver, but need the support from Cl and alkali dopants (mainly Cs) [75]. Their joined effect can be categorized as structure promotion.

In catalysts for nucleophilic oxidation, promoters can serve various purposes. They can help tuning the nucleophilicity of surface O atoms to the stability of the reactant, stabilizing redox cations in desired oxidation states or stabilizing the mixed oxides in desired phase structures. They can also favor the reduction of gas-phase oxygen by the reoxidation sites. Oxygen anion mobility required to conduct O^{2-} ions from reoxidation to reaction sites (cf. Section 4.3.3) can be enhanced by doping oxides with foreign ions of different valence. Notably, oxygen ion conductivity must be complemented by appropriate electron conductivity for charge compensation, which can also be achieved by adding suitable dopants (promoters). The complexity of catalyst formulations does not always allow a clear definition of the effects caused by the individual additives.

4.4 Stability, regeneration, and reactivation

Catalysts change during steady-state operation, although they should not, by definition. Activities decrease, and selectivities deteriorate. While usually starting at sizeable rates, changes may become very gradual with increasing time-on-stream. In plant operation, productivity losses by deactivation are made up for by gradually increasing the temperature, which may, however, affect selectivity. Reasons for deactivation are diverse, and so are the strategies for slowing it down. Reasons depend on the nature of catalysts and feeds, and on the reaction conditions – strategies on the range of permissible changes on them. Often, deactivated catalysts can be regenerated. Without such perspective, catalyst ingredients can be recycled. For materials not worth recycling, uses in other industries can be sometimes found to avoid disposal, which can be very costly due to the toxicity of some ingredients.

Table 4.1: Running lives of technical catalysts and related technologies for designing continuous processes.

Running life of the order of	Process	Process structure
Seconds	Fluid catalytic cracking	Riser with regenerator
Minutes	Propane dehydrogenation (Cr_2O_3/Al_2O_3)	Batteries of packed beds switched between reaction and regeneration phases
Hours/days	Low-pressure version of naphtha reforming	Moving-bed reactor + regenerator section[1]
Weeks	Friedel–Crafts alkylation (benzene with ethene)	Multiple packed beds, periodical regeneration of individual beds
Months	High-pressure version of naphtha reforming	
Years	Selective hydrocarbon oxidations	Multitubular or fluid-bed reactors[2]; off-site treatment
	NH_3 synthesis	Staged packed beds; no regeneration
	SCR of nitrogen oxides with NH_3 (stationary sources)	Monolith reactors; off-site treatment

[1]Batch or continuous
[2]Catalysts employed in fluidized beds may have shorter life times

Where regeneration is feasible, stable processes can be established with suitable engineering solutions, even for catalysts with extremely short running life. In Table 4.1, some processes are listed to indicate the options available for handling different frequencies of reaction/regeneration cycles. Running lives range from seconds to years; the technological options include continuous regeneration of fluid-bed catalysts, continuous or batch regeneration of catalysts from moving beds, periodic regeneration of fixed beds, and off-site treatments.

The following discussion is focused on catalysis in the gas phase, where high temperatures are a major issue. The topic has been comprehensively reviewed in ref. [90], from which much of the information given later has been taken. More details on the relations between catalyst deactivation and process design, as briefly illustrated in Table 4.1, can be found in refs. [80, 91]. Specific aspects of catalyst deactivation in the liquid phase and in electrocatalysis are addressed in Sections 4.7 and 4.9.

4.4.1 Mechanisms of catalyst deactivation

Catalyst deactivation can be caused by components from the feed, by thermal effects, by losses, and by mechanical degradation. Losses are related to the thermal effects in gas-phase reactions, but are caused by leaching in the liquid phase. Different effects may superimpose in real deactivation phenomena.

4.4.1.1 Poisoning and fouling

Poisoning and fouling result almost exclusively from species delivered by the feed.[13] Poisons compete for adsorption sites required for the surface reaction, or initiate damage of the solid structure. Fouling relates to "dirt" on the catalyst, bound to the surface by physical rather than by chemical interactions. It may, for instance, affect catalyst performance by blocking porosity. Differentiation between poisoning and fouling becomes intricate when fouling species cover active sites. Precursors of carbonaceous deposits, which are a major reason for catalyst decay, are often formed on active sites of a metal component, and block them. They may, however, also spill over to the support, where the deposits are further converted and stored. In literature, coking is often categorized as fouling, but its effect on performance is designated as poisoning.

Poisons can be reversible or irreversible depending on how strong they are adsorbed by the active sites. A reversible poison can be included in the kinetic rate law, like any other reactant. Its effect disappears at a rate depending on its desorption kinetics, once it has been removed from the feed. As an example, N-bases are reversible poisons for the acid-catalyzed cracking of hydrocarbon molecules (e.g., in FCC). Notably, catalytic surfaces can also be poisoned by strongly adsorbing reactants or products. The limitation of ammonia synthesis rates by strongly adsorbed NH_x species (cf. discussion of Figure 4.1b) can be categorized as inhibition (or poisoning) by the reaction product. Irreversible poisons block sites completely and can only, if at all, be removed by regeneration procedures. As long as there are no additional effects on selectivity, the kinetic influence of irreversible poisoning is accounted for by a decay function multiplied with the rate constant.

On metal surfaces, poisons may also influence the reactivity of the neighboring sites by electronic interactions. Although electronic effects operate only via short distances (a few atomic units), the behavior of the surface may be significantly changed at moderate coverage of the poison. Interactions between the poison and the surface may also result in (adsorbate-induced) surface reconstruction. Finally, the adsorption of a poison may be only the first step to more serious damage resulting from its reaction with the catalyst.

Poisoning effects depend on the interactions offered by the catalytic surface: there are no poisons per se. Acid sites are poisoned by bases, basic sites by acidic molecules. Oxygen can be a reactant or a poison depending on the nature of the catalyst and the catalytic process. H_2S and other sulfur compounds are extremely strong poisons for metal catalysts employed in (de)hydrogenating processes, like naphtha reforming, NH_3 synthesis, FTS, but also in the steam reforming of methane. In the water-gas shift reaction (2.10) applied to condition syngas for NH_3 production (cf. Appendix A1), H_2S is lethal for the more active Cu-based catalyst (low-temperature shift), while Fe-oxide-based

---

13 Poisoning by components introduced during catalyst preparation (e.g., from contaminated chemicals) may be a persistent problem in research. It should be, however, eliminated on the commercial scale.

high-temperature shift catalysts tolerate sizeable sulfur levels. Sulfide surfaces do not only tolerate sulfur, but some reactions (e.g., alkene hydrogenation) proceed on them without being delayed by H_2S.

Oxygen, which is another prominent poison, cannot be tolerated in any form in some processes, but the damage depends on the applied conditions. While ammonia synthesis catalysts are quickly deactivated when surface Fe atoms are oxidized, water is a reaction product and therefore always present in FTS, which proceeds at lower temperatures. Gradual transformation of active Fe carbide into oxide phases indeed belongs to the processes to which deactivation of Fe-based FT catalysts has been ascribed.

Water is extremely unwelcome in many processes, because it accelerates all kinds of transport processes, resulting in sintering and solid-state reactions (Section 4.4.1.3). In naphtha reforming catalysts, it also causes the loss of the chloride promoter, which keeps the acidity of the Al_2O_3 support on the level required for sustaining high rates of the acid-catalyzed reaction steps (Figure 1.6c). On the other hand, in catalysts for end-of-pipe exhaust gas treatment, noble metals and even zeolites operate in environments containing ≈ 10 vol% H_2O for years.

Many catalysts are poisoned by species containing IVA – VIIA group elements. Adsorbed halide may initiate loss of catalyst material via volatile halides. Atoms and ions of heavy metals (Pb, Bi, As, Sn, Cd, etc.) block the surface of metal catalysts, but some of them are deliberately used to poison side reactions (Section 4.3.5). Strongly adsorbed unsaturated molecules like CO or acetylene interfere with reactions involving weaker bonds between the reactants and the surface. The adverse influence of acetylene on ethylene oxide synthesis over Ag catalysts is a major reason for the effort taken to hydrogenate the alkyne in the steam cracker C_2 fraction.

In residue hydrotreatment, designed to obtain environmentally acceptable fuels even from very heavy oil fractions, asphaltenes[14] and aromatic Ni or V compounds are major poisons. On low concentration levels, the latter are also present in heavy distillates, and affect cracking processes like FCC or hydrocracking (cf. Section 4.3.1, Appendix A1). Hydrocracking combines HDS and HDN with a subsequent cracking step. The former is typically catalyzed by a NiMo catalyst, the latter by an acidic catalyst that is capable of activating hydrogen, e.g., Pt in a zeolite. As this combination is sensitive to NH_3 and H_2S released upstream, both were removed between the stages in the past. Meanwhile, activities of cracking catalysts are so high that they perform well even at poison levels established after the NiMo catalyst, which has allowed dropping the scrubber unit and performing the whole process in one reactor, with zoned catalyst filling.

---

14 Asphaltenes are macromolecular compounds (molecular weight between 700 and 1,500) in natural carbonaceous materials like crude oil or coal, which are soluble in toluene, but not in light alkanes. They are aggregated via interactions of their aromatic sections, and are potent coke precursors.

Poisons can be further differentiated with respect to their selectivity and toxicity, and catalysts with respect to their tolerance (residual activity at full coverage of poison). Poisoning selectivity describes the change of activity per adsorbed quantity of the poison. Small amounts of a selective poison kill a large percentage of activity, for instance when a reaction requires very specific sites like the B5 or C7 sites for ammonia synthesis (Section 4.2.3). The quantification of these parameters can be difficult because of pronounced concentration profiles occurring under real conditions. Strongly interacting poisons are never well distributed: they are preferentially deposited at pore mouths and in the initial sections of reactors. More details can be found in ref. [90].

Adsorption of a poison may initiate more substantial changes in the catalyst. An active phase may be transformed into a less active one, e.g., the active χ-Fe_5C_2 phase of FT catalysts into Fe oxides, as mentioned earlier. Cobalt in Co-based FT catalysts may be oxidized as well, which was, however, shown to be relevant only for particles well below the optimum size of ≈ 6 nm [92, 93]. Gas-phase components such as water may accelerate deactivation processes based on solid-state mobility. The fast sintering of Cu particles in methanol synthesis catalysts exposed to Cl-containing feeds is another example of such processes, which will be discussed in Section 4.4.1.3.

Alternatively, forming a volatile compound with catalyst components, gas-phase species may cause sintering via the gas phase or even losses of the active component. During CO methanation over Ni, $Ni(CO)_4$ may be formed, which causes dramatic growth of Ni particles. Depending on the reaction conditions, Ni may be even deposited downstream of the catalyst bed (cf. Section 4.4.1.2). Particle growth via mass transport through the fluid can also occur in the liquid phase, where active component may be lost by leaching and re-deposited on the support as a part of a particle growth mechanism (see Section 4.4.1.2).

Fouling is mostly related to carbon deposition, examples beyond it are rare. One of them is the SCR of nitrogen oxides in SO_2-containing flue gases (e.g., from coal firings). If the catalyst is too active for SO_2 oxidation, $(NH_4)_2SO_4$ formed from SO_3, NH_3, and water is deposited on the catalyst, and fouls downstream equipment (cf. Section 4.3.1). In specific process configurations (high-dust SCR units immediately after the boiler), the catalyst is also fouled by ashes not yet removed from the gas, which subjects it to high mechanical stress (see later).

As mentioned earlier, carbon deposits formed on surfaces may differ strongly in structure, hydrogen content, and reactivity, depending on the reaction conditions and the nature of the catalyst. On metals, decomposition of hydrocarbons in the absence of hydrogen and their dehydrogenation to carbide species can start at rather low temperatures (on Ni, for instance, at ≈ 400 K). These surface carbide species, which are related to the CH_x chain propagators in the FT reaction, are sometimes referred to as C_α. They can polymerize on the surface to amorphous films (C_β) or diffuse into the bulk of some base metals to form bulk carbide phases (C_γ). At higher temperatures, whiskers and fibers can be formed from β- and/or γ-carbon (cf. Section 3.1.1.5), or crystalline graphitic platelets may be segregated. The latter processes can be accompanied by fragmentation of the metal particles. Coke segregated from metal surfaces may completely embed catalyst particles and even block pores.

Carbidization may be reversed by hydrogenation or oxidation of the carbonaceous species, which becomes more demanding when C–C bonds or even aromatic structures have been formed. The coke accumulation rate is the difference between the rates of coke deposition and removal, i.e., of its gasification by reaction with O_2, H_2, or H_2O. Obviously, coking is preferred in feeds of low H_2 or O_2 content. Coking rates depend on the reactivity of the carbon source. On Ni, for instance, coking is more intense with alkynes than with alkenes or alkanes. Weaker bonding forces to surface carbon, higher hydrogenation activity, and low carbon solubility limit problems with carbon formation on noble metals. On acidic support surfaces, they are, however, integrated in bifunctional coking mechanisms (see later).

On oxide and sulfide surfaces, coke formation is typically initiated by cracking of coke precursors, and proceeds via reactions of the resulting carbenium ions (cf. Section 4.5.2) by cyclization and dehydrogenation steps, by oligomerization, aromatization, and substitution. As cracking and dehydrogenation require strong acidity, the rates of coke formation grow with increasing strength and concentration of the acid sites. Rates also depend on the reactivity of the hydrocarbon. Due to preformed structural motifs, aromatic compounds, in particular, polyaromatics, are most powerful coke precursors. The ranking between open-chain hydrocarbons (alkenes > i-alkanes > n-alkanes) reflects their capabilities for forming carbenium ions.

Coking in zeolites is special because of their spatial constraints. Despite the high acidity of their Brønsted sites, carbon amounts deposited are usually smaller than on catalysts with open porosity. Oligomers of n-alkene reactants may be trapped in pores of medium size under rather mild conditions (≈ 430 K). This soft coke with H/C ratios > 1 may be abundant. At higher temperatures, it can be partly removed by cracking while the remaining material is transformed to hard coke (H/C < 1). From feed streams, such coke is formed above 575 K. Due to its aromatic nature, it requires the larger space available in wide-pore zeolites or at the channel crossings of medium-sized pores. Carbonaceous material can, however, be also piled into smaller pores, and block pore mouths at the external crystallite surface. Soft and hard coke may be differentiated by TPO (Section 3.4.2.5).

The impact of coking on intra-zeolite diffusion (and, hence, on reaction rates) depends on the nature of the pore system. It is most dramatic in one-dimensional pores (e.g., in MOR), while diffusion rates are less affected by rising carbon content in pore systems offering diffusion paths in three dimensions. At high coke contents, significant parts of the pore system may be cut off from the gas phase around the crystals. More details about coke in zeolites may be found in ref. [94].

The cooperation between metal surfaces and acid sites in bifunctional catalysts, e.g., of naphtha reforming, opens a new route to coke formation. Dehydrogenating feed molecules with low coking potential, the metal surface provides powerful coke precursors, which can be further reacted and compacted by acid-catalyzed steps on adjacent portions of the support. Further dehydrogenation, which is tedious on acid sites, proceed on the metal, while the emerging coke molecule wobbles around the contact zone between the metal and the support. Integrating more precursor species provided by the metal surface, the coke molecule grows from this region toward the bare support surface, where it is further modified by interactions with the acid sites.

4.4.1.2 Loss of active component

Losses are usually induced by reactions between the active component and components of the fluid phase, although losses into the liquid phase can also result from flaws in catalyst preparation.[15] An active component may become soluble when the redox potential changes, in particular, if its leached form can be stabilized by ligands available in the liquid phase. Such leaching has been observed with CNT-supported Pd in the oxidation of alcohols (e.g., ethanol) to acids. Pd was largely dissolved in the liquid phase as Pd^{II} acetate, but after reduction of the acetate by the alcohol, also in the form of metal clusters [95]. The case is special because the dissolved species were completely trapped by the support at the end of the runs (release and catch mechanism, cf. ref. [96]). In a continuous regime, a trapping stage would be required to recover all Pd from the liquid phase.

In gas-phase catalysis, loss is a relatively rare phenomenon, which can, however, cause great trouble where it occurs. Volatile forms may be halides, carbonyls, oxyhydroxides, or alcoholates. Ruthenium, for instance, is not suited for three-way catalysis, because it forms a volatile oxide (RuO_4) under rather mild conditions. Pt is much more stable toward oxidation, but above 1,000 K, it bleeds from catalysts in the form of PtO_2. This is the reason why getter gauzes, usually made of a Pd alloy, are installed downstream of Pt–Rh gauzes used, for instance, for NH_3 oxidation to NO (Ostwald process, cf. Section 2.8.2). The loss of Ni by $Ni(CO)_4$ formation has been mentioned already in Section 4.4.1.1. Likewise, Ru carbonyls complicate the application of this element in FT synthesis. Cu chlorides can cause the loss of copper, for instance, from catalysts for chlorine synthesis by the oxidation of HCl (Deacon process).

Mo^{VI} oxide species are known to become volatile at elevated temperatures in the presence of water, which causes Mo losses in selective oxidation processes with Mo-based catalysts. The Formox process for the oxidation of methanol to formaldehyde (eq. (2.69)), a competitor to the Ag-catalyzed process mentioned in Section 2.8.2, is the best-known example of such losses. The reaction is catalyzed by $Fe_2(MoO_4)_3$, but the catalyst contains excess MoO_3 to make up for losses during operation, in particular, to prevent the formation of Fe oxides that catalyze combustion. It has been found that volatilization is orders of magnitude more intense with methanol than with water, which suggests volatilization of Mo as an alkoxy compound [97]. For this reason, the Formox process has raised most research activity related to losses among oxidation reactions with Mo-containing catalysts.

---

15 Stability toward leaching is a property that should be routinely examined for each new catalyst for use in the liquid phase, and actually also for known catalysts, upon extension of the reaction conditions.

4.4.1.3 Thermal effects

Table 4.2: Melting temperatures of some catalyst components.

Compound	T_M / K	Compound	T_M / K	Compound	T_M / K	Compound	T_M / K
Pt	2028	Re	3458	Cu	1356	MoS$_2$	1458
PtO$_2$	723	Fe	1808	CuO	1599	(Al$_2$O$_3$[b]	2318)
PtCl$_4$	643[a]	Co	1753	CuCl$_2$	893	(SiO$_2$[c]	1883
Pd	1828	Ni	1725	Ag	1233		
PdO	1023[a]	NiO	2228	Au	1336		
Rh	2258	NiCl$_2$	1281	Mo	2883		
Rh$_2$O$_3$	1373[a]	Ni(CO)$_4$	254	MoO$_3$	1068		

[a]Decomposes; [b]α-Al$_2$O$_3$; [c]quartz.

Thermal effects are caused by the acceleration of solid-state diffusion at high temperatures. The resulting stabilization of the metastable catalytic materials, for instance, by decreasing the external surface area per unit mass (sintering) or by enabling frozen solid-state reactions, impairs their fitness as catalysts. As mentioned in Section 3.1.4, solid-state diffusion is characterized by the Hüttig and Tammann temperatures $T_{Hüt}$ and T_{Tamm}, which specify conditions where surface atoms (e.g., at exposed positions) and bulk atoms, respectively, become mobile. For bulk-sized material, these temperatures are related to the melting points T_M by eqs. (3.12).

In Table 4.2., melting points of important catalyst components are given, from which estimates for $T_{Hüt}$ and T_{Tamm} can be derived via eqs. (3.12). Notably, such data can explain tendencies, but should not be used to predict the onset or even rates of particle coarsening. Tammann and Hüttig temperatures refer to processes within particles or on their surfaces, while sintering of dispersed particles proceeds via mass transport between them. Sintering rates are therefore also influenced by interactions experienced by migrating species en route. In addition, melting temperatures of nano-sized aggregates decrease with decreasing particle size (cf. Section 3.1.4). By analogy, the sintering behavior of metastable supports, like γ-Al$_2$O$_3$ or silica gel, depends on their surface area, which is not covered by equations (3.12).

Table 4.2 illustrates why Cl-contaminated feed can strongly accelerate the sintering of active metal or oxide particles, and why noble metal particles alternatively operating in reducing and oxidizing environments (in three-way catalysis, cf. Appendix A1) usually sinter during excursions into the oxidizing regime. Group 11 metal particles are much more prone to sintering, than particles of group 10 metals, although gold has been found to benefit from surprising stabilization effects on some supports.

Mass transport during particle coarsening mostly proceeds by surface diffusion of atoms. Movement of crystallites is also possible, though at higher temperatures and over short distances. Transport through the gas phase is unlikely at temperatures relevant for catalysis, except for cases mediated by volatile intermediates. The role of atom diffusion and particle coalescence in different stages of particle sintering has been discussed in ref. [98].

Sintering rates of supported metal particles depend on the temperature, the atmosphere, and on interactions offered by other components, e.g., by the support surface or by promoters (Section 4.4.2). Acceleration of metal particle growth by just traces of water in the gas phase is a rather universal phenomenon, and explains severe specifications on water contents in many technical processes. Interactions with gas-phase components may transfer the active phase into a form more prone to sintering, as illustrated by Cl traces favoring Cu particle growth in methanol synthesis catalysts (cf. Section 4.4.1.1), or by the preferred sintering of Pt on the lean[16] excursions of the oscillating reaction regime of three-way catalysis. On the other hand, trapping metal oxide species by strong support interactions is a successful strategy to delay sintering of noble metal particles in oxidizing environments (Section 4.4.2).

Support porosity delays sintering of supported phases because pore walls introduce segregation between particles. Although interactions with the support are weaker for metals than for oxides, there are also support influences in the sintering of metal particles. For rates of particle growth in vacuum, for instance, a ranking $Pt/Al_2O_3 < Pt/SiO_2 < Pt/C$ was reported [90].

Diffusion processes during the sintering of support materials may proceed across the surfaces, in the bulk, or along grain boundaries. Phase transitions, e.g., among the numerous Al_2O_3 modifications (cf. Section 2.3.2.2) or from anatase to rutile, result in strong losses of specific surface area. They may be initiated at the external surface or in the bulk of particles, and involve all types of diffusion mechanisms just mentioned.

Like in particle coarsening, rates of support sintering depend on the temperature, the atmosphere, and on the interactions with other components – the supported species. The adverse influence of water on the stability of most supports is probably related to an increased hydroxylation degree at their surfaces, where dehydroxylation upon temperature excursions may convert contact points between grains to material bridges. The supported species may favor or delay sintering: in V_2O_5-WO_3/TiO_2, the vanadia component stabilizes the support surface, the tungsta component exerts the opposite influence. To withstand the severe conditions in automotive converters, γ-Al_2O_3 is stabilized with La and/or Ba oxide additives. On the contrary, many supports are destabilized by alkali cations.

The fate of the supported species during sintering depends on their miscibility with the support. They may form a mixed phase with it, or even react with the support to a new compound. Alternatively, they may segregate from the sintering support. In V_2O_5-WO_3/TiO_2 catalysts for the SCR of NO by NH_3, WO_3 was found to segregate as a crystalline phase at

---

16 Lean – (slight) oxygen excess; rich – (slight) excess of reducing components (unburnt fuel and CO).

high temperatures, while no crystalline V-containing phase was detected even in samples of very low residual surface area [99].[17]

The dealumination of zeolites (steaming, cf. Sections 2.3.2.3 and 3.1.4) seems to be a related form of thermal damage. Actually, it is a hydrolysis reaction rather than a thermal degradation. Temperatures required for the extraction of Al at Brønsted sites, e.g., during template combustion, can be upshifted hundreds of degrees by strictly excluding or removing moisture. Steaming is also impeded when zeolite protons are replaced by alkali cations, which is of course at the expense of acidity. In NH_3–SCR, even highly exchanged Cu–ZSM-5 or Fe–ZSM-5 catalysts were found to deactivate, when kept in realistic car exhaust at high temperatures. The driving force for their breakdown is the clustering tendency of the redox cations, which is enhanced by moisture. Dealumination at the Brønsted sites left over at the cation sites occurs as a secondary damage. The problem was solved by placing Cu in zeolites of the small-pore CHA (chabazite) topology. The narrow cage windows of these zeolites, which do not affect reactant diffusion, interfere with cation mobility and clustering. The catalysts can be further stabilized by exchanging residual Brønsted sites with alkali.

The intergrowth of supported species with the support is an example of an undesired solid-state reaction gaining momentum at high temperatures. Al_2O_3-supported Cr_2O_3 used for alkane dehydrogenation forms Cr–Al mixed oxides, when overheated. Ni^{2+} or Co^{2+} ions may diffuse into the spinel lattice of transition aluminas, where they are lost for forming active sites in CoMoS/NiMoS structures after sulfidation (Figures 2.43b and 4.15).

Such solid-state reactions may be induced by a previous, sometimes reversible poisoning of the catalyst, for instance, when metal particles are oxidized, and their cations find stable sites within the support. The loss of Rh as highly inert $RhAl_2O_4$ in earlier three-way catalysts was preceded by the oxidation of Rh nanoparticles. A similar deactivation was also found for Ru/Al_2O_3 catalysts [100]. Ni and V species deposited in FCC catalysts from heavy feeds can irreversibly damage the zeolite component by silicide formation, if they are not passivated by additives and/or extracted by off-site treatments (cf. Section 4.4.3).

Amorphization of zeolites due to hydrothermal stress or transition of a metastable support modification to the stable one, define upper temperature limits for the use of catalysts.

4.4.1.4 Mechanical degradation

Mechanical failure causes different types of damage, which may result in loss of catalyst material, or interfere with the operation of the reactor. Pellets at the bottom of fixed beds may be crushed under the bed weight. Catalyst particles may shrink by splitting off fines (attrition), which may be blown out of the reactor or deposited within in zones

---

17 It was supposed in ref. [99] that the surface V oxide species remain at the grain boundaries, which form upon coalescence of support particles.

of calmed flow. At high flow velocities, catalyst particles and monolith coatings may be eroded and removed from the reactor.

Crushed particles increase the pressure drop in fixed beds. This is most unwelcome, in particular, in the narrow tubes of multitubular reactors (Section 2.8.2), where it may cause tube blockages. Attrition is most familiar in reactors with moving catalyst particles, i.e., fluidized beds, moving beds, or slurry reactors. In fluidized beds, it affects not only the catalyst, but also reactor walls and internal elements like cooling coils, which are exposed to a sandblasting regime. In fixed beds, where it may be caused by temperature gradients during upsets in the reaction regime, it can contribute to growing pressure drop in narrow catalyst tubes. Preventing attrition is a major issue for monoliths in automotive converters, which are subject to drastic temperature excursions, depending on the driving regime.

Erosion may happen, for instance, with fluid-bed particles, previously shrunk by attrition or with chips of washcoats, previously destabilized by thermal stress. It is dramatically enhanced when the feed contains solid particles, e.g., in SCR facilities placed in high-dust process configurations. Figure 4.24 shows erosion damage caused by extended exposure of monoliths to abrasive flue gases.

Mechanical properties of catalysts are beyond the scope of this book. The reader may consult ref. [90] for more information on this topic.

4.4.2 Preventing catalyst decay/promoting stability

There are three general strategies to deal with the stability problem, which need to be combined: catalyst lifetimes can be extended by adapting suitable reaction conditions, by choosing suitable engineering solutions, and by improving the catalysts themselves. Deactivation by coking and by some sorts of poisons can be easily removed by regeneration. Therefore, many processes work in a reaction/regeneration mode (cf. Table 4.1). In other cases, recovery of the original performance, if possible at all, requires larger effort. Thus, extending the lifetime of catalysts is a major challenge in technical catalysis, which can be handled successfully only when the reasons underlying catalyst decay are identified.

Adapting reaction conditions first and foremost means taking poisons out of feeds. Before contacting refinery fractions with metal catalysts, sulfur compounds in them are destroyed by hydrodesulfurization. The H_2S released is removed in a sequence of a scrubbing step and a guard reactor with a ZnO bed, which limits residual S contents of <0.1 ppm. Carbon oxides are removed from NH_3 synthesis gas in a sequence of two water-gas shift steps (cf. Appendix A1), a CO_2 scrubbing step, and a methanation step, which converts the residual CO to the inert methane (eq. (4.1) reversed). Alkynes are removed from alkene streams for oxidation or polymerization by selective hydrogenation (Section 4.3.4).

When poisons do not completely annihilate catalyst activity, the problem may be handled by improving the catalyst to an extent that allows a satisfactory residual activity level to remain under the given concentrations of the poison. Hydrocracking over

Figure 4.24: Erosion of monoliths from a high-dust SCR facility. Reproduced from ref. [101].

Pt/zeolite catalysts working at the NH_3 and H_2S levels established in the upstream HDN/HDS stage (cf. Section 4.4.1.1) are examples of such an achievement. Notably, in this case, the main role of Pt is to hydrogenate coke precursors rather than to catalyze an rds. When the poison affects the rds, a trade-off between the cost of its removal and the losses by the decreased activity must be sought.

In the end-of-pipe processes of exhaust-gas treatment, e.g., in automotive converters, the cost of feed conditioning cannot be realized via product prices. Therefore, waste streams are usually processed without any conditioning,[18] i.e., catalysts regularly need to achieve specified performance and durability at the given poison levels. These depend on the source of the feed, but typically include a high load of water.

Feeds raising inacceptable cost for conditioning are also quite common in the refining of heavy oil fractions, which are rich in coke precursors and poisons (e.g., Ni, V porphyrin compounds). The latter are distinguished by low mobility, which provides opportunities to handle them by both catalyst design and reaction engineering using a chromatographic approach.

---

18 The control of the O_2 content in gasoline car exhaust by the λ sensor for complying with the requirements of the three-way catalyst (cf. Appendix A1) is the only case of waste gas conditioning known to the authors.

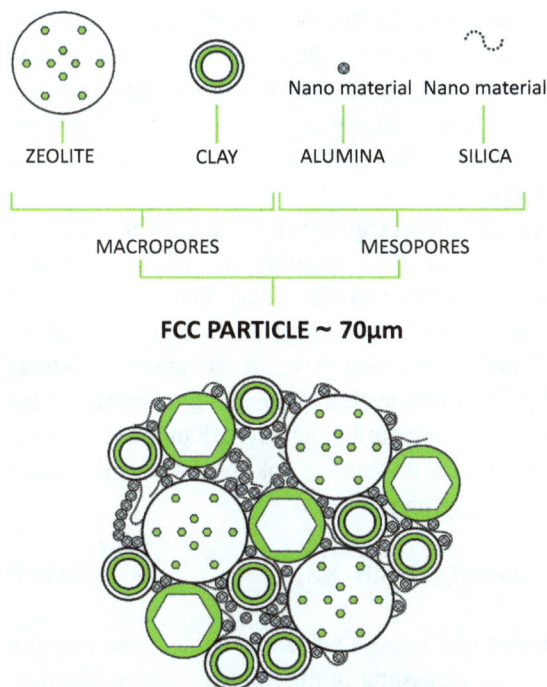

Figure 4.25: Morphology of an FCC catalyst. Reproduced from ref. [102] with permission from the Royal Society of Chemistry.

In FCC catalysts processing vacuum gasoils, zeolites are considered the cracking components, although feed molecules are far too bulky to enter their cavities. Thus, the zeolites are integrated into a matrix containing a number of additional components, e.g., alumina and silica binders, and clays for density and heat capacity regulation (Figure 4.25). Alumina surfaces and Al_2O_3/SiO_2 interfaces provide moderate acidity that allows precracking feed molecules to intermediates fitting into the zeolite micropores. Both coke and metal compounds are deposited in the macropores of the matrix before getting in touch with the zeolite. The matrix also contains various promoters. Passivating agents (Sb, Bi, or Sn compounds) are added to delay deactivation of the zeolite by silicide formation (cf. Section 4.4.1.3). There is also some Pt, which takes care that coke can be completely burnt off, while the catalyst is kept in the regenerator. Finally, the catalyst contains components that make sure that the sulfur introduced by the feed is not released as SO_2 in the regenerator. It evolves in the riser as H_2S instead, which can be separated from the product stream by standard technology.

While the FCC catalyst is a sophisticated piece of catalysis art, the chromatographic principle can also be applied in the simpler form of egg-yolk catalysts. They contain the expensive active component at the center of porous pellets, the external porosity of which serves to trap a poison (cf. Figure 3.17). In such catalysts, pore size optimization is most important to delay clogging of the external pores by trapped species. In residue hydroprocessing, an analogous strategy is realized on the reactor scale: asphaltenes and metal contaminants are trapped in a guard reactor by a cheap wide-pore catalyst, which hydrogenates the phthalocyanine ligands of the latter (hydrodemetallization), before the feed proceeds to the main reactor with a state-of-the-art HDN catalyst. The guard-bed catalyst is periodically substituted.

Competing with the target reaction, carbon deposition can be delayed by improving the selectivity of the latter. Selectivity promoters may, therefore, be stability promoters at the same time. On the other hand, promoters can also be applied to accelerate a coke gasification reaction in order to decrease the net coking rate. In the hydrocracking catalysts mentioned earlier, this is the main purpose of Pt, i.e., it operates as a stability promoter. K^+ ions are known to catalyze coke gasification by steam. In steam reforming of methane (4.1), K^+ delays coke formation by mitigating the strength of the Ni–C bond, but probably also by accelerating gasification of the existing coke by the steam reactant. In the dehydrogenation of ethylbenzene to styrene,

$$C_6H_5-CH_2-CH_3 \rightleftarrows C_6H_5-CH=CH_2 + H_2 \qquad (4.11)$$

acceptable conversions require temperatures around 900 K for equilibrium reasons. The hydrocarbons are diluted with steam. Potassium in the iron oxide-based catalyst, which is actually part of the active $KFeO_2$ phase, helps in keeping coke deposition under control by catalyzing its gasification by steam.

Where coke deposition cannot be prevented with reasonable effort, the catalyst may be eligible for a continuous production process that includes periodic or continuous catalyst regeneration. As mentioned earlier, suitable technology depends on the timescale of deactivation (cf. Table 4.1). More details on catalyst regeneration can be found in Section 4.4.3; for more information on technology, the reader may consult textbooks of industrial chemistry [103–105]. Different handling of the fouling problem has resulted in competing technology for some important industrial processes: naphtha reforming and cracking of vacuum gasoils (cf. Section 4.3.1 and Appendix A1).

Thermal degradation may be obviously delayed by limitations on the reaction temperature, and by meticulous elimination of water from the reaction atmosphere. If this is impractical or insufficient, catalyst design becomes the major tool for delaying particle growth. The underlying mass transport can be influenced via the driving forces, the distances, and the diffusion coefficients. Transport is mostly driven by the difference in the stability of the smaller and the larger particles. Interactions of the metal with a more refractive component decrease this driving force. This is why Pt is stabilized by alloying with Re in naphtha-reforming catalysts. The activity and stability of Co-based FT catalysts are strongly related to the interactions between Co and the support. While the Co precursor is easily reduced on carbon supports (e.g., on CNTs), full reduction is often not achieved, even with reduction promoters, on oxide supports (cf. Section 4.2.4).

On the other hand, stabilizing interactions for the metal particles are weaker on the carbon surfaces. Therefore, the advantage of carbon-supported Co-FT catalysts is limited by stability problems.

The sintering of Co might also be influenced by interactions of the diffusing species with the support: diffusion across the weakly interacting carbon surface should be faster than across an alumina surface. While this idea has not been examined so far, the stabilization of Pt by CeO_2 surfaces is a most striking example of a sintering process suppressed by frustrated diffusion. Ce^{4+}, which is present in three-way catalysts in the form of $CeZrO_x$ mixed oxides, was found to trap cationic Pt on surface sites (Ce^{4+}–O–Pt^{n+}–O–Ce^{4+}) [106]. In the oscillating reaction regime of three-way catalysis, Pt is reduced to extremely small nanoparticles during the rich excursions, but oxidized and trapped in atomic dispersion during the lean excursions, instead of sintering via Pt oxide phases as on other supports. By supporting Pt on the $CeZrO_x$ component, in quantities adapted to its surface capacity, both stability and dispersion of Pt can be strongly improved. Similar stabilization by surface Ce^{4+} oxide species is also known for Pd^{2+}, and has been found to cause dramatic differences in activity, selectivity, and stability of supported bimetallic Pd–Au catalysts in three-way catalysis [107].

Finally, sintering and undesired solid-state reactions may also be impeded by increasing the distances between the particles and enhancing the tortuosity along diffusion paths. For this reason, the stability of supported nanoparticles can be improved by optimization of the support porosity. Spacers preventing coalescence of crystallites in bulk catalysts (e.g., Ca or Si oxides in the Fe catalyst of NH_3 synthesis, cf. Section 4.2.4) are structural and stability promoters at the same time. Currently, research is invested into approaches to separate metal nanoparticles by the debris of a porous precursor structure, e.g., by pyrolysis and reduction of MOFs, or to prevent sintering by encapsulating metal particles in porous shells of a spacer material.

Sintering of support surfaces can be delayed by blocking free OH groups with suitable cations. Alumina surfaces can be stabilized by La^{3+} or Ba^{2+} species as mentioned earlier. Anatase employed in V_2O_5-WO_3/TiO_2 catalysts for NO_x abatement in flue gases is often stabilized by impregnation with a silica additive.

In some cases, undesired solid-state reactions can be eliminated by just separating their reactants in the catalyst. Since the detection of Rh aluminate formation in three-way catalysts, Rh has been supported on ZrO_2, which is washcoated as a separate layer (cf. Section 2.3.4). As alloying of components is undesired in three-way catalysis anyway, washcoats nowadays consist of several layers (up to 6), and are often topped by a getter layer (alumina or a zeolite), which traps corrosion products or poisons from lubricants (Zn and P).

4.4.3 Regeneration of catalysts and the end of their life cycle

Whenever possible and economical, deactivated catalysts are subjected to procedures aimed at the regeneration of their original performance. If there is no reasonable option for regeneration, they can be passed on for other uses, for reclaim and recycle of valuable catalyst components, or for disposal. Due to environmental risks inherent in many catalyst materials, disposal may be costly, and is usually the last resort.

The perspective of regeneration depends on the nature of the deactivation process. Carbon deposits can be removed by gasifying them with air, water, CO_2, or H_2. Relative rates of *noncatalyzed* carbon gasification at 1,070 K (O_2:H_2O:CO_2:H_2 = 100:3:1:0.003) give a flavor of the different efficiencies of these reactants [90]. Redox elements in the catalyst can, however, strongly accelerate the oxidative routes. Although temperatures required for such oxidation depend on the type of carbon present (cf. Section 4.4.1.1), complete combustion of the deposits can usually be achieved by air at 700–900 K in the presence of a redox element. Rates with other oxidants are smaller by orders of magnitude. Reference [90] lists, however, several cases, in which carbon (actually soft carbon) is removed by H_2 or H_2O.

To avoid hotspots due to high oxidation rates, regeneration is usually started with low O_2 contents. Temperatures also need to be strictly controlled during the subsequent increase of the oxygen content because hydrothermal stress increases all risks of solid-state mobility and loss, which are carefully minimized during catalyst operation. Due to the high reaction rates, coke oxidation is typically controlled by heat and mass transfer. The reaction progresses into the pellet, gradually consuming the shrinking core of coke, which subjects the interior of the pellet to particular thermal stress.

In the regeneration of sulfide-poisoned metal catalysts, it is important to avoid the formation of thermally stable sulfates, which keep sulfur at the surface until it is re-reduced to sulfide during catalyst operation. The problem is most relevant for base metals like Ni, which are oxidized together with the sulfide. From steam-reforming catalysts, most sulfides can be removed by steam at ≈970 K, i.e., close to their operation regime [90]. For catalysts that would not survive such conditions, it may be possible to find oxygen-deficient conditions, under which the oxidation of the sulfide may be differentiated from that of the metal. On noble metals, the sulfates are less stable, but the sulfate ion can be trapped by some supports, which may limit the success of the regeneration.

Thermal damage is usually irreversible; sintering of some noble metals being a notable exception. Their redispersion can be achieved by an oxychlorination treatment, charging HCl or a chlorohydrocarbon and dilute O_2 at ≈770 K. It proceeds via volatile intermediates ($Pt(CO)_2Cl_2$ and/or $PtCl_2(AlCl_3)_2$) [90, 108] formed at the metal surfaces, and trapped as surface Pt chloride species elsewhere on the support. Oxychlorination, the details of which may vary with the nature of metals and supports [90], is a standard procedure with naphtha-reforming catalysts, which refreshes the chloride content of the alumina support at the same time. It is also employed to reactivate hydrocracking catalysts.

In the past, regeneration was performed exclusively on site, for fixed beds right in the reactor, for moving catalysts in special sections integrated in the plant. Meanwhile, there is a market for companies remediating more substantial damage by more laborious technologies, for which the catalyst is discharged. Typical steps performed in such a treatment are cleaning/sieving to remove dust and fines, washing in water, washing with reagents, re-impregnation of active components (e.g., Co, Mo, and suitable ligands to attenuate their interactions with the alumina surface in HDS catalysts), drying, and calcination.

Fe, Ni, and V deposited in FCC catalysts are transformed into soluble or volatile forms by routes like sulfidation with subsequent chlorination or oxidation in air [109]. Vanadium can be extracted as vanadate after mere calcination. Metals remaining in the catalyst may be passivated by Bi or Sb reagents. The significant, though not complete, recovery of performance warrants reintegration of demetallized material into the FCC catalyst cycle. In other processes, regenerated catalysts may perform similar to fresh batches, in particular, when active components have been replenished.

The perspective of a catalyst not eligible to regeneration depends on its composition: on the value of its components, on the expenditure for their recovery (including separation!), on their environmental impact, and on the resulting expenditure upon disposal. Precious metals are almost always reclaimed and recycled. Their separation from supports and monoliths can be achieved by methods of hydro- or pyrometallurgy or by chlorination. In wet routes, the noble metals are dissolved, e.g., in aqua regia, and re-reduced after the separation from the support material by the addition of a non-noble metal, e.g., Al. In pyrometallurgical routes, the catalysts are fired with additives (lime, Fe, Cu compounds). Supports melt and react with lime to a low viscous slag while the noble metals adsorb on Fe or Cu additives at the bottom of the reactor. In chlorination, the metals are converted to volatile chlorides by reaction with Cl_2 at high temperatures. More details can be found in ref. [109].

Base metals are reclaimed from bulk catalysts or if metal concentrations are high, but to a smaller extent than precious metals. If they are associated with other metals, procedures may be more laborious and expensive than even their extraction from ores or scrap. In some cases, direct addition of catalysts to raw materials of metallurgical lines is nevertheless possible, for instance, in the production of special steels. Co, Mo, Ni, W, and V in spent hydrotreatment catalysts can be solubilized by leaching processes after roasting in air at high temperatures. Separation of the metals requires a series of extractions or fractionated precipitations [109]. Cu and ZnO from methanol synthesis catalysts can be used for the production of Cu and Zn chemicals.

Alumosilicate-based catalysts and other support materials, even with minor residual metal contents, can be applied as slag-forming additives in metallurgy. Slags from pyrometallurgic noble metal recovery are useful for the production of refractory and abrasive materials. Al salts and oxhydroxides can be obtained as byproducts of the hydrometallurgical route. When complying with pertinent environmental specifications, spent catalysts or residues from metal reclaim operations are employed in pit fills.

4.5 Types of reaction mechanisms

4.5.1 Getting insight into surface processes

Elucidating reaction mechanisms is a major step toward understanding reactive systems all over chemistry. In surface catalysis, it provides the basis for microkinetic models, which describe macroscopic reaction rates as the result of microscopic (surface) processes (Section 4.6). Work on reaction mechanisms is nowadays substantially supported by theoretical studies. Although progress on catalyst development and design of catalytic production processes can also be achieved on a lower level of theoretical maturity, combining atomistic models of catalytic phenomena and their quantum-chemical description with hydrodynamic models bears the promise of an ultimate control on catalytic processes on various time and length scales.

In this book, basic reaction mechanisms have been discussed in Section 2.5.1, and the combination of adsorption, desorption, and surface reaction steps to a catalytic reaction has been exemplified with the synthesis of ammonia (Figure 2.57). Such reactions can combine to reaction networks, comprising consecutive and competitive steps, as shown for the selective oxidation of butane to maleic anhydride in Figure 4.22. Given the large number of catalytic reactions and catalysts, and the ever growing number of hypotheses and speculations on mechanisms in literature, the present section intends to provide the reader with information on the tools available for studying mechanisms, on criteria for the valuation of hypotheses, and with instructive examples highlighting typical mechanisms in thermal catalysis (for photo- and electrocatalysis, see Sections 4.8 and 4.9).

The discussion on reaction mechanisms is driven by the intention to find networks of surface steps that comply with spectroscopic evidence on adsorbates present under reaction conditions, as also with results of studies on reaction kinetics (reaction orders, their dependence on reaction conditions, and rates of candidate steps), on the nature of the active sites, and with evidence from studies with labeled reactants (isotope effects, fate of labeled atoms). The source of such networks has traditionally been chemical intuition, often inspired by assuming analogies to steps in coordination catalysis. Important tools for mechanistic studies like *in situ* or *operando* spectroscopy, transient kinetic methods that might resolve steps faster than the rds (e.g., TAP reactor, SSITKA, cf. Section 3.2.1), or mature methods of theoretical chemistry have become available only in the late twentieth century. Many mechanisms proposed before had to be revised. Successful concepts put forward much earlier, like the Horiuti-Polanyi mechanism of alkene hydrogenation and the Mars–van Krevelen mechanism of oxidation reactions (Section 2.5.1), are all the more admirable.

Despite formidable improvements in the toolbox available for mechanistic studies, many mechanisms more or less accepted by the scientific community at present remain open to challenge by deeper insight or even alternative concepts. The blindness of (*operando*) spectroscopy to the coexistence of reaction intermediates and spectators has been a major problem, which is being tackled by combining spectroscopy with transient kinetic techniques. When, for instance, a catalyst is studied by SSITKA and IR spectroscopy simultaneously, rates of label exchange on different adsorbates can be related to rates of product formation, which differentiates intermediates from spectators.

Modulation excitation spectroscopy (MES) dealing with the response of spectroscopic signals and of the reaction rate to periodic variations of feed composition is a promising tool as well. Time-resolved spectra recorded during a periodic sequence of concentration changes (e.g., steps up and down) are transformed into phase-resolved spectra with the formalisms of phase-sensitive detection (PSD). PSD strongly suppresses noise and hides information related to unchanged parts of the sample. The technique is well described in ref. [110], where it is simultaneously applied to DRIFTS and XAFS measurements of supported metal catalysts during NO reduction by CO. Still, the interpretation of the features detected in this mode of spectroscopy needs more experience and theoretical support, which is, however, likely to change in the future.

Reaction mechanisms put forward should comply with some minimum criteria before being taken seriously. They should describe the whole catalytic cycle returning to the initial situation, as is standard in literature on coordination catalysis. In heterogeneous catalysis, the discussion is often focused on what is considered the rds, because the remaining fast steps are very difficult to elucidate. Such simplification is justified only if the rate-limiting character of the step under scrutiny is beyond doubt. If this cannot be demonstrated by own work, citation of reliable literature supporting the claim is the least one can do to keep credibility. Actually, the identification of the rds may be all but simple. In redox mechanisms, for instance, where the active metal shuttles between two oxidation states, its state in the reaction environment should indicate the starting point of the rds, because it is quickly restored via the remaining fast steps of the cycle. Often, however, oxidation states intermediate between the two "end points" are observed. While this may indicate that rates of reduction and reoxidation steps are comparable, the observed value may also be just an average between the oxidation states of active metal sites and spectator cations. In this situation, the experiment remains inconclusive with respect to the rds.

Reaction mechanisms should also comply with available evidence on the nature of the active site, if this evidence is credible. For supported metal oxide catalysts, for instance, a mechanism proceeding on a single metal cation is at variance with an active site model comprising a pair of metal cations, because it fails to explain the role of the second ion. In such catalysts, reactions requiring sites with more than one metal ion reveal themselves by a disproportionate increase of the rate with the metal content. Rate measurements with series of catalysts containing variable concentrations of the active element are therefore a simple and effective tool of catalysis research.

The very different nature of the reaction types encountered in catalysis suggests an easy classification of mechanisms. In acid–base catalysis, for instance, reactions catalyzed by Brønsted sites are most important; in selective hydrocarbon oxidation – reactions are often initiated by the oxidative activation of a C–H bond, in polymerization – the insertion of a metal atom into a C–H bond is a crucial step. Although differences are dramatic indeed (cf. Section 4.5.2), they are not easily pinned via the nature of bonds between the intermediates and the surface. While the metal-carbon bond is covalent, ionic species (carbenium ions forming an ion pair with negatively charged surface oxygen) are usually assumed as intermediates in reactions catalyzed by Brønsted sites. It has been claimed, however, that the nonsolvated ionic charge is unstable on surfaces exposed to the gas phase, and the intermediate forms a covalent bond with oxygen. The problem is discussed in some detail in Section 4.5.2. In oxidation catalysis, there are a number of mechanisms where intermediates may be suspected to be radicals, but such species are hard to detect unambiguously. A case in which the radical character of a key intermediate has been supported by quantum chemical studies is discussed in Section 4.5.2.

The active site type is another opportunity to categorize mechanisms (Section 4.5.3). Even complex reactions may completely proceed on a well-defined site, similar as in a vacancy of a metal complex catalyst. On a metal surface, however, active sites for complex reactions are not easily delimited, and reaction products may differ strongly when nature and coverage degrees of adsorbates vary depending on temperatures and the gas-phase composition. In bifunctional mechanisms, an intermediate changes between two sites of different types. It may be a stable molecule[19] or a reactive intermediate. While distances between sites involved are often not larger than a few Ångstroms, there is no limit on the range that stable species may cover. Spillover of hydrogen is a special case in which this range strongly depends on the properties of the solid phase (cf. Section 4.5.4).

Section 4.5.5 briefly summarizes mechanisms involving the participation of secondary species. Section 4.5.6 shows how reactions may proceed via different mechanisms, depending on the sites available; sometimes, even with comparable rates.

---

19 In this case, the surface reaction would have to be addressed as a reaction network, according to the conventions of this book. As this is at variance with part of the literature, the convention will be handled flexibly in the case of bi- or multifunctional mechanisms.

4.5.2 Reaction types and catalytic mechanisms: catalysis on Brønsted sites, with nucleophilic oxygen, and via metal-carbon bonds

The following examples are related to reaction types, which are also familiar in homogeneous catalysis. They illustrate the difficulty to specify the nature of the reactive intermediates on the solid surfaces. When reactions involve metal-carbon bonds, as for instance alkene metathesis or polymerization, it is an additional challenge to find out how the metal-organic intermediates are formed on surfaces dominated by metal-oxygen or metal-chloride bonds.

4.5.2.1 Reactions catalyzed by solid Brønsted acids

In liquid, in particular, in aqueous media, there is little doubt that reactions catalyzed by Brønsted acids proceed via carbenium ions. It is unlikely that this changes when the Brønsted sites are located on a solid surface, although there is less space for a solvate shell. At solid–gas interfaces, however, stabilization of ionic charges by solvation is not possible at all.

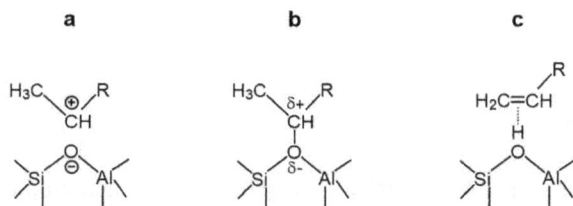

a   **b**   **c**

Scheme 4.5: Interaction of a zeolite Brønsted site with an alkene, resulting in a carbenium ion (a), an alkoxide (b), or a π-complex (c).

The doubts in the stability of carbenium ions go back to the theoretical work of the 1990s, according to which ion pairs, which consist of carbenium ions and the conjugate base of zeolite Brønsted sites (Scheme 4.5a) are unstable, relative to the surface alkoxide structures (Scheme 4.5b) [111]. The carbenium ion was proposed to be a transition state of acid-catalyzed reactions only. In the following years, theoretical work on the interaction of Brønsted sites with various hydrocarbons was refined by using larger models that capture the influence of spatial constraints, and by including van der Waals interactions, e.g., by applying functionals eligible for dispersion forces and by comparing the results of small clusters at the core of QM/MM models with results from wave function-based electron correlation methods.

Energies of carbenium ions always exceeded those of the alkoxide forms or of π-complexes between the Brønsted site and the adsorbate (Scheme 4.5c). The alkoxides, which were often most favorable in terms of energy, suffer an entropic penalty for being more localized than both carbenium ions and π-complexes. Indeed, in a study on species formed in H-FER from C_3–C_5 alkenes [112], π-complexes were found to be most stable, while alkoxides were assigned minority species, if present at all. Although the entropic term favors carbenium ions, in particular, at higher temperatures, they were not expected to become sufficiently stable to be observable, for instance, by spectroscopy. Spectroscopic evidence at solid–gas interfaces is indeed available only for structures in which the charge can be delocalized over several aromatic rings, e.g., triphenyl carbenium ions.

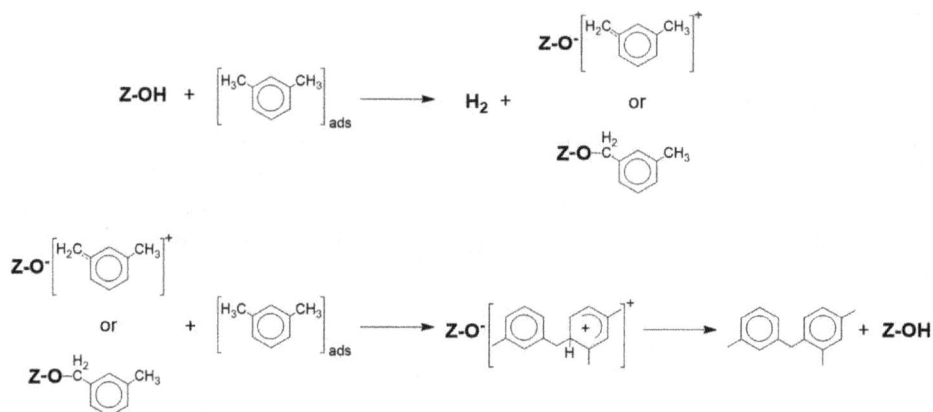

Scheme 4.6: Intermediates during *m*-xylene dimerization.

On the other hand, it was also found that carbenium ions may be local minima on the reaction coordinate. In xylene dimerization, hydride abstraction from adsorbed xylene by the Brønsted site may result in alkoxide species or carbenium ions (Scheme 4.6). In ref. [113], their energies were found to be rather different (Figure 4.26), but all below the energy of the initial reactants (not shown in the figure), i.e., even the carbenium ion is a local minimum. Substitution of the activated *m*-xylene (carbenium ion or alkoxide) into the second *m*-xylene ring results in a benzenium cation, which is a stable intermediate on the reaction coordinate. To complete the reaction, it returns a proton to the zeolite. It is therefore also accessible vice versa by protonation of the adsorbed dimer (Figure 4.26).

In the above-mentioned study on C_3–C_5 adsorbate structures on H-FER [112], tertiary carbenium ions were found to feature local minima on the potential energy surface. Apparently, they can be short-lived intermediates of reactions, which would be sufficient to reconcile the success of reactivity rules derived from the properties of carbenium ions (see further) with theory. On the other hand, as potential minima were not found for nonbranched carbenium ions, secondary carbenium ions as reaction intermediates remain under doubt. Notwithstanding, mechanisms of gas-phase reactions catalyzed by solid Brønsted acids are discussed with carbenium-ion intermediates in the following.

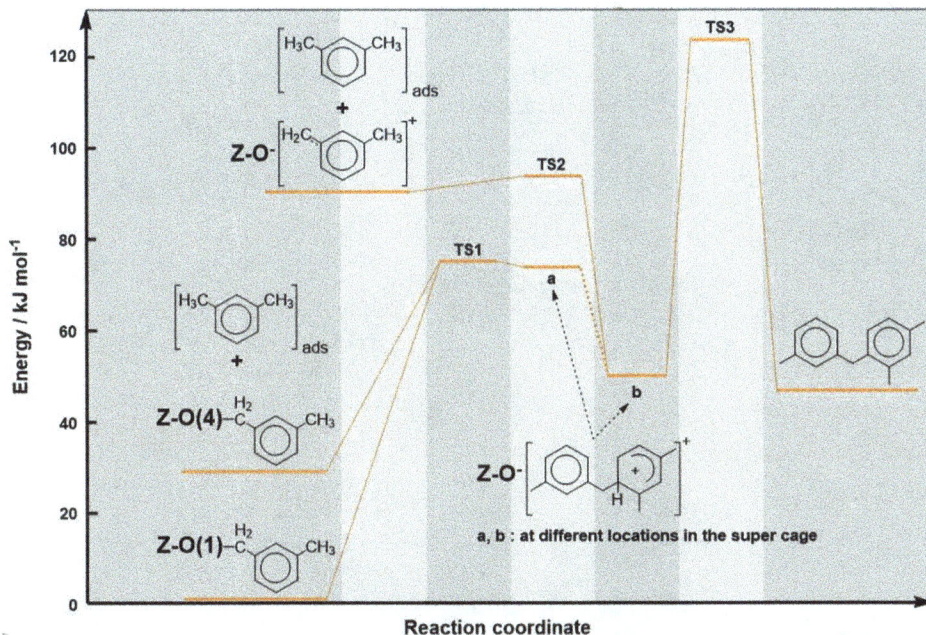

Figure 4.26: Reaction coordinate of *m*-xylene dimerization in the supercage of a H-faujasite (starting with adsorbed reactants). The initial adsorbates (alkoxide or carbenium ion) were evaluated, together with the standby second xylene molecule. O(1) and O(4) designate nonequivalent O atoms in faujasite. Adapted with permission from ref. [113]. Copyright 2003 American Chemical Society.

Catalysis by solid Brønsted acids is most familiar in hydrocarbon chemistry. Alkenes and aromatics can be isomerized by shifting double bonds along chains, by (de)branching the carbon skeleton, or by redistributing alkyl substituents around rings. Brønsted sites can catalyze C–C bond formation (alkene oligomerization, isobutane alkylation, Friedel–Crafts alkylation), and rupture (hydrocarbon cracking, dealkylation/transalkylation of alkyl aromatics). Likewise, many additions at double bonds and the reverse elimination reactions are catalyzed by protons, e.g., the hydration of alkenes and the elimination of water from alcohols. The MTH reaction can be considered as water elimination as well.

The mechanisms of these reactions can be summarized in the simple scheme shown in Figure 4.27, where typical routes to carbenium ions have been labeled f1 through f3: the protonation of an alkene (f1), the dehydration of an alcohol (f2), and hydride abstraction from an alkane (f3, see also Scheme 4.6). Obviously, all these reactions are reversible, and their combination describes mechanisms of real acid-catalyzed reactions. Combining f1 with f2 describes the addition of water to a double bond and the reverse elimination of water from an alcohol. Combinations with f3 are less familiar. Dehydrogenation of alkanes merely on acidic zeolites, which was found in Russia in the times of the former Soviet Union (summarized in ref. [114]), is

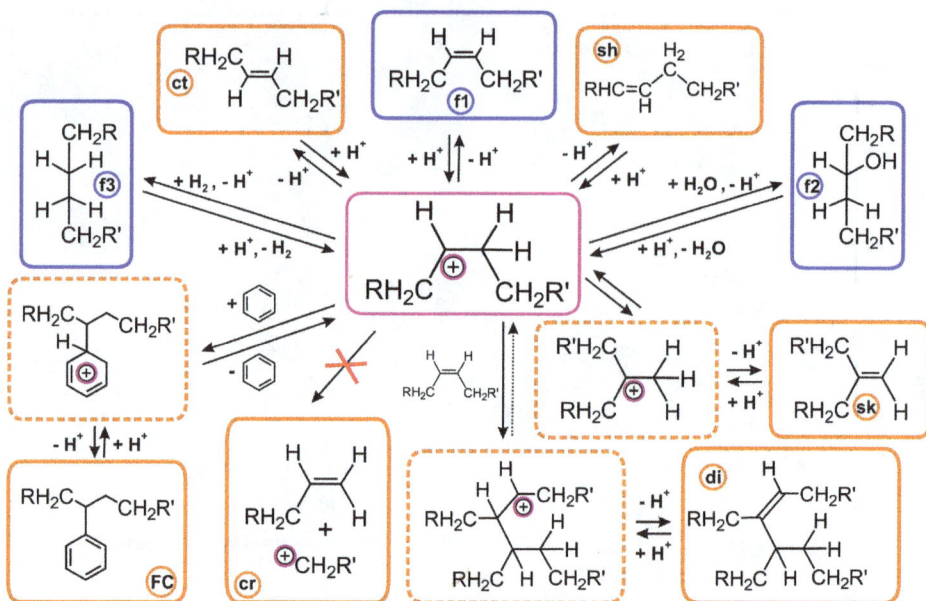

Figure 4.27: Mechanisms of reactions catalyzed by solid Brønsted acids.

meanwhile well established though. Hydride abstraction requires strong acidity and gives access to carbenium-ion chemistry in feeds not containing either double bonds or leaving groups for eliminations.[20]

Figure 4.27 shows a number of reaction channels accessible to the carbenium ion, which can be combined with the formation steps to reaction mechanisms. In the carbenium ion, rotation is possible around what was a double bond in the reactant. If this happens before the proton returns to the catalyst, a cis-trans isomerization occurs (f1 + ct). If the catalyst takes the proton from a different C atom, the double bond is shifted along the chain (f1 + sh). The electrophilic carbenium ion can be substituted at an aromatic ring position, which results in Friedel–Crafts alkylation (f1 + FC). Likewise, the carbenium ion can attack the double bond of another alkene molecule, which results in a dimer after desorption from the Brønsted site (F1 + di). The reverse cracking reaction proceeds by splitting up the C–C bond in β-position to the charge (step cr), and leaves behind an alkene and a carbenium-ion. The version shown in Figure 4.27 is, however, very unfavorable, because it results in the formation of a highly unstable primary carbenium ion.

---

20 The idea to hydrogenate alkenes over Brønsted sites sounds odd now, but according to the principle of microscopic reversibility (Section 1.1), there must be hydrogenation activity at sites capable of dehydrogenating alkanes. It may be small and easily poisoned, e.g., by coking.

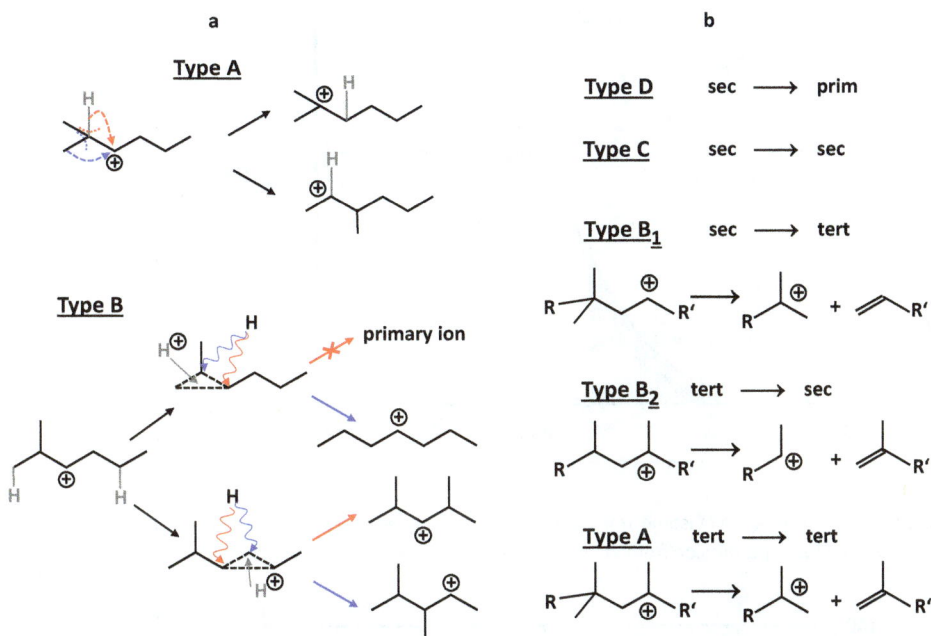

Figure 4.28: Chemistry of carbenium ions: routes for isomerization (a) and splitting (b).

Cracking is, therefore, preceded by skeletal isomerizations of the carbenium ion (Figure 4.27, f1 + sk). There are two types of skeletal rearrangements [115]: those where the number of branching points is preserved (type A) or changed (type B). Isomerizations of type A are thought to proceed via sequences of hydride or alkyl shifts (Figure 4.28a). The former move the charge, while the latter move the branching group along the chain. Their rates are very high, but decrease with the mass of the shifted group. Type B rearrangements proceed via nonclassical protonated cyclopropane structures, as illustrated in Figure 4.28a. The proton coming from one cyclopropane corner may end up at the other two C atoms. This results in two new carbenium ion structures, among which primary structures are penalized for stability reasons.

In cracking, the rates of bond rupture at the β-position depend on the nature of the carbenium ions involved [115]. β-Scissions of type A, in which the tertiary structure of the intermediate is preserved (cf. Figure 4.28b), are even faster than type A rearrangements, and both are faster than desorption [115]. Type B_1 and B_2 β-scissions, which involve tertiary and secondary carbenium ions (Figure 4.28b), proceed at rates comparable to type B isomerizations, while type C and D (the latter shown in Figure 4.27) are the slowest routes. Steps slower than desorption require re-adsorptions of intermediate products to become relevant, which is easily accomplished in zeolite pores.

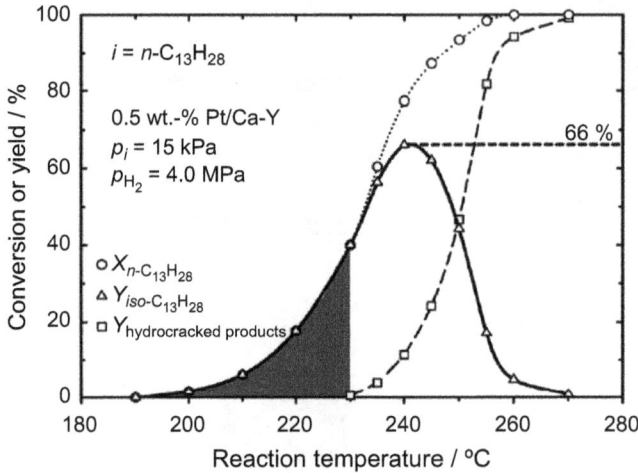

Figure 4.29: The relation of isomerization and hydrocracking: *n*-tridecane over Pt/Ca-Y. Reproduced from ref. [115] with permission from Wiley-VCH Verlag GmbH & Co. KGaA, Weinheim, Germany.

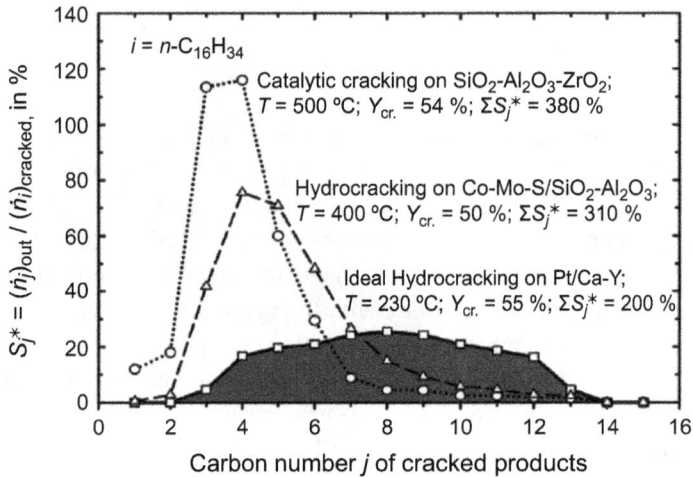

Figure 4.30: The relation of isomerization and hydrocracking: product distributions from *n*-hexadecane under different reaction conditions. Reproduced from ref. [115] with permission from Wiley-VCH Verlag GmbH & Co. KGaA, Weinheim, Germany.

Cracking of the usually saturated hydrocarbon feeds is greatly facilitated if the catalyst contains a component with (de)hydrogenation activity, e.g., a noble metal or a sulfide. Even at high H_2 pressures, the minor alkene concentrations established allow carbenium-ion catalysis under very mild conditions. The example shown in Figure 4.29 illustrates that isomerization indeed precedes cracking: at a H_2 pressure of 40 bar, isomerization products of *n*-tridecane appear at 470 K, while cracking starts at ≈500 K. The yield curves support this conjecture.

Under these mild conditions, the chain length distribution of the cracking products from C_iH_{2i+2} (in Figure 4.30: i = 16) ranges from 3 to (i – 3). There is a broad maximum on a plateau extending from 4 to (i – 4). About 60–90% of the products are mono- and dibranched, while tribranched structures are absent due to the ease of the type A cracking route (Figure 4.28b). In ref. [115], the product distribution of this "ideal hydrocracking" is explained in more detail with the interplay of isomerizations and β-scissions of the different types. Its symmetry and a yield of 200 product molecules per 100 feed molecules (ΣS_j^* in Figure 4.30) suggest that it results from one cracking event per feed molecule, on average, and that the surface reactions are rate-limiting. The chain length plateau between 4 and (i–4) indicates that type A β-scission is the dominant but not exclusive cracking step.

Due to the different degrees of branching in the isomers involved in the various β-scission steps, hydrocracking can be influenced by spatial restrictions. The small channel dimensions of MFI interfere with type A scission; therefore, the central chain length plateau extends between 3 and (i–3) with H-ZSM-5. The relation between the cracking routes also depends on the length of the feed molecules, because β-scissions of types A, B, and C require minimum lengths of 8, 7, and 6 C atoms, respectively. As a result, cracking rates increase drastically from *n*-hexane to n-heptane over Pd/H-ZSM-5 and, slightly less pronounced, between n-heptane and n-octane over Pd/H-Y [115].

At higher reaction temperatures, the chain length distribution of hydrocracking products becomes asymmetric, leaning toward smaller molecules. In Figure 4.30, this is illustrated with a distribution over Co–Mo–S/SiO$_2$–Al$_2$O$_3$ at 723 K, notably at nearly the same conversion as with Pt/Ca–Y at 500 K. The third curve (SiO$_2$–Al$_2$O$_3$–ZrO$_2$) extends the tendency into a range more typical for FCC. An increasing role of rate limitation by desorption allows multiple β-scissions, before molecules are released. Similar distributions would be obtained with Pt/Ca–Y at higher temperatures (at conversions up to 100%), which shows the flexibility of hydrocracking with respect to product composition. The dotted curve in Figure 4.30 is, however, actually from a monofunctional catalyst, where the processes discussed so far are modified by other mechanistic steps: by protolytic cracking and by hydride transfer.

Protolytic cracking, sometimes referred to as Haag–Dessau cracking, is initiated by the protonation of alkanes at their most substituted C atom, as illustrated with 3-methylpentane in Figure 4.31. The resulting carbonium ion is a transition state, which may collapse along different routes, depending on the substituents at the protonated carbon. As a result, H$_2$, methane, and ethene are observed at very low conversions, which cannot be explained by the carbenium-ion mechanism. With longer feed molecules, the carbenium ions resulting from protolytic cracking undergo further isomerization/β-scission steps along the classical mechanism. Haag–Dessau cracking is favored by high temperatures and narrow pores. As it produces alkenes, from which carbenium ions are easily formed, it is relevant only at low conversions [116].

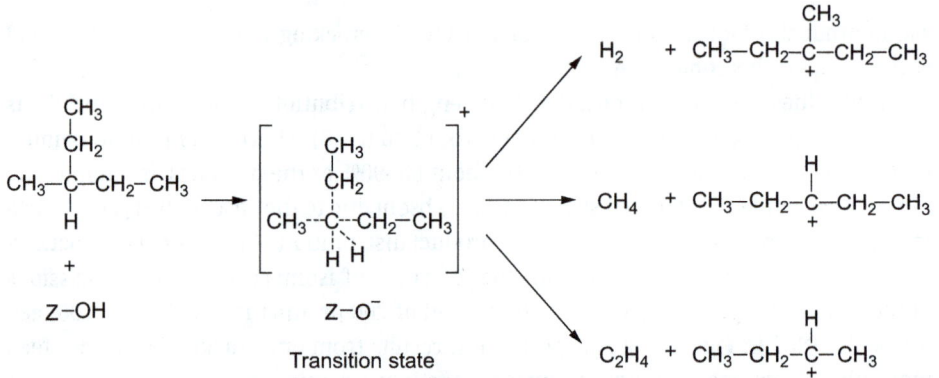

Figure 4.31: The Haag–Dessau mechanism of carbenium ion formation. Adapted from ref. [116] with permission from Elsevier.

In hydride transfer, a carbenium ion abstracts a hydride ion from a saturated reactant molecule (eq. (4.12)). The former desorbs as an alkane, while the new carbenium ion carries on with the cracking reaction. This adds a feature of a chain reaction to hydrocracking.

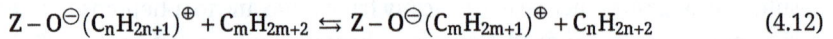

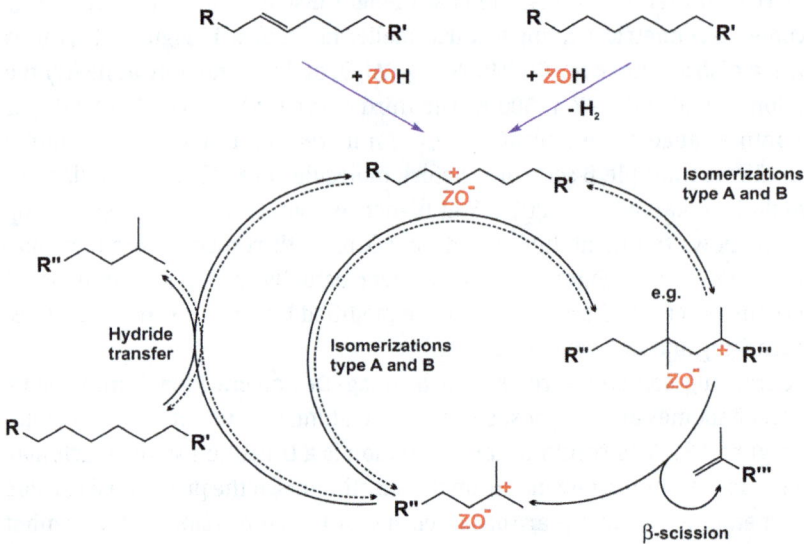

$$Z - O^{\ominus}(C_nH_{2n+1})^{\oplus} + C_mH_{2m+2} \leftrightharpoons Z - O^{\ominus}(C_mH_{2m+1})^{\oplus} + C_nH_{2n+2} \qquad (4.12)$$

Figure 4.32: Hydride transfer and the cyclic mechanism of catalytic cracking.

The complete cycle is illustrated in Figure 4.32. The initial carbenium ion undergoes isomerizations as described earlier, until it has a shape eligible for β-scission. The resulting shorter carbenium ion has the choice between proceeding with isomerization/cracking or abstracting a hydride ion from another feed molecule,[21] which restores the original situation. Therefore, cracking can be sustained with very low concentrations of carbenium ion precursors when hydride transfer rates are high.

The competition between hydride transfer and consecutive cracking is an opportunity to influence the composition of cracking products via the nature of the catalyst (see later). Without hydride transfer, short-chain alkenes from consecutive cracking dominate already at low conversions; without consecutive cracking, a new feed molecule becomes involved after each cracking step (similar as in ideal hydrocracking, cf. Figure 4.30). Chain-length distributions remain broad up to high conversions, with up to 50% alkanes in the product. Generally, hydride transfer favors saturated branched products over (branched) alkenes and aromatics. The latter can be formed by acid-catalyzed cyclization of alkenes (cf. Section 4.5.4).

Hydride transfer is favored by high feed concentrations around the active site, as also by a high concentration of active (and adsorption) sites in the catalyst. In FCC catalysts, the major cracking component is either ultra-stabilized Y zeolite (US-Y) with a very low residual Al content, or rare-earth (RE) exchanged Y, in which most of the active sites are formed from the RE-cations by the Hirschler-Plank mechanism (cf. Scheme 2.1c). Hydride transfer is much more intense in RE-Y. The complex structure of FCC catalysts (cf. Figure 4.25) allows mixing both components in any desired ratio.

In the presence of water, Brønsted sites dissociate into a fluxional ion pair consisting of a hydrated H_3O^+ ion and the (hydrated) anionic surface charge left behind. In a study based on ab initio molecular dynamics, the protonation of H_2O by Brønsted sites of different zeolites was found to be complete when three water molecules are available, while detachment of the cluster from the wall surface, starting with four water molecules, is almost complete with eight water molecules [117]. In the clusters, the positive charge is extremely mobile. Their size is limited because the gain in enthalpy by further growth (relative to condensation in unconstrained space) does no longer pay off the loss in entropy due to the fixation in a constrained situation. In zeolite pores, the proximity of Brønsted sites can cause very high local H_3O^+ concentrations and, correspondingly, high ionic strengths (e.g., 5 mol/L for H-ZSM-5 with Si/Al = 15) [118]. The resulting high polarity of the medium causes dramatic effects on rates of reactions catalyzed by the hydronium ions, which may be superimposed with the impact of repulsive interactions when reactants contain hydrophobic groups. The situation is illustrated with an example in Section 4.7.

---

21 Only hydride transfer between species of different chain lengths propagates the cracking reaction. Transfers between species of equal chain lengths are just isomerizations.

4.5.2.2 Allylic oxidation of propene

Allylic oxidation is a prominent class of selective oxidation reactions, mostly performed with propene, and sometimes with isobutene. It gives access to important intermediates (acrolein, eq. (4.2)) and monomers (acrylonitrile, eq. (4.8), acrylic acid via acrolein). As mentioned in Section 4.2.3, commercial propene (amm)oxidation processes work with Bi-promoted molybdate catalysts, which combine a great number of additional components in a microstructure adapted to the underlying Mars–van Krevelen mechanism (cf. Section 4.3.3).

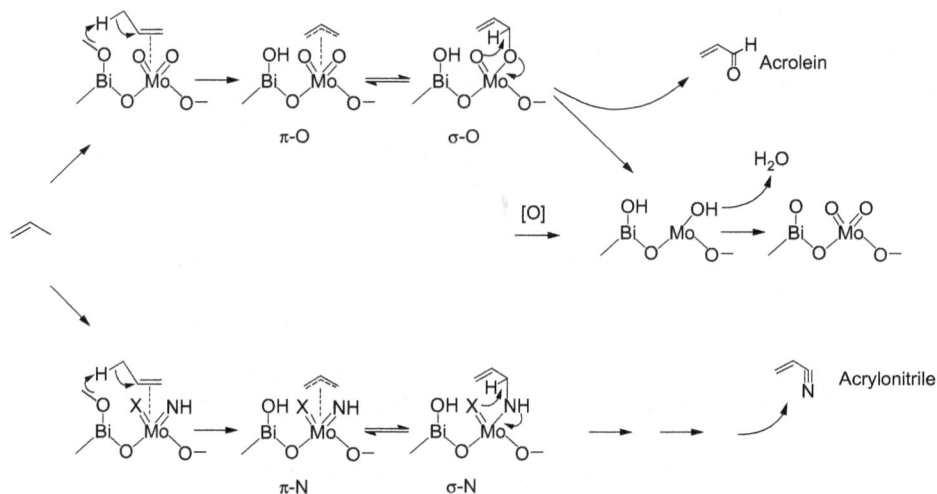

Figure 4.33: Mechanism of propene (amm)oxidation according to Grasselli et al. Adapted from ref. [76] with permission from Springer Nature.

Figure 4.33 shows mechanisms for allylic (amm) oxidation put forward in the early 1980s by R. K. Grasselli [27], who also significantly contributed to the development of the technical processes utilized up to now. They integrate early insight obtained in tracer studies: the abstraction of an allylic hydrogen by nucleophilic oxygen is rate-limiting; the reaction proceeds via an allylic intermediate, which levels the difference between the chain ends; and the aldehyde oxygen in the product originates from the catalyst surface.

In the mechanism, propene is (amm)oxidized at sites comprising a binary Mo^{VI}-O-Bi^{III} structure. The rate-determining abstraction of an allylic H atom is achieved by an O atom bridging a Bi and an adjacent Mo atom (which misses in Figure 4.33). The resulting allyl species is coordinated to the Mo^{VI} ion of the binary site. To proceed to acrolein (upper part of the figure), the allyl species is inserted into a Mo=O bond. This creates a surface-alkoxide species, from which the end product is accessed by another H abstraction. The electrons released in these oxidations stay in the active site, which is reduced after the desorption of acrolein. To close the cycle, water is split off from the OH groups formed; oxygen is replenished and the electrons are drained by the transport steps, characteristic for the Mars–van Krevelen mechanism (Section 2.5.1).

The proposed ammoxidation mechanism (lower part of Figure 4.33) is based on the assumption that the molybdenyl oxygen is partly exchanged by nitrogen under the reaction conditions. By insertion of the allyl species into the Mo=NH bond, an allylamine adsorbate is formed, which is converted into acrylonitrile via three successive H abstraction steps.

It took almost 30 years before a modified mechanism was put forward that reconciled the contradictions between the more recent experimental evidence (cf. Section 4.2.3) and the features of the schemes shown in Figure 4.33. The mechanism is based on DFT model calculations, which were made for model surfaces that expose the binary Mo^{VI}-O-Bi^{III} binary structure [29] (Figure 4.34). Regarding the route to acrolein, the major changes, relative to the earlier mechanism, are that the first H atom is not abstracted by an O atom bound to Bi, but by one of the molybdenyl oxygens, and that reoxidation is very fast and proceeds after each reductive step [29]. The influence of the adjacent Bi atom is indirect: its lone pair favors H abstraction from propene by destabilizing the HOMO, and stabilizing the LUMO of the molybdate unit, thus facilitating the accommodation of an electron in a Mo-O π^* orbital during the reaction. Reoxidation was not studied; the step is indicated in Figure 4.34 by just restoring the original coordination of the Mo^{VI} site. The radical character of the allyl intermediate was confirmed by calculations.

In Figure 4.34, the allyl radical is adsorbed at an adjacent Bi ion until it is inserted into a nearby Mo=O bond. The following H abstraction from the allyl alkoxide intermediate is rather demanding,[22] but not rate-limiting, because it starts well below zero on the energy scale. In the resulting adsorbed acrolein, the unpaired electrons on the adsorbate and the adsorption site (Mo) pair in an extended resonance structure: the lowest spin state of the structure is a doublet. After reoxidation of the Mo site holding the H atom just abstracted (Figure 4.34), acrolein is split off from its Mo site, which reduces the latter to the +4 state. Its reoxidation releases most of the reaction enthalpy before acrolein is desorbed into the gas phase.

The mechanism proposed by A. T. Bell et al. for ammoxidation, more recently on the basis of experimental evidence and DFT studies [120, 121], differs more strongly from its predecessor (Figure 4.33) than that of acrolein formation. Mo=NH groups, for which experimental evidence had not been found in the meantime, were shown to be unstable in the presence of water. On the reaction coordinate depicted in Figure 4.35, NH_3 chemisorption happens only after the formation of the allyl radical, probably to emphasize the paral-

---

22 There are slight differences in energies and details of conclusions because of different exchange correlation functionals used in the studies published in ref. [29] and cited in ref. [119].

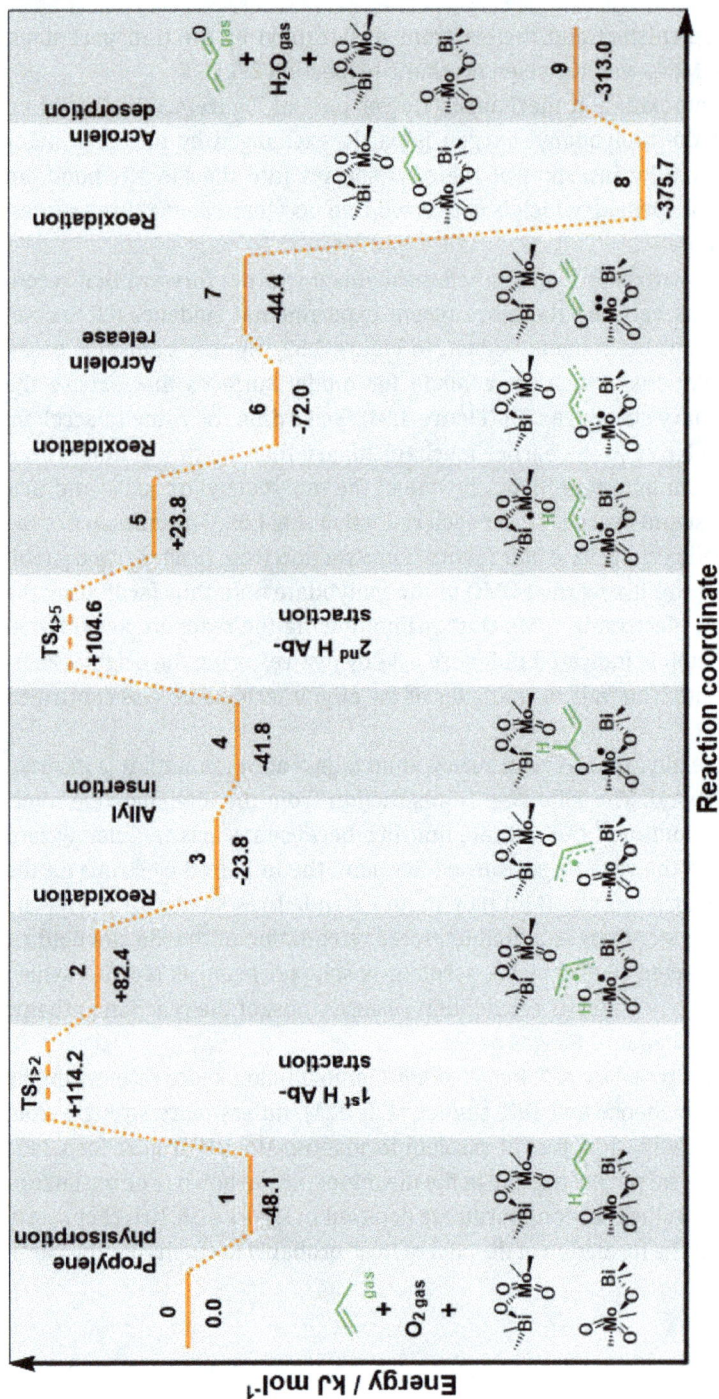

Figure 4.34: Propene oxidation according to Bell et al. (adapted with permission from ref. [119], copyright 2017 American Chemical Society). Energies correspond to those published in ref. [120], which differ slightly from those of ref. [29]. Gas-phase O_2 ends up in water and in molybdenyl-O, the latter indirectly via O^{2-} diffusion from the reoxidation site.

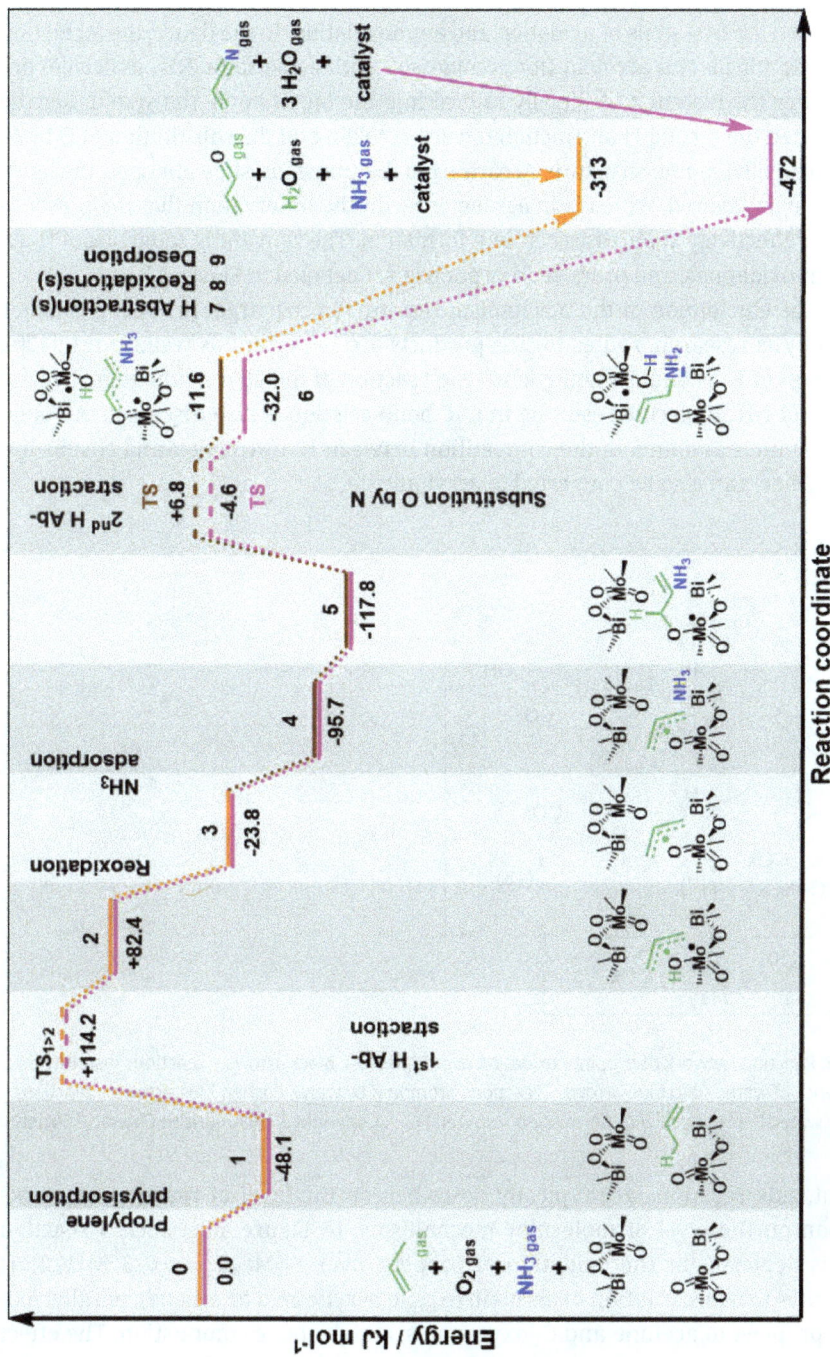

Figure 4.35: Propene ammoxidation according to Bell et al. Adapted with permission from ref. [120], copyright 2017 American Chemical Society.

lelism between the first steps of oxidation and ammoxidation. In the figure, the formation of both acrylo nitrile and acrolein (for comparison with a spectator NH_3 molecule) are depicted. After the insertion of the ally radical into the Mo=O bond, the system has the choice between the second H abstraction toward acrolein and the substitution of O by N, resulting in an allyl amine structure. According to the transition state energies, the latter route is strongly favored, which is in agreement with the observation that the presence of ammonia effectively suppresses acrolein formation. The remaining sequence of H-abstractions, reoxidations, and product desorption is not detailed in Figure 4.35.

While the elucidation of the mechanism resulting in the target product is a major goal of catalysis research, routes to side products are important as well. In ref. [120], the formation of HCN and acetonitrile via the reaction of the allyl amine intermediate with a second NH_3 molecule resulting in C–C bond scission is also discussed. Acrolein obtained in minor amounts in the competition between H abstraction and N substitution (see earlier) can also be converted to acrylonitrile.

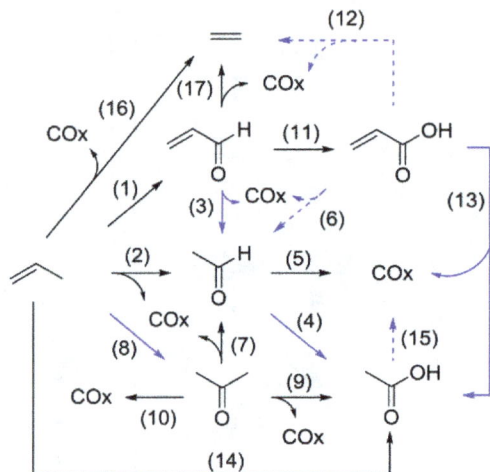

Figure 4.36: Reaction network of propene oxidation over $Bi_2Mo_3O_{12}$; black arrows – reactions insensitive to the presence of water, solid blue arrows – reactions promoted by water, dashed blue arrows – reactions inhibited by water. Reproduced with permission from ref. [122]. Copyright 2016 American Chemical Society.

At present, side reactions are typically described on the level of reaction networks rather than on the level of molecular mechanisms. In Figure 4.36, such a reaction network is depicted for the oxidation of propene over $Bi_2Mo_3O_{12}$ at 623 K [122]. It reveals routes to overoxidation of acrolein (e.g., to acrylic acid or ethene), parallel oxidation of propene to acetone and C_2 oxygenates, and propene combustion. The effect of water on these reactions is quite different. The reaction network operating during the interaction of propene with $Bi_2Mo_3O_{12}$ is, however, still under discussion, as summarized in ref. [119].

4.5.2.3 Mechanisms involving metal-carbon bonds

Intermediates and catalysts with single or double bonds between metal and carbon atoms, are very familiar in mechanisms of molecular catalysis, e.g., for the hydrogenation, hydroformylation, metathesis, or polymerization of alkenes. All these reactions can also be catalyzed by heterogeneous catalysts. Therefore, the presence of similar intermediates on the solid surfaces, and even the identities of reaction mechanisms are quite likely.

Due to the complexity of real surfaces – on which single-site catalysts are much more difficult to achieve than in the realm of molecular catalysis – and to the complications raised for many characterization methods by the anisotropies of solid materials and their surfaces, insight into the molecular processes is much more limited for the heterogeneous than for the molecular versions of these reactions. In polymerization, signals of polymer material suppress the signature of reactive species after very short reaction times. In addition, DFT studies related to polymerization catalysis have only recently achieved a satisfactory level of accuracy because the influence of van der Waals interactions turned out to be significant even where the growing chain was represented by an only short hydrocarbon group [123]. Despite these problems, the present discussion on reaction mechanisms is briefly summarized in the following for two alkene reactions: metathesis and polymerization. The former involves metal carbene species, and the latter (with the exception of ROMP, see later), metal-carbon σ bonds.

During alkene metathesis, the alkylidene groups forming the participating alkene molecules are redistributed until equilibrium is achieved:

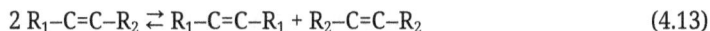

$$2\, R_1{-}C{=}C{-}R_2 \rightleftarrows R_1{-}C{=}C{-}R_1 + R_2{-}C{=}C{-}R_2 \qquad (4.13)$$

(R_1, R_2 = H or alkyl groups). For symmetric molecules ($R_1 = R_2$), the reaction is nonproductive. Metathesis with a cyclic alkene results in ring enlargement, which is the basis of Ring opening metathesis polymerization (ROMP). There are a large number of (mostly molecular) systems, which catalyze reaction (4.13) also with functionalized alkenes. Metathesis is, therefore, an important tool of organic synthesis. Several industrial-scale processes include metathesis steps, mostly on heterogeneous catalysts (for a brief review see ref. [124]). Re, Mo, W, and Ru are the active elements in molecular metathesis catalysts. All but Ru catalyze alkene metathesis also when supported on SiO_2, Al_2O_3, or alumosilicates. Meanwhile, organometallic metathesis catalysts have also been grafted on solid supports (cf. Section 3.1.3.4). As shown further, their reactivity by far exceeds that of sites originating from surface oxide species of the same element. This has shaken the widespread belief in a close analogy between metathesis sites and mechanisms over molecular and traditional heterogeneous metathesis catalysts (see later) [124].

In molecular catalysis, the active sites of alkene metathesis are metal carbenes. By reacting with a feed molecule, they form a metallacyclobutane intermediate, which is

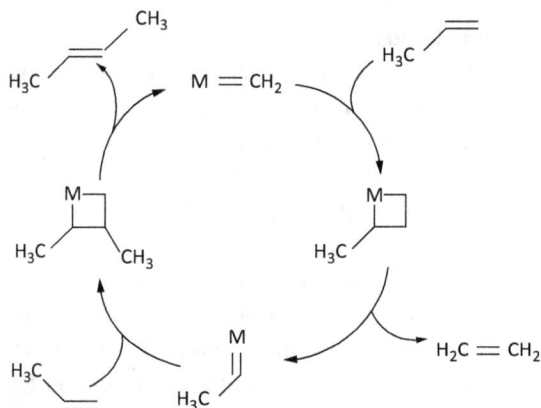

Figure 4.37: Mechanism of metathesis on molecular catalysts. Adapted from ref. [124].

cleaved into a product alkene and a second carbene species (Figure 4.37). By analogous reaction with another feed molecule, the latter produces the second alkene molecule and restores the initial carbene. All steps are reversible. Both carbene sites and metallacyclobutane intermediates are well-documented. Molecular catalysts may contain the Me=C bond. The concentration of active complexes can be determined by titrating the carbenes with a suitable (e.g., deuterated) alkene.

For catalysts derived from supported oxide species, the validity of the Chauvin mechanism (Figure 4.37) has been assumed from early on, but direct evidence has been scarce. The applicability of the alkene titration approach for counting the active species to alumina-supported ReO_x sites [125] was an early indirect support for the analogy between mechanisms operating in solution and on surfaces. IR and UV-Vis spectroscopic evidence for the presence of carbenes and of a metallacyclobutane structure on a MoO_3/SiO_2 catalyst was reported. Although spectra were not recorded *in operando*, they were obtained after an activation procedure, resulting in a dramatic increase of metathesis activity. The analogy of metathesis mechanisms between molecular and heterogeneous catalysts was supported to some extent by DFT studies. In alkene titration studies, metathesis sites were typically found to form only from a small minority of the metal cations exposed (between <1% and a few %, for more detail – see ref. [124]).

Recent work of I.E. Wachs et al. has significantly expanded the knowledge on the active sites, challenging at the same time, the validity of the Chauvin mechanism (Figure 4.37) for supported oxide catalysts. Combining spectroscopic techniques that target the structures of surface oxide species (Raman, XAFS) and of adsorbates (IR) with an improved alkene titration approach and kinetic studies in stationary and instationary regimes, the authors found the surface carbene species to differ strongly in abundance and thermal stability. Over a ReO_x/Al_2O_3 catalyst, for instance, the ratio between

Re=CH$_2$ and Re=CH–CH$_3$ species during self-metathesis of propene ((4.13) with R$_1$ = CH$_3$, R$_2$ = H) was 7 at 303 K, but increased to >20 above 373 K [126].

The low stability of carbenes on supported Re catalysts, very different from carbene complexes, was a pivotal observation in these studies. Most of them decompose under mild conditions, which should have caused major errors in earlier alkene titration work, where catalysts were subjected to intermediate flushing or even evacuation, prior to the titration. With an improved method, percentages of metal (Re) atoms forming carbenes under mild conditions exceeded those reported earlier by 1–2 orders of magnitude. At the same time, the low stability of carbene sites caused their concentration to decrease dramatically with growing temperatures, while reaction rates increase as with any other reaction.

With their novel titration tool, Wachs et al. studied the dependence of metathesis rates on active site (carbene) concentrations, employing a series of catalysts with different Re contents. Their conclusions were quite surprising: at 343 K, turnover frequencies correlated with the square of the carbene concentration, i.e., the rds involves two carbene sites, clearly at variance with the mechanism of Figure 4.37. Combined with a reaction order of 1 for propene under these conditions, this observation suggests that the rds is a dissociative adsorption of propene onto two adjacent carbene sites. The structure of the resulting intermediates remained unresolved. Above 343 K, the propene reaction order increased to 2. On this basis, a reaction between two allyl species adsorbed on adjacent carbene sites was proposed as rds – also different from Figure 4.37 [126].

The access to turnover frequencies of metathesis over supported catalysts is a step forward toward a fair comparison of their performance with that of molecular catalysts. The problem of their different application ranges remains, however: the former are typically employed to convert short alkenes in the gas phase, while the latter are used in the liquid phase. Comparison can be made instead with metal-organic complexes grafted to solid surfaces by SOMC (cf. Section 3.1.3.4). According to ref. [127], Re is by two orders of magnitude more active for the metathesis of propene, when grafted on Al$_2$O$_3$ as methyltrioxorhenium than in sites derived from isolated alumina-supported surface ReO$_4$ species. This estimate was considered a lower limit. As mentioned earlier, such drastic difference in reactivities lends credibility to the proposed divergence of metathesis mechanisms between the catalyst types compared. It would not be the only example of a reaction proceeding via different mechanisms on different sites presenting the same catalytic element (see Section 4.5.6).

As mentioned earlier, a satisfactory model for heterogeneous metathesis catalysts should also explain, how the organometallic sites (carbenes) operating the mechanism are formed from the oxide precursors. In carbene complexes, metal atoms are below their highest oxidation state. It was, therefore, proposed that this should also hold for their oxidation state in the oxide precursors, which was supported by a report on the inability of Mo^{VI}/Al$_2$O$_3$ to develop metathesis activity (cf. [124]). Meanwhile, it is known that metathesis sites may originate from the supported metal ions in their highest oxidation state, e.g., Al_2O_3-supported W^{VI} [128], Mo^{VI} [129], and Re^{VII} [127]. They can, however, be obtained also from lower oxidation states, e.g., from Mo^{IV}/Al_2O_3 [129], which has been largely disregarded in recent research, because the reduction required to access this state is often difficult to control.

Reactivity data suggest that the replacement of oxygen by a carbene ligand proceeds swiftly even on catalysts that are extremely difficult to reduce in H_2 (e.g., WO_3/Al_2O_3 [128]). The problem appears, however, not yet settled. Information on the models discussed may be found in refs. [130, 131].

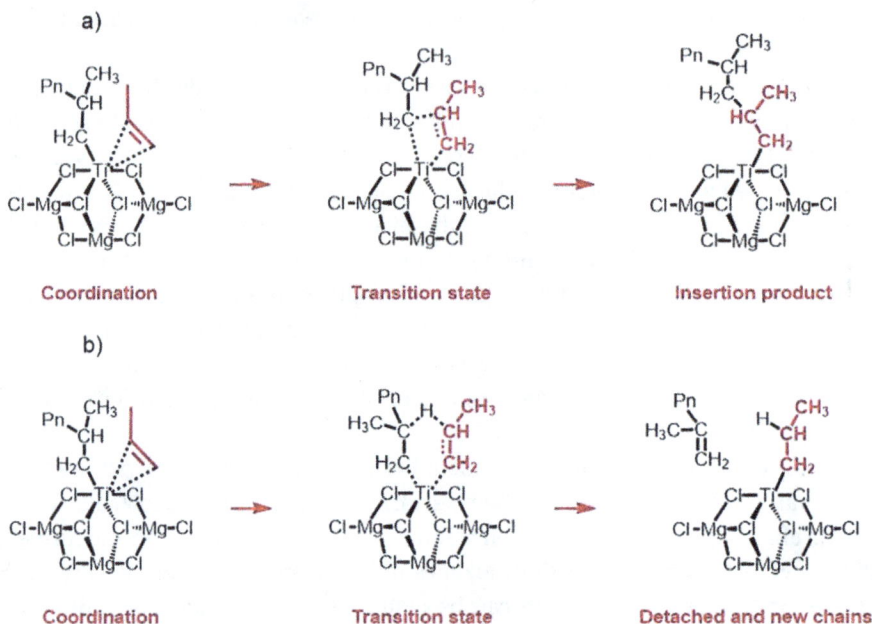

Figure 4.38: Mechanisms relevant in Ziegler–Natta polymerization: (a) Cossee–Arlman mechanism of chain growth and (b) chain transfer to monomer (adapted from ref. [123] with permission from Elsevier).

Catalytic alkene polymerization is by far the most important reaction involving metal-carbon bonds. Although a mechanism via carbene intermediates was discussed in the past,[23] the Cossee–Arlman mechanism (Figure 4.38a) is generally accepted for alkene polymerization with Ziegler–Natta catalysts (see below). In this mechanism, the growing chain is linked to the metal site via a σ bond. The alkene is inserted via a metallacyclobutane structure as well, but this is a transition state rather than an intermediate. Notably, chain and adsorbed monomer exchange their coordination sites after each insertion step.

---

23 Green–Rooney mechanism; the following discussion disregards ROMP, which proceeds via the Chauvin mechanism (Figure 4.37).

With the exception of polyethylene, the repetitive units in polyolefins $(CH(R)-CH_2)_n$ comprise an asymmetric carbon atom. The degree of regularity in the chains[24] strongly influences the material properties of the polymer. Ziegler–Natta catalysts achieve high degrees of tacticity, but the reasons for this are beyond the scope of this book and cannot be derived from the simplified representation of the surface in Figure 4.38a. For more details, the reader may consult ref. [123].

Polymerization is a chain reaction in which the Cossee–Arlman mechanism describes only the propagation step. Chain initiation, i.e., the formation of the active site from the precursor, chain termination, and chain transfer are other important ingredients of the reaction system.

Ziegler–Natta catalysts comprise a compound of an electropositive active metal (groups 4–10, e.g., Ti, Zr, V, or Co), which is combined with a co-catalyst: an organometallic compound of early main-group elements (groups 1, 2, 13), e.g., an Al alkyl. In a typical version, $TiCl_4$ is impregnated onto the surface of nano-sized $MgCl_2$ crystallites previously activated, for instance, by a mechanochemical treatment. In addition, the catalysts contain electron donors, e.g., esters (phthalates, benzoates), alkoxysilanes, or 1,3-diethers [123], which can be considered as promoters.

During activation by the co-catalyst, the catalytic metal becomes reduced (e.g., Ti^{IV} to Ti^{III}, (Ti^{II})) and alkylated, which creates a metal-carbon bond, suited for insertion of monomer molecules. The order of reduction and alkylation, reasons for the very different efficiency of co-catalysts, and the question if a Cl ligand needs to dissociate from the fully coordinated metal in the precursor (e.g., $TiCl_4/MgCl_2$) before reduction and alkylation can take place are issues that have been dealt with in computational studies. While the results are conclusive with respect to co-catalyst efficiency, they permit rejecting some routes rather than specifying a dominating mechanism.

Chain transfer terminates the growing chain, replacing it by another component of the reaction system, which may or may not start a new chain. The latter case is actually a termination. As a result, the polymerization rate decreases. If the new chain propagates without delay, the rate remains unaffected, but the polymerization degree decreases because there are now more chains growing from the same amount of monomer.

Transfer reactions reveal themselves by specific modifications at the chain ends (see Figures 4.38b, 4.39). While the major transfer routes have been known for long, the assessment of their importance was complicated by the heterogeneous site structure of solid catalysts. Meanwhile, their detailed study has strongly benefited from the availability of single-site molecular catalysts (metallocene catalysts). Under realistic reaction conditions, transfer to the monomer (Figure 4.38b) was found to be most important. It happens by the transition of a H atom from the β-C atom in the chain to the monomer. As

---

24 Stereoregularity or tacticity – C atoms all of identical, of alternating, or of arbitrary configuration, regioregularity – 1–2 insertion dominating over 2–1 insertion.

a)

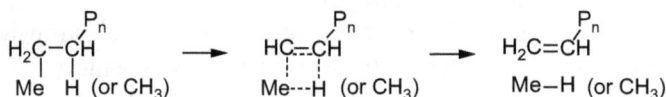

b)

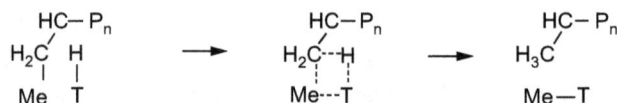

c)

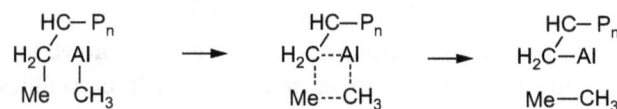

Figure 4.39: Transfer routes to other than monomer species: (a) β-hydrogen (methyl) elimination; (b) transfer to hydrogen donor; and (c) transfer to co-catalyst residues.

its six-center transition state requires more space than the four-center transition state of insertion, it can be delayed by placing bulkier substituents in the coordination sphere of the catalytic element, which is easier for molecular than for solid catalysts. When the vacancy is not immediately filled by a monomer, e.g., at low monomer concentrations, a number of other channels become relevant, e.g., transfer of hydrogen from the chain or from a hydrogen donor to the metal (Figure 4.39). Depending on the transfer agent, the chain end may be saturated or unsaturated, or even contain an Al atom. The H-donor H_2 (T = H) can be deliberately added to tune the polymerization degree.

Nearly half the world supply of high-density polyethylene is produced using Phillips catalysts (Cr/SiO_2). These are prepared by impregnating SiO_2 with CrO_3, followed by severe calcination, or by reacting a calcined SiO_2 with an organochromium compound, e.g., chromocene. When starting from Cr^{VI}, the catalyst is reduced either by a treatment in CO or by the monomer in an initial induction period. Phillips catalysts are rather special because they need no co-catalyst to become active. This has created the "missing hydrogen problem" because unlike the monomer ethene, the growing chain has an uneven number of H atoms.

The mechanism of catalyst activation and the oxidation state of Cr in the active sites (Cr^{II} or Cr^{III}) have been discussed over decades. Regarding the latter question, strong evidence in favor of Cr^{III} has been recently obtained by comparing the polymerization activity of well-defined (dinuclear) SiO_2–supported Cr^{II} and Cr^{III} sites prepared by methods of SOMC (cf. Section 3.4.1.4), and characterized by XAFS [132].

A recent attempt to solve the missing hydrogen problem is highlighted in Figure 4.40a: a H atom abstracted by the catalyst from the monomer adsorbs at an oxygen bridge between Si and Cr. Although the whole catalytic cycle was examined by DFT

a)

b)

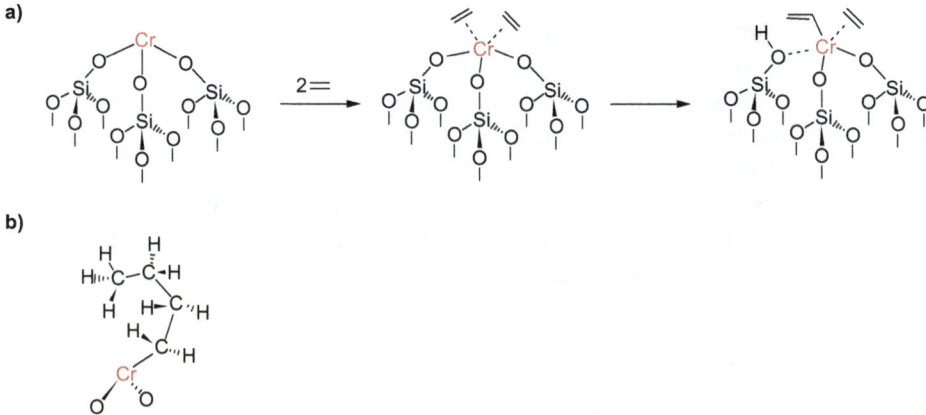

Figure 4.40: Active sites of the Phillips catalyst for ethene polymerization: (a) hypothetic activation route featuring donation of a H atom by the monomer and (b) active site model allowing the calculation of realistic polymerization rates and molecular weights. Adapted from ref. 133 (a) and 134 (b). Copyright 2015 American Chemical Society (both figures).

calculations [135], the model was called into question by a more recent study [134] in which the reaction rates were evaluated for 373 K (instead of 298 K in [135]) and an additional option for proton (back) transfer was included. The authors found that the site shown in Figure 4.40a produces only oligomers for several reasons, not the least being a fast back transfer of the proton, resulting in desorption of the chain. Having also examined a number of other approaches for solving the missing hydrogen problem, they reported that barriers of the mechanistic steps (initiation, Cossee–Arlman propagation, and transfer), which would allow the formation of polymer chains at a rate compatible with experimental data, were obtained only with the model shown in Figure 4.40b. The authors stated explicitly that the origin of the active site (Figure 4.40b) remained unresolved.

4.5.3 Site isolation and catalytic mechanism

Reaction channels available on catalytic surfaces may depend on the degree of separation between the active sites. On metal surfaces, reaction sites are adjacent to each other, and adsorbate mobility also allows interactions across longer distances. This situation allows very direct responses of product selectivity to changes in feed composition, which are mediated by the corresponding changes in surface coverages of intermediates. On the other hand, isolated redox cations in zeolites or on support surfaces offer just one or two vacancies for adsorbates, and maybe for cooperation with an adjacent support site. In most cases, such site responds to variations of feed composition and reaction conditions by changing the rate of "its" characteristic reaction. Side products are formed by these catalysts on sites of different structure (e.g., clustered oxide phases) rather than by switches in the operation mode of the major sites.

In the following, catalysis on isolated sites is exemplified by the NH_3-SCR (equation (3.136)) over redox zeolites and, to get a more complete picture, over supported V_2O_5(-WO_3)/TiO_2 catalysts. N_2 formation from NO and NH_3 is most likely to proceed via a comproportionation, which needs to be prepared by (partially) oxidizing either NH_3 or NO. Both were proposed in several variations in the past and have become part of the widely accepted reaction mechanisms.

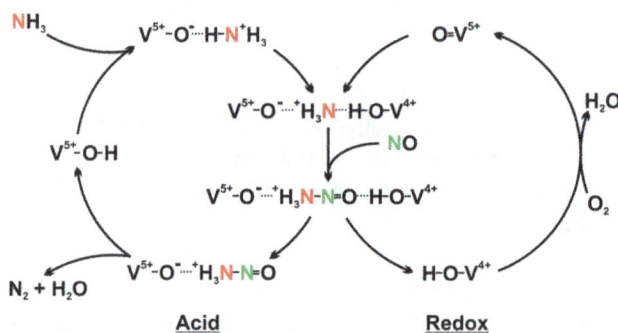

Figure 4.41: Mechanism of NH_3-SCR over V_2O_5/TiO_2 catalysts according to Topsøe et al. Adapted from ref. [136] with permission from Elsevier.

As mentioned in Section 4.2.3, the mechanism proposed by Topsøe et al. for NH_3-SCR over V_2O_5/TiO_2 catalysts proceeds on binary V-O-V sites, the V atoms of which operate as Brønsted and redox sites, respectively (Figure 4.41) [136]. NH_3 adsorbed on the Brønsted site is oxidatively activated by the adjacent V^V ion and reacts with gas-phase (or weakly adsorbed) NO[25] to N_2 and H_2O. The V^V-OH site is ready for the next NH_3 molecule, the V^{IV}-OH site is reoxidized by gas-phase oxygen.

The mechanism was supported by spectroscopic evidence, but in particular, by its successful application as a basis for a microkinetic model of NH_3-SCR over V_2O_5/TiO_2 catalysts. It neglects critical points in steps that are not considered rate-limiting, such as reoxidation or water release, which requires two adjacent (?) V-OH groups. Recently, Ferri et al. concluded from modulation excitation IR spectroscopy (cf. Section 4.5.1) that NH_3 is activated by a Lewis site rather than by a Brønsted site [137]. Moreover, a central piece of IR-spectroscopic evidence supporting the mechanism depicted in Figure 4.41 has been challenged [138]. While this shows that the problem is all but settled, the relevance of V-O-V pairs, suggested by the dependence of rates on V contents, is unquestioned.

---

25 The rate law over these catalysts has a form typical for the Eley–Rideal mechanism.

Mechanistic proposals for NH₃-SCR over redox zeolites almost exclusively assume oxidation of NO, prior to comproportionation. Based on IR studies that showed the signals of nitrite and NH_4^+ and the formation of N_2 responding to changes in the reaction conditions synchronously, the formation of N_2 from NO_2^- and NH_4^+ ions was proposed early. Alternative views, according to which some NO is oxidized to NO_2 to allow fast SCR (3.137) to proceed, met rejection because negative correlations between rates of NO_2 formation and of SCR were observed. As the nature of the active sites was under debate, mechanisms discussed were often written with molecules and ions, omitting their relation to the surface.

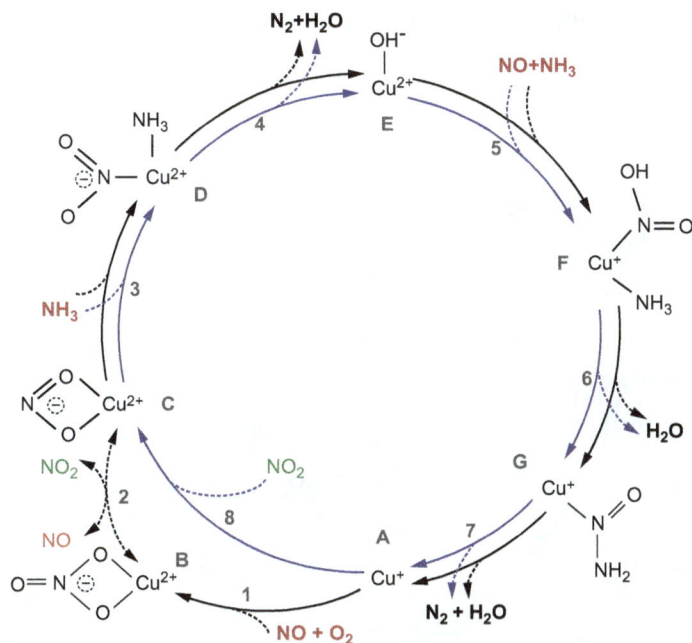

Figure 4.42: Mechanism of NH₃–SCR over single Cu ions in Cu–CHA. Reproduced with permission from ref. [139]. Copyright 2015 American Chemical Society.

Based on relevant literature data related to Fe and Cu zeolites, on the evidence of Cu–CHA being a single-site catalyst (cf. Section 4.2.3), and on own *in situ* XAFS and DFT work, Janssen et al. proposed a full redox cycle, which covers reduction and reoxidation of the active Cu sites (Figure 4.42). In this cycle, N_2 is produced during both the reduction of Cu^{2+} and the oxidation of Cu^+. Oxidative adsorption of NO on Cu^{2+}, which is coordinated with an OH group (E) to keep charge-neutrality at a single framework Al site (cf. Section 3.1.3.1), results in the formation of adsorbed HNO_2. Simultaneous adsorption of NH₃ at the same Cu^{2+} ion creates the starting point (F) for the first comproportionation via an NH₂-NO intermediate (G), which is decomposed into N_2 and H_2O (reaction 7). Different routes are available for the reoxidation of the Cu^+ ion left behind

(A). If there is NO_2 in the gas phase, Cu^+ can be immediately oxidized to the Cu(II) nitrite species C (reaction 8). Otherwise, nitrate is formed by reaction with NO and O_2, which is subsequently reduced to species C by NO. NH_3 adsorption onto species C again brings ammonia and nitrite species close together, which results in the release of N_2 and the restoration of the Cu^{2+}–OH site.

The mechanism is somewhat more complex than it seems at first sight. Only fast SCR, which is described by the inner (blue) circle, proceeds on one single site. The stoichiometry of the outer cycle does not comply with that of standard SCR, which does not cover the NO_2 molecule released in step 2. Only when this molecule reacts with more NH_3 and NO on an isolated site (maybe a different one) via the fast SCR cycle, the standard SCR cycle becomes closed. Thus, standard SCR indeed requires gas-phase NO_2 in this mechanism, which is, however, not formed by simple NO oxidation.

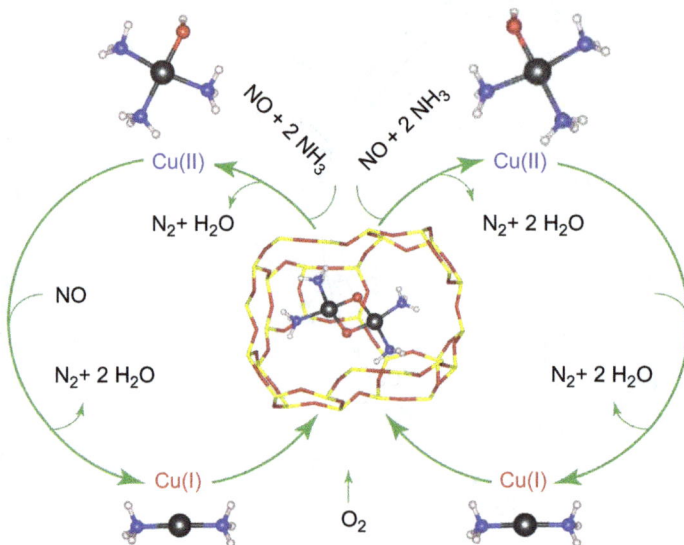

Figure 4.43: Transient Cu dimer in the mechanism of NH_3–SCR over Cu–CHA. From ref. [35], adapted with permission from AAAS. The original figure shows the mechanism for different types of Cu sites in the adjacent cages.

The mechanism in Figure 4.42 neglects that ammonia tends to coordinate at intra-zeolite Cu ions as long as it is present in the gas phase. It is at variance with all findings on the nature of active sites in Fe zeolites where fast SCR was found to proceed on a minority of isolated Fe sites [140], clearly different from the oligomers identified for standard SCR. While such comparison with another catalytic element cannot reject a mechanism, it has been meanwhile shown that reoxidation of isolated $Cu^+(NH_3)_2$ species is very slow, but proceeds swiftly if they can form temporary dimers, which open a different reaction channel (Figure 4.43). As pointed out in Section 4.2.3, the mobility required for this arises from the dielectric properties of the NH_3 ligands, by which the electrostatic forces between the framework and the Cu ions are attenuated, but not annihilated. With increasing average distance between the Cu ions, the oxidation half-cycle becomes rate-limiting; the rate becomes dependent on the square of the Cu concentration, and an increasing part of the Cu ions becomes suspended from the catalytic cycle.

Figure 4.43 depicts the SCR mechanism for two Cu sites in adjacent cages, though without much detail. The reoxidized dimer is split up by two NO molecules into nitrite species similar to D in Figure 4.42, which allow comproportionation. NH_3 molecules are coordinated to form the the $Cu^{2+}OH(NH_3)_3$ ions depicted, which proceed analogous to reactions 5 through 7 in Figure 4.42, before they need to pair up again for reoxdidation. Fast SCR differs from standard SCR by its more facile reoxidation reaction (cf. Figure 4.42), which also allows remote Cu ions contributing to the reaction. In standard SCR, rate limitation by reoxidation and quiescence of remote Cu sites are relevant only at low Cu contents. Therefore, the advantage of fast over standard SCR should be marginal for Cu-CHA with high Cu contents, which is indeed the case. Based on this advance, discussion is going on to clarify more retails of the reaction (e.g., in ref. [141]).

In reactions on unconfined surfaces, mobility of intermediates plays a major role. They are formed by adsorption or by initiating steps, and combine according to reactivities and abundances. This will be exemplified by two reactions in which the sites used in consecutive steps are of the same type: the oxidation of ammonia to NO and the Fischer-Tropsch synthesis. Bifunctional mechanisms, which combine steps proceeding on sites of different nature, are discussed in Section 4.5.4.

Mechanistic models for NH_3 oxidation (2.67) summarize results from kinetic studies with polycrystalline metal surfaces, from surface science work, but mostly from studies in the TAP reactor, some of which have already been highlighted in Section 3.2.1, and from related DFT calculations. Notably, on bare Pt surfaces, NH_3 is quite stable even at high reaction temperatures, while its hydrogen is easily stripped off when surface oxygen is available. Assuming the presence of molecular NH_3 and atomic O adsorbates, this can be written as [142]

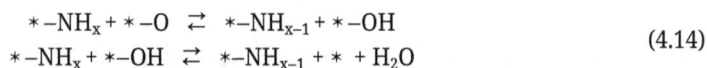

$$*-NH_x + *-O \rightleftarrows *-NH_{x-1} + *-OH$$
$$*-NH_x + *-OH \rightleftarrows *-NH_{x-1} + * + H_2O \tag{4.14}$$

where $*$ is a metal site. The combination of adsorbed N atoms provided by (4.14) with adsorbed oxygen results in NO (4.15). Its desorption is rate-determining and practically irreversible in the presence of O_2:

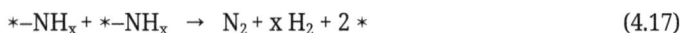

$$*-N + *-O \rightarrow *-NO + * \tag{4.15}$$

$$*-N + *-N \rightarrow N_2 + 2* \tag{4.16}$$

$$*-NH_x + *-NH_x \rightarrow N_2 + x\,H_2 + 2* \tag{4.17}$$

$$*-N + *-NO \;\rightarrow\; *-N_2O + * \tag{4.18}$$

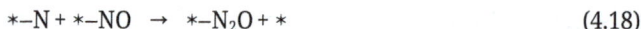

The undesired but thermodynamically preferred N_2 molecule is formed from adsorbed N atoms (4.16), but it may also result from reactions among adsorbed NH_x species ($x < 3$) exemplified by (4.17). The also undesired side product N_2O originates from the reaction between atomic nitrogen and the adsorbed NO (4.18).

DFT studies of the reactants interacting with a Pt(100) facet revealed a somewhat unusual reason for product selectivity [143]: while the barriers for reactions (4.15) and (4.16) are very similar and small (<10 kJ/mol) if the reactants are adsorbed at adjacent sites, the energy required for the reactants to achieve this proximity is different. Smaller repulsion between N and O than between two N species (6 kJ/mol to be overcome for N/O, 26 kJ/mol for N/N) favors the pairing of N with O atoms, which react swiftly to NO.

Beyond this effect, product selectivity is dramatically influenced by the gas phase composition and the nature of the metal [143]. At very low O_2 pressures, even H_2 can be desorbed from Pt surfaces, probably formed via reactions similar to (4.17). A high coverage of the adsorbed O atoms is crucial for good NO selectivity. If their coverage is low, most of the surface oxygen is consumed for the dehydrogenation of NH_3 (4.14) and water formation (from *–OH adsorbates). The shortage of *–O species favors the combination of *–N species to N_2 (4.16), and even the NO already formed can be trapped by *–N as N_2O (4.18).

At sufficiently high O_2 excess, nearly 100% NO selectivity can also be achieved with Rh, which is, however, more prone to the undesired side reactions at lower P_{O_2} than Pt. Reasons for this behavior, which are related to the energetics of adsorbates on Rh (here Rh(100)) can be found in ref. [142].

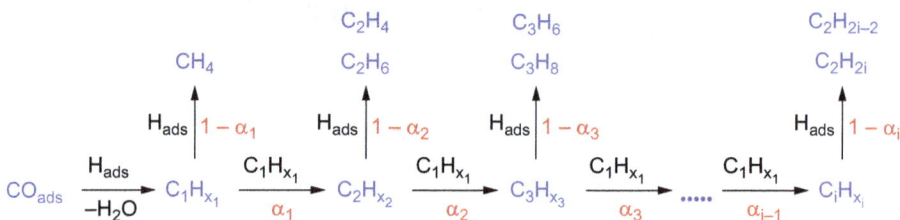

Figure 4.44: The carbide mechanism of Fischer–Tropsch synthesis.

The FT reaction is among the most complicated reactions in heterogeneous catalysis. It is a polymerization of C_1 adsorbates, which results in hydrocarbon molecules rather than carbonaceous deposits (cf. Sections 3.1.5 and 4.3.1). As a chain reaction, it consists of an initiation step, followed by chain growth steps, and stopped by a termination step. There is still debate on the nature of species involved in these steps. The following treatment is focused on the carbide mechanism (Figure 4.44) favored by many researchers, while the competing CO insertion mechanism is only briefly mentioned.

Both mechanisms are initiated by the dissociation of the adsorbed CO, either by mere interaction with the surface or aided by adsorbed hydrogen. The resulting C atoms are hydrogenated to CH_x adsorbates (x<4); adsorbed O atoms are ideally desorbed as water. A C_1H_{x1} species starting a chain has two options: extend the chain by reacting with a suitable "monomer" (in the carbide mechanism – C_1H_{x1}, in the CO insertion mechanism – CO) or terminate the chain by reacting with more adsorbed hydrogen; in this case, to methane. The related probabilities are $α_1$ and $1 – α_1$, respectively (Figure 4.44). The C_2H_{x2} adsorbate obtained by the first extension, and all further links of the emerging hydrocarbon chain, have the same choice between chain extension (with probability $α_i$) and termination by desorption as hydrocarbon molecules $(1 – α_i)$. Such combination of C_1 building blocks to longer units is no exclusive property of FT catalysts, but can be best exploited on them (cf. ref. [144]).

The distribution of chain lengths in FT products has been derived assuming that the chain growth probabilities $α_i$ do not depend on the chain length of the intermediates ($α$ = const). The resulting Anderson–Schulz–Flory distribution

$$W_n = n\left(1 - α^2\right) α^{n-1} \tag{4.19}$$

(W_n – mass fraction of products with n carbon atoms) is indeed observed in a wide range of n (typically between 4 and 12). Deviations at higher and lower carbon numbers have been assigned to the insertion of re-adsorbed light $α$–alkenes into growing chains at high conversions. In promoted catalysts, distributions can sometimes be fitted with two chain growth probabilities.

According to the carbide mechanism, there is no C atom in the products that does not arise from CO dissociation. To obtain chains, the rate of CO splitting must strongly exceed the rates of the other steps, in particular, of methanation which is actually favored by thermodynamics, and of the other termination reactions (Figure 4.44). In addition, the chain extension steps must be faster than methane formation. The barriers for these routes crucially depend on the strength of the metal-carbon bond.

In the competition between methanation and the formation of a first C–C bond, a strong Me-C bond disfavors methane formation where it is broken completely, while it is only modified when a C_1 fragment is inserted. According to DFT studies, barriers for converting a CH_x species to methane are between 100 and 140 kJ/mol on relevant surfaces of Co, Rh, and Ru, while those of typical chain growth steps are of the order of 70 kJ/mol [144]. Due to its higher activation energy, however, methane formation becomes dominant at higher temperatures. More noble metals have higher barriers for splitting CO and lower barriers for the hydrogenation of CH_x to methane. For this reason, Ni is a good methanation catalyst, but exhibits poor selectivities for heavier products. On the other hand, Fe surfaces bind all species, in particular, C, stronger than the other metals mentioned. Therefore, FT catalysis needs higher temperatures on Fe catalysts, but can proceed with remarkably low methane selectivities even under these conditions.

In the CO insertion mechanism, the activated chain end binds a CO molecule, forming an aldehyde intermediate. From it, the next member of the growing chain is obtained by hydrogenation and dehydroxylation. The mechanism responds to the observation of oxygenates among FT products over suitable catalysts (Rh, Ru) under mild conditions. The role of CO insertion was investigated in a recent study with (simplified) microkinetic models representing the competing mechanisms, which were fed with rate constants of elementary steps from DFT calculations and BEP correlations [144]. From the comparison between modeled and experimental tendencies of activity and selectivity at varied energies of the Me-C bond, it was concluded that chain growth via the CO insertion route is not likely. Instead, CO insertion is an additional option for the termination of the carbide route. In Figure 4.45, the relation between carbide and CO insertion routes is depicted according to [144].

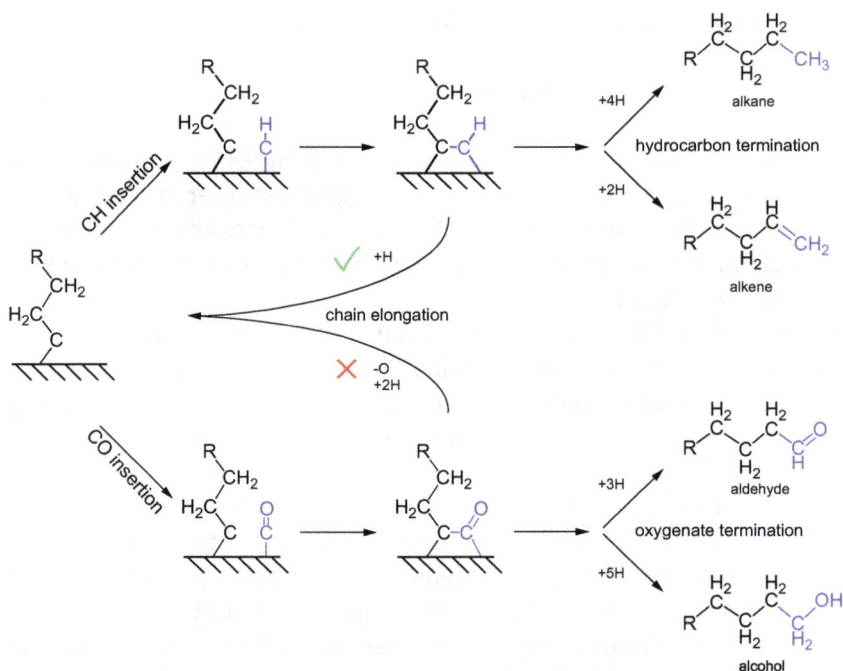

Figure 4.45: A recent view on the role of CO insertion in Fischer–Tropsch chemistry. Adapted from ref. [144] with permission from the Royal Society of Chemistry. The monomer of the growing chain is specified as CH, although insertion of CH_2 is assumed in much of the literature. About the reasons for preferring the CH building block instead, the reader may consult ref. [144].

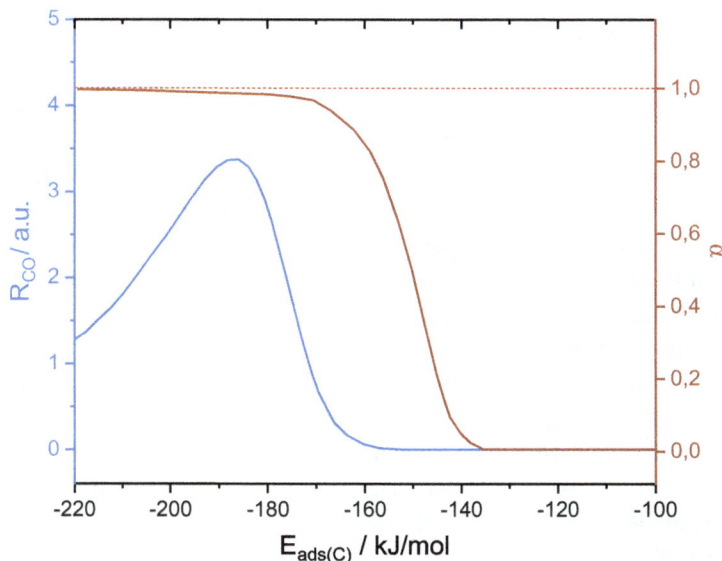

Figure 4.46: Typical dependence of rate and chain growth probability on the metal-carbon bond strength in Fischer–Tropsch catalysis. Adapted from ref. [144] with permission from the Royal Society of Chemistry.

Figure 4.46 highlights the influence of the Me-C bond strength ($E_{ads(C)}$) on key parameters of the FT reaction: its rate R_{CO} and the chain growth probability α. The data were obtained by microkinetic modeling with parameters from BEP correlations and DFT work [144] (see earlier). With increasing strength of the Me-C bond, the rds changes from the splitting of CO to the termination steps (desorption of products), but the discussion is complicated by the undesired methanation route.

When the metal binds CO weakly (right-hand side in Figure 4.46), the activation energy of CO splitting, $E_{a(CO)}$, is very high: the surface is crowded with CO, but there is no CO splitting. With growing Me-C binding energy, $E_{a(CO)}$ decreases, but while CO splitting starts at very low rates, the resulting C_1 species are completely hydrogenated to methane. Further enhanced Me-C interaction creates chain growth at first. As long as the rate of CO splitting, R_{CO}, does not exceed those of methane formation and chain extension, the chain growth probability α remains low. This changes when R_{CO} exceeds the other two rates, and α increases strongly, although R_{CO} and, hence, the total reaction rate may be still relatively small. When CO splitting becomes fast, the chain growth probability may be already pretty close to 1. R_{CO} peaks when it exceeds the rates of the chain termination reactions, which marks the transition between ranges limited by monomer kinetics and by chain growth. In the latter, the growing chains accumulate on the surface and poison it, because the Me-C bond is now very tight.

Notably, the relations between R_{CO} or α and $E_{ads(C)}$ depend on the experimental conditions. At higher temperatures, the rise of α requires stronger Me-C bonds because of the stronger temperature dependence of methanation rates. From their data, the authors of ref. [144] concluded that Co-based FT catalysts work in the monomer kinetics range, right from the R_{CO} peak.

As mentioned in Section 3.5.3, CO dissociation proceeds most easily on sites that allow several metal atoms interacting with the molecule, e.g., the B5 sites of Co or Rh. Assistance of surface H atoms allows CO splitting at rather low Me-C energies (not included in Figure 4.46 because of different site structure), which are, however, not sufficient to suppress methane formation. Therefore, chain growth probabilities α become satisfactory only when B5 sites compete with this route successfully on particles large enough to stabilize this kind of sites (cf. Section 4.2.3).

Despite considerable progress in the understanding of the Fischer–Tropsch reaction, many observations remain to be explained, in particular, with the more complicated Fe catalysts. They exhibit conspicuous differences to Co-based systems [48], being, for instance able to operate under very low CO partial pressures, while the H_2/CO ratio must be kept below critical limits with Co. Other differences are related to the influence of water (promoting chain growth with Co, inhibiting with Fe), and of alkali (an essential promoter with Fe, useless with Co).

4.5.4 Mechanisms depending on transport steps

Reactions on unconfined surfaces involve transport of adsorbed intermediates over lengths of a few atomic distances (cf. previous section). In this section, the nature of intermediates changing between sites allows their desorption, and transport may proceed by diffusion across the surface or via the fluid phase, overcoming much longer distances. Transported species are stable molecules in most cases, although there are now well-supported examples where a short-lived intermediate crosses over between sites (e.g., an allyl radical, cf. Figure 4.34), or where gas-phase reactions are initiated by species activated at a catalyst surface. Stable intermediates that change sites are actually parts of reaction networks. In the following, such networks are, nevertheless, sometimes referred to as "mechanisms," as is customary in much of the literature.

If the site visited by a desorbed species is of the same type as the one just left, the nature of the consecutive step is the same as well. An alkane molecule, for instance, may be dehydrogenated step-by-step at different sites of the same type. This may compete with a sequence of the same steps without desorption of the intermediates, and it may be difficult to differentiate these routes when the readsorption probability is high. In the gas-phase hydrogenation of benzene, the unsaturated intermediates, which are less stable than the end product, cyclohexane, are not detected under most reaction conditions. However, in a specific approach combining the catalyst with two liquid phases (an aqueous and an oily one), the performance of Ru in the hydrogenation of benzene by H_2 even allowed creating a successful industrial process for the production of cyclohexene [145, 146], which is an intermediate for caprolactam monomer.

If site types and the nature of the consecutive steps differ, the catalyst is bifunctional (or multifunctional). Mechanisms combining redox steps with transformations on Brønsted sites are classic examples of bifunctional catalysis, but the composite nature of mechanisms is also familiar in selective oxidation, as highlighted earlier for the oxidation of butane to maleic anhydride (cf. Section 4.3.3, Figure 4.22). Inappropriate combination of sites may provide undesired results, e.g., when a desorbed intermediate becomes cleaved or even gasified at a site offering oxygen of excessive electrophilicity.

The major role of the cooperation between redox and Brønsted sites arises from the fact that important conversions of alkanes, like isomerization, cyclization, or cracking, which are also catalyzed by noble metal surfaces, proceed much more swiftly via the carbenium-ion mechanisms of acid catalysis (cf. Section 4.5.2.1). The rather sluggish hydride abstraction from alkanes to carbenium ions can be avoided if a redox component (usually Pt) makes available alkenes. Under typical reaction conditions that include sizable H_2 pressures, their concentration may be extremely low. The same applies to the product (alkenes, desorbed from the Brønsted sites) which is immediately saturated over adjacent Pt surfaces. Such bifunctional cooperation is the basis for ideal hydrocracking (Section 4.5.2.1, Figure 4.30), as also for light naphtha isomerization, in which the octane number of straight-run gasoline consisting largely of unbranched alkanes is improved from ≈70 to >90.

Naphtha reforming, in which the octane number of the heavy gasoline fraction is enhanced by aromatization of the ingredients, rather than by isomerization, is a prime example for bifunctional catalysis. In Figure 1.6c, the description of naphtha reforming sometimes encountered in textbooks is highlighted in grey: substituted cyclopentane molecules are converted to aromatics by ring enlargement and dehydrogenation.

While not being wrong, this cuts off much of the complexity of this process, which also includes the isomerization of alkanes, their cyclization, probably first to C_5 rings, and the undesired cracking and coking reactions. All steps that change hydrocarbon skeletons proceed on acid sites. The respective carbenium ions arise from unsaturated intermediates formed from saturated precursors on the metal surface (for the nature of the catalysts – see Section 4.3.5). While equilibrium concentrations of these intermediates are small, the (highly endothermal) aromatization step, which releases three H_2 molecules, is favored and provides the driving force for the whole process. The cracking reactions proceed via mechanisms discussed in Section 4.5.2.1. Among the molecules shown in Figure 1.6, cyclopentadiene is known as a particularly potential coke precursor (cf. Section 4.3.1).

FCC catalysts (cf. Figure 4.25) are multifunctional in a different sense. Although cracking in the zeolite pores is monofunctional under severe process conditions (cf. Section 4.5.2.1), most feed hydrocarbons are too large to enter these pores. They need to be precracked, which is achieved on the less-acidic sites of the macroporous matrix (cf. Section 4.4.2). Distances between sites visited by the reactants are in μm rather than in nm range. Zeolite-coated $Cu/ZnO–Al_2O_3$ catalyst particles in which the shell converts methanol formed from syngas over the core material to dimethyl ether (cf. Section

2.3.4), feature another example of this kind of bifunctionality, in which the different nature of the reactions involved is more emphasized. As described in Section 4.4.2, the FCC catalyst offers a number of additional functions to delay its deactivation and speed up its regeneration. These functions are usually not included in the meaning of multi-functionality.

When distances between sites are small and covered by surface diffusion, it is hardly possible to identify the nature of the migrating species experimentally. The evidence for the radical nature of the first intermediate in allylic oxidation, for instance, relies completely on theory (cf. Section 4.5.2.2). On the other hand, there are cases, where reactive intermediates desorbing from the surface have been hypothesized to initiate consecutive gas-phase reactions, e.g., under conditions of ultra-short contact time catalysis (cf. Section 4.3.3). The idea was examined for the ODH of ethane over Al_2O_3-supported Pt by sampling the effluent behind the short catalyst bed via side ports along a distance that allows covering a sizable range of residence times [147]. From their observations, the authors concluded that the role of the catalyst was mainly to combust part of the ethane. This heats the catalyst and the gas stream to temperatures at which the remaining ethane is pyrolyzed to ethene in the gas phase. By adding H_2, which is oxidized prior to the hydrocarbon, ethane conversion was almost completely shifted to the gas phase.

While this example casts doubt on the relevance of reactive intermediates triggering gas-phase reactions, there are oxides known to release methyl radicals when interacting with methane at elevated temperatures. In the effluent of CH_4 pulses charged to beds of MgO or Sm_2O_3 above 800 K in the TAP reactor (cf. Section 3.2.1), methyl radicals were detected under conditions that exclude gas-phase collisions (Knudsen regime) [148]. Both oxides catalyze the oxidative coupling of methane (OCM, eq. (3.18)). In the molecular diffusion regime allowing gas-phase collisions, ethane was obtained from methane pulses at 1,073 K. These observations support a mechanism, according to which ethane is formed by recombination of methyl radicals near the surface on which they are generated. The obstinate limitations of C_2 selectivity were assigned to the competition of O_2 for the methyl radicals in the gas phase. Notably, according to ref. [148], the nature of the oxygen involved in the activation step may depend on the oxide: on MgO, reaction of methane with the exposed lattice oxygen results in methyl radicals, with adsorbed oxygen species in total oxidation instead. On Sm_2O_3, methyl radicals also originated from the interaction of methane with adsorbed oxygen.

Recent work with MgO revealed an additional, more selective route to ethane via the coupling of methyl radicals on the surface [149]. The sites required for this are, however, unstable under reaction conditions, and the steady-state performance is dominated by radical coupling in the gas phase. Notably, on freshly activated catalysts, formation and desorption of methyl radicals seems to be possible even at room temperature, but the sites are immediately poisoned by the coupled product water, and can be kept active only at the high temperatures characteristic of the OCM reaction [150].

Meanwhile, a transport of a reactive intermediate over macroscopic distances has also been observed in a catalyst type that provides competitive activities for NH_3-SCR.

These catalysts were originally designed to exploit the fact that standard SCR (3.136) is the sum of NO oxidation (4.20)

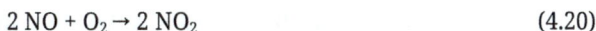

$$2\,NO + O_2 \rightarrow 2\,NO_2 \tag{4.20}$$

and of fast SCR (3.137) in stoichiometric terms [151].[26] This was achieved by just mixing oxides of high activity for NO oxidation (e.g., $MnO_x/CeZrO_x$) with catalysts well suited for fast SCR, e.g., Fe zeolites [152]. It was, however, soon found that these mixtures work equally well if the Fe zeolite is replaced by a H-zeolite, the fast-SCR activity of which is by orders of magnitude smaller.

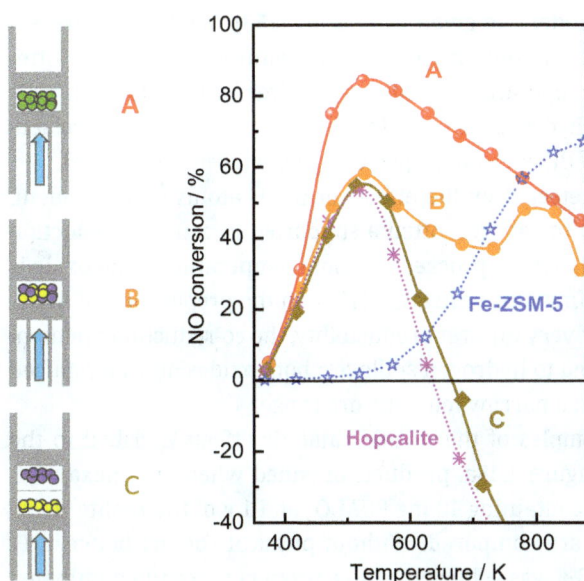

Figure 4.47: Influence of component separation on the performance of SCR catalysts obtained by mixing a redox oxide (hopcalite) with a zeolite (Fe-ZSM-5). Conditions: 0.1% NO, 0.1% NH_3, 2% O_2 in He, 300,000 h^{-1}; negative NO conversions result from NO formation by NH_3 oxidation. Adapted from ref. [152] with permission from Elsevier.

In Figure 4.47, an experiment designed to examine the stability of the species conveying the cooperation between the components (here: Fe-ZSM-5 and hopcalite, a Cu-Mn mixed oxide) is depicted. To vary the distance between them, they were combined in three different modes: by pressing them together into pellets (A), by pressing them into indi-

---

[26] In diesel exhaust gas treatment, this stoichiometric relation is utilized by oxidizing some NO to NO_2 over the Diesel oxidation catalyst, which substantially accelerates the downstream reduction of nitrogen oxides, in particular, over Fe zeolites or V_2O_5-WO_3/TiO_2 catalysts.

vidual pellets and mixing these (B), or by arranging them in two layers (C). In the well-mixed state A, the catalyst exhibits the typical synergism: its NO conversion exceeds those measured with the individual components at almost all temperatures. In mixture B, the synergy breaks down below 550 K, although some improvement is left between 600 and 700 K. The layered arrangement C behaves like the pure hopcalite. Obviously, the synergy is mediated by a short-lived intermediate: a stable mediator, e.g., NO_2, would definitely reach the adjacent Fe-ZSM-5 pellets in B or the downstream Fe-ZSM-5 layer in C. The identity of the intermediate has not yet been established.

Spillover is a transport phenomenon, which has been controversial in the catalysis community over decades. In a wide definition, the term spillover designates the transport of species formed or adsorbed on a surface onto another surface, where they cannot be adsorbed or formed under the given conditions [153]. While this covers a number of transport phenomena without any miraculous features, e.g., O^{2-} ion diffusion in multicomponent catalysts operating the Mars–van Krevelen mechanism (Sections 2.5.1 and 4.3.3), spillover of hydrogen is considered the reason for some spectacular effects, although the nature of the migrating species is not yet well understood.

Hydrogen spillover can be detected by the effects of the H atoms on a recipient, which is in physical contact with the source, often a supported Pt catalyst. Reduction of yellow WO_3 to blue W_4O_{11} by H_2, which proceeds at room temperature if the oxide is mixed with a Pt catalyst, was its first demonstration [153]. In the preparation of bimetallic catalysts from precursors of very different reducibility, the co-reduction phenomenon (cf. Section 3.4.2.5) is ascribed to hydrogen spillover: both oxides may be reduced at the same low temperature or in a narrow temperature range.

Figure 4.48 summarizes examples of the drastic catalytic effects ascribed to the spillover of hydrogen [154]. In Figure 4.48a, products obtained when an n-hexane/H_2 mixture was charged to reactors containing 10 mg Pt/Al_2O_3 and 1 g of the zeolite H-erionite in different arrangements are compared. Without physical contact between Pt and zeolite (A), a conversion of 48% was achieved, with a product spectrum dominated by C_3 and C_4 cracking products. Physical contact between zeolite and Pt (B) changed the catalytic behavior drastically: isomerization became the dominating route, though at a lower reaction rate. The identical result with Pt behind the zeolite layer (C) shows that the effect does not depend on the flow direction. Figure 4.48b illustrates the influence of hydrogen spillover on coke formation during the hydroconversion of n-butane over 400 mg H-MOR under more severe conditions. Conversion and selectivity effects were similar to those reported in Figure 4.48a. In the absence of Pt (A), the zeolite became black due to coking. 40 mg Pt/Al_2O_3 placed before or below the zeolite (B and C, respectively) caused part of the bed to remain white. Arranging the Pt catalyst in four layers alternating with zeolite layers, completely suppressed coke deposition on the latter (D).

Atomic hydrogen provided by spillover has been proposed to cause the dramatic changes in hydrocarbon conversion and selectivity (Figure 4.48a) and in the extent of coking (Figure 4.48b) by two effects: It hydrogenates intermediates relevant for coke formation and for the rupture of C–C bonds, and it opens a new route to the carbenium

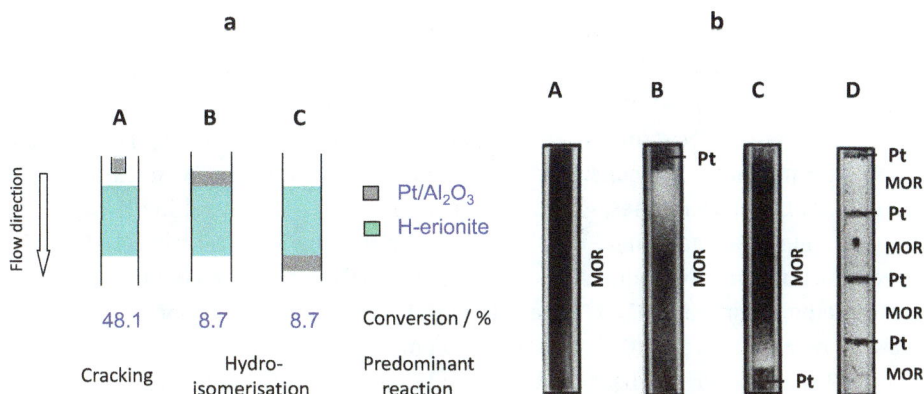

Figure 4.48: Switching of reaction mechanisms in the hydroconversion of alkanes by spilt-over hydrogen: (a) interaction of *n*-hexane with H-erionite with or without contact to Pt/Al$_2$O$_3$, from experiments in closed cycle (cf. Fig. 3.21b) at 500 K, samples taken after 7 h; (b) photographs showing the influence of Pt/Al$_2$O$_3$ ("Pt") on coke formation on H-MOR ("MOR") during the hydroconversion of *n*-butane (flow reactor, 633 K). Adapted from ref. [154] with permission from Elsevier.

Metal

$$H_2 \leftrightharpoons 2H^* \tag{1}$$

$$\updownarrow$$
$$\text{support}$$

Support

$$RH_2 + H^* \rightarrow RH^* + H_2 \tag{2}$$

$$RH^* + Z{-}OH \leftrightharpoons H^* + Z{-}O^-RH^+ \tag{3}$$

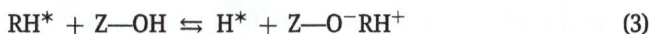
$$Z{-}O^-RH^+ \leftrightharpoons Z{-}O^-i{-}RH^+ \tag{4}$$

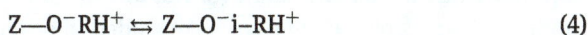
$$H^* + Z{-}O^-i{-}RH^+ \leftrightharpoons i{-}RH^* + Z{-}OH \tag{5}$$

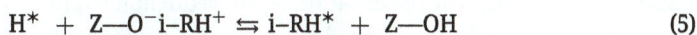
$$i{-}RH^* + H^* \rightarrow i{-}RH_2 \tag{6}$$

(2) through (6)

$$RH_2 + 2H^* \rightarrow i{-}RH_2 + H_2$$

Scheme 4.7: Alkane isomerization with spillover hydrogen.

ion. The latter is depicted in Scheme 4.7 (adapted from ref. [155]). Activated hydrogen is generated on the metal surface ((1) in Scheme 4.7) and fed to the support into a pool, the lateral extension of which depends on conditions and support properties.

On the support, H abstraction from the alkane RH_2 by the spillover hydrogen results in a radical RH^* (1). Its stabilization as a carbenium ion (3) returns the H atom invested in (2). After the isomerization step (4), the isomer ion is first desorbed as a radical ((5), reverting (3)), which recombines to the isoalkane with another H^* (6). In (5) and (6), the H atoms formed in (1) are consumed, though at a different place. The spent H atoms are replaced by adjacent ones, which initiate a migration of the resulting H vacancies toward the metal surface where they are filled with H atoms from reaction (1). Notably, the equilibrium between spillover H atoms and H_2 can be established only by the metal surface: to be desorbed as H_2, H^* needs to march back from the support to the metal. Likewise, reactions (2) and (6) can be reverted only by metal sites.

Spillover hydrogen has been controversial because the migrating species cannot be detected by the traditional characterization techniques for H atoms (e.g., ^{1}H-NMR), and more convincing evidence, e.g., by Inelastic neutron scattering (INS) has been rather recent (cf. ref. [153]). Promising hypotheses with respect to its nature are available only for semiconducting supports (see later). On the other hand, there are no alternative explanations for phenomena like co-reduction or the "remote reactivity control," highlighted in Figure 4.48, without transport of other species. It has been argued, for instance, that alkene intermediates of the classical bifunctional isomerization mechanism mentioned earlier can migrate up to 100 μm to find Brønsted sites and return to the metal for hydrogenation. Distances involved in the example of Figure 4.48 are, however, by two orders of magnitude longer (up to 15 mm [154]). Co-reduction was shown to operate even through carbon layers, which would block any transport of metal atoms from one to the other side (cf. ref. [153]).

Rates and ranges of hydrogen spillover critically depend on the support. In a study with model samples prepared by electron beam lithography, Karim et al. compared spillover from Pt to Fe oxide particles deposited on thin TiO_2 or Al_2O_3 films at different distances from each other, using spatially resolved XAS. At 343 K in H_2, the range was beyond experimental limitations on TiO_2, where surface reduction of Ti^{4+} to Ti^{3+} could be detected, while the maximum range was only around 15 nm on Al_2O_3, which remained unreduced. Rates of spillover were concluded to differ by ten orders of magnitude on these supports [156]. The observation of Ti^{4+} reduction supports a mechanism in which H atoms formed on the metal deposit an electron on an oxide cation at the metal-oxide perimeter and form OH groups with nearby surface oxygen. Spillover proceeds by concerted hopping of the electrons to adjacent cations and of protons to adjacent O^{2-} ions. Explanations for spillover on nonreducible surfaces, including zeolites, typically rely on the presence of structural defects. For more details on spillover, instructive examples, and potential applications, the reader may consult ref. [153].

4.5.5 Mechanisms involving pools of secondary species

While catalysts usually activate reactants by directly interacting with them, there are a few cases where the catalytic effect is substantially conveyed or modified by secondary species formed from feed molecules. The best-known examples of this catalytic operation mode, which have been introduced earlier, are briefly recalled and differentiated here.

In the methanol-to-hydrocarbon process (MTH, cf. Sections 4.2.3 and 4.3.2), the Brønsted sites of the zeolite are instrumental for both the formation of the intraporous hydrocarbon pool and the release of product molecules from it, but the nature of the pool, and, therefore, also of the products, is determined by the spatial confinements imposed by the pore system. The residence time of C atoms in the pool, from their incorporation by a methylation event to their release in a product molecule, is likely to exhibit a wide distribution. While the active sites are busy creating and splitting C–C bonds, the result of their effort is modified by the shape selectivity of the pore system imprinted on the hydrocarbon pool.

Spillover hydrogen is another type of secondary species that conveys a catalytic effect from a primary site (the metal dissociating the H_2 molecule) to other parts of the catalyst, e.g., the support. On the support surface, the H atoms provide a specific reactivity pattern, but they are not able to equilibrate with molecular hydrogen (cf. Scheme 4.7). The reactivity pattern includes the hydrogenation of coke precursors that would otherwise accumulate on the Brønsted sites in hydrocarbon conversions.

As shown in Section 4.2.3, carbon deposits *in situ* grown on surfaces of oxides or carbons during ODH reactions are catalysts doing completely without the sites of the primary surfaces below. This does not exclude influences of the latter on structure, reactivity, and stability of the supported carbon layer [157]. Such materials compete with carbons deliberately prepared and conditioned for use as catalysts. Carbonaceous deposits were also reported to catalyze the direct dehydrogenation of *n*-butane to butenes and butadiene under reaction conditions severe enough to allow developing ordered graphitic carbon structures [158].

4.5.6 The multiplicity of sites and mechanisms

This section deals with the observation that reactions may proceed via different mechanisms, depending on the properties of catalysts. Discussing this idea requires differentiating between mechanistic proposals (or speculations) and real surface processes, which is intricate: while it does not need a textbook chapter to find competing mechanisms for many reactions in literature, committing oneself to one of the options may not be safe at the present state of knowledge. The following discussion about mechanisms of the (standard) SCR of NO by NH_3 (eq. 3.136) illustrates how this dilemma may be handled. Subsequently, the multiplicity of mechanisms will be exemplified with methanol synthesis and with the oxidation of CO over supported gold.

As indicated in Section 4.5.3, the multitude of mechanistic proposals with respect to standard SCR can be differentiated according to the way for achieving N oxidation states suitable for comproportionation – by oxidation of either N^{3-} to N^{2-} or of N^{2+} to N^{3+}.

In addition, mechanisms differ in their requirements to the active site – they may get along with a single transition metal ion or require a pair of them.

As mentioned earlier, there is evidence for the superiority of paired sites for V-based catalysts [34] and for Fe zeolites [33]. The capability also of isolated V sites to catalyze standard SCR was demonstrated in a study with (VO)-ZSM-5 in which a correlation between activity and the abundance of VO cations characterized by EPR was obtained with samples that did not contain detectable amounts of clustered VO_x species [159]. This is a significant result because in studies with coexisting species of very different intrinsic activity, it is difficult to clarify doubts on the contribution of the less active sites, which may be obscured by inaccuracies in the specification of the majority contribution (cf. discussion in ref. [33]). Indisputable evidence for the SCR activity also of isolated Fe oxo sites was presented only recently [160]. It was observed with a Fe-ZSM-5 sample containing such sites almost exclusively, though probably rather at the walls of mesopores created during preparation than in the zeolite micropores.

The ideas put forward in ref. [35] for Cu-CHA support the view that standard SCR can proceed on both isolated and paired cations, though with higher rates on the latter site type. The reduction half cycle, starting with an oxidation of N^{+2} to N^{+3} (cf. Figure 4.42), is very fast with isolated Cu ions, but rates of standard SCR on them are limited by a very sluggish reoxidation. If temporary dimers (Figure 4.43) can be formed, reoxidation is no longer rate-determining; the SCR reaction becomes fast. There are no mechanistic proposals with analogous detail for Fe zeolites, but much of the evidence available is compatible with these ideas, save that there are no transient dimers in Fe zeolites; the reaction is operated by permanent dimers (oligomers), and a large part of the isolated Fe sites is just reduced to Fe^{2+}. The Topsøe mechanism for V_2O_5/TiO_2 (Figure 4.41), with an initial oxidation of N^{-3} to N^{-2}, is completely different. As it has been seriously challenged recently (cf. Section 4.5.3), it may be no longer a credible witness for the statement that reactions may proceed with similar rates using different mechanisms, although there is as yet no evidence that would support a completely different model, e.g., the cycle discussed for redox zeolites. The existence of yet another SCR mechanism allowing reaction rates similar to those of commercial catalysts is, however, strongly suggested by the performance obtained with physical mixtures of NO oxidation catalysts and zeolites in their H form (Section 4.5.4).[27]

The analogy between Cu and Fe zeolites ends in the ideas on mechanisms of fast SCR (3.137), although isolated cations were assigned as active sites for both catalyst types (the capability of $(VO)^{2+}$ ions in ZSM-5 to catalyze fast SCR was unfortunately not examined in ref. [159]). In Cu-CHA, the merit of NO_2 is to accelerate reoxidation and to also reoxidize sizes too remote to find partners for pairing [35], but as the rds is in

---

27 The data in ref. [160] suggest that high SCR rates may be achieved also with isolated Fe oxo sites. These have been so far considered responsible rather for the rates at low temperatures [31, 161], which implies that they are operating with a lower activation energy than the binary sites dominating at higher temperatures. Thus, future work may reveal yet another SCR mechanism operating on an improved catalyst type.

the reduction half cycle, the gain in activity is incremental. The activity of Fe zeolites, however, increases by orders of magnitude when NO_2 is added to the feed, and fast SCR has been shown to proceed on a small minority of isolated Fe sites only [140, 162]. It has not yet been elucidated, why they stand out by a very high activity.

The different modes of H_2 dissociation (homo- or heterolytic) are also familiar in surface catalysis. They cause differences in mechanisms of hydrogenations, though not so much of simple molecules, like alkenes. Dissociation of H_2 is homolytic on metals, but heterolytic on oxides, where the proton binds to surface oxygen and the hydride ion to the cation. One may presume that H atoms on metal surfaces prefer hydrogenating non-polar bonds of alkenes or alkynes, while H ions may be better suited for polar bonds, like in C=O or C≡O, but reality is more complex than that. As illustrated in Section 4.3.4, metal particles can be made to preferentially hydrogenate C=O bonds close to C=C bonds, and there are oxide catalysts capable of hydrogenating alkenes and of catalyzing the H_2/D_2 exchange (e.g., ZrO_2, cf. [163]). On the other hand, oxide catalysts like modified ZrO_2, Cu chromite ("Adkins catalyst"), or Zn chromite are indeed suited to hydrogenate rather stable carboxyl groups in acids or esters to OH groups. Promoted ZnO was the first catalyst used in industry for the synthesis of methanol by the hydrogenation of CO:

$$CO + 2\,H_2 \;\rightleftarrows\; CH_3OH \tag{4.21}$$

$$CO_2 + 3\,H_2 \;\rightleftarrows\; CH_3OH + H_2O \tag{4.22}$$

Methanol can be obtained by the hydrogenation of CO (4.21) or of CO_2 (4.22). In the latter case, the water released converts CO to CO_2 and H_2, i.e., the stoichiometry of eq. (4.21) is restored by the water-gas shift reaction (2.10), which is at equilibrium under typical reaction conditions. While the metals typically used for methanol synthesis (Cu or, when suitably promoted, Pd) are capable of producing methanol from both sources, ZnO is poisoned by CO_2, which blocks the oxygen vacancies catalyzing the reaction (see later).

Figure 4.49a shows typical intermediates discussed for the hydrogenation of CO and of CO_2 on Cu surfaces. In the rds, CO is hydrogenated to a formyl species, which is notably bound to the surface via its C atom. While being hydrogenated to the methoxy species via adsorbed formaldehyde, the adsorbate turns around and becomes adsorbed at the O end. The final step to adsorbed methanol is (generally) omitted in the figure. In the hydrogenation of CO_2, the first intermediate formate is further hydrogenated to methoxy via a methylenebis(oxy) species and adsorbed formaldehyde (not shown). All intermediates are bound via oxygen; the hydrogenation of the formate species is rate-limiting.

The complex interaction between these reaction mechanisms in real catalysis has been illustrated in a study of Studt et al. [164]. While all preparation routes except for some involving Zn fail to produce very small Cu particles (cf. Section 3.1.1.1), which is one of the reasons for the use of the $Cu/ZnO–Al_2O_3$ system in industry, the authors took advantage of a recent achievement that gave access to Cu particles of similar dispersion

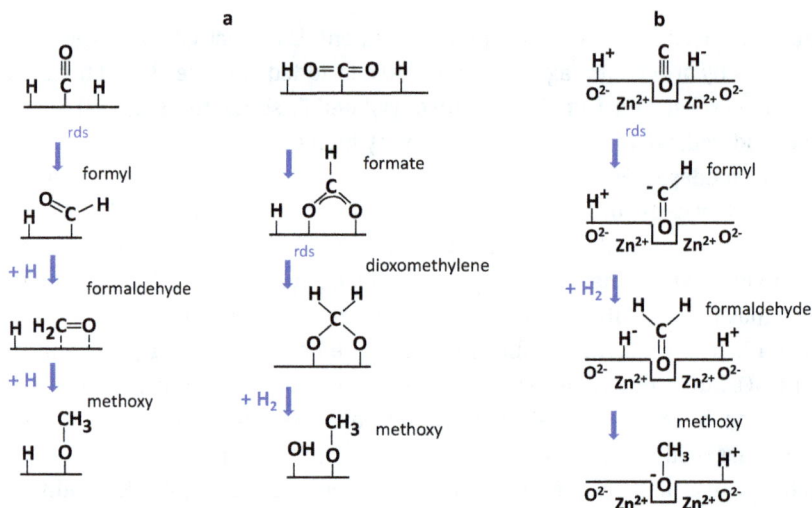

Figure 4.49: Different mechanisms of methanol synthesis: (a) hydrogenation of CO or of CO_2 on Cu surfaces and (b) hydrogenation of CO in an oxygen vacancy of ZnO.

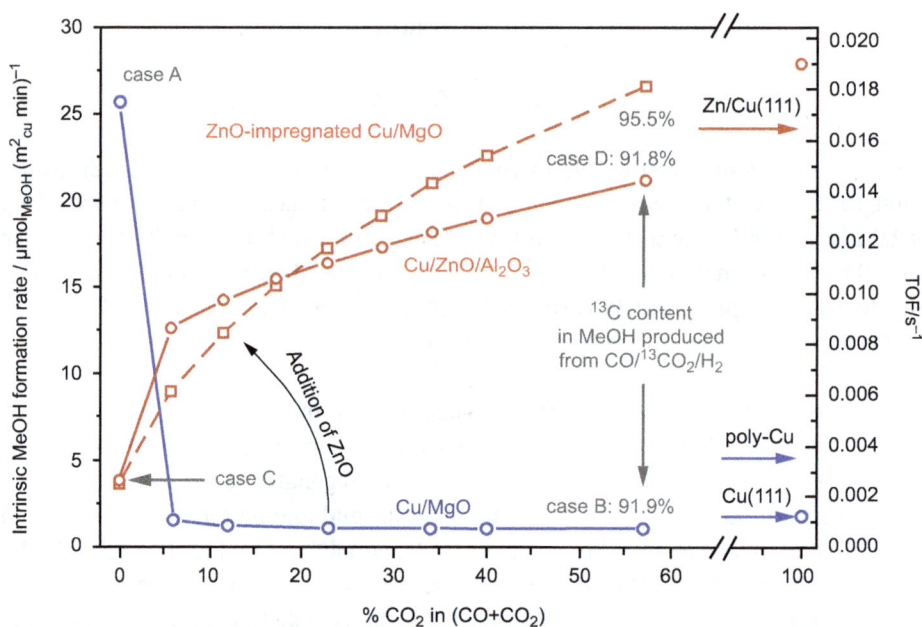

Figure 4.50: Response of unpromoted and of Zn-promoted Cu to the presence of CO_2 in methanol synthesis feed. Reproduced from ref. [164] with permission from Wiley-VCH Verlag GmbH & Co. KGaA, Weinheim, Germany.

on MgO. This allowed comparing monometallic and Zn-promoted Cu on the level of industrially relevant reaction rates.

The influence of CO_2 on the rates for methanol formation over both catalyst types is shown in Figure 4.50. Notably, the data were measured under differential conditions that eliminate the poisoning effect of water present in the reaction atmosphere at higher CO_2 contents under integral conditions. Therefore, the monotonic increase of the TOF over Cu/ZnO–Al_2O_3 with growing CO_2 content, steeply at low concentrations, but monotonically increasing toward 100% CO_2, is according to expectations because of the absence of water in the gas phase. Indeed, the TOF measured with pure CO_2 is close to a value reported in literature for a Zn-doped Cu(111) facet. An experiment with ^{13}C-labeled CO_2 (at 57% CO_2) showed more than 90% of methanol to originate from CO_2, which is in agreement with earlier works.

Notably, unpromoted Cu provides the same activity for methanol formation as the promoted system – but only in pure CO. Just 5 vol% CO_2 is sufficient to quench this activity almost completely. The residual TOF is on the same order of magnitude as the activity achieved with pure CO_2 on a Cu(111) surface, and the more than 90% [13]CH_3OH in the effluent obtained with [13]CO_2 in the feed (at 57% CO_2) confirms that hydrogenation of CO is almost completely suppressed in the presence of CO_2. Obviously, the Zn promoter completely reverses the behavior of Cu with respect to CO_2 in the feed. According to ref. [164], an impregnation of Cu/MgO with just 5 wt.% Zn is sufficient to also make this catalyst behave like the Cu/ZnO–Al_2O_3 system.

These observations can be explained by the results of DFT calculations that model the hydrogenation of CO and CO_2 on a (stepped) Cu(211) facet – either bare or with the steps decorated by Zn [164]. On the bare surface, a key difference between the reactions relates to the stability of intermediates: in CO hydrogenation, they are all less stable than the initial state (bare surface + reactants),[28] in CO_2 hydrogenation, the formate species is more stable. Hence, formate, which accumulates on the Cu surface even at low CO_2 concentrations, poisons the hydrogenation of CO. As even the interaction between Cu and the formate ion is relatively weak, the latter is only moderately activated for the next steps. Therefore, the CO_2 hydrogenation rate remains low.

The Zn promoter enhances the interaction between the surface and adsorbates binding via O atoms, in particular for the polar formate species. The increased rates of CO_2 hydrogenation have been ascribed to the availability of mixed Cu–(O–)Zn sites in literature. Surface interactions of species binding via a C atom are destabilized by Zn, which explains the poor performance of Cu/ZnO–Al_2O_3 in the hydrogenation of CO.

In the presence of Zn, the reaction zone is confined to the perimeter between ZnO and the Cu surface. The role of Cu is mainly to supply H atoms to the intermediates bound in this region, but it is most likely also part of the mixed adsorption sites for

---

28 That is, stationary coverages of intermediates are low, but as the final product is more stable than the reactants, the reaction can proceed, even with high rates.

the latter. ZrO_2 is another component exerting a strong promoting influence on Cu in methanol synthesis. Studies on Zr-promoted Cu/SiO_2 catalysts showed that the reaction is likewise confined to the perimeter between Cu and ZrO_2. In this case, the adsorbates (regardless of whether derived from CO or CO_2) appear to be bound exclusively to ZrO_2, which reduces the role of Cu exclusively to the supply of activated hydrogen [165].

If there is no metal supplying H atoms, CO can be also hydrogenated by H ions from heterolytic H_2 dissociation (Figure 4.49b). The active site is an O vacancy. Under reducing conditions, such vacancies are permanently formed in the surfaces of ZnO, in particular, when the oxide is structurally perturbed by altervalent dopants like Al_2O_3 or Cr_2O_3. CO adsorbs in these vacancies only weakly, without a preferential orientation to the surface according to DFT calculations [166]. The study showed an alternating addition of hydride ions and protons to the adsorbate, resulting in methanol via formyl, formaldehyde, and methoxy intermediates (Figure 4.49b) to be feasible. Methanol formation rates achievable with ZnO-based catalysts are on the same order of magnitude as those obtained with Cu-based catalysts, but at more than 100 K higher temperatures.

The extraordinary career of supported Au nanoparticles in catalysis has been mentioned several times in this book. The discovery of CO oxidation proceeding with high rates at just 200 K ([167], see also [168]) initiated decades of intensive research effort, which resulted (inter alia!) in hundreds of papers dealing with the reaction mechanism, the nature of the active sites, and of the metal-support interaction. Without going into detail with respect to the plethora of proposals published, this section will be closed by introducing reaction mechanisms very likely operating at different temperatures, at particular feed compositions, on different supports, or over different Au sites.

As mentioned earlier (Sections 2.3.3 and 3.4.8.4), supported noble metal particles facilitate the reduction of the support along the three-phase contact line, designated as perimeter. This is the basis of the "gold-assisted Mars–van Krevelen mechanism," which was inspired by the titration of reactive TiO_2 surface oxygen with CO pulses in the TAP reactor. It has been further supported by theoretical calculations, and defined with respect to its scope of validity recently [169]. Its reduction step was shown to be activated: with decreasing temperature, the quantity of surface oxygen titrated by the CO pulses decreased as well until the reaction was completely inhibited below 250 K. Opposed to this, surface oxygen removed at a high temperature could be completely replenished with nearly identical time dependence in the whole temperature range studied.

According to the theoretical model, a CO molecule adsorbed on a perimeter gold atom captures an adjacent O atom from the support in an activated (and rate-determining) step, and is subsequently desorbed as CO_2. The remaining vacancy is stabilized by charge transfer to nearby Au atoms, which are displaced toward the vacancy at the same time, in particular, when the particle is very small [169]. Due to this stabilization, the first step (CO oxidation combined with TiO_2 reduction) is nearly thermoneutral.

The fast reoxidation step was likewise explored with a number of models describing potential routes, e.g., O_2 dissociation in a single vacancy (with different fates of the second O atom) or O_2 dissociation in two adjacent vacancies. Barrier-less reoxidation could, however, be obtained only over sites with a double vacancy, which is difficult to reconcile with the substantial O_2 contents in typical feeds [169]. While this suggests that the reoxidation process requires more attention, the study is remarkable in covering the whole catalytic cycle, including parts that are not rate-limiting.

High CO oxidation rates over Au supported on reducible oxides even below 150 K require, however, explanations beyond the Au-assisted Mars–van Krevelen mechanism; likewise, activities over gold supported on some nonreducible oxides, which are smaller, but still remarkable. In these mechanisms, the activation of O_2 for the reaction with adsorbed CO is the crucial step.

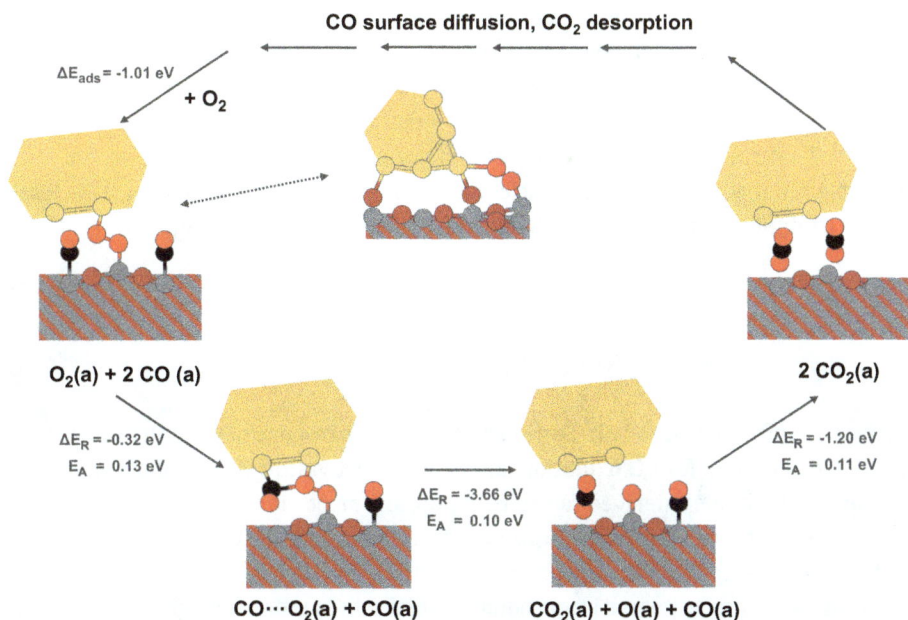

Figure 4.51: CO oxidation with molecular oxygen on the surface of Au/TiO_2 catalysts. The central structure relates to the periodic model evaluated in ref. [170], which consisted of a three-layered Au nanorod bound to a four-layered TiO_2 cluster. The remaining structures allude to the real, more extended Au/TiO_2 perimeter; bonds between support O and Au are omitted. Support O - dark red, O from O_2 - bright red. Adapted from ref. [170].

A typical mechanism discussed for reducible oxides is illustrated in Figure 4.51. O_2, which cannot be strongly adsorbed either on TiO_2 or on extended Au surfaces (see, however, later), is stabilized in an $Au–O–O–Ti^{4+}$ peroxo structure with a formidable adsorption energy of -1 eV. Charge transfer from Au toward Ti^{4+} and from both into the antibonding $2\pi^*$ orbital of O_2 contributes to the elongation of the O–O bond and the activation of the molecule. The interaction of an adjacent CO adsorbate with this peroxide results in a $CO \cdot O_2$ complex from which CO_2 is formed. The remaining O atom reacts with the next CO molecule (Figure 4.51). All these steps have low activation energies. CO_2 desorption and CO diffusion that complete the cycle (cf. ref. [170]) are omitted in the figure. Their kinetic barriers are small as well. The overall activation energy of 0.16 eV predicted for the reaction was in good agreement with the result of rate measurements between 110 and 130 K, performed by tracking the IR intensity of CO bands. Notably, this is well below the activation energies measured at higher temperatures, which scatter around 25 kJ/mol.

The lower CO oxidation activity of Au/Al_2O_3 in dry media, as compared to Au/TiO_2, has been ascribed to a failure of Al^{3+} to allow forming an analogous peroxo species, but convincing mechanistic models are lacking. Lopez et al. have pointed out that the adsorption energy of O_2 on Au and its tendency to dissociate increase dramatically at very small particle dimensions due to the upshift of the 5d band at highly exposed sites (cf. Section 4.2.2) [171]. As a result, most of CO oxidation proceeds on the Au surface when particles are sufficiently small or rough, while perimeter sites are only responsible for the finer ranking of catalysts made with different supports. While this definitely describes an aspect important for the discussion, there have been doubts that it gets to the heart of the problem. When, for instance, TiO_2-supported Au^0 clusters of extreme roughness and irregularity were converted to well-shaped Au particles of 2–3 nm size by calcination in air, a decrease of the CO oxidation activity was found, but within the same order of magnitude [172]:

$$*CO + *OOH \rightarrow *COOH + *O \tag{4.23}$$

$$*COOH + SUP–(OH)(H_2O)_{n-1} \rightarrow CO_2 + * + SUP–(H_2O)_n \tag{4.24}$$

It has been mentioned already in Section 4.2.4 that the presence of moisture opens up a new reaction channel for CO oxidation, allowing the activation of O_2 without involving the support. The models derived for these conditions meet the challenge to explain the support influence, which has been reported to remain significant for Au on TiO_2 and also Al_2O_3 in moist feed ([173], at variance with [61]). In ref. [173], this was ascribed to the poisoning of the perimeter by carbonate, which binds more strongly on Al_2O_3 than on TiO_2. Such explanation raises the question – why does the mechanism (or its rds) need the perimeter and why it cannot proceed elsewhere on the Au particle. The rds proposed in ref. [173] is the decomposition of a carboxyl species previously formed from the adsorbed hydroperoxide and CO ((4.23), to be compared with (4.5)) to CO_2, a free site and a proton (4.24). The latter is delivered to the water layer on the support in return for the proton consumed to form the hydroperoxy species. It is plausible that this transfer may be delayed on Au sites far from the support surface.

Irrespective of the details, it is very likely that CO oxidation proceeds along different routes in the presence and in the absence of water; and insufficient control of the water content has been suggested to have caused large scatter of rate data reported in literature for room temperature [61]. The conditions for the transition between the mechanisms remain, however, to be defined. At low temperatures, layers of liquid or solid water may interfere with stable performance; at high temperatures and low H_2O partial pressures, the "dry" routes should take over under conditions that are difficult to specify.

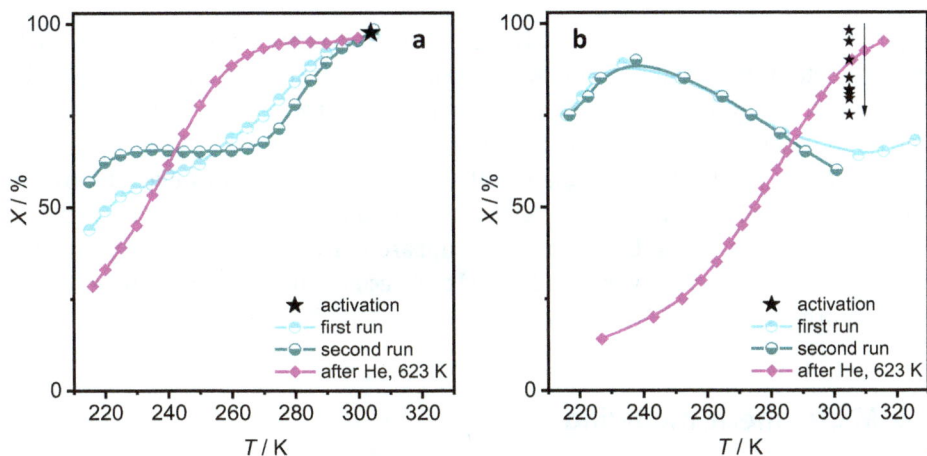

Figure 4.52: CO conversions in successive runs over Au/TiO$_2$, prepared by deposition/precipitation (dp) and freeze-drying (i.e., avoiding thermal stress), measured in 1% CO, 20% O$_2$, balance He: (a) 4% Au, dp at pH ≈ 7, 315,000 mL/g s, and (b) 2% Au, dp at pH = 8.7, 80,000 mL/g s. Adapted from ref. [172] with permission from Wiley-VCH Verlag GmbH & Co. KGaA, Weinheim, Germany.

The experiments summarized in Figure 4.52 suggest the existence of yet another type of Au site capable of catalyzing CO oxidation at high rates and along a different route. The Au/TiO$_2$ samples employed were prepared by combining deposition-precipitation (cf. Section 3.1.3.3) and freeze-drying, with the intention to avoid thermal reduction of initial Au^{3+} species to Au^0. With such a catalyst, for which the absence of Au^0 was confirmed by XANES, catalytic CO oxidation was detected by IR spectroscopy at 90 K [172]. After their activation by exposure to the feed at ≈300 K,[29] the CO conversion X exhibited an unusual dependence on temperature; it remained constant or even decreased in an intermediate temperature range before resuming its expected raise toward 100% (second run in Figure 4.52a, first run in Figure 4.52b). After the second run, the catalysts were treated in He at 623 K, which converts all Au^{3+} to Au^0. This transformed the X–T curves to their usual shape, with quite different activities achieved by the two samples compared in Figure 4.52.

The X–T characteristics prior to thermal treatment reveal contributions of different sites to reaction rates. According to XANES data, the catalyst used for Figure 4.52a contained 87% Au^{3+} and 13% Au^0 after the second run [172] (the one of Figure 4.52b was not studied). In Figure 4.52a, the conversion curves, before and after thermal treatment, appear to be just displaced above 280 K. At least for the second run, the conversion in this temperature range can be assigned to Au^0 sites. The plateau below arises from a superposition of this contribution (with E_A ≈25 kJ/mol) with a process characterized by

---

[29] Black stars in Figure 4.52 indicate CO conversions at the end of the activation (a) or a decreasing tendency during activation of an instable sample (b).

negative apparent activation energy. The latter can be nicely observed in Figure 4.52b, showing data from a catalyst batch, which deactivated during its first contact with the feed (see arrow). As a result, the contribution of Au^0 sites was smaller, and CO conversion clearly decreased between 240 and 280 K.

In ref. [172], catalytic CO oxidation at the lowest temperatures over the freeze-dried catalysts was assigned to Au^{3+} sites. There is no evidence on the reaction mechanism operating on them, but the negative apparent activation energy suggests that one of the reactants is only weakly adsorbed and becomes unavailable with increasing temperature.

4.6 Microkinetic modeling

Microkinetic modeling (MKM) targets one of the ultimate goals of surface catalysis: to describe catalytic processes by models that represent the elementary steps of the reaction mechanism with parameters predicted by theoretical calculations (cf. Section 1.6). Its development, tools and achievements, and its perspectives are briefly summarized in this section.

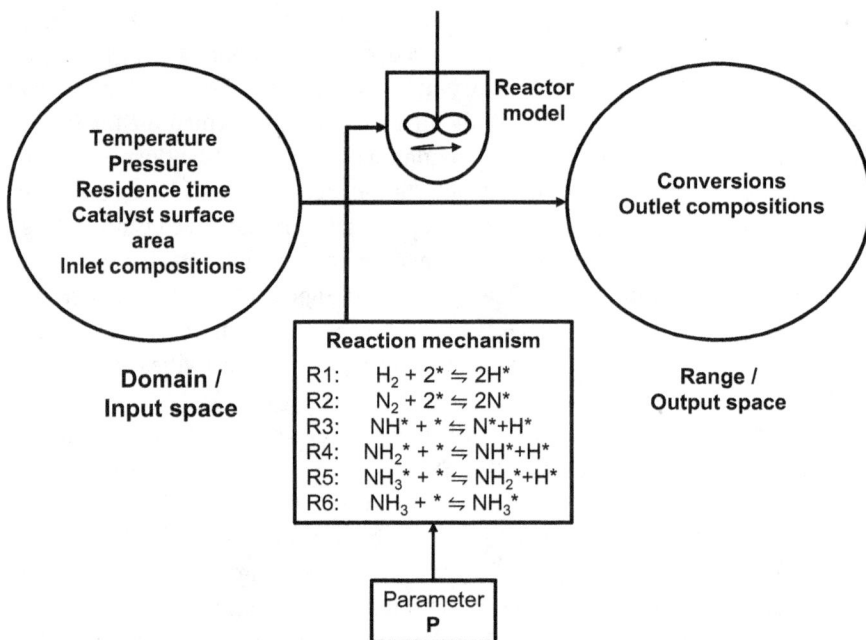

Figure 4.53: Modeling a catalytic reactor with reaction rate information contributed by a microkinetic model. Adapted with permission from ref. [174]. Copyright 2011 American Chemical Society.

Its task is highlighted in Figure 4.53 in its relation to reaction engineering. The output of a catalytic reactor (conversion and outlet concentrations) is to be determined by applying the input variables to the kinetic rate model of the relevant reaction(s) under the hydrodynamic conditions of the reactor (here – a CSTR). Instead of a rate model, a reaction mechanism labeled with the specification "Parameters *P*" is depicted in Figure 4.53, which stands for a rate equation of the type shown in eq. (2.27) or (2.28) for each elementary step *i*. The parameters *P* are the metrics describing the temperature dependence of the rate constants k_i – either the pre-exponential factor A_0, a correction term β, and the activation energy E_A of a modified Arrhenius equation (4.25) or the activation enthalpy $\Delta H^{\ddagger}$ and the activation entropy $\Delta S^{\ddagger}$ of the Eyring equation (2.26):

$$k_i = A_{0,i} \left(\frac{T}{T_{ref}} \right)^{\beta_i} exp\left(-E_{A,i}/RT \right) \tag{4.25}$$

As mentioned in Section 2.6.1, simple power rate laws were used in the kinetic terms of early models for catalytic reactors. They, however, hold only in a narrow range of reaction conditions. Meanwhile, kinetic models based on hypothetical reaction mechanisms, such as Hougen-Watson type rate laws ("global" rate expressions, cf. Section 2.6.1), have become standard in this field. The complementary information used to derive these rate laws (specification of rds, of most abundant surface intermediate, validity of Bodenstein principle) often originates from chemical intuition, but may also result from experimental observations (e.g., identification of rds and of masi from spectroscopic evidence). Kinetic parameters are derived from fitting model predictions to experimental data, which can also be used to discard models derived with inappropriate assumptions (model discrimination).

As global kinetic models depend on hypotheses (not the least, the mean-field assumption, see later), they cannot ideally reproduce the behavior of the system, even in the range of conditions covered by the experiments. Therefore, the potential of experimental data to allow model discrimination depends on their statistics. It is not unusual that the data may be described by different models or by different parameter sets within one model with similar deviations (multiplicity of rate expressions or of model parameters, respectively). Due to the approximate nature of the models, even an acceptable fit with physically consistent parameters is no proof for the underlying model hypothesis (cf. ref. [175]). Although global kinetic models represent the experimental data in wider ranges than power rate laws, they should not be extrapolated either. They definitely fail with a change of the rds, which is not included in any model hypothesis.

The principles of microkinetic modeling, which were first applied in an analysis of reactions of CO with O_2 or NO over single-crystal or supported Rh catalysts [176], were systemized and generalized by Dumesic in 1993 [177]. The promise of this approach is to make available quantitative data on all surface processes during catalytic reactions, most of which are not accessible experimentally. Surface concentrations of intermediates and of vacant sites can be calculated for any set of conditions. Rates and their temperature dependence are accessible for all steps of the mechanism. In networks allowing more than one path from reactants to products, the contribution of these paths to observed conversions can be assessed (reaction path analysis). Sensitivity analysis differentiates kinetically irrelevant (nonsensitive) steps, which are partly or almost completely equilibrated, from sensitive, i.e., relevant, steps. If the sensitivity of one step stands out over the remaining ones, it is rate-determining. Last but not least, the model allows specifying the complete reaction coordinate.

However, to pick up this treasure, model parameters must be determined and models must be validated with the background of experimental rate data. In the 1990s, this was possible only by fitting model predictions to experimental data, as mentioned earlier. The high number of model parameters requires a more extended experimental data base. Fitting must be constrained by conditions, ensuring thermodynamic consistency of the results, e.g., by keeping appropriate relations between activation energies of forward and back reactions and reaction enthalpies, and between pre-exponential factors in both directions and reaction entropies (cf. ref. [175]).

Models are often highly nonlinear. Parameter estimation is hampered by an abundance of local optima featuring, for instance, solutions in which the net rate becomes limited by steps, which are very fast in reality. Inserting kinetic parameters from independent studies of reaction steps reduces the degrees of freedom in the fit, decreases the risk of getting lost in local optima, and enhances the prospect of meaningful results. For solving a microkinetic model for NH_3 synthesis over Cs-promoted Ru/MgO catalysts, for instance, Arrhenius parameters for N_2 desorption were derived by modeling nitrogen TPD[30] from the sample, and for N_2 adsorption by modeling the $^{28}N_2/^{30}N_2$ isotope exchange. Parameters for the hydrogenation of N adsorbates were obtained via TPSR of atomic nitrogen pre adsorbed on the catalyst [178]. TPA[30] and SSITKA (Section 3.2.1) are additional tools for independent kinetic studies in support of MKM.

The importance of kinetic parameters from independent studies was emphasized even in 2009 on the basis of a study dealing with a relatively simple reaction system: NH_3 decomposition over Ru/Al$_2$O$_3$ [179]. The authors pointed out a number of tools for adapting the structure of the data set to the nature of the model in order to avoid unnecessary experiments. In their informatics-based Design of experiments (DoE), the operating space was searched with a Monte-Carlo approach to identify conditions optimal for parameter estimation, e.g., for examining the rate-limiting quality of individual steps. Sensitivity analysis with proposed kinetic models allows distinguishing critical from noncritical steps on the basis of information available prior to experiments. In the fit procedure, noncritical kinetic parameters were fixed to values estimated by semi-empirical methods (see later). Even with these constraints, more than hundred local optima were observed in the fitting space when the model was fitted to the data from the informatics-based DoE without inserting kinetic parameters from independent sources.

In the evaluation of MKM, kinetic parameters estimated by semi-empirical approaches may be employed as fixed parameters (see earlier) or as initial guesses for parameters

---

30 cf. Sections 3.4.2.3 and 3.4.2.4.

to be fitted. Using easily accessible data like thermodynamic properties of components in the gas phase or atomic binding energies to the catalytic elements, semi-empirical methods offer inexpensive estimates of activation energies, pre-exponential factors, and of chemisorption energies of adsorbates. The group additivity concept, for instance, developed for predicting thermochemical properties of large molecules, from increments assigned to their subgroups long ago, was transferred from the gas-phase to surface applications. As an example, the formation enthalpy ΔH_f of an n-propyl group adsorbed on a metal Me can be estimated by adding the increments of three carbon atoms in different bonding configurations with C, H, and Me atoms [175]. In addition to adsorption energies, the "bond order conservation" (boc) method also provides estimates of activation energies by modeling energetic changes along the reaction coordinate with Morse potentials. It covers even influences of surface coverage on activation energies [175]. Estimates for pre-exponential factors are available in literature as well as in approaches to forecast adsorption entropies [175]. Meanwhile, semiempirical methods benefit from the opportunity of benchmarking their estimates against parameters from first-principle calculations.

As mentioned in Section 3.5, realistic activation energies and rate constants can be indeed obtained from first-principle calculations for some types of surfaces nowadays. However, due to the high computational effort needed, this approach is too laborious and expensive to provide more than a frame of benchmarks when large networks are to be treated, e.g., the conversion of glycerol on a Pt surface, exemplified in ref. [175], where ≈ 100 relevant dehydrogenations and ≈ 80 C–C bond cleavage reactions were identified. On the other hand, while "chemical accuracy" targeted for the prediction of individual kinetic parameters tolerates substantial deviations (cf. Section 3.5), relations between parameters of reactions proceeding on the same surface may be captured much more accurately than that. This is the basis for the success of the linear scaling relations among thermochemical properties or between activation energies and reaction enthalpies, which have turned microkinetic modeling into a key tool for the interpretation of activity trends in the periodic table, thanks to the work of J. K. Nørskov and his group. Examples for the use of the BEP relation (3.156), which relates activation energies to reaction enthalpies under the condition of an unchanged nature of the transition state have been given in Sections 3.5.3, 4.2.1, 4.2.4, and 4.5.3. Equation (4.26) describes an analogous scaling relation between adsorption energies $\Delta H_{ads,\ AH_x}$ of hydrogen-containing groups, e.g., CH_3 or NH_2 and the binding energy of the element A on a surface, :

$$\Delta H_{ads, AH_X} = a\Delta H_{ads, A} + \beta \tag{4.26}$$

In eq. (4.26), just the intercept β needs to be calibrated with the data from one metal, while the slope a can be determined from the valence of the element A (cf. ref. [175]). By using large databases comprising metals that exhibit reaction rates spreading by many orders of magnitude, the application of the BEP relation and of (4.26) has provided a favorable environment for validating microkinetic models or drawing appropriate conclusions from failing attempts.

For handling large networks, e.g., the conversion of glycerol on metal surfaces, a hierarchical strategy was recommended in ref. [175]. Using estimated adsorption energies, the stability of potential intermediates may be probed, and by using estimated reaction barriers, irrelevant reactions may be rejected. Among the relevant ones, those with the lowest barrier may be refined for more accuracy with an expensive tool, i.e., DFT.

Microkinetic modeling is based on the mean-field assumption, i.e., it assumes a uniform distribution of identical sites on the catalyst surface, which is characterized by unique values of thermochemical properties and kinetic parameters. Adsorbate-adsorbate interactions, multiplicity of sites, e.g., on different facets of particles, size-dependent properties of particles and clusters, or dynamic catalyst behavior (e.g., spatiotemporal patterns mentioned in Section 2.5.1) are not covered by this approach. These aspects can be investigated with kinetic Monte Carlo methods, however, at the expense of large computational demand. How to include these phenomena also into the mean-field treatments of MKM has been discussed in ref. [175].

The progress of computational chemistry is turning microkinetic modeling into a powerful tool for the prediction of catalytic reaction rates. At present, the reliability of models depends mostly on the completeness of reaction pathways implemented and on the precision of their rate constants (cf. tutorial reviews in refs. [180, 181]). The straightforward implementation of derived parameters into chemical kinetics codes like CHEMKIN or MKMCXX [182] ultimately allows the further utilization of MKM for process optimization. This has already been realized for the analysis of important, but simple reactions involving small molecules such as CO oxidation, water-gas shift, and NH_3 and CH_3OH synthesis.

4.7 Heterogeneous catalysis in liquids: what is special?

Although the fluid delivering reactants to a catalytic surface and recovering the products may be a gas or a liquid, principles of catalysis are explained and exemplified mostly with gas-phase reactions in this book. Simply speaking, this is because processes at the gas–solid interface are better understood than circumstances in the liquid phase. Characterization of surface structures, and of the nature and the dynamics of adsorbates is easier at the gas–solid than at the liquid–solid interface, for which promising techniques (e.g., ATR-IR or SFG, cf. Sections 3.4.8.3 and 3.4.8.2) are of much more recent origin. Moreover, there are no influences of solvation at the gas–solid interface where processes may be fairly complex even without those as exemplified throughout this book.

While application of heterogeneous catalysis for commodities is dominated by gas–solid reactions,[31] catalysis in liquid–solid systems has long been driven by the intention to substitute stoichiometric by catalytic processes in fine chemistry (e.g., oxidations with

---

31 Hydrotreatment of heavy oil fractions, the HPPO process for propene epoxidation with H_2O_2 and, on a smaller scale, fat hardening by hydrogenation being notable exceptions.

chromate/permanganate by catalytic oxidations with H_2O_2 or even air). Meanwhile, the incentive to substitute fossil sources for fuels and materials by renewable sources (cf. Chapter 5) has dramatically increased the attention to this field, for instance, for biomass valorization. The quest for a solar economy is also behind the recent upsurge of photocatalysis, which proceeds at the liquid–solid interface in most cases; likewise of electrocatalysis, where the liquid–solid configuration is exclusive. For this reason, peculiarities of heterogeneous catalysis in liquid phases are summarized here.

The differences introduced by the liquid nature of the fluid originate from both its physical and chemical specifics. Temperatures and pressures are limited to ranges in which the liquid is stable. The high density of liquids decreases diffusion rates of dissolved species, but increases heat capacities and rates of heat transfer. Solvation by solvents or liquid reactants affects adsorption equilibria and transition states, but may also favor damage of catalysts, e.g., by leaching. Solvents may not be innocent and form by-products with reactants, which may influence the main reaction or the state of the catalyst (for more detail on this point see ref. [183]).

Low temperatures required for phase stability favor higher selectivities, similar as in homogeneous catalysis (cf. Section 1.2). The larger heat capacities of liquids and the option of solvent refluxing facilitate the handling of highly exothermal reactions. Due to the low diffusivities, mass transfer limitations are much more serious at the interfaces of solids with liquids, than with gases. This calls for high turbulence at liquid–solid interfaces. At the same time, it cancels the benefit of pellet porosity because the extension of the layer participating in the reaction can shrink to μm size (cf. Section 2.6.2). Violent agitation of very small particles in slurry reactors is the preferred engineering approach for solid catalysts in liquid media. Where this is impractical and a fixed-bed arrangement is preferred (e.g., in trickle beds for three-phase reactions), expensive active components may be confined to the external shell of the pellets (cf. Sections 2.3.1 and 2.8.2).

As the reaction conditions for homogeneous and heterogeneous liquid-phase catalysis are similar, the competition between these modes of catalysis is rather direct. Together with other advantages (cf. Section 1.2), the higher density and mobility of active sites is clearly in favor of homogeneous catalysis, and makes it the only realistic choice in some situations (e.g., catalytic approaches for digesting solid (bio)materials). However, these drawbacks of solid catalysts, which are even enhanced when rates are limited by mass transfer, can still be compensated by the ease of separating the catalyst from the reaction mixture (cf. ref. [184] for a discussion related to biomass conversion). Therefore, research with heterogeneous catalysts can also make sense for processes where selectivity demands seem to favor homogeneous catalysis. Regarding pore-diffusion limitations, the introduction of transport pores into microporous materials (hierarchical pore systems, cf. Section 2.3.1) is a promising strategy to enhance the accessibility of active sites.

Due to the importance of catalyst stability for technical applications, leaching is a most critical issue in heterogeneous liquid-phase catalysis. Solvation of cations by solvents or reactants favors their dissolution, which may be further enhanced by pH excursions, in particular, toward acidic conditions. The dissolved ions are not necessarily inert toward the reaction of interest: ref. [185] presents examples where oxidation activity observed with redox zeolites predominantly originated from leached cations. These cations may be reduced to colloidal metal clusters by a reactant [95]. There are cases where the leached component returns completely to the support at the end of a batch run (release and catch principle [96], cf. Section 4.4.1.2). Where this does not happen, leaching results in loss of the active component, which may contaminate the reaction product. Rigorous proof of heterogeneity can be obtained by a combination of recycling experiments, hot-filtration tests, and trace analyses of the filtrate. The problem is discussed instructively in ref. [185].

Heterogeneous catalysis in liquids comprises hydrogenations, hydrotreatment, and selective oxidations, but also acid- and base-catalyzed reactions. In reactions with H_2 or O_2, the system is complicated by the additional phase boundary between the liquid and the gas. Depending on the engineering solution, the gas contacts the solid only via the liquid (slurry reactor), or it also has direct access to the surface (trickle bed, Section 2.8.2). In any case, such reactions are run at elevated pressures to increase the concentration of the dissolved gaseous reactant.

For selective liquid-phase oxidations, various oxidants are available, among which, reactants with peroxo groups (hydroperoxides, H_2O_2) facilitate achieving high selectivities, while air bears more promise for cost reduction. Oxidations with air may proceed via radical mechanisms and result in wide distributions of reaction products, e.g., alkane oxidation with dissolved (or leached) Co or Mn salts, resulting in a wide range of oxygenates with shorter chain lengths.

H_2O_2, which is typically applied in aqueous solution, is most effectively activated by Ti-substituted siliceous zeolites like TS-1. Competitive adsorption of water at the Ti sites interferes with H_2O_2 activation. To maintain the hydrophobicity of the pore system, catalysts are synthesized via routes allowing for very low defect densities (Si-O-Ti instead of Si-OH + HO-Ti) and for small crystallite sizes, in order to escape mass transfer limitations.

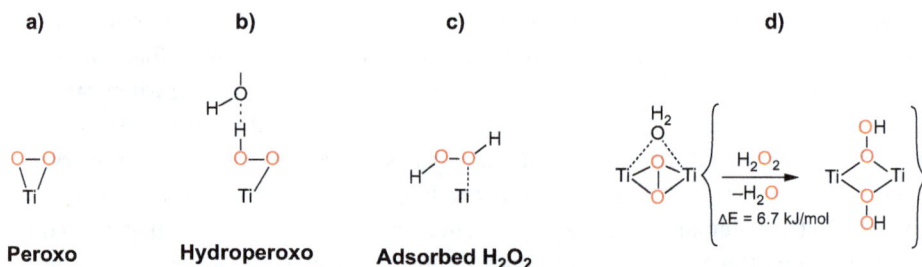

Scheme 4.8: Models for the interaction of H_2O_2 with Ti in TS-1: (a) peroxo complex; (b) hydroperoxo structure; (c) adsorbed H_2O_2; and (d) interaction with binary Ti site.

Among the models for the interaction of framework Ti with H_2O_2 proposed in literature (Scheme 4.8), none has received a clear preference so far. The peroxo model (a) assumes the existence of a titanyl group Ti=O, which releases its oxygen as H_2O to accommodate the peroxo species and is restored, after an O atom has been donated to the reactant [186]. For obtaining the hydroperoxo structure (b), H_2O_2 is thought to cleave a Ti–O–Si bridge, the O atom of which accommodates the remaining H atom. The model was supported by the observation of a characteristic IR band in the OH range [187]. Alternatively, H_2O_2 was assumed to be activated by mere chemisorption (c). The recent proposal of a binary Ti site operating in Ti silicalites (cf. Section 4.2.3, where the pros and cons of this idea are discussed) resulted in model (d) for H_2O_2 activation. The first molecule forms a peroxo structure, delivering its hydrogen to the O atom originally bridging the Ti sites. With a second H_2O_2 molecule, a symmetric structure with two hydroperoxo groups is obtained, which drives the catalytic cycle [36].

To react larger molecules over this kind of site, there has been tremendous effort to substitute Ti into siliceous (or Al-containing) wide-pore zeolites or into the walls of ordered mesoporous materials (cf. Section 3.1.1.1). While some of these catalysts turned out to be useful, none can compete with well-prepared TS-1 in terms of hydrophobicity and stability. The same holds for many siliceous or AlPO materials substituted with small amounts of V, Cr, Co, or other elements, which are also able to catalyze selective oxidations with H_2O_2.

Supported Pt and Pd, monometallic or combined, often promoted with Bi, dominated research on liquid-phase oxidation with O_2 till the end of the twentieth century. While capable of achieving high activities and promising selectivities, they can be damaged by their significant susceptibility for leaching, particularly in aqueous media. Supported Au nanoparticles exhibit smaller activities, but due to the higher inertness of gold, they are more stable. They are much more specific and allow high or even complete selectivities [188], which has been ascribed to the very specific reactivity around the perimeter between Au particles and support (cf. Section 4.5.6).

Due to its importance for the valorization of biogenic raw materials like glycerol or sugars (cf. Chapter 5), the selective oxidation of alcohols and of other oxygenates by supported gold has been the subject of a great exploratory effort. Strong influences of solvents and supports were noted. The oxidation of alcohols often requires the presence of a base, e.g., to activate the resulting aldehyde or neutralize the resulting acid. Meanwhile conditions for base-free operation have been found for many feeds (e.g., for glycerol [189]). As Au is miscible with both Pt and Pd, bimetallic combinations were tested to enhance its activity without unacceptable losses in selectivity and stability. Success achieved along these lines has been ascribed to both electronic and geometric effects [188, 189], but the details are often not yet well-understood.

The "labyrinth" [188] of experimental data collected around Au-catalyzed alcohol oxidation has, so far, defied generalizing interpretations. Regarding reaction mechanisms, it has been proposed that radical chemistry dominates in organic phases, while aqueous media favor ionic reactions [188]. The former has been supported by a model for the solvent-free epoxidation of alkenes (Figure 4.54), which was inspired by the observation that a small amount of radical initiator favors this reaction – either by improving epoxide selectivity or, for terminal alkenes, by making it possible at all[32] (summarized in [190]).

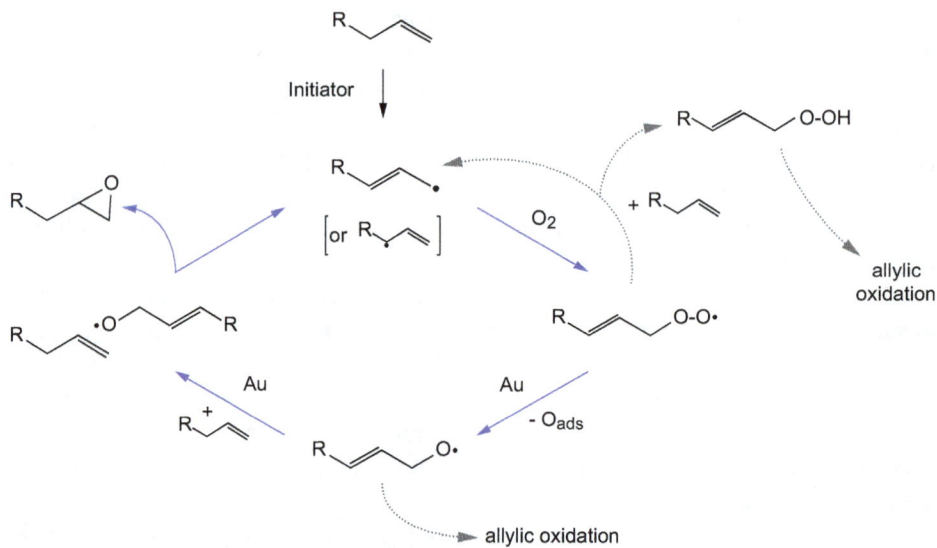

Figure 4.54: Simplified mechanism of alkene epoxidation over supported Au catalysts (the same cycle can be accessed by the isomeric radical indicated in parentheses).

The mechanism runs along the typical routes of autoxidation, i.e., via an allyl radical (Figure 4.54), the formation of which is favored by the initiator. After its oxidation to a peroxy radical, the latter would be converted to the typical products of allylic oxidation without a catalyst. On the Au surface, however, it is cleaved to an en-oxy radical, which is the epoxidizing agent. From delivering its O atom to an alkene molecule, it emerges as the initial allyl radical ready to start the next cycle. In this (simplified) mechanism, which has been assembled from proposals in refs. [190, 191], the role of Au is mainly to suppress the allylic routes by cleaving the peroxy radical and by favoring the epoxidation step. Regarding the role of gold in the formation of the allyl radical, refs. [190, 191] remain indifferent. Such role was assumed in a mechanism for the initiator-free epoxidation of cyclohexene [192]. In this study, however, epoxidation was observed only when Au was combined with a suitable co-catalyst (WO$_3$).

The previous discussion illustrates the problems in describing processes at the liquid–solid interface in more detail than provided by heuristic concepts. In the following,

---

32 The initiator has the side job of oxidizing stabilizers coming with the alkenes, which is neglected here.

two studies pioneering research on the fundamentals of adsorption and catalysis in such systems are briefly highlighted to illustrate the complexity of the situation and the level of understanding achieved at this point of time.

It has been mentioned earlier that water significantly affects the properties of adsorbates on metal surfaces (cf. Section 3.4.8.4, Figure 3.122). Liquid-phase adsorption comprises a number of steps that are absent in gas-phase adsorption. H_2, for instance, replaces water at the metal surface and becomes hydrated by water molecules. The corresponding equations (written with D_2O) are:

$$H_2(g) + 2\,{}^*(Pt) \rightleftarrows 2\,H^*(Pt) \tag{4.27}$$

$$D_2O^*(Pt) \rightleftarrows D_2O(l) + {}^*(Pt) \tag{4.28}$$

$$H^*(Pt) + D_2O(l) \rightleftarrows H^*(Pt)\cdots D_2O \tag{4.29}$$

which adds to:

$$H_2(g) + 2\,D_2O^*(Pt) \rightleftarrows 2\,H^*(Pt)\cdots D_2O \tag{4.30}$$

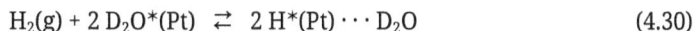

These reactions are superimposed by H-D exchange at the Pt surface resulting in HD, D_2, and HDO. Ref. [193] shows how rate constants of hydrogen adsorption and desorption, and hence, the equilibrium constant of reaction (4.30) are accessible by tracking the kinetics of the whole process. By studying it with a Pt/silicalite-1 catalyst at different temperatures and pressures and performing DFT-based ab initio molecular dynamics simulations of the situation, the authors obtained thermodynamic data, which are briefly discussed in the following.

Hydrogen adsorption was found to be much weaker in the aqueous than in the gas phase. Coverage with H was complete at 0.02 bar in gas-phase adsorption, but just around 50% at 50 bar H_2 in the presence of water. The destabilization of the adsorbate results from both a decreased adsorption enthalpy (−50 kJ/mol for (4.30) versus −72 kJ/mol for (4.27)) and a more negative adsorption entropy (−174 J/mol K vs. −125 J/mol K). The losses in adsorption enthalpy seem to suggest that water interacts with Pt stronger than with adsorbed hydrogen. The changed adsorption entropy reflects a smaller mobility of adsorbed H atoms in the presence of water: in the simulations, the surface diffusion coefficient of H atoms decreased by as much as 80%. The smaller mobility of H atoms corresponds with a higher ordering degree in the adjacent water layer, which contributes to the lower stability via both enthalpy and entropy effects. It was found that the adsorption enthalpy can also be influenced by the concentration of H_3O^+ ions (varied, for instance, via the Si/Al ratio of MFI-type zeolites): it decreases in highly acidic environment.

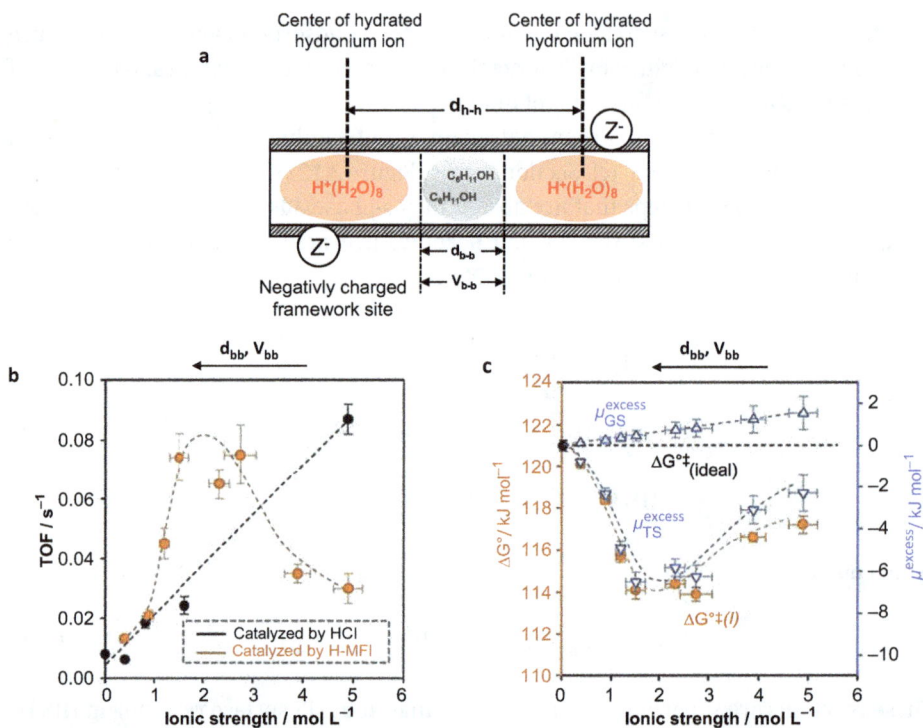

Figure 4.55: The influence of ionic strength and of spatial constraints on the rate of cyclohexanol dehydration in MFI zeolites: (a) model: water clusters hydrating hydronium ions dissociated from Brønsted sites limit pore space available for the hydrophobic reagent cyclohexanol; (b) dependence of the turnover frequency (rate per H^+ available) on the ionic strength: comparison of liquid phase constrained by zeolite pore with unconstrained case (where ionic strength was varied by dissolving LiCl); and (c) influence of the ionic strength in the zeolite pore on the chemical potentials of ground state and transition state and the resulting effect on the free Gibbs activation energy. From ref. [118], reproduced with permission from AAAS.

Careful examination of activation parameters changing under the influence of the active site concentration in zeolites has resulted in a remarkable level of under-standing of a liquid-phase reaction catalyzed by Brønsted sites: the dehydration of cyclohexanol. As outlined in Section 4.5.2.1, the interaction of zeolite Brønsted sites with water results in clusters of up to eight water molecules that stabilize the fluctuat-ing protonic charge and remain spatially related to the anionic framework charge. In a study of cyclohexanol adsorption from aqueous solutions [194], it was found that the alcohol occupies only the pore space left over by these clusters (Figure 4.55a): unlike short-chain alcohols, cyclohexanol cannot replace water in the clusters, which, there-fore, feature an additional steric constraint. Beyond this steric effect, adsorption is also influenced by the overall repulsive interactions between cyclohexanol and the aqueous clusters. With increasing Al content, their importance grows at the expense of the dispersive interactions among the cyclohexanol molecules, which results in an increasing cyclohexanol activity coefficient. This can be quantified within the ionic strength concept, which describes the changes of the chemical potential under the influence of ionic charges in the system. The effects are dramatic because of the high density of charged species in the pores of Al-rich zeolites.

The influence of the ionic strength on the thermodynamic activity of the hydronium ions is similar and causes a strong increase also of their catalytic activity in cyclohexanol dehydration. A comparison of turnover frequencies observed in aqueous HCl and in MFI pores at identical ion strengths (Figure 4.55b) suggests, however, that circumstances in the pore are more complex [118]. They need to be discussed regarding the influence of the ionic strength on the chemical potential of both the ground and the transition state (μ_{GS}, μ_{TS}) at varying distances, d_{BB}, between water clusters. The former is a cyclohexanol molecule adjacent to a hydrated hydronium ion; the latter a hydrated cyclohexyl carbenium ion, which is obviously more polar. As mentioned earlier, cyclohexanol is destabilized at high ionic strengths; its chemical (excess) potential grows with decreasing d_{BB} (Figure 4.55c). At low ionic strengths, this is outbalanced by a strong stabilization of the transition state by the more polar environment, which results in a decrease of $\Delta G^{\ddagger}$, i.e., higher reaction rates. With further decreasing d_{BB}, however, the space also becomes constricted for the TS, which results in its destabilization and, hence, decreasing reaction rates.

The work of Lercher's group gives a taste of how scientific concepts may help understanding seemingly disparate observations in catalysis at liquid–solid interfaces, by revealing hidden correlations.

4.8 Structure and performance in photocatalysis

The first and foremost requirement of a photocatalyst is the availability of a suitable bandgap allowing charge separation by light, preferably by sunlight. As mentioned in Sections 1.4.3 and 2.5.2, the resulting charge carriers may get lost by recombination on their way to the surface or thereat, if the electron transfer to or from the reactants is slow. The relation between structure and performance of photocatalysts has, therefore, various aspects: finding or engineering materials with suitable bandgaps, impeding recombination processes, accelerating charge transfer to the reactants at the surface, and, last but not least, the stability of the catalytic phase under reaction conditions. These aspects are discussed in the following with reference to photocatalytic water splitting.

Photocatalysts for water splitting are typically metal oxide, nitride, or oxynitride phases. With a few exceptions, sulfide-containing materials fail for stability reasons [195]. An alternative development has been started by a report on metal-free graphitic carbon nitride materials effectively catalyzing H_2 evolution from water[33] in visible light [196].

Cations in oxide compounds with suitable bandgaps have either a d^0 or a d^{10} electronic configuration. Ti^{4+}, Zr^{4+}, V^{5+}, Nb^{5+}, Ta^{5+}, W^{6+}, and Ce^{4+} are examples of the former and Ga^{3+}, In^{3+}, Ge^{4+}, Sn^{4+}, and Sb^{5+} of the latter group. d^n cations with $0 < n < 10$ were found to exhibit inefficient photoresponses [195]. Promising materials include binary oxides like TiO_2, ZrO_2, CeO_2, β-Ga_2O_3, ternary oxides ($SrTiO_3$ and other perovskites, niobates, tantalates, tungstates, etc.), and more complex phases, which may contain both d^0 and d^{10} cations. Trends in efficiency, though complex in detail, can mostly be explained well on the basis of electron structure models [195]. The more dispersed character of the conduction band in d^{10} photocatalysts (derived from s- and p-orbitals) compared with d^0 catalysts (derived from d orbitals) causes a higher mobility of charge carriers in the former.

The oxides introduced so far have bandgaps in the UV region and are not active in visible light. Some of them, which are quite efficient for splitting water in UV light, have been optimized along strategies generally pursued in photocatalysis. Their targets — high surface area, short distances for charge carrier diffusion, and high carrier preservation and mobility are conflicting goals calling for a trade-off. Small particles allowing high surface areas and short diffusion lengths are often rich in structural defects, which impede carrier mobility and facilitate recombination processes. Improving the structure by thermal treatments, however, causes sintering. Reference [195] highlights some instructive examples, where diffusion lengths were minimized keeping acceptable structural integrity, e.g., by exfoliation and recalcination of layered oxide structures, or by firing the amorphous 1.8 nm walls of mesoporous Ta_2O_5 under a temporary protective silicate layer.

It has been also reported that the coexistence of (well-crystalline) phases may favor the photocatalytic performance of semiconducting materials. Such superiority of mixed over pure phases has been assigned to a more stable charge separation, when electrons or holes can be trapped by a second phase, prior to recombination. The remarkable performance of the mixed-phase titania P25®, cf. Section 2.3.2.5, has been often explained by such phase-cooperation (cf. ref. [197]). The problem, however, does not seem to be settled; in ref. [198], synergy by phase cooperation in P25® was rejected for four photocatalytic reactions.

---

33 Activity of photocatalysts for half-reactions can be studied separately, when the other half-reaction is replaced by a redox process proceeding more easily (cf. Section 2.7). H_2 formation can be isolated, for instance, in the presence of sacrificial reductants like alcohols, thiosulfate, or (as in ref. 196) triethanolamine, which are easier oxidized than the O atom in water. Sacrificial oxidants like silver nitrate are used to investigate the O_2 formation step, because they are easier reduced than the H atom in water.

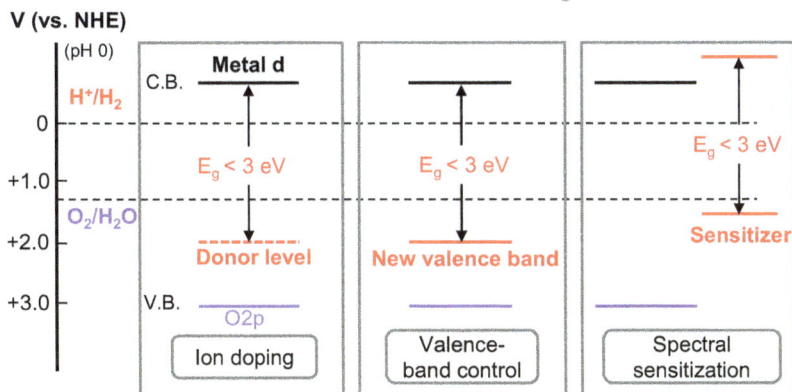

Figure 4.56: Strategies for the development of catalysts for water splitting with visible light. Adapted from ref. [195] with permission from Elsevier.

In oxides, the top of the O 2p valence band is typically around +3.0 eV versus the normal hydrogen electrode. As H_2O can be reduced only at <0 eV (cf. Figure 4.56), H_2 evolution, and hence, total water splitting can be achieved with them only in UV light. Figure 4.56 highlights the strategies applied, so far, to develop catalysts capable of splitting water with visible light [195].

Doping the host lattice, e.g., TiO_2, with a suitable element introduces levels at potentials <3 eV, which are suited for charge separation by visible light (Figure 4.56). Dopants may be d^n metal ions ($0 < n < 10$) or alternative anions: S^{2-}, N^{3-}, C^{4-}. However, the impurity atoms affect charge carrier migration in the material: to make up for charge imbalances, the structure usually contains vacancies, which favor recombination of holes and electrons.

Although this problem may be solved by a co-doping approach (e.g., TiO_2 with Cr^{3+} and Sb^{5+}), the valence-band control strategy, which operates with mixed compounds that form bands instead of discrete levels in the relevant energy range (Figure 4.56), has been more successful. In these materials, the top of the valence band may be formed by s states of heavy metal cations (e.g., Bi^{3+} or Sn^{2+} in materials like $BiVO_3$ or $SnNb_2O_6$) or by p states of anions like N^{3-} coordinated around d^0 or d^{10} cations. Oxynitrides combined with suitable co-catalysts are among the best catalysts for water splitting with sunlight, known to date. Because of the more dispersed conduction band, oxynitrides of d^{10} cations bear most promise for this purpose, e.g., solid solutions of GaN and ZnO $((Ga_{1-x}Zn_x)(N_{1-x}O_x))^{34}$ or of $ZnGeN_2$ and ZnO $((Zn_{1+x}Ge)(N_2O_x))$. The bandgap of the Ga-Zn-N-O materials can be further decreased by including indium [195].

Sensitizers can be organic dyes or inorganic semiconductors with narrow bandgap and an excited state below 0 eV. From this state, the excited sensitizer injects elec-

---

34 Ga/Zn = 1.4–20. Their band structure deviates from the simple model shown in Figure 4.56b (cf. ref. [195] and literature cited therein).

trons into the conduction band of the catalyst, which consumes them, typically via a co-catalyst, for H_2 evolution. For total water splitting, re-reduction of the sensitizer by water needs to win competition with the transfer of electrons from the co-catalyst to the sensitizer. So far, sensitized photocatalysts have been mostly studied for H_2 evolution, but total water splitting was also achieved [199].

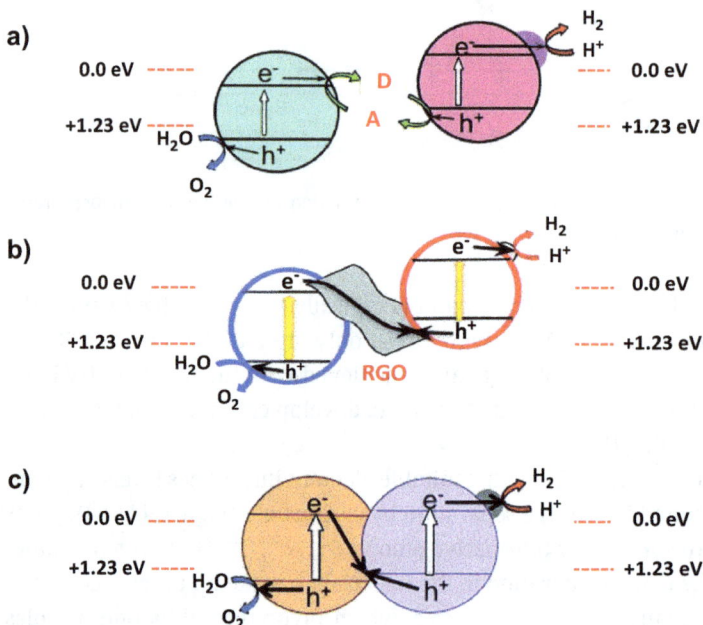

Figure 4.57: Z-scheme photocatalysts with different electron mediators: (a) dissolved electron mediator; (b) solid electron mediator; and (c) electron mediation by contact. Adapted from ref. [200] with permission from the Royal Society of Chemistry.

Water splitting can be also performed with systems in which two semiconductors with narrow band gap are coupled by an electron mediator or just by contact (Figure 4.57). Absorption of photons by these "z-scheme photocatalysts" results in evolution of H_2 and O_2 on the respective semiconducting components (or on their co-catalysts), while the mediator helps the electrons left over after excitation in one component to the holes left over in the other one. This dual principle, which mimics biological photosynthesis, has been proposed as early as 1979 [201]. Most realizations operate with dissolved electron mediators (Figure 4.57a), e.g., Fe^{3+}/Fe^{2+}, IO_3^-/I^-. The use of reduced graphene oxide (RGO) to establish contact between the components (b) and the identification of z-scheme pairs operating with direct contact (c) are rather recent achievements [200, 202]. Deposition of the two components on separate electrodes that are connected by an external circuit results in a photoelectrochemical cell.

The z-scheme bears the promise to allow combining the most promising catalysts for H_2 and O_2 evolution to water splitting systems with unprecedented efficiency. However, to realize this promise with dissolved redox mediators, characteristic side reactions need to be suppressed: the reduced mediator (D in Figure 4.57a) may compete with water for holes in the O_2 evolution catalyst, its oxidized form (A) with H^+ for hot electrons in the H_2 evolution catalyst. The state of the art in water splitting with z-scheme systems is discussed in refs. [200, 202].

As mentioned earlier, co-catalysts are crucial components of photocatalysts, many of which do not exhibit any activity at all, without them. Their performance depends on the efficiency of two electron transfer steps: the transition through their interface with the semiconductor and the exchange with reactants adsorbed on their surface. They influence the fate of charge carriers by their trapping effect (cf. Section 2.5.2), but some of them also catalyze the back reaction. As an easy-to-handle co-catalyst for H_2 evolution, Pt, for instance, is often employed in proof-of-principle experiments that demonstrate other functions of the photocatalyst, but owing to its high activity for water formation from H_2 and O_2, it is inferior to many other reduction co-catalysts. Due to the complex interactions superimposing in the operation of co-catalysts, their development is still based on heuristic strategies, to a large extent.

With respect to H_2 evolution, the intention to escape the back reaction has shifted the focus from noble metals to oxides like RuO_x, and later to Cr-containing mixed oxides, e.g., of Rh or Cu. Using a Rh–Cr mixed oxide instead of RuO_2 as a reduction co-catalyst for $(Ga_{1-x}Zn_x)(N_{1-x}O_x)$, the apparent quantum yield during water splitting at 410 nm was increased by an order of magnitude [195]. Similar performance can be achieved with noble metal nanoparticles encapsulated by a thin Cr_2O_3 layer, which allows permeation of H_2 while O_2 is kept out. Co-catalysts for water oxidation are mostly oxides like MnO_x, CrO_x, or CoO_x.

Co-catalysts are applied in very low amounts. With increasing loading, their efficiency exhibits a volcano-type behavior with a peak typically well below 1 wt.%. The decrease beyond has been often ascribed to an optical shielding of the semiconductor surface. For many systems with metal co-catalysts, this drop starts at metal contents that are too low to justify such interpretation. An alternative explanation, which operates with the effect of fields set up between electrons trapped in the metal particles and the semiconductor, can be found in ref. [203].

Co-catalysts can be deposited by impregnation (cf. Section 3.1.3.2), but photodeposition techniques have been found to be much more specific. They even allow depositing oxidation and reduction co-catalysts separately from each other, when crystals have anisotropic facets [204]. Selective deposition of dual co-catalysts (Pt, Co_3O_4) on the facets of anisotropic crystals of $SrTiO_3$ (bandgap – 3.2 eV) resulted in strong improvements of water splitting rates compared with the same system prepared with isotropic (cubic) crystals. The effect was assigned to a smaller influence of recombination [205]. With anisotropic Al-doped $SrTiO_3$ crystals loaded with Rh/Cr_2O_3 and CoOOH co-catalysts on different facets, quantum efficiencies near 100% were achieved during water splitting in visible light, indicating a nearly complete absence of charge carrier recombination [206].

Photocatalysts are dynamic systems and exhibit transient behavior that may include activation periods, but, in particular, deactivation. The latter may be caused by photocorrosion or, in reactions other than water splitting, the deposition of intermediates or byproducts on the surface. Corrosion may be enhanced when photoabsorber or co-catalyst change oxidation states due to imbalances of charges generated during irradiation. Research on these phenomena is in an early stage. Reference [207] gives a flavor of the problems encountered in this field.

4.9 Structure and performance in electrocatalysis

4.9.1 Electrocatalytic reaction mechanisms

Proceeding exclusively at liquid–solid interfaces, electrocatalysis resists the elucidation of reaction mechanisms as does heterogeneous liquid-phase catalysis (Section 4.7), but offers additional obstacles related to the charge transfer processes involved. The challenges concern both experiment and theory. In the experiment, the potential driving the surface reaction needs to be scrupulously differentiated from potentials caused by diffusion processes or by ohmic resistances in the catalyst, between catalyst and electrode, in the electrolyte, etc. (cf. Section 3.2.3). Concepts of the surface reaction need to take into account that charge transfer and bond rupture or formation are not necessarily concerted, although they are usually lumped together in the equations describing elementary steps (e.g., eqs. (4.38) through (4.40) below). Moreover, there are fewer opportunities for *in situ* spectroscopy of surface processes than in thermal catalysis. Together with the general complexity of processes at real surfaces (heterogeneity of site structure, adaptation of the surface state to the chemical potential, coexistence of intermediates and spectators), this explains why less issues on reaction mechanisms can be considered settled in electrocatalysis than in other versions of surface catalysis, particularly in catalysis of gas-phase reactions.

In this situation, the challenge to theory is threefold. Its importance for the elucidation of reaction mechanisms is even higher than in thermal catalysis, but at the same time the phenomena to be modeled are much more complex, and experimental data suitable for benchmarking theoretical results, e.g., binding energies of adsorbates, are less accessible. Often, comparison is possible only with rates and product selectivities resulting from a sequence or even a network of elementary steps [208].

Quantum chemistry is able to supply binding energies of intermediates with catalytic surfaces, e.g., of $*O$, $*H$, $*O_2$, $*OH$, $*OOH$, $*H_2O_2$, and $*H_2O$ ($*$ – surface site), and to describe their dependence on the applied potential, e.g., during the reduction or the evolution of O_2. Kinetic barriers between them are evaluated, which allows estimating rate constants of assumed elementary steps [208]. However, a full understanding of electrocatalytic processes goes beyond arranging intermediates in suitable sequences, as in eqs. (4.38) through (4.40). It specifies also the type of sites involved and their contributions; it considers the effects of solvation, of changing surface coverage, of electron and mass transfer [208].

The following discussion highlights how insight in real reaction mechanisms can be obtained and illustrates the methods with examples from water electrolysis (oxygen evolution reaction (OER), hydrogen evolution reaction (HER)) and from the hydrogen-fueled PEMFC (ORR and hydrogen oxidation reaction (HOR)).

As mentioned in Section 2.7, the Tafel slope (cf. eq. (2.64)) is related to the catalytic reaction mechanism, but conclusions from this macrokinetic level on microkinetics are rarely unambiguous. The interpretation of the Tafel slope via the Butler–Volmer equation (cf. eq. (2.63)) holds for simple reactions at low coverages of reactants and spectators. To extend to the slightly more complex schemes of proton reduction, we express the influence of the overpotential η on the current density j or the rate constant k by reverting eq. (2.63):

$$j/j_0 = k/k_0 = exp(-a\,f\,\eta) \tag{4.31}$$

where $f = F/RT$ and k_0 is the rate constant, k, at zero potential. Further, three different options for the reduction of H^+ in acid medium are specified. Two of them (the Volmer step (4.32) and the Heyrovsky step (4.33)) assume a simultaneous transfer of proton and electron (proton-coupled electron transfer, PCET):

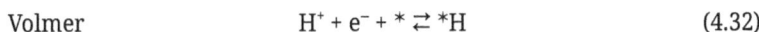

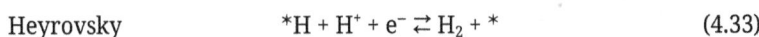

Volmer	$H^+ + e^- + * \rightleftarrows {}^*H$	(4.32)
Heyrovsky	${}^*H + H^+ + e^- \rightleftarrows H_2 + {}^*$	(4.33)
Tafel	$2\,{}^*H \rightleftarrows H_2 + 2\,{}^*$	(4.34)

When the Volmer step (4.32) is rate-determining, the coverage of sites by hydrogen θ_H is low because H_2 evolution via either (4.33) or (4.34) is fast. Neglecting the back reaction and expressing activities as a, the rate r is:

$$r = r_{Vo} = k_{Vo}\,a_{H^+}\,(1-\theta_H)$$

and the current density j_{Vo} is accessible via

$$j_{Vo}/j_{Vo,0} = k_{Vo}/k_{Vo,0} = exp(-a_{Vo}\,f\,\eta_{Vo}) \tag{4.35}$$

i.e., the Butler–Volmer equation is valid in this case. With a rate-determining Heyrovsky step (4.33), r becomes:

$$r = r_{He} = k_{He}\, a_{H^+}\, \theta_H$$

As the Volmer step is now quasi-equilibrated, θ_H can be derived by inserting r_{Vo} (see earlier) and $r_{-Vo} = k_{-Vo,0}\, \theta_H$ into the treatment used to derive the Langmuir isotherm (Section 2.4.3). From this,

$$\theta_H = \frac{K_{Vo,0}\, a_{H^+}}{exp(f\,\eta_{Vo}) + K_{Vo,0}\, a_{H^+}}$$

and, with , $K_{Vo,0} = k_{Vo,0}/k_{-Vo,0}$

$$j_{He}/j_{He,0} = k_{He}/k_{He,0} = \frac{K_{Vo,0}\, a_{H^+}^2\, exp(-a_{He}\, f\,\eta_{He})}{exp(f\,\eta_{Vo}) + K_{Vo,0}\, a_{H^+}} \tag{4.36}$$

which differs dramatically from (4.35). For a rate-determining Tafel step (4.34), $r = r_{Ta} = k_{Ta,0}\, \theta_H^2$, and therefore,

$$j_{Ta}/j_{Ta,0} = k_{Ta}/k_{Ta,0} = \left(\frac{K_{Vo,0}\, a_{H^+}}{exp(f\,\eta_{Vo}) + K_{Vo,0}\, a_{H^+}}\right)^2 \tag{4.37}$$

In (4.37), the potential influence operates only via the hydrogen coverage, θ_H. Simulations of eqs. (4.35)–(4.37) using realistic relations between rate constants resulted in Tafel slopes of 120, 40, and 30 mV/dec as usually assigned to rate-determining Volmer, Heyrovsky, and Tafel steps [209]. However, when θ_H grew above 0.6, the Tafel slope increased to 120 mV/dec also for the Heyrovsky step, which limits the diagnostic potential of Tafel analysis, even for a reaction as simple as the HER.

Although the HOR uses the same steps as the HER ((4.32) through (4.34)), their influence on the kinetics of hydrogen oxidation is quite different. This may be easily demonstrated with the case of a rate-determining Tafel step. This step (reverse (4.34)) can be rate-limiting only, when the reverse Heyrovsky step (4.33) is too slow to compete. At the same time, the consumption of *H (reverse (4.32)) must be fast. Therefore, θ_H remains small, which cancels any potential influence for the HOR, quite in contrast to the HER (cf. eq. (4.37)): the HOR rate with rate-determining Tafel step is simply $r_{-Ta} = k_{-Ta,0}\, p_{H2}$. With a rate-determining Heyrovsky step, the HOR rate complies with Butler–Volmer kinetics, while a rate-determining Volmer step results in a complex kinetic expression, because its reactant *H can be formed in two reactions (reverse (4.33) and (4.32)).

For the reduction and evolution of O_2 with up to four electron transfer steps (see later), the relation between molecular mechanisms and resulting macrokinetics is much more complicated. Details may be found in ref. [209], where a large body of experimental information is compiled. Similar to thermal catalysis, kinetic rate laws allow reliable conclusions on molecular mechanisms only in rare cases, but they can be used to reject certain mechanistic options. Ref. [209] shows many cases where the Tafel slope is coverage dependent. Deviations of the Tafel slope from ideal behavior have been also attributed to other reasons, e.g., to a change of the rds or to an influence of the potential on the number of active sites [210]. Another possible reason is discussed later.

As for H_2 evolution and oxidation, different mechanistic options are discussed for the ORR and the OER and may operate on different catalysts. A peroxide species is an intermediate in the associative mechanism of the ORR, which proceeds in four PCETs to H_2O (4.38) or stops at H_2O_2 after two electron transfers (4.39). On metals with high O adsorption energies and low kinetic barriers to dissociation, a dissociative mechanism (4.40) might be preferred. It can be differentiated from the associative mechanism by the number, z, of electrons transferred (eq. 2.57), which is always 4, in contrast to the associative mechanism (4.38), where the parallel peroxide formation often results in $z < 4$. Which mechanism operates on a given catalyst, which step if any is rate-limiting, and how the bond between adsorbates and catalyst is, are matters of ongoing research:

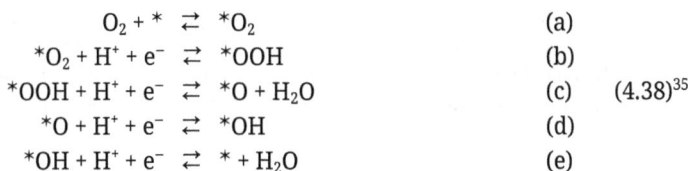

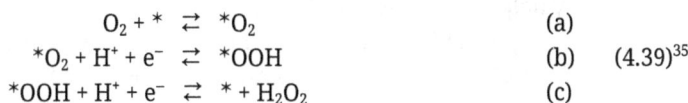

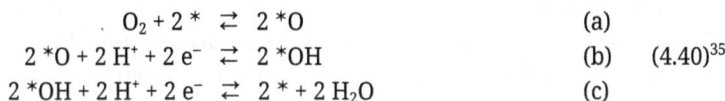

$$O_2 + * \rightleftarrows *O_2 \qquad \text{(a)}$$
$$*O_2 + H^+ + e^- \rightleftarrows *OOH \qquad \text{(b)}$$
$$*OOH + H^+ + e^- \rightleftarrows *O + H_2O \qquad \text{(c)} \qquad (4.38)^{35}$$
$$*O + H^+ + e^- \rightleftarrows *OH \qquad \text{(d)}$$
$$*OH + H^+ + e^- \rightleftarrows * + H_2O \qquad \text{(e)}$$

$$O_2 + * \rightleftarrows *O_2 \qquad \text{(a)}$$
$$*O_2 + H^+ + e^- \rightleftarrows *OOH \qquad \text{(b)} \qquad (4.39)^{35}$$
$$*OOH + H^+ + e^- \rightleftarrows * + H_2O_2 \qquad \text{(c)}$$

$$O_2 + 2* \rightleftarrows 2*O \qquad \text{(a)}$$
$$2*O + 2H^+ + 2e^- \rightleftarrows 2*OH \qquad \text{(b)} \qquad (4.40)^{35}$$
$$2*OH + 2H^+ + 2e^- \rightleftarrows 2* + 2H_2O \qquad \text{(c)}$$

Beyond the discussion of Tafel slopes, information on reaction mechanisms may also arise from product analyses, from spectroscopy with working electrodes, and from theoretical studies. The relevance of peroxo species for the ORR has been first concluded from the observation of H_2O_2 on some catalysts, where it may be even the major product. More recently, the differentiation of O isotopes by DEMS (Differential electrochemical mass spectrometry) has been applied, for instance, to find out if O_2 released in the OER contains oxygen from the electrode surface (see later). Spectroscopy with working electrodes is easier with techniques addressing the catalytic material than the adsorbates. Detection of surface species has been achieved by ATR-IR (e.g., peroxide and

---

35 Analogous mechanisms may be written also for basic media.

superoxide during the ORR on Ge(100) [211]), with APPES, and with NEXAFS [212] (for the techniques, see Sections 3.4.8.3, 3.4.5.2, and 3.4.4.1).

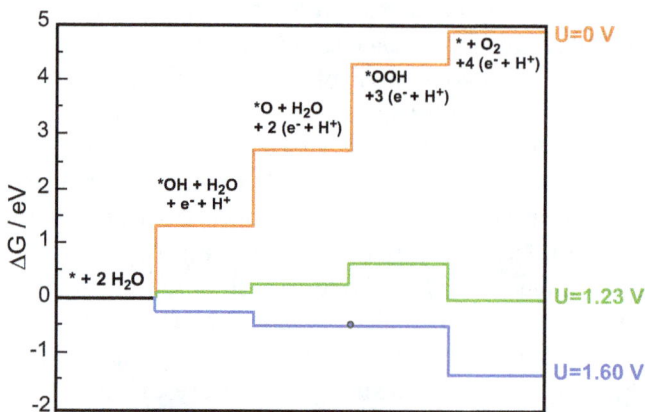

Figure 4.58: Thermochemical analysis of an electrocatalytic reaction mechanism. The free energies of the intermediates on RuO_2 (110) at three different potentials ($U = 0$, $U = 1.23$ V, and $U = 1.60$ V) are depicted. At the equilibrium potential ($U = 1.23$ V), most reaction steps are still uphill in free energy. At 1.60 V, there are no more uphill steps. Adapted from ref. [213] with permission from Elsevier.

Supplying estimates for binding energies of adsorbates and kinetic barriers between them, theory helps sorting out most likely reaction paths and specifying the bottleneck among them. Figure 4.58 illustrates a method to achieve this on a merely thermochemical basis. For a RuO_2(110) facet, the diagram shows how the free Gibbs energies of intermediates along an OER mechanism equivalent to the reverse of (4.38) respond to the potentials [213]. The reaction can proceed only if none of its steps is uphill. As the equilibrium potential of 1.23 V is not sufficient to achieve this, overpotential is required. Its magnitude is adequate when ΔG of the last uphill step (in Figure 4.58, the formation of *OOH) becomes zero. The corresponding step is referred to as potential-determining step (pds); it is most likely, but not necessarily also the rds. An agreement between predicted and observed overpotential supports such assignment. The method does not make use of kinetic barriers.

In an alternative approach, kinetic barriers between intermediates and their dependence on changing potentials were taken into account [208]. Studying the ORR on Pt(111), the authors allowed for various mechanisms ((4.38),(4.40), (4.39) to H_2O_2, which decomposes to H_2O). Due to different influences of the potential on rate constants, the mechanisms were found to compete. Contributions of routes to the observed rates and even the nature of the rds depended on the potential.

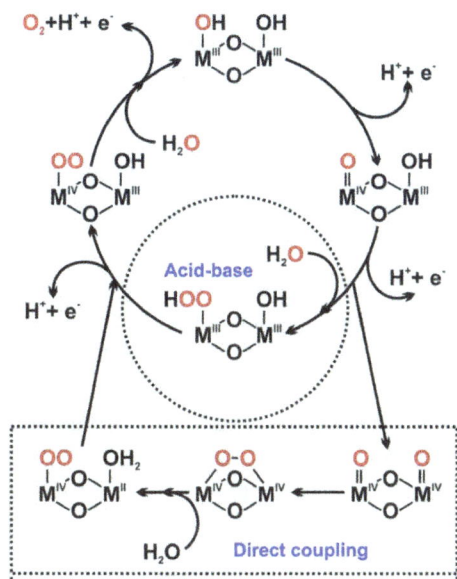

Figure 4.59: Mechanistic proposals for the OER with and without participation of lattice oxygen. Reproduced from ref. [214] with permission from John Wiley & Sons.

The OER requires potentials, which cause the oxidation of most noble metal surfaces, e.g., those of Ru or Ir. To find out if this surface oxygen participates in the OER, the reaction is run with $H_2^{16}O$ using anodes prepared with ^{18}O,[36] and the isotopic composition of the released O_2 is studied by DEMS [214]. There are examples for the initial evolution of either $^{18}O_2$ or $^{16}O_2$. The mechanisms proposed to explain these observations are presented in Figure 4.59. The acid–base mechanism is analogous to (4.38) and requires only one of the two metal atoms depicted. It produces O_2 from H_2O exclusively. In the direct coupling mechanism, the O–O bond is formed from two adjacent oxyl groups, which originate from surface OH groups, i.e., a detection of initial $^{18}O_2$ formation identifies this mechanism. It remains, however, to be elucidated why a catalyst operates along one of these mechanisms. RuO_2, for instance, was found to operate the OER via the acid–base mechanism when prepared thermally, but via the direct coupling mechanism when prepared electrochemically [214].

For the OER over IrO_2, relations between electrode charging and bond rearrangement were investigated with a combination of pulse voltammetry and *in situ* XANES at the Ir L_2/L_3 and O K edges with DFT modeling of the mechanism (identified as (4.38)) [215]. While the influence of capacitive charging on the reaction was negligible, the OER rates correlated with hole concentrations, mainly in the Ir L levels, created by the changing bias (pseudocapacitive charging). Above a critical potential, the Tafel slope of the OER increased. This was related to the observation that the hole concentration became less dependent on the potential in this range, i.e., not to a change of the rds. Activation

---

36 By oxidizing the metal surface with $^{18}O_2$ or by electrochemical oxidation in $H_2^{18}O$.

energies and reaction enthalpies remained linearly correlated at different hole coverages, as predicted by the BEP relation (3.156). Apparently, the influence of double-layer effects set up by the capacitive charge is negligible in this reaction.

The complexity of the OER, the tools available for getting insight in the surface processes involved, and the level of understanding achieved so far have been illustrated in an short review, recently [216]. The effort has been compared with climbing a ladder of complexity. Its two lowest rungs are related to thermodynamic aspects. On the third one, the thermodynamic effects of candidate steps are evaluated with models that assume low surface coverages and neglect the impact of the potential on surface properties (cf. Figure 4.58). On the fourth rung, the influence of the solvent on surface and intermediates is added; on the fifth rung the influence of the potential on surface conditions (adsorbate coverages, oxidation states, e.g., pseudocapacitive charging, see earlier). Only on the sixth rung, the actual kinetic barriers come into play (cf. ref. [208]), as also the relation between bond reorganization and electron transfer. Still, influences of anions and even of cations on the OER, which are known from experiment, are not included on this rung, i.e., the ladder is not yet long enough.

4.9.2 Recent topics of electrocatalysis

The broad application of electrochemical devices like low-temperature fuel cells or electrolyzers is strongly penalized by the price of the catalytic materials best suited for the most important reactions: Pt for HER, HOR, ORR, and methanol oxidation (MOR), IrO_2 and RuO_2 for OER. Although competitive and even improved performance in the ORR was often achieved when Pt was alloyed with cheaper metals, e.g., Cu or Ni, it still remained a major catalyst component. Alternatives based on new material classes may become competitive with the traditional catalysts, e.g., heteroatom-doped carbon nanomaterials for the ORR (see later). The recent attention to high-entropy alloys (HEA) bears promise for creating new catalysts for all reactions mentioned earlier.

According to an early definition, HEAs are alloys of five or more principal elements, with the concentration of each element being between 35 and 5 at% [217]. Their particular stability arises, to a large extent, from high mixing entropy contributions. This suggests that synergistic effects expected from the cooperation of various different metal atoms may also be stable under the severe conditions of electrochemistry. Due to the multiplicity of coexisting site configurations, adsorption energies of reactants exhibit a near-continuum distribution, instead of one or a few discrete values. The peak of the distribution can be shifted along the energy scale via the alloy composition [218]. Optimization of a catalyst for a simple reaction aims in creating surfaces, where the energy distribution of key reactant adsorption peaks at the optimum energy obtained from an analysis of the reaction according to the Sabatier principle (cf. Section 4.2.1).

While this suggests a paramount importance of theory for identifying promising compositions, the task is very much complicated by the great number of candidate elements and configurations (cf. ref. [219] for a review of this field). It is all the more hardly possible to get an overview over structures and resulting performances, using conventional methodology. Therefore, HEAs in general, and in particular, for the use as electrocatalysts are a field with a specific need for support by techniques of high-throughput experimentation (cf. Section 5).

The variety of site configurations on HEA surfaces, which allows mimicking even Pt with non-noble metals [218], may also raise challenges. It favors reactions requiring multifunctionality or proceeding via cascade mechanisms, but it may, at the same time, offer active sites for undesired side reactions. Electrochemical reduction of CO or CO_2, for instance, typically competes with the HER from electrolyte water. Moreover, due to the distribution of adsorption energies, only a fraction of the available sites catalyzes the reaction with high rates, unlike in some traditional materials, where this fraction can be high when the adsorption energy marks the top of the Sabatier volcano curve. There are also challenges in the synthesis of HEAs, which may be looked up in refs. [218, 220].

Despite these problems, promising new systems have already been identified, e.g., for hydrogen evolution during water electrolysis, where research targets three types of HER catalysts: Pt-based, Pd-based, and nonprecious metal-based ones. Unlike pure metals, HEAs are not easily deposited on supports as nanoparticles, which is no disadvantage for compositions without noble metals. In HEA electrodes, electron transport is even facilitated by the absence of support and binders and of their interfaces with metal components (catalyst particles, connections). This may be exemplified by a recent study on the HEA CuAlNiMoFe, which was prepared with an excess of Al ($\approx$82 at%) to allow creating mesoporosity by subsequent leaching, and of Cu ($\approx$12 at%), which remained as a conductive skeleton [221]. In the HER in basic or neutral medium, this HEA outperformed all nonprecious metal catalysts known so far. Although its specific activity (i.e., mA/cm^2 ECSA, cf. eq. (1.10)) remained well below that of Pt, the HEA-based cathode achieved reference current densities even at lower overpotentials than a commercial Pt/C catalyst, which was attributed to its superior internal conductivity. Based on DFT work, the authors assigned the HER activity to Ni-Mo or Fe-Mo pairs, where Mo atoms interact with OH species, while H atoms bind to Ni or Fe atoms before they recombine.

As a process of converting a greenhouse gas to useful products, CO_2 reduction (CO_2RR) receives much attention, despite its moderate energy efficiency (30–40% depending on the product [222]). Which of these products (CO, formic acid/formate, CH_4, CH_3OH, or ethene) dominates, depends strongly on the catalyst. As a rule of thumb, formate is the main product over Bi, Cd, Hg, In, Pb, Sn, and Tl, whereas Ag, Au, Ga, Pd, and Zn favor CO formation. Tendencies are also influenced by electrolytes and concentrations (see later). Due to the competition between CO_2RR and HER, metals efficient for HER are less suited for CO_2RR. Copper has attracted much attention, because of its variable product spectrum comprising CH_4, C_2H_4, and various alcohols [222]. It has been pointed out that many observations with respect to product distributions in CO_2RR can be explained by considering the binding energies of four noncoupled intermediates: H*, COOH*, CO*, and CH_3O^* [223].

$$CO_2 + H_2O + 2\,e^- \rightarrow CO + 2\,OH^- \tag{4.41}$$

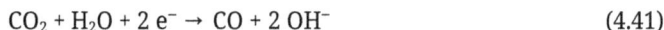

Among the quantities determining the nature of reduction products over Cu, the potential is most important. In a KCl electrolyte, CO and formate were reported to be released at less negative potentials (>−1.50 V vs. Ag/AgCl) than ethene and other C–C coupled products, which were formed between −1.65 and −1.70 V. The fully hydrogenated methane appeared at −1.70 V, before HER became predominant [222]. CO_2 reduction causes the pH value near the electrode to increase (cf. eq. (4.41)), which tends to suppress the competing HER. To benefit from this, anions with buffer capacity like HPO_4^{2-} should be avoided. Keeping the anion constant, pronounced changes in selectivity were observed also when the (alkali) cations were varied. The effects were related to ionic radii, the softness of the electron shell, and to hydration properties [222].

In an XAFS-based study supported by DFT work, roles of Cu species in the CO_2RR were explored [224]. Cu^{2+} is inactive, impeding charge transport either as oxide or as carbonate. Pristine Cu is not able to split the C–O bonds either, but becomes active when it contains subsurface O atoms. Cu_2O can also split CO_2, but only at oxygen vacancies. During a cathodic scan, the reaction starts after Cu^{2+} is reduced and vacancies appear in Cu_2O, where CO is formed from CO_2. When the HER potential is reached, Cu_2O is reduced to Cu^0 containing subsurface oxygen. As the interface is rich in protons under these conditions, the second C–O bond is also split, and C_2H_4, and at higher potential, even CH_4 are formed.

Improved behavior of Cu in CO_2RR activity has been achieved also by nanoengineering its morphology with respect to surface roughness, defect structures, or shapes (nanowires, hollow fibers, etc., summarized in ref. [224]). The impact of the reaction conditions on the Cu surface is another important aspect of CO_2RR catalysis, which has been elucidated in ref. [225]. For more general review on CO_2RR over Cu catalysts, see ref. [226].

Electrocatalytic CO_2 reduction can also produce H_2/CO mixtures, i.e., syngas. Au and Ag are best suited for this purpose, because they do not catalyze hydrogenation and C–C coupling, even at potentials driving the HER. Ag is preferred because of its price. As the catalyst operates under mass transport limitations because of limited CO_2 solubilities in electrolytes, the CO/H_2 ratio obtained depends on the rate of CO_2 supply. Working with a commercially available gas diffusion electrode[37] and with gaseous, instead of dissolved, CO_2 (Figure 4.60), 1,200 h of continuous operation with a current density customary in

---

37 The type is employed also as oxygen depolarization cathode in modern chloralkali technology, cf. Appendix A1.

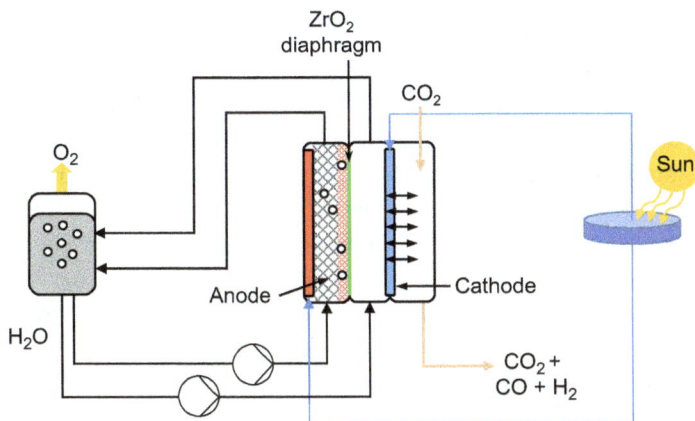

Figure 4.60: Production of syngas by electrocatalytic reduction of CO_2. Adapted from ref. [227] with permission from Springer Nature.

technical electrolyzers (300 mA/cm^2) were demonstrated, which brings syngas production from water, CO_2, and green energy closer to an industrial perspective [227].

As mentioned earlier, the multielectron transfer processes of oxygen electrocatalysis are bottlenecks in devices processing oxygen, such as fuel cells, electrolyzers, metal-air batteries. Even with the best catalysts available to date, overpotentials are ≈0.4 V at current densities of 1.5 A/cm^2(geo). Alternative catalysts based on nonprecious transition metals can be usually applied only in basic solutions, where they achieve current densities of 0.1 A/cm^2(geo) at overpotentials of ≈0.4 V [228].

In early experimental work, Ru was found on the top of a volcano plot of activities vesus enthalpies of transition from a lower to a higher oxidation state, with Ir and Pt quite nearby [228]. Ru and Ir oxides (further abbreviated as RuO_2 and IrO_2) are the best OER catalysts known to date, which can also operate in acidic solutions. They have been prepared via various routes like thermal and electrochemical oxidation or sputtering, from various precursors, and with different thermal conditioning steps. Leaching has always remained a problem: dissolution interferes even with studies on single crystal facets, which has complicated basic research and the identification of structure–activity relations in this field [214]. Not only OER activities of RuO_2 and IrO_2 electrodes depend strongly on the preparation mode, but also their stability, which is often inversely correlated to activity. For more detail, the reader may consult ref. [214].

In alkaline medium, promising OER performance can be achieved also by nonprecious transition metal oxides, e.g., electrodeposited mixed oxides like $NiFeO_x$ or $CoFeO_x$ [229], perovskites [228] or layered double hydroxides [230]. Theory-based search for optimum surface compositions has been impeded by the complexity of surface processes, so far (see Section 4.9.1). Mechanisms may differ between catalytic elements involved, which complicates the identification of descriptors covering wide ranges of catalysts [214, 228]. How to cope with this problem has been discussed for perovskite-based OER catalysts in ref. [228].

As mentioned earlier, heteroatom-doped carbon nanomaterials are being extensively studied as alternative ORR catalysts. B, N, P, S, and/or F are heteroatoms, which have been introduced into hosts like CNTs, graphene, graphite, and even fullerenes [231]. In another recently discovered class of materials, 3d metals like Fe or Co are dispersed on N-modified carbon nanomaterials [232]. Their activity has been, meanwhile, attributed to the heteroatom-doped graphene structures. The catalyst type bears the promise to provide materials that are resistant to corrosion, friendly to the environment, and stable against typical poisons for metal catalysts. High activity and stability have been shown for them in basic and neutral media, and for some systems, even at the technologically most important low pH values, where, however, Pt-based catalysts have, so far, remained superior [231].

Promising performance was reported, for instance, for N-CNTs, where pyridinic nitrogen was identified as the active N site by a correlation of its abundance measured by XPS with the ORR activity [233]. The activity of heteroatom-doped graphene systems has been ascribed to the perturbation of the aromatic system resulting in modified charge distributions and to missing bonds at defect sites [231].

References

[1]   Pecoraro, T. A.; Chianelli, R. R., Hydrodesulfurization catalysis by transition metal sulfides. *Journal of Catalysis* **1981**, *67* (2), 430–445. doi.org/10.1016/0021-9517(81)90303-1

[2]   Jacobsen, C. J. H.; Dahl, S.; Clausen, B. S.; Bahn, S.; Logadottir, A.; Nørskov, J. K., Catalyst design by interpolation in the periodic table: Bimetallic ammonia synthesis catalysts. *Journal of the American Chemical Society* **2001**, *123* (34), 8404–8405. doi 10.1021/ja010963d

[3]   Moses, P. G.; Grabow, L. C.; Fernandez, E. M.; Hinnemann, B.; Topsøe, H.; Knudsen, K. G.; Nørskov, J. K., Trends in hydrodesulfurization catalysis based on realistic surface models. *Catalysis Letters* **2014**, *144* (8), 1425–1432. doi 10.1007/s10562-014-1279-4

[4]   Medford, A. J.; Vojvodic, A.; Hummelshøj, J. S.; Voss, J.; Abild-Pedersen, F.; Studt, F.; Bligaard, T.; Nilsson, A.; Nørskov, J. K., From the Sabatier principle to a predictive theory of transition-metal heterogeneous catalysis. *Journal of Catalysis* **2015**, *328*, 36–42. doi.org/10.1016/j.jcat.2014.12.033

[5]   Rodriguez, J. A.; Goodman, D. W., Surface science studies of the electronic and chemical properties of bimetallic systems. *The Journal of Physical Chemistry* **1991**, *95* (11), 4196–4206. doi 10.1021/j100164a008

[6]   Petrova, O.; Kulp, C.; Pohl, M.-M.; ter Veen, R.; Veith, L.; Grehl, T.; van den Berg, M. W. E.; Brongersma, H.; Bron, M.; Grünert, W., Chemical leaching of Pt–Cu/C catalysts for electrochemical oxygen reduction: Activity, particle structure, and relation to electrochemical leaching. *ChemElectroChem* **2016**, *3* (11), 1768–1780. doi 10.1002/celc.201600468

[7]   Hammer, B.; Nørskov, J. K., Theoretical surface science and catalysis – calculations and concepts. *Advances in Catalysis.* **2000**, *45*, 71–129. doi.org/10.1016/S0360-0564(02)45013-4

[8]   Chorkendorff, I.; Niemantsverdriet, J. W., *Concepts of modern catalysis and kinetics*. 2 ed.; VCH: Weinheim, 2006. ISBN: 9783527605644

[9]   Hammer, B.; Nielsen, O. H.; Nørskov, J. K., Structure sensitivity in adsorption: CO interaction with stepped and reconstructed Pt surfaces. *Catalysis Letters* **1997**, *46* (1), 31–35. doi 10.1023/A:1019073208575

[10]   Brodén, G.; Rhodin, T. N.; Brucker, C.; Benbow, R.; Hurych, Z., Synchrotron radiation study of chemisorptive bonding of CO on transition metals – polarization effect on Ir(100). *Surface Science* **1976**, *59* (2), 593–611. doi.org/10.1016/0039-6028(76)90038-8

[11]   Vojvodic, A.; Nørskov, J. K.; Abild-Pedersen, F., Electronic structure effects in transition metal surface chemistry. *Topics in Catalysis* **2014**, *57* (1), 25–32. doi 10.1007/s11244-013-0159-2

[12]   Somorjai, G. A.; Blakely, D. W., Mechanism of catalysis of hydrocarbon reactions by platinum surfaces. *Nature* **1975**, *258*, 580–583. doi: 10.1038/258580a0

[13]   Van Santen, R. A., Complementary structure sensitive and insensitive catalytic relationships. *Accounts of Chemical Research* **2009**, *42* (1), 57–66. doi 10.1021/ar800022m

[14]   Mortensen, P. M.; Dybkjær, I., Industrial scale experience on steam reforming of CO_2-rich gas. *Applied Catalysis A: General* **2015**, *495*, 141–151. doi.org/10.1016/j.apcata.2015.02.022

[15]   Somorjai, G. A.; Materer, N., Surface structures in ammonia synthesis. *Topics in Catalysis* **1994**, *1* (3), 215–231. doi 10.1007/BF01492277

[16]   Honkala, K.; Hellman, A.; Remediakis, I. N.; Logadottir, A.; Carlsson, A.; Dahl, S.; Christensen, C. H.; Nørskov, J. K., Ammonia synthesis from first-principles calculations. *Science* **2005**, *307* (5709), 555–558. doi 10.1126/science.1106435

[17]   Tanaka, K.-I.; Okuhara, T., Regulation of intermediates on sulfided nickel and MoS_2 catalysts. *Catalysis Reviews - Science and Engineering* **1977**, *15*, 249–292. doi: 10.1080/03602457708081726

[18]   Nørskov, J. K.; Clausen, B. S.; Topsøe, H., Understanding the trends in the hydrodesulfurization activity of the transition metal sulfides. *Catalysis Letters* **1992**, *13* (1), 1–8. doi 10.1007/BF00770941

[19]   Hinnemann, B.; Nørskov, J. K.; Topsøe, H., A Density Functional Study of the chemical differences between type I and type II MoS_2-based structures in hydrotreating catalysts. *The Journal of Physical Chemistry B* **2005**, *109* (6), 2245–2253. doi 10.1021/jp048842y

[20]   Vogelaar, B. M.; Kagami, N.; van der Zijden, T. F.; van Langeveld, A. D.; Eijsbouts, S.; Moulijn, J. A., Relation between sulfur coordination of active sites and HDS activity for Mo and NiMo catalysts. *Journal of Molecular Catalysis A: Chemical* **2009**, *309* (1–2), 79–88. doi 10.1016/j.molcata.2009.04.018

[21]   Drescher, T.; Niefindt, F.; Bensch, W.; Grünert, W., Sulfide catalysis without coordinatively unsaturated sites: Hydrogenation, cis–trans isomerization, and H_2/D_2 scrambling over MoS_2 and WS_2. *Journal of the American Chemical Society* **2012**, *134*, 18896–18899. doi 10.1021/ja3074903

[22]   Daage, M.; Chianelli, R. R., Structure-function relationships in molybdenum sulfide catalysts – the rim-edge model. *Journal of Catalysis* **1994**, *149* (2), 414–427. doi 10.1006/jcat.1994.1308

[23]   Lauritsen, J. V.; Besenbacher, F., Model catalyst surfaces investigated by scanning tunneling microscopy. In *Advances in Catalysis* **2006**, 50, 97–147. doi.org/10.1016/S0360-0564(06)50003-3

[24]   Jaramillo, T. F.; Jørgensen, K. P.; Bonde, J.; Nielsen, J. H.; Horch, S.; Chorkendorff, I., Identification of active edge sites for electrochemical H_2 evolution from MoS_2 nanocatalysts. *Science* **2007**, *317* (5834), 100. doi 10.1126/science.1141483

[25]   Védrine, J. C.; Hutchings, G. J.; Kiely, C. J., Molybdenum oxide model catalysts and vanadium phosphates as actual catalysts for understanding heterogeneous catalytic partial oxidation reactions: A contribution by Jean-Claude Volta. *Catalysis Today* **2013**, *217*, 57–64. doi.org/10.1016/j.cattod.2013.01.004

[26]   Volta, J. C.; Portefaix, J. L., Structure sensitivity of mild oxidation reactions on oxide catalysts – a review. *Applied Catalysis* **1985**, *18* (1), 1–32. doi.org/10.1016/S0166-9834(00)80296-1

[27]   Grasselli, R. K.; Burrington, J. D., Molecular probes for the mechanism of selective oxidation and ammoxidation catalysis. *Industrial & Engineering Chemistry Product Research and Development* **1984**, *23* (3), 393–404. doi 10.1021/i300015a014

[28]   Merzlikin, S. V.; Tolkachev, N. N.; Briand, L. E.; Strunskus, T.; Wöll, C.; Wachs, I. E.; Grünert, W., Anomalous surface compositions of stoichiometric mixed oxide compounds. *Angewandte Chemie International Edition* **2010**, *49* (43), 8037–8041. doi 10.1002/anie.201001804

[29]   Getsoian, A. B.; Shapovalov, V.; Bell, A. T., DFT+U investigation of propene oxidation over bismuth molybdate: Active sites, reaction intermediates, and the role of bismuth. *The Journal of Physical Chemistry C* **2013**, *117* (14), 7123–7137. doi 10.1021/jp400440p

[30]   Voskoboinikov, T. V.; Chen, H.-Y.; Sachtler, W. M. H., On the nature of active sites in Fe/ZSM-5 catalysts for NOx abatement. *Applied Catalysis B: Environmental* **1998**, *19* (3–4), 279–287. doi 10.1016/S0926-3373(98)00082-4

[31]   Schwidder, M.; Santhosh Kumar, M.; Klementiev, K. V.; Pohl, M. M.; Brückner, A.; Grünert, W., Selective reduction of NO with Fe-ZSM-5 catalysts of low Fe content I. Relations between active site structure and catalytic performance. *Journal of Catalysis.* **2005**, *231* (2), 314–330. doi 10.1016/j.jcat.2005.01.031

[32]   Brandenberger, S.; Kröcher, O., Estimation of the fractions of different nuclear iron species in uniformly metal-exchanged Fe-ZSM-5 samples based on a Poisson distribution. *Applied Catalysis B* **2010**, *373* (1–2), 168–175. doi 10.1016/j.apcata.2009.11.012

[33]   Grünert, W.; Ganesha, P. K.; Ellmers, I.; Velez, R. P.; Huang, H. M.; Bentrup, U.; Schünemann, V.; Brückner, A., Active sites of the selective catalytic reduction of NO by NH$_3$ over Fe-ZSM-5: Combining reaction kinetics with postcatalytic Mössbauer spectroscopy at cryogenic temperatures. *ACS Catalysis* **2020**, *10* (5), 3119–3130. doi 10.1021/acscatal.9b04627

[34]   Wachs, I. E.; Deo, G.; Weckhuysen, B. M.; Andreini, A.; Vuurman, M. A.; de Boer, M.; Amiridis, M. D., Selective catalytic reduction of NO with NH$_3$ over supported vanadia catalysts. *Journal of Catalysis* **1996**, *161* (1), 211–221. doi 10.1006/jcat.1996.0179

[35]   Paolucci, C.; Khurana, I.; Parekh, A. A.; Li, S. C.; Shih, A. J.; Li, H.; Di Iorio, J. R.; Albarracin-Caballero, J. D.; Yezerets, A.; Miller, J. T.; Delgass, W. N.; Ribeiro, F. H.; Schneider, W. F.; Gounder, R., Dynamic multinuclear sites formed by mobilized copper ions in NOx selective catalytic reduction. *Science* **2017**, *357* (6354), 898–903. doi 10.1126/science.aan5630

[36]   Gordon, C. P.; Engler, H.; Tragl, A. S.; Plodinec, M.; Lunkenbein, T.; Berkessel, A.; Teles, J. H.; Parvulescu, A.-N.; Copéret, C., Efficient epoxidation over dinuclear sites in titanium silicalite-1. *Nature* **2020**, *586* (7831), 708–713. doi 10.1038/s41586-020-2826-3

[37]   Weckhuysen, B. M., Titanium atoms pair up in industrial catalyst. *Nature* **2020**, *586* (7831), 678–679. doi 10.1038/s41586-020-2826-3

[38]   Kattel, S.; Ramirez, P. J.; Chen, J. G.; Rodriguez, J. A.; Liu, P., Active sites for CO$_2$ hydrogenation to methanol on Cu/ZnO catalysts. *Science* **2017**, *355* (6331), 1296–1299. doi 10.1126/science.aal3573.

[39]   Grossmann, D.; Klementiev, K.; Sinev, I.; Grünert, W., Surface alloy or metal-cation interaction-the state of Zn promoting the active Cu sites in methanol synthesis catalysts. *Chemcatchem* **2017**, *9* (2), 365–372. doi 10.1002/cctc.201601102

[40]   Kuld, S.; Thorhauge, M.; Falsig, H.; Elkjaer, C. F.; Helveg, S.; Chorkendorff, I.; Sehested, J., Quantifying the promotion of Cu catalysts by ZnO for methanol synthesis. *Science* **2016**, *352* (6288), 969–974. doi 10.1126/science.aaf0718

[41]   Laudenschleger, D.; Ruland, H.; Muhler, M., Identifying the nature of the active sites in methanol synthesis over Cu/ZnO/Al$_2$O$_3$ catalysts. *Nature Communications* **2020**, *11* (1), 3898. doi 10.1038/s41467-020-17631-5

[42]   Lauritsen, J. V.; Kibsgaard, J.; Olesen, G. H.; Moses, P. G.; Hinnemann, B.; Helveg, S.; Nørskov, J. K.; Clausen, B. S.; Topsøe, H.; Laegsgaard, E.; Besenbacher, F., Location and coordination of promoter atoms in Co- and Ni-promoted MoS$_2$-based hydrotreating catalysts. *Journal of Catalysis* **2007**, *249* (2), 220–233. doi 10.1016/j.jcat.2007.04.013

[43]   Moses, P. G.; Hinnemann, B.; Topsøe, H.; Nørskov, J. K., The effect of Co-promotion on MoS$_2$ catalysts for hydrodesulfurization of thiophene: A density functional study. *Journal of Catalysis* **2009**, *268* (2), 201–208. doi.org/10.1016/j.jcat.2009.09.016

[44]   Guerrero-Ruiz, A.; Rodríguez-Ramos, I., Oxydehydrogenation of ethylbenzene to styrene catalyzed by graphites and activated carbons. *Carbon* **1994**, *32* (1), 23–29. doi.org/10.1016/0008-6223(94)90005-1

[45]   Dahl, I. M.; Kolboe, S., On the reaction mechanism for hydrocarbon formation from methanol over SAPO-34: 2. isotopic labeling studies of the co-reaction of propene and methanol. *Journal of Catalysis* **1996**, *161* (1), 304–309. doi.org/10.1016/0008-6223(94)90005-1

[46]   Olsbye, U.; Svelle, S.; Bjørgen, M.; Beato, P.; Janssens, T. V. W.; Joensen, F.; Bordiga, S.; Lillerud, K. P., Conversion of methanol to hydrocarbons: How zeolite cavity and pore size controls product selectivity. **2012**, *51* (24), 5810–5831. doi.org/10.1002/anie.201103657

[47]   Dahl, S.; Logadottir, A.; Jacobsen, C. J. H.; Nørskov, J. K., Electronic factors in catalysis: The volcano curve and the effect of promotion in catalytic ammonia synthesis. *Applied Catalysis A: General* **2001**, *222* (1), 19–29. doi.org/10.1016/S0926-860X(01)00826-2

[48]   Schulz, H.; Riedel, T.; Schaub, G., Fischer–Tropsch principles of co-hydrogenation on iron catalysts. *Topics in Catalysis* **2005**, *32* (3), 117–124. doi 10.1007/s11244-005-2883-8

[49]   Graf, B.; Muhler, M., The influence of the potassium promoter on the kinetics and thermodynamics of CO adsorption on a bulk iron catalyst applied in Fischer–Tropsch synthesis: a quantitative adsorption calorimetry, temperature-programmed desorption, and surface hydrogenation study. *Physical Chemistry Chemical Physics* **2011**, *13* (9), 3701–3710. doi 10.1039/C0CP01875A

[50]   Jacobs, G.; Das, T. K.; Zhang, Y. Q.; Li, J. L.; Racoillet, G.; Davis, B. H., Fischer-Tropsch synthesis: Support, loading, and promoter effects on the reducibility of cobalt catalysts. *Applied Catalysis A: General* **2002**, *233* (1–2), 263–281. doi 10.1016/s0926-860x(02)00195-3

[51]   Johnson, G. R.; Bell, A. T., Effects of Lewis acidity of metal oxide promoters on the activity and selectivity of Co-based Fischer–Tropsch synthesis catalysts. *Journal of Catalysis* **2016**, *338*, 250–264. doi.org/10.1016/j.jcat.2016.03.022

[52]   Schlögl, R., Preparation and activation of the technical ammonia synthesis catalyst. In *Catalytic Ammonia Synthesis. Fundamentals and Practice*, Jennings, J. R., Ed. Springer: 1991; pp 19–108. ISBN: 978-0-306-43628-4

[53]   Strongin, D. R.; Somorjai, G. A., A surface science and catalytic study on the effects of aluminium oxide and potassium on the ammonia synthesis over iron single-crystal surfaces. In *Catalytic Ammonia Synthesis. Fundamentals and Practice*, Jennings, J. R., Ed. Plenum: New York and London, 1991; pp 133–178. ISBN: 978-0-306-43628-4

[54]   Kurtz, M.; Bauer, N.; Büscher, C.; Wilmer, H.; Hinrichsen, O.; Becker, R.; Rabe, S.; Merz, K.; Driess, M.; Fischer, R. A.; Muhler, M., New synthetic routes to more active Cu/ZnO catalysts used for methanol synthesis. *Catalysis Letters* **2004**, *92* (1–2), 49–52. doi 10.1023/B:CATL.0000011085.88267.a6

[55]   Grunwaldt, J.-D.; Molenbroek, A. M.; Topsøe, N.-Y.; Topsøe, H.; Clausen, B. S., *In-situ* investigation of structural changes in Cu/ZnO catalysts. *J. Catal.* **2000**, *194* (2), 452–460. doi 10.1006/jcat.2000.2930

[56]   Lunkenbein, T.; Schumann, J.; Behrens, M.; Schlögl, R.; Willinger, M. G., Formation of a ZnO overlayer in industrial Cu/ZnO/Al₂O₃ catalysts induced by strong metal-support interactions. *Angewandte Chemie-International Edition* **2015**, *54* (15), 4544–4548. doi 10.1002/anie.201411581

[57]   Muhler, M.; Törnqvist, E.; Nielsen, L. P.; Clausen, B. S.; Topsøe, H., On the role of adsorbed atomic oxygen and CO₂ in copper based methanol synthesis catalysts. *Catalysis Letters* **1994**, *25* (1), 1–10. doi 10.1007/BF00815409

[58]   Behrens, M.; Zander, S.; Kurr, P.; Jacobsen, N.; Senker, J.; Koch, G.; Ressler, T.; Fischer, R. W.; Schlögl, R., Performance improvement of nanocatalysts by promoter-induced defects in the support material: Methanol synthesis over Cu/ZnO:Al. *Journal of the American Chemical Society* **2013**, *135* (16), 6061–6068. doi 10.1021/ja310456f

[59]   Chinchen, G. C.; Davies, P. S., R. J., The historical development of catalytic oxidation processes. In *Catalyis: Science and Technology*, Springer: Berlin, Heidelberg, 1987; pp 1–67. ISBN: 9783642932786

[60]   Zhang, Y.; Zhao, R.; Sanchez-Sanchez, M.; Haller, G. L.; Hu, J.; Bermejo-Deval, R.; Liu, Y.; Lercher, J. A., Promotion of protolytic pentane conversion on H-MFI zeolite by proximity of extra-framework aluminum oxide and Brønsted acid sites. *Journal of Catalysis* **2019**, *370*, 424–433. doi.org/10.1016/j.jcat.2019.01.006

[61] Ojeda, M.; Zhan, B.-Z.; Iglesia, E., Mechanistic interpretation of CO oxidation turnover rates on supported Au clusters. *Journal of Catalysis* **2012**, *285* (1), 92–102. doi.org/10.1016/j.jcat.2011.09.015

[62] Murzin, D. Y.; Mäki-Arvela, P.; Toukoniitty, E.; Salmi, T., Asymmetric heterogeneous catalysis: Science and engineering. *Catalysis Reviews* **2005**, *47* (2), 175–256. doi 10.1081/CR-200057461

[63] Liu, Y.; Xuan, W. M.; Cui, Y., Engineering homochiral metal-organic frameworks for heterogeneous asymmetric catalysis and enantioselective separation. *Advanced Materials* **2010**, *22* (37), 4112–4135. doi 10.1002/adma.201000197

[64] Harris, J. W.; Bates, J. S.; Bukowski, B. C.; Greeley, J.; Gounder, R., Opportunities in catalysis over metal-zeotypes enabled by descriptions of active centers beyond their binding site. *ACS Catalysis* **2020**, *10* (16), 9476–9495. doi 10.1021/acscatal.0c02102

[65] Shifa, T. A.; Vomiero, A., Confined catalysis: Progress and prospects in energy conversion. *Advanced Energy Materials* **2019**, *9* (40). Art. No. 1902307. doi 10.1002/aenm.201902307

[66] Albonetti, S.; Cavani, F.; Trifirò, F., Key aspects of catalyst design for the selective oxidation of paraffins. *Catalysis Reviews* **1996**, *38* (4), 413–438. doi.org/10.1006/jcat.1996.0123

[67] Okuhara, T.; Mizuno, N.; Misono, M., Catalysis by heteropoly compounds—recent developments. *Applied Catalysis A: General* **2001**, *222* (1), 63–77. doi.org/10.1016/S0926-860X(01)00830-4

[68] Khajavi, H.; Stil, H. A.; Kuipers, H.; Gascon, J.; Kapteijn, F., Shape and transition state selective hydrogenations using Egg-Shell Pt-MIL-101(Cr) Catalyst. *ACS Catalysis* **2013**, *3* (11), 2617–2626. doi 10.1021/cs400681s

[69] Zheng, S. R.; Jentys, A.; Lercher, J. A., Xylene isomerization with surface-modified HZSM-5 zeolite catalysts: An *in situ* IR study. *Journal of Catalysis* **2006**, *241* (2), 304–311. doi 10.1016/j.jcat.2006.04.026

[70] Clark, L. A.; Sierka, M.; Sauer, J., Computational elucidation of the transition state shape selectivity phenomenon. *Journal of the American Chemical Society* **2004**, *126* (3), 936–947. doi 10.1021/ja0381712

[71] Tada, M.; Iwasawa, Y., Advanced design of catalytically active reaction space at surfaces for selective catalysis. *Coordination Chemistry Reviews* **2007**, *251* (21), 2702–2716. doi.org/10.1016/j.ccr.2007.06.008

[72] Batiot, C.; Hodnett, B. K., The role of reactant and product bond energies in determining limitations to selective catalytic oxidations. *Applied Catalysis A: General* **1996**, *137* (1), 179–191. doi.org/10.1016/09 26-860X(95)00322-3

[73] Deshlahra, P.; Iglesia, E., Reactivity and selectivity descriptors for the activation of C–H bonds in hydrocarbons and oxygenates on metal oxides. *The Journal of Physical Chemistry C* **2016**, *120* (30), 16741–16760. doi 10.1021/acs.jpcc.6b04604

[74] Wachs, I. E., Recent conceptual advances in the catalysis science of mixed metal oxide catalytic materials. *Catalysis Today* **2005**, *100* (1–2), 79–94. 10.1016/j.cattod.2004.12.019

[75] Özbek, M. O.; van Santen, R. A., The mechanism of ethylene epoxidation catalysis. *Catalysis Letters* **2013**, *143* (2), 131–141. doi 10.1007/s10562-012-0957-3

[76] Grasselli, R. K., Fundamental principles of selective heterogeneous oxidation catalysis. *Topics in Catalysis* **2002**, *21* (1–2), 79–88. 10.1023/A:1020556131984

[77] Bodke, A. S.; Olschki, D. A.; Schmidt, L. D.; Ranzi, E., High selectivities to ethylene by partial oxidation of ethane. *Science* **1999**, *285* (5428), 712. doi 10.1126/science.285.5428.712

[78] Dragomirova, R.; Wohlrab, S., Zeolite Membranes in catalysis-from separate units to particle coatings. *Catalysts* **2015**, *5* (4), 2161–2222. doi 10.3390/catal5042161

[79] Lintz, H. G.; Reitzmann, A., Alternative reaction engineering concepts in partial oxidations on oxidic catalysts. *Catalysis Reviews* **2007**, *49* (1), 1–32. doi 10.1080/01614940600983467

[80] Moulijn, J. A.; van Diepen, A. E.; Kapteijn, F., Catalyst deactivation: Is it predictable?: What to do? *Applied Catalysis A: General* **2001**, *212* (1), 3–16. doi.org/10.1016/S0926-860X(00)00842-5

[81] Teschner, D.; Vass, E.; Hävecker, M.; Zafeiratos, S.; Schnörch, P.; Sauer, H.; Knop-Gericke, A.; Schlögl, R.; Chamam, M.; Wootsch, A.; Canning, A. S.; Gamman, J. J.; Jackson, S. D.; McGregor, J.; Gladden, L. F., Alkyne hydrogenation over Pd catalysts: A new paradigm. *Journal of Catalysis* **2006**, *242* (1), 26–37. doi 10.1016/j.jcat.2006.05.030

[82]   García-Mota, M.; Bridier, B.; Pérez-Ramírez, J.; López, N., Interplay between carbon monoxide, hydrides, and carbides in selective alkyne hydrogenation on palladium. *Journal of Catalysis* **2010**, *273* (2), 92–102. doi.org/10.1016/j.jcat.2010.04.018

[83]   García-Mota, M.; Gómez-Díaz, J.; Novell-Leruth, G.; Vargas-Fuentes, C.; Bellarosa, L.; Bridier, B.; Pérez-Ramírez, J.; López, N., A density functional theory study of the 'mythic' Lindlar hydrogenation catalyst. *Theoretical Chemistry Accounts* **2011**, *128* (4), 663–673. doi 10.1007/s00214-010-0800-0

[84]   Luneau, M.; Lim, J. S.; Patel, D. A.; Sykes, E. C. H.; Friend, C. M.; Sautet, P., Guidelines to achieving high selectivity for the hydrogenation of α,β-unsaturated aldehydes with bimetallic and dilute alloy catalysts: A review. *Chemical Reviews* **2020**, *120* (23), 12834–12872. doi 10.1021/acs.chemrev.0c00582

[85]   Lan, X.; Wang, T., Highly selective catalysts for the hydrogenation of unsaturated aldehydes: A review. *ACS Catalysis* **2020**, *10* (4), 2764–2790. doi 10.1021/acscatal.9b04331

[86]   Kahsar, K. R.; Schwartz, D. K.; Medlin, J. W., Control of metal catalyst selectivity through specific noncovalent molecular interactions. *Journal of the American Chemical Society* **2014**, *136* (1), 520–526. doi 10.1021/ja411973p

[87]   Chiu, M. E.; Watson, D. J.; Kyriakou, G.; Tikhov, M. S.; Lambert, R. M., Tilt the molecule and change the chemistry: Mechanism of S-promoted chemoselective catalytic hydrogenation of crotonaldehyde on Cu(111). *Angewandte Chemie-International Edition* **2006**, *45* (45), 7530–7534. doi 10.1002/anie.200603408

[88]   Boronat, M.; May, M.; Illas, F., Origin of chemoselective behavior of S-covered Cu(111) towards catalytic hydrogenation of unsaturated aldehydes. *Surface Science* **2008**, *602* (21), 3284–3290. doi 10.1016/j.susc.2008.08.023

[89]   Tsang, S. C.; Cailuo, N.; Oduro, W.; Kong, A. T. S.; Clifton, L.; Yu, K. M. K.; Thiebaut, B.; Cookson, J.; Bishop, P., Engineering preformed cobalt-doped platinum nanocatalysts for ultraselective hydrogenation. *ACS Nano* **2008**, *2* (12), 2547–2553. doi 10.1021/nn800400u

[90]   Argyle, M. D.; Bartholomew, C. H., Heterogeneous catalyst deactivation and regeneration: A review. *Catalysts* **2015**, *5* (1), 145–269. doi 10.3390/catal5010145

[91]   Sie, S. T., Consequences of catalyst deactivation for process design and operation. *Applied Catalysis A: General* **2001**, *212* (1), 129–151. doi.org/10.1016/S0926-860X(00)00851-6

[92]   Tsakoumis, N. E.; Rønning, M.; Borg, Ø.; Rytter, E.; Holmen, A., Deactivation of cobalt based Fischer–Tropsch catalysts: A review. *Catalysis Today* **2010**, *154* (3), 162–182. doi.org/10.1016/j.cattod.2010.02.077

[93]   Saib, A. M.; Moodley, D. J.; Ciobîcă, I. M.; Hauman, M. M.; Sigwebela, B. H.; Weststrate, C. J.; Niemantsverdriet, J. W.; van de Loosdrecht, J., Fundamental understanding of deactivation and regeneration of cobalt Fischer–Tropsch synthesis catalysts. *Catalysis Today* **2010**, *154* (3), 271–282. doi.org/10.1016/j.cattod.2010.02.008

[94]   Karge, H. G., Coke formation on zeolites. *Studies in Surface Science and Catalysis* **2001**, *137*, 707–746. doi.org/10.1016/S0167-2991(01)80258-5

[95]   Dong, W.; Chen, P.; Xia, W.; Weide, P.; Ruland, H.; Kostka, A.; Köhler, K.; Muhler, M., Palladium nanoparticles supported on nitrogen-doped carbon nanotubes as a release-and-catch catalytic system in aerobic liquid-phase ethanol oxidation. *ChemCatChem* **2016**, *8* (7), 1269–1273. doi.org/10.1016/j.cattod.2010.02.008

[96]   Gruttadauria, M.; Giacalone, F.; Noto, R., "Release and catch" catalytic systems. *Green Chemistry* **2013**, *15* (10), 2608–2618. doi 10.1039/C3GC41132J

[97]   Andersson, A.; Hernelind, M.; Augustsson, O., A study of the ageing and deactivation phenomena occurring during operation of an iron molybdate catalyst in formaldehyde production. *Catalysis Today* **2006**, *112* (1), 40–44. doi.org/10.1016/j.cattod.2005.11.052

[98]   Hansen, T. W.; DeLaRiva, A. T.; Challa, S. R.; Datye, A. K., Sintering of catalytic nanoparticles: Particle migration or ostwald ripening? *Accounts of Chemical Research* **2013**, *46* (8), 1720–1730. doi 10.1021/ar3002427

[99]   Kompio, P. G. W. A.; Brückner, A.; Hipler, F.; Manoylova, O.; Auer, G.; Mestl, G.; Grünert, W., V_2O_5-WO_3/TiO_2 catalysts under thermal stress: Responses of structure and catalytic behavior in the selective catalytic reduction of NO by NH_3. *Applied Catalysis. B* **2017**, *217*, 365–377. doi 10.1016/j.apcatb.2017.06.006

[100]   Elmasides, C.; Kontarides, D. I.; Grünert, W.; Verykios, X. E., XPS and FTIR study of Ru/Al_2O_3 and Ru/TiO_2 catalysts: Reduction characteristics and interaction with a methane-oxygen mixture. *Journal of Physical Chemistry B* **1999**, *103* (2), 5227–5239. 10.1021/jp9842291

[101]   Zeng, H.; Yuan, J.; Wang, J., Numerical and Experimental Investigations on Reducing Particle Accumulation for SCR-deNOx Facilities. *Applied Sciences* **2019**, *9* (19), 4158. doi 10.3390/app9194158

[102]   Vogt, E. T. C.; Weckhuysen, B. M., Fluid catalytic cracking: Recent developments on the grand old lady of zeolite catalysis. *Chemical Society Reviews* **2015**, *44* (20), 7342–7370. dx.doi.org/10.1039/C5CS00376H

[103]   Baerns, M.; Behr, A.; Brehm, A.; Gmeling, J.; Hofmann, H.; Onken, U.; Renken, A., *Technische Chemie*. Wiley-VCH: Weinheim, 2006. ISBN: 978-3527310005

[104]   Satterfield, C. N., *Heterogeneous Catalysis in Industrial Practice*. 2nd ed.; McGraw Hill: 1991. ISBN: 978-0070548862

[105]   Murzin, D. Y., *Engineering Catalysis*. de Gruyter: Berlin, 2020. doi 10.1515/9783110614435

[106]   Nagai, Y.; Hirabayashi, T.; Dohmae, K.; Takagi, N.; Minami, T.; Shinjoh, H.; Matsumoto, S., Sintering inhibition mechanism of platinum supported on ceria-based oxide and Pt-oxide – support interaction. *Journal of Catalysis.* **2006**, *242*, 103–109. doi 10.1016/j.jcat.2006.06.002

[107]   Ulrich, V.; Moroz, B.; Pyrjaev, P.; Sinev, I.; Bukhtiyarov, A.; Gerasimov, E.; Bukhtiyarov, V.; Roldan Cuenya, B.; Grünert, W., Three-way catalysis with bimetallic supported Pd-Au catalysts: Gold as a poison and as a promotor. *Applied Catalysis B: Environmental* **2021**, *282*, 119614. doi.org/10.1016/j.apcatb.2020.119614

[108]   Birke, P.; Engels, S.; Kranz, M.; Meiler, F., Feststoffreaktionen in Katalysatoren und Katalysatorkomponenten. VI. Untersuchungen an chlormodifizierten Pt/γ-Al_2O_3-Katalysatoren. *Zeitschrift für anorganische und allgemeine Chemie* **1980**, *467* (1), 7–16. doi.org/10.1002/zaac.19804670102

[109]   Yoo, J. S., Metal recovery and rejuvenation of metal-loaded spent catalysts. *Catalysis Today* **1998**, *44* (1), 27–46. doi.org/10.1016/S0920-5861(98)00171-0

[110]   Ferri, D.; Kumar, M. S.; Wirz, R.; Eyssler, A.; Korsak, O.; Hug, P.; Weidenkaff, A.; Newton, M. A., First steps in combining modulation excitation spectroscopy with synchronous dispersive EXAFS/DRIFTS/mass spectrometry for *in situ* time resolved study of heterogeneous catalysts. *Physical Chemistry Chemical Physics* **2010**, *12* (21), 5634–5646. doi 10.1039/B926886C

[111]   Kazansky, V. B., Adsorbed carbocations and transition states in heterogeneous acid catalyzed transformations of hydrocarbons. *Catalysis Today* **1999**, *51* (3–4), 419–434. doi 10.1016/S0920-5861(99)00031-0

[112]   Ren, Q.; Rybicki, M.; Sauer, J., Interaction of C_3–C_5 alkenes with zeolitic Brønsted sites: π-Complexes, alkoxides, and carbenium ions in H-FER. *The Journal of Physical Chemistry C* **2020**, *124* (18), 10067–10078. doi 10.1021/acs.jpcc.0c03061

[113]   Clark, L. A.; Sierka, M.; Sauer, J., Stable mechanistically-relevant aromatic-based carbenium ions in zeolite catalysts. *Journal of the American Chemical Society* **2003**, *125* (8), 2136–2141. doi 10.1021/ja0283302

[114]   Kharlamov, V. V.; Minachev, K. M., *Okislitel'no-Vosstanovitel'nyi Kataliz na Zeolitakh (Redox Catalysis by Zeolites)*. Nauka: Moscow, 1999; p 199.

[115]   Weitkamp, J., Catalytic hydrocracking-mechanisms and versatility of the process. *ChemCatChem* **2012**, *4* (3), 292–306. doi 10.1002/cctc.201100315

[116]   Kotrel, S.; Knözinger, H.; Gates, B. C., The Haag–Dessau mechanism of protolytic cracking of alkanes. *Microporous and Mesoporous Materials* **2000**, *35–36*, 11–20. doi.org/10.1016/S1387-1811(99)00204-8

[117] Grifoni, E.; Piccini, G.; Lercher, J. A.; Glezakou, V. A.; Rousseau, R.; Parrinello, M., Confinement effects and acid strength in zeolites. *Nature Communications* **2021**, *12* (1), Art No. 2630. doi 10.1038/s41467-021-22936-0

[118] Pfriem, N.; Hintermeier, P. H.; Eckstein, S.; Kim, S.; Liu, Q.; Shi, H.; Milakovic, L.; Liu, Y.; Haller, G. L.; Baráth, E.; Liu, Y.; Lercher, J. A., Role of the ionic environment in enhancing the activity of reacting molecules in zeolite pores. *Science* **2021**, *372* (6545), 952–957. doi:10.1126/science.abh3418

[119] Sprenger, P.; Kleist, W.; Grunwaldt, J.-D., Recent advances in selective propylene oxidation over bismuth molybdate based catalysts: Synthetic, spectroscopic, and theoretical approaches. *ACS Catalysis* **2017**, *7* (9), 5628–5642. doi 10.1021/acscatal.7b01149

[120] Licht, R. B.; Bell, A. T., A DFT Investigation of the mechanism of propene ammoxidation over α-bismuth molybdate. *ACS Catalysis* **2017**, *7* (1), 161–176. doi 10.1021/acscatal.6b02523

[121] Licht, R. B.; Vogt, D.; Bell, A. T., The mechanism and kinetics of propene ammoxidation over α-bismuth molybdate. *Journal of Catalysis* **2016**, *339*, 228–241. doi.org/10.1016/j.jcat.2016.04.012

[122] Bui, L.; Chakrabarti, R.; Bhan, A., Mechanistic origins of unselective oxidation products in the conversion of propylene to acrolein on $Bi_2Mo_3O_{12}$. *ACS Catalysis* **2016**, *6* (10), 6567–6580. doi 10.1021/acscatal.6b01830

[123] Bahri-Laleh, N.; Hanifpour, A.; Mirmohammadi, S. A.; Poater, A.; Nekoomanesh-Haghighi, M.; Talarico, G.; Cavallo, L., Computational modeling of heterogeneous Ziegler-Natta catalysts for olefins polymerization. *Progress in Polymer Science* **2018**, *84*, 89–114. doi.org/10.1016/j.progpolymsci.2018.06.005

[124] Lwin, S.; Wachs, I. E., Olefin metathesis by supported metal oxide catalysts. *ACS Catalysis* **2014**, *4* (8), 2505–2520. doi 10.1021/cs500528h

[125] Chauvin, Y.; Commereuc, D., Chemical counting and characterization of the active sites in the rhenium oxide/alumina metathesis catalyst. *Journal of the Chemical Society, Chemical Communications* **1992**,(6), 462–464. doi 10.1039/C39920000462

[126] Lwin, S.; Wachs, I. E., Reaction mechanism and kinetics of olefin metathesis by supported $ReOx$/Al_2O_3 Catalysts. *ACS Catalysis* **2016**, *6* (1), 272–278. doi 10.1021/acscatal.5b02233

[127] Zhang, B.; Wachs, I. E., Identifying the catalytic active site for propylene metathesis by supported $ReOx$ catalysts. *ACS Catalysis* **2021**, *11* (4), 1962–1976. doi 10.1021/acscatal.0c04773

[128] Grünert, W.; Feldhaus, R.; Anders, K.; Shpiro, E. S.; Minachev, K. M., Reduction behavior and metathesis activity of WO_3/Al_2O_3 Catalysts III. The activation of WO_3/Al_2O_3 catalysts. *Journal of Catalysis* **1989**, *120*, 444–456 doi 10.1016/0021-9517(89)90284-4

[129] Grünert, W.; Stakheev, A. Y.; Mörke, W.; Feldhaus, R.; Anders, K.; Shpiro, E. S.; Minachev, K. M., Reduction and metathesis activity of MoO_3/Al_2O_3 catalysts II. The activation of MoO_3/Al_2O_3 catalysts. *Journal of Catalysis* **1992**, *135*, 287–299. doi 10.1016/0021-9517(92)90286-Q

[130] Amakawa, K.; Wrabetz, S.; Kröhnert, J.; Tzolova-Müller, G.; Schlögl, R.; Trunschke, A., *In situ* generation of active sites in olefin metathesis. *Journal of the American Chemical Society* **2012**, *134* (28), 11462–11473. doi 10.1021/ja3011989

[131] Chakrabarti, A.; Wachs, I. E., Activation mechanism and surface intermediates during olefin metathesis by supported $MoOx$/Al_2O_3 catalysts. *The Journal of Physical Chemistry C* **2019**, *123* (19), 12367–12375. doi 10.1021/acs.jpcc.9b02426

[132] Conley, M. P.; Delley, M. F.; Siddiqi, G.; Lapadula, G.; Norsic, S.; Monteil, V.; Safonova, O. V.; Copéret, C., Polymerization of ethylene by silica-supported dinuclear Cr^{III} Sites through an initiation step involving C–H bond activation. *Angewandte Chemie International Edition* **2014**, *53* (7), 1872–1876. doi.org/10.1002/anie.201308983

[133] Conley, M. P.; Delley, M. F.; Núñez-Zarur, F.; Comas-Vives, A.; Copéret, C., Heterolytic activation of C–H bonds on Cr^{III}–O surface sites is a key step in catalytic polymerization of ethylene and dehydrogenation of propane. *Inorganic Chemistry* **2015**, *54* (11), 5065–5078. doi 10.1021/ic502696n

[134] Fong, A.; Yuan, Y.; Ivry, S. L.; Scott, S. L.; Peters, B., Computational kinetic discrimination of ethylene polymerization mechanisms for the Phillips (Cr/SiO$_2$) catalyst. *ACS Catalysis* **2015**, *5* (6), 3360–3374. doi 10.1021/acscatal.5b00016

[135] Delley, M. F.; Núñez-Zarur, F.; Conley, M. P.; Comas-Vives, A.; Siddiqi, G.; Norsic, S.; Monteil, V.; Safonova, O. V.; Copéret, C., Proton transfers are key elementary steps in ethylene polymerization on isolated chromium(III) silicates. *Proceedings of the National Academy of Sciences* **2014**, *111* (32), 11624.

[136] Topsøe, N.-Y.; Dumesic, J. A.; Topsøe, H., Vanadia-titania catalysts for selective catalytic reduction of nitric-oxide by ammonia .2. Studies of active-sites and formulation of catalytic cycles. *Journal of Catalysis* **1995**, *151*, 241–252. doi 10.1006/jcat.1995.1025

[137] Marberger, A.; Ferri, D.; Elsener, M.; Kröcher, O., The significance of Lewis acid sites for the selective catalytic reduction of nitric oxide on vanadium-based catalysts. *Angewandte Chemie International Edition* **2016**, *55* (39), 11989–11994. doi 10.1002/anie.201605397

[138] Lai, J.-K.; Wachs, I. E., A perspective on the Selective Catalytic Reduction (SCR) of NO with NH$_3$ by supported V$_2$O$_5$–WO$_3$/TiO$_2$ catalysts. *ACS Catalysis* **2018**, *8* (7), 6537–6551. doi 10.1021/acscatal.8b01357

[139] Janssens, T. V. W.; Falsig, H.; Lundegaard, L. F.; Venneström, P. N. R.; Rasmussen, S. B.; Moses, P. G.; Giordanino, F.; Borfecchia, E.; Lomachenko, K. A.; Lamberti, C.; Bordiga, S.; Godiksen, A.; Mossin, S.; Beato, P., A consistent reaction scheme for the selective catalytic reduction of nitrogen oxides with ammonia. *ACS Catalysis* **2015**, *5* (5), 2832–2845. doi 10.1021/acscatal.8b01357

[140] Pérez Vélez, R.; Ellmers, I.; Huang, H.; Bentrup, U.; Schünemann, V.; Grünert, W.; Brückner, A., Identifying active sites for fast NH$_3$-SCR of NO/NO$_2$ mixtures over Fe-ZSM-5 by operando EPR and UV-vis spectroscopy. *Journal of Catalysis* **2014**, *316*, 103–111. doi 10.1016/j.jcat.2014.05.001

[141] Chen, L.; Janssens, T. V. W.; Venneström, P. N. R.; Jansson, J.; Skoglundh, M.; Grönbeck, H., A complete multisite reaction mechanism for low-temperature NH$_3$-SCR over Cu-CHA. *ACS Catalysis* **2020**, *10* (10), 5646–5656. doi 10.1021/acscatal.0c00440

[142] Pérez-Ramírez, J.; Kondratenko, E. V.; Kondratenko, V. A.; Baerns, M., Selectivity-directing factors of ammonia oxidation over PGM gauzes in the Temporal Analysis of Products reactor: Primary interactions of NH$_3$ and O$_2$. *Journal of Catalysis* **2004**, *227* (1), 90–100. doi 10.1016/j.jcat.2004.06.023

[143] Novell-Leruth, G.; Ricart, J. M.; Pérez-Ramírez, J., Pt(100)-catalyzed ammonia oxidation studied by DFT: Mechanism and microkinetics. *The Journal of Physical Chemistry C* **2008**, *112* (35), 13554–13562. doi 10.1021/jp802489y

[144] van Santen, R. A.; Markvoort, A. J.; Filot, I. A. W.; Ghouri, M. M.; Hensen, E. J. M., Mechanism and microkinetics of the Fischer–Tropsch reaction. *Physical Chemistry Chemical Physics* **2013**, *15* (40), 17038–17063. doi 10.1039/C3CP52506F

[145] Nagahara, H.; Ono, M.; Konishi, M.; Fukuoka, Y., Partial hydrogenation of benzene to cyclohexene. *Appl. Surf. Sci.* **1997**, *121–122*, 448–451. doi.org/10.1016/S0169-4332(97)00325-5

[146] Chen, Z.; Sun, H.; Peng, Z.; Gao, J.; Li, B.; Liu, Z.; Liu, S., Selective hydrogenation of benzene: Progress of understanding for the Ru-based catalytic system design. *Industrial & Engineering Chemistry Research* **2019**, *58* (31), 13794–13803. doi 10.1021/acs.iecr.9b01475

[147] Henning, A. D.; Schmidt, L. D., Oxidative dehydrogenation of ethane at short contact times: Species and temperature profiles within and after the catalyst. *Chemical Engineering Science* **2002**, *57* (14), 2615–2625. doi.org/10.1016/S0009-2509(02)00155-0

[148] Buyevskaya, O. V.; Rothaemel, M.; Zanthoff, H. W.; Baerns, M., Transient studies on reaction steps in the oxidative coupling of methane over catalytic surfaces of MgO and Sm$_2$O$_3$. *Journal of Catalysis* **1994**, *146* (2), 346–357. doi 10.1006/jcat.1994.1073

[149] Schwach, P.; Frandsen, W.; Willinger, M.-G.; Schlögl, R.; Trunschke, A., Structure sensitivity of the oxidative activation of methane over MgO model catalysts: I. Kinetic study. *Journal of Catalysis* **2015**, *329*, 560–573. doi.org/10.1016/j.jcat.2015.05.007

[150] Schwach, P.; Hamilton, N.; Eichelbaum, M.; Thum, L.; Lunkenbein, T.; Schlögl, R.; Trunschke, A., Structure sensitivity of the oxidative activation of methane over MgO model catalysts: II. Nature of active sites and reaction mechanism. *Journal of Catalysis* **2015**, *329*, 574–587. doi.org/10.1016/j. jcat.2015.05.008

[151] Stakheev, A. Y.; Baeva, G. N.; Bragina, G. O.; Teleguina, N. S.; Kustov, A. L.; Grill, M.; Thogersen, J. R., Integrated DeNO$_x$-DeSoot catalytic systems with improved low-temperature performance. *Topics in Catalysis* **2013**, *56* (1–8), 427–433. doi 10.1007/s11244-013-9991-7

[152] Salazar, M.; Hoffmann, S.; Tkachenko, O. P.; Becker, R.; Grünert, W., Hybrid catalysts for the selective catalytic reduction of NO by NH$_3$: The influence of component separation on the performance of hybrid systems. *Applied Catalysis B* **2016**, *182*, 213–219. doi 10.1016/j.apcatb.2015.09.028

[153] Xiong, M.; Gao, Z.; Qin, Y., Spillover in heterogeneous catalysis: New insights and opportunities. *ACS Catalysis* **2021**, *11* (5), 3159–3172. doi 10.1021/acscatal.0c05567

[154] Steinberg, K. H.; Mroczek, U.; Rössner, F., Hydrogen spillover in the conversion of n-alkanes on zeolites. *Studies in Surface Science and Catalysis* **1989**, 46, 81–90. doi.org/10.1016/ S0167-2991(08)60969-6

[155] Prins, R., Hydrogen spillover. Facts and fiction. *Chemical Reviews* **2012**, *112* (5), 2714–2738. doi 10.1021/cr200346z

[156] Karim, W.; Spreafico, C.; Kleibert, A.; Gobrecht, J.; VandeVondele, J.; Ekinci, Y.; van Bokhoven, J. A., Catalyst support effects on hydrogen spillover. *Nature* **2017**, *541* (7635), 68–71. doi 10.1038/ nature20782

[157] Melián-Cabrera, I.; Zarubina, V.; Nederlof, C.; Kapteijn, F.; Makkee, M., An *in situ* reactivation study reveals the supreme stability of γ-alumina for the oxidative dehydrogenation of ethylbenzene to styrene. *Catalysis Science & Technology* **2018**, *8* (15), 3733–3736. doi 10.1039/C8CY00748A

[158] McGregor, J.; Huang, Z. Y.; Parrott, E. P. J.; Zeitler, J. A.; Nguyen, K. L.; Rawson, J. M.; Carley, A.; Hansen, T. W.; Tessonnier, J. P.; Su, D. S.; Teschner, D.; Vass, E. M.; Knop-Gericke, A.; Schlögl, R.; Gladden, L. F., Active coke: Carbonaceous materials as catalysts for alkane dehydrogenation. *Journal of Catalysis* **2010**, *269* (2), 329–339. doi 10.1016/j.jcat.2009.11.016

[159] Wark, M.; Brückner, A.; Liese, T.; Grünert, W., Selective catalytic reduction of NO by NH$_3$ over vanadium- modified zeolites. *Journal of Catalysis* **1998**, *175* (1), 48–61. doi 10.1006/jcat.1998.1974

[160] Wierzbicki, D.; Clark, A. H.; Kröcher, O.; Ferri, D.; Nachtegaal, M., Monomeric Fe Species in Square Planar Geometry Active for Low Temperature NH$_3$-SCR of NO. *The Journal of Physical Chemistry C* **2022**, *126* (41), 17510–17519. doi 10.1021/acs.jpcc.2c03480

[161] Brandenberger, S.; Kröcher, O.; Tissler, A.; Althoff, R., The determination of the activities of different iron species in Fe-ZSM-5 for SCR of NO by NH$_3$. *Applied Catalysis B: Environmental* **2010**, *95* (3–4), 348–357. doi.org/10.1016/j.apcatb.2010.01.013

[162] Ellmers, I.; Pérez Vélez, R.; Bentrup, U.; Schwieger, W.; Brückner, A.; Grünert, W., SCR and NO oxidation over Fe-ZSM-5 – the influence of the Fe content. *Catalysis Today* **2015**, *258*, 337–346. doi 10.1016/j.cattod.2014.12.017

[163] Iglesia, E.; Barton, D. G.; Biscardi, J. A.; Gines, M. J. L.; Soled, S. L., Bifunctional pathways in catalysis by solid acids and bases. *Catalysis Today* **1997**, *38* (3), 339–360. doi.org/10.1016/ S0920-5861(97)81503-7

[164] Studt, F.; Behrens, M.; Kunkes, E. L.; Thomas, N.; Zander, S.; Tarasov, A.; Schumann, J.; Frei, E.; Varley, J. B.; Abild-Pedersen, F.; Nørskov, J. K.; Schlögl, R., The mechanism of CO and CO$_2$ hydrogenation to methanol over Cu-based catalysts. *ChemCatChem* **2015**, *7* (7), 1105–1111. doi.org/10.1002/ cctc.201500123

[165]   Fisher, I. A.; Woo, H. C.; Bell, A. T., Effects of zirconia promotion on the activity of Cu/SiO$_2$
         for methanol synthesis from CO/H$_2$ and CO$_2$/H$_2$. *Catalysis Letters* **1997,** *44* (1), 11–17.
         doi 10.1023/A:1018916806816

[166]   Kurtz, M.; Strunk, J.; Hinrichsen, O.; Muhler, M.; Fink, K.; Meyer, B.; Wöll, C., Active sites on oxide
         surfaces: ZnO-catalyzed synthesis of methanol from CO and H$_2$. **2005,** *44* (18), 2790–2794.
         doi.org/10.1002/anie.200462374

[167]   Haruta, M.; Kobayashi, T.; Sano, H.; Yamada, N., Novel gold catalysts for the oxidation of carbon
         monoxide at a temperature far below 0°C. *Chemistry Letters* **1987,**(2), 405–408. doi 10.1246/
         cl.1987.405

[168]   Haruta, M., Catalysis of gold nanoparticles deposited on metal oxides. *Cattech* **2002,** *6* (3), 102–115.
         doi 10.1023/A:1020181423055

[169]   Schlexer, P.; Widmann, D.; Behm, R. J.; Pacchioni, G., CO Oxidation on a Au/TiO$_2$ nanoparticle
         catalyst via the Au-assisted Mars–van Krevelen mechanism. *ACS Catalysis* **2018,** *8* (7), 6513–6525.
         doi 10.1021/acscatal.8b01751

[170]   Green, I. X.; Tang, W.; Neurock, M.; Yates, J. T., Spectroscopic observation of dual catalytic sites
         during oxidation of CO on a Au/TiO$_2$ catalyst. *Science* **2011,** *333* (6043), 736–739. doi 10.1126/
         science.1207272

[171]   Lopez, N.; Janssens, T. V. W.; Clausen, B. S.; Xu, Y.; Mavrikakis, M.; Bligaard, T.; Nørskov, J. K., On the
         origin of the catalytic activity of gold nanoparticles for low-temperature CO oxidation. *Journal of
         Catalysis* **2004,** *223* (1), 232–235. doi.org/10.1016/j.jcat.2004.01.001

[172]   Grünert, W.; Großmann, D.; Noei, H.; Pohl, M. M.; Sinev, I.; De Toni, A.; Wang, Y.; Muhler, M.,
         Low-temperature CO oxidation with TiO$_2$-supported Au^{3+} ions. *Angewandte Chemie International
         Edition.* **2014,** *53,* 3245–3249.

[173]   Saavedra, J.; Pursell, C. J.; Chandler, B. D., CO Oxidation kinetics over Au/TiO$_2$ and Au/Al$_2$O$_3$ catalysts:
         Evidence for a common water-assisted mechanism. *Journal of the American Chemical Society* **2018,**
         140 (10), 3712–3723. doi 10.1021/jacs.7b12758

[174]   Prasad, V.; Vlachos, D. G., Multiscale Model and informatics-based optimal design of experiments:
         Application to the catalytic Decomposition of Ammonia on Ruthenium. *Industrial & Engineering
         Chemistry Research* **2008,** *47* (17), 6555–6567. doi 10.1021/ie800343s

[175]   Salciccioli, M.; Stamatakis, M.; Caratzoulas, S.; Vlachos, D. G., A review of multiscale modeling
         of metal-catalyzed reactions: Mechanism development for complexity and emergent behavior.
         Chemical Engineering Science **2011,** *66* (19), 4319–4355. doi.org/10.1016/j.ces.2011.05.050

[176]   Oh, S. H.; Fisher, G. B.; Carpenter, J. E.; Goodman, D. W., Comparative kinetic studies of CO/O$_2$ and
         CO/NO reactions over single crystal and supported rhodium catalysts. *Journal of Catalysis* **1986,** *100*
         (2), 360–376. doi.org/10.1016/0021-9517(86)90103-X

[177]   Dumesic, J. A.; Rudd, D. F.; Aparicio, L. M., *The Microkinetics of Heterogeneous Catalysis*. American
         Chemical Society: 1993. doi 10.1002/aic.690400620

[178]   Hinrichsen, O.; Rosowski, F.; Muhler, M.; Ertl, G., The microkinetics of ammonia synthesis catalyzed
         by cesium-promoted supported ruthenium. *Chemical Engineering Science* **1996,** *51* (10), 1683–1690.
         doi.org/10.1016/0009-2509(96)00027-9

[179]   Prasad, V.; Karim, A. M.; Arya, A.; Vlachos, D. G., Assessment of Overall Rate Expressions
         and multiscale, microkinetic model uniqueness via experimental data injection: Ammonia
         decomposition on Ru/γ-Al$_2$O$_3$ for hydrogen production. *Industrial & Engineering Chemistry Research*
         2009, *48* (11), 5255–5265. doi 10.1021/ie900144x

[180]   Motagamwala, A. H.; Dumesic, J. A., Microkinetic modeling: A tool for rational catalyst design.
         Chemical Reviews **2021,** *121* (2), 1049–1076. doi 10.1021/acs.chemrev.0c00394

[181]   Wittreich, G. R.; Alexopoulos, K.; Vlachos, D. G., Microkinetic modeling of surface catalysis. In
         Handbook of Materials Modeling: Applications: Current and Emerging Materials, Andreoni, W.; Yip,

S., Eds. Springer International Publishing: Cham, 2020; pp 1377–1404. doi 10.1007/978-3-319-44680-6_5

[182] Filot, I. A. W.; van Santen, R. A.; Hensen, E. J. M., The optimally performing Fischer-Tropsch catalyst. *Angewandte Chemie-International Edition* **2014**, *53* (47), 12746–12750. doi 10.1002/anie.201406521

[183] Mallat, T.; Baiker, A., Reactions in "sacrificial" solvents. *Catalysis Science & Technology* **2011**, *1* (9), 1572–1583. doi 10.1039/C1CY00207D

[184] Rinaldi, R.; Schüth, F., Design of solid catalysts for the conversion of biomass. *Energy & Environmental Science* **2009**, *2* (6), 610–626. doi 10.1039/B902668A

[185] Sheldon, R. A.; Wallau, M.; Arends, I. W. C. E.; Schuchardt, U., Heterogeneous catalysts for liquid-phase oxidations: Philosophers' Stones or Trojan Horses? *Accounts of Chemical Research* **1998**, *31* (8), 485–493. doi10.1021/ar9700163

[186] Huybrechts, D. R. C.; Debruycker, L.; Jacobs, P. A., Oxyfunctionalization of alkanes with hydrogen-peroxide on titanium silicalite. *Nature* **1990**, *345* (6272), 240–242. doi 10.1038/345240a0

[187] Lin, W.; Frei, H., Photochemical and FT-IR Probing of the active site of hydrogen peroxide in Ti silicalite sieve. *Journal of the American Chemical Society* **2002**, *124* (31), 9292–9298. doi 10.1021/ja012477w

[188] Della Pina, C.; Falletta, E.; Prati, L.; Rossi, M., Selective oxidation using gold. *Chemical Society Reviews* **2008**, *37* (9), 2077–2095. doi 10.1039/b707319b

[189] Villa, A.; Dimitratos, N.; Chan-Thaw, C. E.; Hammond, C.; Prati, L.; Hutchings, G. J., Glycerol oxidation using gold-containing catalysts. *Accounts of Chemical Research* **2015**, *48* (5), 1403–1412. doi 10.1021/ar500426g

[190] Gupta, U. N.; Dummer, N. F.; Pattisson, S.; Jenkins, R. L.; Knight, D. W.; Bethell, D.; Hutchings, G. J., Solvent-free aerobic epoxidation of Dec-1-ene using gold/graphite as a catalyst. *Catalysis Letters* **2015**, *145* (2), 689–696. doi 10.1007/s10562-014-1425-z

[191] Bawaked, S.; Dummer, N. F.; Bethell, D.; Knight, D. W.; Hutchings, G. J., Solvent-free selective epoxidation of cyclooctene using supported gold catalysts: An investigation of catalyst re-use. *Green Chemistry* **2011**, *13* (1), 127–134. doi 10.1039/C0GC00550A

[192] Ovoshchnikov, D. S.; Donoeva, B. G.; Williamson, B. E.; Golovko, V. B., Tuning the selectivity of a supported gold catalyst in solvent- and radical initiator-free aerobic oxidation of cyclohexene. *Catalysis Science & Technology* **2014**, *4* (3), 752–757. doi 10.1039/C3CY01011B

[193] Yang, G.; Akhade, S. A.; Chen, X.; Liu, Y.; Lee, M.-S.; Glezakou, V.-A.; Rousseau, R.; Lercher, J. A., The nature of hydrogen adsorption on platinum in the aqueous phase. *Angewandte Chemie International Edition* **2019**, *58* (11), 3527–3532. doi.org/10.1002/anie.201813958

[194] Eckstein, S.; Hintermeier, P. H.; Zhao, R.; Baráth, E.; Shi, H.; Liu, Y.; Lercher, J. A., Influence of hydronium ions in zeolites on sorption. *Angewandte Chemie International Edition* **2019**, *58* (11), 3450–3455. doi.org/10.1002/anie.201812184

[195] Maeda, K., Photocatalytic water splitting using semiconductor particles: History and recent developments. *Journal of Photochemistry and Photobiology C: Photochemistry Reviews* **2011**, *12* (4), 237–268. doi.org/10.1016/j.jphotochemrev.2011.07.001

[196] Wang, X. C.; Maeda, K.; Thomas, A.; Takanabe, K.; Xin, G.; Carlsson, J. M.; Domen, K.; Antonietti, M., A metal-free polymeric photocatalyst for hydrogen production from water under visible light. *Nature Materials* **2009**, *8* (1), 76–80. doi 10.1038/nmat2317

[197] Hurum, D. C.; Agrios, A. G.; Gray, K. A.; Rajh, T.; Thurnauer, M. C., Explaining the enhanced photocatalytic activity of Degussa P25 mixed-phase TiO_2 using EPR. *The Journal of Physical Chemistry B* **2003**, *107* (19), 4545–4549. doi 10.1021/jp0273934

[198] Ohtani, B.; Prieto-Mahaney, O. O.; Li, D.; Abe, R., What is Degussa (Evonik) P25? Crystalline composition analysis, reconstruction from isolated pure particles and photocatalytic activity test. *Journal of Photochemistry and Photobiology A-Chemistry* **2010**, *216* (2–3), 179–182. doi 10.1016/j.jphotochem.2010.07.024

[199]   Youngblood, W. J.; Lee, S.-H. A.; Kobayashi, Y.; Hernandez-Pagan, E. A.; Hoertz, P. G.; Moore, T. A.; Moore, A. L.; Gust, D.; Mallouk, T. E., Photoassisted overall water splitting in a visible light-absorbing dye-sensitized photoelectrochemical cell. *Journal of the American Chemical Society* **2009**, *131* (3), 926–927. doi 10.1021/ja809108y

[200]   Kudo, A.; Yoshino, S.; Tsuchiya, T.; Udagawa, Y.; Takahashi, Y.; Yamaguchi, M.; Ogasawara, I.; Matsumoto, H.; Iwase, A., Z-scheme photocatalyst systems employing Rh- and Ir-doped metal oxide materials for water splitting under visible light irradiation. *Faraday Discussion* **2019**, *215* (0), 313–328. doi 10.1039/C8FD00209F

[201]   Bard, A. J., Photoelectrochemistry and heterogeneous photo-catalysis at semiconductors. *Journal of Photochemistry* **1979**, *10* (1), 59–75. doi.org/10.1016/0047-2670(79)80037-4

[202]   Maeda, K., Z-Scheme water splitting using two different semiconductor photocatalysts. *ACS Catalysis* **2013**, *3* (7), 1486–1503. doi 10.1021/cs4002089

[203]   Mills, A.; Bingham, M.; O'Rourke, C.; Bowker, M., Modelled kinetics of the rate of hydrogen evolution as a function of metal catalyst loading in the photocatalysed reforming of methanol by Pt (or Pd)/TiO$_2$. *Journal of Photochemistry and Photobiology A: Chemistry* **2019**, *373*, 122–130. doi. org/10.1016/j.jphotochem.2018.12.039

[204]   Li, R.; Han, H.; Zhang, F.; Wang, D.; Li, C., Highly efficient photocatalysts constructed by rational assembly of dual-cocatalysts separately on different facets of BiVO$_4$. *Energy & Environmental Science* **2014**, *7* (4), 1369–1376. doi 10.1039/C3EE43304H

[205]   Mu, L.; Zhao, Y.; Li, A.; Wang, S.; Wang, Z.; Yang, J.; Wang, Y.; Liu, T.; Chen, R.; Zhu, J.; Fan, F.; Li, R.; Li, C., Enhancing charge separation on high symmetry SrTiO$_3$ exposed with anisotropic facets for photocatalytic water splitting. *Energy & Environmental Science* **2016**, *9* (7), 2463–2469. doi 10.1039/C6EE00526H

[206]   Takata, T.; Jiang, J.; Sakata, Y.; Nakabayashi, M.; Shibata, N.; Nandal, V.; Seki, K.; Hisatomi, T.; Domen, K., Photocatalytic water splitting with a quantum efficiency of almost unity. *Nature* **2020**, *581* (7809), 411–414. doi 10.1038/s41586-020-2278-9

[207]   Mei, B.; Han, K.; Mul, G., Driving surface redox reactions in heterogeneous photocatalysis: The active state of illuminated semiconductor-supported nanoparticles during overall water-splitting. *ACS Catalysis* **2018**, *8* (10), 9154–9164. doi 10.1038/s41586-020-2278-9

[208]   Keith, J. A.; Jacob, T., Theoretical studies of potential-dependent and competing mechanisms of the electrocatalytic oxygen reduction reaction on Pt(111) **2010**, *49* (49), 9521–9525. doi.org/10.1002/anie.201004794

[209]   Shinagawa, T.; Garcia-Esparza, A. T.; Takanabe, K., Insight on Tafel slopes from a microkinetic analysis of aqueous electrocatalysis for energy conversion. *Scientific Reports* **2015**, *5* (1), 13801. doi 10.1038/srep13801

[210]   Holewinski, A.; Linic, S., Elementary mechanisms in electrocatalysis: Revisiting the ORR tafel slope. *Journal of the Electrochemical Society* **2012**, *159* (11), H864–H870. doi 10.1149/2.022211jes

[211]   Nayak, S.; Biedermann, P. U.; Stratmann, M.; Erbe, A., *In situ* infrared spectroscopic investigation of intermediates in the electrochemical oxygen reduction on n-Ge(100) in alkaline perchlorate and chloride electrolyte. *Electrochimica Acta* **2013**, *106*, 472–482. doi.org/10.1016/j.electacta.2013.05.133

[212]   Choi, Y. W.; Mistry, H.; Roldan Cuenya, B., New insights into working nanostructured electrocatalysts through operando spectroscopy and microscopy. *Current Opinion in Electrochemistry* **2017**, *1* (1), 95–103. doi.org/10.1016/j.coelec.2017.01.004

[213]   Rossmeisl, J.; Qu, Z. W.; Zhu, H.; Kroes, G. J.; Nørskov, J. K., Electrolysis of water on oxide surfaces. *Journal of Electroanalytical Chemistry* **2007**, *607* (1), 83–89. doi.org/10.1016/j.jelechem.2006.11.008

[214]   Reier, T.; Nong, H. N.; Teschner, D.; Schlögl, R.; Strasser, P., Electrocatalytic oxygen evolution reaction in acidic environments – reaction mechanisms and catalysts. *Advanced Energy Materials* **2017**, *7* (1), 1601275. doi.org/10.1002/aenm.201601275

[215]   Nong, H. N.; Falling, L. J.; Bergmann, A.; Klingenhof, M.; Tran, H. P.; Spöri, C.; Mom, R.; Timoshenko, J.; Zichittella, G.; Knop-Gericke, A.; Piccinin, S.; Pérez-Ramírez, J.; Roldan Cuenya, B.; Schlögl, R.; Strasser, P.; Teschner, D.; Jones, T. E., Key role of chemistry versus bias in electrocatalytic oxygen evolution. *Nature* **2020,** *587* (7834), 408–413. doi 10.1038/s41586-020-2908-2

[216]   Falling, L. J.; Velasco-Vélez, J. J.; Mom, R. V.; Knop-Gericke, A.; Schlögl, R.; Teschner, D.; Jones, T. E., The ladder towards understanding the oxygen evolution reaction. *Current Opinion in Electrochemistry* **2021,** *30,* 100842. doi.org/10.1016/j.coelec.2021.100842

[217]   Miracle, D. B.; Senkov, O. N., A critical review of high entropy alloys and related concepts. *Acta Materialia* **2017,** *122,* 448–511. doi.org/10.1016/j.coelec.2021.100842

[218]   Löffler, T.; Ludwig, A.; Rossmeisl, J.; Schuhmann, W., What makes high-entropy alloys exceptional electrocatalysts? *Angewandte Chemie-International Edition* **2021,** *60* (52), 26894–26903. doi 10.1002/anie.202109212

[219]   Batchelor, T. A. A.; Pedersen, J. K.; Winther, S. H.; Castelli, I. E.; Jacobsen, K. W.; Rossmeisl, J., High-entropy alloys as a discovery platform for electrocatalysis. *Joule* **2019,** *3* (3), 834–845. doi.org/10.1016/j.joule.2018.12.015

[220]   Li, W.; Liu, P.; Liaw, P. K., Microstructures and properties of high-entropy alloy films and coatings: A review. *Materials Research Letters* **2018,** *6* (4), 199–229. doi 10.1080/21663831.2018.1434248

[221]   Yao, R.-Q.; Zhou, Y.-T.; Shi, H.; Wan, W.-B.; Zhang, Q.-H.; Gu, L.; Zhu, Y.-F.; Wen, Z.; Lang, X.-Y.; Jiang, Q., Nanoporous surface high-entropy alloys as highly efficient multisite electrocatalysts for nonacidic hydrogen evolution reaction. *Advanced Functional Materials* **2021,** *31* (10), 2009613. doi.org/10.1002/adfm.202009613

[222]   Hori, Y., Electrochemical CO_2 reduction on metal electrodes. In *Modern Aspects of Electrochemistry,* Vayenas, C. G.; White, R. E.; Gamboa-Aldeco, M. E., Eds. Springer New York: New York, NY, 2008; pp 89–189. doi 10.1007/978-0-387-49489-0_3

[223]   Bagger, A.; Ju, W.; Varela, A. S.; Strasser, P.; Rossmeisl, J., Electrochemical CO_2 reduction: A classification problem. **2017,** *18* (22), 3266–3273. doi.org/10.1002/cphc.201700736

[224]   Velasco-Vélez, J.-J.; Jones, T.; Gao, D.; Carbonio, E.; Arrigo, R.; Hsu, C.-J.; Huang, Y.-C.; Dong, C.-L.; Chen, J.-M.; Lee, J.-F.; Strasser, P.; Roldan Cuenya, B.; Schlögl, R.; Knop-Gericke, A.; Chuang, C.-H., The role of the copper oxidation state in the electrocatalytic reduction of CO_2 into valuable hydrocarbons. *ACS Sustainable Chemistry & Engineering* **2019,** *7* (1), 1485–1492. doi 10.1021/acssuschemeng.8b05106

[225]   Möller, T.; Scholten, F.; Thanh, T. N.; Sinev, I.; Timoshenko, J.; Wang, X.; Jovanov, Z.; Gliech, M.; Roldan Cuenya, B.; Varela, A. S.; Strasser, P., Electrocatalytic CO_2 reduction on CuO_x nanocubes: Tracking the evolution of chemical state, geometric structure, and catalytic selectivity using operando spectroscopy **2020,** *59* (41), 17974–17983. doi.org/10.1002/anie.202007136

[226]   Nitopi, S.; Bertheussen, E.; Scott, S. B.; Liu, X.; Engstfeld, A. K.; Horch, S.; Seger, B.; Stephens, I. E. L.; Chan, K.; Hahn, C.; Nørskov, J. K.; Jaramillo, T. F.; Chorkendorff, I., Progress and perspectives of electrochemical CO_2 reduction on copper in aqueous electrolyte. *Chemical Reviews* **2019,** *119* (12), 7610–7672. doi 10.1021/acs.chemrev.8b00705

[227]   Haas, T.; Krause, R.; Weber, R.; Demler, M.; Schmid, G., Technical photosynthesis involving CO_2 electrolysis and fermentation. *Nature Catalysis* **2018,** *1* (1), 32–39.

[228]   Hong, W. T.; Risch, M.; Stoerzinger, K. A.; Grimaud, A.; Suntivich, J.; Shao-Horn, Y., Toward the rational design of non-precious transition metal oxides for oxygen electrocatalysis. *Energy & Environmental Science* **2015,** *8* (5), 1404–1427. doi 10.1039/C4EE03869J

[229]   McCrory, C. C. L.; Jung, S.; Peters, J. C.; Jaramillo, T. F., Benchmarking heterogeneous electrocatalysts for the oxygen evolution reaction. *Journal of the American Chemical Society* **2013,** *135* (45), 16977–16987. doi 10.1021/ja407115p

[230]   Yang, F.; Sliozberg, K.; Sinev, I.; Antoni, H.; Bähr, A.; Ollegott, K.; Xia, W.; Masa, J.; Grünert, W.; Roldan Cuenya, B.; Schuhmann, W.; Muhler, M., Synergistic effect of cobalt and iron in layered double

hydroxide catalysts for the oxygen evolution reaction. *ChemSusChem* **2017**, *10* (1), 156–165. doi.org/10.1002/cssc.201601272

[231] Dai, L.; Xue, Y.; Qu, L.; Choi, H.-J.; Baek, J.-B., Metal-free catalysts for oxygen reduction reaction. *Chemical Reviews* **2015**, *115* (11), 4823–4892. doi 10.1021/cr5003563

[232] Hu, J. W.; Liu, W.; Xin, C. C.; Guo, J. Y.; Cheng, X. S.; Wei, J. Z.; Hao, C.; Zhang, G. F.; Shi, Y. T., Carbon-based single atom catalysts for tailoring the ORR pathway: A concise review. *Journal of Materials Chemistry A* **2021**, *9* (44), 24803–24829. doi 10.1039/d1ta06144e

[233] Kundu, S.; Nagaiah, T. C.; Xia, W.; Wang, Y.; Dommele, S. V.; Bitter, J. H.; Santa, M.; Grundmeier, G.; Bron, M.; Schuhmann, W.; Muhler, M., Electrocatalytic activity and stability of nitrogen-containing carbon nanotubes in the oxygen reduction reaction. *The Journal of Physical Chemistry C* **2009**, *113* (32), 14302–14310. doi 10.1021/jp811320d

5 With catalysis into the Anthropocene age: strategies and a look ahead

In this book, basics and special aspects of surface catalysis have been discussed and tools for the development of catalysts and catalytic processes have been described. Surface catalysis has been closely linked with industrial practice from early on. Its industrial breakthrough in the Haber–Bosch process resulted from a combination of basic research (Fritz Haber's insight into the role of pressure in the ammonia equilibrium, actually an early application of a physicochemical principle to a catalytic reaction), of industrial research (a huge screening effort to find the most suitable catalyst, designed and driven by Alwin Mittasch), and of engineering innovation (the creation of high-pressure technology guided by Carl Bosch).

Present catalysis research has come from these roots, in particular from the first two of them. Catalytic reaction engineering creates environments allowing an optimum output from the interaction between catalyst and reactants on industrial scales. It is crucial for the practical application of catalysts and can give important impulses for their improvement, but it is beyond the scope of this book, because the catalyst is a constant in this field. Engineering models of catalytic processes include mass and heat transfer in catalyst beds and in catalyst particles as well as the kinetics of surface reactions, which may be covered even by microkinetic approaches nowadays. However, physics and chemistry of surface processes and the competence to make catalysts better, which are in the focus of this book, are beyond the scope of catalytic reaction engineering. For more information on state and recent development of this field, the reader may consult refs. [1, 2]. Textbooks that allow accessing it from the basics have been cited elsewhere in this book.

Catalysis research has always included the two directions epitomized by Haber's and Mittasch's work: the effort to understand the catalytic process in terms of physical chemistry, and the empirical search for new catalytic materials. Although the agenda of science is actually to make empiricism redundant, both directions benefited from each other: results of empirical work raised problems to be dealt with in fundamental investigations, and insights from the latter enhanced the design of screening work. The efficiency of basic research has been continuously advanced by the improvement of characterization tools and the extension of computing capacities. Catalyst screening saw an upsurge in the 1990s, which was inspired by the development of combinatorial drug design.

The term "combinatorial chemistry" designates techniques that allow (i) parallel preparation of a large number of compounds or samples (so-called libraries), (ii) a largely parallelized assessment of their merits with respect to a target property, and (iii) the automated processing of the data obtained in order to provide a background for human input into the process or to create the basis for an automated design of subsequent optimization steps. The method is best developed in the synthesis of peptides,

https://doi.org/10.1515/9783110632484-005

of which a huge number of individuals can be obtained by varying the sequence of well-known building blocks, for instance, by the split-and-pool principle. Due to the high throughputs achievable for synthesis and assessment of samples, combinatorial chemistry is also referred to as high-throughput screening or experimentation (HTS or HTE). In the development of solid materials, in particular of catalysts, where combinatorial library design is much more difficult to realize, the terms HTS or HTE are preferred.

Preparation and assay establish a relation between the target property of library members and their molecular structure or composition.[1] The accuracy of assays is limited because of the often enormous size of libraries to be analyzed. Therefore, only a selection of promising candidates (the leads) results from such effort, which is referred to as stage I study in HTS of catalysts. The leads are subsequently examined again in parallelized methodology, but now under better defined conditions and with more accurate analytical tools. From this stage II, a few candidates proceed to the traditional program of process development that includes lab reactor studies, examination of alternative preparations, physicochemical characterization, and work in pilot plant and full plant scales.

Equipment for the HTS of solid catalysts comprises facilities for automated preparation and testing of samples and an IT environment that allows control of the process, data tracking and mining, and library design. Arrays for catalyst synthesis are realized in microwell or titer plates, the points of which are supplied with precursor quantities according to the intended library structure by synthesis robots.

Having studied Section 3.1 of this book, the reader may doubt that the various techniques available for catalyst preparation, often with specific tricks and intricacies, are all eligible to automatization and provide useful results when performed under very simple, standardized circumstances. Preparation routes lend themselves to parallelization in different extents. The best results are obtained where a solid matrix submitted (e.g., a support pellet) is modified by liquid additives, i.e., in impregnation, ion exchange, deposition-precipitation, etc. Extended libraries of multi-metal samples for catalytic electrodes can be created by physical vapor deposition or sputtering techniques, e.g., by sequential deposition of metals on conducting substrates in suitable concentration profiles across the array surface.

Parallelization has been achieved even for protocols of zeolite synthesis and of sol–gel processes, which include washing and thermal conditioning steps. The formation of solid phases from large liquid volumes (e.g., precipitation, co-precipitation) or from the gas phase (e.g., spray drying) is, however, difficult to integrate into parallelized

---

1 In solid-state HTS, sample composition resulting from the preparation protocol is usually encoded in the location of samples in a sample array. In the most powerful split-pool techniques, traceability of polymer beads, on the surface of which library molecules are synthesized, would be lost during pooling unless special encoding strategies were applied.

techniques. Instead, bulk (mixed) metal oxide catalysts, which are typically synthesized by these methods, can be obtained by impregnating suitable precursors into porous carbon beads, which are then removed by combustion [3]. Regarding the presumably suboptimal microstructure of materials from robotic syntheses, it should be kept in mind that stage I HTS targets only the identification of leads, which are examined in more detail or abandoned in later stages.

The boom of HT materials development has resulted in new technology allowing the fast characterization of samples with respect to structure and reactivity. Diffracto-meters or spectrometers are equipped with movable sample holders that scan the sample array relative to the incident beam for sequential analyses. For special purposes, HT characterization can be performed under controlled atmospheres.

Catalytic testing requires suitable reactors: facilities for parallelized contact between samples and reactants in an appropriate hydrodynamic regime and for observing the resulting reaction rates by measuring output concentrations. Early proofs of concept, in which the activity of metal-containing catalyst beads in the oxyhydrogen reaction was assessed by just flowing the O_2/H_2 mixture across the titer plate and observing the colors of glowing beads, did not comply with any of these requirements. Hydrodynamics and product analyses have remained a permanent point of concern in equipment for stage I screening. On the other hand, in new equipment for type II studies, the quality of rate measurements is often comparable with that familiar from traditional methodology.

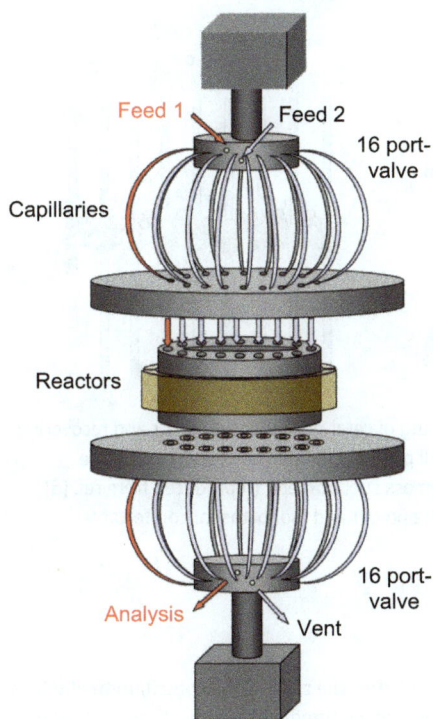

Figure 5.1: Parallel reactor for stage II high-throughput testing of catalysts. In normal operation, the upper valve just operates as (or may be replaced by) a gas distributor, the lower valve sequentially selects effluent streams for analysis. Reproduced from ref. [4] with permission from Elsevier.

In a stage II test setup, catalyst samples are placed in parallel reactor tubes heated by a multichannel oven as exemplified in Figure 5.1. They are charged with the same feeds at the same rates via a gas distributor (cf. legend to Figure 5.1). Effluents proceed to a selector valve that picks one stream for analysis and vents the remaining ones. In systems of this type, only the total gas flow needs to be regulated, its uniform distribution to the reactors is achieved by tuning the respective pressure drops.[2] Being fast in the analysis of effluent compositions, mass spectrometry is employed for concentration analysis in phase II HTS wherever possible. However, when selectivity tracking is mandatory, there may be no way around gas chromatography, where the choice of the separation method may limit the productivity of the whole system. Stage II HTS is also familiar for liquid-phase reactions, which are performed in sets of miniature reactors or autoclaves kept under identical conditions.

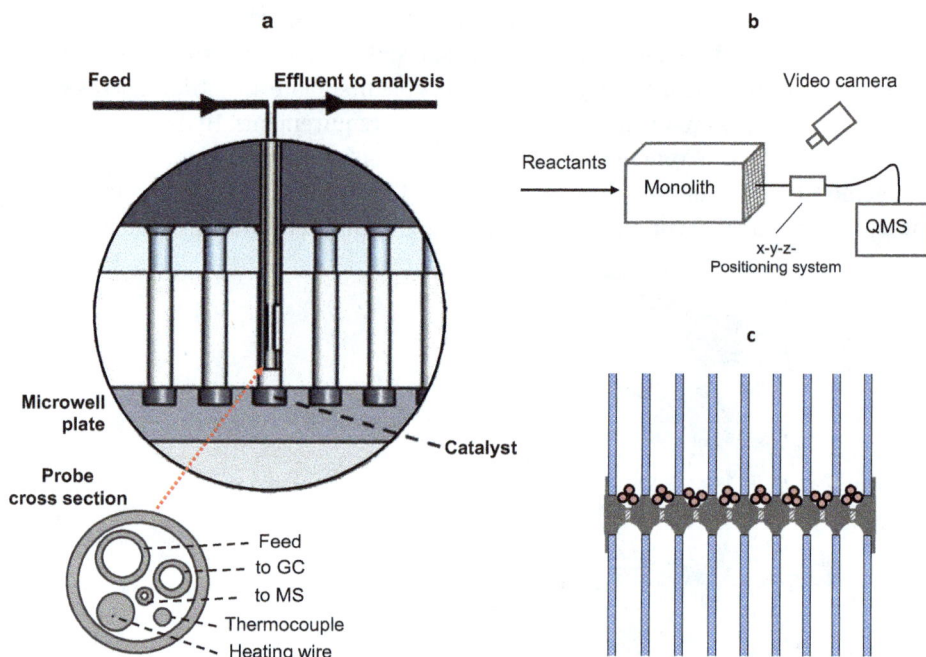

Figure 5.2: Approaches for stage I high-throughput screening of catalysts: (a) dosing feeds and recovering effluents by a probe approaching to the wells of a microwell plate via suitable bores in a cover plate (adapted from ref. [5]); (b) monolith reactor with scanning mass spectrometer (reproduced from ref. [6] with permission from Elsevier and (c) combination of sample holder and monoliths for comfortable handling and testing of samples (see text).

---

2 Which should be dominated by the capillaries before and after the reactor, the contribution of which should be negligible. In high-end equipment, each gas line may be fitted with an own flow controller.

Schemes with individual reactors as in Figure 5.1 are far too laborious for stage I testing of catalyst performance. Filling hundreds of tubes and reintegrating them into the setup would be the bottleneck of the procedure. On the other hand, the advantage of testing catalysts right on the microwell plate from preparation is bought by ill-defined gas flow around the pellets. Figure 5.2a shows an attempt to solve these problems. The feed is dosed and the effluent is withdrawn by probes approaching the samples through a mask, which is placed on the microwell plate and forms cylindrical channels above them.

Monoliths also allow establishing appropriate hydrodynamic conditions for catalyst testing. Other than in washcoated monoliths, the catalyst beds, which need to be kept in place, e.g., by glass wool plugs, use the whole channel cross sections. The effluents can be effectively sampled by a scanning mass spectrometer, which combines a sampling capillary capable of visiting the monolith channels with a mass analyzer (Figure 5.2b). The major challenge in this approach is the uniform distribution of the gas flow on the channels, its major retard the filling procedure. Figure 5.2c depicts an approach, in which two identical monolith sections are combined by a sample holder plate, which is easy to fill.[3] The system was used for testing more than 500 catalyst samples in one run employing a color reaction of the key reactant NO for performance assessment [3].

Stage I HTS has been also reported for heterogeneous catalysis in the liquid phase and for homogeneous catalysis. It is typically performed in 96-well titer plates familiar from pharmacy or in analogous arrangements of vials that can be heated or shaken.

After 30 years of HTS, hopes to discover completely new breakthrough compositions for catalysts or electrodes seem to have failed [7]. On the other hand, ideas and equipment developed during this effort are utilized everywhere in industry and academia. While highly automated systems are far too expensive for research institutes and serve purposes less typical for academic research, rate measurements parallelized on stage II level and HT equipment for structure analysis and spectroscopy also save valuable time in academia. In industrial catalyst research, fast optimization across parameter fields is a frequent task. Large companies tend to keep own HT equipment for this purpose, e.g., customer-tailored robotic laboratories, which are available in the market [7]. On the other hand, companies with suitable know-how offer HT services to customers. Catalysis-related projects have been successful also on the commercial scale, but the results do not appear in the open literature. Overviews about HTS for materials development including catalysis research can be found in refs. [4, 7, 8].

Other than suggested by some early claims, HTS has not replaced, but rather accelerated research on catalytic surface processes. Above all, it has revolutionized catalyst screening, which had, however, never ceased contributing to the body of knowledge fol-

---

3 When geometries of sample holder and preparation plate correlate, catalyst transfer from the latter to the former may be achieved by a simple head-over-heels turn.

lowing programs inspired also by the available insight into fundamentals. At the doorstep of an age facing the challenge of converting the energy and raw materials supply of mankind from fossil to renewable resources, HTS complements the choice of strategies available to catalysis for contributing to this development. Future will require progress of catalysis research on all levels of understanding, from accelerated screening of new materials and treatments via the creation and utilization of heuristic models for complex processes to the description of catalytic reaction systems by atomistic models. Likewise, alternative routes to catalytic materials and improved control over existing ones are urgently needed. An analysis of the present situation characterized by isolated players exchanging data just via publications and a strategy how to turn this into a unified research data infrastructure can be found in ref. [9].

The present epoch is sometimes referred to as Anthropocene age, because the activities of mankind exert a significant (negative) influence on the state of our planet, which is most obvious in the climate crisis. To cope with these effects, technologies are required that fulfill the material needs of mankind without using energy or raw materials from fossil sources. This may finally exclude also nuclear energy because of the radioactive waste disposal problem. Paths and intensities of the transition process depend strongly on political decisions on national or regional scales like ambitions in greenhouse gas reduction rates, attitudes to bridging technologies like nuclear power or CO_2 capture and storage, market interventions in favor or to the disadvantage of specific technologies, etc. Therefore, only some basic aspects will be outlined in the following discussion of opportunities and challenges for surface catalysis.

Running economy on renewable sources means relying almost completely on solar-driven processes for energy extraction[4] and on biogenic feedstocks.[5] Solar power makes large contributions to power consumption in many countries already now. With the exception of hydroelectricity, however, the output of solar-driven power generation is subject to fluctuations, e.g., the day–night cycle of photovoltaics or meteorological influences on wind energy. It may, therefore, temporarily fail to provide the load required by the related economy. In economies with high contributions of intermittent energy sources, this is compensated by electricity from permanent sources at present, e.g., power stations using fossil or nuclear fuels. The problem might be solved if excess electricity frequently delivered by the intermittent sources could be stored. Unfortunately, battery storage of electricity at capacities scaling with the needs of an economy is a problem far from any realistic solution. Therefore, methods are sought to store the energy into molecules or materials, from which it may be recovered when or wherever required.

---

4 Being actually not renewable, geothermal energy may contribute to climate protection if greenhouse gas savings by its utilization exceed expenses for installation and disposal of equipment.

5 This excludes inorganic materials produced from mineral resources like metals, or P- and N-fertilizers. Possible limitations of these sources are usually disregarded in the discussion of the more urgent climate problem.

At present, H_2 generated by water electrolysis is the primary storage molecule for electrical energy. If produced with solar electricity, it is labeled green. To avoid disadvantages due to low volumetric energy density, diffusivity, and corrosivity of H_2, it can be stored in suitable media by adsorption, absorption, or by hydrogenation of suitable molecules, but efficiencies of these processes need to be improved. Alternatively, H_2 may be employed to convert unreactive molecules like CO_2 or N_2 into fuels of high energy density, which may be easily stored and shipped (power-to-gas concept).

Reactions suitable for this purpose like ammonia synthesis or hydrogenation of CO_2 to methane (Sabatier reaction) or to methanol are well known, but improvements of catalysts and processes are highly welcome for their exploitation on capacity scales dictated by national energy systems. Methane can be distributed via existing pipeline nets. Handling of methanol or ammonia is feasible despite significant health risks. Both can be employed either as hydrogen storage media or as fuels. Electricity can be recovered directly from methanol provided efficiency and stability of the direct methanol fuel cell (DMFC) can be improved by further development. Being also an intermediate to chemicals like alkenes (via MTH), acetic acid, or formaldehyde, or to alternative fuels like dimethyl ether or polyoxymethylene dimethyl ethers, methanol is the most versatile energy storage molecule and may become an important platform chemical in the future.

These options are, however, penalized by the expenditure for providing N_2 or CO_2 at acceptable purities and concentrations. Concentrated CO_2 is available only from a few industrial sources as lime production[6] and fermentation processes, its separation from the ambient atmosphere is very costly. It can be, however, recovered after combustion processes, in particular if these are run with (almost) pure oxygen (oxyfuel processes). CO_2 cycles combining CO_2 hydrogenation with oxyfuel combustion of the resulting fuels are conceivable in a sustainable economy. They need air separation capacity to supply the oxygen released later from the cycle as H_2O. The coupled product N_2 can by hydrogenated to NH_3, which is a versatile hydrogen storage molecule as well. Therefore, CO_2 and N_2 hydrogenation may be favorably coupled in chemical energy storage technology.

Replacing oil and gas for feeding national economies by biomass meets very different challenges. Oil and gas are relatively easy to handle. They can be pumped; oil can be distilled into fractions suitable for intended uses. Originating from huge deposits, they can be supplied with constant properties over long periods of time, although composition and properties may differ substantially depending on the deposit tapped. Oil and gas consist almost completely of hydrocarbons; their oxygen content is very low, even below the optimum composition for some important uses, e.g., gasoline. Other minor heteroatom components (S-, N-, metal compounds) adverse for equipment and environment are removed during refining (cf. Appendix A1).

---

6 Lime and related products like cement are important fossil CO_2 sources, which need to be phased out like coal, oil, and gas. Hydrogenating CO_2 from lime burning can be, therefore, only a bridging technology.

Biomass can be of plant, animal, or microbic origin (ca. 80%, 5%, and 15%, respectively, of total biomass). Being highly polar polymers, most biomaterials are solids, not meltable, and poorly soluble in almost all solvents, and need to be digested for further chemical processing. Fats, oils, or fragrances are notable exceptions from this rule. Plant biomass for the most part consists of carbohydrates, e.g., cellulose and hemicellulose (ca. 40% and 25%, respectively, of total biomass [10]), starch, and saccharose. Chitin and chitosane, which are important types of animal biomass, are polyamino saccharides. Lignin, a major constituent of wood, features a complex structure, which can be conceived as resulting from random couplings of p-hydroxy cinnamyl alcohols. Its contribution to total biomass is estimated as 20–30% [10], while other natural compounds like fats, oils, and proteins make up for ca. 5%. More information about the nature and abundance of biomaterials can be found in ref. [10].

The oxygen content of biomass is much higher than that of oil or gas, which have gone through a geologic processing. The O/C ratio of carbohydrates like (hemi)cellulose is ≈0.8 (at H/C ≈ 1.5), that of lignin is still around 0.4. Therefore, deoxygenation is an important step in many approaches for biomass valorization. Biomaterials contain also mineral components (ashes), which may be potent catalyst poisons.

Different from oil and gas, the availability of many biomaterials (e.g., crops) is subject to seasonal variations. Their sources (fields and forests) are spread over wide territories, from which they must be collected for processing. This favors a decentralized structure of many small, flexible plants operating mostly in batch mode opposite to oil refining, which is run in a few large plants that benefit from the advantages of continuous operation and from economy of scale. With the exception of some thermal processes, biomass is digested with liquid reactants. Most consecutive steps proceed in the liquid phase as well. This suggests a preference for homogeneous and biocatalysts in biomass valorization, but the classical advantage of heterogeneous catalysts, their easy separation from the reaction medium, calls them into play also in this field.

In biomass processing, requirements to solid catalysts differ significantly from those to catalysts used in oil refining or chemical industry. Catalysts should be stable against leaching and poisoning in polar media in a wide pH range, but also in unusual environments (e.g., ionic liquids, water under subcritical or supercritical conditions). As discussed in ref. [10], this suggests a new role for polymer-based catalysts. Indeed, depolymerization of cellulose by an ion-exchange resin was achieved in a ionic liquid, which is able to dissolve even wood. Reference [10] highlights also some other aspects important for the application of solid catalysts in biomass valorization.

Catalysis with selected biomass feedstocks has been known for ages, e.g., the alcoholic fermentation of sugar or soap-making from oils and fats. Oleochemistry, the study of oils and fats, is a traditional field of chemistry and technology, which includes many catalytic reactions. Various uses of cellulose, e.g., the production of cellulose fibers, involve homogeneous catalysis at some stages.

Early approaches for a broader use of biomass were focused on thermal procedures like pyrolysis and gasification. Wood pyrolysis has been exercised for centuries to obtain charcoal. The spectrum of its byproducts,[7] however, does not recommend this approach for a broader use. Biomass (forest litter, crop straws, etc.) can be converted to syngas via modified coal gasification technologies, which attracted attention to this field far before the chemical storage of energy became a major incentive. However, the aspect that delicate chemical structures provided by nature could be better utilized than by dismantling them in order to make simpler molecules confines the role of this approach to biomass of low value.

Meanwhile, production of biofuels from energy crops, such as bioethanol by fermentation of glucose from sugar cane, maize or even cellulose, or biodiesel by transesterification of vegetable oils, has grown to industrial scales (for a discussion on sustainability of related processes see ref. [10]). This has raised concern about competition between biofuels and food production for land and feedstock, resulting in enhanced deforestation. Three generations of biomass feedstock are now differentiated (cf. ref. [11]). Materials rich in sugars, starches, or oils are designated first-generation feedstock. The focus of research has shifted to second-generation feedstock like lignocellulose from various sources, organic wastes, or glycerol, which is an excess co-product of biodiesel production, while work with the third-generation feedstock algae is on an exploratory level.

Notably, the ultimate goal of biomass valorization is far beyond biofuel production because, finally, all materials made nowadays on a petrochemical basis must be either accessible starting from biomass or CO_2, or substituted by other materials accessible on this basis. Scenarios on how to achieve this transition feature a huge variety of opportunities also for surface catalysis. These are easier to take than the digestion steps, but include serious challenges in particular with respect to selectivity. In ref. [11], the reader can find information on the state of research in this field and about primary treatment of lignocellulose biomass.

Today, biofuels are admixed to conventional fuels in quantities tolerated by existing engines. Their further role in the competition between advanced internal combustion engines and electric cars appears open. In a similar way, new products from bio-based feedstock and related processes are likely to be introduced as drop-in products or processes into existing infrastructure first [12], while routes to a sustainable economy that allow keeping societies stable on the way are sorted out.

A forecast on likely developments of surface catalysis will, therefore, combine recent tendencies in research and technology of fossil-based processes with the challenges of technology transition. The everlasting strive to improve activity, selectivity, and stability of catalysts in order to make better use of energy and raw materials, to allow transition to cheaper feedstock, and to improve properties of products will continue on both fossil and biogenic platforms. Improved environmental protection will remain urgent as long as lives of people are affected by pollutants, and greenhouse gases are handled

---

7 Methane, H_2, CO, CO_2, a bio-oil of often low stability and caloric value.

and emitted. A greater role of (photo)catalysis in wastewater treatment, consequent abatement of methane emissions from various sources, compact NO and soot abatement in diesel exhaust, and exhaust treatment for ships of all tonnages remain important incentives. Combining catalysis and separation as in catalytic membranes or catalytic distillation in more industrial-scale processes, replacing more radical polymerization processes by catalytic technology (e.g., for copolymerization of alkenes with polar monomers), and saving waste in fine chemicals production by finding more catalytic routes, are examples for goals important in industry [12].

The quest to achieve high performance with earth-abundant and environmentally harmless catalysts is an overriding goal in these efforts, be it in mature technology like three-way catalysis or in emerging processes like photo- or electrocatalytic water splitting. A breakthrough to low-temperature fuel cells allowing efficient and stable operation at low Pt contents would, for instance, dramatically change competition among automotive drive technologies.

The challenges of technology transition can be met only when all kinds of energy inputs into the catalytic process are taken into account [12]. This explains the recent renaissance of electrocatalysis and the upsurge of photo- and photoelectrocatalysis. Ammonia synthesis not via the Haber–Bosch process, for instance, but at room temperature on catalytic electrodes, has become a hot topic. Photocatalysts allow forming NH_3 (and O_2) from N_2 and water [13]. Recently, catalytic NH_3 formation from N_2 and H_2 has been observed to be induced even by a mechanochemical procedure at room temperature [14]. By analogy, H_2, which is now predominantly produced by steam reforming of natural gas or, to improve sustainability, by water electrolysis , can be also made by steam reforming of low-grade bio-oils or glycerol. Photocatalytic processes with solid or molecular catalysts allow direct splitting of water into H_2 and O_2, or supply H_2 while the oxygen reacts with organic contaminants in the (waste) water feed. In the future, competition between different fields of catalysis will be complemented by more cooperation, which may result in solutions that combine different versions of catalysis in cascade mechanisms or even in one catalyst [12]. Solutions allowing direct utilization of sunlight without the detour via electricity or high temperatures will be preferred. They are, however, most challenging. The artificial leaf, for instance, which produces sugars (or at least hydrocarbons) from CO_2, H_2O, and visible light is a vision at present rather than a realistic goal.

As mentioned earlier, the increasing biogenic origin of material flows will require more catalysts stable in highly polar liquid media, toward pH excursions and poisons. They should tolerate variations in feed properties and transient process regimes. Catalysts should be identified that can be applied in unusual media like supercritical fluids and in plasma. In electrocatalysis, three-dimensional electrodes provide higher surface areas for the interaction with the liquid phase, and, hence, higher rates in production processes and improved sensitivity in analytic applications. Further improvement of their performance can be expected from the optimization of flow conditions in them using advanced microfabrication techniques. The development has inspired visions of nanostructured catalytic devices made, for instance, by 3d-printing and by tools of nanomanipulation, in which molecular traffic control guides liquid reactants through a set of nanoreactors providing the catalytic functions necessary to operate a complex reaction network, which would nowadays require several process steps.

Although "prognoses are difficult, in particular if they deal with the future,"[8] it is certainly safe to predict that despite a foreseeable decay of some branches, surface catalysis will remain a key element of chemical, energy, and environmental technologies during the transition of economies to a sustainable basis.

References

[1] Salciccioli, M.; Stamatakis, M.; Caratzoulas, S.; Vlachos, D. G., A review of multiscale modeling of metal-catalyzed reactions: Mechanism development for complexity and emergent behavior. *Chemical Engineering Science* **2011**, *66* (19), 4319–4355. doi.org/10.1016/j.ces.2011.05.050

[2] Deutschmann, O., *Modeling and Simulation of Heterogeneous Catalytic Reactions: from the Molecular Process to the Technical System*. Wiley-VCH: Weinheim, 2013. ISBN: 978-3-527-63988-5

[3] Schüth, F.; Baumes, L.; Clerc, F.; Demuth, D.; Farrusseng, D.; Llamas-Galilea, J.; Klanner, C.; Klein, J.; Martinez-Joaristi, A.; Procelewska, J.; Saupe, M.; Schunk, S.; Schwickardi, M.; Strehlau, W.; Zech, T., High throughput experimentation in oxidation catalysis: Higher integration and "intelligent" software. *Catalysis Today* **2006**, *117* (1), 284–290. doi.org/10.1016/j.cattod.2006.05.038

[4] Farrusseng, D., High-throughput heterogeneous catalysis. *Surface Science Reports* **2008**, *63* (11), 487–513. doi.org/10.1016/j.surfrep.2008.09.001

[5] Kim, D. K.; Maier, W. F., Combinatorial discovery of new autoreduction catalysts for the CO_2 reforming of methane. *Journal of Catalysis* **2006**, *238* (1), 142–152. doi.org/10.1016/j.jcat.2005.12.001

[6] Claus, P.; Hönicke, D.; Zech, T., Miniaturization of screening devices for the combinatorial development of heterogeneous catalysts. *Catalysis Today* **2001**, *67* (4), 319–339. doi.org/10.1016/S0920-5861(01)00326-1

[7] Maier, W. F., Early years of high-throughput experimentation and combinatorial approaches in catalysis and materials science. *ACS Combinatorial Science* **2019**, *21* (6), 437–444. doi 10.1021/acscombsci.8b00189

[8] Potyrailo, R.; Rajan, K.; Stoewe, K.; Takeuchi, I.; Chisholm, B.; Lam, H., Combinatorial and high-throughput screening of materials libraries: Review of state of the art. *ACS Combinatorial Science* **2011**, *13* (6), 579–633. doi 10.1021/co200007w

[9] Wulf, C.; Beller, M.; Bönisch, T.; Deutschmann, O.; Hanf, S.; Kockmann, N.; Krähnert, R.; Özaslan, M.; Palkovits, S.; Schimmler, S.; Schunk, S. A.; Wagemann, K.; Linke, D., A Unified research data infrastructure for catalysis research – Challenges and concepts. *ChemCatChem* **2021**, *13* (14), 3223–3236. doi.org/10.1002/cctc.202001974

[10] Rinaldi, R.; Schüth, F., Design of solid catalysts for the conversion of biomass. *Energy & Environmental Science* **2009**, *2* (6), 610–626. doi 10.1039/B902668A

[11] Ning, P.; Yang, G.; Hu, L.; Sun, J.; Shi, L.; Zhou, Y.; Wang, Z.; Yang, J., Recent advances in the valorization of plant biomass. *Biotechnology for Biofuels* **2021**, *14* (1), 102. doi 10.1186/s13068-021-01949-3

---

8 Karl Valentin (1882–1948), Bavarian humorist.

[12] Centi, G.; Perathoner, S., *Science and Technology Roadmap on Catalysis for Europe*. European Research Intstitute on Catalysis a.i.s.b.l.: 2016. http://gecats.org/gecats_media/ Science+and+Technology+Roadmap+on+Catalysis+for+Europe+2016.pdf, accessed Mar 12, 2023

[13] Medford, A. J.; Hatzell, M. C., Photon-driven nitrogen fixation: Current progress, thermodynamic considerations, and future outlook. *ACS Catalysis* **2017**, *7* (4), 2624–2643. doi.org/10.1021/ acscatal.7b00439

[14] Reichle, S.; Felderhoff, M.; Schüth, F., Mechanocatalytic room-temperature synthesis of ammonia from its elements down to atmospheric pressure. *Angewandte Chemie-International Edition* **2021**, *60* (50), 26385–26389. doi 10.1002/anie.202112095

Appendix

A1 Basic information on applications of (thermal) catalysis, photocatalysis, and electrocatalysis

Ammonia synthesis (Haber–Bosch process)

Purpose: Fixation of nitrogen from air to get a starting point for the production of N fertilizers, of explosives, etc. Practically, all nitrogen in synthetic materials originates from NH_3. With solar hydrogen, NH_3 might be a future energy storage molecule.
Principle/chemistry: See eq. (2.19).
Catalysts: Unsupported Fe or supported Ru, both multipromoted.
Reaction mechanism: Figure 2.57.
Details on catalysts: Preparation – Sections 3.1.1 and 3.1.1.4; microstructure – Figure 4.17; promoters – Section 4.2.4, Figures 4.16 and 4.17; rationale behind activity trends – Sections 4.2.1, 4.2.3, and Figure 4.12

Ammonia oxidation (Ostwald process)

Purpose: Oxidation of NH_3 to NO, with subsequent process stages to obtain NO_2 and HNO_3.
Principle/chemistry: See eq. (2.67).
Catalysts: Pt-Rh gauzes (Figure 2.67).
Reaction mechanism: See Section 4.5.3.
Miscellaneous: To achieve kinetic control required to avoid N_2 formation, the reaction is performed at high temperatures (1,100–1,200 K) and contact times on the order of 1 ms (ultra-short contact time catalysis, Section 2.8.2). Pronounced initial transients of the catalyst gauzes [1], deactivation by Pt loss (Section 4.4.1.2).

Methanol synthesis

Purpose: Production of methanol from syngas. As syngas can be obtained completely from renewable sources, methanol may become an important future platform molecule.
Principle/chemistry: Equations (4.21) or (4.22) combined with eq. (2.10), depending on the catalyst, cf. Section 4.5.6.
Catalysts: Commercial: $Cu/ZnO–Al_2O_3$, other Cu-based systems under research, not commercialized - supported Pd, in history: $ZnO–Cr_2O_3$.
Reaction mechanism: Section 4.5.6, Figure 4.49.

https://doi.org/10.1515/9783110632484-006

Details on (Cu) catalysts: Preparation – Sections 3.1.1.1, and 3.1.4; role of Zn – Section 4.2.3; promoters – Section 4.2.4; response to redox potential – Figures 4.18, 4.49, and 3.156.

Selective oxidation of methanol

Purpose: Production of formaldehyde from methanol.
Principle/chemistry: See eq. (2.69). The reaction is commercialized in fundamentally different processes: the Formox process operating between 520 and 670 K (a) and silver-based processes (900–1,000 K (b)).
Catalysts: (a) $Fe_2(MoO_4)_3$ and (b) Ag.
Reaction mechanism: Process (a) is typical oxidative dehydrogenation (cf. Section 4.3.3); in (b), eq. (2.69) may be superimposed by nonoxidative dehydrogenation ($CH_3OH \rightarrow CH_2O + H_2$), in particular at higher temperatures. H_2 can be oxidized to water or appear in the effluent.
Miscellaneous: (a) To make up for Mo losses, catalysts are prepared with excess MoO_3 (cf. Section 4.4.1.2).

(b) An example of ultra-short contact time catalysis (Section 2.8.2). To stabilize the Ag granules, the reaction is performed at high partial pressure of water. This limits the formaldehyde concentration in the end product to ca. 37% (formalin). Higher concentrations are accessible via the Formox process (a).

Recent reviews: (a) – ref. [2] and (b) – ref. [3].

Methanol to gasoline (MTG), Methanol to olefins (MTO), Methanol to propylene (MTP) - see Methanol to hydrocarbons (MTH)

Methanol to hydrocarbons (MTH), also designated as Methanol to gasoline (MTG), Methanol to olefins (MTO), or Methanol to propylene (MTP)

Purpose: Synthesis of highly aromatic hydrocarbon fractions (MTG) or of hydrocarbon fractions rich in alkenes (MTO and MTP) from methanol.
Principle/chemistry: See Scheme 4.1.
Catalysts: H zeolites or H-SAPOs of different pore systems (cf. Section 4.3.2).
Reaction mechanism: Sections 4.2.3, 4.3.2, and 4.5.5.

Fischer–Tropsch (FT) reaction

Purpose: Synthesis of hydrocarbons of wide length distribution from syngas. Products may be *n*-alkanes and/or *n*-alkenes, with some catalyst types, significant selectivities for oxygenates may be obtained. With syngas from solar sources, FTS may become a major source of sustainable fuels and hydrocarbons for material synthesis.

Principle/chemistry/mechanism: Surface polymerization of adsorbed C_1 monomers; see Section 4.5.3, Figures 4.44 and 4.45.

Catalysts: Commercial – multipromoted unsupported Fe catalysts (high temperature) or supported Co catalysts (low temperature); not commercialized – Ru, Rh.

Details on catalysts: Promoters – Sections 4.2.4 and 4.3.5; poisoning – Section 4.4.1.1; rationale behind activity/selectivity trends – Section 4.5.3, Figure 4.46.

Higher alcohol synthesis (HAS)

Purpose: Synthesis of alcohols $C_nH_{2n+1}OH$ ($n > 1$) from syngas.

Principle: Combination of C_1 monomers with and without bond to oxygen.

Catalysts: There are four approaches to arrange the coexistence of C_1 monomers with and without C–O bond on the catalyst surface:
- Modification of FT catalysts (in particular Co catalysts) with Cu, which hydrogenates CO without splitting the C–O bond.
- Enhancing the C–O splitting tendency in methanol synthesis catalysts by adding suitable promoters, e.g., Cs.
- Ru- or Rh-based FT catalysts: Due to an only moderate C–O splitting tendency, they form the required mixture of C_1 monomers per se.
- (Promoted) Mo-based catalysts, e.g., MoS_2. Their hydrogenation activity is modified by promoters to enhance their CO-splitting capabilities

Miscellaneous: Not commercialized, but of interest for a sustainable economy; recent reviews: refs. [4, 5], the latter with a focus on CO_2 hydrogenation.

Steam reforming of methane (SRM)

Purpose: Production of syngas (or hydrogen) from methane and water.

Principle/chemistry: Equation (4.1); highly endothermal (ΔH_R = +206 kJ/mol)

Catalysts: (Promoted) Ni/Al_2O_3 (commercial), Ru, Rh, Ir

Reaction mechanism: Stepwise dehydrogenation of CH_4 without or with assistance of O_{ads} (#1, #2, respectively, * – Ni site) provided by the dissociation of water (#3). CO is formed from CH_x ($x = 1$ or 0) and O_{ads} (#4), H_2 from recombination of 2 H_{ads} (not shown). Methane dissociation is rate determining.

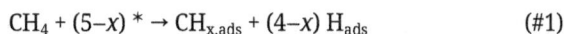

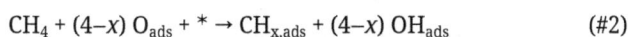

$$CH_4 + (5{-}x) \, {}^* \rightarrow CH_{x,ads} + (4{-}x) \, H_{ads} \qquad (\#1)$$

$$CH_4 + (4{-}x) \, O_{ads} + {}^* \rightarrow CH_{x,ads} + (4{-}x) \, OH_{ads} \qquad (\#2)$$

$$H_2O + 3 \, {}^* \leftrightarrows 2\,H_{ads} + O_{ads} \tag{#3}$$

$$CH_{x.ads} + O_{ads} + (x{-}1)^* \rightarrow CO_{ads} + x\,H_{ads} \tag{#4}$$

Details on catalysts: Preparation – Section 3.1.1.1; poisoning by coking and counter-measures – 4.2.2, 4.2.3, 4.3.1, and 4.4.2.

Dry reforming of methane (DRM)

Purpose: Production of syngas from methane and CO_2.
Principle/chemistry: $CH_4 + CO_2 \rightarrow 2\,CO + 2\,H_2$; highly endothermal ($\Delta H_R$ = +247 kJ/mol).
Catalysts: Similar to SRM.
Reaction mechanism: Similar to SRM, (#3) replaced by (stepwise) dissociation of CO_2.
Miscellaneous: Not commercialized as an individual process, but part of SRM, which is often run with admixture of CO_2. Deactivation by coke formation via endothermal methane pyrolysis ($CH_4 \rightarrow C + 2\,H_2$) or the exothermal Boudouart equilibrium ($2\,CO \leftrightarrows C + CO_2$) is a key problem.

Catalytic partial oxidation (POX) of methane

Purpose: Production of syngas (or hydrogen) from methane and oxygen.
Principle/chemistry: $2\,CH_4 + O_2 \rightarrow 2\,CO + 4\,H_2$.
Catalysts: (Supported) Ni, Co, or noble metals like Pt, Rh.
Reaction mechanism: Pronounced hot spots at the entrance of fixed beds suggest a two-step scheme with initial total oxidation and subsequent conversion of H_2O and CO_2 with residual methane via SRM and DRM, respectively. Measurements of axial concentration profiles with a Rh catalyst revealed formation of CO, H_2, and H_2O right from the reactor entrance, the latter being converted via SRM further in the bed. CO_2 formation was negligible (cf. ref. [6]).
Miscellaneous: Not commercialized.

Methanation

Purpose: Conversion of CO (a) or CO_2 (b) to methane for reducing the CO content of ammonia syngas to ppm levels (a) or (with solar hydrogen) as a step for storage of solar energy or of CO_2 in cycles in a sustainable economy (b).
Principle/chemistry: (a) Reverse of (4.1).

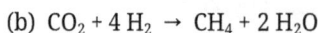

$$\text{(b)} \quad CO_2 + 4\,H_2 \rightarrow CH_4 + 2\,H_2O \qquad \text{(Sabatier reaction)}$$

Catalysts: (Promoted) Ni/Al_2O_3, Ru (active, but expensive).
Miscellaneous: (b) is part of power-to-gas projects (cf. Chapter 5).

Oxidative coupling of methane (OCM)

Purpose: Production of ethane/ethene from methane.
Principle/chemistry: Equation (3.18).
Catalysts: Mostly promoted metal oxides that activate CH_4 by creating surface defects rather than by a reduction of the metal cation, e.g., Li-doped MgO, Sr/La_2O_3, but also $(SiO_2$- or MgO-) supported Mn-doped Na_2WO_4.
Reaction mechanism: Under stationary conditions, methyl radicals formed on the surface recombine in the adjacent gas-phase layers (cf. Section 4.5.4).
Miscellaneous: Not commercialized. At the high temperatures required for methane activation by known catalysts (950–1,200° K), C_2 selectivity strongly decreasing with conversion renders C_2 yields >30% inaccessible under stationary conditions.
For more processes of methane activation, see ref. [6].

Water-gas shift

Purpose: Tuning the CO/H_2 ratio in the production of syngas; in syngas for NH_3 or in H_2 production – transformation of CO to CO_2, which is subsequently removed by scrubbing, with co-production of H_2
Principle/chemistry: Equation (2.10).
Catalysts: In large-scale applications - high-temperature shift (HTS) – $Fe_2O_3(80–90\%)$/ Cr_2O_3, low-temperature shift (LTS) – $Cu/ZnO–Al_2O_3$; for H_2 generation from hydrocarbons in compact devices, e.g., for fuel-cell-driven cars (not commercial) – Pt, Au, or Cu on reducible supports (often containing Ce).
Reaction mechanism: HTS – redox cycle with active phase Fe_3O_4: CO reduces Fe^{3+}, H_2O reoxidizes Fe^{2+}; LTS and noble metal-based catalysts – competing hypotheses with formate or carboxylate intermediates, cf. ref. [7].
Miscellaneous: HTS catalysts tolerate more S in feeds than LTS catalysts. At the higher reaction temperatures required by them (>620° K), residual CO contents are limited to 3–5% by thermodynamic equilibrium. For H_2 or NH_3 syngas, LTS catalysts operating at ≈470° K leave behind ≈0.3% CO, which is converted to methane subsequently (see "methanation"). See also ref. [7].

Hydrogenation

Hydrogenation catalysis is an extended, multifaceted field. Substrates range from simple alkenes, alkynes, and aromatic rings via aldehydes, ketones, carboxylic acids, and esters to nitro groups, nitriles, and imines. Applications include tasks of selective hydrogenation (see later) and of hydrogenation in the presence of sulfide poisons. Basic knowledge on catalysts suitable for various substrates and on typical complications (e.g., side reactions as in the hydrogenation of nitriles or nitroaromatics), which is not very recent, is summarized in ref. [8]. Hydrogenation in the presence of sulfur can be achieved using Co–Mo, Ni–Mo, and Ni–W catalysts applied in hydrotreatment (cf. Sections 2.3.3, 4.2.3), but also with zeolite-supported noble metals (Pt–Pd, see, e.g., ref. [9]).

Selective hydrogenation of unsaturated hydrocarbons

Purpose: Hydrogenation of triple bonds or of conjugate double bonds to isolated double bonds.
Chemistry: $R-C{\equiv}C-R' + H_2 \rightarrow R-CH{=}CH-R'$, e.g., ethyne and propyne
$\qquad\qquad R-CH{=}CH-CH{=}CH-R' + H_2 \rightarrow R-CH_2-CH{=}CH-CH_2-R'$, e.g., butadiene
Catalysts: Promoted Pd on macroporous supports or deposited in eggshell profiles; for strategies to enhance selectivity, see Section 4.3.4.
Miscellaneous: Several applications in the processing of steam cracker products: hydrogenation of ethyne or propyne in an excess of ethene or propene in the C_2 or C_3 cut, respectively, hydrogenation of residual butadiene after butadiene extraction from C_4 cut.

Selective hydrogenation of unsaturated aldehydes/ketones

Purpose: Hydrogenation of C=O in the presence of C=C bond (reverse order is straightforward with catalysts for the hydrogenation of C=C bonds or of aromatics, e.g., Ni, Pd, and Pt).
Chemistry: $R-CH{=}CH-(CH_2)_n-C(O)-R' + H_2 \rightarrow R-CH{=}CH-(CH_2)_n-CHOH-R'$ $\qquad$ $(n \geq 0)$
Catalysts: (1) Group 8–10 metals, in particular Pt and Ru, but also Ni and Co; (2) group 11 metals: Au, Ag, and Cu. Group (1) prefers C=C relative to C=O if not modified by promoters or by a second metal (for strategies of modification, see Section 4.3.4.), group (2) achieves selectivity also unmodified, but at low activity.
Miscellaneous: Selectivity easier to achieve with $n > 0$. For details, see Section 4.3.4, refs. [10, 11].

Hydrotreatment

The term "hydrotreatment" summarizes treatments applied to oil fractions in order to remove heteroatoms (HDS, HDN, HDMe for S, N, metals, respectively)[1] and aromatics from them. Heteroatoms are released in reactions, which are parallel, but not independent. Hydrotreatment of very heavy fractions (residue hydrotreatment) involves also mild cracking and removal of asphaltenes (cf. Section 4.4.1.1).

Hydrodesulfurization (HDS)

Purpose: Splitting of S from S-containing contaminants[2] of oil fractions without or with concomitant hydrogenation of aromatic ingredients depending on intended uses in gasoline or diesel, respectively.

Chemistry/principle: Direct desulfurization of the aromatic system (DDS) or hydrogenation with subsequent splitting of the C–S single bond (HYD), cf. **Scheme A1a**. As DDS requires S vacancies, the relation between these routes depends on catalysts and reaction conditions.

Catalysts: For gasoline – Al_2O_3-supported Co–Mo-sulfide, keeping aromaticity, for diesel and heavier fractions – Al_2O_3-supported Ni–Mo-sulfide, favoring hydrogenation, for heavy fractions Al_2O_3-supported Ni–W-sulfide, but also unsupported mixed sulfides [12], cf. Sections 2.3.3, 4.2.1, and 4.2.3.

Miscellaneous: Reaction conditions and reactors depend strongly on the boiling range of feeds. HDS of heavier fractions is affected by a poisoning influence of most N-containing contaminants (cf. refs. [13, 14] for the interactions between HDS and HDN). For information on the activation of HDS catalysts see Section 3.1.5, on off-site regeneration – Section 4.4.3.

Hydrodenitrogenation (HDN)

Purpose: Splitting of N from N-containing, mostly aromatic contaminants of oil fractions, with concomitant hydrogenation of aromatic ingredients.

---

1 HDO is of little relevance for oil fractions, because their O-containing contaminants are relatively unstable. It is, however, becoming a key technology for biofuel refining.

2 For stability reasons, only aromatic structures are relevant, thiophene being the least, substituted dibenzothiophenes the most challenging substrates.

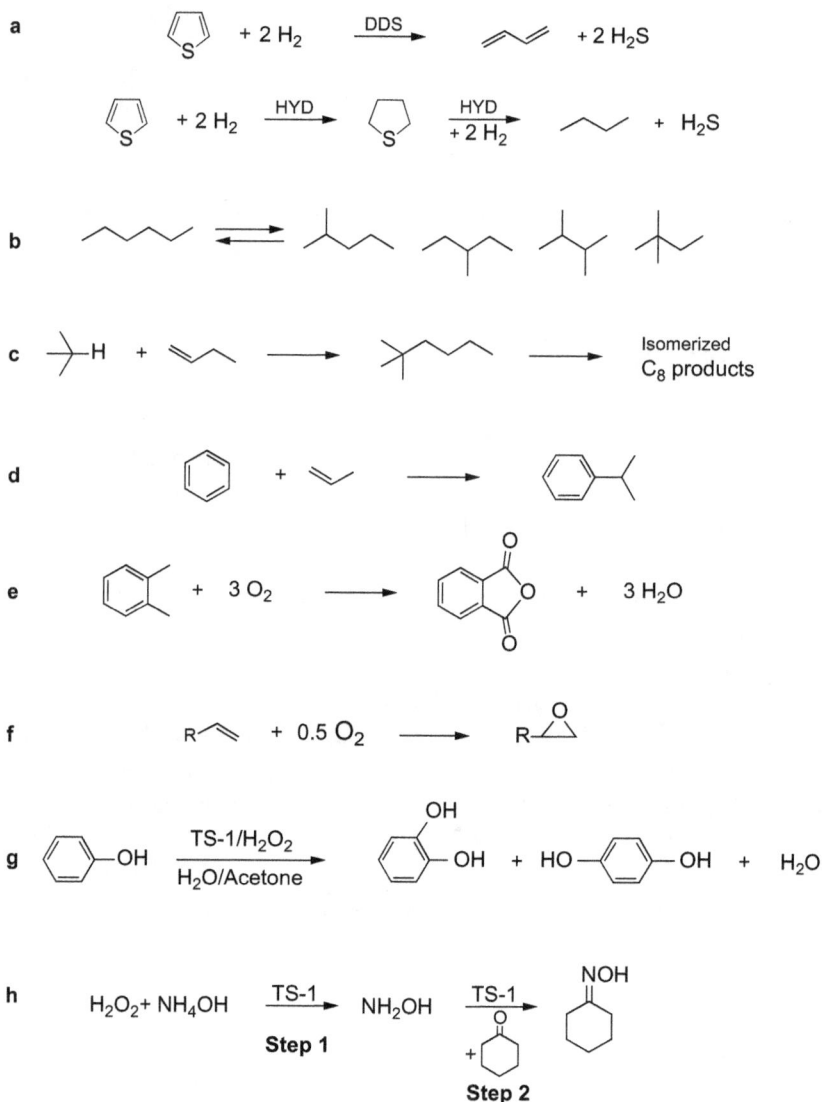

Scheme A1: Catalytic reactions treated in Appendix A1 (a) DDS and HYD routes of HDS exemplified with thiophene; (b) light naphtha isomerization exemplified with *n*-hexane; (c) isobutane alkylation exemplified with 1-butene; (d) Friedel–Crafts alkylation of benzene with propene; (e) oxidation of *o*-xylene to phthalic anhydride; (f) epoxidation of ethene (R = H) or propene (R = CH_3); (g) hydroxylation of phenol; and (h) ammoximation of cyclohexanol.

Chemistry/principle: Hydrogenation of rings containing N atoms (HYD route), subsequent hydrogenolysis (ring opening and NH_3 elimination).
Catalysts: Al_2O_3-supported Ni–Mo-sulfide, for very heavy fractions Al_2O_3-supported Ni–W-sulfide. But also unsupported mixed sulfides [12].

Miscellaneous: HDN may be favored by H_2S (not by S-containing contaminants) for reasons that are discussed in refs. [14, 15].

Hydrodemetallization (HDMe)

Purpose: Removal of metal compounds from heavy oil fractions, typically residues.
Chemistry/principle: Destabilization of the metal complex by hydrogenation and hydrogenolysis of its ligands (e.g., phthalocyanine structures). The metal is deposited where the reaction occurs.
Catalysts: Ni–Mo on wide-pore Al_2O_3.
Miscellaneous: Catalyst deactivation is irreversible; therefore, HDMe is often performed in guard beds with cheap, disposable catalysts that precede the HDN/HDS stage (cf. Section 4.4.2).

Hydrocracking

Purpose: Enhancing yields of light transportation fuels (in particular diesel) from oil by cracking vacuum distillates, low-value gas oils, or residues.
Principle: Catalytic cracking at high H_2 pressures with a catalyst containing a metal (or sulfide) component to delay deactivation of acid sites by coking (cf. Section 4.3.1).
Catalysts: Al_2O_3-supported Ni–Mo or Ni–W catalyst for HDS/HDN in front of cracking catalyst (noble metal/acidic support), cf. Sections 4.4.1.1 and 4.4.2.
Mechanism: Sections 4.5.2.1 and 4.5.4.
Miscellaneous: Expensive, but highly versatile, suited for very heavy feeds; in trickle beds; for hydrocracking residues with high coking contents slurry processes are considered.

Fluid catalytic cracking (FCC)

Purpose: Enhancing yields of light transportation fuels (in particular gasoline) from oil by cracking vacuum distillates and low-value gas oils from other processes.
Principle: Very fast catalytic cracking under drastic conditions in the absence of H_2, reactor-regenerator scheme: combustion of coke during regeneration in stationary fluid bed introduces reaction heat required for cracking in a riser (cf. Sections 2.8.2 and 4.3.1).
Mechanism: Sections 4.5.2.1 and 4.5.4.
Catalysts: Polyfunctional catalyst combining several components in a complex architecture (cf. Section 4.4.2, Figure 4.25); cracking components can be US-Y, rare-earth-modified Y and H-ZSM-5 (cf. Sections 4.3.2 and 4.5.2.1).

Mechanism: Section 4.5.2.1.
Miscellaneous: Catalyst deactivation – Sections 4.4.1.1 and 4.4.1.3; FCC is cheaper than hydrocracking, but due to the need to match heats recovered from coke combustion and required for the cracking reaction less flexible.

Naphtha reforming

Purpose: Enhancing octane numbers of straight-run heavy naphthas by synthesis of C_{7+} aromatics from saturated cyclic and open-chain C_{7+} hydrocarbons.
Chemistry: Section 4.5.4, Figure 1.6c.
Catalysts, promoters: Sections 4.2.4, 4.3.5, and 4.4.2.
Miscellaneous: Initial catalyst stabilization – Section 3.1.5; catalyst deactivation and countermeasures – Sections 4.4.1.1 and 4.4.3; process versions – Section 4.3.1.

Light naphtha isomerization

Purpose: Enhancing octane numbers of straight-run light naphthas by isomerization (isohexanes instead of benzene).
Chemistry: Scheme A1b.
Catalysts: Pt on acidic supports: chlorided γ-Al_2O_3 or mordenite.
Mechanism: See hydrocracking (Section 4.5.2.1), due to lower temperatures (420–520 K), isomerization dominates over cracking.
Miscellaneous: Residual n-alkanes are recycled after n-/i-alkane separation by molecular sieves.

Isobutane alkylation

Purpose: Synthesis of high-octane fuel components from isobutane and alkenes (mixed alkenes in dilute streams).
Chemistry: Scheme A1c.
Catalysts: Commercial processes work with corrosive, toxic homogeneous catalysts like H_2SO_4 or HF. Tremendous effort to replace them by solid acids has not yet succeeded in delaying coking to an extent that would allow for competitive processes in reaction/regeneration mode. For more details, see ref. [16].
Miscellaneous: Alkylation is now also studied with ionic liquids, in the liquid phase or heterogenized (SILP catalysts, cf. Section 2.3.3).

Friedel–Crafts alkylation

Purpose: Synthesis of alkyl aromatic intermediates from benzene and alkenes, in particular of ethylbenzene (a, for styrene) and of cumene (b, for phenol).
Chemistry: (a) eq. (4.9).
   (b) **Scheme A1d**, parallel formation of n-propylbenzene undesired.
Catalysts: H-zeolites, cf. Section 4.3.1, choice of zeolite – Sections 2.3.2.3 and 4.3.1.
Mechanism: Section 4.5.2.1.
Miscellaneous: Importance of transalkylation activity to handle problem of consecutive substitutions – Section 4.3.1.

Synthesis of alkyl *tert*-butyl ethers

Purpose: Synthesis of octane boosters for gasoline fuels.
Chemistry: Addition of suitable alcohols to the isobutene double bond – methanol for MTBE, ethanol for ETBE.
Catalysts: Acidic ion exchange resins.
Miscellaneous: On the commercial scale, reactions are realized as catalytic distillation.

Xylene isomerization

Purpose: Isomerization of m-xylene into the equilibrium mixture. After isolating p- and o-xylene (the treatment of EB is beyond the scope of this book), the remaining m-xylene is recycled.
Chemistry: m-Xy $\leftrightarrows$ o-Xy (20), m-Xy (50), p-Xy (20), EB (10)
   In parentheses – equilibrium percentages at ≈ 700 K.
Catalysts: Pt on acidic support: alumosilicate or ZSM-5.
Mechanism: The reaction can proceed via an intra- or an intermolecular mechanism. For more details, see ref. [17].
Miscellaneous: The p-selectivity can be enhanced by the shape selectivity of ZSM-5 (cf. Section 4.3.2).

Dehydrogenation of propane

Purpose: An option to handle propene shortage on markets.
Chemistry: $C_3H_8 \leftrightarrows C_3H_6 + H_2$
   for thermodynamic reasons at high temperatures
Catalysts: K-promoted Cr_2O_3/Al_2O_3 (a) or supported Pt promoted by Sn or Ga (b), also Ga- or V-based catalysts, for more details, see ref. [18].

Miscellaneous: The reaction may be performed in the presence of H_2 to delay coking. Reaction/regeneration cycles in known commercial processes last between 2 min and 10 days [18].

Dehydrogenation of ethylbenzene

Purpose: Synthesis of styrene monomer.
Chemistry: Equation (4.11); for thermodynamic reasons at high temperatures.
Catalysts: K-promoted Fe oxide (cf. Section 4.4.2).
Miscellaneous: Recent research targets alternative processes like the ODH of ethylbenzene with O_2 (eq. (4.3), cf. Sections 4.2.3 and 4.5.5) or with the soft oxidant CO_2 instead of the technical direct dehydrogenation process, which remains, however, dominant.

Metathesis of propene

Purpose: Flexibility in alkene markets.
Chemistry: (4.13) with $R_1 = CH_3$, $R_2 = H$.
Catalysts: Section 4.5.2.3.
Reaction mechanism: Section 4.5.2.3.
Miscellaneous: Producing more ethene and butene from excess propene was the intention behind the first commercial plant. Meanwhile, the reaction is rather applied to make propene from ethene (some of which is dimerized before).

For metathesis of alkanes see Section 3.1.3.4.

Polymerization with Ziegler–Natta or Phillips catalysts

Purpose: Synthesis of (polyolefin) polymers.
Catalysts: Section 4.5.2.3.
Reaction mechanism: Section 4.5.2.3.
Miscellaneous: For information on the development of heterogeneous Ziegler–Natta and of metallocene catalysts and on their use in industry, see refs. [19, 20].

Allylic oxidation of propene

Purpose: Synthesis of acrolein.
Chemistry: Equation (4.2).

Catalysts: In industry: "Bi–Mo" mixed oxides (actually Ni–Mo mixed oxides promoted with Bi and other components, see Section 4.3.3); also suitable according to early research: U–Sb, Fe–Sb mixed oxides, mixed oxides containing Cu.
Reaction mechanism: Section 4.5.2.2.
Active sites: Section 4.2.3.
Miscellaneous: Most of the acrolein is subsequently oxidized to acrylic acid (see later). The two sequential oxidation processes have withstood all effort to run them with a single catalyst or a catalyst combination owing to diverse requirements to catalyst and conditions.

Electrophilic oxidation of acrolein

Purpose: Synthesis of acrylic acid.
Chemistry: 2 CH$_2$=CH–CHO + O$_2$ → 2 CH$_2$=CH–COOH.
Catalysts: Cu-promoted V–Mo mixed oxides (with W as structural promoter, see also ref. [21]).
Miscellaneous: The reaction has long been out of focus of academic research because the technical process is rather mature. For an example of recent research, see ref. [22].

Ammoxidation of propene

Purpose: Synthesis of acrylo nitrile.
Chemistry: Equation (4.8).
Catalysts: "Bi–Mo" mixed oxides (actually Co–Mo mixed oxides promoted with Bi and other components, see Section 4.3.3).
Reaction mechanism: Section 4.5.2.2.
Active sites: Section 4.2.3.
Miscellaneous: Aromatic nitriles are obtained by analogous ammoxidation of alkylaromatics, however, with catalysts typically containing V. For a recent piece of research, see ref. [23].

Oxidative dehydrogenation (ODH) and (amm)oxidation of propane

Purpose: Energy-saving access to propene (by ODH, a), access to typical propene oxidation products (in particular acrylic acid, b) or to acrylo nitrile (c) from propane instead of propene.
Chemistry: (a) 2 CH$_3$-CH$_2$-CH$_3$ + O$_2$ → 2 CH$_2$=CH-CH$_3$ + 2 H$_2$O
(b) CH$_3$-CH$_2$-CH$_3$ + 2 O$_2$ → CH$_2$=CH-COOH + 2 H$_2$O
(c) CH$_3$-CH$_2$-CH$_3$ + 2 O$_2$ + NH$_3$ → CH$_2$=CH-CN + 4 H$_2$O

Catalysts: (a) Isolated VO_x sites in several systems [24]; (b) MoVTeNbO, a complex crystalline mixed-oxide system; and (c) MoVSbNbO, a system analogous to MoVTeNbO.

For more details on (a), see ref. [18].

Miscellaneous: Due to the relatively strong C–H bonds in propane, high selectivities are not easily achieved (Section 4.3.3). Only propane ammoxidation with MoVSbNbO has so far been developed to a level that suggests commercialization. Although used in industry, it does not seem to challenge the existing technology.

ODH and (amm)oxidation of ethane

Purpose: Energy-saving access to ethylene, acetic acid, and acetonitrile (here confined to the ODH version).

Chemistry: $2\ CH_3–CH_3 + O_2 \rightarrow 2\ CH_2=CH_2 + 2\ H_2O$.

Catalysts: Nb-promoted NiO, MoVTeNbO, crystalline Mo–V mixed oxides [22, 25].

Miscellaneous: ODH of ethane is being commercialized by Linde Engineering, with reaction products ethene and acetic acid.

Oxidation of *n*-butane to maleic anhydride (MA)

Chemistry: $2\ C_4H_{10} + 7\ O_2 \rightarrow 2\ C_4H_2O_3 + 8\ H_2O$

Catalysts: vanadium phosphates, active phase – $(VO)_2P_2O_7$ (VPP).

Reaction network: Section 4.3.3, Figure 4.22.

Miscellaneous: MA production from *n*-butane replaced earlier processes based on the oxidation of *n*-butene or of benzene;

Conditioning of initial catalyst – Section 3.1.5.

VPP catalysts lose phosphorous, which may be continuously replaced from the gas phase [26].

For a good summary on VPP catalysts and a recent piece of basic research, see ref. [27].

Oxidation of *o*-xylene to phthalic anhydride (PA)

Chemistry: Scheme A1e.

Catalysts: Promoted V_2O_5/TiO_2 catalysts; typical promoters: Sb_2O_3, Cs, P.

Reaction network: See Figure A1; selectivity for PA up to 83 mol%; beyond *o*-xylene, maleic anhydride (MA) and intermediates on the route to MA are also sources of carbon oxides.

Miscellaneous: Multitubular reactors with up to 30,000 tubes; catalysts with different compositions are arranged in 4–6 sequential layers of different lengths; the choice of

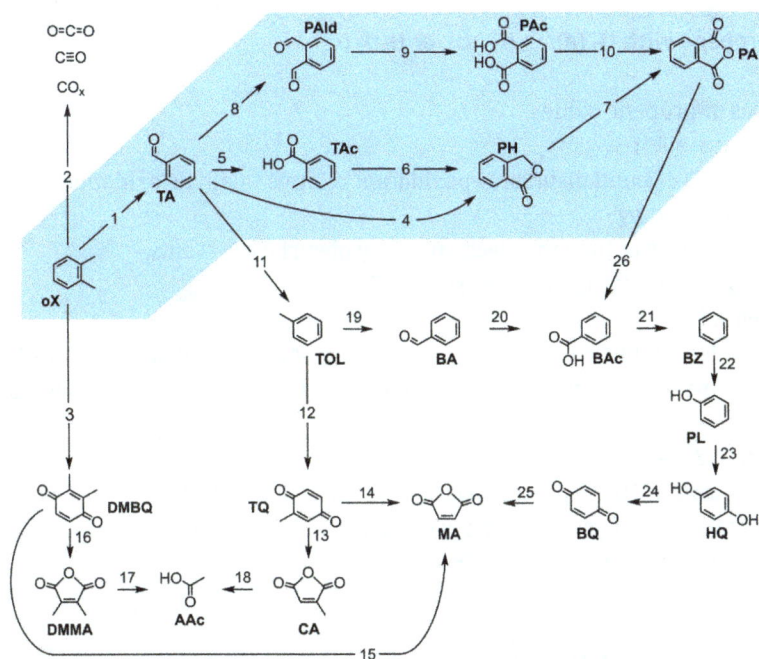

Figure A1: Reaction network of *o*-xylene oxidation; routes from intermediates to carbon oxides omitted for the sake of simplicity. Abbreviations: AAc – acetic acid; BA – benzaldehyde; BAc – benzoic acid; BQ – benzoquinone; BZ – benzene; CA – citraconic anhydride; DMBQ – 2,3-dimethyl-*p*-benzoquinone; DMMA – 2,3-dimethyl maleic anhydride; HQ – hydroquinone; MA – maleic anhydride; oX – *o*-xylene; PA – phthalic anhydride; PAld – phthalaldehyde; PH – phthalide; PL – phenol; TA – tolualdehyde; TAc – toluic acid; TOL – toluene; TQ – toluquinone. Adapted from ref. [28] with permission from Elsevier.

promoters depends on the position along the tube, i.e., on the stage of the reaction such as activation of reactants, main conversion, or clean-up of byproducts. In between, there may be buffer layers to extend the lifetime of the whole arrangement. For more details, see refs. [28–30].

Naphthalene can be converted to PA over the same type of catalyst. Where it is available at reasonable prices, it is processed in mixed feeds with *o*-xylene.

Epoxidation of ethene

Purpose: Synthesis of ethylene oxide.
Chemistry: See **Scheme A1f**.
Catalysts: Ag/α-Al$_2$O$_3$, promoted mainly with alkali (per synthesis) and Cl (via organo-chlorine compounds added to the feed), cf. Section 4.3.5.
Reaction mechanism: Electrophilic oxidation, cf. Sections 4.3.3 and 4.3.5.

Epoxidation of propene with O₂ (a), O₂/H₂ (b), or H₂O₂ (c)

Purpose: Synthesis of propene oxide.
Chemistry: See **Scheme A1f**.
The reactivity of allyl C–H bond disfavors epoxidation relative to allylic oxidation. The problem has been targeted by
(a) promoting Ag catalysts to tune the electrophilicity of surface O atoms;
(b) adding H_2 to O_2 for modifying available surface O sites
(c) epoxidizing with H_2O_2.
Catalysts: (a) Promoted supported Ag, (b) TiO_2- or TS-1-supported Au or Ag, and (c) TS-1.
Reaction mechanism: (b) Intermediate formation of H_2O_2 (and Ti-OOH species) is likely, cf. ref. [31];
(c) See Sections 4.2.3 and 4.7.
Miscellaneous: The recent HPPO process based on (c) had a market share of 5% with growing tendency as of 2009. The remaining propylene oxide is produced without solid catalysts (cf. ref. [32]).

Other selective oxidations with H₂O₂

Hydroxylation of phenol: Scheme A1g
Ammoximation of cyclohexanone: Scheme A1h
(both commercially realized with TS-1, cf. ref. [33])

Total oxidation. . .

. . . of CO

Purpose: As a seemingly simple, prototypical reaction, a drosophila of surface catalysis, and at the same time an important part of catalytic air purification technology, CO oxidation has been subject of an enormous number of studies from various points of view.
Chemistry: $2\,CO + O_2\ (\leftrightarrows)\ 2\,CO_2$
For CO oxidation in the presence of H_2 (PROX), see further.
Catalysts: Supported noble metals (Pd, Pt, and Au), in particular on reducible supports, some binary oxides (CuO and Co_3O_4), mixed oxides, in particular materials containing Cu, such as hopcalite (Cu–Mn(–Co)) or Cu chromites, and perovskites. For a brief survey on suitable catalysts, the reader may consult ref. [34], for a more elaborate treatment, see refs. [35, 36].
Reaction mechanism: Depends on catalysts and conditions; Langmuir–Hinshelwood versus Mars–van Krevelen mechanism on noble metals: Section 2.5.1; multiplicity of mechanisms over supported Au: Section 4.5.6 (see also Section 4.2.4).

Miscellaneous: Oscillations of concentrations and rates, spatiotemporal patterning cf. Section 2.5.1; CO oxidation in the presence of other reactants (propene) – Section 3.4.8.4.

... of hydrocarbons

Purpose: Hydrocarbon combustion has been studied for two very different applications: (a) for avoiding formation of nitrogen oxides in combustion processes by controlling flame temperatures in heat generation units, where both temperature T and hydrocarbon concentrations c_{hc} are high, and (b) removal of volatile organic compounds (VOCs) from flue gases (T, c_{hc} – low).

Chemistry: $C_xH_y + (x + y/4)O_2 \rightarrow x\, CO_2 + y/2\, H_2O$

(a) Catalytic combustion

Catalysts: In the staged combustion process, first-stage catalysts are designed for high activity, in the second stage, stability at high temperatures is crucial [37].
- First-stage catalysts: Supported Pd (preferred for methane), Pt (preferred for higher hydrocarbons), oxides and mixed-oxides containing Cu, Co, Mn, and Ce, cf. [38].
- Second-stage catalysts: Spinels, perovskites, and hexaaluminates [38].

Miscellaneous: The second stage may feed a final homogeneous combustion zone [37].

(b) Catalytic oxidation of VOCs

Catalysts: Supported Pt, Pd, Ru (Au); Co, Ni, V, Cu, Mn, Cr, Ce, Fe oxides and mixed oxides, perovskites; for a comprehensive review, see ref. [39].

Miscellaneous: Compared with the first stage of catalytic combustion, catalysts are stressed by poisons in feeds rather than by excessive evolution of reaction heat.

... of chlorohydrocarbons

Purpose: Detoxification of waste gas streams.

Catalysts: VO_x and CrO_x on various supports and in mixed oxides, also Co, Mn, Ce (mixed) oxide catalysts; among noble metals only Ru, other metals like Pt or Pd suffering from losses (chloride formation), see ref. [39].

... of SO₂

Purpose: First step in the synthesis of H_2SO_4 from SO_2.
Chemistry: $2\ SO_2 + O_2 \leftrightarrows 2\ SO_3$
Catalysts: Noble metals, in particular Pt, commercial: K-promoted V_2O_5/SiO_2.
Miscellaneous: The K-promoted V_2O_5/SiO_2 catalyst as an SLP system: Sections 2.3.3 and 4.2.4.

Preferential oxidation of CO in the presence of H₂ (PROX)

Purpose: Removal of residual CO from H_2 for feeding fuel cells.
Chemistry: $2\ CO + O_2 + H_2 \rightarrow 2\ CO_2 + H_2$
Catalysts: Supported Pt, Ru, Ir, or Rh, promoted with reducible oxide species, e.g., Pt–Fe, Ir–Ce, also with Cu, Co promotors, (cf. ref. [40]).
Supported Au catalysts, Cu–Ce mixed oxides.
Miscellaneous: Supported Au catalysts tend to deactivate and lose selectivity with increasing temperature, Cu–Ce mixed oxides are poisoned by the reaction products (CO_2 and H_2O).

Oxychlorination of ethene

Purpose: Synthesis of 1,2-dichloroethane as an intermediate in vinyl chloride monomer synthesis.
Chemistry: $2\ C_2H_4 + 4\ HCl + O_2 \rightarrow 2\ C_2H_4Cl_2 + 2\ H_2O$
Catalysts: $CuCl_2/\gamma\text{-}Al_2O_3$.
Miscellaneous: For recent research on the topic, see ref. [41].

Deacon process

Purpose: Recycling of the HCl byproduct from chlorination processes.
Chemistry: $4\ HCl + O_2 \rightarrow 2\ Cl_2 + 2\ H_2O$
Catalysts: Traditionally $CuO/CuCl_2$ (cf. Section 4.4.1.2), more recently RuO_2/TiO_2.
Miscellaneous: Low volatility of Ru chlorides allows for a successful commercial process. For an analysis of the reaction, see ref. [42].

Three-way catalysis

Purpose: Simultaneous conversion of CO, unburnt hydrocarbons, and NO/NO_2 to CO_2, H_2O, and N_2 in the exhaust of gasoline cars.

Chemistry: $2\,CO + O_2 \rightarrow 2\,CO_2$

$\qquad\qquad 2\,C_mH_n + (2m + n/2)\,O_2 \rightarrow 2m\,CO_2 + n\,H_2O$

$\qquad\qquad 2\,NO + 2\,CO \rightarrow N_2 + 2\,CO_2$

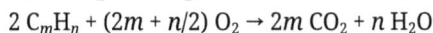

(NO may be actually reduced by any oxidizable mixture component)

As high conversions can be achieved simultaneously only at a stoichiometric air–fuel ratio, the latter is controlled using a λ sensor. Due to the on–off nature of the control, the reaction regime is transient, changing between rich and lean regions a few times per second (cf. Section 4.4.1.3). The catalyst needs to be capable of sequentially storing oxygen and hydrocarbons for sustaining oxidation under rich conditions and reduction under lean conditions, respectively.

Catalysts: Pt, Rh, Pd supported on (La-modified) Al_2O_3, ZrO_2, $CeZrO_x$; oxygen storage functions provided by Ce compounds (Section 2.3.2.5), hydrocarbon storage by noble metals.

Reaction mechanism: Dissociation of O_2, hydrocarbons, and NO on the metal surfaces (NO best on Rh), recombinations and reaction with adsorbed CO to final products; for a typical surface reaction mechanism used for kinetic models, see ref. [43].

Miscellaneous: For questions related to catalyst stability, see Sections 4.4.1.3 and 4.4.2; layered structure of catalysts, see Sections 2.3.4 and 4.4.2.

Diesel oxidation catalysts (DOC)

Purpose: Oxidation of CO, unburnt fuel, and soluble organic fraction of particulate matter in diesel exhaust, partial oxidation of NO to NO_2 to be used in downstream SCR or particle traps (see there).

Chemistry: See total oxidation reactions.

Catalysts: Pt and/or Pd, typically supported on Al_2O_3.

Miscellaneous: For details, see ref. [44].

Selective catalytic reduction (SCR) of NO with NH₃

Purpose: Reduction of NO/NO_2 to N_2 in the presence of oxygen for cleaning flue gases from combustion processes or lean automobile exhaust (diesel or lean gasoline engines).

Chemistry: Equations (3.136) and (3.137).

Catalysts: W- or Mo-promoted V_2O_5/TiO_2 (stationary sources), Cu-zeolites (Cu-CHA), Fe-zeolites (Fe-BEA) in passenger diesel cars, W-promoted V_2O_5/TiO_2 in commercial vehicles; in research, great attention also to Ce-containing catalysts.

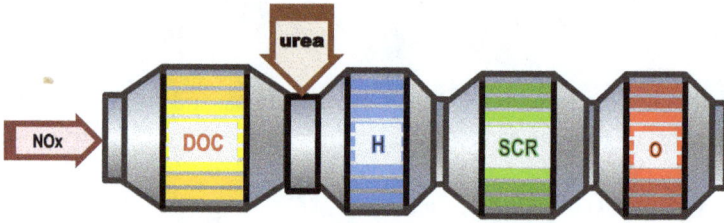

Figure A2: Diesel exhaust duct with abatement of nitrogen oxides by urea-SCR; DOC, diesel (particle) oxidation; H, urea hydrolysis; SCR, NH$_3$-SCR; O, oxidation of breakthrough NH$_3$ to N$_2$.

Reaction mechanism: cf. Sections 4.5.3 and 4.5.6.

Active sites: cf. Section 4.2.3.

Miscellaneous: In cars, NH$_3$ is stored as urea (in aqueous solution), which is injected into the exhaust and hydrolyzed on a catalyst prior to the SCR step (H in Figure A2). Performance on Fe- or V-based catalysts can be improved to compete with Cu-CHA by oxidizing part of NO to NO$_2$ on the preceding diesel oxidation catalyst (DOC in Figure A2), which shifts chemistry on the SCR catalyst from (3.136) to (3.137). Finally, excess NH$_3$ is oxidized to N$_2$ over a Pt-based catalyst (O in Figure A2). Industry tends to integrate SCR into other treatment steps, e.g., into particle filters.

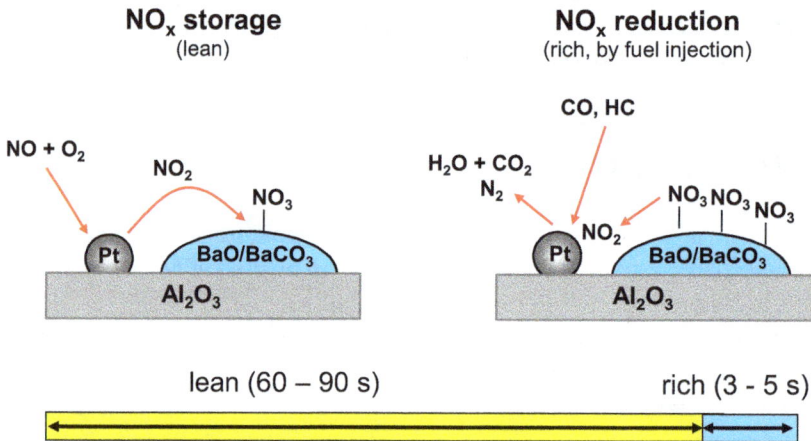

Figure A3: Principle of the nitrogen storage and reduction (NSR) technology.

Nitrogen storage and reduction (NSR)

Purpose: Reduction of NO/NO$_2$ to N$_2$ in the presence of oxygen in lean automobile exhaust (diesel or lean gasoline engines), competes with NH$_3$-SCR in these applications.

Catalysts: A supported noble metal (Pt or Pt–Rh) combined with a component storing N as nitrate or nitrite, typically BaO.

Principle: In a reaction/regeneration scheme, NO is oxidized to NO_2 and stored as nitrate under lean conditions (Figure A3). For regeneration, fuel is injected. Its combustion raises the temperature, which decomposes stored nitrate/nitrite. Released nitrogen oxides are reduced by excess fuel (Figure A3).

Miscellaneous: Due to stability of $BaSO_4$ strongly poisoned by S in fuel; requires, therefore, ultra-low sulfur diesel qualities. Commercialized in particular for small passenger cars.

Treatment of diesel soot

Purpose: Trapping soot particles escaping from the DOC by diesel particle filters (DPF), which are regenerated by incineration of combustible soot constituents (all except for ashes).

Principle: Wall-flow monolith (Figure 2.65b), often coated with oxidation catalysts. Regeneration either passive (parallel to filtering) or active (in a regeneration phase).

Passive regeneration:

- Without catalyst: $C + 2\,NO_2 \rightarrow CO_2 + 2\,NO$,
  (NO_2 produced from NO, e.g., on DOC)
- With catalyst on DPF: In addition, $C + O_2 \rightarrow CO_2$.
  Catalyst components range from Pt, Pd to base metal (mixed) oxides containing Ce/Zr, Mn, Fe, etc; challenging catalytic regime as contact between reactant (soot) and catalyst is usually loose.
- With fuel-borne catalyst: A fuel-soluble compound of a catalytic metal is dissolved in the fuel (e.g., Fe or Ce, concentration in modern versions <10 ppm); included in the soot, i.e., in close contact, it catalyzes soot incineration effectively.

In most applications, passive regeneration technologies are not sufficiently flexible to cover all driving regimes. They are, therefore, mostly complemented by active regeneration phases.

Active regeneration: To achieve regeneration, the DPF temperature is raised by fuel injection (in-cylinder or post-cylinder), by dedicated fuel burners, or by electrical heaters.

For more detail, see ref. [45].

Photocatalytic water splitting

Purpose: Production of H_2 from water using exclusively light (in particular sunlight).
Chemistry: Equation (2.23).
Catalysts: Depend on wavelength range of light applied:
- UV light: d^0 or d^{10} binary or higher oxides (Section 4.8).
- Sunlight: oxynitrides of d^{10} cations, z-scheme photocatalysts, with suitable co-catalyst, cf. Section 4.8.

Reaction mechanism: cf. Sections 2.5.2 and 4.8.
Miscellaneous: A commercial realization (still far ahead) would have to deal with the immediate separation of the resulting H_2/O_2 mixture. Hydrogen oxidation and oxygen reduction can be studied separately in arrangements replacing the complementary step by sacrificial oxidation/reduction processes, cf. ref. [46].

H$_2$-fueled PEM fuel cell

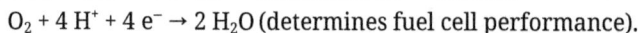

Purpose: Production of electricity from hydrogen and oxygen.
Chemistry: $2 H_2 \rightarrow 4 H^+ + 4 e^-$ ($H^+ \rightarrow$ membrane, $e^- \rightarrow$ circuit)
$O_2 + 4 H^+ + 4 e^- \rightarrow 2 H_2O$ (determines fuel cell performance).
Catalysts: HOR – Pt, ORR – see Section 4.9.2.
Reaction mechanisms: Section 4.9.1.
Miscellaneous: For a recent review, see ref. [47].

Direct methanol fuel cell (DMFC)

Purpose: Production of electricity from methanol and oxygen.
Chemistry: $CH_3OH + H_2O \rightarrow CO_2 + 6 H^+ + 6 e^-$
($H^+ \rightarrow$ membrane, $e^- \rightarrow$ circuit)
$3/2 O_2 + 6 H^+ + 6 e^- \rightarrow 3 H_2O$
Catalysts: See ref. [48].
Miscellaneous: In the DMFC, self-poisoning of the anode by intermediates (CO_{ads}) adds to the well-known challenges with the ORR.

Water electrolysis

Purpose: Production of hydrogen and oxygen from electricity.
Chemistry: Equations (2.59) through (2.61).
Catalysts: HER – traditionally Ni, for recent development cf. Section 3.5.3. OER: cf. Section 4.9.2.

Reaction mechanisms: Section 4.9.1.

Miscellaneous: Traditionally performed in alkaline electrolytes using diaphragm cells; more recent development: PEM electrolyzers, which require, however, catalysts stable in acidic environment; for more details see refs. [49, 50],

Electrocatalytic CO_2 reduction

Purpose: Conversion of CO_2 and H_2O to value-added products,

Chemistry: $x\, CO_2 + n\, H^+ + n\, e^- \rightarrow$ product $+ y\, H_2O$.

$$2\, H_2O \rightarrow O_2 + 4\, H^+ + 4\, e^-.$$

"Product" (E_0 increasing): $(COOH)_2/HCOOH/CO/CH_3OH/CH_3\text{-}CHO/C_2H_4/C_2H_5OH/C_3H_7OH/C_3H_6O/CH_3COOH/C_2H_6/CH_4/$graphite.

Catalysts: See Section 4.9.2.

Reaction mechanisms: See ref. [51].

Miscellaneous: Despite encouraging developments in some applications not yet competitive with respect to costs and energy demand. For a recent assessment, see ref. [51].

Chloralkali process

Purpose: Synthesis of chlorine from brine solutions.

Chemistry: $4\, Cl^- \rightarrow 2\, Cl_2 + 4\, e^-$ (Cl_2 evolution, CER)

$$4\, H_2O + 4\, e^- \rightarrow 4\, OH^- + 2\, H_2 \ \text{(HER)}$$

In total: $4\, H_2O + 4\, Cl^- \rightarrow 2\, Cl_2 + 2\, H_2 + 4\, OH^-$

Catalysts: Cathode – Ni or steel; anode – $RuO_2\text{-}TiO_2/Ti$.

Miscellaneous: Actual anodes catalyze the competing OER, suffer from instability and overpotentials with respect to the CER. For a review on recent research on CER, see ref. [52].

To decrease the cell voltage, the HER at the cathode can be replaced by the ORR performed at an Ag-based gas diffusion electrode (oxygen depolarized cathode, ODC). This shifts the cathode reaction to

$$O_2 + 2\, H_2O + 4\, e^- \rightarrow 4\, OH^-,$$

and the total reaction to

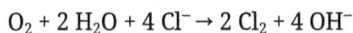

$$O_2 + 2\, H_2O + 4\, Cl^- \rightarrow 2\, Cl_2 + 4\, OH^-$$

Brine electrolysis with ODC saves ca. 25% of electrical energy, which is only partly balanced by capital expenses, loss of H_2 value and expenses for O_2 [53]. For an analysis of life cycle impact on environment, see ref. [54].

A2 Derivation of some equations presented in this book

A2.1 BET equation (2.16)

Equations (2.15) allow expressing the dependence of all coverages θ_j on the adsorptive pressure p, which is then lumped with ratios of rate constants into the reduced pressures α and β:

$$\theta_1/\theta_0 = p\,{}^{k_{ads,1}}/_{k_{des,1}} = \alpha \text{ and } \theta_j/\theta_{j-1} = p\,{}^{k_{ads,j}}/_{k_{des,j}} = \beta_j$$

Because of the last condition in Box 2.2, $k_{ads,j}/k_{des,j}$ is the same (β) for all layers but the first one for which the ratio is designated α. All θ_j can now be evaluated by simple recursive formulae: $\theta_2 = \beta\theta_1 = \beta\,\alpha\theta_0$ and $\theta_j = \beta\theta_{j-1} = \cdots = \beta^{j-1}\alpha\theta_0$. When α/β is abbreviated by c, the coverages θ_j can be expressed as

$$\theta_j = \beta^j c\theta_0, \text{ with } c = \alpha/\beta \quad (*)$$

Next, the adsorbed quantity n_{ads} is evaluated and related to the monolayer capacity n_{mono}: $n_{ads} = 1n_1 + 2n_2 + \cdots + jn_j = \Sigma_{j=1}^{\infty} j\,n_j$ (cf. model in Figure 2.53). The relation between n_j and n_{mono} is $n_j = n_{mono}\,\theta_j$; therefore, $n_{ads} = \Sigma_{j=1}^{\infty} j\,n_j = n_{mono}\,c\,\theta_0\,\Sigma_{j=1}^{\infty} j\beta^j$. The series $\Sigma_{j=1}^{\infty} j\beta^j$ can be replaced by $\beta/(1-\beta)^2$, therefore

$$n_{ads} = n_{mono}\,c\,\theta_0\,\beta/(1-\beta)^2 \quad (**)$$

The balance of coverages is now employed to eliminate θ_0:
$\theta_0 = 1 - \Sigma_{j=1}^{\infty} \theta_j = 1 - c\,\theta_0\,\Sigma_{j=1}^{\infty} \beta^j$, where $\Sigma_{j=1}^{\infty} \beta^j$ can be replaced by $\beta/(1-\beta)$ to get $\theta_0 = 1 - \Sigma_{j=1}^{\infty} \theta_j = 1 - c\,\theta_0\beta/(1-\beta)$.

Isolating θ_0 and substituting it into (**) gives

$$n_{ads} = n_{mono}c\beta/(1-\beta)(1-\beta(c-1)) \quad (***)$$

Now, what is β? Per definition, it is $\beta = pk_{ads,j}/k_{des,j}$. When $p \rightarrow p_0$, the adsorptive condenses, i.e., $n_{ads} \rightarrow \infty$. According to (***), this happens at $\beta(= p_0 k_{ads,j}/k_{des,j}) = 1$. This relation allows replacing $k_{ads,j}/k_{des,j}$, by $1/p_0$, i.e., $\beta = p/p_0$, and

$$n_{ads} = \frac{n_{mono}\,c\,\frac{p}{p_0}}{\left(1 - \frac{p}{p_0}\right)\left(1 - \frac{p(c-1)}{p_0}\right)} \tag{2.16}$$

A2.2 Derivation of a Hougen–Watson rate law – First order reaction with rate-limiting adsorption of *A*

From eq. (2.32b) and $r_{surf} = 0$: $K_2 = \theta_P/\theta_A$, $\theta_P = K_2\theta_A$

From eq. (2.32c) and $r_{des,P} = 0$:

$$K_p = \theta_P/p_P(1 - \theta_A - \theta_P) = K_2\theta_A/(1 - \theta_A - K_2\theta_A)$$
$$K_P p_P = \theta_A(K_2 + K_P p_P + K_P K_2 p_P)$$
$$\theta_A = K_P p_P/(K_P p_P + K_2(1 + K_P p_P)) *$$

Rearrange (2.32a)

$$r = k_1 p_A (1 - \theta_A - K_2\theta_A) - k_{-1} \theta_A = k_1 p_A - \theta_A (k_1 K_2 p_A + k_1 p_A + k_{-1})$$

Insert *:

$$r = k_1 p_A - \frac{K_P p_P(k_1 K_2 p_A + k_1 p_A + k_{-1})}{K_P p_P + K_2(1 + K_P p_P)}$$

Multiply and cancel...

$$r = \cdots = \frac{k_1 p_A - k_{-1} p_P K_P/K_2}{1 + K_P p_P + p_P K_P/K_2} = \frac{k_1(p_A - p_P k_{-1} K_P/k_1 K_2)}{1 + K_P p_P + p_P K_P/K_2}$$

with $k_{-1}/k_1 = K_A^{-1}$ and $K_P/K_A K_2 = K^{-1} \rightarrow$ eq. (2.33)

A2.3 Solution of eq. (2.42)

Inserting $C_i = a \exp(bZ)$ into (2.42) as a first guess shows that b can be $+\Phi_0$ or $-\Phi_0$.

Therefore, an ansatz using a linear combination of the exponential functions is most promising:

$$C_i = a_1 \exp(\Phi_0 Z) + a_2 \exp(-\Phi_0 Z) \; (*).$$

a_1 and a_2 are found with the boundary conditions. $C_i(0) = 1$ results in $a_1 + a_2 = 1$ (**).
Inserting C_i in $(dC_i/dZ)_{Z=1} = 0$ gives $a_1 \Phi_0 \exp(\Phi_0) - a_2 \Phi_0 \exp(-\Phi_0) = 0$, and finally $a_2 = a_2 \exp(-2\Phi_0)$ (***)

From (**) and (***), and using cosh $x = (\exp(x) + \exp(-x))/2$, a_1 and a_2 are obtained:
$a_1 = \exp(-\Phi_0)/2 \cosh \Phi_0$; $a_2 = \exp(\Phi_0)/2 \cosh \Phi_0$.
a_1 and a_2 are then inserted into the ansatz (*) to obtain (2.43).

A2.4 Derivation of eq. (2.51) within the porous sphere model

The definition (2.50) is expanded by the sphere volume to compare reactant quantities converted by the diffusion-limited and the unlimited reactions: $\eta_p = V\,r_{\text{eff}}/V\,r$ (note: the Thiele modulus is discussed within a homogeneous model). For a first-order reaction, the denominator is $V\,r = 4\,\pi\,r_{\text{sph}}^3\,k\,c_{S,i}/3$.

The diffusion-limited reaction rate is represented by the diffusive flow into the pellet(!): $V\,r_{\text{eff}} = D_{\text{eff}}\,4\,\pi\,r_{\text{sph}}^2\,(d\,c_i/d\,r_{\text{rad}})_{r_{\text{sph}}}$. For the meaning of r_{rad} see footnote 26 in Section 2.6.2. Using the reduced coordinates C_i and R specified there, the derivative can be written as

$$\frac{dc_i}{dr_{\text{rad}}} = \frac{c_{S,i}}{r_{\text{sph}}}\frac{dC_i}{dR}.$$

dC_i/dR is obtained by differentiating (2.48) under the condition $R = 1$ (because of $r_{\text{rad}} = r_{\text{sph}}$, see earlier):

$$\left(\frac{dc_i}{dr_{\text{rad}}}\right)_{r_{\text{sph}}} = \frac{c_{S,i}}{r_{\text{sph}}}\frac{\Phi_o\cosh\Phi_o - \sinh\Phi_o}{\sinh\Phi_o}$$

and

$$V r_{\text{eff}} = 4\,\pi\,r_{\text{sph}}\,D_{\text{eff}}\,c_{S,i}\frac{\Phi_o\cosh\Phi_o - \sinh\Phi_o}{\sinh\Phi_o}.$$

Therefore,

$$\eta_P = \frac{V\,r_{\text{eff}}}{V\,r} = \frac{3\,D_{\textit{eff}}\,\Phi_0}{r_{\text{sph}}^2\,k}\left(\frac{1}{\tanh\Phi_0} - \frac{1}{\Phi_0}\right)$$

Replacing $r_{\text{sph}}^2\,k/D_{\textit{eff}}$ by Φ_0^2, eq. (2.51) is obtained.

References

[1]   Hannevold, L.; Nilsen, O.; Kjekshus, A.; Fjellvåg, H., Reconstruction of platinum–rhodium catalysts during oxidation of ammonia. *Applied Catalysis A: General* **2005**, *284* (1), 163–176. doi.org/10.1016/j. apcata.2005.01.033

[2]   Thrane, J.; Mentzel, U. V.; Thorhauge, M.; Høj, M.; Jensen, A. D., A review and experimental revisit of alternative catalysts for selective oxidation of methanol to formaldehyde. *Catalysts* **2021**, *11* (11), 1329. doi 10.3390/catal11111329

[3]   Millar, G. J.; Collins, M., Industrial production of formaldehyde using polycrystalline silver catalyst. *Industrial & Engineering Chemistry Research* **2017**, *56* (33), 9247–9265. doi 10.1021/acs.iecr.7b02388

[4]   Luk, H. T.; Mondelli, C.; Ferré, D. C.; Stewart, J. A.; Pérez-Ramírez, J., Status and prospects in higher alcohols synthesis from syngas. *Chemical Society Reviews* **2017**, *46* (5), 1358–1426. doi 10.1039/C6CS00324A

[5]   Zeng, F.; Mebrahtu, C.; Xi, X. Y.; Liao, L. F.; Ren, J.; Xie, J. X.; Heeres, H. J.; Palkovits, R., Catalysts design for higher alcohols synthesis by CO_2 hydrogenation: Trends and future perspectives. *Applied Catalysis B: Environmental* **2021**, *291*, Art. No. 120073, doi 10.1016/j.apcatb.2021.120073

[6]   Horn, R.; Schlögl, R., Methane activation by heterogeneous catalysis. *Catalysis Letters* **2015**, *145* (1), 23–39. doi 10.1007/s10562-014-1417-z

[7]   Ratnasamy, C.; Wagner, J. P., Water gas shift catalysis. *Catalysis Reviews* **2009**, *51* (3), 325–440. doi 10.1080/01614940903048661

[8]   Chen, B.; Dingerdissen, U.; Krauter, J. G. E.; Lansink Rotgerink, H. G. J.; Möbus, K.; Ostgard, D. J.; Panster, P.; Riermeier, T. H.; Seebald, S.; Tacke, T.; Trauthwein, H., New developments in hydrogenation catalysis particularly in synthesis of fine and intermediate chemicals. *Applied Catalysis A: General* **2005**, *280* (1), 17–46. doi.org/10.1016/j.apcata.2004.08.025

[9]   Yoshimura, Y.; Toba, M.; Matsui, T.; Harada, M.; Ichihashi, Y.; Bando, K. K.; Yasuda, H.; Ishihara, H.; Morita, Y.; Kameoka, T., Active phases and sulfur tolerance of bimetallic Pd-Pt catalysts used for hydrotreatment. *Applied Catalysis A: General* **2007**, *322*, 152–171. doi 10.1016/j.apcata.2007.01.009

[10] Luneau, M.; Lim, J. S.; Patel, D. A.; Sykes, E. C. H.; Friend, C. M.; Sautet, P., Guidelines to achieving high selectivity for the hydrogenation of α,β-unsaturated aldehydes with bimetallic and dilute alloy catalysts: A review. *Chemical Reviews* **2020**, *120* (23), 12834–12872. doi 10.1021/acs.chemrev.0c00582

[11] Lan, X.; Wang, T., Highly selective catalysts for the hydrogenation of unsaturated aldehydes: A review. *ACS Catalysis* **2020**, *10* (4), 2764–2790. doi 10.1021/acscatal.9b04331

[12] Eijsbouts, S.; Mayo, S. W.; Fujita, K., Unsupported transition metal sulfide catalysts: From fundamentals to industrial application. *Applied Catalysis A* **2007**, *322* (1), 58–66. doi 10.1016/j.apcata.2007.01.008

[13] Egorova, M.; Prins, R., The role of Ni and Co promoters in the simultaneous HDS of dibenzothiophene and HDN of amines over Mo/γ-Al_2O_3 catalysts. *Journal of Catalysis* **2006**, *241* (1), 162–172. doi 10.1016/j.jcat.2006.04.011

[14] Bello, S. S.; Wang, C.; Zhang, M.; Gao, H.; Han, Z.; Shi, L.; Su, F.; Xu, G., A review on the reaction mechanism of hydrodesulfurization and hydrodenitrogenation in heavy oil upgrading. *Energy & Fuels* **2021**, *35* (14), 10998–11016. doi 10.1021/acs.energyfuels.1c01015

[15] Prins, R.; Egorova, M.; Röthlisberger, A.; Zhao, Y.; Sivasankar, N.; Kukula, P., Mechanisms of hydrodesulfurization and hydrodenitrogenation. *Catalysis Today* **2006**, *111* (1), 84–93. doi.org/10.1016/j.cattod.2005.10.008

[16] Díaz Velázquez, H.; Likhanova, N.; Aljammal, N.; Verpoort, F.; Martínez-Palou, R., New insights into the progress on the isobutane/butene alkylation reaction and related processes for high-quality fuel production. A critical review. *Energy & Fuels* **2020**, *34* (12), 15525–15556. doi 10.1021/acs.energyfuels.0c02962

[17] Guisnet, M.; Gnep, N. S.; Morin, S., Mechanisms of xylene isomerization over acidic solid catalysts. *Microporous and Mesoporous Materials* **2000**, *35–6*, 47–59. doi 10.1016/s1387-1811(99)00207-3

[18] Carter, J. H.; Bere, T.; Pitchers, J. R.; Hewes, D. G.; Vandegehuchte, B. D.; Kiely, C. J.; Taylor, S. H.; Hutchings, G. J., Direct and oxidative dehydrogenation of propane: From catalyst design to industrial application. *Green Chemistry* **2021**, *23* (24), 9747–9799. doi 10.1039/d1gc03700e

[19] Claverie, J. R.; Schaper, F., Ziegler-Natta catalysis: 50 years after the Nobel Prize. *MRS Bulletin* **2013**, *38* (3), 213–218. doi 10.1557/mrs.2013.52

[20] Gahleitner, M.; Resconi, L.; Doshev, P., Heterogeneous Ziegler-Natta, metallocene, and post-metallocene catalysis: Successes and challenges in industrial application. *MRS Bulletin* **2013**, *38* (3), 229–233. doi 10.1557/mrs.2013.47

[21] Kampe, P.; Giebeler, L.; Samuelis, D.; Kunert, J.; Drochner, A.; Haaß, F.; Adams, A. H.; Ott, J.; Endres, S.; Schimanke, G.; Buhrmester, T.; Martin, M.; Fuess, H.; Vogel, H., Heterogeneously catalysed partial oxidation of acrolein to acrylic acid—structure, function and dynamics of the V–Mo–W mixed oxides. *Physical Chemistry Chemical Physics* **2007**, *9* (27), 3577–3589. doi 10.1039/B700098G

[22] Ishikawa, S.; Ueda, W., Microporous crystalline Mo–V mixed oxides for selective oxidations. *Catalysis Science & Technology* **2016**, *6* (3), 617–629. doi 10.1039/C5CY01435B

[23] Goto, Y.; Shimizu, K.-i.; Kon, K.; Toyao, T.; Murayama, T.; Ueda, W., NH_3-efficient ammoxidation of toluene by hydrothermally synthesized layered tungsten-vanadium complex metal oxides. *Journal of Catalysis* **2016**, *344*, 346–353. doi.org/10.1016/j.jcat.2016.10.013

[24] Albonetti, S.; Cavani, F.; Trifirò, F., Key aspects of catalyst design for the selective oxidation of paraffins. *Catalysis Reviews* **1996**, *38* (4), 413–438. doi.org/10.1006/jcat.1996.0123

[25] Védrine, J. C., Heterogeneous partial (amm)oxidation and oxidative dehydrogenation catalysis on mixed metal oxides. *Catalysts* **2016**, *6* (2), 22. doi 10.3390/catal6020022

[26] Lesser, D.; Mestl, G.; Turek, T., Transient behavior of vanadyl pyrophosphate catalysts during the partial oxidation of n-butane in industrial-sized, fixed bed reactors. *Applied Catalysis A: General* **2016**, *510*, 1–10. doi.org/10.1016/j.apcata.2015.11.002

[27] Kappel, I.; Böcklein, S.; Park, S.; Wharmby, M.; Mestl, G.; Schmahl, W. W., Crystal imperfections of industrial vanadium phosphorous oxide catalysts. *Catalysts* **2021**, *11* (11), 1325. doi 10.3390/catal11111325

[28] Marx, R.; Wölk, H.-J.; Mestl, G.; Turek, T., Reaction scheme of o-xylene oxidation on vanadia catalyst. *Applied Catalysis A: General* **2011**, *398* (1), 37–43. doi.org/10.1016/j.apcata.2011.03.006

[29] Richter, O.; Eberle, H.-J.; de Munck, N. A.; Marx, R.; Turek, T.; Mestl, G., Oxidation of o-xylene and naphthalene to phthalic anhydride - Catalyst development. In *Industrial Arene Chemistry*, Mortier, J., Ed. Wiley-VCH: Weinheim, 2023. ISBN: 978-3-527-34784-1

[30] de Munck, N. A.; Richter, O.; Eberle, H.-J.; Mestl, G., Oxidation of o-xylene and naphthalene to phthalic anhydride. In *Industrial Arene Chemistry*, Mortier, J., Ed. Wiley-VCH: Weinheim: 2023. ISBN: 978-3-527-34784-1

[31] Huang, J.; Haruta, M., Gas-phase propene epoxidation over coinage metal catalysts. *Research on Chemical Intermediates* **2012**, *38* (1), 1–24. doi 10.1007/s11164-011-0424-6

[32] Russo, V.; Tesser, R.; Santacesaria, E.; Di Serio, M., Chemical and technical aspects of propene oxide production via hydrogen peroxide (HPPO process). *Industrial & Engineering Chemistry Research* **2013**, *52* (3), 1168–1178. doi 10.1021/ie3023862

[33] Perego, C.; Carati, A.; Ingallina, P.; Mantegazza, M. A.; Bellussi, G., Production of titanium containing molecular sieves and their application in catalysis. *Applied Catalysis A: General* **2001**, *221* (1), 63–72. doi.org/10.1016/S0926-860X(01)00797-9

[34] Prasad, R.; Singh, P., A review on CO oxidation over copper chromite catalyst. *Catalysis Reviews – Science and Engineering* **2012**, *54*, 224–279. doi 10.1080/01614940.2012.648494a

[35] Freund, H. J.; Meijer, G.; Scheffler, M.; Schlögl, R.; Wolf, M., CO oxidation as a prototypical reaction for heterogeneous processes. *Angewandte Chemie-International Edition* **2011**, *50* (43), 10064–10094. doi 10.1002/anie.201101378

[36] Royer, S.; Duprez, D., Catalytic oxidation of carbon monoxide over transition metal oxides. *ChemCatChem* **2011**, *3* (1), 24–65. doi 10.1002/cctc.201000378

[37] Forzatti, P., Status and perspectives of catalytic combustion for gas turbines. *Catalysis Today* **2003**, *83* (1–4), 3–18. doi 10.1016/S0920-5861(03)00211-6

[38] Chen, J. H.; Arandiyan, H.; Gao, X.; Li, J. H., Recent advances in catalysts for methane combustion. *Catalysis Surveys from Asia* **2015**, *19* (3), 140–171. doi 10.1007/s10563-015-9191-5

[39] He, C.; Cheng, J.; Zhang, X.; Douthwaite, M.; Pattisson, S.; Hao, Z. P., Recent advances in the catalytic oxidation of volatile organic compounds: A review based on pollutant sorts and sources. *Chemical Reviews* **2019**, *119* (7), 4471–4568. doi 10.1021/acs.chemrev.8b00408

[40] Liu, K.; Wang, A. Q.; Zhang, T., Recent advances in preferential oxidation of CO reaction over platinum group metal catalysts. *ACS Catalysis* **2012**, *2* (6), 1165–1178. doi 10.1021/acs.chemrev.8b00408

[41] Muddada, N. B.; Fuglerud, T.; Lamberti, C.; Olsbye, U., Tuning the activity and selectivity of CuCl$_2$/gamma-Al$_2$O$_3$ ethene oxychlorination catalyst by selective promotion. *Topics in Catalysis* **2014**, *57* (6–9), 741–756. doi 10.1007/s11244-013-0231-y

[42] Over, H.; Schomäcker, R., What makes a good catalyst for the Deacon process? *ACS Catalysis* **2013**, *3* (5), 1034–1046. doi 10.1021/cs300735e

[43] Zeng, F.; Hohn, K. L., Modeling of three-way catalytic converter performance with exhaust mixture from natural gas-fueled engines. *Applied Catalysis B: Environmental* **2016**, *182*, 570–579. doi 10.1016/j.apcatb.2015.10.004

[44] Russell, A.; Epling, W. S., Diesel oxidation catalysts. *Catalysis Reviews* **2011**, *53* (4), 337–423. doi 10.1080/01614940.2011.596429

[45] Guan, B.; Zhan, R.; Lin, H.; Huang, Z., Review of the state-of-the-art of exhaust particulate filter technology in internal combustion engines. *Journal of Environmental Management* **2015**, *154*, 225–258. doi.org/10.1016/j.jenvman.2015.02.027

[46] Kong, D.; Zheng, Y.; Kobielusz, M.; Wang, Y.; Bai, Z.; Macyk, W.; Wang, X.; Tang, J., Recent advances in visible light-driven water oxidation and reduction in suspension systems. *Materials Today* **2018**, *21* (8), 897–924. doi.org/10.1016/j.jenvman.2015.02.027

[47] Xie, M.; Chu, T. K.; Wang, T. T.; Wan, K. C.; Yang, D. J.; Li, B.; Ming, P. W.; Zhang, C. M., Preparation, performance and challenges of catalyst layer for proton exchange membrane fuel cell. *Membranes* **2021**, *11* (11). DOI 10.3390/membranes11110879

[48] Tiwari, J. N.; Tiwari, R. N.; Singh, G.; Kim, K. S., Recent progress in the development of anode and cathode catalysts for direct methanol fuel cells. *Nano Energy* **2013**, *2* (5), 553–578. Doi 10.1016/j.nanoen.2013.06.009

[49] Zeng, K.; Zhang, D., Recent progress in alkaline water electrolysis for hydrogen production and applications. *Progress in Energy and Combustion Science* **2010**, *36* (3), 307–326. doi.org/10.1016/j.pecs.2009.11.002

[50] Carmo, M.; Fritz, D. L.; Mergel, J.; Stolten, D., A comprehensive review on PEM water electrolysis. *International Journal of Hydrogen Energy* **2013**, *38* (12), 4901–4934. doi.org/10.1016/j.ijhydene.2013.01.151

[51] Nitopi, S.; Bertheussen, E.; Scott, S. B.; Liu, X.; Engstfeld, A. K.; Horch, S.; Seger, B.; Stephens, I. E. L.; Chan, K.; Hahn, C.; Nørskov, J. K.; Jaramillo, T. F.; Chorkendorff, I., Progress and perspectives of electrochemical CO$_2$ reduction on copper in aqueous electrolyte. *Chemical Reviews* **2019**, *119* (12), 7610–7672. doi 10.1021/acs.chemrev.8b00705

[52] Wang, Y.; Liu, Y.; Wiley, D.; Zhao, S.; Tang, Z., Recent advances in electrocatalytic chloride oxidation for chlorine gas production. *Journal of Materials Chemistry A* **2021**, *9* (35), 18974–18993. doi 10.1039/D1TA02745J

[53] Kintrup, J.; Millaruelo, M.; Trieu, V.; Bulan, A.; Mojica, E. S., Gas diffusion electrodes for efficient manufacturing of chlorine and other chemicals. *The Electrochemical Society Interface* **2017**, *26* (2), 73–76. doi 10.1149/2.f07172if

[54] Jung, J.; Postels, S.; Bardow, A., Cleaner chlorine production using oxygen depolarized cathodes? A life cycle assessment. *Journal of Cleaner Production* **2014**, *80*, 46–56. doi.org/10.1016/j.jclepro.2014.05.086

Index